# VINYL CATIONS

# VINYL CATIONS

PETER J. STANG
Chemistry Department
The University of Utah
Salt Lake City, Utah

ZVI RAPPOPORT
Department of Organic Chemistry
The Hebrew University of Jerusalem
Jerusalem, Israel

MICHAEL HANACK
L. R. SUBRAMANIAN
Institut für Organische Chemie
University of Tübingen
Tübingen, West Germany

1979

ACADEMIC PRESS
A Subsidiary of Harcourt Brace Jovanovich, Publishers
New York   London   Toronto   Sydney   San Francisco

*6352 – 4983*

CHEMISTRY

ACADEMIC PRESS, INC.
111 Fifth Avenue, New York, New York 10003

*United Kingdom Edition published by*
ACADEMIC PRESS, INC. (LONDON) LTD.
24/28 Oval Road, London NW1 7DX

Library of Congress Cataloging in Publication Data
Main entry under title:

Vinyl cations.

Includes bibliographical references.
1. Cations. 2. Vinyl compounds. I. Stang,
Peter J.
QD561.V56    547'.8432'36    79–21330
ISBN 0–12–663780–6

PRINTED IN THE UNITED STATES OF AMERICA

79 80 81 82    9 8 7 6 5 4 3 2 1

*Eine Geschichte zu schreiben,*
*ist immer eine bedenkliche Sache.*
*Denn bei dem redlichsten Vorsatz*
*kommt man in Gefahr, unredlich*
*zu sein; ja, wer eine solche*
*Darstellung unternimmt, erklärt*
*im Voraus, daß er manches ins*
*Licht, manches in Schatten*
*setzen werde.*

Johann Wolfgang von Goethe

# CONTENTS

# PREFACE

Until the mid 1960s, vinyl cations were generally regarded as either extremely difficult to produce or nonexistent. Interest in these positively charged, but still electron rich, intermediates has greatly increased in recent years, as can be seen by a perusal of the literature. In 1964, for example, one of us reviewed "Nucleophilic Reactions on Carbon–Carbon Double Bonds." Of the 115 pages, 20 were devoted to nucleophilic vinylic substitution; the possibility of a vinylic $S_N1$ reaction was but briefly mentioned. In 1969, the same author in an extensive 114-page review on nucleophilic vinylic substitution could cover vinylic $S_N1$ reactions in four pages. By 1973, the field had sufficiently advanced so that another of the present authors could write a 120-page review in the general area of vinyl cations alone with about a third of it devoted to the direct solvolytic generation of vinyl cations. Other general and specialized reviews have appeared since; and when in 1977 we extrapolated these data, it turned out that every 4 to 5 years each subchapter grows four- to fivefold! Consequently, precisely in 1979, "Vinyl Cations" will have matured enough to deserve coverage in an advanced monograph devoted in its entirety to this special field of research. Those of the present authors who have entered or plan to enter other fields of research think of it as a farewell, while those of us who remain in the area of vinyl cations feel that it is a chance for reflection and consolidation. Therefore, after consultation with Academic Press, the present book came into existence.

The book was written by authors from three continents, with overall coordination by P. J. Stang. Chapters 1 and 2, and parts of 5, 7, and 9, were written by P. J. Stang in Salt Lake City; Chapter 8 and part of 5 were written by M. Hanack and Chapters 3 and 4 by L. R. Subramanian, both in Tübingen; Z. Rappoport in Jerusalem contributed Chapter 6 and parts of 5, 7, and 9. Each author was attracted to the field from different starting points, e.g., solvolyses, nucleophilic reactions at double bonds, and homoallenyl rearrangements. The authors are understandably not in complete agreement on all aspects of the field; one author prefers to emphasize the differences between trisubstituted and vinyl cations, while the others are more interested in the similarities of these two species. Although we strived for unification, we also tried to make each chapter self-contained. We attempted to avoid overlap, but this was not always possible, and some was left when this seemed desirable for the sake of completeness in individual chapters.

Peter J. Stang
Zvi Rappoport
Michael Hanack
L. R. Subramanian

# ACKNOWLEDGMENTS

We are pleased to acknowledge the exchange of data and ideas with several colleagues in the field. Professors C. J. Collins, W. Drenth, L. Ghosez, C. A. Grob, C. C. Lee, G. Modena, M. D. Schiavelli, P. v. R. Schleyer, W. M. Schubert, H. Schwarz, H. Taniguchi and Drs. Y. Apeloig, M. Kaftory, S. Kobayashi, and G. Lodder contributed published and unpublished results and commented on specific topics.

We are indebted to our students and co-workers mentioned in the numerous references, who shared our enthusiasm and sometimes our frustrations in conducting research in the field. Research support came from the Volkswagen Foundation for a joint program of M. Hanack and Z. Rappoport; the Deutsche Forschungsgemeinschaft and the Fonds der Chemischen Industrie to M. Hanack; and the Research Corporation, the Petroleum Research Fund administered by the American Chemical Society and, partially, the NIH Cancer Institute to P. J. Stang are gratefully acknowledged. P. J. Stang thanks the Alexander von Humboldt Foundation for a Senior U. S. Scientist Award enabling him to spend the academic year 1977–1978 and part of 1979 in Tübingen, where a major portion of the book was written. Zvi Rappoport extends his thanks to his colleagues and friends in the Department of Chemistry, University of Victoria in Canada for their hospitality during a sabbatical stay in 1976–1977, where he started his work on the book.

Last, but not least, we want to express our appreciation to Academic Press and to Miss F. B. Sundermann and Miss M. Sautter in Tübingen for the artwork and to Mrs. C. Sauer in Salt Lake City for her expert typing of the final manuscript.

Peter J. Stang
Zvi Rappoport
Michael Hanack
L. R. Subramanian

# VINYL CATIONS

# 1

# INTRODUCTION AND HISTORICAL BACKGROUND

Few ideas have proved to be as valuable in organic chemistry as the concept of reactive intermediates. Thousands of seemingly unrelated organic transformations can be correlated, unified, and understood through the involvement of a few reaction intermediaries. Reactive intermediates and their concomitant transition states are the cornerstone of mechanistic organic chemistry as practiced by current physical–organic as well as synthetic chemists.

Among the numerous reactive intermediates in organic chemistry there are essentially four major intermediates involving simple nontetracoordinated carbon species: carbanions, radicals, carbenes, and carbenium ions (Table 1.1). It is interesting that of these four species the corresponding

**TABLE 1.1**

**Simple Reactive Intermediates of Carbon**

| Intermediate | Structure | Corresponding unsaturated species | |
|---|---|---|---|
| | | Structure | Name |
| Carbanion | $-\overset{\mid}{\underset{\mid}{C}}{}^{-}$ | $\text{C}=\overset{-}{\text{C}}-$ | Vinyl anion |
| Radical | $-\overset{\mid}{\underset{\mid}{C}}\cdot$ | $\text{C}=\overset{\cdot}{\text{C}}-$ | Vinyl radical |
| Carbene | $\overset{}{\underset{}{C}}:$ | $\text{C}=\text{C}:$ | Alkylidene carbene |
| Cation | $-\overset{\mid}{\underset{\mid}{C}}{}^{+}$ | $\text{C}=\overset{+}{\text{C}}-$ | Vinyl cation |

unsaturated analogues of carbanions [1] and radicals [2] have been known and investigated and their chemistry understood almost as well and for about as long as the simple intermediates themselves. On the other hand, the unsaturated analogues of carbenes and carbenium ions, namely, alkylidene carbenes and vinyl cations, have been investigated much less extensively, and an understanding of their chemistry has been acquired much more recently.

The term "vinyl cation" refers to a reactive intermediate (**1**) in which the electron-deficient carbon is an integral part of a $\pi$ unsaturation. In **1**, the formally positive carbon atom is bonded to only two substituents, and hence it is a member of the disubstituted carbenium ion family, in contrast to trisubstituted carbenium ions (**2**) derived from saturated precursors. Species **1** has also been called vinylic cation, as well as carbynium [3] and vinyl carbenium ion. The latter two terms are clearly incorrect since carbynium implies a triple bond and vinyl carbenium ion is properly considered to be an ion of type **3**.

$$\begin{array}{ccc}
\underset{R}{\overset{R}{\diagdown}}C{=}\overset{+}{C}{-}R & R{-}\underset{R}{\overset{R}{\underset{|}{\overset{|}{C^+}}}} & \underset{R}{\overset{R}{\diagdown}}C{=}C\underset{R}{\overset{C^+}{\diagup}} \\
\mathbf{1} & \mathbf{2} & \mathbf{3}
\end{array}$$

Carbenium ions have been known [4] since about 1900 and in the intervening three-quarters of a century have been investigated more extensively than any other reactive intermediate, yet until the 1960's only trisubstituted cations were considered acceptable electron-deficient carbon intermediates, and next to nothing was known about vinyl cations. This was the case despite the fact that other related members of the disubstituted cation group, namely, the resonance-stabilized acylium (**4**) and nitrilium cations (**5**), were by then well established and accepted.

$$-\overset{+}{C}{=}O \longleftrightarrow -C{\equiv}\overset{+}{O} \qquad -\overset{+}{C}{=}N{-} \longleftrightarrow -C{\equiv}\overset{+}{N}{-}$$

$$\mathbf{4} \qquad\qquad\qquad \mathbf{5}$$

Among the numerous reasons why this was the case, two stand out: (a) vinyl cations were for a long time regarded as unattractive reaction intermediates because of their alleged high energy, particularly in relation to their trisubstituted analogues; and, perhaps more important, (b) until recently there were no ready general precursors for the solvolytic generation of simple alkylvinyl cations. The somewhat erroneous notion of the unusually high energy of vinyl cations arose from the experimental observations that simple alkylvinyl halides do not precipitate silver halides, even under forcing conditions or conditions that readily give silver halide from primary halides [5]. This notion persisted despite the fact that at least certain stabilized vinyl cations were undoubtedly involved in electrophilic additions to alkynes.

The fact that propyne and allene both give acetone upon hydration in aqueous sulfuric acid has been known since the 1870's [6]. However, vinyl cations were first postulated as possible reaction intermediates by Jacobs and Searles [7] in 1944 to explain the formation of alkyl acetate from the acid-catalyzed hydration of alkoxyacetylenes. This suggestion of Jacobs and Searles and the existence of vinyl cations were not generally recognized and accepted until the 1960's, when detailed mechanistic investigations [8] of the acid-catalyzed hydration, first of alkynyl ethers and thio ethers and then of arylacetylenes, provided compelling evidence for their actual existence (see Chapter 3).

Parallel to the investigations of electrophilic additions to alkynes were the studies of Hanack and co-workers [9] on the generation of vinyl cations by multiple-bond participation in the reaction of homopropargyl and homoallenyl substrates and of Bertrand and Santelli [10] on homoallenyl substrates, which established the possibility of solvolytic generation, albeit indirectly, of such species.

A major step in the development of vinyl cation chemistry occurred in 1964 when Grob and Cseh [11] reported the first direct solvolytic generation of vinyl cations by solvolysis of ring-substituted arylvinyl halides. Consequently, aryl-substituted vinyl cations have been extensively investigated, primarily by Rappoport and co-workers [12], and in the case of $\beta$-heteroatom substituted systems by Modena and collaborators, and have played an important role in the elucidation of the exact mechanisms of vinylic substitutions.

Further progress in the field was made possible by the development and use of the fluorosulfonate "super" leaving groups, such as fluorosulfonates, trifluoromethanesulfonates (triflates), and more recently nonafluorobutanesulfonates (nonaflates). Due to their superior leaving ability, these groups react some $10^6$ to $10^8$ times faster than chlorides or bromides and made possible the direct solvolytic generation of simple alkylvinyl cations as first reported by Stang and Summerville [13] and even the direct generation of the strained cyclic vinyl cations [14]. Hence, by the late 1960's, the direct solvolytic generation of the gamut of vinyl cations, from simple alkyl-substituted ones to the highly stabilized allenyl cations, was accomplished. The ready solvolytic generation of vinyl cations in turn provided the means for a detailed investigation of their exact nature and chemical behavior, such that by the 1970's vinyl cations were respectable members of the establishment of reactive intermediates.

In the late 1960's and early 1970's, vinyl cations also attracted theoretical attention, and calculations were carried out that provided further insight into the nature of these novel intermediates. More recently, they have been generated in the gas phase and studied by ion cyclotron resonance spectroscopy, as well as investigated by nuclear magnetic resonance (nmr) in strongly acidic media.

Besides vinyl cations, considerable attention has recently been focused on two related species: aryl (6) and allenyl (7) cations; on the other hand, little is known about monosubstituted carbenium ions, namely, ethynyl cations (8) [15].

$$
\underset{\textbf{6}}{\overset{+}{\bigcirc}} \qquad \underset{\textbf{7}}{\overset{}{\underset{/}{\diagdown}} C{=}C{=}\overset{+}{C}{-} \longleftrightarrow {-}C{\equiv}C{-}\underset{|}{\overset{+}{C}}{-}} \qquad \underset{\textbf{8}}{R{-}C{\equiv}\overset{+}{C}}
$$

This volume provides a comprehensive and detailed treatment of vinyl cations. It establishes that reaction via vinyl cations is a viable pathway among the multitude of mechanistic routes for vinylic substitution. It shows that our present understanding of the nature and behavior of these reactive intermediates is comparable to our knowledge of any of the other simple carbon intermediates in Table 1.1. Finally, it attempts to provide a brief prognosis of future studies involving vinyl cations.

Besides vinyl cations, this volume briefly touches on aryl (6) and ethynyl (8) cations and considers allenyl cations (7) from the viewpoint of direct solvolytic generation only from appropriate allenyl precursors. It does not cover the well-established acylium (4) and nitrilium (5) cations, nor does it discuss keteneiminium ions (9) derived from α-haloenamines (10) since they are discussed extensively in a recent monograph [16].

$$
\underset{\textbf{9}}{(R)_2C{=}\overset{+}{C}{-}\overset{..}{N}(R)_2} \longleftrightarrow (R)_2C{=}C{=}\overset{+}{N}(R)_2 \qquad \underset{\textbf{10}}{(R)_2C{=}\overset{\overset{\displaystyle X}{|}}{C}{-}N(R)_2}
$$

As indicated in the preface, various aspects of vinyl cation chemistry have been reviewed in the last decade during the evolution of this field. Reference is made to these reviews in appropriate places in the text, and, for convenience, they are also listed in chronological order at the end of this chapter [17].

## REFERENCES

1. D. J. Cram, "Fundamentals of Carbanion Chemistry." Academic Press, New York, 1965.
2. C. Walling, "Free Radicals in Solution." Wiley, New York, 1957.
3. J. J. Jennen, *Chimia* **20**, 309 (1966).

4. C. D. Nenitzescu, *in* "Carbonium Ions" (G. A. Olah and P. von R. Schleyer, eds.), Vol. I, pp. 1–75. Wiley (Interscience), New York, 1968.
5. R. L. Shriner, R. C. Fuson, and D. Y. Curtin, "The Systematic Identification of Organic Compounds," 4th Ed. Wiley, New York, 1964.
6. R. Fittig and A. Schrohe, *Chem. Ber.* **8**, 367 (1875); G. Gustavson and N. Demjanoff, *J. Prakt. Chem.* [2], **38**, 201 (1888).
7. T. L. Jacobs and S. Searles, Jr., *J. Am. Chem. Soc.* **66**, 686 (1944).
8. W. Drenth *et al.*, *Recl. Trav. Chim. Pays-Bas* **79**, 1002 (1960); **80**, 797, 1285 (1961); **82**, 375, 385, 394, 405, 410 (1963); D. S. Noyce, M. A. Matesich, M. D. Schiavelli, and P. E. Peterson, *J. Am. Chem. Soc.* **87**, 2295 (1965); R. W. Bott, C. E. Eaborn, and D. R. M. Walton, *J. Chem. Soc.* p. 384 (1965); P. E. Peterson and J. E. Duddey, *J. Am. Chem. Soc.* **85**, 2865 (1963); **88**, 4990 (1966).
9. M. Hanack *et al.*, *Tetrahedron Lett.* p. 2191 (1964); p. 875 (1965); *Chem. Ber.* **99**, 1077 (1966); *Acc. Chem. Res.* **3**, 209 (1970).
10. M. Bertrand and M. Santelli, *C. R. Acad. Sci., Ser. C* **259**, 2251 (1964).
11. C. A. Grob and G. Cseh, *Helv. Chim. Acta* **47**, 194 (1964); C. A. Grob, *Chimia* **25**, 87 (1971).
12. Z. Rappoport and A. Gal, *J. Am. Chem. Soc.* **91**, 5246 (1969); Z. Rappoport, *Acc. Chem. Res.* **9**, 265 (1976).
13. P. J. Stang and R. H. Summerville, *J. Am. Chem. Soc.* **91**, 4600 (1969); P. J. Stang, *Acc. Chem. Res.* **11**, 107 (1978).
14. W. D. Pfeifer, C. A. Bahn, P. von R. Schleyer, S. Bocher, C. E. Harding, K. Hummel, M. Hanack, and P. J. Stang, *J. Am. Chem. Soc.* **93**, 1513 (1971).
15. S. Miller and J. Dickstein, *Acc. Chem. Res.* **9**, 358 (1976).
16. L. Ghosez and J. Marchand-Brynaert, *in* "Iminium Salts in Organic Chemistry" (H. Bohme and H. G. Viehe, eds.), Part I, pp. 422–528. Wiley, New York, 1976.
17. W. Drenth, *in* "The Chemistry of Organic Compounds (N. Kharasch, ed.), Vol. 2, Chap. 7. Pergamon, Oxford, 1966; R. C. Fahey, *Top. Stereochem.* **3**, 237 (1968); H. G. Richey and J. M. Richey, *Carbonium Ions* **2**, 899 (1970); M. Hanack, *Acc. Chem. Res.* **3**, 209 (1970); G. Modena and U. Tonellato, *Adv. Phys. Org. Chem.* **9**, 185 (1971); C. A. Grob, *Chimia* **25**, 87 (1971); P. J. Stang, *Prog. Phys. Org. Chem.* **10**, 205 (1973); G. Modena and U. Tonellato, *Chim. Ind. (Milan)* **56**, 80 (1974); L. R. Subramanian and M. Hanack, *J. Chem. Educ.*, **52**, 80 (1975); Z. Rappoport, *Acc. Chem. Res.* **9**, 265 (1976); M. Hanack, *Acc. Chem. Res.* **9**, 364 (1976); H. Taniguchi, S. Kobayashi, and T. Sonoda, *Kagaku No Ryoiki*, **30**, 689 (1976); P. J. Stang, *Acc. Chem. Res.* **11**, 107 (1978); M. Hanack, *Angew Chem., Int. Ed. Engl.* **17**, 333 (1978); G. H. Schmid, *in* "The Chemistry of the Carbon–Carbon Triple Bond" Pt. 1 (S. Patai, ed.), Wiley (Interscience), New York, 1978.

# 2

# THERMODYNAMICS AND
# THEORETICAL CALCULATIONS

## I. THERMODYNAMIC CONSIDERATIONS

Among the most interesting aspects of reaction intermediates are their inherent stabilities, their stabilities in relation to each other as well as to their progenitors, and their exact structures. Experimentally these properties can, of course, be directly investigated only if the intermediates under consideration have sufficiently long lifetimes for physical or at least spectral observation and measurement. To date, with few exceptions, vinyl cations have proved to be too elusive for such direct observation and measurement.

Nevertheless, a limited but valuable amount of thermodynamic data are available for the gas phase, as obtained by mass spectroscopy or ion cyclotron resonance (ICR) techniques, and are summarized in Table 2.1, together with data for related trisubstituted carbenium ions and neutral hydrocarbons.

$$R_1^+ + R_2H \longrightarrow R_1H + R_2^+ \tag{1}$$

The data in Table 2.1 and the isodesmic hydride transfer of Eq. (1) can be used to calculate the relative stabilities of two ions. Such a calculation shows that in the gas phase the parent vinyl cation is 25 kcal/mole more stable than the methyl cation but 15 kcal/mole less stable than the ethyl cation. Similarly, the 2-propenyl cation is comparable in stability to the 1-propyl cation but 15 kcal/mole less stable than its saturated counterpart, the isopropyl cation. These data also show that the 3-methyl-2-buten-2-yl cation is comparable in stability to the isopropyl cation. Since energy differences of ions tend to be compressed in the liquid phase compared to the gas phase, due to solvation [6], the differences in stabilities between carbenium ions and vinyl cations in solution are likely to be even smaller.

**TABLE 2.1**

**Gas-Phase Thermodynamic Data for Cations and Hydrocarbons**

| Species | $\Delta H_f^\circ$ (kcal/mole) | Reference |
|---|---|---|
| $CH_3^+$ | 261 | 1, 5 |
| $CH_3CH_2^+$ | 219 | 1, 5 |
| $CH_3CH_2CH_2^+$ | 208 | 1, 5 |
| $CH_3\overset{+}{C}HCH_3$ | 192 | 1, 5 |
| $CH_2{=}CH\overset{+}{C}H_2$ | 226 | 2 |
| $CH_2{=}\overset{+}{C}H$ | 266 | 2 |
| $CH_2{=}\overset{+}{C}CH_3$ | 237 | 3a |
| $CH_3CH{=}\overset{+}{C}CH_3$ | 218 | 3b |
| $(CH_3)_2C{=}\overset{+}{C}CH_3$ | 202 | 3b |
| $H_2C{=}\overset{+}{C}CH(CH_3)_2$ | 207 | 3b |
| $CH_3CH{=}\overset{+}{C}CH_2CH_3$ | 204 | 3b |
| $CH_4$ | $-18$ | 4a |
| $CH_3CH_3$ | $-20$ | 4a |
| $CH_3CH_2CH_3$ | $-25$ | 4a |
| $CH_2{=}CH_2$ | 12 | 4a |
| $CH_3CH{=}CH_2$ | 5 | 4a |
| $CH_3CH{=}CHCH_3$ | $-2$ | 4b |
| $CH_3CH{=}CHCH_2CH_3$ | $-7$ | 4b |
| $(CH_3)_2C{=}CHCH_3$ | $-10$ | 4b |
| $CH{\equiv}CH$ | 54 | 4a |
| $CH_3C{\equiv}CH$ | 44 | 4a |

Since primary $sp^2$-hybridized carbenium ions or carbenium ion-like species are readily accessible by solvolysis [7], it is not surprising that, contrary to earlier considerations [8] and as will be discussed in later chapters, even simple alkylvinyl cations may be generated under appropriate solvolytic conditions. On the other hand, since few primary carbenium ions have been observed with sufficiently long lifetimes for detailed spectral observation, even in strongly acidic media [9], the reason for the paucity of accurate physical and spectral measurements on vinyl cations to date is self-evident.

Equations (2) and (3) in conjunction with the data in Table 2.1 also show

$$CH_3CH_2{}^+ + HC\equiv CH \longrightarrow CH_2{=}CH_2 + CH_2{=}\overset{+}{C}H \qquad (2)$$

$$CH_3CH_2{}^+ + CH_3C\equiv CH \longrightarrow CH_2{=}CH_2 + CH_3\overset{+}{C}{=}CH_2 \qquad (3)$$

that proton addition to alkynes is energetically comparable to addition to alkenes. This perhaps surprising result is the consequence of the higher ground state energy (due to strain) of alkynes compared to alkenes, off-setting the higher energy of vinyl cations relative to alkylcarbenium ions and thereby indirectly making vinyl cations more accessible. This accounts for the fact, at least in retrospect, that vinyl cations were first observed and established as viable reaction intermediaries in the acid-catalyzed hydration of certain alkynes (see Chapters 1 and 3).

## II. THEORETICAL CALCULATIONS

Fortunately, in cases in which experimental data are difficult to obtain and hence are lacking, theoretical calculations are becoming increasingly available. With the advent of high-speed computers, it has, at least in principle, become possible to calculate the exact geometry and energy of molecules from first principles alone. Such exact calculations rely on quantum mechanics and can be carried out only by solving the appropriate Schrödinger equation for the desired molecular system. The difficulties and complexities involved in such detailed calculations for even the simplest organic molecules are well known, and hence alternative approaches have been employed [10]. All of these techniques involve the variational method whereby the total energy is minimized as a function of certain parameters. Variational calculations can be carried out without recourse to experimental data other than the fundamental constants (*ab initio* calculations) or with the aid of some empirical data (semiempirical methods). Since there are a wide variety of empirical parameters and a large choice of determinants and basis functions, there are numerous semiempirical and *ab initio* techniques [10].

As valuable as these calculations are in giving both new insights and guidance to organic chemistry, they must be approached with caution. For one thing, different types of calculations with different approximations and approaches often give different, and sometimes contradictory, results for the same molecule. Hence, care must be exercised in the cross-comparison of results that have been obtained by different calculational methods. Second, almost all calculations are carried out for isolated molecules in the gas phase. The temptation is always great to compare *directly* the results of these theoretical calculations in the gas phase with experimental data in the condensed phase. However, such direct and quantitative comparisons are seldom valid and are particularly dangerous for ionic species such as carbocations, because solvation reduces the magnitude of electronic and polarization effects and may even change the relative stabilities of ions with different sizes and charge distributions [11].

Although numerous calculations have been carried out on carbocations [12], the available theoretical data on vinyl cations are still limited. Furthermore, since even the simplified approaches are time-consuming and relatively expensive, particularly for larger molecules, most calculations on vinyl cations have been done on the parent $C_2H_3^+$ molecule and more recently on a few substituted vinyl cations.

## A. Parent Vinyl Cation

The first calculations on the parent vinyl cation were reported by Hoffmann and employed the extended Hückel MO method [13]. The open form **1** was predicted to be more stable than the bridged structure **2**. A number of

subsequent calculations have appeared, at both the semiempirical and *ab initio* levels, and are summarized in Table 2.2. It is evident from these data that different calculations predict a different structure for the parent vinyl cation, with semiempirical methods [15, 16, 18] favoring the bridged form **2** and single-determinant *ab initio* calculations [17, 19, 20, 21] favoring the classic open ion **1**. This divergence of results for the two different levels of calculation have been ascribed to the overemphasis of the stability of two-electron three-center bonds (and hence the bridged form **2**) by the semiempirical methods. Hence, for a while it appeared that vinyl cations had the open classic structure **1**. However, once correlation was included in the *ab initio* methods, they again predicted the bridged form **2** to be the more stable [21] by some 7 kcal/mole. The latest calculations [22] indicate that the stability difference between **1** and **2** is very small; there is an energy difference of only 1 to 2 kcal/mole, with the "bridged structure probably having the lower energy" [22a,b] (see, however, Dewar and Rzepa [22c]). Furthermore, according to these calculations [22a,b], the barrier to interconversion between **1** and **2** is very small, with an activation energy of only 1 to 3 kcal/mole, as depicted in Fig. 2.1.

**TABLE 2.2**

**Theoretical Calculations on $C_2H_3^+$**

| Method of calculation[a] | $E(2) - E(1)$, kcal/mole[b] | Reference |
|---|---|---|
| EHT | >0 | 13 |
| INDO | +14 | 14 |
| NDDO | −32 | 15 |
| CNDO | −6 | 16 |
| EGTO | +19 | 17 |
| CNDO/FK | −6 | 18 |
| STO-3G | +18 | 19 |
| 4-31G | +19 | 20 |
| 6-31G | +6 | 20 |
| PEGTO | +5 | 21 |
| C-PEGTO | −7 | 21 |
| *Ab initio* + CI | −1 to −2 | 22a,b |

[a] See original literature for definition and details.
[b] A positive difference means that the open ion **1** is more stable than the bridged structure **2**.

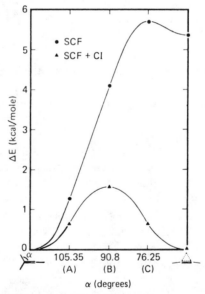

**Fig. 2.1.** Section of the potential surface corresponding to the optimal transformation path of the classic to the bridged vinyl cation. For both cations, the difference $\Delta E = E - E$ (classic form) is plotted against the reaction coordinate $\alpha$. [Reproduced by permission of the publisher from *J. Chem. Phys.* **64**, 4159 (1976).]

**TABLE 2.3**

**Calculated [22a] Bond Lengths and Bond Angles for the Open (1) and Bridged (2) Parent Vinyl Cation**

| Structural feature | 1 | 2 |
|---|---|---|
| $C_1=C_2$ | 1.263 Å | 1.210 Å |
| $C_1-H_1$ | 1.075 Å | 1.276 Å |
| $C_2-H_2$ | 1.086 Å | 1.074 Å |
| $\alpha$ | 180° | 179.1° |
| $\beta$ | 119.9° | 61.7° |

**TABLE 2.4**

**Calculated [22a] Charge Distributions for the Open (1) and Bridged (2) Parent Vinyl Cation**

| Charge/atom | 1 | 2 |
|---|---|---|
| $C_1$ | 0.307 | 0.050 |
| $C_2$ | -0.237 | 0.050 |
| $H_1$ | 0.308 | 0.271 |
| $H_2$ | 0.311 | 0.314 |

Weber and McLean [22a,b] calculated a singlet–singlet vertical first excitation energy $^1A_1 \rightarrow {}^1A_2$ ($\pi \rightarrow \sigma^*$) of 2.37 eV for the open ion **1** and 6.53 eV for the bridged form **2**. The important structural features of ions **1** and **2** as determined by Weber and McLean (22) are summarized in Table 2.3. The respective charge distributions for **1** and **2** are summarized in Table 2.4.

Both forms of $C_2H_3{}^+$ (**1** and **2**) are planar, with $C_{2v}$ symmetry in their most stable conformation. The calculated C=C bond length for the open ion **1** is between the experimental values of 1.330 and 1.203 Å for ethylene [23] and acetylene [24], respectively. The calculated $C_1-H_1$ bond lengths of ion **1** are comparable to the experimental C–H of 1.076 Å in ethylene, whereas the $C_2-H_2$ bond lengths are somewhat longer. The calculated C=C bond length for **2** approaches the experimental value of 1.203 Å for acetylene, whereas the calculated C–H$_1$ distances of 1.276 Å are considerably larger than the normal C–H distances. This suggests that ion **2** resembles a $\pi$-complexed protonated acetylene.

The calculated charge distributions (Table 2.4) are instructive and indicate that, in both ionic forms **1** and **2**, the unit positive charge is largely distributed on the hydrogens, with in fact a slight negative charge on $C_2$ of ion **1**.

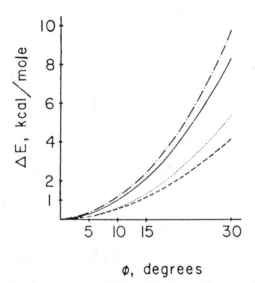

**3**

All of the theoretical calculations [15–22] predict a linear sp-hybridized geometry (**1**) for vinyl cations rather than a trigonal sp²-hybridized geometry (**3**). Figure 2.2 shows a plot of the energy required to bend the third hydrogen out of the HCH plane in the methyl cation as determined by calculations [25]. These calculations confirm the organic chemist's intuitive and qualitative feeling that carbenium ions prefer a trigonal planar geometry with an empty p orbital and that it takes energy to bend the methyl cation from its low-energy planar form. Figure 2.3 shows a similar plot for the energy re-

**Fig. 2.2.** *Ab initio* and MINDO energies for the methyl cation, all referenced to the same zero of energy (at $\varphi = 0$ for each $\theta$): (—·—) *ab initio*, $\theta = 90°$; (—) *ab initio*, $\theta = 120°$; (· · ·) MINDO, $\theta = 90°$; (– – –) MINDO, $\theta = 120°$. [Reproduced by permission of the publisher from *J. Am. Chem. Soc.*, **91**, 1037 (1969).]

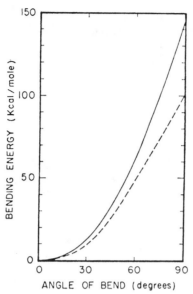

**Fig. 2.3.** INDO energy for bending methine CH bond in vinyl cation **1** (—), in-plane bending; (–––) out-of-plane bending. [Reproduced by permission of the publisher from *J. Am. Chem. Soc.* **91**, 5350 (1969).]

quired, as determined by INDO calculations [15], for bending the methine CH bond out of the linear arrangement **1** toward the trigonal geometry **3**. The energy required to deform a linear vinyl cation is even greater than that required to deform the planar methyl cation, in fact, the bent $sp^2$-hybridized structure **3** with a C=C–H bond angle of 120° is some 50 kcal/mole higher [15, 17] in energy than the linear sp-hybridized form **1**. These predictions on the preferred geometry of vinyl cations, as derived from calculations on the parent ion, in general are in accord with indirect experimental evidence derived from solvolysis data, as discussed in later chapters.

## B. Substituted Vinyl Cations

The energies and structures of a number of α- and β-substituted vinyl cations have been calculated by *ab initio* methods at the STO-3G and 4-31G level. In particular, methyl- (**4**), ethenyl- (**5**), ethynyl- (**6**), cyclopropyl- (**7**),

**TABLE 2.5**

**Calculated Energies[a] [26] for Reaction (4)**

| Ion | Substituent | RHF/STO-3G | RHF/4-31G |
|-----|-------------|------------|-----------|
| 1 | H | −25.9 | −25.2 |
| 4 | $CH_3$ | 0.0 | 0.0 |
| 5 | $CH=CH_2$ | 15.1 | 14.3 |
| 6 | $C\equiv CH$ | 2.0 | −0.1 |
| 7 | c-$C_3H_5$ | 15.8 | 17.3 |
| 8 | $C_6H_5$ | 34.2 | — |

[a] In kilocalories per mole.

and phenyl-substituted (8) vinyl cations have been investigated [26]. The relative energies of these ions were compared by means of the isodesmic [27] hydride transfer reaction (4) and are summarized in Table 2.5. A

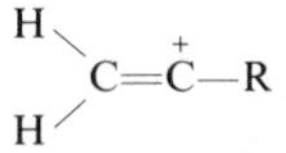

$$H_2C = \overset{+}{C} - R$$

4  R = $CH_3$
5  R = $CH=CH_2$
6  R = $C\equiv CH$
7  R = cyclopropyl, c-$C_3H_5$
8  R = phenyl, $C_6H_5$

$$H_2C=\overset{+}{C}R + H_2C=CHCH_3 \longrightarrow H_2C=\overset{+}{C}CH_3 + H_2C=CHR \qquad (4)$$

positive energy indicates a greater stabilization by the substituent (relative to methyl) in the cation than in the corresponding neutral molecule and hence gives an indication of the relative stabilities of the corresponding vinyl cations. The resulting energies of such isodesmic reactions are well described by these methods, and the estimated error limits are of the order of 2 to 5 kcal/mole [27, 28].

The data of Table 2.5 indicate that the effectiveness of $\alpha$ substituents in stabilizing a vinyl cation follow the order $C_6H_5 \gg$ c-$C_3H_5 \cong CH=CH_2 \gg C\equiv CH \cong CH_3 \gg$ H. Specifically, an $\alpha$-methyl substituent in 4 makes the 2-propenyl cation some 26 kcal/mole more stable than the parent ion 1, and an $\alpha$-phenyl group (8) provides an additional stabilization of 34 kcal/mole.

Ethenyl and ethynyl groups possess $\pi$ electrons that can stabilize a cationic center by allylic-type conjugation, as represented by 5 ↔ 5′ and 6 ↔ 6′. Such delocalization in 5, however, is possible only in the perpendicular conformation 5a, where the interaction between the "empty" cationic p orbital and the HOMO of the substituent is maximized. Indeed,

$$CH_2 = \overset{+}{C} - CH = CH_2 \longleftrightarrow CH_2 = C = CH - \overset{+}{C}H_2$$

**5**           **5′**

$$CH_2 = \overset{+}{C} - C \equiv CH \longleftrightarrow CH_2 = C = C = \overset{+}{C}H$$

**6**           **6′**

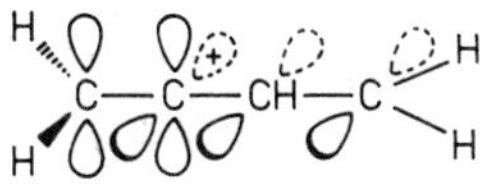

**5a** 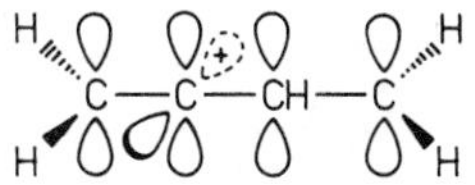           **5b**

conformation **5a** is found to be 22 kcal/mole more stable than the planar form **5b** (amounting to the barrier of rotation around the $\overset{+}{C}$–C substituent bond). The lower stability of the planar conformation **5b** has been experimentally established (see Chapter 5) by Grob and Pfaendler [29].

Interestingly, a triple bond as in **6** provides a stabilization comparable only to that of a methyl group and much lower than the stabilization provided by an α double bond. This low stabilization by an α triple bond has been attributed to a cancellation between a stabilizing π conjugation and a destabilizing σ withdrawal by the acetylenic group [26].

The ability of cyclopropyl and phenyl rings to stabilize a carbenium ion is well known [30]. However, there has been some dispute as to the relative efficiences of the stabilization provided by the two rings; recent ICR data [31] indicate that in the *gas phase* a phenyl substituent is superior to a cyclopropyl ring in stabilizing primary and secondary carbenium ions, whereas in solution different experimental probes give differing results [32]. As the data in Table 2.5 indicate, calculations show that in the gas phase a phenyl group is 18 kcal/mole more effective in stabilizing a vinyl cation than is a cyclopropyl group. In solution, the reverse is true, with α-cyclopropylvinyl cations (**7**) being formed faster than α-phenylvinyl ions (**8**) [33]. The discrepancy between the solvolysis and gas-phase data is probably due to the aforementioned differences in media effects [6] and, in particular, to the preferential solvation of the smaller and less polarizable ion **7** compared to **8** [11, 12].

The stabilizing effect of the phenyl and cyclopropyl rings once again arises from the interaction between the empty p orbital and the HOMO of the ring. For these interactions to be effective, the rings in **7** and **8** must be properly oriented. Both simple Hückel MO [34] and *ab initio* [26] calculations predict a favored bisected conformation (**7a**) rather than the perpendicular conformation **7b** for the α-cyclopropylvinyl system. An analogous

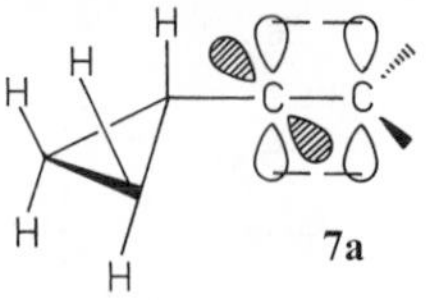

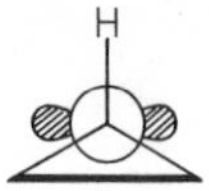

7a        bisected

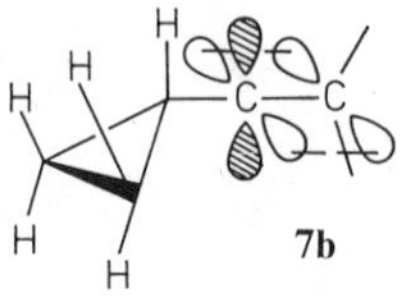

7b        perpendicular

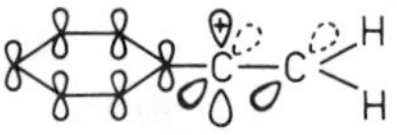

8a

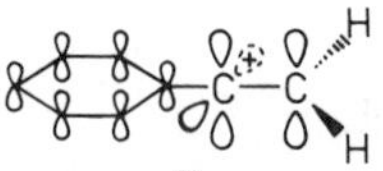

8b

situation holds in the saturated counterpart of **7**, the cyclopropylcarbinyl system. The energy difference between **7a** and **7b** is computed to be 16 kcal/mole in favor of **7a** [26]. Similarly, the conformation **8a**, in which the vinylic $\pi$ bond and the aromatic ring are deconjugated (i.e., perpendicular), is favored by 25 kcal/mole over the conformation **8b** [26] (see also Chapter 6). Calculations also indicate [26] that similarly substituted ethyl cations are 12 to 17 kcal/mole more stable than their vinyl cation counterparts (**4–8**), indicating that the stabilizing effects of the substituents examined are inherently similar for the alkenyl and alkyl cations.

In contrast to $\alpha$ substitution on **1**, $\beta$ substitutions as expected provide considerably less stabilization. That is, the 1-propenyl cation **9**, although some 11 kcal/mole more stable than **1**, is 14 to 16 kcal/mole less stable than **4** and is predicted to rearrange to **4** with little or no activation energy [35].

$$CH_3CH{=}\overset{+}{CH}$$

**9**

Calculations [35] also predict, in accord with recent gas-phase [3] and other experimental data (see Chapter 4), that protonation of allene should occur on the terminal carbon, giving vinyl cation **4**, rather than on the central carbon, which would ultimately result in the allyl cation **10**. The preferential formation of the less stable vinyl cation **4** (*vide supra*) over the

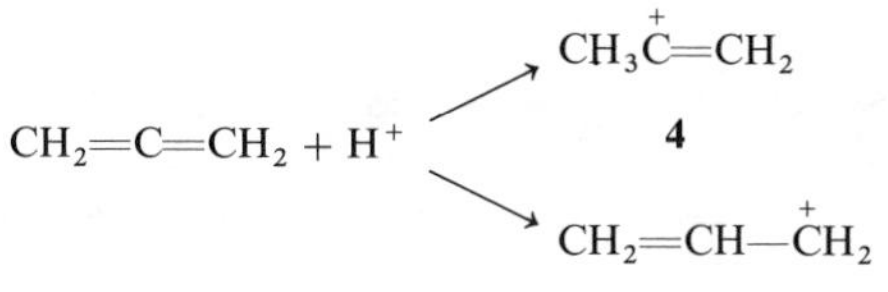

$$CH_2=C=CH_2 + H^+ \begin{cases} \nearrow & CH_3\overset{+}{C}=CH_2 \\ & \mathbf{4} \\ \searrow & CH_2=CH-\overset{+}{C}H_2 \\ & \mathbf{10} \end{cases}$$

allyl cation **10** can be ascribed to the instability of the initially formed intermediate, a perpendicular allyl cation (**10b**), in the central protonation. Ion **10b** has been calculated to be some 35 to 42 kcal/mole less stable than the fully conjugated **10a**, and its lower stability has been ascribed primarily to the destabilizing inductive effect of a perpendicular $\pi$ bond [35, 36]. Indirect experimental evidence [37] indicates a barrier of 38 to 43 kcal/mole for the interconversion of **10a** and **10b**.

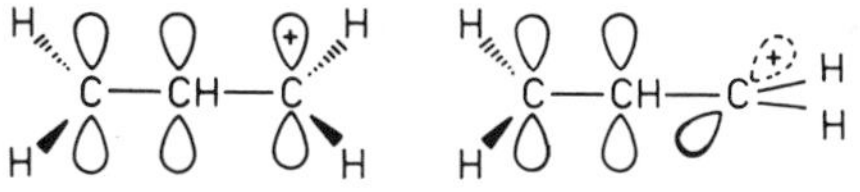

**10a**                                          **10b**

## C. Miscellaneous Calculations

Calculations have also appeared on the effects of $\alpha$- [38] and $\beta$-electropositive [39] substituents (first short-period groups) on the stabilities of alkyl and vinyl cations.

The stabilities of the $\alpha$-substituted cations **11** and **12** were compared with those of the parent ethyl and vinyl (**1**) cations, respectively, by means of the isodesmic reactions (5) and (6). The results are summarized in Table 2.6. A number of interesting trends emerge from the data in Table 2.6. One is that the two different levels of calculations do not always give the same results. Since the 4-31G values are more reliable, they are more valuable and will be discussed further. It is evident from the data that with the

$$CH_3\overset{+}{C}HX + CH_3CH_3 \longrightarrow CH_3CH_2{}^+ + CH_3CH_2X \tag{5}$$

**11**

$$CH_2=\overset{+}{C}X + CH_2=CH_2 \longrightarrow CH_2=\overset{+}{C}H + CH_2=CHX \tag{6}$$

**12**

exception of F, all first-row elements are capable of cation stabilization when substituted $\alpha$ to the cationic center. The stabilizing effects of most substituents are similar for the ethyl and vinyl cations but are of a larger magnitude in the latter case. Surprisingly, the best stabilization is provided

TABLE 2.6

Calculated Stabilization Energies[a] for α-Substituted Ethyl and Vinyl Cations

|  | Equation (5) | | Equation (6) | |
| --- | --- | --- | --- | --- |
| Substituent | STO-3G[b] | 4-31G[b] | STO-3G[b] | 4-31G[b] |
| Li | 80.2 | 69.4 | 95.1 | 91.9 |
| BeH | 22.9 | 17.1 | 31.5 | 31.7 |
| Planar BH$_2$[b] | 8.6 | 3.6 | 28.6 | 26.7 |
| Perpendicular BH$_2$[b] | 25.1 | 20.8 | 9.0 | 5.7 |
| CH$_3$ | 24.1 | 22.5 | 25.9 | 25.2 |
| Perpendicular NH$_2$[b] | −0.6 | 10.4 | 68.8 | 61.5 |
| Planar NH$_2$[b] | 79.7 | 74.7 | 13.5 | 11.8 |
| OH | 55.2 | 32.3 | 37.7 | 19.7 |
| F | 29.0 | −0.2 | 6.9 | −20.8 |

[a] In kilocalories per mole.
[b] See Apeloig et al. [38] for exact definition and further details.

by Li, despite the lack of p or d electrons, and has been ascribed to α donation [38]. The least stabilization is afforded by F, despite the availability of nonbonding electrons for possible π donation. Apparently, the stronger σ withdrawal of F due to its high electronegativity outweighs its π-donor ability by the nonbonding lone pairs. Similarly, OH is less effective in stabilizing **1** than is a methyl group, once again due to electronegativity and σ withdrawal.

Likewise, for the β substituents the energies derived from the isodesmic Eq. (7) provide estimates of the cation stabilities relative to the parent vinyl

$$XCH{=}\overset{+}{C}H + CH_2{=}CH_2 \longrightarrow XCH{=}CH_2 + CH_2{=}\overset{+}{C}H \qquad (7)$$

cation **1** [39]. The results are summarized in Table 2.7. As the data indicate, once again Li provides the greatest stabilization, and F the least. The stabilizing influence of these β substituents has been ascribed to inductive and hyperconjugative effects, with the latter being dominant [39]. However, in contrast to the behavior of the other first-row elements, both F and OH are destabilizing.

In accord with the suggestions that certain metals such as Li might provide considerable stabilization to cations, platinum-stabilized vinyl cations (**13**) have been experimentally invoked in the reaction of certain alkynyl- and alkenylplatinum(II) compounds [40] (see Chapter 9).

$$Pt\overset{+}{C}{=}CHR$$

**13**

**TABLE 2.7**

**Calculated Stabilization Energies$^a$ for β-Substituted Vinyl Cations**

| Substituent (X) | STO-3G$^b$ | 4-31G$^b$ |
|---|---|---|
| H | 0.0 | 0.0 |
| Li | 92.2 | 89.0 |
| BeH | 28.9 | 26.4 |
| BH$_2$ perpendicular$^b$ | 17.5 | 13.9 |
| BH$_2$ planar$^b$ | 9.6 | 6.6 |
| CH$_3$ | 12.8 | 10.9 |
| NH$_2$ perpendicular$^b$ | 3.0 | 1.6 |
| NH$_2$ planar$^b$ | 10.9 | 5.4 |
| OH | −2.0 | −10.8 |
| F | −17.9 | −32.0 |

$^a$ In kilocalories per mole.
$^b$ See Apeloig et al. [39] for definition and details.

These calculations [38, 39] also predict that fluorovinyl cation **14** is more stable than **15**, in accord with previous calculations [41] indicating that

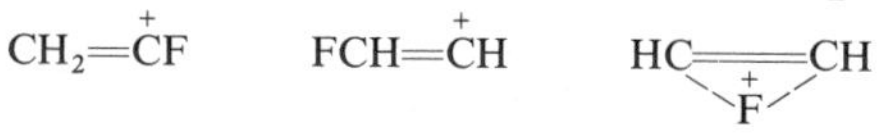

$$CH_2{=}\overset{+}{C}F \qquad FCH{=}\overset{+}{C}H \qquad HC{\equiv\!\!\equiv}CH$$

$$\overset{}{\underset{F}{\diagdown\!\diagup}}$$

**14**          **15**          **16**

proton approach to acetylene is symmetric but that the asymmetric approach resulting in **14** rather than **15** is favored for fluoroacetylene. Another theoretical study showed that open fluorovinyl cation **15** is 31 kcal/mole more stable than the bridged ion **16** [42].

Calculations have also been carried out on the open and bridged oxygen-(**17**) [43] and sulfur-substituted (**18**) [44] vinyl cations. In the former, the open ion **17a** was found to be 18 kcal/mole more stable than the bridged oxirenium ion **17b**. On the other hand, the thiirenium ion **18b** was found to be 3 kcal/mole more stable than the open ion **18a**, in agreement with the

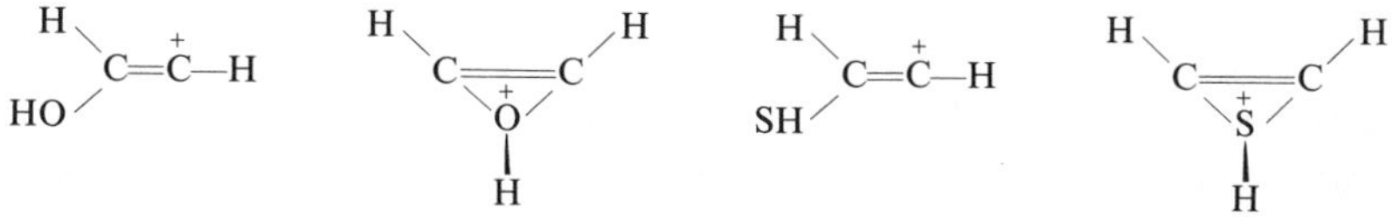

**17a**          **17b**          **18a**          **18b**

recent nmr observation [45] of thiirenium ions as well as a large body of experimental data on both the addition of sulfenyl halides to alkynes and the solvolyses of β-thiovinyl sulfonates (see Chapters 3 and 7).

Phenyl bridging has also been theoretically investigated. Preliminary results indicate [46a] that the bridged vinylidenephenonium ion **19** is some 9 kcal/mole more stable than the $\beta$-phenylvinyl cation **20**. The hydrogen-bridged isomer **21** is some 7 kcal/mole higher in energy than the open ion

CH=CH

$C_6H_5$—CH=CH$^+$          HC=CC$_6$H$_5$
                                               H

19                    20                    21

**20** [46a]. The 9 kcal/mole stabilization energy provided by the bridging phenyl in **19** is considerably less than the 35 kcal/mole difference in the analogous saturated systems **22** and **23** [46b]. This difference is probably

CH$_2$—CH$_2$

$C_6H_5CH_2CH_2^+$

22                    23

due to the higher ring strain of **19** compared to that of **22**. Furthermore, no activation energy for bridging was found in **22**, but an activation energy of 7 kcal/mole was calculated for bridging to occur in the vinylidenephenonium ion **19** [46a]. This activation energy was ascribed to the need to rotate the phenyl group out of the ground state $\pi$–$\pi$ conjugation before collapse to the bridged form can occur [46a].

Semiempirical [47] as well as *ab initio* [48] (STO-3G level) calculations have also been carried out on the isomeric cyclopropylidenemethyl (**24**) and cyclobutenyl (**25**) cations. Cation **24** is an exceptionally stable primary vinyl

24          25          26

cation due to favorable interaction between the empty p orbital and the antisymmetric Walsh orbital of the cyclopropyl ring. Calculations indicate that **24** is 8 kcal/mole more stable than the secondary 2-propenyl cation **4** [48]. On the other hand, cyclobutenyl ion in its classic form (**25**) should be a highly strained and hence unstable intermediate. Yet semiempirical calculations [47] predict a nonclassic stabilization in the form of a nonclassic ion (**26**) for the cyclobutenyl system in accord with a large body of experimental data [33] (see Chapter 5). On the other hand, *ab initio* calculations

[48] at the 4-31G level predict **24** to be more stable than **26** by 5 kcal/mole. However, when correlation interaction is included, the nonclassical ion **26** is predicted to be some 9 kcal/mole more stable than **24**. Hence, once again, theory and experimental results are in harmony with each other.

EHMO and MINDO/2 calculations have been carried out on the mechanism of olefin [49], acetylene [50], and allene [50] cycloadditions to vinyl cations (Scheme 2.1). These calculations predict a nonconcerted dipolar addition pathway via intermediate **27** for olefins and **28** for acetylene and allene rather than a concerted process as the energetically favored mechanism. They further predict [50] that vinyl cations should react fastest with acetylene, least with ethylene, and at an intermediate rate with allene.

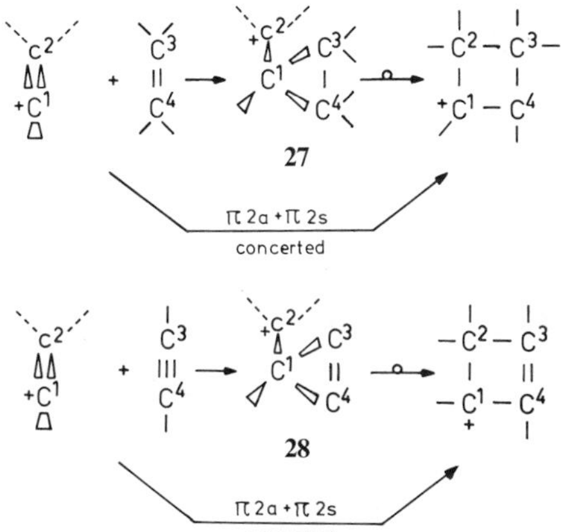

**Scheme 2.1**

Finally, calculations have been carried out on the allenyl cations **29–31** [51]. For both the parent (**29**) and the methyl-substituted systems, the results indicate a greater charge on the propargylic position than on the allenyl carbon [51]. These results are in agreement with the preferential capture of these ions at the propargylic position in solvolytic displacement reactions (see Chapter 5).

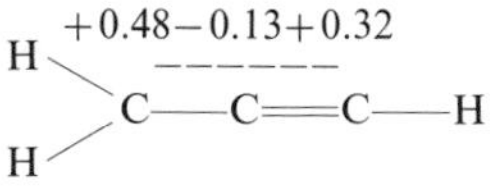

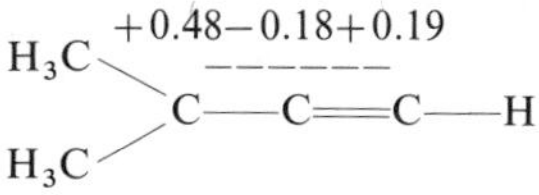

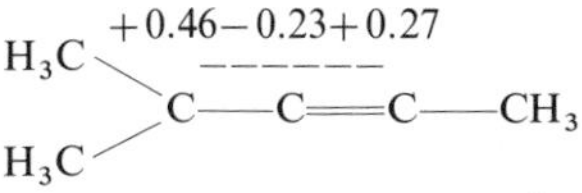

# REFERENCES

1. F. P. Lossing and G. P. Semeluk, *Can. J. Chem.* **48**, 955 (1970).
2. F. P. Lossing, *Can. J. Chem.* **49**, 357 (1971); **50**, 3973 (1972).
3. (a) D. H. Aue, W. R. Davidson, and M. T. Bowers, *J. Am. Chem. Soc.* **98**, 6700 (1976); (b) D. H. Aue, personal communication.
4. (a) L. Radom, W. J. Hehre, and J. A. Pople, *J. Am. Chem. Soc.* **93**, 289 (1971), and references therein; (b) S. W. Benson, F. R. Cruickshank, D. M. Golden, G. R. Haugen, H. E. O'Neal, A. S. Rodgers, R. Shaw, and R. Walsh, *Chem. Rev.* **69**, 279 (1969).
5. See also J. L. Franklin, J. G. Dillard, H. M. Rosenstock, J. T. Herron, and K. Draxl, "Ionization Potentials, Appearance Potentials and Heats of Formation of Gaseous Positive Ions," NSRDS-NBS-26. Natl. Bur. Stand., Washington, D.C., 1969.
6. E. S. Amis and J. F. Hinton, "Solvent Effects on Chemical Phenomena," Vol. 1. Academic Press, New York, 1973.
7. A. Streitwieser, Jr., "Solvolytic Displacement Reactions." McGraw-Hill, New York, 1962; J. M. Harris, *Prog. Phys. Org. Chem.* **11**, 89 (1974).
8. R. L. Shriner, R. C. Fuson, and D. Y. Curtin, "The Systematic Identification of Organic Compounds," 4th Ed. Wiley, New York, 1964.
9. G. Fraenkel and D. G. Farnum, *in* "Carbonium Ions" (G. A. Olah and P. von R. Schleyer, eds.), Vol. 1, pp. 237–255. Wiley (Interscience), New York, 1968; G. A. Olah, *Angew. Chem., Int. Ed. Engl.* **12**, 173 (1973).
10. See, e.g., H. F. Schaefer, III, ed., "Modern Theoretical Chemistry." Plenum. New York, 1976; W. L. Jorgensen and L. Salem, "The Organic Chemist's Book of Orbitals." Academic Press, New York, 1973; J. A. Pople and D. L. Beveridge, "Approximate Molecular Orbital Theory." McGraw-Hill, New York, 1970; M. J. S. Dewar, "The Molecular Orbital Theory of Organic Chemistry." McGraw-Hill, New York, 1969; R. G. Parr, "The Quantum Theory of Molecular Electronic Structure." Benjamin, New York, 1963; J. C. Slater, "Quantum Theory of Molecules and Solids." McGraw-Hill, New York. 1963; A. Streitwieser, Jr., "Molecular Orbital Theory for Organic Chemists," Wiley, New York, 1961.
11. See, e.g., D. H. Aue, H. M. Webb, and M. T. Bowers, *J. Am. Chem. Soc.* **97**, 4137 (1976); **98**, 318 (1976); J. F. Wolf, P. G. Harch, and R. W. Taft, *J. Am. Chem. Soc.* **97**, 2904 (1975); R. W. Taft and E. M. Arnett, *in* "Proton Transfer Reactions" (E. F. Caldin and V. Gold, eds.), pp. 31–101. Chapman & Hall, London, 1975.
12. For a review and leading references, see L. Radom, D. Poppinger, and R. C. Haddon, *in* "Carbonium Ions" (G. A. Olah and P. von R. Schleyer, eds.), Vol. 5, pp. 2303–2426. Wiley (Interscience), New York, 1976.
13. R. Hoffmann, *J. Chem. Phys.* **40**, 2480 (1964).
14. T. Yonezawa, H. Nakatsuji, and H. Kato, *J. Am. Chem. Soc.* **90**, 1239 (1968).
15. R. Sustmann, J. E. Williams, Jr., M. J. S. Dewar, L. C. Allen, and P. von R. Schleyer, *J. Am. Chem. Soc.* **91**, 5350 (1969).
16. J. E. Williams, Jr., V. Buss, L. C. Allen, P. von R. Schleyer, W. A. Lathan, W. J. Hehre, and J. A. Pople, *J. Am. Chem. Soc.* **92**, 2141 (1970).
17. A. C. Hopkins, K. Yates, and I. G. Csizmadia, *J. Chem. Phys.* **55**, 3835 (1971).
18. H. Kollmar and H. O. Smith, *Theor. Chim. Acta* **20**, 65 (1971).
19. W. A. Lathan, W. J. Hehre, and J. A. Pople, *J. Am. Chem. Soc.* **93**, 808 (1971).
20. P. C. Hariharan, W. A. Lathan, and J. A. Pople, *Chem. Phys. Lett.* **14**, 385 (1972).
21. B. Zurawski, R. Ahlirchs, and W. Kutzelnigg, *Chem. Phys. Lett.* **21**, 309 (1973).
22. (a) J. Weber, M. Yoshimine, and A. D. McLean, *J. Chem. Phys.* **64**, 4159 (1976); (b) J. Weber and A. D. McLean, *J. Am. Chem. Soc.* **98**, 875 (1976).
23. K. Kuchitsu, *J. Chem. Phys.* **44**, 906 (1966).

24. W. J. Lafferty and R. J. Thibault, *J. Mol. Spectrosc.* **14**, 79 (1964); E. H. Plyler and E. D. Tidwell, *J. Opt. Soc. Am.* **53**, 589 (1963).
25. J. E. Williams, R. Sustmann, L. C. Allen, and P. von R. Schleyer, *J. Am. Chem. Soc.* **91**, 1037 (1969).
26. Y. Apeloig, P. von R. Schleyer, and J. A. Pople, *J. Org. Chem.* **42**, 3004 (1977).
27. W. J. Hehre, R. Ditchfield, L. Radom, and J. H. Pople, *J. Am. Chem. Soc.* **92**, 4796 (1970).
28. W. J. Hehre and J. A. Pople, *J. Am. Chem. Soc.* **97**, 6941 (1975), and references therein.
29. C. A. Grob and H. R. Pfaendler, *Helv. Chim. Acta* **53**, 2130 (1970).
30. See, e.g., W. J. Le Noble, "Highlights of Organic Chemistry," Chap. 20. Dekker, New York, 1974.
31. J. F. Wolf, P. G. Harch, R. W. Taft, and W. J. Hehre, *J. Am. Chem. Soc.* **97**, 2902 (1975).
32. See references 26 and 31 for further discussion and additional references.
33. M. Hanack, *Acc. Chem. Res.* **9**, 364 (1976).
34. D. R. Kelsey and R. G. Bergman, *J. Am. Chem. Soc.* **93**, 1953 (1971).
35. L. Radom, P. C. Hariharan, J. A. Pople, and P. von R. Schleyer, *J. Am. Chem. Soc.* **95**, 6531 (1973).
36. S. D. Peyerimhoff and R. J. Buenker, *J. Chem. Phys.* **51**, 2528 (1969).
37. V. Buss, R. Gleiter, and P. von R. Schleyer, *J. Am. Chem. Soc.* **93**, 3927 (1971).
38. Y. Apeloig, P. von R. Schleyer, and J. A. Pople, *J. Am. Chem. Soc.* **99**, 1291 (1977).
39. Y. Apeloig, P. von R. Schleyer, and J. A. Pople, *J. Am. Chem. Soc.* **99**, 5901 (1977).
40. R. A. Bell and M. H. Chisholm, *Inorg. Chem.* **16**, 687 (1977); R. A. Bell, M. H. Chisholm, D. A. Cough, and L. A. Rankel, *Inorg. Chem.* **16**, 677 (1977).
41. D. T. Clark and D. B. Adams, *Tetrahedron* **29**, 1887 (1973); C. U. Pittman, *J. Phys. Chem.* **77**, 494 (1973).
42. I. G. Csizmadia, V. Lucchini, and G. Modena, *Theor. Chim. Acta* **39**, 51 (1975).
43. I. G. Csizmadia, F. Bernardi, V. Lucchini, and G. Modena, *J. Chem. Soc., Perkin Trans. II* p. 542 (1977).
44. I. G. Csizmadia, A. J. Duke, V. Lucchini, and G. Modena, *J. Chem. Soc., Perkin Trans. II* p. 1808 (1974).
45. G. Capozzi, O. Delucchi, V. Lucchini, and G. Modena, *Chem. Commun.* p. 248 (1975).
46. (a) Y. Apeloig, personal communication; (b) W. J. Hehre, *J. Am. Chem. Soc.* **94**, 5919 (1972).
47. H. Fischer, K. Hummel, and M. Hanack, *Tetrahedron Lett.* p. 2169 (1969).
48. Y. Apeloig, J. B. Collins, P. von R. Schleyer, J. A. Pople, and D. Cremer, submitted for publication; Y. Apeloig, personal communication.
49. H. U. Wagner and R. Gompper, *Tetrahedron Lett.* p. 4061 (1971).
50. H. U. Wagner and R. Gompper, *Tetrahedron Lett.* p. 4065 (1971).
51. H. Mayr, personal communication; see also D. Mirejovsky, W. Drenth and F. B. van Duijneveldt, *J. Org. Chem.* **43**, 763 (1978).

# 3

# ELECTROPHILIC ADDITIONS TO ALKYNES AND PARTICIPATION OF THE TRIPLE BOND IN SOLVOLYSIS

Electrophilic addition to alkynes is a well-known industrial process for the preparation of technically important vinyl derivatives. This reaction was used in the industry for a long time without a detailed knowledge of its mechanistic aspects. The addition of several different electrophiles to alkynes has been studied only sporadically in the past from a mechanistic viewpoint. However, in recent years the interest in electrophilic additions to alkynes has greatly increased, and a number of reviews have appeared [1]. This may be attributed, at least partially, to the identification of vinyl cations as intermediates and the formation of interesting products in addition to the usual adducts in some of these reactions. Indeed, the early convincing evidence for vinyl cation intermediates was provided by the acid-catalyzed hydration of various alkynes. Nevertheless, work in the field of electrophilic addition to alkynes made rapid progress only when the existence of vinyl cations as intermediates in the solvolysis of vinylic substrates (Chapters 5, 6, and 7) was established.

Since the triple bond of the alkynes is electron-rich, one can expect that these compounds will undergo electrophilic additions rather easily. The protonation energies of alkynes are only slightly higher than those of alkenes (Chapter 2). Consequently, alkynes can undergo electrophilic additions as well as, and often better than alkenes depending on the structure of the substrate, the electrophile, and the solvent used. The influence of solvent in these reactions is very marked, and it frequently changes the mechanism.

Electrophilic addition to alkynes has been studied by several groups from the mechanistic and, independently, from the preparative viewpoints. Accordingly, this chapter considers these two aspects in different sections. Section I discusses the multitude of mechanisms possible for electrophilic addition to alkynes and the ways in which these are probed kinetically and stereochemically to prove the existence of the intermediate vinyl cations. Although many similarities in the electrophilic addition to alkenes and alkynes exist, the reaction rates observed under identical conditions are very different. The prospects and limitations in explaining these results in the light of the stability of vinyl cations versus alkyl cations are treated in a subsection. Section II presents the preparation of different products formed via the intermediate vinyl cations in the electrophilic additions to alkynes.

Participation of triple bonds with a developing cationic center during solvolysis reactions is an extended example of electrophilic addition to triple bonds involving vinyl cations and is therefore treated here as well. Reactions of other electrophilic reagents, such as peracids, mercuric salts, and electron-deficient nitrogen, proceeding through a noncationic transition state are beyond the scope of this chapter.

## I. MECHANISMS OF ELECTROPHILIC ADDITIONS TO ALKYNES

### A. Addition of Protic Electrophiles

#### 1. Water

Under normal conditions simple alkynes do not react with water. In 1875 Fittig and Schrohe [2] found acetone in a reaction mixture obtained by pouring a concentrated sulfuric acid solution saturated with propyne into a large amount of cold water. This was the first definite report on the hydration of alkynes. Later Kutscheroff [3] generalized the addition of water to triple bonds in the presence of mercuric ions as a catalyst. This useful method is still successfully employed in many syntheses [4].

The addition of water to alkynes can also be carried out by the use of an acid catalyst. The mechanism of this reaction has only recently been clearly established. In analogy with alkenes, this reaction is general acid-catalyzed and involves a proton transfer to the triple bond as the rate-determining step. As will be shown later, kinetic studies involving activation parameters, kinetic isotope effects, acidity function dependence, isotopic labeling, substituent effects, etc., have established that the acid-catalyzed hydration of alkynes takes place via the $A-S_E2$ mechanism. This mechanism includes neither a preequilibrium nor a $\pi$ complex as an intermediate [5].

a. *Alkynyl Ethers, Thio Ethers, and Amines.*   The mechanism of acid-catalyzed hydration of ethynyl alkyl ethers **1** was first studied by Jacobs and Searles [6]. The hydration proceeds easily and quantitatively simply by shaking with dilute mineral acids to give esters as the main products. The rate of hydration of **1** in aqueous ethanol at 25° is first order with respect to both the alkynyl ether and the hydronium ion. The vinyl- and phenyl-substituted

$$HC\equiv COR + H_2O \xrightarrow{\ H^+\ } MeCOR$$
$$\underset{O}{\overset{\parallel}{}}$$

**1a** R = Et
**1b** R = $n$—Bu
**1c** R = CH=CH$_2$
**1d** R = Ph

alkynyl ethers **1c** and **1d**, respectively, reacted ca. $10^3$ times slower than the alkyl-substituted compounds **1a** and **1b**. It is suggested that the contributing resonance structure **2** explains both the ease of the acid-catalyzed protonation and the direction (regiospecificity) of addition to the alkynyl ethers. Hence,

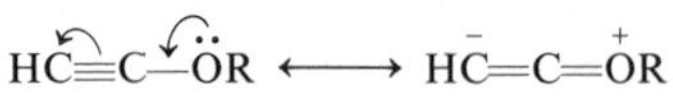

$$HC\equiv C{-}\overset{..}{O}R \longleftrightarrow H\overset{-}{C}{=}C{=}\overset{+}{O}R$$

**2**

the decreased contribution of **2** when R = vinyl or phenyl accounts for the low reactivity of **1c** and **1d**. The mechanism given in Scheme 3.1, in which the first step of the reaction is rate determining, was suggested [6].

$$RC\equiv CXR_1 + H_3O^+ \xrightarrow{\ slow\ } RHC{=}\overset{+}{C}XR_1 + H_2O \qquad (1)$$

$$\mathbf{3},\ X = O, S \qquad\qquad\qquad \mathbf{9}$$

$$RHC{=}\overset{+}{C}XR_1 + H_2O \xrightarrow{\ fast\ } RHC{=}\underset{\displaystyle CXR_1}{\overset{\displaystyle{}^+OH_2}{\mid}} \qquad (2)$$

$$RHC{=}\underset{\displaystyle CXR_1}{\overset{\displaystyle{}^+OH_2}{\mid}} + H_2O \rightleftarrows RHC{=}\underset{\displaystyle CXR_1}{\overset{\displaystyle OH}{\mid}} + H_3O^+$$

$$RHC{=}\underset{\displaystyle CXR_1}{\overset{\displaystyle OH}{\mid}} \rightleftarrows RCH_2\overset{O}{\overset{\parallel}{C}}XR_1$$

Scheme 3.1

Drenth and co-workers conducted much more detailed mechanistic studies on the addition to many alkynyl ethers and thio ethers. Compounds **3**, which were substituted both at the alkynyl carbon atom and at the hetero-atom, yielded the corresponding esters or thio esters as the main product of hydration. The results are summarized in Table 3.1.

**TABLE 3.1**

**Acid-Catalyzed Hydration of Alkynyl Ethers, Thio Ethers, and Amines in Aqueous Buffers at 25°**

| Compound | Main product | $\rho^*$ | $k_{H_2O}/k_{D_2O}$ | Reference |
|---|---|---|---|---|
| HC≡CSR (**4**, R = Me, Et, i-Pr, t-Bu, CH=CH₂) | MeCSR $\overset{\parallel}{O}$ | −0.73 | 2.13[a] | [7–9] |
| RC≡CSEt (**5**, R = Me, Et, n-Pr, i-Pr, t-Bu, D) | RCH₂CSEt $\overset{\parallel}{O}$ | −3.76 | 1.9[b] | [8, 9, 10] |
| HC≡COR (**6**, R = Me, Et, n-Pr, i-Pr, t-Bu) | MeCOR $\overset{\parallel}{O}$ | −7 | | [11, 12] |
| RC≡COEt (**7**, R = Me, Et, n-Bu, i-Pr) | RCH₂COEt $\overset{\parallel}{O}$ | −2.57 | | [14] |
| cis-MeC≡COCH=CHMe (**8**) | MeCH₂COCH=CHMe $\overset{\parallel}{O}$ | | 1.7 | [13] |

[a] For R = Et.

[b] For R = Me₂C— $\overset{\mid}{OH}$

$$RC≡CXR_1 + H_2O \xrightarrow{H^+} RCH_2COXR_1$$

**3**

$$X = O, S$$

The rates of hydration of **3** in aqueous buffers were first order with respect to the alkyne and the hydronium ion. The rates in aqueous acetic acid–sodium acetate buffers depend on the concentration of undissociated acetic acid, indicating that the hydration is general acid-catalyzed and that the mechanism involves a slow proton transfer. This is corroborated by the high value of the solvent isotope effect, $k_{H_2O}/k_{D_2O}$ (Table 3.1), which rules out a preequilibrium step for the hydration. In the case of **4** (R = Et) no deuterium was incorporated at the alkynyl position when this compound was isolated after partial hydration in D₂O [9], in line with a rate-determining protonation of the triple bond. The A–S$_E$2 mechanism of Scheme 3.1, which involves the intermediate vinyl cation **9**, fits the data.

Further support for this mechanism is given by the large negative Taft's $\rho^*$ values (Table 3.1), which show the development of a substantial positive charge on the carbon substituted by the R group in the transition state. The negative $\rho^*$ values for a change of the substituent on the heteroatom are discussed below.

The rates of hydration of tertiary propargyl alcohols **10** were also studied by Drenth *et al.* These were found to react 10 to $10^3$ times faster than calculated on the basis of the Taft's equation [10, 14]. These reactions were also subject to general acid catalysis, indicating that the hydration of **10** also proceeds via a vinyl cation. The discrepancy of the calculated and the observed rates for the reaction may be due to anchimeric assistance by the hydroxyl group in the transition state for the protonation as depicted in **11**.

$$R_2CC{\equiv}CXEt \xrightarrow{\ H^+\ } R_2CCH{=}\overset{+}{C}XEt \xrightarrow[\ 2.\ -H_2O\ ]{1.\ OH^-} R_2C{=}CHCXEt$$

with OH on first carbon, OH on second structure, and O (double bond) on product

**10**

$$\left[ R_2C\underset{CH}{\overset{OH}{\diagup\diagdown}} CXEt \right]^+$$

**11**

In the hydration reaction the addition of water can occur either simultaneously with the proton transfer or in a subsequent step, which could be rate determining (A-2 process). The latter possibility is ruled out by the following data. If step 2 of Scheme 3.1 were rate determining, a bulkier substituent $R_1$ would hinder it and slow the hydration. However, *tert*-butylthioethyne (**4**, $R = t$-Bu) underwent hydration faster than ethylthioethyne (**4**, $R = Et$) under the same conditions, indicating that a water molecule is not involved in the rate-determining step [9]. The generally good Taft correlation for alkyl substituents R and $R_1$ in **3** shows that the steric influence of these substituents on the hydration reaction is negligible.

The negative $\Delta S^{\ddagger}$ values, which are in the range of $-1$ to $-6$ eu, were considered to be in agreement with this mechanism [9, 13]. However, it has been argued that $\Delta S^{\ddagger}$ values are not very useful [15] and should be treated with caution in the interpretation of reaction mechanisms [16].

It is interesting that the alkynyl vinyl ether **8** is protonated preferably at the triple bond and not at the double bond, as seen from the product obtained (Table 3.1).

A rate-determining proton transfer is supported by the acid dependency. The Grunwald plot [17] of (log $k_1$ − log $f_{H^+}$) against $Y_0$ was linear for compounds **7a** and **12** [9]. The rate of hydration of vinylthioethyne **4a** could be followed conveniently in perchloric acid solutions at 25°, and the plot of log $k_1$ against $-H_0$ was linear with a slope of 1.07 [9]. Although this Zucker–Hammett criterion [18] has been applied mainly to acid-catalyzed reactions involving a rapid preequilibrium, Long and Paul [5] reasoned that for slow proton transfers such as the A–$S_E2$ mechanism, $\log k_1$ will be linear in $-H_0$ as long as water is not covalently bound in the rate-determining step.

$$MeC{\equiv}COEt \qquad MeC{\equiv}CSt\text{-Bu} \qquad HC{\equiv}CSCH{=}CH_2$$

$$7a \qquad\qquad\qquad 12 \qquad\qquad\qquad 4a$$

According to Bunnett's treatment of acidity functions [19], a plot of ($\log k_1 + H_0$) against $\log a_{H_2O}$ should be linear with a slope $w$, which is characteristic of the type of the reaction and is related to the question of participation of water in the slow step. In the early work, the fast hydration rates of heterosubstituted alkynes prevented Drenth and co-workers from extending the measurements to high acidities except for the slow-reacting vinylthioethyne **4a**. More recently, Verhelst and Drenth were able to measure the rates of several 1-alkynyl thio ethers at different perchloric acid concentrations at constant ionic strength by using a stopped-flow technique [20]. The plots of $\log k_1$ against the acidity function $-H_0$ were linear with unit slope (Table 3.2). These data and the negative values of Bunnett's $w$ parameters (Table 3.2), compared to the positive values of 1.2 to 3.3 postulated for reactions involving rate-determining addition of water, are clear evidence for the A–$S_E2$ mechanism.

**TABLE 3.2**

**Slopes $z$ of Zucker–Hammett Plots and Bunnett's $w$ Parameters [20]**

| Compound | $z$ | $w$ |
|---|---|---|
| HC≡CSEt | $1.16 \pm 0.04$ | $-2.4 \pm 0.3$ |
| MeC≡CSEt | $1.31 \pm 0.04$ | $-3.4 \pm 0.5$ |
| EtC≡CSEt | $1.16 \pm 0.04$ | $-2.4 \pm 0.5$ |
| MeC≡CSt-Bu | $1.03 \pm 0.03$ | $-0.2 \pm 0.5$ |

The regiospecificity of the addition argues strongly for the intermediacy of vinyl cations in the acid-catalyzed hydration of alkynes. The charge distribution in the ground state of the alkynes is strongly determined by the contribution of the dipolar structure **13b**. This will direct the reaction into a Markownikoff addition in which the product ion and the transition state leading to it are stabilized by the resonance structure **14b**.

$$HC{\equiv}CXR \longleftrightarrow H\overset{-}{C}{=}C{=}\overset{+}{X}R \xrightarrow{H^+} H_2C{=}\overset{+}{C}XR \longleftrightarrow H_2C{=}C{=}\overset{+}{X}R$$

$$13a \qquad\qquad 13b \qquad\qquad 14a \qquad\qquad 14b$$

Among the heterosubstituted alkynes studied, the alkynyl thio ethers underwent hydration $10^3$ to $10^4$ times slower than alkynyl ethers. The relatively poor transmittance of polar effects by sulfur in the transition state, as seen by the less negative $\rho^*$ values for the thio ethers (Table 3.1), is accounted for by the diffuse 3d orbitals of sulfur. This behavior is borne out by the fragmentation reactions observed in the hydration of alkynyl ethers **6** when R = i-Pr or t-Bu [12]. The fragmentation products, acetic acid and an alkene, were derived from the mesomeric vinyl cation structure **15** and accounted for 20% in the case of **6a** (R = i-Pr) and for 60% in the case of **6b** (R = t-Bu) [12]. The stability of the alkyl cation formed plays an obvious role in increasing the extent of fragmentation.

$$HC{\equiv}C{\overset{\overset{\displaystyle Me}{|}}{\underset{\underset{\displaystyle Me}{|}}{O}}CR} \xrightarrow{H^+} CH_2{=}C{\overset{\overset{\displaystyle Me}{|}}{\underset{\underset{\displaystyle Me}{|}}{\overset{+}{O}}CR}} \longleftrightarrow CH_2{=}C{\overset{+}{{-}}}O{-}{\overset{\overset{\displaystyle Me}{|}}{\underset{\underset{\displaystyle Me}{|}}{CR}}} \longrightarrow$$

**6a** R = H                                    **15**
**6b** R = Me

$$R{\overset{\overset{\displaystyle Me}{|}}{\underset{\underset{\displaystyle Me}{|}}{C}}}{+} \quad + \quad CH_2{=}C{=}O$$

$$\downarrow {-H^+} \qquad\qquad \downarrow {H_2O}$$

$$R{\overset{}{C}}{=}CH_2 \qquad\qquad MeC{\overset{\displaystyle O}{\underset{}{\|}}}OH$$
$$\underset{\displaystyle Me}{|}$$

An interesting result of this investigation is the observed higher rate of hydration for the monosubstituted alkyne **5** (R = H) as compared with that of the substituted analogue (**5**, R $\neq$ H) [8]. The observed rate constant for compound **5** (R = H) was about $2 \times 10^3$ times higher than the value calculated on the basis of Taft's equation. Three possible explanations for this discrepancy were offered: (1) a hyperconjugation in the transition state resembling the ion; which is more important when R = H; (2) hyperconjugation in the ground state of the methyl-substituted derivative, which is not present when R = H; and (3) steric hindrance to solvation of the transition state, which is more important for the reaction when R $\neq$ H [8]. The secondary isotope effect in aqueous tartaric acid buffers was inverse, $k_{DC{\equiv}CSEt}/k_{HC{\equiv}CSEt} = 1.03$. Calculations of the isotope effect expected for this reaction gave much higher values (1.54, 1.23, 1.21) on the basis of different models, and an explanation for the discrepancy was not given. It should be noted that, in the generation of vinyl cations by solvolysis, the isotope effect depends on the positive charge dispersal ability of the substituent attached to the vinylic carbon substituted by the leaving group (Chapter 5), and the charge dispersal ability of the sulfur may reduce the isotope effect due to hyperconjugation.

The acid-catalyzed hydration of alkynylamines (ynamines) was recently investigated [21]. The hydration of 1,1-dialkylamino-2-phenylethynes (16) in weakly basic aqueous buffers and in aqueous dioxane was found to be general acid-catalyzed, with a rate-determining proton addition to the triple bond [21]. The products were the amides, indicating that the proton attacks the phenyl-substituted carbon. Products derived from the alternative mode of addition, which generates the $\alpha$-aryl-stabilized vinyl cation 18, were not observed. Consequently, the stabilizing ability of the lone-pair electrons of the nitrogen in 17 exceeds the stabilizing effect of the phenyl group in the ion 18. A high solvent deuterium isotope effect of $k_{H_2O}/k_{D_2O} = 4$ was observed for the hydration of ynamines. Analysis of the effect of alkyl substituents R on nitrogen by the Taft relation gave a $\rho^*$ value of $-2.2$. Therefore, an $A-S_E2$ mechanism with rate-determining proton transfer to the triple bond to give the vinyl cation 17 as an intermediate is reasonable to assume here (see Scheme 3.1).

$$PhC\equiv CNR_2 + H_2O \xrightarrow{H^+} PhCH=\overset{+}{C}NR_2 \xrightarrow{^-OH} PhCH_2\underset{\underset{O}{\|}}{C}NHR_2$$

$$\quad\; 16 \qquad\qquad\qquad\quad 17 \qquad\qquad\quad$$

$$Ph\overset{+}{C}=CHNR_2$$

$$18$$

When the addition of water to ynamines 16 was carried out in acidic acetic acid–sodium acetate buffer solutions, the reaction proceeded approximately $10^4$ times slower than expected from a Brönsted extrapolation, indicating that the mechanism is different from that in a basic medium. From careful kinetic studies the authors suggested the mechanism given in Scheme 3.2. Steps (a) and (b) represent the formation of a complex between acetic acid and the ynamine, which then is protonated at the nitrogen atom. The complex formation was indicated by the red shift observed in the ultraviolet absorption maxima of ynamines when acetic acid was present. Reactions (c) and (d) proceed by the transfer of the proton from nitrogen to carbon in a reaction catalyzed by acetic acid or hydroxyl ions with the slow formation of vinyl cations. A concerted process in which proton removal and donation occur in one step involving the complexes 19 and 19a with acetic acid has been proposed. In addition, pathways (e) and (f) may take place.

A comparison of the data for the $H_2PO_4^-$-catalyzed hydration reactions reveals that ynamines react $10^6$ times faster than alkynyl ethers. The reason

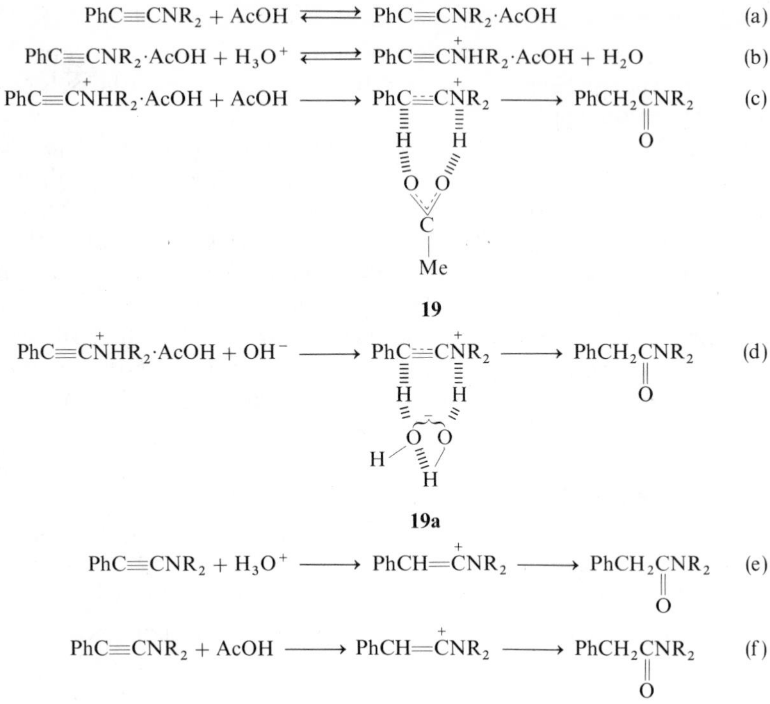

$$PhC\equiv CNR_2 + AcOH \;\rightleftharpoons\; PhC\equiv CNR_2 \cdot AcOH \qquad\qquad (a)$$

$$PhC\equiv CNR_2 \cdot AcOH + H_3O^+ \;\rightleftharpoons\; PhC\equiv \overset{+}{C}NHR_2 \cdot AcOH + H_2O \qquad (b)$$

$$PhC\equiv \overset{+}{C}NHR_2 \cdot AcOH + AcOH \longrightarrow PhC\equiv CNR_2 \longrightarrow PhCH_2CNR_2 \qquad (c)$$

**19**

$$PhC\equiv \overset{+}{C}NHR_2 \cdot AcOH + OH^- \longrightarrow PhC\equiv CNR_2 \longrightarrow PhCH_2CNR_2 \qquad (d)$$

**19a**

$$PhC\equiv CNR_2 + H_3O^+ \longrightarrow PhCH=\overset{+}{C}NR_2 \longrightarrow PhCH_2CNR_2 \qquad (e)$$

$$PhC\equiv CNR_2 + AcOH \longrightarrow PhCH=\overset{+}{C}NR_2 \longrightarrow PhCH_2CNR_2 \qquad (f)$$

Scheme 3.2

might be a better stabilization of the vinyl cation **14** (**X** = *N*) because of the decreased electronegativity of nitrogen relative to oxygen [21]. Therefore, the reactivity among heterosubstituted alkynes for the hydration follows the order alkynylamines ≫ alkynyl ethers ≫ alkynyl thio ethers.

   b. *Aryl Alkynes.*   Another class of alkynes in which the derived vinyl cations are stabilized by the overlap of the vacant *p* orbital with the occupied aromatic π orbitals are the aryl-substituted alkynes (see also Chapter 2). Although the reaction of phenylacetylene and ethyl phenylpropiolate with cold concentrated sulfuric acid followed by water to give acetophenone and benzoylacetic acid has been known [22, 23] since the 1880's, the mechanistic aspects have been studied only lately. These have been very useful in establishing the intermediacy of vinyl cations in the acid-catalyzed hydration of aryl alkynes, which generate α-aryl-substituted vinyl cations by obeying the Markownikoff's addition rule.

$$ArC\equiv CH \xrightarrow{\;H^+\;} Ar\overset{+}{C}=CH_2$$

**TABLE 3.3**

**Acid-Catalyzed Hydration of Aryl Alkynes**

| Compound | Reaction medium | Products | Temp (°C) | $\rho$ | References |
|---|---|---|---|---|---|
| $XC_6H_4C{\equiv}CGeEt_3{}^a$ **20** | Aqueous MeOH–HClO$_4$ | $XC_6H_4C{\equiv}CH + Et_3GeOH$ | 29 | −3.3 | [24] |
| $XC_6H_4C{\equiv}CH^a$ **21** | Aqueous H$_2$SO$_4$–HOAc | $XC_6H_4\overset{\displaystyle O}{\overset{\|}{C}}Me$ | 50.2 | −4.3 | [25] |
| $XC_6H_4C{\equiv}CH^b$ **21** | Aqueous H$_2$SO$_4$ | $XC_6H_4\overset{\displaystyle O}{\overset{\|}{C}}Me$ | 25 | −3.84 | [27] |
| $PhC{\equiv}CMe$ **22** | Aqueous H$_2$SO$_4$ | $Ph\overset{\displaystyle O}{\overset{\|}{C}}CH_2Me$ | 25 | | [28] |
| $XC_6H_4C{\equiv}C\overset{\displaystyle O}{\overset{\|}{C}}OH^b$ **23** | Aqueous H$_2$SO$_4$ | $CO_2 + Me\overset{\displaystyle O}{\overset{\|}{C}}C_6H_4X$ | 25 | −4.77 | [29] |
| $XC_6H_4C{\equiv}C\overset{\displaystyle O}{\overset{\|}{C}}Ph^b$ **24** | Aqueous H$_2$SO$_4$ | $XC_6H_4CH_2\overset{\displaystyle O}{\overset{\|}{C}}Ph$ | 25 | −4.2 | [30] |

$^a$ See Table 3.4 for X.

$^b$ X = p-OMe, p-Me, p-Cl.

Early work of Eaborn et al. [24] on the acid cleavage of phenylethynyl-triethylgermanium compounds 20 led to more detailed studies of the mechanism of hydration of aryl alkynes. These authors analyzed the rates of acid cleavage of several meta- and para-substituted phenylethynyltriethyl-germanes in aqueous methanolic perchloric acid by the Yukawa–Tsuno equation [26], in which $\log k$ is plotted against $\rho[\sigma + r(\sigma^+ - \sigma)]$. The fairly negative $\rho$ value of $-3.3$ (Table 3.3) is indicative of a dependence of the hydration on the electron demands of the substituents. A rate-determining proton transfer to the triple bond was proposed for this acid cleavage reaction. The intermediate vinyl cation 25 is attacked by nucleophiles more rapidly at the germanium atom than at the carbon atom to give alkynes as products.

$$XC_6H_4C\equiv CGeEt_3 + YOH \xrightarrow{H^+} XC_6H_4\overset{+}{C}=CHGeEt_3 \longrightarrow XC_6H_4C\equiv CH + Et_3GeOY$$

$$\qquad\quad 20 \qquad\qquad\qquad\qquad\qquad\qquad 25$$

The relative rates of hydration of several meta- and para-substituted phenylacetylenes to the corresponding acetophenones were also investigated by Bott et al. [25]. As the data in Table 3.4 indicate, large rate accelerations were caused by electron-releasing substituents. The very large negative $\rho$ value of $-4.3$ (Table 3.3) obtained by applying the Yukawa–Tsuno equation [26] is indicative of a developing positive charge next to the aryl group in the transition state, i.e., proton transfer to the triple bond. The

**TABLE 3.4**

Relative Rates of Hydration of
$XC_6H_4C\equiv CH$ in Acetic Acid–Water–
Sulfuric Acid at 50.2° [25]

| X | $k_{rel}$ |
|---|---|
| $m$-CF$_3$ | 0.0047 |
| $m$-Br | 0.018 |
| $m$-Cl | 0.021 |
| $p$-Br | 0.22 |
| $p$-I | 0.24 |
| $p$-Cl | 0.28 |
| $m$-MeO | 0.52 |
| H | 1.00 |
| $m$-Me | 1.92 |
| $p$-$t$-Bu | 10.5 |
| $p$-Me | 15.6 |
| $p$-MeO | 950 |

magnitude of the values is remniscent of the $\rho^+$ values obtained in the solvolysis of $\alpha$-arylvinyl derivatives (Table 6.7). A preequilibrium protonation is ruled out since the tritium label in $PhC{\equiv}CT$ was not lost after partial hydration, and a mechanism analogous to Scheme 3.1 (via **26**) is applicable.

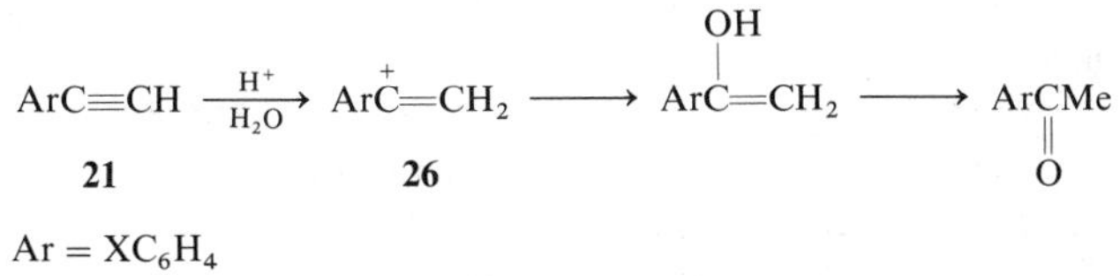

$$ArC{\equiv}CH \xrightarrow[H_2O]{H^+} Ar\overset{+}{C}{=}CH_2 \longrightarrow Ar\overset{\overset{\displaystyle OH}{|}}{C}{=}CH_2 \longrightarrow Ar\overset{\overset{\displaystyle O}{\|}}{C}Me$$

$$\quad\quad\;\, \textbf{21} \quad\quad\quad\quad\quad\;\; \textbf{26}$$

$Ar = XC_6H_4$

The smaller negative $\rho$ value found for phenylethynyltriethylgermanes **20** relative to alkynyl compounds **21** (Table 3.3) is possibly due to the attenuation of the influence of the aryl group in **25** by electron release from the triethylgermanium group.

Comprehensive mechanistic studies on the hydration of phenylacetylene and substituted phenylacetylenes **21** were conducted by Noyce *et al.* The general acid-catalyzed nature of the hydration was established by following the rate of hydration of *p*-methoxyphenylacetylene (**21**, X = *p*-OMe) in formic acid–formate buffers. The reaction gave a negative $\rho$ value of $-3.84$ (Table 3.3) [27]. A linear correlation to the acidity function was noticed for phenylacetylenes **21** when $\log k_{obsd}$ was plotted against $-H_0$ with a slope close to 1.2.

**TABLE 3.5**

**Solvent Deuterium Isotope Effects for the Hydration of $XC_6H_4 C{\equiv}CH$ (21) at 25° [28]**

| X | $-H_0$ range | $k_{H_2SO_4}/k_{D_2SO_4}$[a] | $k_{H_2SO_4}/k_{D_2SO_4}$[b] |
|---|---|---|---|
| H | 2.34–3.24 | 2.67–3.32 | 2.46 |
| Me | 0.85–2.12 | 2.39–2.74 | 2.7 |
| MeO | $-0.73$–0.6 | 2.67–3.18 | 3.82 |
| Cl | 1.91–3.21 | 2.97–3.1 | 2.98 |

[a] Range of values observed over the acidity range.
[b] Isotope effect at $H_0 = -2.0$.

The rates of hydration of **21** were also measured in deuterosulfuric acid [28] and were found to be smaller than in sulfuric acid. The solvent deuterium isotope effects (Table 3.5) calculated from the rate data are fairly high, and this is in accord with a rate-limiting proton transfer to the triple bond for the hydration reaction. As has been discussed for the hydration of alkynyl ethers and thio ethers, these data agree with an A-$S_E2$ mechanism for the hydration of phenylacetylenes **21** with vinyl cation **26** as intermediate.

The presence of electron-attracting substituents on the alkynyl carbon atom did not divert the course of the mechanism in the acid-catalyzed hydration of arylpropiolic acids and arylbenzoyl alkynes [29, 30]. Rate-determining protonation to give the vinyl cations **27** and **28** rather than 1,4 addition of water was found for the hydration of **23** and **24**, respectively, in aqueous sulfuric acid. The rates of hydration of both compounds gave a linear correlation when plotted against $-H_0$ with a slope of near unity. The $\rho$ values were highly negative (Table 3.3). This fact coupled with the high solvent deuterium isotope effect observed ($k_{H_2SO_4}/k_{D_2SO_4}$ = 2 to 5 at $H_0 = -3.38$) and the linear correlation of rate against $-H_0$ is consistent with an A-$S_E2$ mechanism. Although arylpropiolic acids do not differ in their mechanism of hydration from their unsaturated analogues, arylbenzoyl alkynes differ markedly in their behavior from $\alpha,\beta$-unsaturated ketones, where a 1,4 addition of water with protonation of oxygen occurs.

$$ArC{\equiv}CCOH \xrightarrow{H^+} Ar\overset{+}{C}{=}CHCOH \xrightarrow{H_2O} ArCMe + CO_2$$
$$\quad\quad\|\quad\quad\quad\quad\quad\quad\|\quad\quad\quad\quad\quad\quad\|$$
$$\quad\quad O\quad\quad\quad\quad\quad\quad O\quad\quad\quad\quad\quad\quad O$$

**23**                         **27**

$$ArC{\equiv}CCPh \xrightarrow{H^+} Ar\overset{+}{C}{=}CHCPh \xrightarrow{H_2O} ArCCH_2CPh$$
$$\quad\quad\|\quad\quad\quad\quad\quad\quad\|\quad\quad\quad\quad\quad\|\quad\|$$
$$\quad\quad O\quad\quad\quad\quad\quad\quad O\quad\quad\quad\quad\quad O\quad O$$

**24**                         **28**

Noyce and Schiavelli [28] found that unsubstituted phenylacetylene reacted 30 times faster than the methyl compound 1-phenylpropyne (**22**), analogous to the alkynyl thio ethers (see also p. 30). Also, a secondary kinetic isotope effect $k_{HC{\equiv}CPh}/k_{DC{\equiv}CPh}$ of 1.1 was observed. The reason for this effect is not very clear; however, hyperconjugative overlap of the C–H bond with the developing vacant $p$ orbital of the cation in the transition state can play a role. The hyperconjugation is more effective for the unsubstituted phenylacetylene, the transition state of which contains two hydrogen atoms, for effective stabilization as depicted in **29**. According to Streitwieser's theory, the secondary isotope effect arising from hyperconjugation should show some dependence on the relative amount of positive charge at the atom adjacent to the aryl group [31]. The observed secondary kinetic isotope effect $k_H/k_D$ for 4-ethynyl-$d$-anisole (MeOC$_6$H$_4$C≡CD) is closer to unity (1.07), showing that the importance of the hyperconjugative effect decreases as electron-releasing substituents are placed in the ring [28].

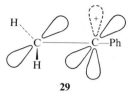

**29**

The formation of vinyl cations as the rate-determining step in the acid-catalyzed hydration of alkynes has thus been established by several mechanistic parameters. Mention must also be made of the Brönsted α values of 0.5 and 0.74 found for the hydration of alkynyl ethers, thio ethers, and amines [21]. The values mentioned above indicate that proton transfer from the medium is more than 50% complete in the transition state of the rate-determining step [32].

In most cases, the intermediate vinyl cations are attacked by water to give enols, which convert to the keto form. Therefore, stereochemical investiga-

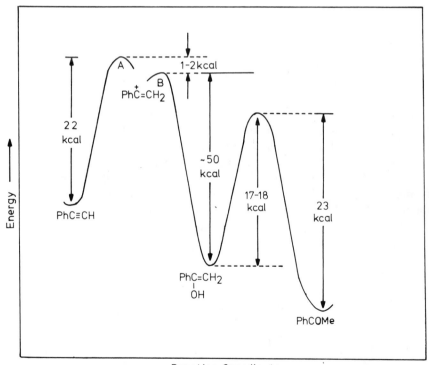

Reaction Coordinate

**Fig. 3.1.** Energy–reaction coordinate diagram for the acid-catalyzed hydration of phenylacetylene. The ordinate is not to scale [27].

tions were not carried out. The possibility that vinyl cations (26) formed can lose a proton to give back alkynes was observed by Noyce and Schiavelli [27]. Thus when phenylacetylene was hydrated in 40% tritium labeled sulfuric acid (up to 50% completion of the reaction), 0.6% exchange was noted in the recovered unreacted material.

A good representation of the energy–reaction coordinate diagram for the acid-catalyzed hydration of phenylacetylene is given in Fig. 3.1. Constructed by Noyce and Schiavelli [27], it takes into consideration the heats of combustion of these compounds and the rate of enolization of acetophenone [18].

c. *Tertiary Propargyl Alcohols.* Tertiary propargyl alcohols (10) used for the study of hydration by Drenth *et al.* [14] (see p. 28) contained a heteroatom (O or S) on the acetylenic carbon atom. The influence of this substituent is not available for tertiary propargyl alcohols containing an alkyl or aryl substituent on the acetylenic carbon atom. The acid-catalyzed reaction of the latter compounds is known as the Meyer–Schuster and Rupe rearrangement, in which mesomeric cations 30 and 31 are involved as intermediates [33]. Here the vinyl cation known as the allenyl cation (see Chapter 6) is stabilized by resonance with the neighboring double bond to form the alkynyl cations.

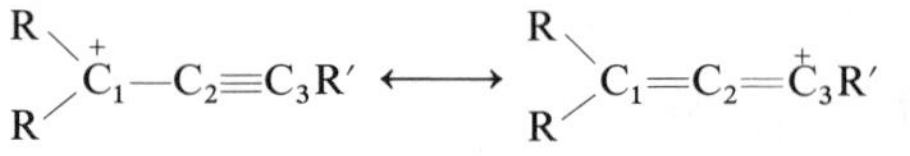

$$\begin{array}{c} R \\ \diagdown \overset{+}{C_1}{-}C_2{\equiv}C_3R' \end{array} \longleftrightarrow \begin{array}{c} R \\ \diagdown C_1{=}C_2{=}\overset{+}{C_3}R' \end{array}$$

$$\begin{array}{c} R \diagup \end{array} \qquad\qquad \begin{array}{c} R \diagup \end{array}$$

30                              31

Acid-catalyzed reactions of the α-tertiary alcohols 32 and 33 have been carried out in order to elucidate the mechanism of this reaction [34]. Treatment of the acetylenic alcohols 32 and 33 with 5% $H_2SO_4$/50% HOAc at 80° gave the α,β-unsaturated ketones 35 and 36 in 70 and 95% yield, respectively. The cyclohexenyl derivative 33 reacted via the Rupe rearrangement as indicated by the formation of 1-(1-propynyl)cyclohexene (34), which could be detected even at elevated temperatures. From compound 32

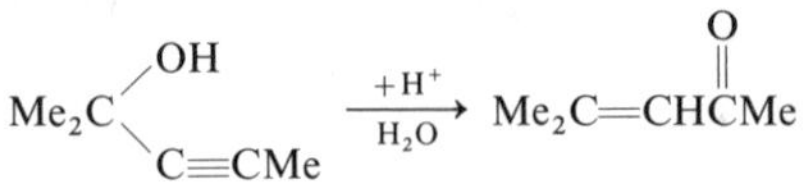

32                              35

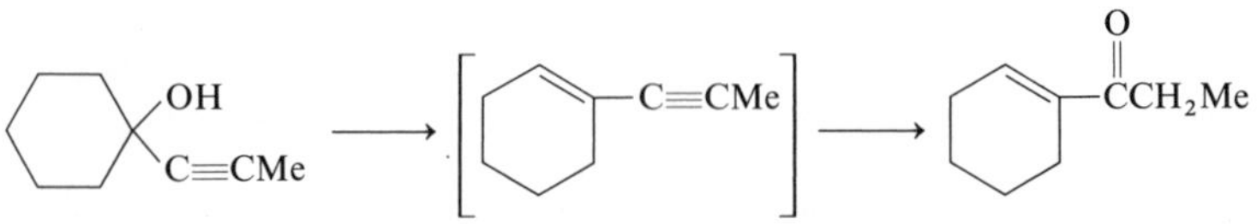

33              34              36

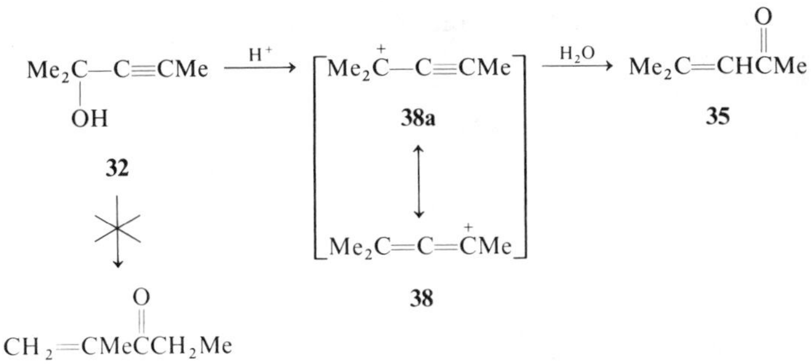

$$
\underset{\substack{\big| \\ \text{OH} \\[2pt] \textbf{32}}}{\text{Me}_2\text{C}}\text{—C}\equiv\text{CMe} \xrightarrow{\ \text{H}^+\ } \left[ \overset{+}{\text{Me}_2\text{C}}\text{—C}\equiv\text{CMe} \right]_{\textbf{38a}} \xrightarrow{\ \text{H}_2\text{O}\ } \text{Me}_2\text{C}=\text{CHC}\overset{\text{O}}{\overset{\|}{\text{C}}}\text{Me}
$$

**38a** ⇅ **38**

$$\text{Me}_2\text{C}=\text{C}=\overset{+}{\text{C}}\text{Me}$$

**35**

$$\overset{\text{O}}{\overset{\|}{\text{CH}_2=\text{CMeCCH}_2\text{Me}}}$$

**37**

neither the corresponding intermediate hydrocarbon nor the ketone **37** could be found, and hence it reacts via a 1,3-hydroxy shift (Meyer–Schuster rearrangement) to give the ketone **35**. The results obtained show that tertiary propargyl alcohols can rearrange according to two different mechanisms depending on the structure of the substrate. The fact that the dimethyl-hydroxy compound **32** did not react in the same manner as the cyclohexanol derivative **33** may be due to the low tendency of the carbenium ion **38a** to expel a proton to form a terminal double bond [34].

Edens *et al.* [35] studied in detail the mechanism of Meyer–Schuster rearrangement of tertiary propargyl alcohols. The rates of hydration of eight triaryl and diaryl tertiary propargyl alcohols (**39a–h**) were investigated in 40% dioxane/60% aqueous sulfuric acid. The alcohols **39–h** were found to give the rearrangement products, $\alpha,\beta$-unsaturated ketones **40a–h**, exclusively and quantitatively. The hydration was not found to be general acid-catalyzed as in the case of the thio ethers (**10**) observed by Drenth *et al.* [14] (see p. 28). Any mechanism involving a proton transfer in the rate-determining step was ruled out by the inverse solvent isotope effects ($k_{\text{H}_2\text{O}}/k_{\text{D}_2\text{O}} = <1$) observed for **39a** and **39g**. The other mechanism (Scheme 3.3) involves a rapid pre-equilibrium step followed by slow rearrangement of the conjugate acid.

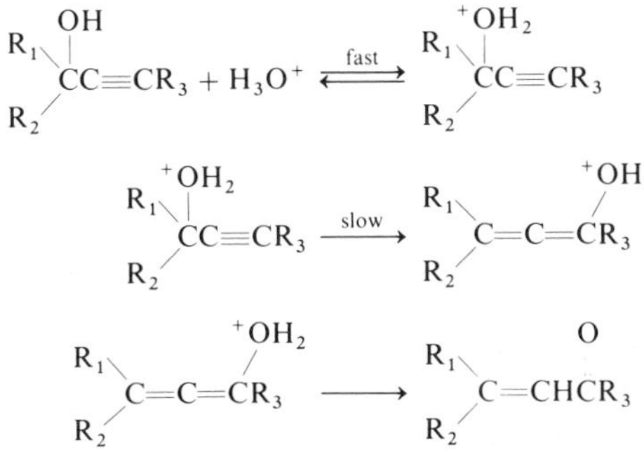

$$
\underset{R_2}{\overset{R_1}{\phantom{.}}}\!\!\!\!\!\underset{\phantom{.}}{\overset{\text{OH}}{\text{CC}}}\equiv\text{CR}_3 + \text{H}_3\text{O}^+ \underset{\ }{\overset{\text{fast}}{\rightleftharpoons}} \underset{R_2}{\overset{R_1}{\phantom{.}}}\!\!\!\!\!\underset{\phantom{.}}{\overset{^+\text{OH}_2}{\text{CC}}}\equiv\text{CR}_3
$$

$$
\underset{R_2}{\overset{R_1}{\phantom{.}}}\!\!\!\!\!\underset{\phantom{.}}{\overset{^+\text{OH}_2}{\text{CC}}}\equiv\text{CR}_3 \xrightarrow{\ \text{slow}\ } \underset{R_2}{\overset{R_1}{\phantom{.}}}\!\!\!\!\!\underset{\phantom{.}}{\overset{^+\text{OH}_2}{\text{C}}}=\text{C}=\text{CR}_3
$$

$$
\underset{R_2}{\overset{R_1}{\phantom{.}}}\!\!\!\!\!\underset{\phantom{.}}{\overset{^+\text{OH}_2}{\text{C}}}=\text{C}=\text{CR}_3 \xrightarrow{\ \ } \underset{R_2}{\overset{R_1}{\phantom{.}}}\!\!\!\!\!\underset{\phantom{.}}{\overset{\text{O}}{\text{C}}}=\text{CHCR}_3
$$

Scheme 3.3

The formation of the ketone from the allenol is considered to take place rapidly.

$$\underset{R_2}{\overset{R_1}{\diagdown}} C \underset{OH}{\overset{C \equiv CR_3}{\diagdown}} \xrightarrow[H_2O]{H^+} \underset{R_2}{\overset{R_1}{\diagdown}} C = CH\overset{O}{\overset{\|}{C}}R_3$$

$$\quad\quad\quad\quad\quad 39 \quad\quad\quad\quad\quad\quad\quad\quad\quad\quad 40$$

**a** $R_1 = R_2 = R_3 = Ph$
**b** $R_1 = R_2 = Ph; R_3 = p\text{-}ClC_6H_4$
**c** $R_1 = R_2 = Ph; R_3 = p\text{-}MeC_6H_4$
**d** $R_1 = R_2 = Ph; R_3 = p\text{-}MeOC_6H_4$
**e** $R_1 = R_3 = Ph; R_2 = p\text{-}ClC_6H_4$
**f** $R_1 = R_3 = Ph; R_2 = p\text{-}MeOC_6H_4$
**g** $R_1 = R_2 = Ph; R_3 = H$
**h** $R_1 = R_2 = Ph; R_3 = D$

From the kinetic data on the hydration of the several substituted propargyl alcohols **39**, values of $-2.3$ and $-1.6$ were calculated for substituent changes at C-1 and C-2 (Scheme 3.3), respectively. Linear plots of $\log k_1$ against $-H_0$ also point to a charge delocalization from the reaction center to the rearrangement terminus. The degree and nature of involvement of water in the transition state are not unequivocally known, and three ion–dipole transition state structures (**41** to **43**) were suggested. The calculated $k_{H_2O}/k_{D_2O}$ values of ca. 0.4 fit any one of these transition states.

$$R_1 \overset{\delta^+}{-} \overset{\overset{\overset{OH_2}{\vdots}}{}}{C} = C = C - R_3 \quad\quad R_1 \overset{\delta^+}{-} \overset{\overset{\overset{H\diagdown\quad\diagup H}{O}}{\vdots}}{C} = C = C \overset{\delta^+}{-} R_3 \quad\quad R_1 - \overset{\overset{\overset{OH_2}{\vdots}}{}}{C} = C \overset{\delta^+}{-} C - R_3$$
$$\overset{|}{R_2} \quad\quad\quad\quad\quad\quad\quad\quad\quad \overset{|}{R_2} \quad\quad\quad\quad\quad\quad\quad\quad R_2 \quad H_2O$$

$$\quad\quad 41 \quad\quad\quad\quad\quad\quad\quad\quad\quad 42 \quad\quad\quad\quad\quad\quad\quad\quad 43$$

From the foregoing data on the acid-catalyzed hydration of hetero-substituted alkynes and the related reaction of arylacetylenes, there is good evidence that a vinyl cation intermediate is formed in a rate-determining protonation step of the A–$S_E2$ mechanism operative for these alkynes.

## 2. Addition of Carboxylic Acids

The stereochemistry of the products of the addition of electrophiles to alkynes is a useful probe, in addition to the kinetic evidence, in specifying the details of the mechanism. In the acid-catalyzed hydration of alkynes, the stereochemistry of the reaction could not be examined due to ketonization of the intermediate enols. In contrast, in the addition of carboxylic acids and other electrophiles, which will be described later, the vinyl esters and other derivatives formed retained their stereochemical identity on isolation.

The addition may be a concerted or a multistep process. In synchronous reaction, the electrophilic (E) and nucleophilic (Nu) parts of the same molecule approach the substrate from one side only. The stereochemical implication of this process is syn addition, as represented below for alkynes. Concerted syn additions to alkenes were suggested to be symmetry-forbidden [36]; most of the additions occur via a multistep mechanism. However,

$$R_1—C\equiv C—R_2 + ENu \longrightarrow \left[ \begin{array}{c} E^{""}Nu \\ R_1—C{=}C—R_2 \end{array} \right]^{\ddagger} \longrightarrow \begin{array}{c} E \qquad\quad Nu \\ C{=}C \\ R_1 \qquad\quad R_2 \end{array}$$

recent theoretical calculations [37] on the electrophilic additions to alkenes predict that the symmetry-forbidden syn addition can take place in the case of polar $2_\pi + 2_\sigma$ cycloadditions. Concerted syn additions have been observed for bicyclic olefins [38], in which anti additions are seriously disfavored for steric reasons. Charge separation in the transition state, seen from the effect of electron-withdrawing substituents on the rate [38], indicates that the process is electrophilic.

When the addition reaction to alkynes proceeds stepwise, the carbenium ion **44** is generated as the intermediate. This can give *syn*- and *anti*-vinylic adducts according to a syn and an anti periplanar arrangement of the entering groups. Equal amounts of syn and anti adducts are usually not expected in the attack of the nucleophile on the cation **44** since both sides of the cation are not equivalent. Electronic, steric, and conformational effects may play a role and can provoke a preferential attack at the vacant p orbital of the vinyl cation **44** from one side to give predominantly the syn or anti adduct. With excess of the electrophilic reagent, secondary reactions such as addition of a second mole of the reagent to the already formed vinyl derivative are possible.

$$R_1—C\equiv C—R_2 + ENu \xrightarrow{\text{slow}} \begin{array}{c} E \\ C{=}\overset{+}{C}—R_2 + Nu^- \\ R_1 \end{array}$$

**44**

$$\begin{array}{c} E \\ \overset{+}{C}{=}C—R_2 + Nu^- \\ R_1 \end{array} \xrightarrow{\text{fast}} \begin{array}{c} E \qquad\quad R_2 \\ C{=}C \\ R_1 \qquad\quad Nu \end{array} + \begin{array}{c} E \qquad\quad Nu \\ C{=}C \\ R_1 \qquad\quad R_2 \end{array}$$

anti adduct            syn adduct

Both concerted and multistep mechanisms are bimolecular in nature. The attachment of the electrophile to the alkynyl triple bond is the rate-determining step in the case of the nonconcerted mechanism described above, whereas the C–E and C–Nu bonds form simultaneously in the transition state of the synchronous process. Using Ingold's terminology, these reactions are denoted as $Ad_E2$ mechanisms. The $Ad_E2$ mechanism does not differ in principle from the $A-S_E2$ mechanism valid for the acid-catalyzed hydration discussed above. However, the term $A-S_E2$ is preferable for addition reactions occurring via rate-determining proton transfer to the substrate in aqueous medium and having no covalent attachment of water in the transition state.

    a. *Acetylenic Ethers.*   The addition of carboxylic acids to 1-ethoxy-1-alkynes (45) proceeds with initial formation of vinyl esters (46), which consequently add an additional mole of carboxylic acid to give ethyl esters and the acid anhydride. The kinetics of the first stage of the reaction of 45 with acetic acid ($R_1$ = Me) to give the vinyl ester 46 were investigated in

$$RC{\equiv}COEt + R_1COOH \longrightarrow RHC{=}C(OEt)OCR_1 \xrightarrow{\;R_1COOH\;}$$

$$\quad\quad\quad\text{45}\quad\quad\quad\quad\quad\quad\quad\quad\quad\quad\quad\quad\overset{\|}{O}$$

$$\text{46}$$

$$H_2RCC(OEt)(OCOR_1)_2 \longrightarrow H_2RCCO_2Et + (R_1CO)_2O$$

nonaqueous solvents [39]. It was deduced from the kinetic data that four carboxylic acid molecules are involved in benzene solution and at least two acid molecules are involved in dioxane and sulfolane solution due to solvation effects. A $\rho^*$ value of $-3.24$ was obtained for the addition reaction in benzene, showing a considerable sensitivity to the electron-releasing power of the substituent R in 45. In benzene and sulfolane, kinetic isotope effects $k_H/k_D$ of 3.1 and 2.8, respectively, were observed for 45 (R = H). These results and the normal linear salt effect found for the reaction between ethoxyethyne (45, R = H) and chloroacetic acid in dioxane are evidence that the carboxylic addition reaction to alkynyl ethers proceeds by an ionic mechanism.

    From these results it was concluded that proton transfer occurs from the carboxylic acid to the triple bond of the substrate in the rate-determining step involving a vinyl cation–ion pair (47). The complex formation with acetic acid is analogous to the addition of water to ynamines under acidic conditions [see step (c) in Scheme 3.2].

$$RC{\equiv}COEt + R_1CO_2H \xrightarrow{\;slow\;} RHC{=}\overset{+}{C}{-}OEt \xrightarrow{\;fast\;} RHC{=}C(OEt)OCR_1$$

47 (Solvated ion pair)

    b. *Alkyl and Aryl Alkynes.*   The reaction of trifluoroacetic acid (TFA) with 3-hexyne was investigated by Peterson *et al.* [40]. The addition of 0.1

mole of 3-hexyne to an excess of refluxing TFA afforded in 89% yield a mixture of equivalent amounts of E- and Z-3-hexen-3-yl trifluoroacetates (49) (50:50) along with a small amount of 3-hexanone and ca. 2% of the trimeric product hexaethylbenzene (50).

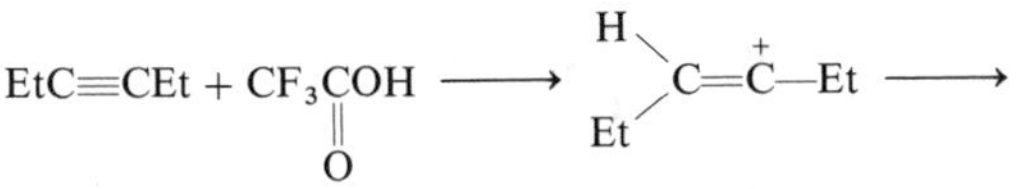

48

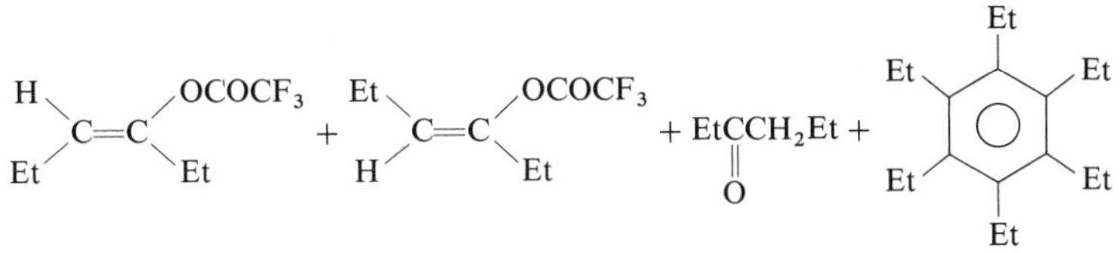

E-49                Z-49                        50

The formation of both syn and anti adducts E- and Z-49 is in line with the stepwise mechanism for this reaction. The above-mentioned reaction appears to be a simple case in which no selectivity was observed, although the nucleophile can react with cation 48 stereoselectively from the least hindered side. With higher concentrations of 3-hexyne (2 moles), the yield of the trimeric product increased up to 30% relative to the vinyl trifluoroacetates 49. The formation of hexaethylbenzene (50) may again involve the vinyl cation 48, reacting with excess 3-hexyne to form the homoallylic vinyl cation 51 and subsequently the trimeric product. 3-Hexanone is formed by a secondary reaction of TFA with the vinyl trifluoroacetates 49.

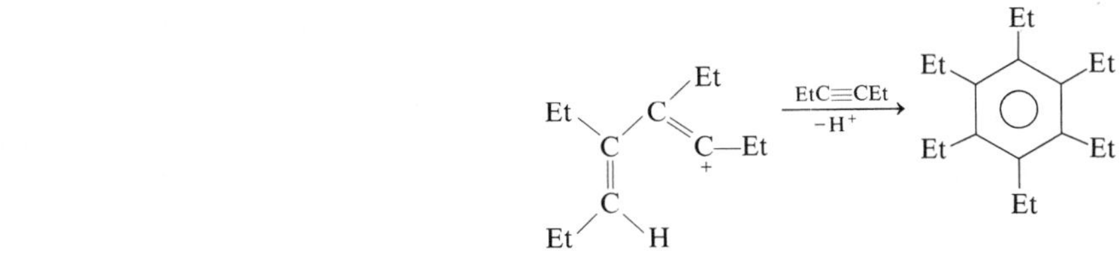

51                              50

The observed nonstereoselective addition of TFA to 3-hexyne raises several questions: Why does the nucleophilic attack on the unsymmetric vinyl cation 48 not take place preferentially from the less hindered side of the β-hydrogen to give the kinetically controlled product E-49? Does any syn/anti equilibration take place during the reaction? Some answers to these questions have been reported.

The high acidity, low nucleophilicity, and high ionizing power makes TFA a suitable medium for the generation of free vinyl cations in which stereospecific interaction with the counterion is minimal and for the study of their steriochemical behavior. In their work on the stereochemical outcome of electrophilic addition to alkynes and solvolyses of vinyl triflates (see Chapter 5), Summerville and Schleyer [41] investigated the reaction of TFA with alkynes 52, 53, and 54. Both anti and syn additions were observed in different proportions, as listed in Table 3.6.

**TABLE 3.6**

**Products of Reaction of Alkynes with Trifluoroacetic Acid[a]**

| Alkyne | Temp, °C | Trifluoroacetate Z isomer (anti), isomer ratio | Trifluoroacetate E isomer (syn), isomer ratio | Z/E ratio |
|---|---|---|---|---|
| MeC≡CMe | 75.4 | 77 | 23 | 3.33 |
|  | 49.8 | 80.5 | 19.5 | 4.13 |
|  | 35.0 | 84 | 16 | 5.11 |
| EtC≡CEt | 75.4 | 48 | 52 | 0.91 |
|  | 60.1 | 49.8 | 50.2 | 0.99 |
| n-PrC≡C—Pr-n | 75.4 | 44 | 56 | 0.77 |
| n-PrC≡CH | Reflux | 50 | 25[b] | 2 |
| n-BuC≡CH | Reflux | 62 | 30[b] | 2.06 |

[a] Data collected from Summerville and Schleyer [41].
[b] Yield of the isomerized products.

Scrutiny of the data in Table 3.6 indicates that 2-butyne (52) gives major amounts of the Z isomer (anti) of vinyl trifluoroacetate (Z-56). The result obtained for 3-hexyne is identical with that of Peterson's value of $Z/E = 1$, whereas a small tendency to form the syn adduct (E isomer) in greater quantity is observed for 4-octyne. Compound Z-56 is indeed the thermodynamically more stable product, but the explanation, thermodynamic instead of kinetic control, was ruled out by the fact that the product ratio Z-56/E-56 of 3.33 (at 75.4°) significantly exceeds the equilibrium ratio Z-56/E-56 of 2.36.

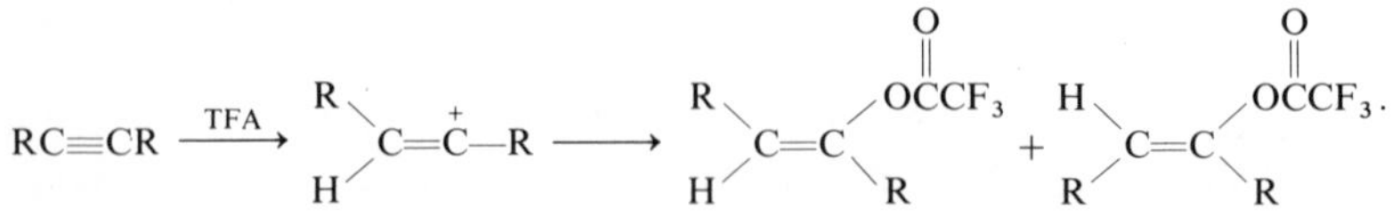

| | | |
|---|---|---|
| 52 R = Me | 55 | Z-56 | E-56 |
| 53 R = Et | | Z-57 | E-57 |
| 54 R = n-Pr | | Z-58 | E-58 |

The product distribution (Table 3.6) from differently substituted alkynes shows that steric effects are possibly involved in determining the product ratios. Thus, the $Z/E$ values at 75.4° for the alkyl groups in the alkynes decrease from 3.33 for methyl to 0.91 for ethyl and 0.77 for $n$-propyl. The main change is found only in going from methyl to ethyl, whereas the small difference in the effective bulk of ethyl and $n$-propyl does not result in a major change in the product ratio.

The high $Z/E$ ratio favoring an anti addition for the symmetric alkyne **52** is intriguing. A syn addition would be expected to predominate on steric grounds, since solvent attack cis to the smaller $\beta$-hydrogen rather than cis to the larger methyl would appear to be more favorable for the intermediate vinyl cation **55** (R = Me). The explanation probably lies in the electronic effects in the transition state. An attraction between the electron-deficient $\beta$-methyl group (due to hyperconjugation) and the attacking nucleophile, as depicted in **59**, was postulated to explain the preference for the $Z$ product.

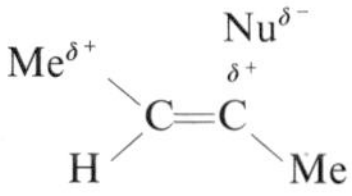

$$
\text{Me}^{\delta+} \diagdown \quad \begin{array}{c} \text{Nu}^{\delta-} \\ {\scriptstyle \delta+} \end{array}
$$
$$
\underset{\text{H}}{\overset{}{\diagup}} \text{C}{=}\text{C} \underset{\text{Me}}{\overset{}{\diagdown}}
$$

**59**

The addition of trifluoroacetic acid to 5-substituted 1-pentynes of type **60** produced a different picture. This is indicated by the observation of rearranged products deriving from a 1,4-X shift (where X = Cl, F, OMe, or OAc; see Scheme 3.4). Kinetic data indicate that the rates are retarded by

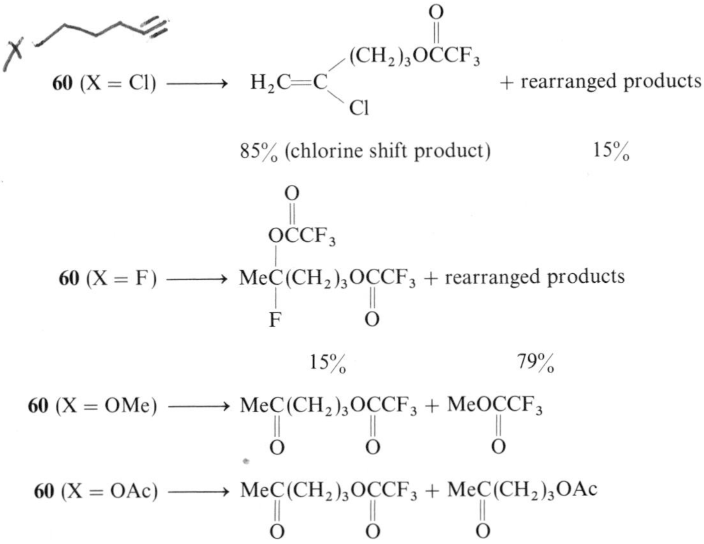

$$ X{\sim}{\sim}{\equiv} $$

$$ \textbf{60 (X = Cl)} \longrightarrow \text{H}_2\text{C}{=}\text{C} \underset{\text{Cl}}{\overset{(\text{CH}_2)_3\text{OCCF}_3}{\diagup}} \quad + \text{ rearranged products} $$

$$ O \atop \| $$ above $(\text{CH}_2)_3\text{OCCF}_3$

85% (chlorine shift product)              15%

$$ \textbf{60 (X = F)} \longrightarrow \text{Me}\overset{\text{OCCF}_3}{\underset{\text{F}}{\text{C}}}(\text{CH}_2)_3\text{OCCF}_3 + \text{ rearranged products} $$

15%                        79%

$$ \textbf{60 (X = OMe)} \longrightarrow \text{MeC}(\text{CH}_2)_3\text{OCCF}_3 + \text{MeOCCF}_3 $$

$$ \textbf{60 (X = OAc)} \longrightarrow \text{MeC}(\text{CH}_2)_3\text{OCCF}_3 + \text{MeC}(\text{CH}_2)_3\text{OAc} $$

Scheme 3.4

electron-withdrawing substituents. The observed extent of this decrease was less than expected if only the inductive effect of the substituents had contributed to the overall rate of addition. Therefore, anchimeric assistance from the 5-X group in the addition of TFA to **60** was suggested [40]. In combination with the studies of the products of the reaction, these data are more consistent with a five-membered ring cation of type **61** than with a linear vinyl cation.

$$HC{\equiv}C(CH_2)_3X$$

**60**                                **61**

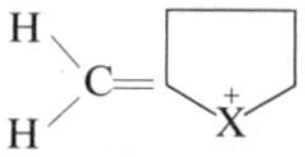

In the case of unsubstituted terminal alkynes, the primary adducts **62** apparently isomerized to give the vinyl trifluoroacetates **63**. The $Z/E$ ratio was about 2, indicating that the thermodynamically more stable $Z$ isomer is the major product in the isomerization process. This is also an indication that the $Z/E$ values obtained by Peterson, Schleyer, and co-workers derive from a kinetically controlled reaction without any isomerization.

$$Me(CH_2)_n C{\equiv}CH \longrightarrow Me(CH_2)_n \overset{+}{C}{=}CH_2 \longrightarrow Me(CH_2)_n C{=}CH_2 \longrightarrow$$

$n = 2, 3$                                                    $OCOCF_3$

**62**

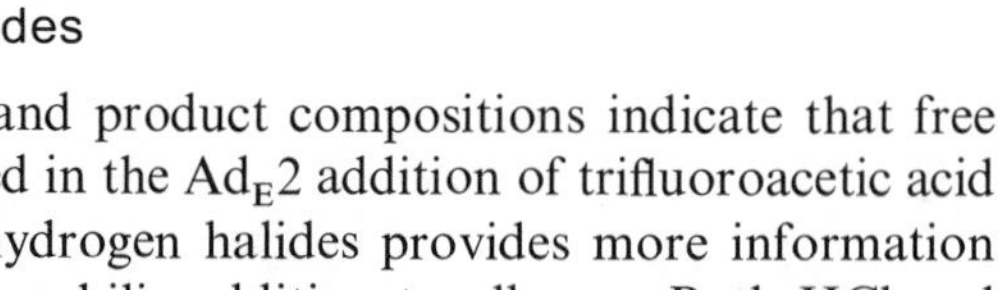

*Z*-**63**                                *E*-**63**

## 3. Addition of Hydrogen Halides

The available kinetic data and product compositions indicate that free linear vinyl cations are involved in the $Ad_E2$ addition of trifluoroacetic acid to alkynes. The addition of hydrogen halides provides more information about the mechanism of electrophilic addition to alkynes. Both HCl and HBr were used for the addition reactions, which were carried out in acetic acid as well as in other common solvents. Analogous to the electrophilic addition to alkenes, both $Ad_E2$ and $Ad_E3$ mechanisms are known, and they are distinguished by kinetic studies and product analyses.

a. *$Ad_E2$ Mechanism.* This mechanism was discussed in Section I,A,2 for the case involving a free vinyl cation. In weakly dissociating solvents, such as acetic acid, the intermediate vinyl cation formed by addition of HCl can be closely associated with the counterion and exist as the ion pair **64**. If ion pairs are indeed involved, a mixture of syn and anti isomers of vinyl chlorides and acetates can be formed. The proportion of syn and anti adducts depends on several factors, which are described below.

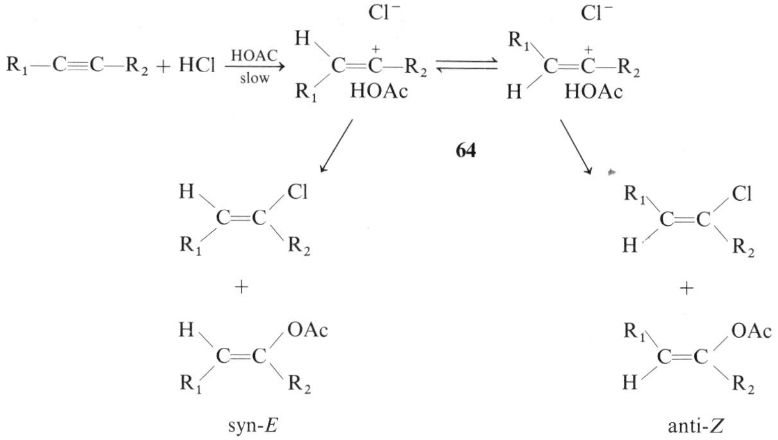

syn-*E*                 anti-*Z*

b. *Ad$_E$3 Mechanism.* This synchronous mechanism is characterized by a simultaneous attack of nucleophile and electrophile from different molecules on the alkynes. It is symbolically denoted as Ad$_E$3 and is classified as termolecular since the transition state is composed of three distinct reactants. The addition of the nucleophile and electrophile can occur either syn via transition state **65** or anti via transition state **66** to give either *syn*- or *anti*-products. This process does not involve a vinyl cation intermediate.

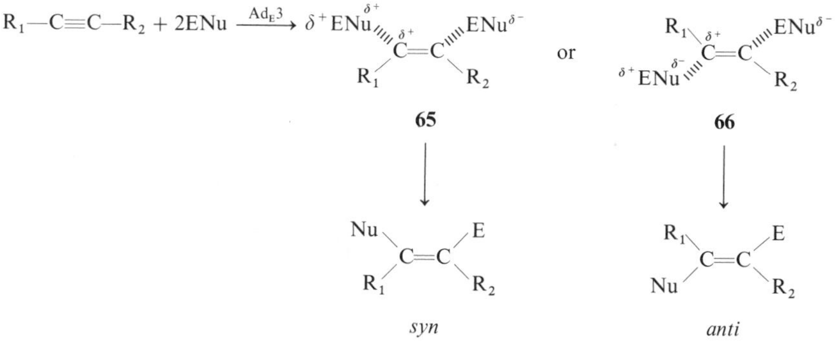

The Ad$_E$3 reactions show third-order kinetics but, if the nucleophile Nu is derived from the original electrophilic reagent ENu, the kinetics may be second order. Similarly, if the nucleophile arises from the solvent, pseudo-second-order kinetics may result. There are several factors affecting these two mechanisms, and there is competition as well between Ad$_E$2 and Ad$_E$3 addition under the same conditions. Thus, the mechanism can be changed from $Ad_E2$ to $Ad_E3$ by adding salts of the acid to the reaction medium.

c. *Alkyl and Aryl Alkynes.* Fahey *et al.* extensively investigated the mechanism of addition of hydrogen chloride in acetic acid to several alkynes. 1-Phenylpropyne added HCl in HOAc to give predominantly the syn adduct, *E*-1-chloro-1-phenylpropene (*E*-**68**) [42], together with small amounts of *Z*-**68** and phenyl ethyl ketone (Scheme 3.5). Control experiments demonstrated that the ketone arose from a secondary reaction of the intermediate vinyl acetates.

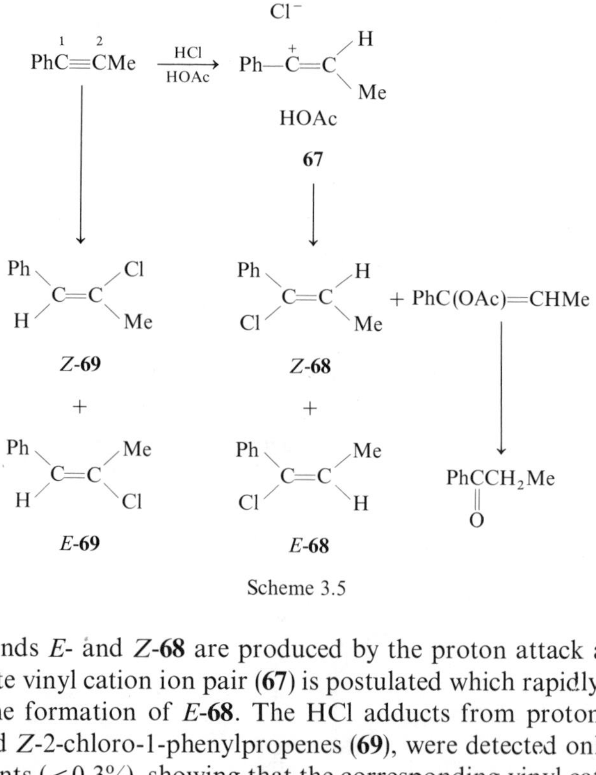

Scheme 3.5

Compounds *E*- and *Z*-**68** are produced by the proton attack at C-2. An intermediate vinyl cation ion pair (**67**) is postulated which rapidly collapses, favoring the formation of *E*-**68**. The HCl adducts from proton attack at C-1, *E*- and *Z*-2-chloro-1-phenylpropenes (**69**), were detected only in negligible amounts ( $< 0.3\%$ ), showing that the corresponding vinyl cation in this case is less stable. When the reaction was carried out in the presence of tetramethylammonium chloride (TMAC), an increase in the rate of reaction and significant amounts of both the anti adducts from proton attack at C-1 and C-2, *Z*-**68** and *Z*-**69**, were observed [43]. These products are therefore derived from a competing $Ad_E3$ process, as shown in Scheme 3.6.

It was calculated that the addition via an $Ad_E2$ mechanism occurs with proton attack at C-2 300 times faster than attack at C-1, i.e., which points to

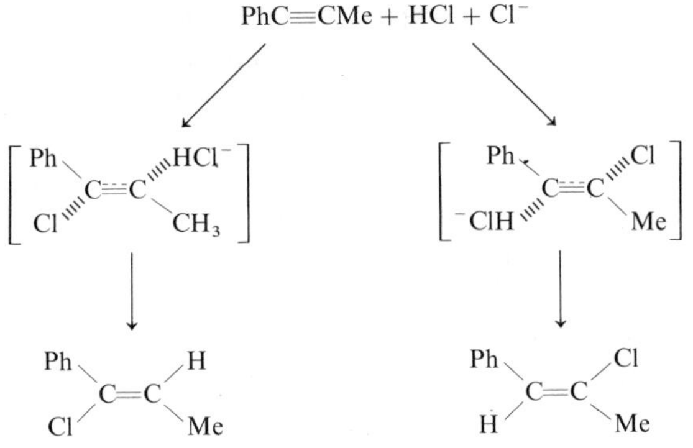

$$PhC{\equiv}CMe + HCl + Cl^{-}$$

Scheme 3.6

the generation of a stabilized vinyl cation ion pair (67). The corresponding ratio for attack in the $Ad_E3$ process is only about 5 (see below). These results with 1-phenylpropyne demonstrate an important difference in the regiospecificity as well as the stereospecificity of the $Ad_E2$ and $Ad_E3$ addition processes. This has potential importance in organic synthesis because one can possibly direct the preferential formation of one particular product only.

The reaction of 3-hexyne with HCl in HOAc yielded the anti adduct Z-3-chloro-3-hexene (Z-70) in fair amounts [44], a small quantity of the syn adduct E-70, and 3-hexanone as the main product (Scheme 3.7). The yield of Z-70 relative to that of 3-hexanone increased with the concentration of HCl. In the presence of TMAC, the yield of anti adduct Z-70 could be raised to 97%, and only a negligible amount of the syn adduct E-70 was found. These data suit an $Ad_E3$ mechanism well.

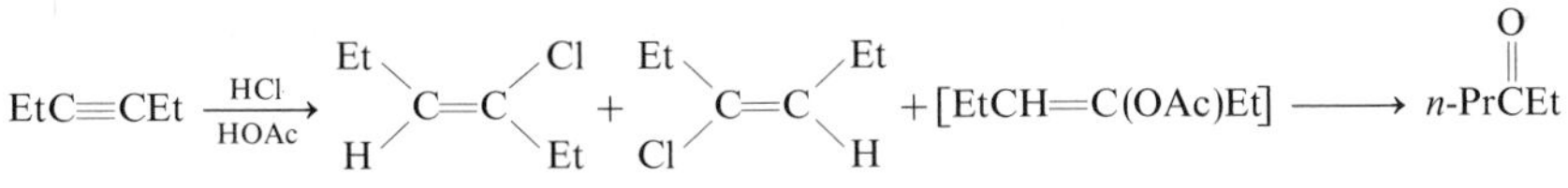

Scheme 3.7

At higher temperatures (50° and 80°), a small increase of the syn adduct E-70 was noted. The formation of syn adduct and its increase with reaction temperature were taken as evidence that a small but significant fraction of the reaction does occur via an $Ad_E2$ mechanism. Hence, both the $Ad_E2$ and the $Ad_E3$ mechanisms are competitive processes for addition of HCl (in HOAc) to 3-hexyne. On comparing the result of HCl (in HOAc) addition to 3-hexyne with that of 1-phenylpropyne, a striking effect of structure in changing the course of reaction can be discerned.

The reaction of 3-hexyne with HCl (in HOAc) differs markedly in mechanism from its reaction with TFA, as discussed in Section I,A,2, where a vinyl cation was shown to be an intermediate. The difference in mechanisms is due to the different nucleophilic and ionizing power of the two solvents employed. The reaction rates and product compositions for the reaction of several alkynes with HCl [43] in acetic acid (at 25°) carried out in the absence and presence of TMAC are summarized in Table 3.7. Addition of HCl in HOAc to phenylacetylene gave $\alpha$-chlorostyrene and acetophenone in the ratio 12:1. Secondary reactions of the initial products were ruled out by showing that the product composition did not vary with time. Addition of 0.2 $M$ TMAC increased the rate of reaction by a factor of only 2 and did not increase the fraction of the $\alpha$-chlorostyrene formed (Table 3.7). The effect of added salt on the rate results from a salt effect on the rate of formation of the tight ion pair. These observations are in agreement with an $Ad_E2$ mechanism for the reaction of phenylacetylene with HCl in acetic acid. This is also probable because the intermediate vinyl cation is stabilized by an aryl group. The formation of less chloride and more ketone in the absence of TMAC and the salt effect observed in the presence of TMAC (Table 3.7) for 1-hexyne and 3,3-dimethyl-1-butyne are suggestive of an $Ad_E3$ mechanism for these alkynes. In the latter case, no rearrangement product was noticed, as would

**TABLE 3.7**

**Initial Rates and Product Composition for the Reaction of Alkynes (0.8 $M$) with HCl (0.75 $M$) in Acetic Acid at 25°[a,b]**

| Alkyne | Tetramethylammonium chloride ($M$) | $10^7 R$ ($M \sec^{-1}$) | Products (%) Chloride | Products (%) Ketone |
|---|---|---|---|---|
| PhC≡CH | 0 | 79 | 92 | 8 |
|  | 0.2 | 174 | 88 | 12 |
| PhC≡CMe | 0 | 19,800 | 89[c] | 11 |
|  | 0.209 | 47,000 | 90[d] | 10 |
| $n$-BuC≡CH | 0 | 0.29 | 35 | 65 |
|  | 0.2 | 2.2 | 82 | 18 |
| $t$-BuC≡CH | 0 | 0.094 | 26 | 74 |
|  | 0.2 | 0.48 | 75 | 25 |
| EtC≡CEt | 0 | 0.87 | 47 | 53 |
|  | 0.2 | 16.7 | 96 | 4 |

[a] Data obtained from Fahey et al. [43].
[b] See Schemes 3.5–3.8 for the reaction products.
[c] Mixture of E-**68** (77%) and Z-**68** (12%).
[d] Mixture of E-**68** (51%), Z-**68** (35%), and Z-**69** (4%).

be expected for a cationic intermediate. The mechanisms and the products obtained from these alkynes are depicted in Scheme 3.8. For 1-hexyne, the competitive $Ad_E2$ mechanism was also observed at higher temperatures, as in the case of 3-hexyne. At 50° and in the absence of TMAC, 40% of 2-chloro-1-hexene (71) was formed by syn addition from the vinyl cation ion pair by an $Ad_E2$ mechanism. The substitution of the terminal hydrogen atom of the triple bond by deuterium in 71 makes it possible to determine the stereochemistry of the addition. Addition of HCl to 1-hexyne-1-$d$ at 50° takes place also via an $Ad_E2$ process, giving 25 to 50% of the deutero analogue of 71 as a mixture of syn and anti adducts, the former predominating to an extent of 60%. The highly competitive nature of the $Ad_E3$ process in the addition reaction to alkyl-substituted alkynes is seen from the fact that the presence of 0.2 $M$ TMAC changed the course of the reaction, giving 90% of the anti adduct.

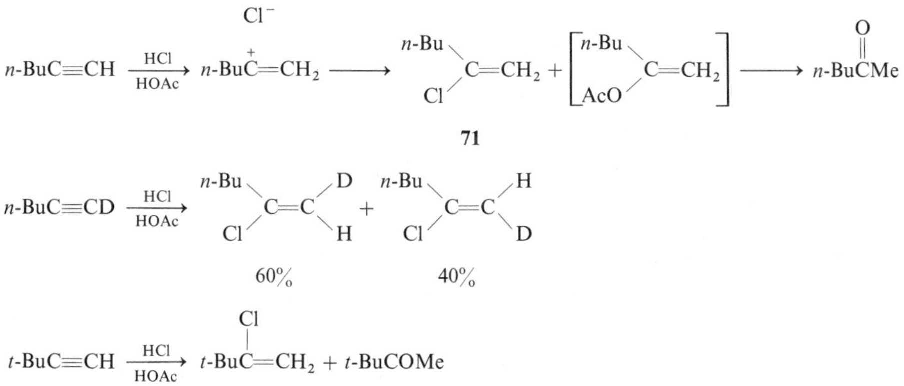

Scheme 3.8

By analyzing the reaction rates and product compositions of the alkynes investigated, Fahey *et al.* assigned relative rates for reactions occurring via $Ad_E2$ and $Ad_E3$ mechanism [43]. An inspection of these values, given in Table 3.8, reveals that the relative rates vary greatly by a factor of $10^4$, and substitution of alkyl by phenyl increases the rate 400- to 800-fold, demonstrating the ability of the phenyl group to stabilize the intermediate vinyl cations. The anomalous rate increase that occurs when H is replaced by a methyl group was also seen in this investigation. The explanation given on p. 36 can also be applied here.

For the $Ad_E3$ addition reaction, the relative rates vary by a factor of less than $10^4$. It can be seen from Table 3.8 that substitution of alkyl by phenyl at the site of chloride attachment results in a decrease in rate, in sharp contrast to the large increase found for $Ad_E2$ addition. Similarly, an effect

TABLE 3.8

Relative Rates in $Ad_E2$ and $Ad_E3$ Addition
Reactions for the Proton Attack at the
Indicated Positions [43]

| Alkynes | $Ad_E2$ | $Ad_E3$ |
|---|---|---|
| $EtCH_2CH_2C\overset{*}{\equiv}CH$ | 1 | 1 |
| $Et\overset{*}{C}{\equiv}CEt$ | ~0.05 | 5 |
| $Me_3CC\overset{*}{\equiv}CH$ | — | 0.2 |
| $Ph\overset{*}{C}{\equiv}CH$ | 800 | <10 |
| $Ph\overset{*}{C}{\equiv}CMe$ | 20 | 1 |
| $Ph\overset{*}{C}{\equiv}CMe$ | <0.1 | 0.1 |

opposite to that found for $Ad_E2$ addition was observed by comparing the relative rates of 1-hexyne and 3-hexyne. Replacement of H by an alkyl group at the site of proton attack increases the rate by a modest factor of 5. These data show that the $Ad_E3$ transition state involves little rehybridization at the site of proton attack and does not closely resemble a vinyl cation intermediate.

Another useful observation that can be gleaned from the work of Fahey [43] is the scant steric effect shown in $Ad_E3$ transition states. For example, 2-hexyne gave a 50:50 mixture of the *anti*-HCl adducts (in HOAc–TMAC) from attack at C-2 and C-3, thereby making no differentiation between the sizes of methyl and *n*-propyl groups. The results discussed above emphasize that alkyl alkynes react predominantly by the $Ad_E3$ mechanism, and aryl alkynes via the $Ad_E2$ mechanism.

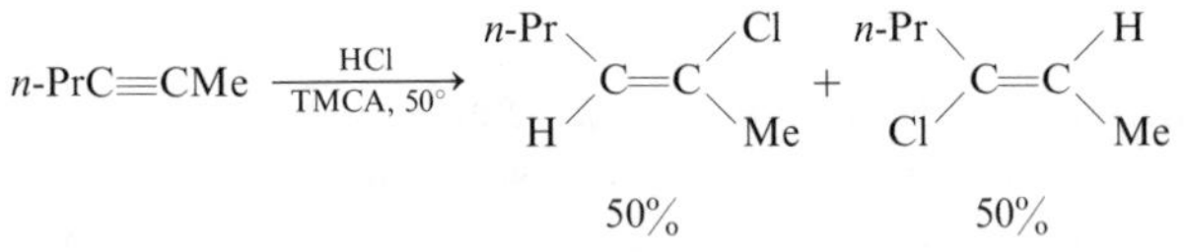

$$n\text{-}PrC{\equiv}CMe \xrightarrow[\text{TMCA, 50}°]{\text{HCl}} \underset{H}{\overset{n\text{-}Pr}{\diagdown}}C{=}C\underset{Me}{\overset{Cl}{\diagup}} + \underset{Cl}{\overset{n\text{-}Pr}{\diagdown}}C{=}C\underset{Me}{\overset{H}{\diagup}}$$

50%                    50%

Addition of HCl to alkynes was carried out in other nonaqueous solvents. Hydrogen chloride adds stereoselectively to 3,3-dimethyl-1-phenyl propyne in methylene chloride in the presence of $ZnCl_2$ to give exclusively the syn adduct, *E*-1-chloro-3,3-dimethyl-1-phenyl-1-butene (*E*-73) [45]. No evidence for the isomeric anti adduct was found.

This reaction demonstrates the influence of substrate structure on the stereochemistry of the products formed. The exclusive formation of the syn adduct is explained by postulating the intermediate linear vinyl cation 72 the empty p orbital of which is in the plane of the molecule (see Chapter 2).

Since the nucleophilic attack in this plane is subjected to the steric and electronic influences of the $\beta$ substituents, the bulky *tert*-butyl group in **72** hinders the attack of the nucleophile from the position trans to the $\beta$-hydrogen atom.

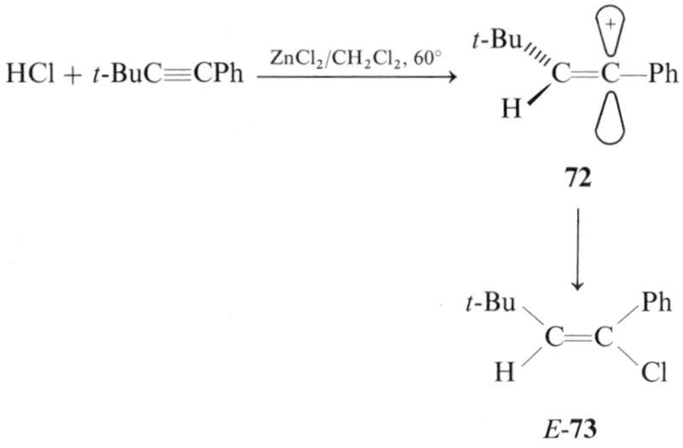

$$HCl + t\text{-}BuC\equiv CPh \xrightarrow{ZnCl_2/CH_2Cl_2, \, 60°}$$

**72**

*E*-**73**

This is one of many examples which demonstrate that the stereoselectivity of the electrophilic addition to alkynes involving a vinyl cation intermediate is determined mainly by its capture by the nucleophile from its less hindered side. Disubstituted alkynes capable of generating an unsymmetric vinyl cation show this effect in many instances, and dominant syn additions can take place, depending mainly on the steric factors outlined above.

**TABLE 3.9**

**Stereochemistry of Addition of HCl to PhC≡CD [46]**

| Solvent | Catalyst | *E*-**75** (syn) isomer ratio | *Z*-**75** (anti) isomer ratio |
|---|---|---|---|
| Methylene chloride | $ZnCl_2$ | 70 | 30 |
| | — | 65 | 35 |
| Acetic acid | — | 60 | 40 |
| Nitromethane | $ZnCl_2$ | 55 | 45 |
| Sulfolane | $ZnCl_2$ | 50 | 50 |
| | — | 75 | 25 |

A nonselective addition via free ion can be distinguished from a syn addition, either concerted or via an ion pair, by studying the product distribution from vinyl cations substituted by chemically identical but isotopically different $\beta$ substituents. The results of the addition of HCl to 1-deutero-2-phenylacetylene (**74**) in different solvents are listed in Table 3.9. A mixture of *E*- and *Z*-$\beta$-deutero-$\alpha$-chlorostyrene was obtained, with the syn adduct *E*-**75** predominating in solvents of low dielectric constants [46]. Addition of DCl to phenylacetylene also gives mostly (70%) the syn adduct Z-75. In the absence of kinetic data an unequivocal answer concerning the mechanism is not possible in most cases. The reaction may occur mainly via a concerted

$$\begin{array}{c} HCl + DC\equiv CPh \\ \mathbf{74} \\ DCl + HC\equiv CPh \end{array} \Bigg\rangle \longrightarrow \quad \underset{H}{\overset{D}{>}}C=C\underset{Cl}{\overset{Ph}{<}} + \underset{H}{\overset{D}{>}}C=C\underset{Ph}{\overset{Cl}{<}}$$

$$\qquad\qquad\qquad\qquad\qquad E\text{-}75 \qquad\qquad Z\text{-}75$$

syn addition where both the C–H and the C–Cl bonds are formed at the same time, or a tight ion pair that collapses in favor of the syn adduct may be involved. The latter situation is probably the case for most reactions carried out in the presence of zinc chloride. The 50:50 ratio of $E$ and $Z$ products obtained in sulfolane/$ZnCl_2$ is the expected ratio for a polar addition via a symmetrically substituted free vinyl cation. Formation of a free ion is due to the efficient solvation of the chloride ion in this solvent.

Similar dominant syn addition was also observed in the reaction of HCl with 1-anisyl-2-deuteroacetylene (**76**) in methylene chloride [47]. The yield of syn adduct E-**77** increased to 66% when the reaction was carried out in the nonpolar petroleum ether solvent, as well as when tetra-$n$-butylammonium chloride was added to the reaction carried out in methylene chloride.

$$AnC\equiv CD + HCl \xrightarrow[0°]{CH_2Cl_2} \underset{H}{\overset{An}{>}}C=C\underset{Cl}{\overset{D}{<}} + \underset{H}{\overset{An}{>}}C=C\underset{D}{\overset{Cl}{<}}$$

$$\quad\;\; \mathbf{76} \qquad\qquad\qquad\qquad E\text{-}77\ (56\%) \qquad Z\text{-}77\ (44\%)$$

Rappoport et al. utilized the addition of HBr to alkynes to prepare the arylvinyl bromides used in their solvolytic studies (see Chapter 6). The effect of $\beta$ groups was felt in the addition of HBr to bis(4-methoxyphenyl)acetylene (**78**) in acetic acid [48]. The relatively bulky bromide ion was captured by the intermediate vinyl cation **79** from the less hindered side to give predominantly the syn adduct, E-α-bromo-4,4′-dimethoxystilbene (**80**). It was noticed that E-**80** isomerized to a mixture richer in the thermodynamically

$$An-C\equiv C-An + HBr \xrightarrow[r.t.]{HOAC} \underset{H}{\overset{An}{>}}C=\overset{+}{C}-An \longrightarrow \underset{H}{\overset{An}{>}}C=C\underset{Br}{\overset{An}{<}} + \underset{H}{\overset{An}{>}}C=C\underset{An}{\overset{Br}{<}}$$

$$\quad\; \mathbf{78} \qquad\qquad\qquad\qquad \mathbf{79} \qquad\qquad\qquad E\text{-}\mathbf{80} \qquad\qquad Z\text{-}\mathbf{80}$$

more stable Z-**80**. Under kinetic control (in the absence of excess HBr), the addition was stereoselective, leading to more than 93% of the syn isomer E-**80**. Exclusive stereoselectivity with the formation of E-**82** was observed [49] when HBr was added to 1-(4-methoxyphenyl)-3,3-dimethylpropyne

(81), analogous to the addition of HCl to 3,3-dimethyl-1-butyne discussed earlier.

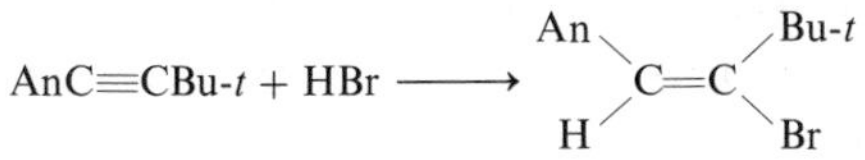

$$AnC{\equiv}CBu\text{-}t + HBr \longrightarrow \begin{matrix} An \\ \diagup \\ H \end{matrix} C{=}C \begin{matrix} Bu\text{-}t \\ \diagdown \\ Br \end{matrix}$$

81                         E-82

Addition of HBr to 1-anisyl-2-deuterioacetylene (76) was carried out in different solvent systems. An excess of 76 was used to avoid any rapid isomerization of the products. The ratios of $E$ and $Z$ isomers obtained are given in Table 3.10. In solvents of low dielectric constants and in the absence of salt, little or no stereoselectivity of the products formed is observed [47]. It was suggested that the formation of a 1:1 mixture of E-84 and Z-84 in acetic acid is indicative of a free vinyl cation (83) as an intermediate in this solvent. The E-84 ⇌ Z-84 isomerization in AcOH is slower than the addition rate [47], indicating that the observed product distribution is kinetically controlled.

**TABLE 3.10**

**Addition of HBr (0.049 $M$) to AnC≡CD (0.05 $M$) [47]**

| Solvent | Added salt | Concentration, $M$ | Temp, °C | E-83 (syn) isomer ratio | Z-83 (anti) isomer ratio |
|---|---|---|---|---|---|
| Petroleum ether | — | — | 0 | 55 | 45 |
|  | — | — | −20 | 60 | 40 |
|  | Et₄NBr | 0.15 | 0 | 65 | 35 |
| Benzol | — | — | 20 | 55 | 45 |
| Methylene chloride | — | — | 20 | 52 | 48 |
|  | Et₄NBr | 0.2 | 20 | 63 | 37 |
| Carbon tetrachloride | — | — | 0 | 52 | 48 |
|  | Bu₄NBr | 0.1 | −20 | 70 | 30 |
| Chloroform | — | — | 0 | 56 | 44 |
|  | Et₄NBr | 0.25 | 0 | 70 | 30 |
|  | Bu₄NBr | 0.5 | 0 | 73 | 27 |
|  | HgBr₂ | 0.002 | −20 | 80 | 20 |
| Acetic acid | — | — | 20 | 50 | 50 |

However, addition of tetraethylammonium bromide to the reaction (in chloroform) reduced the rate of reaction and gave more of the syn adduct E-84. This salt effect is in contrast to the increased rate and anti stereoselectivity caused by added chloride ion in the addition of HCl to 1-phenylpropyne in AcOH [42], which is shown to take place via an $Ad_E3$ mechanism. Presumably, addition of the ion pair $H^+B^-$ generates the vinyl cation ion pair *E*-**85**, the collapse of which to *E*-**84** competes favorably with isomerization to the ion pair *Z*-**85** or with dissociation to the free cation. The use of small amounts of $HgBr_2$ increased the stereoselectivity of the reaction by giving 80% of the syn adduct. It is not clear whether a concerted syn addition or an ion-pair mechanism is operative in this case.

$$An-C\equiv C-D \xrightarrow{H^+} An-\overset{+}{C}=C\begin{smallmatrix}H\\ \\D\end{smallmatrix} \longrightarrow \begin{smallmatrix}An\\ \\Br\end{smallmatrix}C=C\begin{smallmatrix}D\\ \\H\end{smallmatrix} + \begin{smallmatrix}An\\ \\Br\end{smallmatrix}C=C\begin{smallmatrix}H\\ \\D\end{smallmatrix}$$

**76**                    **83**                    *E*-**84**          *Z*-**84**

$$An-C\equiv C-D + H^+Br^- \longrightarrow An-\overset{+}{C}=C\begin{smallmatrix}D\\ \\H\end{smallmatrix} \longrightarrow \begin{smallmatrix}An\\ \\Br\end{smallmatrix}C=C\begin{smallmatrix}D\\ \\H\end{smallmatrix}$$

Br^-

**76**                              *E*-**85**                    *E*-**84**

$$An-\overset{+}{C}=C\begin{smallmatrix}H\\ \\D\end{smallmatrix} \longrightarrow \begin{smallmatrix}An\\ \\Br\end{smallmatrix}C=C\begin{smallmatrix}H\\ \\D\end{smallmatrix}$$

Br^-

*Z*-**85**                    *Z*-**84**

The data presented in Tables 3.9 and 3.10 allow a comparison of the addition of HCl to 1-deuterio-2-phenylacetylene (**74**) and that of HBr to 1-anisyl-2-deuterioacetylene (**76**). The results show that fewer syn adducts are obtained from **76** in AcOH or methylene chloride than from **74**. This is the expected result if concerted syn addition competes with a nonselective reaction of the free ion. The anisyl-substituted ion **83** is more stable, and reaction via the free ion will be more important than in the phenyl-substituted analogue.

The stereochemistry and kinetics of the addition of hydrogen halides to the unsubstituted and 3-substituted 2,3-alkynoic acids were investigated by Bowden and Price [50]. The rate of addition in water or in aqueous dioxane was first order with respect to both the acid and the halide anion and was dependent on the acidity of the medium. The solvent isotope effects $k_{D_2O}/k_{H_2O}$ for propiolic acid were between 1.3 and 1.6. This was considered to be due to a preequilibrium protonation before the rate-determining step. The addition was found to be predominantly anti; the percentage of the anti product increased with the increased nucleophilicity of the halide ion, i.e., $I^- > Br^- > Cl^-$. Addition of $NaClO_4$ decreased the reaction rate but increased the percentage of the anti product. Different mechanisms were suggested to account for the formation of the syn and anti adducts. The anti adduct was postulated to be formed via an initial protonation on the carbonyl group followed by a slow nucleophilic attack of $X^-$ on the protonated acid.

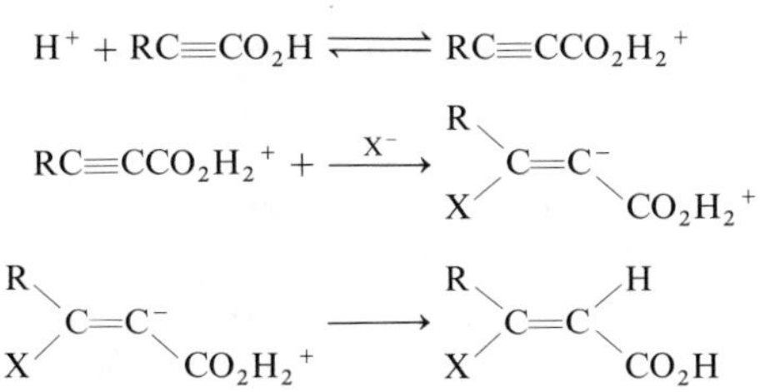

$$H^+ + RC\equiv CO_2H \rightleftharpoons RC\equiv CCO_2H_2{}^+$$

$$RC\equiv CCO_2H_2{}^+ + \xrightarrow{\ X^-\ } \underset{X}{\overset{R}{\diagdown}}C=C\underset{CO_2H_2{}^+}{\diagup}^-$$

$$\underset{X}{\overset{R}{\diagdown}}C=C\underset{CO_2H_2{}^+}{\diagup}^- \longrightarrow \underset{X}{\overset{R}{\diagdown}}C=C\underset{CO_2H}{\overset{H}{\diagup}}$$

In contrast, the syn adduct was suggested to be formed either by a stereo-specific ion pair or molecular hydrogen halide or by a nonspecific vinyl cation mechanism as follows:

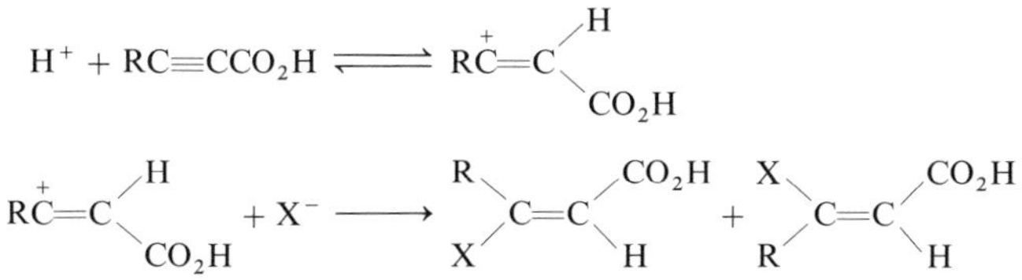

$$H^+ + RC\equiv CCO_2H \rightleftharpoons R\overset{+}{C}=C\underset{CO_2H}{\overset{H}{\diagup}}$$

$$R\overset{+}{C}=C\underset{CO_2H}{\overset{H}{\diagup}} + X^- \longrightarrow \underset{X}{\overset{R}{\diagdown}}C=C\underset{H}{\overset{CO_2H}{\diagup}} + \underset{R}{\overset{X}{\diagdown}}C=C\underset{H}{\overset{CO_2H}{\diagup}}$$

d. *Miscellaneous.* Only very few studies are available on the mechanisms of addition of hydrogen halides to alkynyl systems substituted at the $\omega$ position. Ethyl 3-chloro-3-butenoate (**87**) and its 2,4-dideuterio analogue (**88**) were obtained, respectively, from the additions of HCl and DCl to ethyl 3-butynoate (**86**) [51]. It gas concluded that addition to the triple bond of the enol ester **89** takes place with the formation of the vinyl cation **90**, which is stabilized by the neighboring double bond (Scheme 3.9). Mercuric salts

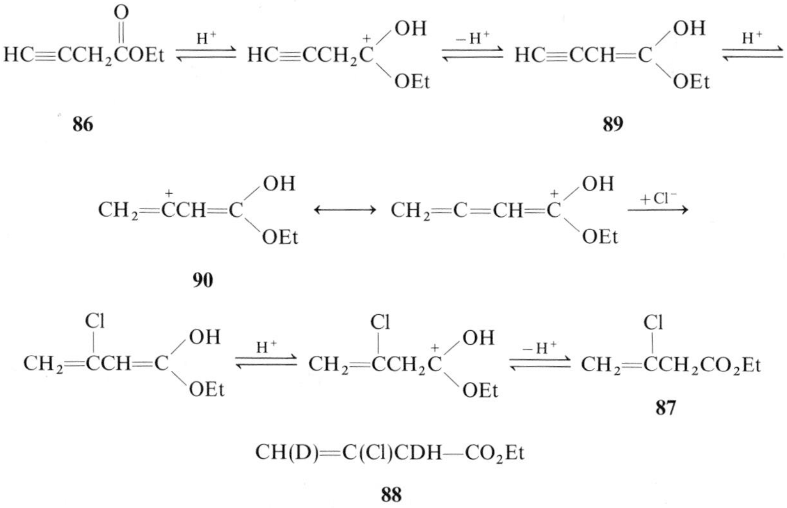

$$HC\equiv CCH_2\overset{O}{\overset{\|}{C}}OEt \xrightleftharpoons{H^+} HC\equiv CCH_2\overset{+}{C}\underset{OEt}{\overset{OH}{\diagup}} \xrightleftharpoons{-H^+} HC\equiv CCH=C\underset{OEt}{\overset{OH}{\diagup}} \rightleftharpoons{\ H^+\ }$$

**86**     **89**

$$CH_2=\overset{+}{C}CH=C\underset{OEt}{\overset{OH}{\diagup}} \longleftrightarrow CH_2=C=CH=\overset{+}{C}\underset{OEt}{\overset{OH}{\diagup}} \xrightarrow{+Cl^-}$$

**90**

$$\underset{CH_2=\overset{Cl}{|}{CCH}=C}{}\underset{OEt}{\overset{OH}{\diagup}} \xrightleftharpoons{H^+} CH_2=\overset{Cl}{\overset{|}{C}}CH_2\overset{+}{C}\underset{OEt}{\overset{OH}{\diagup}} \xrightleftharpoons{-H^+} CH_2=\overset{Cl}{\overset{|}{C}}CH_2CO_2Et$$

**87**

$$CH(D)=C(Cl)CDH-CO_2Et$$

**88**

Scheme 3.9

were found to catalyze the enolization process. This mechanism is supported by the fact that nonenolizable $\beta$-alkynoic esters do not react under these conditions.

The presence of a $\beta$-amino group on the alkynes gave different results. Addition of HCl to alkynyl tertiary amines gave the dihydrogen chloride salt **91** as the initial product [52]. When the dihydrogen chloride salt of 1-di-$n$-propylamino-2-propyne (**91**, R = $n$-Pr) was heated, either alone or in dipolar aprotic solvents, two chloro products (**92** and **93**) were obtained in the ratio 35:65.

$$R_2NCH_2C\equiv CH \xrightarrow{2HCl} R_2\overset{+}{N}HCH_2C\equiv CH \cdot HCl_2^- \xrightarrow[2.\ OH^-]{1.\ \Delta}$$

R = $n$-Pr　　　　　　　　**91**

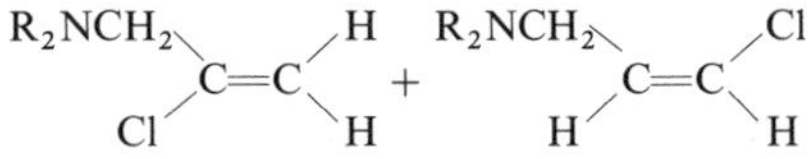

　　　　　　　**92**　　　　　**93**

An $Ad_E3$ reaction with nucleophilic chloride ion attacking both carbon atoms of the triple bond is probably the mechanism operating here. Thus, an $Ad_E3$ mechanism can be easily distinguished by the products obtained containing both the Markownikoff and anti Markownikoff addition products. An $Ad_E2$ mechanism obeys mostly Markownikoff's addition rule and prefers to form a stabilized vinyl cation.

$$R_2\overset{+}{N}HCH_2C\equiv CH \cdot HCl_2^- \longrightarrow R_2NCH_2\overset{+}{C}=CH_2 \cdot HCl_2^- \longrightarrow \mathbf{92}$$

1-Alkynylphosphines **94** were found to add hydrogen halides to the triple bond, although all attempts to hydrate them failed [53]. From the kinetic studies of the reaction, a vinyl cation mechanism was ruled out. The rate-determining step involved the addition of halide to protonated 1-alkynylphosphine, which stereospecifically adds a proton to form the anti adduct **95**.

$$R_2\overset{+}{P}(H)C\equiv CR' + X^- \longrightarrow R_2\overset{+}{P}(H)\overset{-}{C}=C(X)R' \quad (slow)$$

**94**

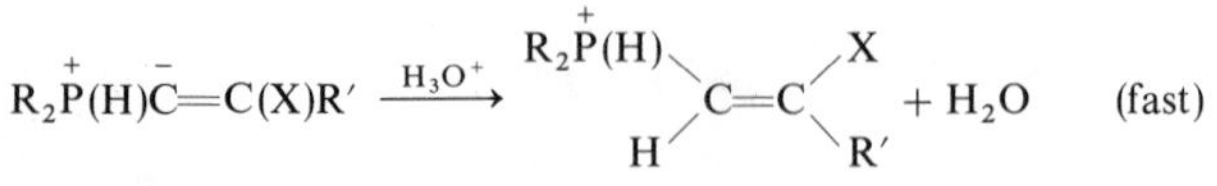

$$R_2\overset{+}{P}(H)\overset{-}{C}=C(X)R' \xrightarrow{H_3O^+} \underset{H}{\overset{R_2\overset{+}{P}(H)}{\diagdown}}C=C\underset{R'}{\overset{X}{\diagup}} + H_2O \quad (fast)$$

X = Cl, Br, I　　　　　　**95**

Addition of HBr to $o$-bis(phenylethynyl)benzene gave an indene derivative [54]. It was proposed that an $Ad_E3$ addition of the electrophile to one triple bond occurs while the other serves concurrently as the nucleophile. Two other reactions to dialkynes were reported. One involved the addition of HBr in HOAc to 2,2'-diethynyldiphenyl to give a phenenthrene derivative [55]. In contrast, in the other case no cyclization took place when HBr was added to 1,8-bis(phenylethynyl)naphthalene, and only the bis adduct was obtained [56].

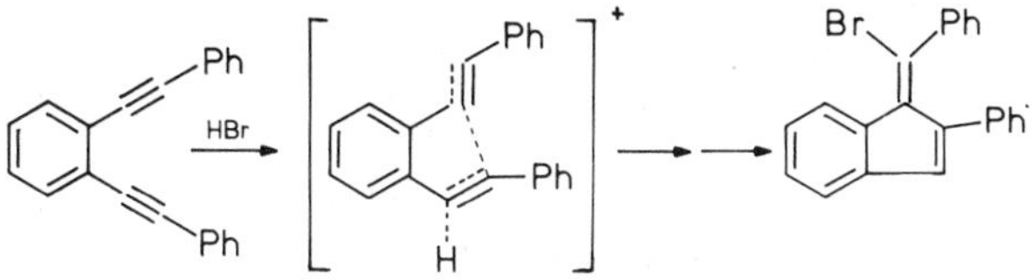

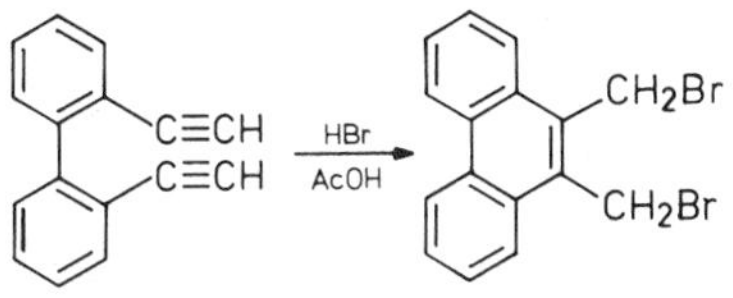

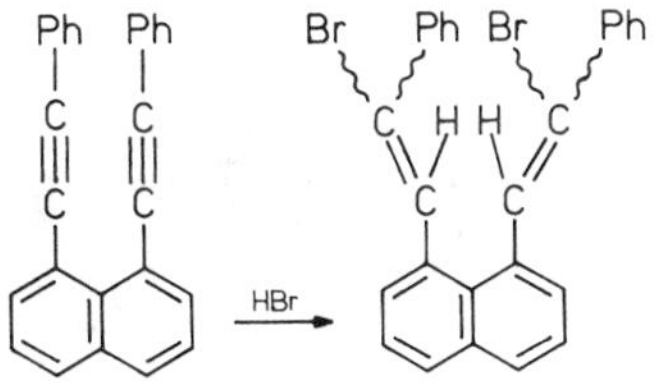

## 4. Perfluorosulfonic Acids

Perfluorosulfonic acids are strong acids [57], and their anions are very weak nucleophiles. Therefore, preferential formation of ionic intermediates rather than an $Ad_E3$ process is more likely to occur in the addition of perfluorosulfonic acids to alkynes than in the addition of other acids. In order to prevent polymerization the reaction should be carried out at low temperatures in inert solvents.

The ratios of the syn and anti adducts obtained [58] by the addition of several mono- and disubstituted alkynes (20% in $SO_2ClF$ at $-120°$) to excess $FSO_3H$ are summarized in Table 3.11, along with other available data in this field. The stereochemistry of addition for the terminal alkynes (propyne, 1-butyne, and 1-hexyne) was determined by using $FSO_3D$. The 80:20 ratio found for the syn and anti adducts did not change over an extended period of time, ruling out any acid-catalyzed isomerization by excess acid. This result is similar to that found for the addition of $CF_3SO_3D$ to 1-hexyne and $CF_3SO_3H$ to 1-hexyne-1-$d$ [41] (Table 3.11).

TABLE 3.11
Addition of Perfluorosulfonic Acids to Alkynes

| Alkyne | Perfluorosulfonic acid/solvent | Temp (°C) | Product | syn isomer ratio | anti isomer ratio | Reference |
|---|---|---|---|---|---|---|
| $MeC\equiv CH$ | $FSO_3D/SO_2ClF$ | $-120$ | $Me(OSO_2F)C=CHD$ | 80 | 20 | [58] |
| $EtC\equiv CH$ | $FSO_3D/SO_2ClF$ | $-120$ | $Et(OSO_2F)C=CHD$ | 80 | 20 | [58] |
| $n\text{-}BuC\equiv CH$ | $FSO_3D/SO_2ClF$ | $-120$ | $n\text{-}Bu(OSO_2F)C=CHD$ | 80 | 20 | [58] |
| $MeC\equiv CMe$ | $FSO_3H/SO_2ClF$ | $-120$ | $Me(OSO_2F)C=CHMe$ | 13 | 87 | [58] |
| $EtC\equiv CEt$ | $FSO_3H/SO_2ClF$ | $-120$ | $Et(OSO_2F)C=CHEt$ | 51 | 49 | [58] |
| $EtC\equiv CEt$ | $FSO_3H/SO_2ClF +$ pyridinium fluorosulfonate | $-78$ | $Et(OSO_2F)C=CHEt$ | 40 | 60 | [58] |
| $ClCH_2C\equiv CCH_2Cl$ | $FSO_3H$ | 0 | $ClCH_2(FO_2SO)C=CHCH_2Cl$ | 17 | 83 | [58] |
| $MeC\equiv CCOOH$ | $FSO_3H$ | 20 | $Me(OSO_2F)C=CHCOH$ (=O) | 85 | 15 | [58] |
| $PhC\equiv CD$ | $FSO_3H/$methylene chloride | $-78$ | $Ph(OSO_2F)C=CHD$ | 50 | 50 | [60] |
| $MeC\equiv CMe$ | $CF_3SO_3H/$pentane | $-20$ | $Me(OSO_2CF_3)C=CHMe$ | 35 | 65 | [59] |
| $n\text{-}BuC\equiv CH$ | $CF_3SO_3D/$pentane | $-20$ | $n\text{-}Bu(OSO_2CF_3)C=CHD$ | 74 | 26 | [41] |
| $n\text{-}BuC\equiv CD$ | $CF_3SO_3H/$pentane | $-20$ | $n\text{-}Bu(OSO_2CF_3)C=CHD$ | 77 | 23 | [41] |

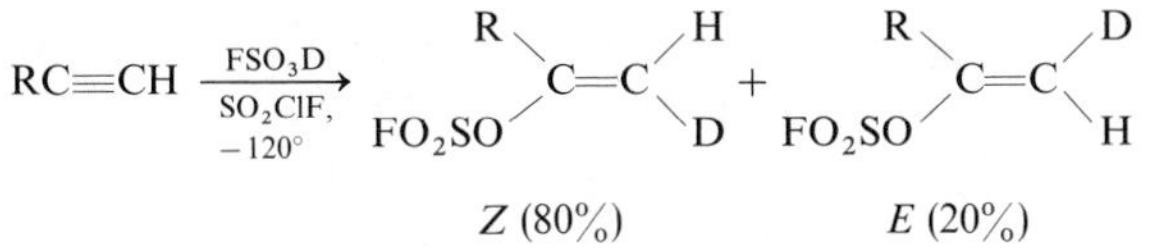

$$RC{\equiv}CH \xrightarrow[\substack{SO_2ClF, \\ -120°}]{FSO_3D}$$

$$\underset{Z\ (80\%)}{\underset{FO_2SO}{R}{\diagdown}C{=}C{\diagup}\underset{D}{\overset{H}{\diagup}}} + \underset{E\ (20\%)}{\underset{FO_2SO}{R}{\diagdown}C{=}C{\diagup}\underset{H}{\overset{D}{\diagup}}}$$

$$R = Me,\ Et,\ n\text{-}Bu$$

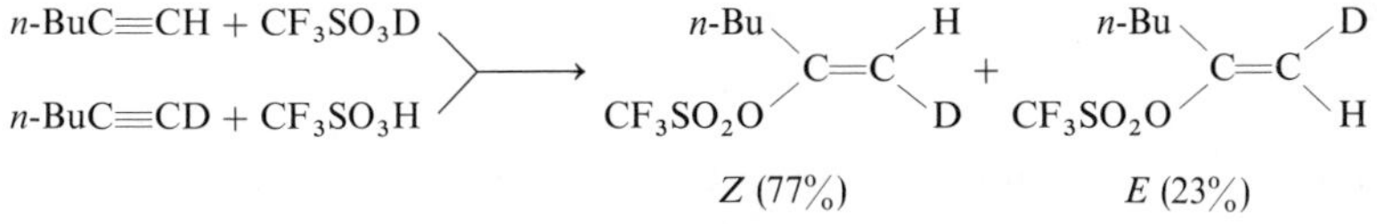

$$\left.\begin{array}{l} n\text{-}BuC{\equiv}CH + CF_3SO_3D \\ n\text{-}BuC{\equiv}CD + CF_3SO_3H \end{array}\right\rangle \longrightarrow \underset{Z\ (77\%)}{\underset{CF_3SO_2O}{n\text{-}Bu}{\diagdown}C{=}C{\diagup}\underset{D}{\overset{H}{\diagup}}} + \underset{E\ (23\%)}{\underset{CF_3SO_2O}{n\text{-}Bu}{\diagdown}C{=}C{\diagup}\underset{H}{\overset{D}{\diagup}}}$$

The mechanism of the dominant syn addition observed for the terminal alkynes is interpreted as initial protonation of the alkyne to form a linear vinyl cation–$FSO_3^-$ ion pair (**96**), which subsequently collapses in favor of syn product in 60% of the cases. The rest (40%) is assumed to escape the solvent cage to give free linear vinyl cations, which undergo nonstereo-specific nucleophilic attack, giving equal amounts of syn and anti addition (Scheme 3.10). However, one cannot rule out the possiblity that the 20% anti addition observed is caused by external attack on the initially formed ion pair **96**.

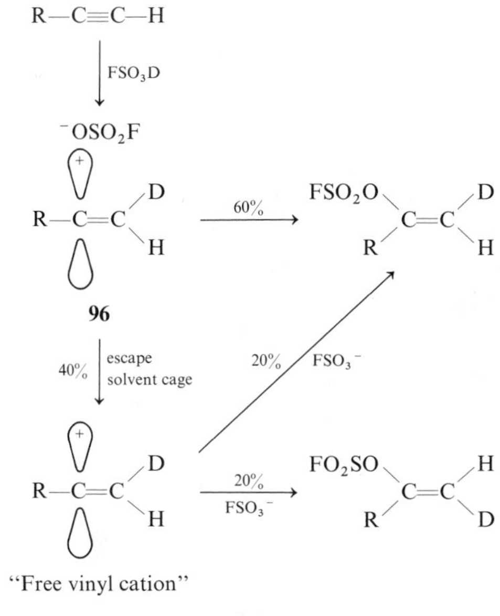

Scheme 3.10

The syn/anti ratios of the vinyl perfluorosulfonates obtained from 2-butyne and 3-hexyne are similar to those found for the trifluoroacetic acid addition [40, 41] to these alkynes (Tables 3.6 and 3.11). In contrast, the anti adducts are preferentially formed in the reaction of trifluoromethanesulfonic acid with 2-butyne and of 1,4-dichloro-2-butyne with $FSO_3H$. Added pyridinium fluorosulfonate increases the formation of anti isomer in the reaction of 3-hexyne with $FSO_3H$. Thus, different values were found for syn/anti ratios of these products from the addition reaction. Although vinyl cations are involved as intermediates, the explanation for this phenomenon is complicated.

Olah and Spear suggested an alternative to Schleyer's proposal for the nonconformity of 2-butyne and 3-hexyne in the electrophilic addition studied [58]. According to these authors, the data for the addition of $FSO_3H$ to 2-butyne, 3-hexyne, and 1,4-dichloro-2-butyne are most consistent with the initial formation of hydrogen-bridged vinyl cations (**97**). The hydrogen-bridged vinyl cations from 3-hexyne probably collapse to the open vinyl cation before nucleophilic attack, giving the capture products of this cation, whereas 2-butyne and 1,4-dichloro-2-butyne react predominantly via the hydrogen-bridged ion, giving mainly the anti adducts. However, these authors do not explain why the very similar bridged vinyl cations **98** and **99** behave so differently. Generally, it appears that the additions to 2-butyne take a course that is different from those to other alkynes. The reason for this is not clear.

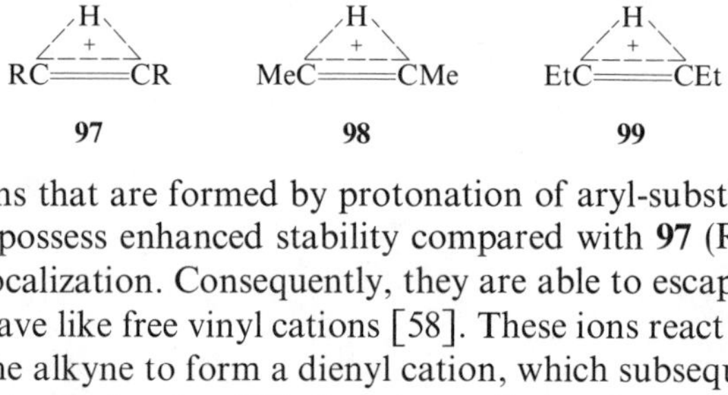

Vinyl cations that are formed by protonation of aryl-substituted alkynes with $FSO_3H$ possess enhanced stability compared with **97** (R = alkyl) due to charge delocalization. Consequently, they are able to escape the ion-pair stage and behave like free vinyl cations [58]. These ions react with a second molecule of the alkyne to form a dienyl cation, which subsequently cyclizes to the cyclobuten-3-yl cation **100**. Cyclobuten-3-yl cations generated by this method have been identified spectroscopically [61].

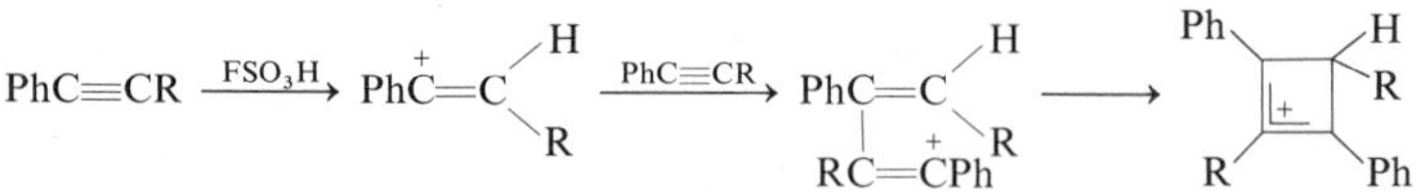

R = Me, Ph, $t$-Bu                                                      **100**

The 1:1 ratio of syn and anti adducts formed in the addition of $FSO_3H$ to 1-deuterio-2-phenylacetylene [60] points to a linear symmetric vinyl cation, which is captured with equal probability from both sides (Table 3.11).

Alkynoic acids and $FSO_3H$ gave mostly the thermodynamically more stable syn adducts due to the isomerization of the initially formed products [58].

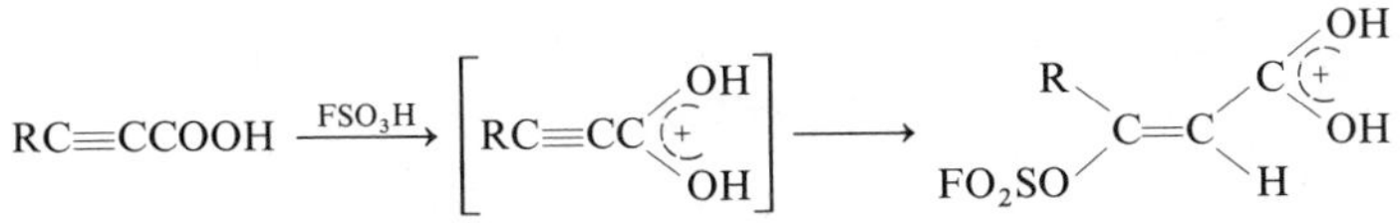

$$RC\equiv CCOOH \xrightarrow{FSO_3H} \left[ RC\equiv CC \overset{OH}{\underset{OH}{\diagup}}{}^{(+)} \right] \longrightarrow \underset{FO_2SO}{\overset{R}{\diagdown}}C=C\overset{C\overset{OH}{\underset{OH}{\diagup}}{}^{(+)}}{\underset{H}{\diagdown}}$$

The protonation of alkynes was also carried out with the combination $FSO_3H$–$SbF_5$ in $SO_2$ solution, which is normally used for observing stable ions spectroscopically. At $-78°$ only complex mixtures of unidentified products were obtained, but, at higher temperatures ($-20°$), signals corresponding to the protons of allyl cations were observed in the nmr spectra [62]. The formation of an allyl cation can take place either synchronously (A) or via a vinyl cation intermediate (B), as shown below.

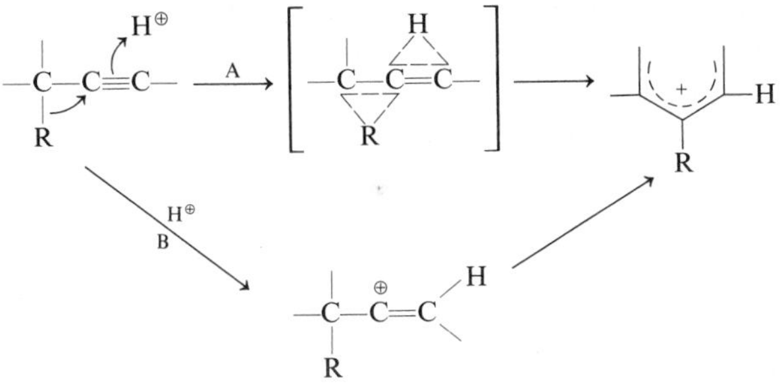

Several alkynes were protonated with $FSO_3H$–$SbF_5$ systems, and the nmr spectra of the allyl cations were studied (see Chapter 8). That the protonation involves a vinyl cation can be shown by conducting the experiment in $FSO_3D$. A synchronous mechanism (path A) would give an allyl cation with deuterium exclusively in the endo position. However, the nmr spectra of the allyl cation from 101 ($R_1 = H$; $R_2 = Me$) showed deuterium scrambling with formation of a 1:1 mixture of E- and Z-D-103. This fact can be taken as evidence for the intermediacy of vinyl cation 102. The incorporation of only one deuterium in the allyl cation also demonstrates that the protonation of alkynes with $FSO_3H$–$SbF_5$ is not an equilibrium process. Dialkyl alkynes and terminal alkynes, not branched at the $\alpha$ position

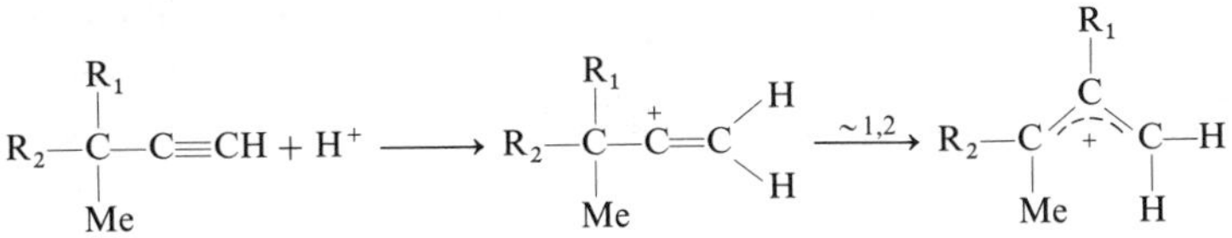

| $R_1$ | $R_2$ |
|-------|-------|
| Me | Me |
| H | Me |
| H | Et |

like **101**, also give allyl cations on protonation with $FSO_3H$–$SbF_5$, and they were also identified by nmr spectroscopy.

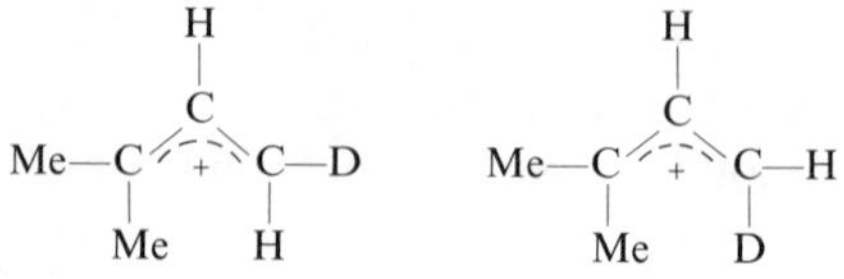

*E*-D-**103**                    *Z*-D-**103**

Tertiary propargyl alcohols were also treated with $FSO_3H$–$SbF_5$, and the resulting mesomeric alkynylcarbenium ions were investigated by $^{13}C$ and $^1H$ nmr spectroscopy [63]. The positive charge was found to be extensively delocalized, with the mesomeric vinyl cations (allenyl cations) contributing extensively to the total ion structure (see Chapter 8).

The data presented on the foregoing pages provide concrete evidence for the intermediacy of vinyl cations in the addition of electrophiles to alkynes. The vinyl cations have been shown to be linear; sometimes they are found intimately associated with the counterion, and sometimes they occur as free vinyl cations. The nature of the substituents and the reaction conditions employed determine the fate of the vinyl cations thus generated.

## B. Addition of Carbenium Ions to Alkynes

Addition of carbenium ions to alkynes can be carried out in two different ways: (1) by generating the cationic species separately and reacting it with an an alkyne and (2) by generating a cationic center at a particular position of the alkyne itself so that an intramolecular electrophilic addition of the cation to the triple bound is feasible. The latter method is discussed in Section III.

### 1. Adamantyl Systems

Early work carried out in this field with adamantyl systems gave evidence for vinyl cation intermediates, as shown by the occurrence of rearranged products. Thus, reaction of acetylene with 1-bromoadamantane in 98% sulfuric acid gave 1-adamantyl methyl ketone (**104**) [64]. On the other hand, 1-hydroxyadamantane reacted with acetylene in 90 to 95% sulfuric acid to

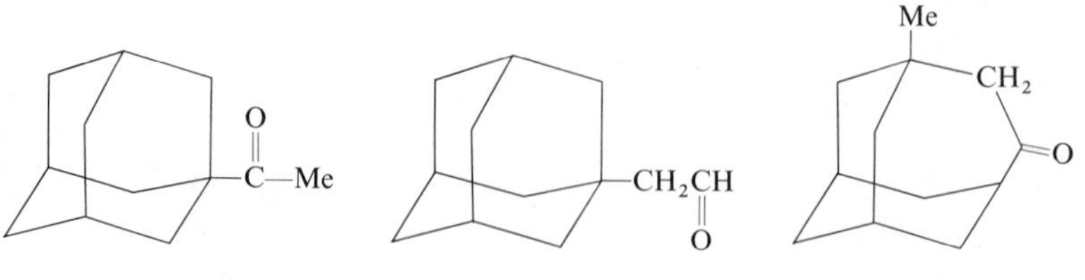

**104**                    **105**                    **106**

yield a mixture of 1-adamantylacetaldehyde (105) and 1-methylhomoadam-antanone (106) [65].

If one invokes the formation of 1-adamantyl cation 107 as the first step, its reaction with acetylene can give rise to the vinyl cation 108, which is trapped by water (present in 90 to 95% $H_2SO_4$) to give mainly adamantyl-acetaldehyde. However, this is an energetically unfavorable primary vinyl cation, and, in 95% sulfuric acid, a 1,2-hydride shift is probably the preferred reaction to give the more stable cation 109 and subsequent formation of 1-adamantyl methyl ketone (104). Part of the vinyl cation rearranges to form the allyl cation 110, which gives the diol 111 by capture of $^-$OH ion and addition of water to the double bond. Pinacol rearrangement of the diol 111 under the strongly acidic conditions employed gives the 1-methylhomo-adamantone 106. These reaction pathways are depicted in Scheme 3.11. It appears that the reaction products depend on the source of the adamantyl cation used. The results indicate that the rearrangement of the vinyl cation 109 to the allyl cation 110 is fast relative to capture of the nucleophile.

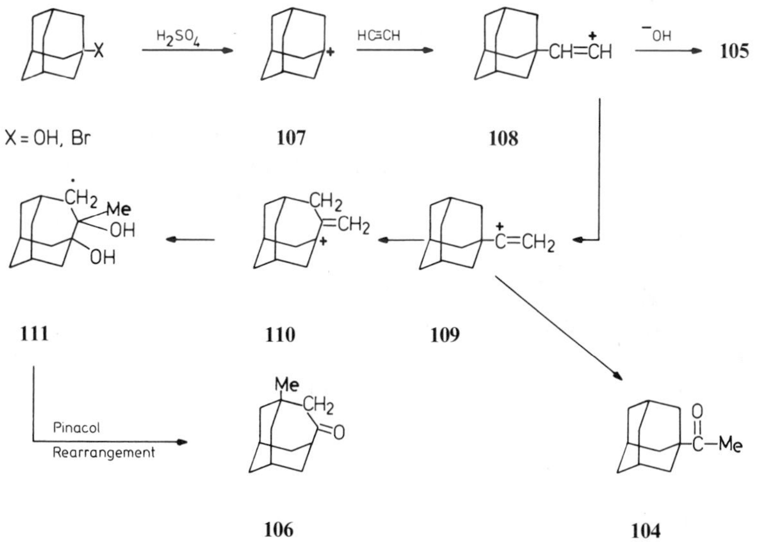

Scheme 3.11

When the reaction of 1-adamantanol with acetylene was carried out at various concentrations of sulfuric acid, the percentage of 1-adamantyl-acetaldehyde (105) decreased at low acid concentration relative to the increase of 1-methylhomoadamantanone (106); i.e., the formation of homoadamantone is favored by the presence of water in the medium [66]. Probably the vinyl sulfate 112 is involved in the reaction and adds 1 mole of sulfuric acid to give the adduct 113. Indeed, a sulfate adduct could be isolated by carrying out the reaction in methanesulfonic acid. Reaction carried out in $D_2SO_4$ gave the deuteromethylene analogue of **105**, indicating that both methylene protons of **105** appear to be derived from the acidic medium.

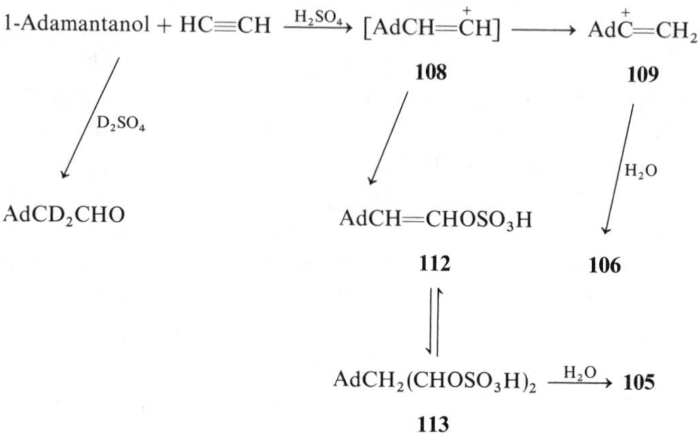

1-Adamantanol $+$ HC$\equiv$CH $\xrightarrow{H_2SO_4}$ [AdCH=$\overset{+}{C}$H] $\longrightarrow$ Ad$\overset{+}{C}$=CH$_2$

**108**

**109**

AdCD$_2$CHO          AdCH=CHOSO$_3$H

**112**            **106**

AdCH$_2$(CHOSO$_3$H)$_2$ $\xrightarrow{H_2O}$ **105**

**113**

Methylhomoadamantanone (**106**) was found as the sole product in the reaction of 1-adamantanol and acetylene in boron trifluoride–etherate complex as solvent. Here the primary cation **108** could not be trapped by the less nucleophilic catalyst and rearranged to give the stable vinyl cation **109**.

The slow reaction of 1-adamantyl bromide with acetylene in sulfuric acid relative to the fast reaction of 1-adamantanol led Kell and Mc Quillin [66] to suppose that the main product, adamantyl methyl ketone, in the former reaction must arise from slow rearrangement of the sulfate intermediate **112**. To test this, the reaction of 1-adamantanol with acetylene in $H_2SO_4$ was carried out for some hours. The formation of adamantyl methyl ketone (**104**) was found to increase at the expense of adamantylacetaldehyde (**105**), the amount of methylhomoadamantanone remaining constant. The re-

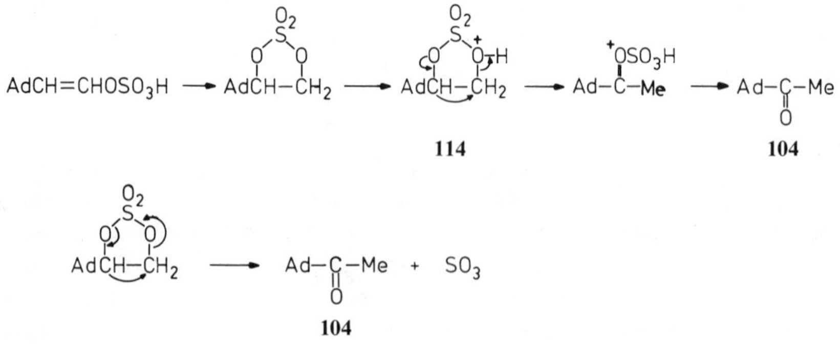

AdCH=CHOSO$_3$H $\longrightarrow$ AdCH–CH$_2$ $\longrightarrow$ AdCH–CH$_2$ $\longrightarrow$ Ad–$\overset{+}{C}$–Me $\longrightarrow$ Ad–C–Me

**114**              **104**

AdCH–CH$_2$ $\longrightarrow$ Ad–C–Me $+$ SO$_3$

**104**

Scheme 3.12

arrangement of the intermediate was rationalized by the authors [66] as shown in Scheme 3.12. The intermediate **114** can also fragment, with loss of an $SO_3$ residue.

From these data, it was concluded that the cation first formed (**108**) either was trapped by sulfuric acid or gave the homoadamantyl intermediate **110** via the vinyl cation **109**, which rearranged too rapidly to be trapped by nucleophiles. The extension of the reaction of 1-adamantanol with acetylene in $H_2SO_4$ to other alkynes gave the corresponding products **115**. Products involving vinyl cation intermediates were also obtained from the reaction of diphenylmethanol with 1-octyne in $H_2SO_4$ and norbornene and acetylene in $H_2SO_4$.

$$Ad—CH_2COR$$

**115** R = Ph, *n*-Pr, *n*-Hex

Addition of the 1-adamantyl cation derived from 1-bromoadamantane with propyn-3-ol in concentrated sulfuric acid gave homoadamantyl methyl ketone **119** [67]. The reaction involves the intermediate vinyl cation **116**, which rearranges to the homoadamantyl cation **118** via the epoxide **117**. An intermolecular hydride shift is assumed to neutralize the positive charge in the cation **118** and to lead to the final product (Scheme 3.13).

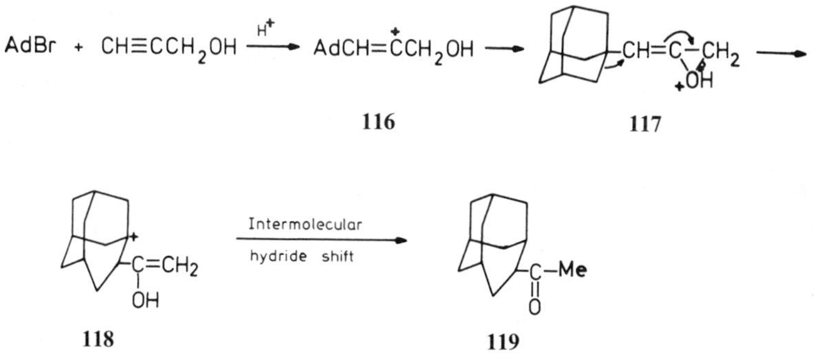

116            117

118            119

Scheme 3.13

## 2. Comparative Study of the Addition of RX and HX to Alkynes

α-Aryl-substituted vinyl halides **120** have been shown to undergo $S_N1$ solvolysis via vinyl cations **121** as intermediates. The phenomenon of ion-pair return and the stereochemistry of the products formed in these solvolysis reactions have been discussed in view of the steric bulk of the β groups in the intermediate vinyl cation–ion pairs involved. An exhaustive treatment of α-arylvinyl cations generated by solvolysis of arylvinyl halides is given in Chapter 6.

$$\underset{R_1}{\overset{R}{>}}C{=}C(X)Ar \xrightarrow{HOS} \underset{R_1}{\overset{R}{>}}C{=}\overset{+}{\underset{\alpha}{C}}{-}Ph + X^- \longrightarrow \text{solvolysis products}$$

$$\underset{\beta}{\qquad}$$

    120                              121

R = R$_1$ = H, Alk, Ar
X = Br, I

When similar vinyl cations are generated in less nucleophilic solvents by electrophilic additions to arylacetylenes, a study of the stereochemical outcome of the products will aid in understanding the nature of vinyl cations. Some data regarding this aspect have been presented in Section I,A,3. However, the main results on the steric effects of substituents in the intermediate vinyl cations have been obtained by Mareuzzi and Melloni [68, 69, 72] and Maroni *et al.* [70, 71]. and these are presented below.

Melloni *et al.* investigated electrophilic addition to alkynes by generating the same vinyl cation by two different methods: (1) by treating alkyl halides, capable of producing stable carbenium ions, with alkynes in the presence of a Friedel–Craft catalyst such as zinc chloride in methylene chloride as solvent, and (2) by reacting hydrogen halides with alkynes using the same catalyst/solvent system. If a linear α-phenylvinyl cation carrying two different β substituents is involved, this offers two diastereotopic sides for reaction with the nucleophile. As discussed above, the ease of attack of the nucleophile depends on the size of the β substituents. A bulkier substituent will hinder attack trans to the hydrogen in **122** and therefore can provoke stereoselective formation of the more crowded syn adduct under kinetic control. Hence, both methods of generating the vinyl cation will afford identical product distributions if the same free ion is involved. All the reactions were conducted under conditions in which no isomerization of the E and Z products was observed.

$$\begin{array}{l} RX + HC{\equiv}CPh \\ HX + RC{\equiv}CPh \end{array} \bigg\rangle \xrightarrow{ZnCl_2} \underset{H}{\overset{R}{>}}C{=}\overset{+}{C}{-}Ph \longrightarrow \underset{H}{\overset{R}{>}}C{=}C\underset{X}{\overset{Ph}{<}} + \underset{H}{\overset{R}{>}}C{=}C\underset{Ph}{\overset{X}{<}}$$

                    122

The steric effect of the β substituent in the cationic intermediate was immediately seen in the reaction of 3,3-dimethyl-1-phenylpropyne with HCl (p. 53). The formation of only the syn adduct in the HCl addition was explained on steric grounds whereby the nucleophile (Cl$^-$) attacks cation **72** from the least hindered side. Thus, adduct **73** corresponds to 100% syn addition of HCl to 3,3-dimethyl-1-phenylpropyne and 100% anti addition of *tert*-butyl chloride to phenyacetylene. Apparently, cation **72** is therefore involved in both processes.

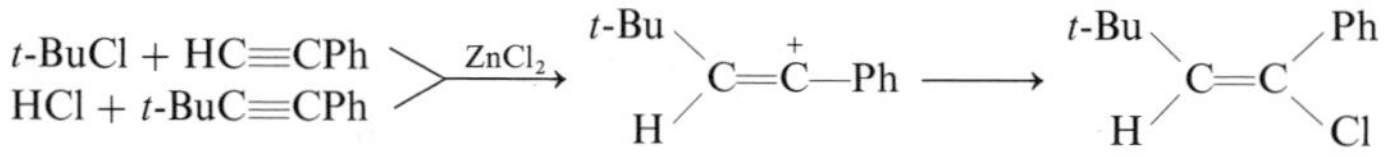

$$t\text{-BuCl} + HC\equiv CPh \atop HCl + t\text{-BuC}\equiv CPh \quad \overset{ZnCl_2}{\searrow\!\!\!\nearrow} \quad {t\text{-Bu}\atop H}\!\!>\!\!C\!\!=\!\!\overset{+}{C}\!-\!Ph \longrightarrow {t\text{-Bu}\atop H}\!\!>\!\!C\!\!=\!\!C\!\!<\!\!{Ph\atop Cl}$$

72         73

In the reaction of *tert*-butyl halides with substituted phenylacetylenes in the presence of zinc chloride, products arising from the addition of HX to the alkynes were also observed [68–70] besides the 1:1 addition products formed from *tert*-butyl halides and alkynes: Table 3.12 lists only the stereo-

**TABLE 3.12**

**Stereochemistry of the Addition of *tert*-Butyl Halides to Alkynes**

| Alkyne, PhC≡CR | *t*-BuCl addition products | | *t*-BuBr addition products | |
|---|---|---|---|---|
| | syn isomer ratio | anti isomer ratio | syn isomer ratio | anti isomer ratio |
| R = H | — | 100 | — | 100 |
| R = Me | — | 100 | — | 100 |
| R = Et | 5 | 95 | 10 | 90 |
| R = Ph | 5 | 95 | — | — |

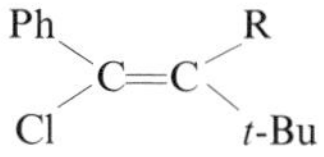

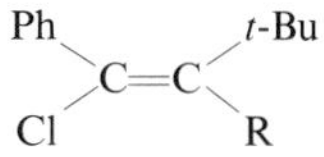

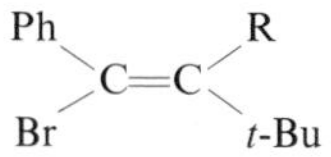

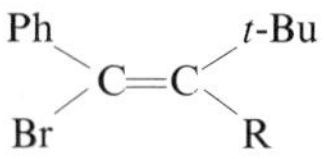

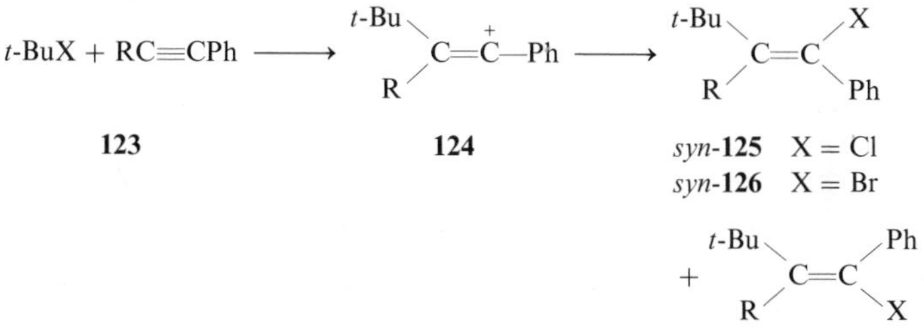

chemical outcome of *tert*-butyl halide addition products. It can be seen from the results given in Table 3.12 that only anti adducts were observed for phenylacetylene and 1-phenylpropyne, and the other two substituted phenyl-acetylenes (**123**, R = Et, Ph) gave predominantly anti addition products.

$$t\text{-BuX} + RC\equiv CPh \longrightarrow {t\text{-Bu}\atop R}\!\!>\!\!C\!\!=\!\!\overset{+}{C}\!-\!Ph \longrightarrow {t\text{-Bu}\atop R}\!\!>\!\!C\!\!=\!\!C\!\!<\!\!{X\atop Ph}$$

123     124     syn-**125** X = Cl<br>syn-**126** X = Br

$$+ \quad {t\text{-Bu}\atop R}\!\!>\!\!C\!\!=\!\!C\!\!<\!\!{Ph\atop X} \quad + \text{ other products}$$

X = Cl, Br
R = Me, Et, *i*-Pr, *t*-Bu, Ph

anti-**125** X = Cl<br>anti-**126** X = Br

Small amounts (5 to 10%) of syn isomers were also found for 1-phenyl-butyne and diphenylacetylene. This trend of reduced stereoselectivity is ascribed to a lower discrimination of the nucleophile while the difference in

# TABLE 3.13

## Stereochemistry of the Addition of PhCH₂X and Ph₂CHX to Alkynes[a]

| Alkyne, PhC≡CR | PhCH₂X addition products | | Ph₂CHX addition products | |
|---|---|---|---|---|
| | Ph–C=C–R with X, CH₂Ph (syn isomer ratio) | 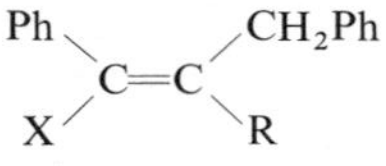 Ph–C=C–R with X, CH₂Ph (anti isomer ratio) | 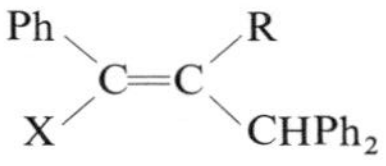 Ph–C=C–R with X, CHPh₂ (syn isomer ratio) | 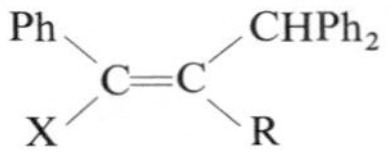 Ph–C=C–R with X, CHPh₂ (anti isomer ratio) |
| R = H | 20 | 80 | 10 | 90 |
| R = Ph | 85 | 15 | No 1:1 adduct was observed | |
| R = CH₂Ph | 50 | 50[b] | — | — |

[a] X = Cl or Br.
[b] Addition product of PhCD₂Cl.

size between the two $\beta$ groups in **124** becomes smaller. Practically no addition products of *tert*-butyl halides with the alkynes **123** (R = *i*-Pr, *t*-Bu) were observed. Instead, only HX adducts were obtained due to the competitive reaction of HX (formed by dehydrohalogenation of *t*-BuX in the presence of Lewis acid) with sterically hindered phenylacetylenes [68–70].

The addition of benzyl chloride and diphenylmethyl chloride to alkynes **123** (R = H, Ph, CH$_2$Ph) provided additional evidence for the intermediacy of vinyl cations in electrophilic additions [70]. Phenylacetylene gave predominantly anti addition product (Table 3.13), but the steric requirement for anti attack by the chloride ion was less pronounced than that for the *tert*-butyl halide addition reactions (cf. Tables 3.12 and 3.13).

$$R_1X + RC{\equiv}CPh \longrightarrow \underset{\substack{\\ \textbf{123}}}{\overset{R}{\underset{R_1}{\diagdown}}C{=}\overset{+}{C}{-}Ph} \longrightarrow \underset{\substack{\\ syn\text{-}\textbf{127}\quad X = Cl \\ syn\text{-}\textbf{128}\quad X = Br}}{\overset{R}{\underset{R_1}{\diagdown}}C{=}C\overset{Ph}{\underset{X}{\diagup}}} + \underset{\substack{\\ anti\text{-}\textbf{127}\quad X = Cl \\ anti\text{-}\textbf{128}\quad X = Br}}{\overset{R}{\underset{R_1}{\diagdown}}C{=}C\overset{X}{\underset{Ph}{\diagup}}}$$

R$_1$ = PhCH$_2$, Ph$_2$CH
R = H, Ph, CH$_2$Ph
X = Cl, Br

Using diphenylacetylene as substrate, Maroni *et al.* were able to demonstrate the nuances in steric effects displayed by similar substituents [71]. While *tert*-butyl chloride and diphenylacetylene reacted to give 95% of the anti product *Z*-**125** (R = Ph), benzyl chloride gave the syn adduct *E*-**127** (R = Ph; R$_1$ = CH$_2$Ph) as the major product (Tables 3.12 and 3.13). Considering the two vinyl cations (**129** and **130**) involved as intermediates, a less difficult approach, cis to the benzyl group, of the nucleophile was perceived for the cation **130**. This gave the following order of steric requirements for the $\beta$ substituents: *t*-Bu > Ph > PhCH$_2$.

$$\overset{Ph}{\underset{t\text{-}Bu}{\diagdown}}C{=}\overset{+}{C}{-}Ph \qquad\qquad \overset{Ph}{\underset{PhCH_2}{\diagdown}}C{=}\overset{+}{C}{-}Ph$$

$$\textbf{129} \qquad\qquad\qquad \textbf{130}$$

A small amount of the cyclization product 2,3-diphenylindene (**132**) was also isolated in the addition reaction of benzyl chloride with diphenylacetylene. The indene derivative **133** became the major product in the reaction of diphenylmethyl chloride with diphenylacetylene. The formation of indene derivatives has been ascribed to the intermal electrophilic attack of the vinyl cation centers in **131** on the $\gamma$-phenyl residue followed by loss of a proton (Scheme 3.14). Internal electrophilic alkylations involving vinyl cations are known (see Chapter 7, Section III,D).

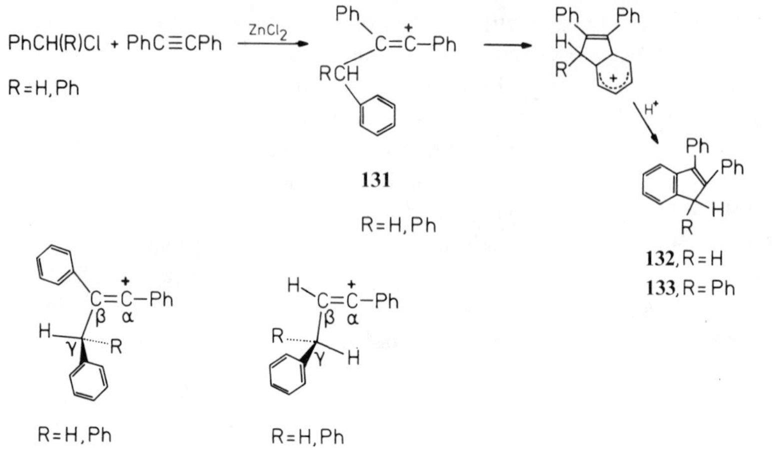

PhCH(R)Cl + PhC≡CPh $\xrightarrow{ZnCl_2}$

R = H, Ph

Ph
\C=C–Ph
RCH
⟨Ph⟩

131

R = H, Ph

Ph   Ph
H⟨...⟩
R'⟨...⟩
$\searrow$ H⁺

Ph
⟨...⟩Ph
⟨...⟩H
R

132, R = H
133, R = Ph

⟨Ph⟩
\C=C–Ph
H–/β  α
Y–R
⟨Ph⟩

R = H, Ph

H\C=C–Ph
R····/β  α
⟨Ph⟩Y–H

R = H, Ph

Scheme 3.14

Examination of models shows that the $\gamma$-phenyl groups are located near the vinyl cation center, favoring cyclization in the case of alkyne substrates having bulkier $\beta$ substituents (Scheme 3.14). In the case of phenylacetylene, the $\gamma$-phenyl group is farther away from the positive center in the favorable low-energy conformation, and therefore cyclization to the indene derivative was not noted. The vinyl cations formed (131) are selective species in the sense that they react exclusively by intramolecular cyclization rather than by nucleophilic attack by an external nucleophile.

Supporting evidence for free vinyl cations in the addition reactions of carbenium ions to alkynes was obtained in the reaction of benzyl-$\alpha,\alpha$-$d_2$ chloride with 1,3-diphenylpropyne [68] in the presence of $ZnCl_2$. The reaction was slow and was therefore carried out in high-boiling 1,2-dichloroethane [68]. The 1:1 ratio of syn and anti adducts obtained corresponds to a symmetric free vinyl cation (136) having two identical substituents at the $\beta$ positions. Apparently, complete equilibration of the intimate syn- and anti-oriented ion pairs 134 and 135 or their dissociation to the free ion 136 is faster than collapse of either 134 or 135 to give preferentially one isomer (Scheme 3.15).

Dominant anti addition of alkyl halides to alkynes, which is ascribed to the steric influence of $\beta$ substituents in the cationic intermediate, corresponds to a syn addition in the case of the reaction of HX with alkynes. The nucleophilic chloride ion prefers to attack from the less hindered side cis to hydrogen. The stereochemical outcomes of the addition of HX to several phenylacetylenes are listed in Table 3.14.

The data given in Table 3.14 show the influence of the bulkier substituents R in directing the attack of the nucleophile on the cation 123. The decreasing trend toward syn adducts follows the order $t$-Bu ≥ Ph > $i$-Pr > Me ≥ D.

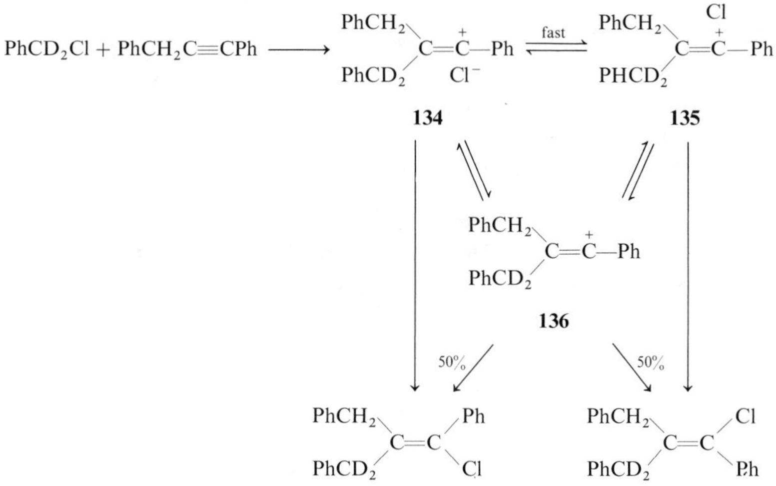

$$PhCD_2Cl + PhCH_2C\equiv CPh \longrightarrow \underset{PhCD_2}{\overset{PhCH_2}{>}}C=\overset{+}{C}-Ph \quad Cl^- \underset{}{\overset{fast}{\rightleftharpoons}} \underset{PHCD_2}{\overset{PhCH_2}{>}}C=\overset{Cl}{\underset{}{C}}-Ph$$

$$\textbf{134} \qquad\qquad\qquad \textbf{135}$$

$$\underset{PhCD_2}{\overset{PhCH_2}{>}}C=\overset{+}{C}-Ph$$

$$\textbf{136}$$

50%　　　　　　　50%

$$\underset{PhCD_2}{\overset{PhCH_2}{>}}C=C\underset{Cl}{\overset{Ph}{<}} \qquad\qquad \underset{PhCD_2}{\overset{PhCH_2}{>}}C=C\underset{Ph}{\overset{Cl}{<}}$$

Scheme 3.15

**TABLE 3.14**

**Stereochemistry of the Addition of HX to Alkynes**

| | HCl addition products | | HBr addition products | |
|---|---|---|---|---|
| | 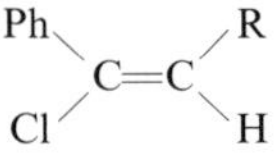 $\underset{Cl}{\overset{Ph}{>}}C=C\underset{H}{\overset{R}{<}}$ | 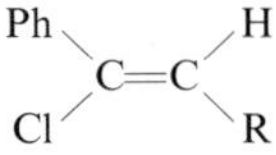 $\underset{Cl}{\overset{Ph}{>}}C=C\underset{R}{\overset{H}{<}}$ | 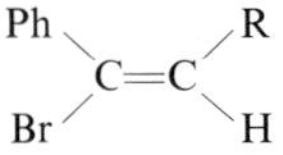 $\underset{Br}{\overset{Ph}{>}}C=C\underset{H}{\overset{R}{<}}$ | 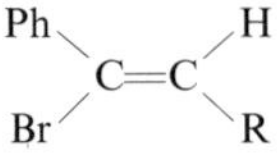 $\underset{Br}{\overset{Ph}{>}}C=C\underset{R}{\overset{H}{<}}$ |
| Alkyne, PhC≡CR | syn-$E$ isomer ratio | anti-$Z$ isomer ratio | syn-$E$ isomer ratio | anti-$Z$ isomer ratio |
| R = D | 70 | 30 | 75 | 25 |
| | 65 | 35[a] | — | — |
| R = Me | 70 | 30 | — | — |
| R = Et | 80 | 20 | — | — |
| R = $i$-Pr | 95 | 5 | — | — |
| R = $t$-Bu | 100 | — | 100 | — |
| R = Ph | 100 | — | — | — |
| R = $CH_2Ph$ | 85 | 15 | 90 | 10 |
| R = $CHPh_2$ | 95 | 5 | 98 | 2 |

[a] Reaction carried out in the absence of catalyst.

The reduced steric effect shown by $CH_2Ph$ in blocking the attack of the nucleophile trans to the hydrogen atom in **123** is similar to that found in the addition of benzyl chloride to diphenylacetylene (Table 3.13).

Experiments to generate the same vinyl cation intermediate were carried out both by the addition of *tert*-butyl, benzyl, and diphenylmethyl halides

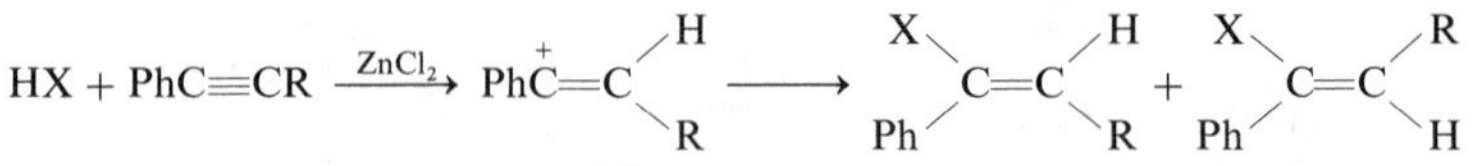

$$HX + PhC\equiv CR \xrightarrow{ZnCl_2} Ph\overset{+}{C}{=}C\overset{H}{\underset{R}{\diagup}} \longrightarrow \underset{Ph}{\overset{X}{\diagdown}}C{=}C\overset{H}{\underset{R}{\diagup}} + \underset{Ph}{\overset{X}{\diagdown}}C{=}C\overset{R}{\underset{H}{\diagup}}$$

| 122 | 123 | syn-$E$ | anti-$Z$ |

X = Cl, Br
R = D, Me, Et, $i$-Pr, $t$-Bu, Ph, PhCH$_2$, Ph$_2$CH

to phenylacetylene and by the reaction of hydrogen halides with **122** (R = $t$-Bu, PhCH$_2$, Ph$_2$CH) [70]. Viewed as a whole, fairly identical product distributions were obtained for both reactions (Scheme 3.16). Therefore, a common cationic intermediate is strongly indicated, regardless of the electrophilic species used for its formation. The observed steric effect of the $\beta$ substituents was similar to that in the addition reactions mentioned above. The major products in both addition reactions are formed under kinetic control, with bulky groups cis to each other. This result rules out any alternative explanation based on product isomerization.

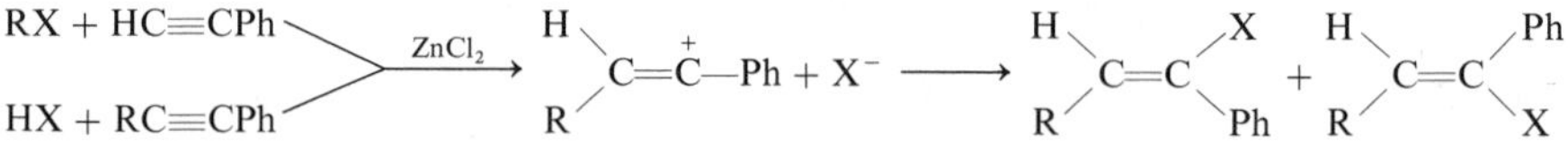

$$\left.\begin{array}{l} RX + HC\equiv CPh \\ HX + RC\equiv CPh \end{array}\right\} \xrightarrow{ZnCl_2} \underset{R}{\overset{H}{\diagdown}}C{=}\overset{+}{C}{-}Ph + X^- \longrightarrow \underset{R}{\overset{H}{\diagdown}}C{=}C\overset{X}{\underset{Ph}{\diagup}} + \underset{R}{\overset{H}{\diagdown}}C{=}C\overset{Ph}{\underset{X}{\diagup}}$$

| | 123 | syn-$E$ | anti-$Z$ |

R = $t$-Bu, PhCH$_2$, Ph$_2$CH
X = Cl, Br

|  | Yield (%), RX addition | | Yield (%), HX addition | | |
|---|---|---|---|---|---|
| R | $E$ | $Z$ | $E$ | $Z$ | |
| PhCH$_2$ | 80 | 20 | 85 | 15 | (X = Cl) |
|  |  |  | 90 | 10 | (X = Br) |
| Ph$_2$CH | 90 | 10 | 95 | 5 | (X = Cl) |
|  |  |  | 98 | 2 | (X = Br) |
| $t$-Bu | 100 | — | 100 | — | (X = Cl, Br) |

Scheme 3.16

Comparison of the two addition reactions (Scheme 3.16) shows a slight trend toward higher extent of formation of syn adducts in the HCl addition to alkynes. This is more pronounced in the addition of HCl to 1-deutero-2-phenylacetylene **(74)**, even though a nearly symmetric vinyl cation is involved as the intermediate in the reaction (p. 53). The stereochemical configuration of the products obtained is therefore slightly different in both cases of addition of alkyl halides and hydrogen halides to alkynes under identical conditions. The different behavior is ascribed to the low dissociation of hydrogen halides in methylene chloride due to the poor solvation.

**TABLE 3.15**

**Stereochemistry of the Addition of Alkyl Chlorides to Alkynes**

| Alkyne, $RC{\equiv}CR_1$ | *t*-BuCl addition products | | PhCH$_2$Cl addition products | |
|---|---|---|---|---|
| | syn-*Z* isomer ratio | anti-*E* isomer ratio | syn-*Z* isomer ratio | anti-*E* isomer ratio |
| R = Me; R$_1$ = H | 20 | 80 | 35 | 65 |
| R = Et; R$_1$ = H | 5 | 95 | — | — |
| R = *n*-Bu; R$_1$ = H | 10 | 90 | 25 | 75 |
| R = Me; R$_1$ = Me | — | 100 | — | — |
| R = Et; R$_1$ = Et | 10 | 90 | — | — |

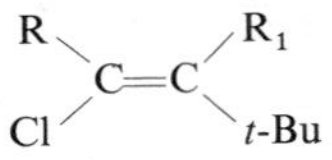

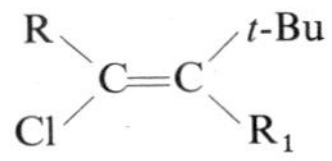

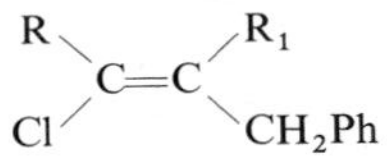

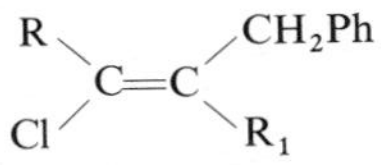

Therefore, the nucleophilic halide ion (here the complexed $ZnCl_3{}^-$) remains very close to the proton during its attachment to the triple bond, giving syn-oriented tight ion pairs. The stable carbenium ions used in the addition reactions are less sensitive to solvation, and consequently less tight ion pairs are formed. The concerted molecular addition mechanism described in Section I,A,2 can be regarded as an alternative explanation for the major syn addition of HCl to alkynes. So far, this possibility has not been unambiguously excluded by experimental means.

The addition of carbenium ions to alkyl-substituted alkynes in the presence of $ZnCl_2$ as catalyst was also investigated by Marcuzzi and Melloni [72]. The vinyl cations generated from these alkynes are stabilized only by inductive and/or hyperconjugative effects. Terminal and symmetrically disubstituted alkynes gave the expected 1 : 1 addition products with *tert*-butyl chloride and benzyl chloride. The stereochemical results are given in Table 3.15.

The data presented in Table 3.15 clearly show that anti adducts were formed in major amounts, in analogy with the addition of alkyl chlorides to phenylacetylenes. The terminal alkynes yielded exclusively Markownikoff adducts, supporting a stepwise mechanism for the reaction (Scheme 3.17). However, compared to the 100% anti addition of *tert*-butyl chloride with phenylacetylene, propyne underwent less stereoselective addition, giving only 80% of the anti adduct. This shows that the relative steric effects of $\beta$ substituents in the cations **137** and **138** also depend on the nature of the $\alpha$ substituent.

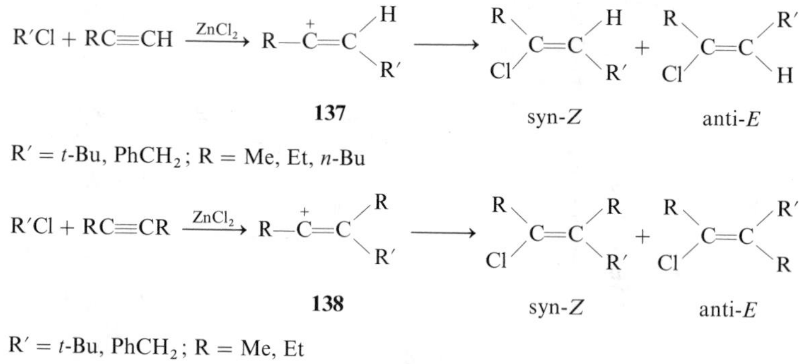

$$R'Cl + RC{\equiv}CH \xrightarrow{ZnCl_2} R-\overset{+}{C}{=}C{\overset{H}{\underset{R'}{\diagdown}}}$$

**137**

R' = *t*-Bu, PhCH$_2$; R = Me, Et, *n*-Bu

syn-*Z*   anti-*E*

$$R'Cl + RC{\equiv}CR \xrightarrow{ZnCl_2} R-\overset{+}{C}{=}C{\overset{R}{\underset{R'}{\diagdown}}}$$

**138**

R' = *t*-Bu, PhCH$_2$; R = Me, Et

syn-*Z*   anti-*E*

Scheme 3.17

## C. Acylation of Alkynes

### 1. By H$^+$/CO

We saw earlier that the stereochemical results of the electrophilic addition to alkyl-substituted alkynes do not allow one to formulate a definite, uniform mechanism. That 2-butyne gave more of the anti adduct was also evidenced in the work of Hogeveen and Roobeek [73], who investigated the reaction of carbon monoxide with vinyl cations. When mixtures of 2-butyne and carbon monoxide were bubbled through a $FSO_3H–SbF_5$ solution at room temperature, the predominant formation of the oxocarbonium ion Z-**140** (Scheme 3.18) at higher concentrations of carbon monoxide and $SbF_5$ relative to that of alkyne (25:10:1, respectively) was detected by $^1H$ nmr spectroscopy. The corresponding isomer (E-**140**) was detected only in very small amounts. An isomerization between E and Z adducts was ruled out by control experiments. The stereoselective anti addition of carbon monoxide to the vinyl cation **139** takes place trans to the $\beta$-hydrogen, although attack from the cis position is less hindered. This observation is similar to that of Summerville and Schleyer [41] and Olah and Spear [58] in the addition of trifluoroacetic acid and fluorosulfonic acid to 2-butyne.

The reaction is proposed to occur as shown in Scheme 3.18 with the formation of the vinyl cation **139**, which is intercepted by carbon monoxide to give the oxocarbonium ion Z-**140**. At higher concentrations of carbon monoxide, but with a 1:1:1 ratio of alkyne, $FSO_3H$, and $SbF_5$, a rearranged

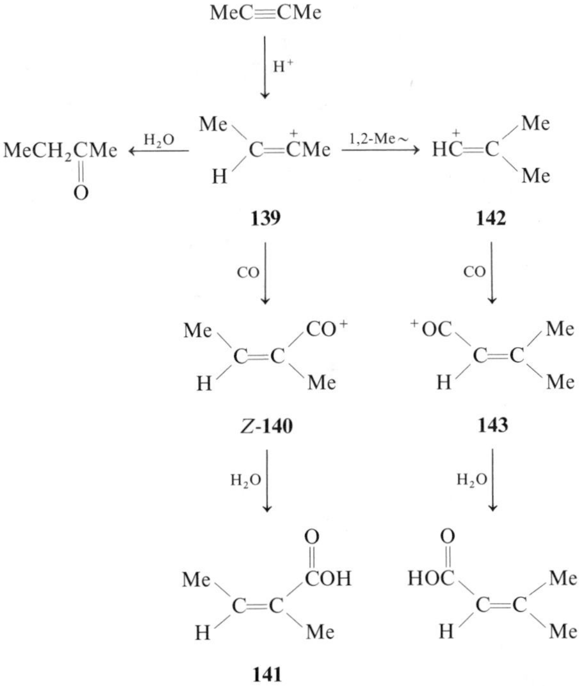

Scheme 3.18

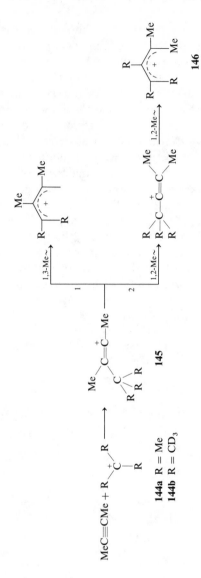

Scheme 3.19

oxocarbonium ion (143) was also detected. A mechanism involving a 1,2-methyl shift via the cation 142 (Scheme 3.18) is discussed in Chapter 7 (Section III,B). A bridged vinyl cation intermediate (98) may be responsible for the major anti adduct found in the reaction.

In an attempt to observe vinyl cations (145) spectroscopically, the addition of tert-butyl cation 144 to 2-butyne was carried out [74]. However, the spectra observed were those of pentamethylallyl cation 146 similar to the spectra obtained on addition of $FSO_3H$–$SbF_5$ to alkynes in $SO_2$ (p. 63). Among the two possible pathways for the formation of vinyl cations (Scheme 3.19), the double 1,2-methyl shift through path 2 was experimentally proved by using the deuterated analogue of the cation 144b. The $^1H$ nmr spectrum showed no signals corresponding to the methyl group in the 2 position, which should have been the case if a 1,3-methyl shift had taken place (see also Chapter 8).

The product compositions obtained in the addition of several electrophilic reagents to different alkynes are summarized in Tables 3.16 and 3.17. In summary, the work of Fahey has shown that alkyl-substituted alkynes prefer to react by the $Ad_E3$ process, although the competing $Ad_E2$ mechanism is observed at higher temperatures (Section 1,A,3) for the addition of HCl (in HOAc). The results are different in the case of addition of strong acids, in which an $Ad_E3$ mechanism can be ruled out due to the less nucleophilic conditions employed. It can be seen from the data in Tables 3.16 and 3.17 that there is a preference for more syn adducts in the addition reaction of simple non-bulky reagents to terminal alkynes, whereas disubstituted alkynes give major amounts of anti adducts. This can be interpreted by the mechanism suggested by Olah and Spear [58] involving symmetric open vinyl cation–ion pairs, with some escaping the solvent cage to give free vinyl cations, which undergo nonstereospecific nucleophilic attack. The effects of solvent and temperature may be seen here, as shown by the result for $CF_3SO_3D$ in pentane, but these are small.

The unsymmetric vinyl cations formed from dialkyl-substituted alkynes probably are more susceptible to electronic effects than sensitive to the steric bulk of $\beta$ substituents. A hydrogen-bridged vinyl cation has also been proposed to explain this phenomenon (p. 62). 1,4-Dichloro-2-butyne has not been included in Table 3.16 because it does not belong to the dialkyl-substituted alkynes due to the chloride functional group. The use of 2-pentyne-2-hexyne, etc., as a substrate in the addition reaction should clarify whether a terminal methyl group increases the tendency for anti addition to occur and, if so, why.

Vinyl cations generated by the bulkier electrophiles hinder the attack of the nucleophile, as shown by the addition of t-BuCl and $PhCh_2Cl$ to 2-butyne and 3-hexyne in the presence of $ZnCl_2$. A 100% anti addition with

**TABLE 3.16**

**Stereochemistry of Electrophilic Additions to Terminal Alkynes**

| Alkyne | Intermediate vinyl cation | Reagent–solvent | Temp (°C) | syn isomer ratio | anti isomer ratio | Reference |
|---|---|---|---|---|---|---|
| MeC≡CH | $MeC^+=C$ (H, D) | $FSO_3D-SO_2ClF$ | −120 | 80 | 20 | [58] |
| | $MeC^+=C$ (H, $CH_2Ph$) | $PhCH_2Cl/ZnCl_2-CH_2Cl_2$ | 60 | 35 | 65 | [72] |
| | $MeC^+=C$ (H, t-BuCl) | $t\text{-}BuCl/ZnCl_2-CH_2Cl_2$ | 40 | 20 | 80 | [72] |
| EtC≡CH | $EtC^+=C$ (H, D) | $FSO_3D-SO_2ClF$ | −120 | 80 | 20 | [58] |
| | $EtC^+=C$ (H, t-Bu) | $t\text{-}BuCl/ZnCl_2-CH_2Cl_2$ | 40 | 5 | 95 | [72] |
| n-BuC≡CH | $n\text{-}BuC^+=C$ (H, D) | $FSO_3D-SO_2ClF$ | −120 | 80 | 20 | [58] |
| | $n\text{-}BuC^+=C$ (H, t-Bu) | $CF_3SO_3D-pentane$ | −20 | 74 | 26 | [41] |
| | $n\text{-}BuC^+=C$ (H, t-Bu) | $t\text{-}BuCl/ZnCl_2-CH_2Cl_2$ | 40 | 10 | 90 | [72] |
| n-BuC≡CD | $n\text{-}BuC^+=C$ (D, H) | $CF_3SO_3H-pentane$ | −20 | 77 | 23 | [41] |

| Substrate | Structure | Conditions | Temp (°C) | | | Ref. |
|---|---|---|---|---|---|---|
| PhC≡CH | $Ph\overset{+}{C}=C\big<^{H}_{D}$ | HCl–HOAc | 50 | 60 | 40 | [43] |
| | | HCl/TMCA–HOAc | 50 | 10 | 90 | [43] |
| | | DCl/ZnCl$_2$–CH$_2$Cl$_2$ | 40 | 70 | 30 | [46] |
| | $Ph\overset{+}{C}=C\big<^{H}_{t\text{-Bu}}$ | $t$-BuX/ZnX$_2$–CH$_2$Cl$_2$ (X = Cl, Br) | 40 | — | 100 | [45, 72] |
| | $Ph\overset{+}{C}=C\big<^{H}_{CH_2Ph}$ | PhCH$_2$Cl/ZnCl$_2$–CH$_2$Cl$_2$ | 40 | 20 | 80 | [68] |
| | $Ph\overset{+}{C}=C\big<^{H}_{CHPh_2}$ | Ph$_2$CHCl/ZnCl$_2$–CH$_2$Cl$_2$ | 40 | 10 | 90 | [68] |
| PhC≡CD | $Ph\overset{+}{C}=C\big<^{D}_{H}$ | HCl/ZnCl$_2$–CH$_2$Cl$_2$ | 40 | 70 | 30 | [46] |
| | | HCl/ZnCl$_2$–sulfolane | 40 | 50 | 50 | [46] |
| | | HCl–HOAc | 40 | 60 | 40 | [46] |
| | | FSO$_3$H–CH$_2$Cl$_2$ | −78 | 50 | 50 | [60] |
| AnC≡CD | $An\overset{+}{C}=C\big<^{D}_{H}$ | HBr–CH$_2$Cl$_2$ | 20 | 52 | 48 | [47] |
| | | HBr/Et$_4{}^{+}$NBr$^{-}$–CH$_2$Cl$_2$ | 20 | 63 | 37 | [47] |
| | | HBr–CHCl$_3$ | 0 | 56 | 44 | [47] |
| | | HBr/Et$_4{}^{+}$NBr$^{-}$–CH$_2$Cl$_2$ | 0 | 70 | 30 | [47] |
| | | HBr/HgBr$_2$–CHCl$_3$ | −20 | 80 | 20 | [47] |

**TABLE 3.17**

**Stereochemistry of Electrophilic Addition to Disubstituted Alkynes**

| Alkyne | Intermediate vinyl cation | Reagent–solvent | Temp (°C) | syn isomer ratio | anti isomer ratio | Reference |
|---|---|---|---|---|---|---|
| MeC≡CMe | $\overset{+}{Me C}=C\overset{Me}{\underset{H}{\big\langle}}$ | CF₃COH ($\overset{\parallel}{O}$) | 50 | 19.5 | 80.5 | [41] |
| | | FSO₃H–SO₂ClF | −120 | 13 | 87 | [58] |
| | | CF₃SO₃H–pentane | −20 | 35 | 65 | [59] |
| | $\overset{+}{Me C}=C\overset{Me}{\underset{t\text{-Bu}}{\big\langle}}$ | $t$-BuCl/ZnCl₂–CH₂Cl₂ | 40 | — | 100 | [72] |
| | $\overset{+}{Me C}=C\overset{Me}{\underset{CH_2Ph}{\big\langle}}$ | PhCH₂Cl/ZnCl₂–CH₂Cl₂ | 60 | 25 | 75 | [72] |
| EtC≡CEt | $\overset{+}{Et C}=C\overset{Et}{\underset{H}{\big\langle}}$ | CF₃COH ($\overset{\parallel}{O}$) | 75.4 | 52 | 48 | [41] |
| | | FSO₃H–SO₂ClF | −120 | 51 | 49 | [58] |
| | | FSO₃H/pyridinium fluorosulfate–SO₂ClF | −78 | 40 | 60 | [58] |
| | | HCl/ZnCl₂–CH₂Cl₂ | 40 | 40 | 60 | [72] |
| | | HCl–HOAc | 60 | <2 | >98 | [43] |
| | $\overset{+}{Et C}=C\overset{Et}{\underset{t\text{-Bu}}{\big\langle}}$ | $t$-BuCl/ZnCl₂–CH₂Cl₂ | 40 | 10 | 90 | [72] |

| Substrate | Ion | Conditions | Temp | | | Ref. |
|---|---|---|---|---|---|---|
| $n\text{-PrC}{\equiv}\text{C}n\text{-Pr}$ | $n\text{-PrC}{\overset{+}{=}}\text{C}(n\text{-Pr})(\text{H})$ | $\text{CF}_3\text{COH}$ ($\overset{\|}{\text{O}}$) | 75.4 | 56 | 44 | [41] |
| $\text{PhC}{\equiv}\text{CMe}$ | $\text{PhC}{\overset{+}{=}}\text{C}(\text{Me})(\text{H})$ | $\text{HCl/ZnCl}_2\text{–CH}_2\text{Cl}_2$ | 40 | 70 | 30 | [68] |
| | | $\text{HCl–HOAc}$ | 25 | 77 | 12 | [43] |
| | | $\text{HCl/TMAC–HOAc}$ | 25 | 51 | 35 | [43] |
| | $\text{PhC}{\overset{+}{=}}\text{C}(\text{Me})(t\text{-Bu})$ | $t\text{-BuX/ZnX}_2\text{–CH}_2\text{Cl}_2$ (X = Cl, Br) | 40 | — | 100 | [68, 69] |
| $\text{PhC}{\equiv}\text{CEt}$ | $\text{PhC}{\overset{+}{=}}\text{C}(\text{Et})(\text{H})$ | $\text{HCl/ZnCl}_2\text{–CH}_2\text{Cl}_2$ | 40 | 80 | 20 | [68] |
| | $\text{PhC}{\overset{+}{=}}\text{C}(\text{Et})(t\text{-Bu})$ | $t\text{-BuX/ZnX}_2\text{–CH}_2\text{Cl}_2$ (X = Cl, Br) | 40 | 10 | 90 | [68, 69] |
| $\text{PhC}{\equiv}\text{C}i\text{-Pr}$ | $\text{PhC}{\overset{+}{=}}\text{C}(i\text{-Pr})(\text{H})$ | $\text{HCl/ZnCl}_2\text{–CH}_2\text{Cl}_2$ | 40 | 95 | 5 | [68] |
| $\text{PhC}{\equiv}\text{C}t\text{-Bu}$ | $\text{PhC}{\overset{+}{=}}\text{C}(t\text{-Bu})(\text{H})$ | $\text{HCl/ZnCl}_2\text{–CH}_2\text{Cl}_2$ | 40 | 100 | — | [68] |
| $\text{PhC}{\equiv}\text{CPh}$ | $\text{PhC}{\overset{+}{=}}\text{C}(\text{Ph})(\text{H})$ | $\text{HCl/ZnCl}_2\text{–CH}_2\text{Cl}_2$ | 40 | 100 | — | [70] |

*(Continued)*

**TABLE 3.17 (continued)**

| Alkyne | Intermediate vinyl cation | Reagent–solvent | Temp (°C) | syn isomer ratio | anti isomer ratio | Reference |
|---|---|---|---|---|---|---|
| | $PhC\overset{+}{=}C$ with Ph, $t$-Bu | $t$-BuCl/ZnCl$_2$–CH$_2$Cl$_2$ | 40 | 5 | 95 | [70] |
| | $PhC\overset{+}{=}C$ with Ph, CH$_2$Ph | PhCH$_2$Cl/ZnCl$_2$–CH$_2$Cl$_2$ | 40 | 85 | 15 | [70] |
| PhC≡CCH$_2$Ph | $PhC\overset{+}{=}C$ with CH$_2$Ph, H | HCl/ZnCl$_2$–CH$_2$Cl$_2$ | 40 | 85 | 15 | [68] |
| | $PhC\overset{+}{=}C$ with CH$_2$Ph, CD$_2$Ph | PhCD$_2$Cl/ZnCl$_2$–CH$_2$Cl$_2$ | 40 | 50 | 50 | [68] |
| PhC≡CCHPh$_2$ | $PhC\overset{+}{=}C$ with CHPh$_2$, H | HCl/ZnCl$_2$–CH$_2$Cl$_2$ | 40 | 95 | 5 | [68] |

the nucleophile attacking trans to the *tert*-butyl group is observed for 2-butyne.

The situation is different for the α-aryl-substituted vinyl cations that are stabilized by charge delocalization into the phenyl ring. Symmetric vinyl cations generated in highly ionizing medium are open free vinyl cations giving a 1:1 product ratio of syn and anti adducts. The major amount of syn addition observed in other media may be due either to preferential syn collapse of the tight ion pair or to concerted molecular syn addition. To distinguish between the two mechanisms, more work must be done.

Steric effects rather than electronic effects of the β substituents in the intermediate vinyl cations are dominant in the addition of carbenium ions to phenylacetylenes. Here the stabilized vinyl cation exists as a free open cation, and the nucleophile chooses to attack it from the less hindered side.

In general, alkyl-substituted vinyl cations are less stabilized than α-aryl-vinyl cations, and so the alkynes behave differently in electrophilic additions. The electrophile and the solvent also play a role in determining the stereo-chemical outcome of the products. The properties of the vinyl cations generated by electrophilic addition to alkynes are not different from those of vinyl cations generated by solvolysis reactions (see Chapters 5, 6, and 7).

## 2. By Acylium Ions

The acylation of alkynes with acid chloride–$AlCl_3$ leads to the formation of β-chlorovinyl ketones, which are useful synthetic intermediates [75]. Vinyl cations were invoked in these reactions to explain the products observed [76]. The stereochemistry and mechanism of acylation of alkynes have been explored by Martens *et al.* [77].

$$R_1C{\equiv}CR_2 + RCX \xrightarrow[-40°]{CH_2Cl_2} \underset{\substack{\| \\ O}}{RC}\diagdown C{=}C\diagup \underset{X}{\overset{R_2}{}} + \underset{\substack{\| \\ O}}{RC}\diagdown C{=}C\diagup \underset{R_2}{\overset{X}{}}$$

Z-147        E-147

$R_1 = R_2 = $ Me, Et
$X = OSO_2CF_3$, $AlCl_4^-$

Acylation was carried out by using both acyl trifluoromethanesulfonates (triflates) and acyl chloride–$AlCl_3$ completes in methylene chloride solution at $-40°$ under kinetically controlled conditions. No isomerization of the products was noted. Both syn and anti adducts (147) were obtained. The ratio of the syn and anti adducts was dependent on the substituent attached to the acylating agent, as shown by the data for 2-butyne in Table 3.18. Acylating agents containing electron-donating substituents were found to

**TABLE 3.18**

**Product Distribution for the Acylation of 2-Butyne [77]**

| Acylating agent | $X = AlCl_4$ | | $X = OSO_2CF_3$ | |
|---|---|---|---|---|
| | syn isomer ratio | anti isomer ratio | syn isomer ratio | anti isomer ratio |
| $p$-AnCOX | — | 100 | 1.5 | 82[a] |
| PhCOX | 5 | 95 | 6.5 | 71.5[a] |
| $p$-NO$_2$C$_6$H$_4$COX | 63 | 37 | 60 | 40 |
| EtCOX | — | 100 | 20 | 80 |
| MeCOX | — | 100 | 30 | 70 |
| CH$_2$ClCOX | 90 | 10 | 15 | 85 |
| CHCl$_2$COX | 100 | — | No reaction | |

[a] Indenone was formed as side product.

favor anti addition, whereas those substituted by electron-withdrawing groups gave more of the syn adducts.

The formation of the anti adducts was explained by postulating a vinyl cation intermediate (**148**) that is formed by the electrophilic attack of the acylium ion on the triple bond (Scheme 3.20). The formation of syn adduct Z-**147** is depicted as involving acyclic transition states similar to **149** (Scheme 3.22). Part of it may also arise by attack of the nucleophile on the vinyl cation (**148**) cis to the acyl group. Which of these pathways will be followed is again dependent on the nature of the electrophilic acylating agent. There-

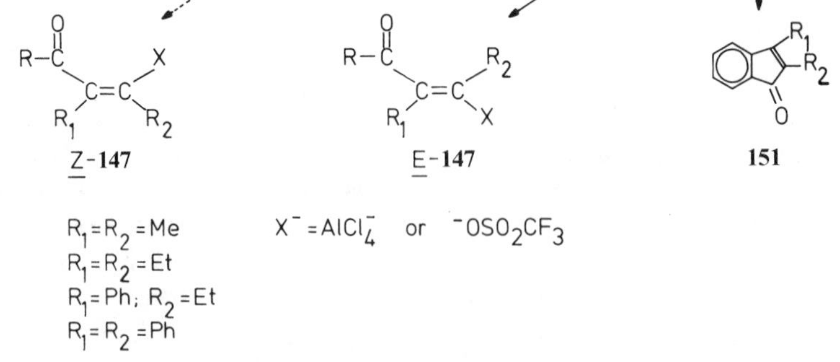

$R_1 = R_2 = Me$      $X^- = AlCl_4^-$   or   $^-OSO_2CF_3$
$R_1 = R_2 = Et$
$R_1 = Ph;\ R_2 = Et$
$R_1 = R_2 = Ph$

Scheme 3.20

fore, the ratio of the syn and anti adducts given in Table 3.18 is very much influenced by electronic factors as well as steric effects, which may be also operative. Hence, the data are not directly comparable with other results on the electrophilic addition discussed so far.

Aroyl triflates were found to yield larger amounts of cyclized products than the acyl chloride–AlCl$_3$ complexes. The formation of these cyclized products, the indenones 151, is ascribed to the intramolecular electrophilic attack of the vinyl cation 150 on the phenyl ring. α-Phenyl alkynes, which generate the vinyl cations 150 (R$_2$ = Ph), give exclusively indenones as products. The smaller amount of indenones obtained in the reaction utilizing acyl chloride–AlCl$_3$ complexes is explained by the higher nucleophilicity of the tetrachloroaluminate anion compared with the triflate anion.

Further evidence for the intermediacy of vinyl cations in the acylation reaction is obtained by the occurrence of a 1,2-methyl shift in the addition of 3,5-dimethoxybenzoyl chloride to 4,4-dimethyl-2-pentyne. The rearranged product 152 was detected in up to 10% yield (Scheme 3.21).

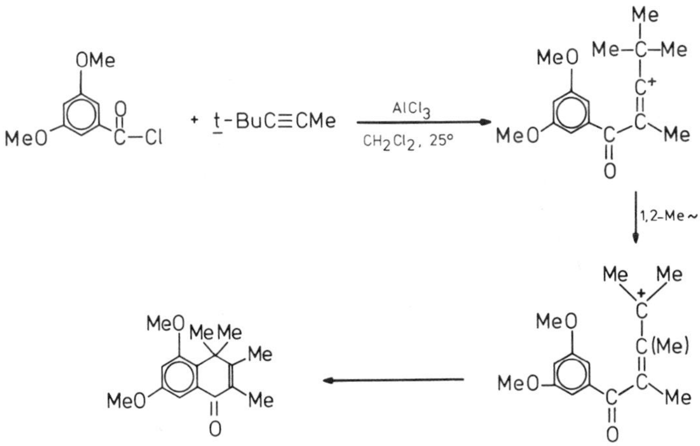

Scheme 3.21

Intramolecular cyclization supporting the intermediacy of vinyl cations was found in the acylation of alkynes with acyl tetrafluoroborates in non-nucleophilic solvents [78]. Both propyne and 2-butyne gave cyclopenten-2-ones (156) on treatment with acyl tetrafluoroborates in methylene chloride–dichloroethane. Two possibilities for the cyclization of the intermediate vinyl cation were considered (Scheme 3.22). The vinyl cation 153 can undergo either an intermolecular electrophilic attack (S$_E$i reaction) via the transition state 154 or a sequence of two steps involving a 1,5-hydride shift to give the cation 155 and subsequent Ad$_E$ cyclization.

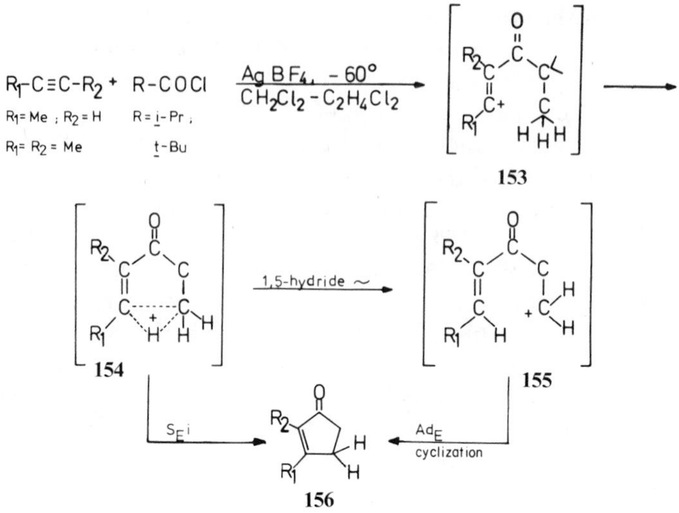

$R_1-C\equiv C-R_2 +$ R-COCl $\xrightarrow[\text{CH}_2\text{Cl}_2-\text{C}_2\text{H}_4\text{Cl}_2]{\text{Ag BF}_4, \ -60°}$

$R_1= Me \ ; R_2 = H$   $R = \underline{i}-Pr$ ,

$R_1= R_2 = Me$   $\underline{t}-Bu$

**153**

1,5-hydride ~

**154**   **155**

$S_Ei$   $Ad_E$ cyclization

**156**

Scheme 3.22

Evidence for a 1,5-hydride shift was obtained by generating cyclo-hexanecarbonyl tetrafluoroborate as the acylating agent [79]. Instead of the expected bicyclic ketone **160**, the fluoride **159** was obtained. In this case, the 1,5-hydride shift in the initially formed vinyl cation **157** gave a secondary carbenium ion (**158**) compared to the less favorable primary cation (**155**) of the acyclic derivative (Scheme 3.23).

1,5-Hydride shifts were also found to occur in the acylation of alkynes with 1-adamantanoyl tetrafluoroborate (**161**) or hexafluoroantimonate (**162**)

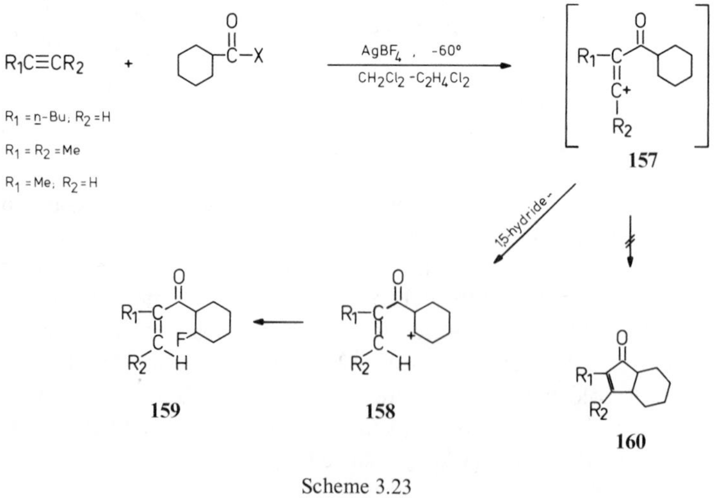

$R_1C\equiv CR_2 +$ [cyclohexane]$\overset{O}{\overset{\|}{C}}-X$ $\xrightarrow[\text{CH}_2\text{Cl}_2 -\text{C}_2\text{H}_4\text{Cl}_2]{\text{AgBF}_4 , \ -60°}$

$R_1 = \underline{n}-Bu, \ R_2 = H$

$R_1 = R_2 = Me$

$R_1 = Me, \ R_2 = H$

**157**

1,5-hydride ~

**159**   **158**   **160**

Scheme 3.23

[80]. The structure of the products formed depended on the nature of the acylium salt and the reaction media employed. Thus, the reaction of alkynes with **161** in methylene chloride–dichloroethane yielded the 2-fluorinated derivatives **165** (Nu = F), which is apparently formed by the reaction of the intermediate **164** with the $BF_4^-$ counteranion. The corresponding chlorides were the main products in the reaction of **162** with alkynes. These were probably formed by the abstraction of chloride ion from the solvent by the cation **164**. When aromatic hydrocarbons were used as solvents, the intermediate cation was intercepted to give compound **165** (Nu = Ar) (Scheme 3.24). Acylation of 2-butyne with **161** gave products derived from a 1,5-hydride shift (**165**, $R_1 = R_2 = Me$; Nu = F) and the cyclized ketone **166** in equal proportions. The latter cannot be formed by direct cyclization and reaction with fluoride ion, and it is apparently formed by a sequence of cyclization proton expulsion and HF addition. Compound **166** was found to be unstable and to eliminate HF to give **167**. All compounds (**165**) obtained from these alkynes had the $E$ configuration at the double bond ($R_2 = H$).

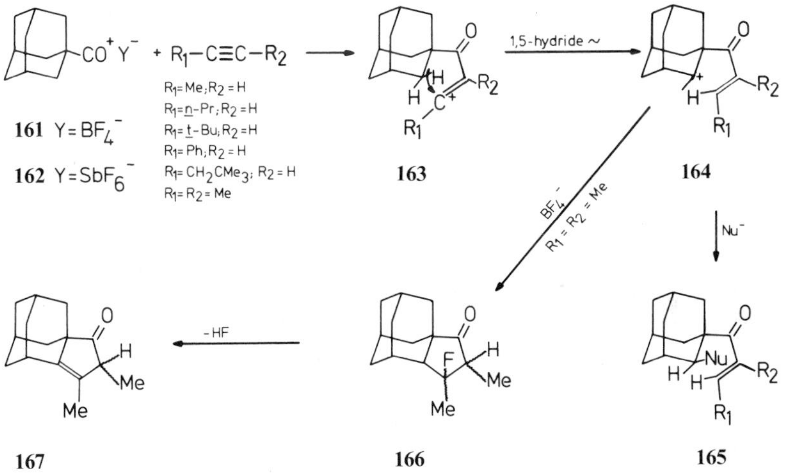

Scheme 3.24

## D. Addition of Halogens

Electrophilic additions of protons or carbenium ions to alkynes, with the probable exception of the addition of $FSO_3H$ to dialkyl-substituted alkynes, proceed via open vinyl cations or via $Ad_E3$ processes. Previous evidence suggesting prior coordination of protons to form $\pi$, complexes is now believed to be unsubstantiated [81].

However, with other electrophiles, and depending on the structure of the substrate, the intermediate vinyl cation may also exist in the bridged structure. The electrophile can form a $\pi$ complex (**168**) with the alkyne before the formation of the bridged structure or the vinyl cation. The bridged structure can be either $\pi$-bridged (169) or $\sigma$-bridged (170). The attack of the nucleophile usually takes place at the more positively charged carbon from a direction opposite the bridged or $\pi$ complex to give anti adducts with high stereoselectivity. The ring opening of three-membered cyclic structures has been analyzed with respect to a planar tetracoordinated structure in the transition state [82].

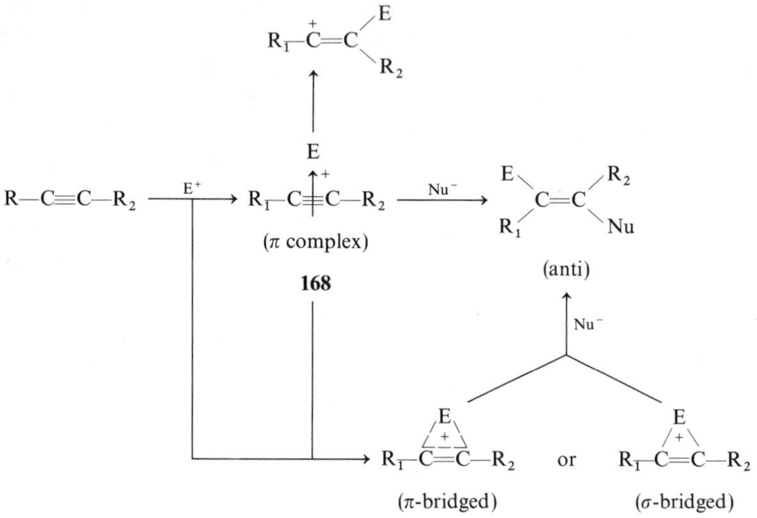

If the bridged intermediate or $\pi$ complex is formed rapidly and reversibly, the transition state for the addition has the structure 171, and the slow step of the reaction is the attack of the nucleophile on 171. This transition state gives a product with the same geometry as that for the $Ad_E3$ anti process, and it is generally difficult to prove that reversible formation of a bridged structure or $\pi$ complex is a necessary step in such a mechanism. Fahey [1b] also considers this variation an $Ad_E3$ anti (trans) addition mechanism.

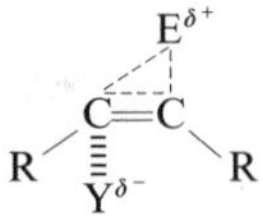

171

In the alkene series, the ionic addition of halogens proceeds via open and bridged ion intermediates [83]. Consequently, the investigation of the halogenation of alkynes is of considerable interest in relation to the possibility of bridging in the addition reaction.

Data on the fluorination of alkynes [1e] are scarce, and no conclusion on the reaction mechanism can be drawn. The additions of the other halogens will be discussed separately, and the reactivity of alkenes and alkynes in halogenation will be compared in Section I,G.

## 1. Addition of Chlorine

Hennion *et al.* investigated the chlorination of 1-hexyne in the late 1930's [84]. Mainly *E*-1,2-dichloro-1-hexene (**172**) was obtained in CCl$_4$, whereas chlorination in hydroxylic solvents gave a mixture of chloro compounds, including solvent-incorporated products. For example, chlorination in methanol at 25° yielded **172** and the solvent-incorporated products **173** and **174** [84]. These early results are of little mechanistic use since no kinetic data are available and formation of the products under kinetic control was not clearly demonstrated.

The addition of chlorine to phenylacetylene in methylene chloride in the presence of oxygen was reported to give equal amounts of *E*- and *Z*-**176** and **177**. The formation of these chlorinated products was ascribed to reactions

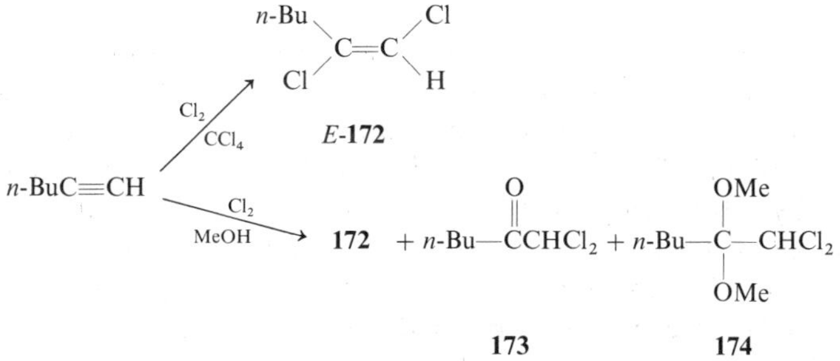

of an intermediate vinyl cation–chloride ion pair (**175**) [1b].

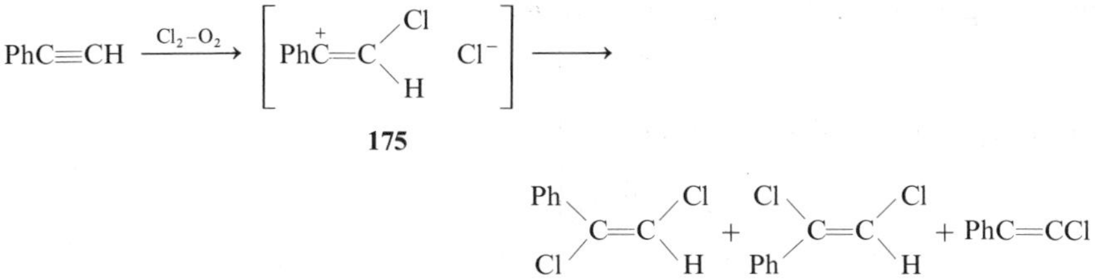

Chlorination of 1-butyne was shown to proceed through a radical mechanism [85], indicating the need for caution before suggesting a cationic mechanism for chlorination of alkynes. The chlorination of cyclopropylacetylene was shown to take place via an ionic mechanism [86]. The products obtained are the dichlorides *E*- and *Z*-**179** along with the rearranged products

**180** and **181** (Scheme 3.25). These are of the same type as the solvolysis products of 1-cyclopropyl-1-iodoethylene, in which a cyclopropyl-stabilized vinyl cation is invoked as an intermediate [87] (Chapter 5, Section IV,B). A radical mechanism can probably be ruled out, since oxygen does not inhibit the reaction or change the product composition significantly, and the vinyl cation **178** is therefore a likely intermediate in the chlorination of cyclopropylacetylene.

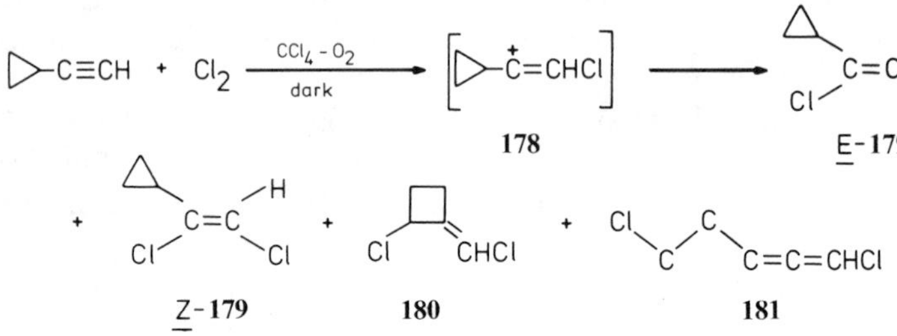

Scheme 3.25

The kinetics of the chlorination of terminal and disubstituted alkynes in acetic acid at 25° have been investigated by Yates and Go [88]. The addition was found to be second order—first order in both alkyne and chlorine. The electrophilic nature of the chlorination is shown by the rate decrease obtained by substituting the alkynes with electron-withdrawing groups. In contrast, alkyl substituents were found to increase the reaction rate. A $\rho^+$ value of $-4.32$ was found when the rates of chlorination of the ring-substituted phenylacetylenes were correlated with $\sigma^+$. This and the formation of solvent-incorporated products in the addition to the *syn*- and *anti*-dichloride with substituted phenylacetylene are in agreement with a mechanism involving a vinyl cation–chloride ion pair.

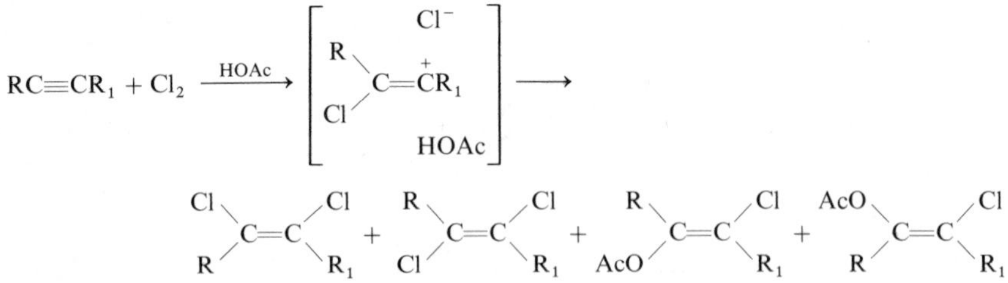

## 2. Addition of Bromine

Compared to chlorination, the addition of bromine to alkynes has been studied in some detail. In early kinetic studies of the bromination of several arylacetylenes in acetic acid, Robertson *et al.* [89] observed that the reaction is electrophilic and is similar to the bromination of alkenes. The reaction rate was found to obey mixed second- and third-order kinetics and was enhanced by electron-donating substituents. In contrast, the work of Sinn et al. [90] on the bromination of diphenylacetylene in bromobenzene is indicative of a nucleophilic reaction. A detailed study clarifying the reaction path of bromination of alkynes was conducted by Pincock and Yates [91].

The rates of addition of bromine in acetic acid to several alkyl- and aryl-substituted alkynes and the structure of the products formed were studied. The rate of addition was found to follow the general equation,

$$-d(Br_2)/dt = [\text{alkyne}]\,(k_2[Br_2] + k_3[Br_2]^2 + k_{Br-}[Br_2][Br^-])$$

At low bromine concentration, the rate constants $k_2$ were determined for all the alkynes studied. These values for a series of ring-substituted phenylacetylenes were found to correlate well with the $\sigma^+$ values, giving a $\rho^+$ value of $-5.2$ [91]. This high negative value indicates the development of extensive positive charge on the aryl-substituted carbon in the transition state and is comparable to the values obtained in the acid-catalyzed hydration of aryl alkynes (Table 3.2). A vinyl cation is therefore a likely intermediate.

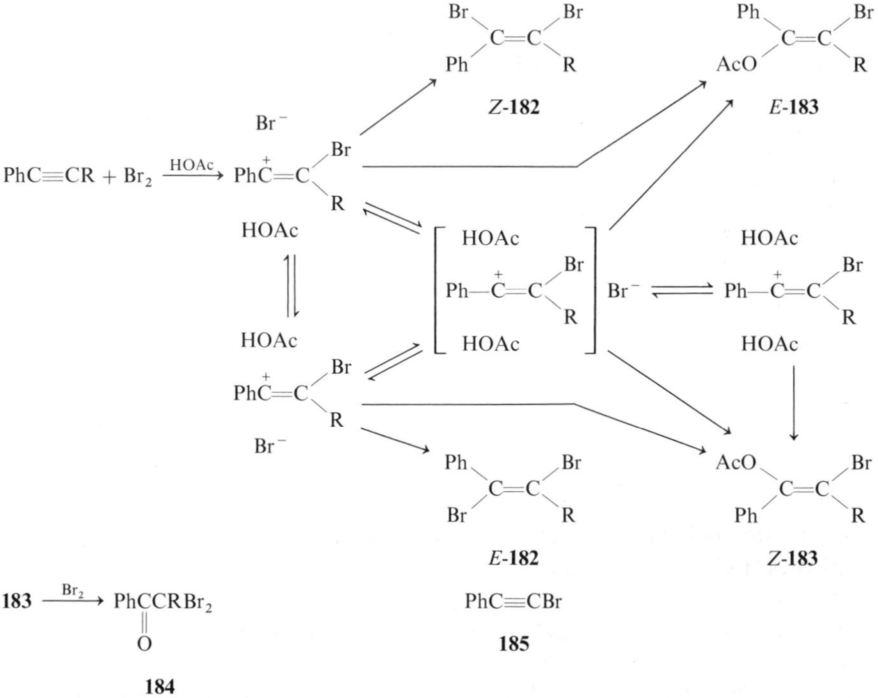

Scheme 3.26

The product distribution from phenylacetylene in the absence of added salt indicates that the bromination is not stereospecific under kinetic control, although the anti adduct predominates over the syn isomer. In addition to the dibromides **182** (R = H), the regiospecific solvent-incorporated products *E*- and *Z*-1-acetoxy-2-bromo-1-phenylpropene (**183**, R = H) and the secondary product **184** (R = H) were formed. Phenylacetylene also gave the bromoalkyne **185**, probably formed by the elimination of HBr from *Z*-**182** (R = H). The formation of both intimate and solvent-separated ion pairs was suggested (Scheme 3.26).

No solvent-incorporated product was found in the bromination of *p*-methylphenylacetylene. The *p*-methylphenylvinyl cation is considered to be more stable than its unsubstituted analogue, and it therefore reacted preferably with the more nucleophilic bromine.

Addition of $0.1 M$ LiBr increased the proportion of the anti isomer in the additions to all the phenylacetylenes, and the anti isomer *E*-**182** constituted more than 99% of the adduct from phenylacetylene itself. It is clear that the bromide ion catalyzed the attack of molecular bromine by the $Ad_E3$ mechanism through a transition state such as **186**.

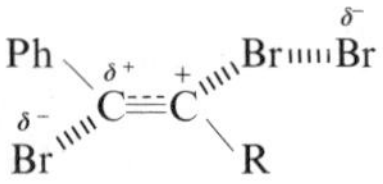

**186**

Alkyl-substituted alkynes gave predominantly *anti*-dibromo adducts, and no syn adduct or solvent-incorporated products were detected [91]. This observation indicates that the electrophilic addition occurs via a cyclic bromonium ion (**187**), which is probably more stable than the linear substituted bromovinyl cation. Ring opening of **187** obviously gives the anti adduct. Bromination in the presence of LiBr again gave the anti adduct almost exclusively via the $Ad_E3$ mechanism.

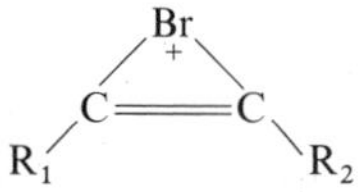

**187**

$$XC_6H_4CH_2C{\equiv}CH$$

**188**   X = H, 4-MeO, 4-Me, 3-CF$_3$

The stereochemistry of the adducts in the addition of bromine to 3-aryl alkynes (**188**) is influenced by the substituent on the phenyl group. The *anti*-dibromide predominated except for the 4-methoxy derivative, which gave more of the syn adduct *Z*-**191** [92]. The product distribution and the $k_2$ values are given in Table 3.19. The 4-methoxy derivative **188** (X = MeO) reacted 2.5 times faster than the unsubstituted derivative. On the basis of the larger amount of syn adduct, the lack of solvent-incorporated products,

**TABLE 3.19**

Product Distribution and $k_2$ Values for the Bromination of 3-Arylpropynes in Acetic Acid at 25° [92]

| | | In AcOH | | In AcOH/0.1 $M$ LiBr | |
|---|---|---|---|---|---|
| $XC_6H_4CH_2C{\equiv}CH$ | $10^3 k_2$ $M^{-1} \sec^{-1}$ | syn ($Z$-191), isomer ratio | anti ($E$-191), isomer ratio | syn ($Z$-191), isomer ratio | anti ($E$-191), isomer ratio |
| X = H | 0.75 | 1 | 99 | 3 | 97 |
| X = $m$-CF$_3$ | 0.64 | 2 | 98 | 2 | 98 |
| X = $p$-Me | 0.82 | 9 | 91 | 2 | 98 |
| X = $p$-MeO | 2.02 | 83 | 17 | 16 | 84 |

and the higher rate for **188** (X = MeO), Pincock and Somawardhana proposed that the cyclic phenonium ion **189** is an intermediate. The phenonium ion formation occurs in the direction anti to the original electrophilic attack by the bromine. Hence, opening by the nucleophilic bromide ion gives only the syn product (Scheme 3.27).

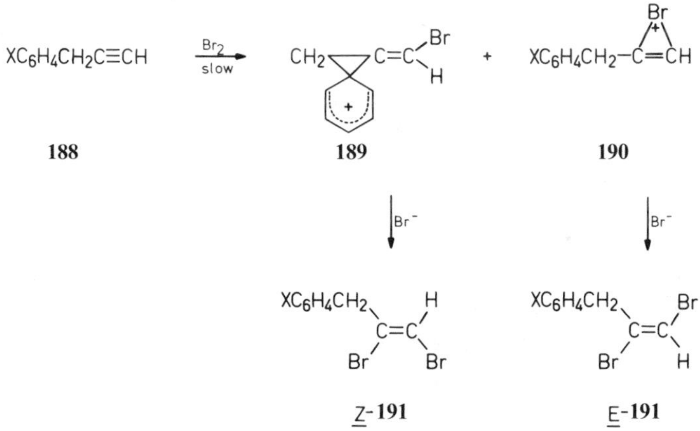

Scheme 3.27

In contrast, the other 3-arylpropynes were suggested to react via the bromonium ion **190**, which opens predominantly with the formation of anti addition products. The Hammett $\rho$ value is only $-0.2$, indicating that only a very small positive charge is developed on C-2 during the bromination. All the 3-arylpropynes reacted via an $Ad_E3$ process in the presence of 0.1 $M$LiBr, as shown by the preferential formation of anti adducts $E$-**191** (Table 3.19).

It is interesting that an $Ad_E3$ mechanism probably explains the preferential formation of the unsubstituted $Z$ isomer $Z$-**191** by anti addition of HBr to the triple bond of 1-bromo-3-phenylpropyne.

$$PhCH_2C{\equiv}CBr \xrightarrow[\text{HOAc}]{\text{HBr}} E\text{-191} + Z\text{-191}$$

$$(11\%) \quad (89\%)$$

The small negative Taft's $\rho^*$ value of $-1.0$ for the $k_3$ process obtained from the study of the bromination of several alkyl-substituted alkynes fits more with the formation of a cyclic bromonium ion than with the formation of an intermediate vinyl cation [93].

Cyclopropylacetylene was found to react with bromine either by an ionic or a radical mechanism [86], depending on the conditions. The radical addition by ultraviolet irradiation gave fewer ring-opened products than bromination under ionic conditions. The latter was similar to the addition of chlorine via a vinyl cation intermediate (Scheme 3.27).

No mechanistic studies have been conducted on the addition of halogens to cyclic alkynes. The bromination of cyclooctyne has been reported to give $76\%$ $Z$-1,2-dibromocyclooctene (192) [94]. The bromination of cyclononyne in $CCl_4$ solution did not yield any transannular product as expected for medium rings. Only $Z$-1,2-dibromocyclononene (193) was isolated and characterized as the bromination product [95], whereas the mechanism was not investigated. An intimate vinyl cation–bromide ion pair and a concerted syn molecular addition are likely possibilities. The eight- and nine-membered cyclic vinyl cations are apparently not very strained, as shown by their solvolytic generation (Chapter 5, Section V,B).

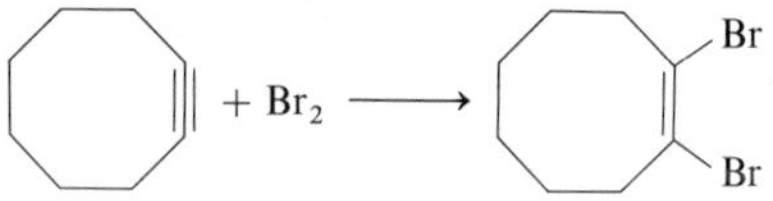

192

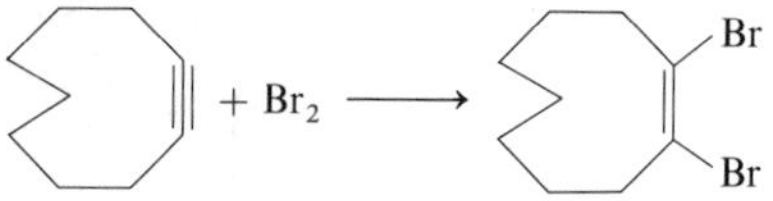

193

The syn/anti ratio of the dibromo adducts obtained by bromination of dianisylacetylene was found to be solvent dependent [48]. The $Z$ isomer $Z$-195 was formed to an extent of 45, 23, 19, and $0\%$ in $CCl_4$, acetic acid, hexane, and acetone, respectively. The stabilizing influence of the anisyl substituent suggests a bromination via an intermediate vinyl cation (194).

194      $Z$-195      $E$-195

## E. Iodination

Berliner and co-workers studied the addition of iodine to substituted sodium phenylpropiolates in water [96]. The reaction was conducted in the presence of iodide ion and was found to be termolecular with a first-order dependency on the substrate, the free iodine, and the iodide ion. Hammett plots using $\sigma$ or $\sigma^+$ were nonlinear, and from the correlation with $\sigma^+$ for the three fast-reacting compounds **196** (X = $p$-OMe, $p$-Me, $m$-Me) a $\rho^+$ value of $-1.66$ was found, whereas a value of $-0.77$ was calculated for the remaining compounds. The low negative $\rho^+$ values show less charge separation in the transition state, which is slightly different for each compound depending on the substituent.

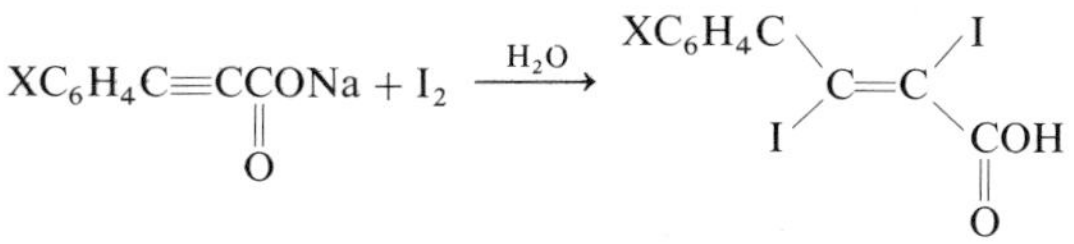

$$XC_6H_4C\equiv CCONa + I_2 \xrightarrow{H_2O} \underset{I}{\overset{XC_6H_4C}{>}}C=C\underset{COH}{\overset{I}{<}}$$

$$O$$

**196**                                **197**

$$\underset{\delta\ I}{\overset{XC_6H_4C}{>}}C\cdots C\underset{CO_2^-}{\overset{I^{\delta+}}{<}}$$

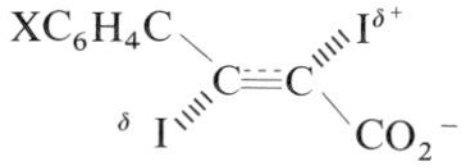

**198**

The products obtained were predominantly the $E$-diiodocinnamic acids **197**, indicating that anti addition is favored. No solvent-incorporated products, i.e., keto acids were formed. These results best fit an $Ad_E3$ termolecular mechanism via transition state **198**. The transition states are assumed to be slightly different for each compound, depending on the substituents, and therefore not completely synchronous. Iodination of propiolic acids also was found to be termolecular and not involving a vinyl cation [97].

In conclusion, electrophilic halogenation of alkynes, at least in the case of bromination, sometimes involves cyclic cationic intermediates. Alkyl-substituted alkynes prefer to react via cyclic onium cations, whereas $\alpha$-phenyl-substituted alkynes, in which the nonbridged ion is more stable, react via the intermediacy of the open ion.

## F. Addition of Sulfenyl and Selenenyl Halides

### 1. Sulfenyl Halides

The addition of sulfenyl halides to alkynyl triple bonds to form 1:1 adducts has been studied both synthetically and mechanistically in a fairly detailed manner. The halogen atom is usually chlorine and sometimes bromine; sulfenyl iodides and fluorides are rarely used in the addition reaction. The addition reaction was first carried out by Kharasch and Assony [98], who showed that a catalytic amount of aluminium chloride was required in the addition of 2,4-dinitrobenzenesulfenyl chloride with acetylene, whereas alkyl-substituted alkynes reacted in the absence of catalyst. Acetylenedicarboxylic acid and its ester did not react with sulfenyl chlorides, and this deactivation by electron-withdrawing groups indicates that the rate-determining step involves an electrophilic attack on the triple bond.

Modena *et al.* showed that the addition of sulfenyl halides to terminal alkynes gives a mixture of two compounds, one having the Markownikoff (M) orientation (199) and the other the anti Markownikoff (aM) orientation (200) [99]. The structure of the alkyne and the solvent affect the orientation

of the products and are interrelated. Moreover, the reaction rate is also dependent on the solvent and on the alkyl substituents at the triple bond but is little affected by substituents in ArSCl, except for the *o*-nitro group, which strongly reduces the rate [99]. Alkanesulfenyl chlorides were found to have a reactivity similar to that of benzenesulfenyl chloride. The structure of the sulfenyl chloride seems to be the least important factor in determining the orientation of the addition [99]. However, recent work in liquid sulfur dioxide [100] showed some effect of the substituent in the sulfenyl chloride on the reactivity. Alkanesulfenyl chlorides reacted slower than arylsulfenyl chlorides in this solvent. The effects of the substituents at the phenyl ring are moderate, giving a $\rho$ value of $+1.1$.

Schmid and Heinola found a nonregiospecific addition of 2,4-dinitrobenzenesulfenyl chloride to 1-phenylpropyne in chloroform [101]. The Markownikoff adduct *E*-202 was formed in 94% yield, and the anti Markownikoff *E*-203 was formed in 6% yield. The stereochemistry of the addition is anti, as judged from the structure of the adducts *E*-202 and *E*-203. On the basis of the second-order kinetics, the anti stereospecificity but nonregiospecific addition, a mechanism involving a bridged rate-determining transition state leading to a thiirenium ion (201), was suggested by Modena [99] and Schmid [101]. Chloride attack at either ring carbon of 201 results in formation of the adduct (Scheme 3.28). The intermediacy of the structurally related thiirenium ions formed by neighboring sulfur participation in solvolysis is discussed in Chapter 7, Section III,D.

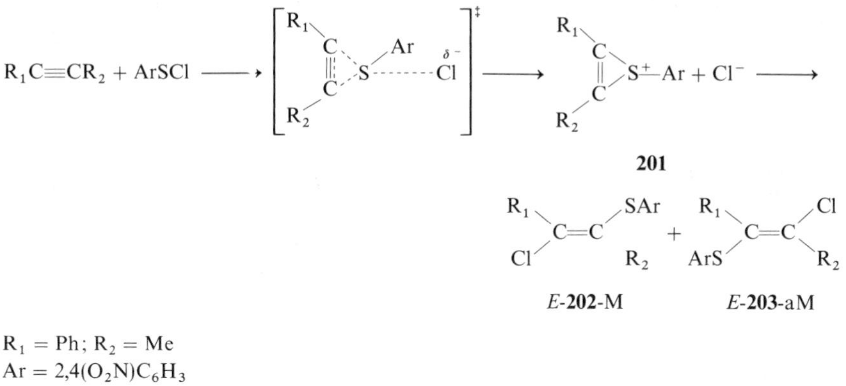

$$R_1C\equiv CR_2 + ArSCl \longrightarrow \left[ \begin{array}{c} R_1 \\ C \\ \| \quad S \cdots\cdots Cl \\ C \\ R_2 \end{array} \overset{Ar}{\underset{\delta^-}{}} \right]^{\ddagger} \longrightarrow \begin{array}{c} R_1 \\ C \\ \| \quad S^{\pm}\!\!-Ar + Cl^- \\ C \\ R_2 \end{array} \longrightarrow$$

**201**

$$\begin{array}{cc} R_1 \quad SAr \\ C=C \\ Cl \quad R_2 \end{array} + \begin{array}{cc} R_1 \quad Cl \\ C=C \\ ArS \quad R_2 \end{array}$$

E-**202**-M          E-**203**-aM

$R_1 = Ph; R_2 = Me$
$Ar = 2,4(O_2N)C_6H_3$

Scheme 3.28

Okuyama et al. [102] provided further evidence for the cyclic thiirenium ion intermediate. They observed a small inverse hydrogen isotope effect $k_H/k_D$ of 0.92 in the addition of 2,4-dinitrobenzenesulfenyl chloride to phenylacetylene and its 1-deutero derivative in acetic acid. This value is smaller than the $k_H/k_D$ value of 1.11 found for the acid-catalyzed hydration of the same substrates [27], which was ascribed to the hyperconjugative effect prevailing over the rehybridization effect. The inverse isotope effect observed suggests a negligible role for the hyperconjugative effect, in line with a thiirenium ion transition state that has a small electron deficiency at the carbon atom. The reaction proceeds with high anti stereospecificity and in the Markownikoff sense, since the product E-1-chloro-2-(2',4'-dinitro-phenylthio)-1-phenylpropene (E-**202**) is formed in >98% yield (in AcOH).

Substitution of the terminal hydrogen by a methyl group enhanced the addition rate of 2,4-dinitrobenzenesulfenyl chloride by a factor of 10 to 20 when the pairs phenylacetylene/1-phenylpropyne and 1-hexyne/2-hexyne were compared [102]. This behavior is the opposite of that found for the acid-catalyzed hydration of alkynes in which a vinyl cation intermediate is involved (Section I,A,1) but is in accord with an intermediate cyclic cation. The $\rho$ value of $-1.46$ for this addition [102] is similar to the value of $-1.25$ obtained by Modena et al. for the solvolysis of 1,2-diaryl-2-arylthiovinyl 2,4,6-trinitrobenzenesulfonates [103], in which an intermediate thiirenium cation was postulated (Chapter 7, Section III,D). Okuyama et al. noted that the value for the sulfenylation reaction is significantly smaller than the values found for the sulfenylation of the olefinic analogues, 1-phenylpropenes

($\rho = -2.64$) and styrenes ($\rho = -2.41$), but close to the values for non-conjugated alkylbenzenes and phenyl allenes ($\rho = -2.0$). Therefore, they suggested that electronic delocalization to the positive sulfur occurs from the $\pi$ bond that is orthogonal to the phenyl group and that the other $\pi$ bond is in conjugation with the phenyl group in the intermediate thiirenium ion, as in **204**.

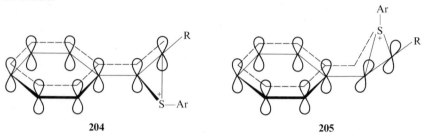

| 204 | 205 |

This proposal of Okuyama was challenged by Schmid *et al.* [104]. They view the thiirenium ion as stabilized by conjugation with the phenyl ring, as in **205**, with the $\pi$ orbitals of the double bond and the benzene ring being orthogonal in the transition state. Reaction of 4-chlorobenzenesulfenyl chloride with ethylene, acetylene, and their phenyl-substituted derivatives supports this kind of stabilization. The kinetics show that substitution of a phenyl group in the acetylene has a large effect on the rate of the reaction, whereas corresponding substitution of ethylene has practically no effect (Table 3.20). This greater effect for alkynes indicates that the phenyl ring stabilizes the transition state more than it stabilizes the ground state. If the

**TABLE 3.20**

**Relative Rate Constants and Product Distribution for the Addition of 4-Chlorobenzenesulfenyl Chloride to Alkenes and Alkynes in 1,1,2,2-Tetrachloroethane at 25° [104]**

| | | | $R_1CH{=}CHR_2$ | | | $R_1C{\equiv}CR_2$ | |
| | | | $M^a$ | $aM^b$ | | $E\text{-}M^a$ | $E\text{-}aM^b$ |
| $R_1$ | $R_2$ | $k_{rel}$ | isomer ratio | isomer ratio | $k_{rel}$ | isomer ratio | isomer ratio |
|---|---|---|---|---|---|---|---|
| H | H | 1.0 | — | — | 1.0 | — | — |
| Ph | H | 0.95 | 100 | — | 113.0 | 100 | — |
| Ph | Me | 1.81 | 100 | — | $3.94 \times 10^3$ | 61 | 39 |
| Ph | Et | — | — | — | $7.71 \times 10^3$ | 51 | 49 |
| Ph | Ph | 0.12 | — | — | 116 | — | — |

$^a$ Markownikoff adduct.
$^b$ anti Markownikoff adduct.

stabilization is that proposed by Okuyama, the inductive effect of the aryl ring would influence the rate of addition, which should correlate with the sum of Taft inductive substituent constants ($\Sigma\sigma^*$), but no such correlation was observed. If similar ground state and transition state stabilizations are present in the case of phenylacetylene, the rate enhancement by phenyl substitution of acetylene would not be large. It would be comparable to that for the ethylene derivatives, in which the stabilization of the ground state and the transition state by the phenyl ring are exactly balanced (Table 3.20). Consequently, structure **205**, in which the charge is delocalized into the phenyl ring, is more compatible with the data.

Schmid et al. state that the rate data of Okuyama et al. [102] correlate better with $\sigma^+$ values (with $\rho^+ = -1.35$) than with $\sigma$ values, showing the importance of conjugation between the substituents and the reaction center. Another convincing argument for the importance of **205** rather than **204** is the formation of 100% E Markownikoff adduct with phenylacetylene. Structure **206**, in which the phenyl ring is in the same plane as the thiirenium ring, would sterically hinder the attack of chloride ion at the α-carbon atom, and preferential attack at $C_\beta$ would give the E anti Markownikoff product for **206** (R ≠ H).

**206**

The addition of 4-chlorobenzenesulfenyl chloride to acetylene and to alkyl-substituted alkynes in 1,1,2,2-tetrachloroethane follows second-order kinetics [105]. Substitution of the terminal hydrogen atoms by alkyl groups enhanced the reaction rate, and the rate constants followed the Taft equation: $\log k_2 = -4.47\,\Sigma\sigma^* + 1.64$. Points for the tert-butyl-substituted alkynes deviated from this relationship. The products of addition to the terminal alkynes gave a mixture of the Markownikoff and the anti Markownikoff adducts (Table 3.21). The overall stereochemistry of sulfenyl chloride addition is anti in nature. These results were interpreted as supporting a cyclic thiirenium ion intermediate. The alkyl groups exert a strong steric effect on the product-determining transition state **207** in which the carbon atom carrying the substituent, the thiirenium ring, and the chloride ion all lie in the same plane (cf. **206**). Consequently, the entering chloride ion is subject to the steric hindrance of the alkyl substituent, and, as the size of the substituent increases from Et to i-Pr to t-Bu, the amount of product with Markownikoff orientation decreases. The reason for the small amount

**TABLE 3.21**

**Stereochemistry of the Addition of 4-Chlorobenzenesulfenyl Chloride to Alkyl-Substituted Alkynes [105]**

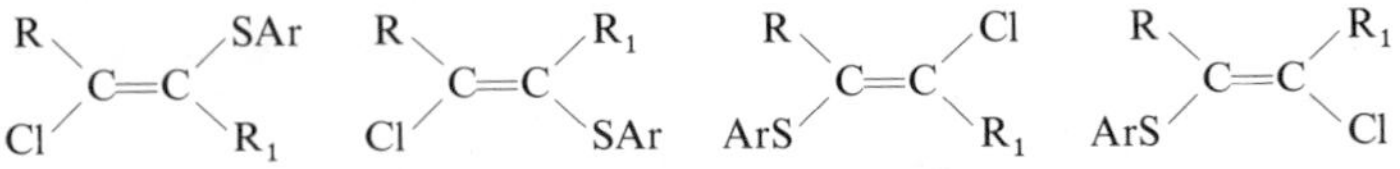

| $RC{\equiv}CR_1$ | | $R{-}C{=}C{-}SAr, Cl, R_1$ | $R{-}C{=}C{-}R_1, Cl, SAr$ | $R{-}C{=}C{-}Cl, ArS, R_1$ | $R{-}C{=}C{-}R_1, ArS, Cl$ |
|------|------|--------------|--------------|---------------|---------------|
| | | *E*-M | *Z*-M | *E*-aM | *Z*-aM |
| R | $R_1$ | isomer ratio | isomer ratio | isomer ratio | isomer ratio |
| Me | H | 14 | 0 | 86 | 0 |
| Et | H | 10 | 0 | 90 | 0 |
| *i*-Pr | H | 27 | 0 | 73 | 0 |
| *t*-Bu | H | 0 | 0 | 100 | 0 |
| *n*-Pr | H | 16 | 0 | 84 | 0 |
| *n*-Bu | H | 20 | 0 | 80 | 0 |
| Me | Et | 60 | 0 | 40 | 0 |
| Me | *i*-Pr | 48 | 0 | 52 | 0 |
| Me | *t*-Bu | 12 | 5 | 76 | 7 |

of Markownikoff- and $Z$ anti Markownikoff-oriented products in the case of 4,4-dimethyl-2-pentyne is not clear [105].

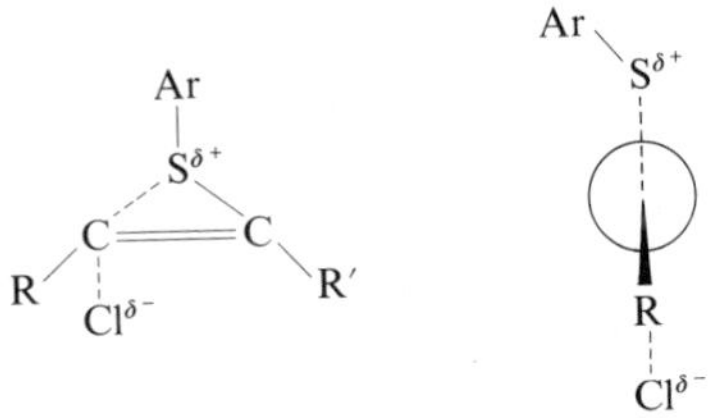

**207**

A thiirenium ion intermediate was proposed to account for the products of addition of disulfur dichloride to phenylacetylenes [106]. Both Markownikoff- and anti Markownikoff-oriented products were obtained in a ratio that depended on the inductive effect of the substituent R (Scheme 3.29). Only $E$ isomers were formed exclusively by anti addition. The intermediate adducts **208** and **209** cannot be isolated and react immediately. When R is an electron-donating group, adduct **208** expels sulfur and reacts with another molecule of alkyne to give the divinyl sulfide **210**. On the other hand, the Markownikoff adduct **209** undergoes intramolecular electrophilic aromatic substitution to form benzothiophene (**211**). The formation of the cyclic products also indicates that the stereochemistry of the addition is anti.

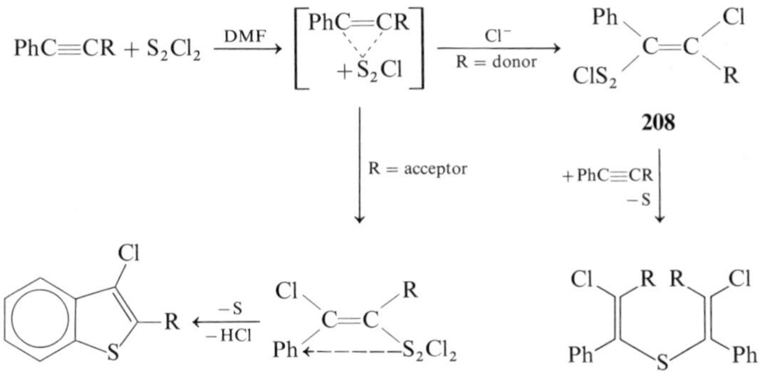

Scheme 3.29

The addition of benzenesulfenyl chloride to ethyl 3-butynoate also proceeds in an anti manner, giving the $E$ anti Markownikoff isomer 212 [107].

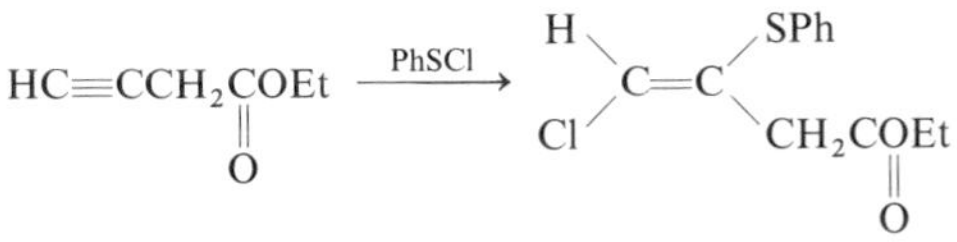

$E$-212-aM

Capozzi et al. [108] were able to observe the trimethylthiirenium ion 214 in liquid sulfur dioxide by nmr (Chapter 8). They were also able to isolate and characterize [109] the stable thiirenium salt 215, which is formed by the addition of 2,2,5,5-tetramethyl-3-hexyne to the salt 213. The high stability of 215 is shown by the fact that the reaction of methanesulfenyl chloride and 2,2,5,5-tetramethyl-3-hexyne in liquid sulfur dioxide gave the chloride 216. When the reaction was carried out in methylene chloride at low temperature, slow opening of 216 gave the $E$ adduct 217. The structure of cyclic thiirenium species has been proved by X-ray analysis of the tetrafluoroborate salt 218 [110]. These results are concrete proof for the existence of thiirenium ion intermediates in the addition of sulfenyl halides to alkynes.

Recent kinetic studies of the addition of several sulfenyl chlorides to 2-butyne and 3-hexyne in $SO_2$ at $-67°$ have given further support to the two-step process involving the thiirenium ion [100]. The effect of the substituents at the sulfur on the rate of the reaction observed in this medium led Capozzi et al. [100] to consider the possibility of a sulfurane structure (219) before the formation of the thiirenium ion in weakly polar solvents. Clearly, more work has to be done to verify this hypothesis.

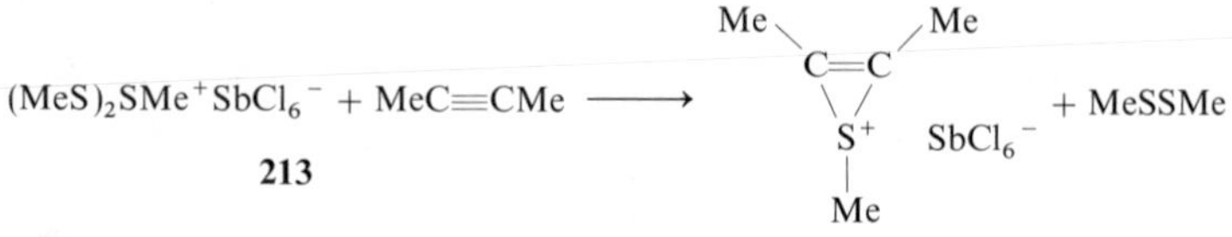

$(MeS)_2SMe^+SbCl_6^- + MeC{\equiv}CMe \longrightarrow$ **213**

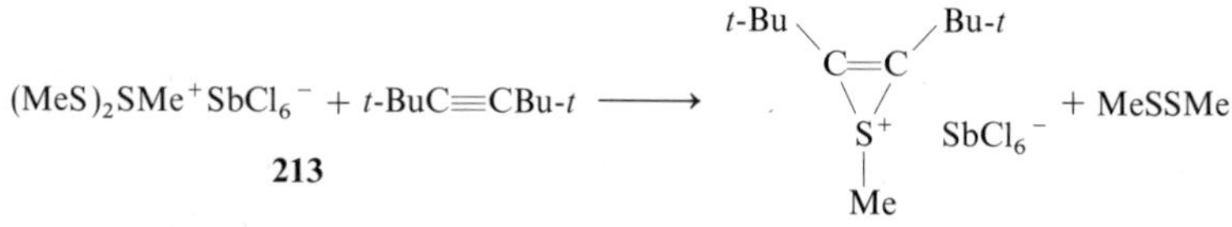

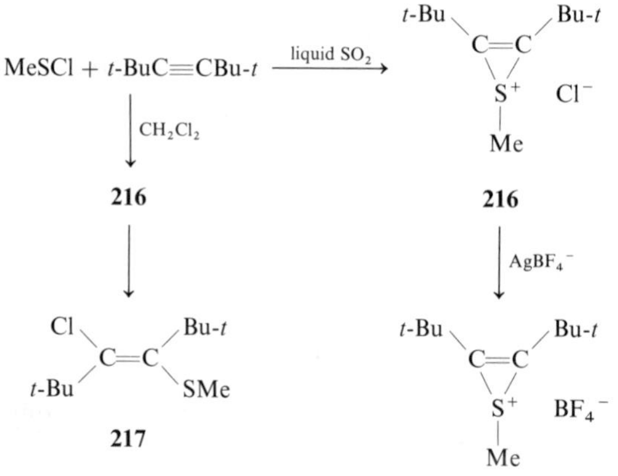

$+ MeSSMe$

**214**

$(MeS)_2SMe^+SbCl_6^- + t\text{-}BuC{\equiv}CBu\text{-}t \longrightarrow$ **213**

$+ MeSSMe$

**215**

$MeSCl + t\text{-}BuC{\equiv}CBu\text{-}t \xrightarrow{\text{liquid } SO_2}$

$\Big\downarrow CH_2Cl_2$

**216**                              **216**

$\Big\downarrow AgBF_4^-$

**217**

**218**

$R{-}C{\equiv}C{-}R' + ArSCl \rightleftharpoons$

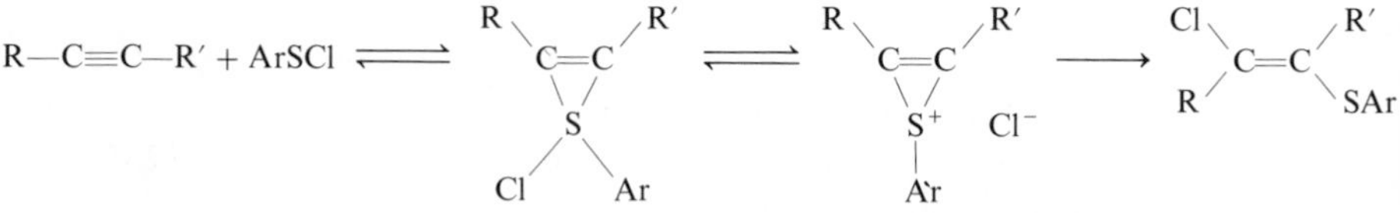

**219**

## 2. Selenenyl Halides

Schmid and Garrat [111] studied the addition of benzeneselenenyl chloride to alkynes. The reaction is overall second order—first order in both the alkyne and the selenenyl chloride. The *E* anti Markownikoff products **221**, which were formed by an anti stereospecific and a non-regiospecific addition, were the major products. The polar effects of alkyl substituents on the triple bond were shown by the $\rho^*$ value of $-2.09$. The steric effect of the alkyl group on the product composition was very similar to that for the addition of 4-chlorobenzenesulfenyl chloride to alkyl-substituted alkynes [105]. A two-step mechanism involving a bridged selinirenium ion (**220**) intermediate was therefore suggested for this reaction [111].

$$\text{PhSeCl} + \text{RC}\equiv\text{CH} \longrightarrow
\underset{\textbf{220}}{
\begin{array}{c} \text{Ph} \\ | \\ \text{Se}^+ \\ / \backslash \\ \text{C}{=}\text{C} \\ R^{\diagup} \quad {}^{\diagdown}\text{H} \end{array}}
\xrightarrow{\text{Cl}^-}
\underset{E\text{-221-aM}}{
\begin{array}{c} \text{Cl}\diagdown \quad \diagup \text{H} \\ \text{C}{=}\text{C} \\ R^{\diagup} \quad {}^{\diagdown}\text{SePh} \end{array}}
+
\underset{E\text{-221-M}}{
\begin{array}{c} \text{PhSe}\diagdown \quad \diagup \text{H} \\ \text{C}{=}\text{C} \\ R^{\diagup} \quad {}^{\diagdown}\text{Cl} \end{array}}
$$

## G. Miscellaneous

The protonation of the excited state of alkynes has been reported [112]. This method of hydration of alkynes was found to depend on the acid concentration. At higher concentration of acid, a higher conversion to ketones occurs in shorter times by regiospecific addition of water to the triple bond in the Markownikoff sense. Although the basicity of alkynes is higher at the excited state, the addition of water could be carried out under milder acidic conditions (0.5 $N$ acid). No addition occurred in the presence of sodium hydroxide, supporting the mechanism involving a vinyl cation intermediate of the type **222**.

$$\text{PhC}\equiv\text{CR} \xrightarrow[\text{H}^+]{hv} \overset{+}{\text{PhC}}{=}\text{CHR} \xrightarrow{\text{H}_2\text{O}} \underset{\textbf{222}}{\overset{\overset{\text{OH}}{|}}{\text{PhC}}{=}\text{CHR}} \longrightarrow \overset{\overset{\text{O}}{\|}}{\text{PhC}}\text{CH}_2\text{R}$$

R = H, Me, Ph

Gas-phase isomerization of alkynyl hydrocarbons on acid catalysts was observed to proceed via vinyl cation intermediates **223** and **224** [113]. 1,2-Hexadiene was the initial product in the conversion of 1-hexyne to 2-hexyne in the presence of $SiO_2-Al_2O_3$ and $H_3PO_4-SiO_2$ catalysts.

$$\text{HC}\equiv\text{CCH}_2\text{Pr-}n \underset{}{\overset{\text{H}^+}{\rightleftharpoons}} \underset{\textbf{223}}{\text{H}_2\text{C}{=}\overset{+}{\text{C}}\text{CH}_2\text{Pr-}n} \underset{}{\overset{-\text{H}^+}{\rightleftharpoons}} \text{H}_2\text{C}{=}\text{C}{=}\text{CPr-}n \overset{\text{H}^+}{\rightleftharpoons}$$

$$\underset{\textbf{224}}{\text{Me}\overset{+}{\text{C}}{=}\text{CPr-}n} \overset{-\text{H}^+}{\rightleftharpoons} \text{MeC}\equiv\text{CPr-}n$$

Rearranged and cyclized products were obtained when 3,3-dimethyl-1-butyne and 3,3-dimethyl-1-phenyl propyne were heated over $H_3PO_4-SiO_2$ catalysts (Scheme 3.30). In both reactions, 1,2-methyl shifts were observed, which in the latter case led to the formation of 1,1,2-trimethylindene (**227**). If the reaction was carried out in the presence of water, there was a competition between hydration and isomerization. These results were explained

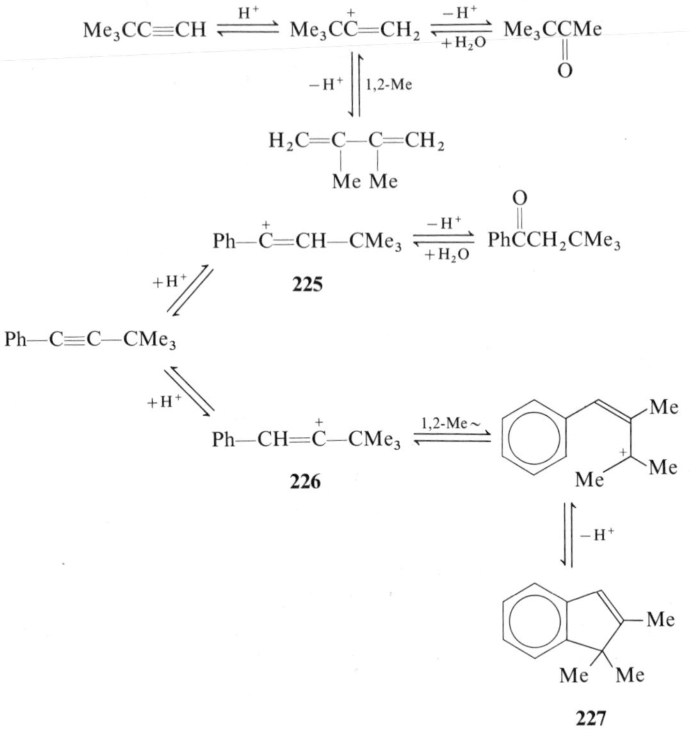

$$Me_3CC\equiv CH \overset{H^+}{\rightleftharpoons} Me_3C\overset{+}{C}=CH_2 \underset{+H_2O}{\overset{-H^+}{\rightleftharpoons}} Me_3CCMe$$

with $\underset{O}{\overset{}{\parallel}}$ on the $Me_3CCMe$ and

$$-H^+ \bigg\Vert\ 1,2\text{-Me}$$

$$\underset{\underset{Me\ Me}{|\quad\ |}}{H_2C=C-C=CH_2}$$

$$Ph-\overset{+}{C}=CH-CMe_3 \underset{+H_2O}{\overset{-H^+}{\rightleftharpoons}} Ph\overset{O}{\overset{\parallel}{C}}CH_2CMe_3$$

**225**

$$+H^+ \nearrow$$

$$Ph-C\equiv C-CMe_3$$

$$+H^+ \searrow$$

$$Ph-CH=\overset{+}{C}-CMe_3 \overset{1,2\text{-Me}\sim}{\rightleftharpoons}$$

**226**

$$\Big\Vert\ -H^+$$

**227**

Scheme 3.30

by postulating vinyl cation intermediates. It is interesting that both types of vinyl cations **225** and **226** are formed from 3,3-dimethyl-1-phenyl propyne. Acid-catalyzed alkyne–allene isomerization in dry sulfolane with $HBF_4$, $HPF_6$, and $H_2SO_4$ also gave similar results [114]. In all the catalytic systems of acids studied, a sequential isomerization of alkyne–allene was noted, with allenes undergoing the isomerization faster than the alkynes. 2-Hexyne was found to have more thermodynamic stability. This isomerization was also explained by postulating vinyl cations (see compounds **223** and **224**) as intermediates with subsequent deprotonation to give alkynes and allenes.

$$HC\equiv CCH_2Pr\text{-}n \rightleftharpoons H_2C=C=CPr\text{-}n \rightleftharpoons MeC\equiv CPr\text{-}n \rightleftharpoons$$

$$MeC=C=CEt \rightleftharpoons EtC\equiv CEt$$

Mention can be made here of the frequently used hydration method in the synthesis of alkynes with Hg(II) ions as catalysts [4]. The mechanism of this reaction is represented as electrophilic attack on the alkyne by Hg(II) to give an organomercury compound (**228**), from which the metal is displaced by a proton. The intermediacy of a mercury $\pi$ complex has not been clearly established. Other metal ions are also known to catalyze the hydration of alkynes, and work has been carried out on the reactions of gold(III) ions with alkynes [115].

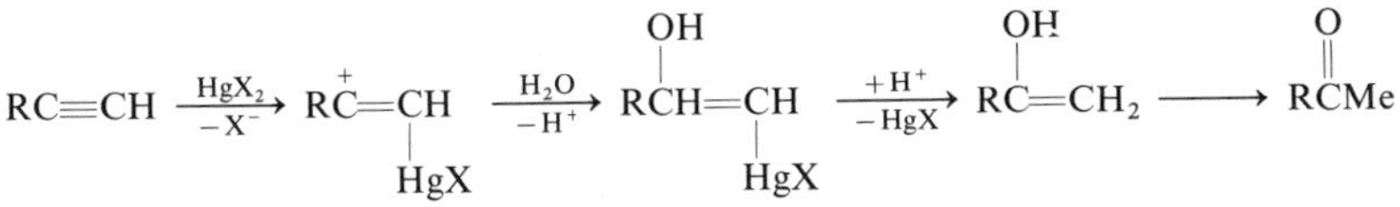

$$RC{\equiv}CH \xrightarrow[-X^-]{HgX_2} \underset{HgX}{R\overset{+}{C}{=}CH} \xrightarrow[-H^+]{H_2O} \underset{HgX}{R\overset{OH}{\overset{|}{C}H}{=}CH} \xrightarrow[-HgX]{+H^+} R\overset{OH}{\overset{|}{C}}{=}CH_2 \longrightarrow R\overset{O}{\overset{\|}{C}}Me$$

**228**

Acetoxy mercuration of 1-aryl-1-propynes proceeds by rearside attack on the intermediate mercurinium ion **229** to give both isomers of the addition product in a ratio decreasing with the ability of the aromatic substituents to stabilize the transition state for an electrophilic addition process [116]. The products are the vinylmercury chloride compounds **230** and **231**, which are formed by the reaction of initially formed vinylmercury acetates with KCl. As shown earlier (p. 55), the presence of $HgBr_2$ increases the stereospecificity of the addition of HBr to alkynes in the syn direction.

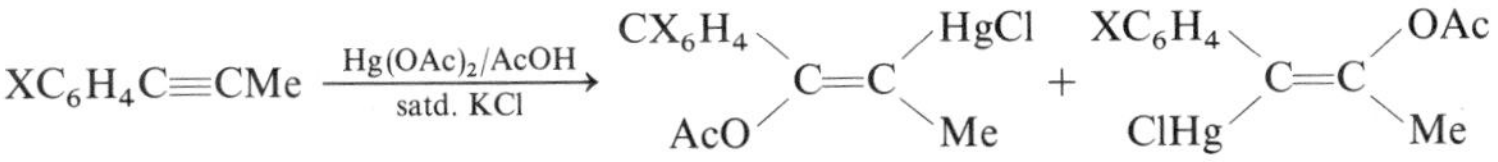

$$XC_6H_4C{\equiv}CMe \xrightarrow[satd. KCl]{Hg(OAc)_2/AcOH} \underset{AcO}{\overset{CX_6H_4}{>}}C{=}C\overset{HgCl}{\underset{Me}{<}} + \underset{ClHg}{\overset{XC_6H_4}{>}}C{=}C\overset{OAc}{\underset{Me}{<}}$$

**230**     **231**

$$\underset{\overset{|}{^-OAc}}{Ar\overset{HgOAC}{\overset{\overset{+}{C}{=}C}{\diagup\ \diagdown}}R}$$

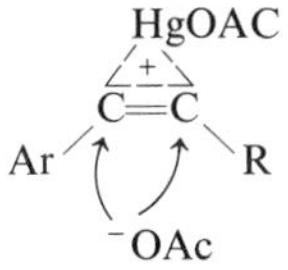

**229**

$X = p\text{-OMe}, p\text{-Me}, H, p\text{-F}, p\text{-Cl}, p\text{-Br}, m\text{-Cl}$

Vinyl cations (**232**) stabilized by platinum metal in the $\alpha$ position have been proposed as intermediates in the protonation of alkynylplatinum(II) acetylides [117].

$$PtC{\equiv}CH + HCl \rightleftharpoons Pt\overset{+}{C}{=}CH_2 \rightleftharpoons PtC(Cl){=}CH_2$$

**232**

Reaction of alkylphenylacetylenes with $CuCl_2$–LiCl or $LiCl$–$I_2$ in acetonitrile (under kinetic control) gave the corresponding dihaloalkenes **233** and **234** with the anti isomer predominating (>90%), except in the chlorination of 3,3-dimethyl-1-phenyl propyne, in which the major amount of the dichloride formed ($\sim$80%) has the syn orientation [118] (Scheme 3.31). The observation that in chlorination the isomer ratio (anti/syn) decreased markedly on changing the alkyl group from primary to secondary

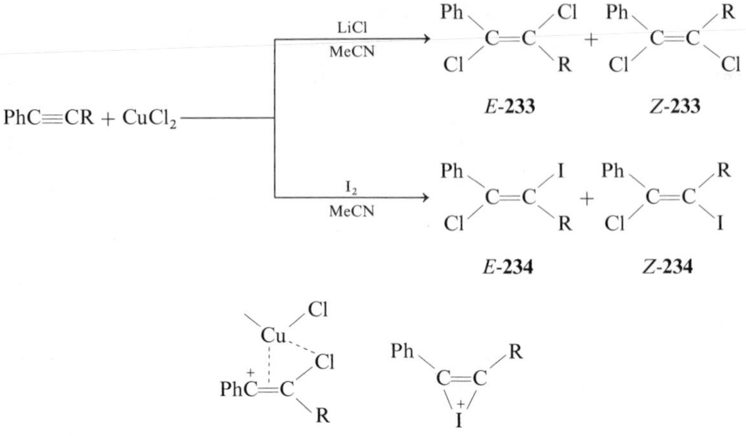

R = H, Me, Et, n-Pr, i-Pr, n-Bu, t-Bu

Scheme 3.31

and then to tertiary points to an open vinyl cation intermediate. This can be best represented as **235**, in which the Cu(I) salt coordinates weakly with the double bond and the chlorine atom. When R becomes sterically large, it can hinder attack on its own side in **235** by CuCl$_3^-$, leading to syn orientation, as in the case of 3,3-dimethyl-1-phenyl propyne. In the chloroiodination of alkynes, a cyclic iodonium ion (**236**) was suggested as the intermediate causing anti stereospecificity (100% for R = H, Me). The assumption of a cyclic intermediate in the case of iodine but not in the case of chlorine parallels the result of solvolysis of β-halogenovinyl sulfonates [119].

In the reaction of alkylphenylacetylenes with SbCl$_5$ in CCl$_4$ yielding the dichloro products E- and Z-**237** under kinetic control, the syn Z isomer was formed predominantly (~90%) irrespective of the alkyl substituent [120]. The very high syn stereospecificity was explained by assuming a concerted or nearly concerted molecular addition of SbCl$_5$ or its dimer to the triple bond, as in the case of alkenes. Chlorination of phenylpropyne and 3,3-dimethyl-1-phenyl propyne with Cl$_2$ in CCl$_4$ showed less stereospecificity for syn addition. The former gave a syn/anti ratio of 32:68, and the latter gave a ratio of 87:13, which can be explained by postulating the intermediacy of corresponding vinyl cations (Section I,D).

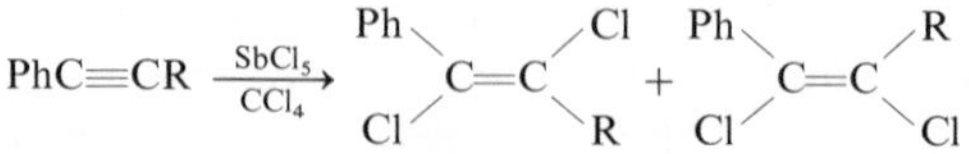

## H. Relative Reactivity of Alkynes and Alkenes toward Electrophilic Reagents

The preceding discussion gave clear evidence that vinyl cations, either free, ion-paired, or bridged, are intermediates in many electrophilic additions to alkynes. There are parallels with the generation of vinyl cations by this method in the formation of trisubstituted carbenium ions by electrophilic additions to alkenes [83]. It is therefore interesting to compare the ease of formation of the two classes of ions as a function of structural parameters in the unsaturated system, of the electrophile, and of the medium. These relative reactivities have been discussed by several authors [121–123].

A survey of the data on the relative reactivities of additions to alkenes and alkynes is given in terms of $k_o/k_a$ ratios in Table 3.22. Although some of the ratios compared were measured at different temperatures, this effects is not taken into account in the discussion below.

An $\alpha$-ferrocenyl group stabilizes a vinyl cation much more than does an $\alpha$-phenyl group. Hydration of ferrocenylacetylene is $10^5$ times faster than hydration of phenylacetylene, and hydration of ferrocenylphenylacetylene (**238**) gives exclusively ferrocenylbenzyl ketone (**239**) [124]. Styrylferrocene, the alkene analogue of **238**, did not react under these mild conditions (25% $H_2SO_4$), and reaction under more vigorous conditions ($CF_3COOH$ or concentrated $H_2SO_4$) gave polymers. Consequently, the addition to the alkyne is faster than the addition to the alkene, with an estimated nine order of magnitude difference between the rates in favor of the alkyne.

$$\text{FcC} \equiv \text{CPh} + \text{H}^+ \longrightarrow \text{Fc}\overset{+}{\text{C}} = \text{CHPh} \xrightarrow{\text{H}_2\text{O}} \text{FcC}\overset{\overset{\text{O}}{\|}}{\text{C}}\text{CH}_2\text{Ph}$$

$$\textbf{238} \qquad\qquad\qquad\qquad\qquad\qquad\qquad \textbf{239}$$

Fc = ferrocene

The relative $k_o/k_a$ ratio for the carbenium ion addition was obtained from a competition experiment of the alkene and the alkyne with diphenyl-methyl chloride in the presence of $ZnCl_2$ in methylene chloride [125].

Inspection of Table 3.22 indicates that, although trisubstituted carbenium ions are estimated to be 10 to 20 kcal/mole more stable than their vinyl cation counterparts (Chapter 2), both species are formed with similar rates, and formation of vinyl cations is sometimes even faster in the acid-catalyzed hydration reactions. Yates et al. [121] ascribed the low $k_o/k_a$ values to the fact that the solvent stabilization of the vinyl cations is much higher than that of the trisubstituted cations. They found a significant solvent effect on the $k_o/k_a$ ratio obtained by the competition method for the bromination of phenylacetylene and styrene: the ratio increased from 0.63 in water to 2590 in acetic acid. However, when the direct $k_o$ and $k_a$ values were measured

**TABLE 3.22**
**Comparative Reactivity of Electrophilic Additions to Alkenes and Alkynes**

| Electrophilic reaction | Substrate | $k_o/k_a{}^a$ | Reference |
|---|---|---|---|
| Acid-catalyzed hydration | $PhCH=CH_2/PhC\equiv CH$ | 0.65 | [121] |
| | $E\text{-}PhCH=CHMe/PhC\equiv CMe$ | 1.5 | [121] |
| | $FcCH=CH_2/FcC\equiv CH$ | 0.11 | [124] |
| | $E\text{-}EtCH=CHEt/EtC\equiv CEt$ | 16.6 | [121] |
| | $Z\text{-}EtCH=CHEt/EtC\equiv CEt$ | 13.9 | [121] |
| | $n\text{-}BuCH=CH_2/n\text{-}BuC\equiv CH$ | 3.6 | [121] |
| | $EtOCH=CH_2/EtOC\equiv CH$ | $5.5 \times 10^{-3}$ | [21] |
| | $PhCH=CHNR_2/PhC\equiv CNR_2{}^b$ | $5.5 \times 10^{-4}$ | [21] |
| Addition of trifluoroacetic acid | $PhCH=CH_2/PhC\equiv CH$ | 0.8 | [122] |
| | $E\text{-}EtCH=CHEt/EtC\equiv CEt$ | 2.6 | [40] |
| | $n\text{-}BuCH=CH_2/n\text{-}BuC\equiv CH$ | 5.2 | [40] |
| Addition of Diphenylmethyl cation | $E\text{-}PhCH=CHPh/PhC\equiv CPh$ | 0.32 | [125] |
| | $Z\text{-}PhCH=CHPh/PhC\equiv CPh$ | $3.3 \times 10^{-3}$ | [125] |
| | $E\text{-}PhCH=CHMe/PhC\equiv CMe$ | 2.22 | [125] |
| | $Z\text{-}PhCH=CHMe/PhC\equiv CMe$ | 0.28 | [125] |
| Bromination$^c$ | $CH_2=CH_2/CH\equiv CH$ | $1.14 \times 10^4$ | [1e] |
| | $PhCH=CH_2/PhC\equiv CH$ | $2.59 \times 10^3$ | [121, 122b] |
| | $Z\text{-}PhCH=CHMe/PhC\equiv CMe$ | $3.6 \times 10^3$ | [121] |
| | $E\text{-}PhCH=CHMe/PhC\equiv CMe$ | $5 \times 10^3$ | [121] |
| | $n\text{-}BuCH=CH_2/n\text{-}BuC\equiv CH$ | $1.82 \times 10^5$ | [121] |
| | $E\text{-}EtCH=CHEt/EtC\equiv CEt$ | $3.36 \times 10^5$ | [121] |
| | $Z\text{-}EtCH=CHEt/EtC\equiv CEt$ | $3.72 \times 10^5$ | [121] |
| | | $3 \times 10^5$ | [123] |
| | | $\sim 10^{4\,d}$ | [130] |
| Chlorination$^c$ | $PhCH=CH_2/PhC\equiv CH$ | $7.2 \times 10^2$ | [121] |
| | $Z\text{-}PhCH=CHMe/PhC\equiv CMe$ | $6.17 \times 10^2$ | [121] |
| | $E\text{-}PhCH=CHMe/PhC\equiv CMe$ | $74 \times 10^3$ | [121] |
| | $n\text{-}BuCH=CH_2/n\text{-}BuC\equiv CH$ | $5.3 \times 10^5$ | [121] |
| | $E\text{-}$ and $Z\text{-}EtCH=CHEt/EtC\equiv CEt$ | $\sim 10^5$ | [121] |
| Addition of 4-chlorobenzenesulfenyl chloride$^d$ | $CH_2=CH_2/CH\equiv CH$ | $2.82 \times 10^4$ | [105] |
| | $PhCH=CH_2/PhC\equiv CH$ | $1.86 \times 10^2$ | [121] |
| | $E\text{-}PhCH=CHMe/PhC\equiv CMe$ | 14 | [121] |
| | $Z\text{-}PhCH=CHMe/PhC\equiv CMe$ | 5 | [121] |
| | $n\text{-}BuCH=CH_2/n\text{-}BuC\equiv CH$ | 84 | [121] |
| | $E\text{-}EtCH=CHEt/EtC\equiv CEt$ | 1.5 | [121] |
| | $Z\text{-}EtCH=CHEt/EtC\equiv CEt$ | 14.0 | [121] |
| Addition of Benzeneselenenyl chloride$^f$ | $CH_2=CH_2/CH\equiv CH$ | $5.6 \times 10^4$ | [1e] |
| | $PhCH=CH_2/PhC\equiv CH$ | $10^2$ | [1e] |
| | $n\text{-}BuCH=CH_2/n\text{-}BuC\equiv CH$ | $6.9 \times 10^3$ | [1e] |
| | $E\text{-}EtCH=CHEt/EtC\equiv CEt$ | $4 \times 10^4$ | [1e] |

$^a$ The $k_{alkene}/k_{alkyne}$ ratio; $k_o$ is used in the literature for $k_{olefin}$.
$^b$ R = piperidino.
$^c$ In acetic acid.
$^d$ In 1,1,2-trichlorotrifluoroethane.
$^e$ In tetrachloroethylene.
$^f$ In chloroform.

[126], the $k_o/k_a$ values were found to be 360 and 2580 in water and AcOH, respectively. These new values are in agreement with those of Ruasse and Dubois [123], who found that solvents do not play a significant role in the bromination of alkenes and alkynes and obtained almost the same $k_{styrene}/k_{phenylacetylene}$ ratios for the bromination in different solvents. Likewise, the bromination of Z-3-hexene and 3-hexyne proceeded with nearly identical rates in acetic acid, methanol, and 50% aqueous methanol [127].

The near solvent independency of the $k_o/k_a$ ratios is explained by Ruasse and Dubois [123] by a mechanism that they designated $Ad_EC$ 1. A rapid preequilibrium between the unsaturated bond and bromine to form a $\pi$ complex is assumed to precede the rate-determining step. Dissociation of the Br–Br bond and the formation of the C–Br bond take place subsequently (Scheme 3.32). The solvent acts both by a general medium effect on the stability of the intermediate and specifically by electrophilic solvation of the leaving bromide ion. The importance of this electrophilic solvation was shown by Dubois et al. in the alkene series by the high solvent isotope effect and by the linear relationship between the bromination rates and the free energies of solvation of the bromide ion in various solvents [128]. Thus, in bromination, the magnitude of the solvent effects depends very little, if at all, on the structure of the cationic part of the transition state.

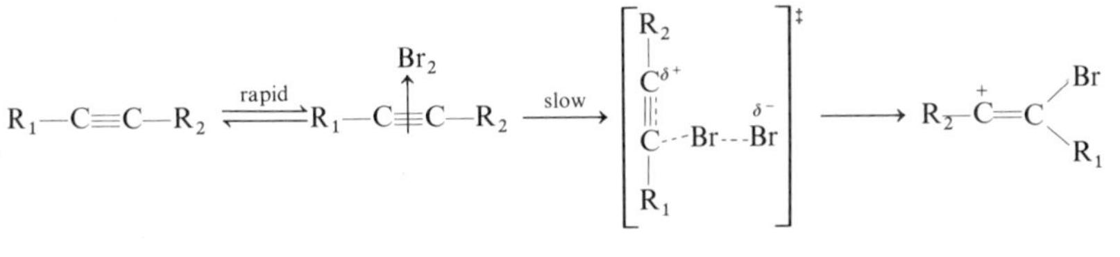

Scheme 3.32

The high $k_o/k_a$ values for the bromination reaction have been ascribed by Ruasse and Dubois [123] to a greater destabilizing effect of the $\beta$-bromine atom on the vinyl cation **240** than on the trisubstituted analogue **241** (Scheme 3.32). A halogen atom in the position $\beta$ to a developing positive charge reduces the rate inductively in the absence of neighboring-group participation, and this destabilization is probably more important in the vinyl cation than in the saturated one (Fig. 3.2). Moreover, ab initio calculations [129]

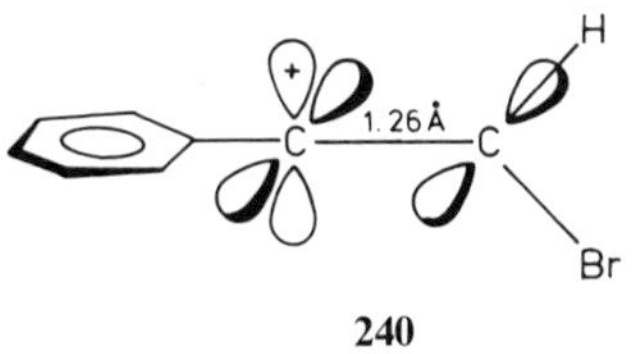

240

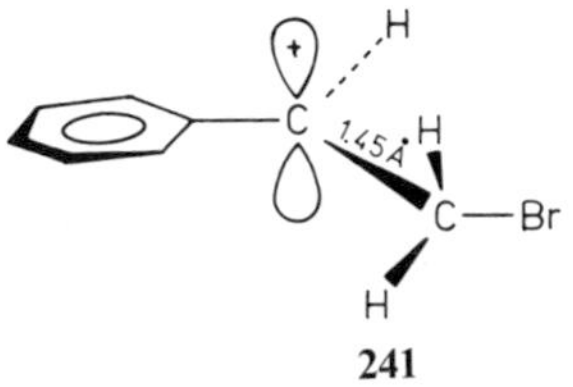

241

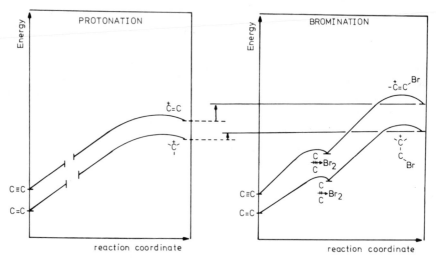

**Fig. 3.2.** Schematic representation of the relative ease of formation of carbenium ions in protonation and bromination [123]. [Reproduced from *J. Org. Chem.* **39**, 1438 (1974) by permission of the American Chemical Society.]

and experimental results (Chapters 5, 6, and 7) indicate that β-substituent effects in vinyl cations are larger than in trisubstituted cations due to the $sp^2-sp^2$ bonds in **240** being shorter than $sp^2-sp^3$ bonds in **241**.

Recent work of Modena *et al.* [122b] also offers a rationale for the widely different reactivity ratios observed for halogenation and acid-catalyzed hydration of alkynes relative to the corresponding alkenes. The reactivity ratio $k_o/k_a$ for bromination for the pair styrene/phenylacetylene was neither sensitive to solvent changes from acetic acid to water nor greatly affected by the presence of substituents in the phenyl ring. On the other hand, a nucleophilically assisted pathway is important in the case of alkyl-substituted alkynes in good nucleophilic solvents, and bridged structures are favored as the intermediate (see below).

Olah and Hakswender [130] relate the decreased reactivity of alkynes compared with alkenes in bromination reactions to the increased energy necessary for the initial π-complex formation of the alkynes. This is due to the cylindrical symmetry of the alkynes, which results in electron delocalization, thereby reducing the total electron availability for π-complex formation. Although one should consider both of the above explanations for the slow reactions in the halogenation of alkynes, more study is necessary to solve this interesting problem.

The $k_o/k_a$ ratio for bromination of alkynes is very high but decreases when the unsaturated system is substituted by phenyl groups. This shows that the cyclic cationic intermediates formed in the bromination of alkyl-substituted alkynes are less stable than those formed in the bromination of alkenes. The $k_o/k_a$ values for chlorination of alkynes and alkenes are somewhat higher than those for bromination. Values for iodination are rare, but available data are in favor of cyclic iodonium ions rather than open $\beta$-iodovinyl cations, as shown in the solvolysis of $\beta$-iodovinyl sulfonates [119].

From a study on the addition of trinitrobenzenesulfonyl hypohalites to alkynes to give $\beta$-halogenovinyl sulfonates, Bassi and Tonellato [131] also came to the conclusion that the geometry of the vinyl cation intermediates may change, depending on both the nature of the halogen and that of the alkyne derivative. In the addition to diaryl alkynes, a gradual shift from a bridged (242) to an open cationic species (243) was apparent on going from iodine to bromine to chlorine, whereas dialkyl alkynes reacted by a bridged species irrespective of the halogen. A similar trend was observed by Pincock and Yates [91] in the bromination of alkynes. Hence, the explanation of the lower $k_o/k_a$ value for aryl-substituted alkynes lies in the fact that an open cation (243) bearing an $\alpha$-aryl residue (R = Ph) is much more effectively stabilized than $\alpha$-alkyl-substituted cation (R = alkyl) of the same geometry, whereas the same change is not expected materially to affect the stability of bridged ions [132]. A poorly effective bridging group such as chlorine facilitates the formation of open cations 243 (X = Cl) rather than the bridged cations (X = Cl), as seen by the lower $k_o/k_a$ value found for chlorination compared to bromination.

242                    243    X = Cl, Br, I

A major role of solvent effect in determining the low $k_o/k_a$ values in the acid-catalyzed hydration reactions for alkenes and alkynes has been ruled out by Modena et al. [122]. The rates of hydration of several alkynes in aqueous sulfuric acid were correlated with the acidity of the medium according to Bunnett and Olsen's [133] free-energy relationship proposed for specific acid-catalyzed reactions [Eq. (1)]. The $\phi_{\ddagger}$ value

$$\log k_\psi + H_0 = \phi_{\ddagger}(H_0 + \log[H^+]) + \log k_0 \tag{1}$$

obtained for the hydration reaction is a measure [Eq. (2)] of the change of the activity coefficient ratio $f_{\ddagger}/f_s$ and hence of the solvation requirements of the transition state:

$$\log f_{H^+} - \log \frac{f_{\ddagger}}{f_s} = (1 - \phi_{\ddagger})\left(\log f_{H^+} - \log \frac{f_{BH^+}}{f_B}\right) \tag{2}$$

TABLE 3.23

Acidity Correlation of the Rates of Hydration of
Alkenes and Alkynes[a] [122a]

| Substrate | $\phi_{\ddagger}{}^{b}$ | $-\log k_0{}^{b}$ |
|---|---|---|
| $XC_6H_4C{\equiv}CH$ | | |
| $X = p\text{-OMe}$ | $-0.67 \pm 0.08$ | $3.67 \pm 0.03$ |
| $p\text{-Me}$ | $-0.58 \pm 0.05$ | $5.25 \pm 0.06$ |
| $H$ | $-0.44 \pm 0.05$ | $6.55 \pm 0.10$ |
| $p\text{-Cl}$ | $-0.29 \pm 0.04$ | $6.85 \pm 0.11$ |
| $XC_6H_4CH{=}CH_2$ | | |
| $X = p\text{-OMe}$ | $-0.54 \pm 0.03$ | $4.38 \pm 0.03$ |
| $p\text{-Me}$ | $-0.34 \pm 0.03$ | $5.50 \pm 0.04$ |
| $m\text{-Me}$ | $-0.22 \pm 0.08$ | $6.10 \pm 0.17$ |
| $H$ | $-0.31 \pm 0.04$ | $6.60 \pm 0.06$ |
| $p\text{-Cl}$ | $-0.28 \pm 0.03$ | $6.95 \pm 0.10$ |
| $p\text{-Br}$ | $-0.20 \pm 0.08$ | $6.80 \pm 0.18$ |
| $m\text{-Cl}$ | $-0.11 \pm 0.06$ | $7.56 \pm 0.12$ |
| $m\text{-Br}$ | $-0.13 \pm 0.07$ | $7.69 \pm 0.24$ |
| $m\text{-NO}_2$ | $-0.05 \pm 0.01$ | $8.67 \pm 0.06$ |
| $EtC{\equiv}CEt$ | $-0.25 \pm 0.04$ | $8.30 \pm 0.06$ |
| $E\text{-EtCH}{=}CHEt$ | $-0.34 \pm 0.02$ | $7.55 \pm 0.08$ |

[a] In aqueous $H_2SO_4$ at 25°.
[b] Here, $\phi_{\ddagger}$ and $\log k_0$ are the slope and intercept, respectively, of the plot $(\log k_\psi + H_0)$ versus $(H_0 + \log[H^+])$.

The lower the $\phi_{\ddagger}$ parameter, the smaller is the solvation requirement of the transition state compared with that of the reactant.

The observed $\phi_{\ddagger}$ values listed in Table 3.23 for the hydration of pairs of analogous alkenes and alkynes are closely similar. Therefore, for each pair considered, the changes in the activity coefficient ratios $f_{\ddagger}/f_s$ accompanying the change in the acidity of the medium are equal. From this it was concluded that the solvation requirements of the transition state for the hydration of analogous alkenes and alkynes are similar. The comparable values obtained for addition of trifluoroacetic acid are also in agreement with this reasoning.

The slope parameters (Table 3.23) decrease steadily in both the substituted styrenes and alkynes on going from electron-withdrawing to electron-donating substituents. The sensitivity of the $\phi_{\ddagger}$ parameters is considered to reflect the differing position of the transition states with (depending on the substituents) different degree of charge dispersal into the transition state structure; i.e., the more fully delocalized, the lower the slope parameters.

By use of both the Bunnett–Olsen and the Hammett equations, the substituent effects for both the styrenes and phenylacetylenes were found to increase with an increase in acid concentration. Also, increasingly larger negative $\rho$ values for phenylacetylenes than for styrenes with increasing acid concentrations were observed. This was explained by a more effective conjugative interaction between the positively charged carbon and the aromatic ring of the vinyl cations compared to trisubstituted cations, due to the shorter $C_{sp}-C_{Ar}$ versus $C_{sp^2}-C_{Ar}$ bond in $\alpha$-aryl-substituted alkenes and alkynes and to mesomeric resonance structures (Section I,A,1,a) in the case of alkenyl and alkynyl ethers and amines. In line with this, the $k_o/k_a$ ratios decrease with the increased conjugative ability of the $\alpha$ substituent on the unsaturated center in the following order: 0.7 (phenyl), 0.1 (anisyl), $5.5 \times 10^{-3}$ (ether), $5 \times 10^{-4}$ (amines) (Table 3.22).

The relative ease of hydration of alkynes was attributed by Pople and Schleyer (Chapter 2) to the higher strain in the alkyne molecule compared with that in the alkene molecule. This ground state difference compensates for the apparently comparable transition state energy differences, which favor the trisubstituted carbenium ions (Fig. 3.2).

The $k_o/k_a$ ratios for the addition of $p$-chlorobenzenesulfenyl chloride are variable and decrease on substitution of the parent compound. This decrease reflects a larger effect of the substituents on the rate of addition to alkynes than on the rate of addition to alkenes [1e]. A similar but smaller effect is evident for the addition of benzeneselenenyl chloride.

In conclusion, the above data suggest that alkynes and alkenes do not differ much in their reaction toward electrophilic reagents. In both cases, the nature of the electrophile and the substituent on the unsaturated system affect the relative reactivities. Solvent effects are important for the individual electrophilic reaction, but they have little influence on the $k_o/k_a$ ratios.

## II. PREPARATIVE ASPECTS OF ELECTROPHILIC ADDITIONS TO ALKYNES

A great variety of interesting products are formed by electrophilic addition reactions to alkynes. Some of these, e.g., ketones, vinyl derivatives, and cyclized products such as indenes, were mentioned in Section I as products obtained during investigations of reaction mechanisms. This section surveys reactions yielding products of preparative interest, which are considered to be formed through vinyl cations as intermediates.

Griesbaum [134] observed the formation of cyclodimerization products in the liquid-phase reactions of anhydrous hydrogen halides with alkynes.

Thus, the reaction of propyne with anhydrous HBr at $-70°$ yielded, besides the normal addition products, the *cis*- and *trans*-cyclobutane derivatives **246** (R = H; X = Br) in about 41% yield [134]. The formation of the four-membered compound is considered to proceed by a $\pi 2_s + \pi 2_a$ cyclo-addition of the initially formed vinyl cation **244** (R = H) to propyne (Chapter 2) to give the resonance-stabilized intermediate cation **245** (R = H; X = Br). Addition of HBr and Br$^-$ then gives the isomeric cyclodimers (Scheme 3.33). The trans product was the major compound among the cyclic products. Propyne and anhydrous HCl did not give any cyclodimer at low temperatures, but a 16% yield of the dimeric product **246** (R = H; X = Cl) was obtained at room temperature [135]. The reaction of HCl with other substituted terminal alkynes also gave dimeric products [135] (Scheme 3.33, X = Cl).

The addition reactions were also conducted with mixtures of different alkynes, which resulted in cross-cyclization products (Scheme 3.33) [136]. The reaction of a 2:1:1 ratio of HCl, propene, and propyne gave products

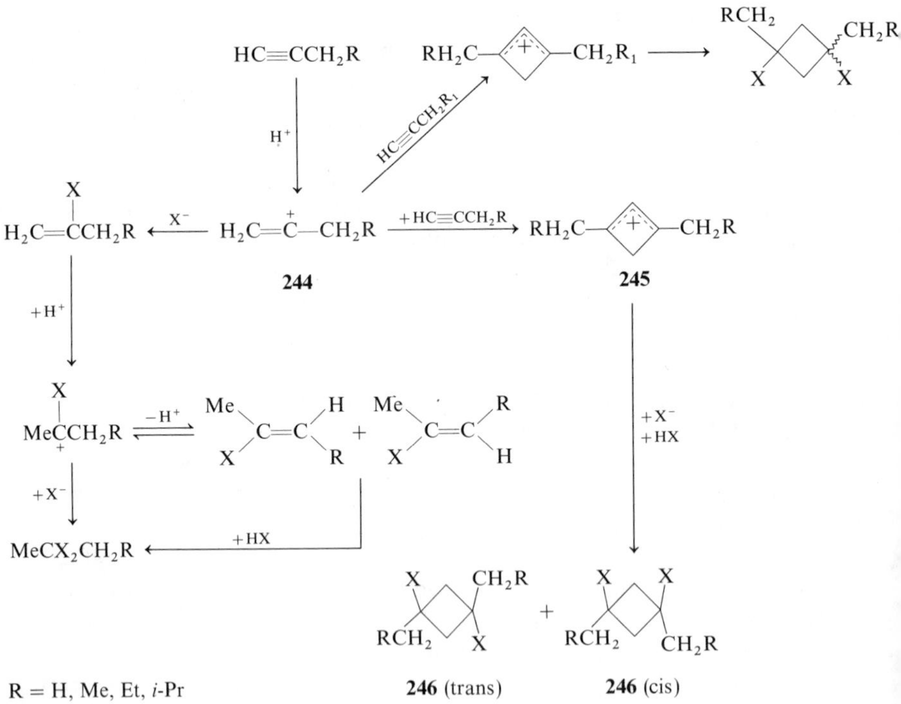

R = H, Me, Et, *i*-Pr          **246** (trans)          **246** (cis)
X = Cl, Br

Scheme 3.33

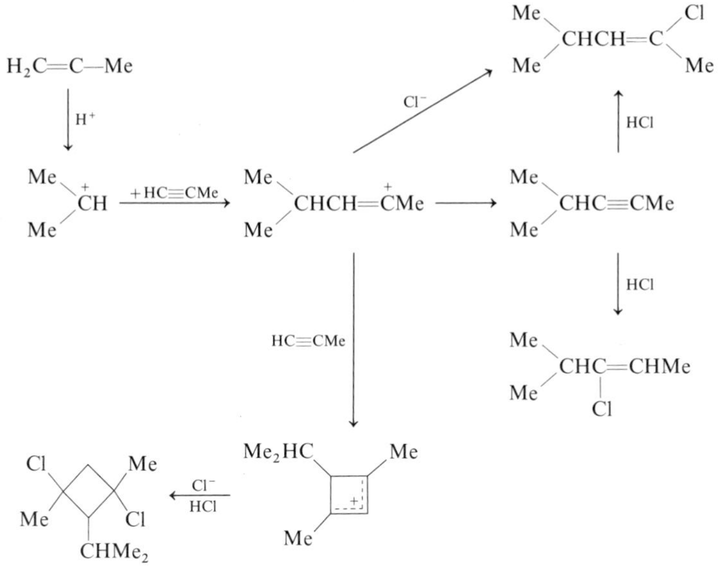

Scheme 3.34

resulting from electrophilic alkylation and subsequent cycloaddition [137] (Scheme 3.34).

More recently, trimerization products were isolated from the reaction of propyne with HCl, which was postulated to involve vinyl cation intermediates [138] (Scheme 3.35). Addition of HX to 2-butyne and 3-methyl-1-butyne gave both of the normal addition products and cyclodimeric products [139, 140], probably via vinyl cations.

Cyclobutene derivatives **248** were also obtained by addition of terminal alkynes to alkenes in the presence of Lewis acid catalysts [141]. The catalytic action of the Lewis acid is probably due to an initial complexation of the alkyne leading to vinyl cation **247**, which then undergoes cycloaddition.

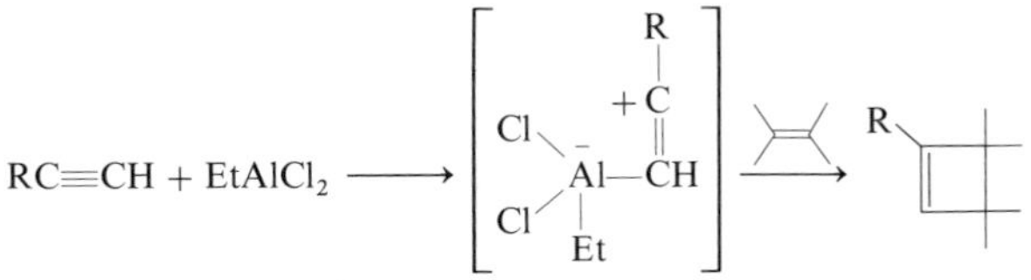

**247**

**248**
(30–60%)

R = alkyl

Ghosez *et al.* showed that the substituted α-dialkylaminovinyl cations (**250**) generated by the action of Lewis acids such as $ZnCl_2$ on α-chloro-enamines (**249**) undergo cycloaddition reaction with alkenes as well as

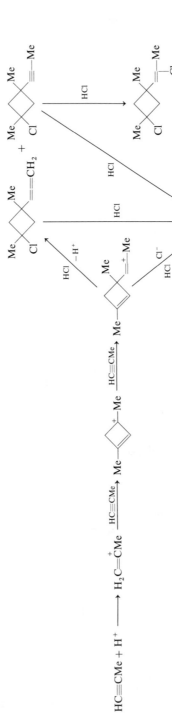

Scheme 3.35

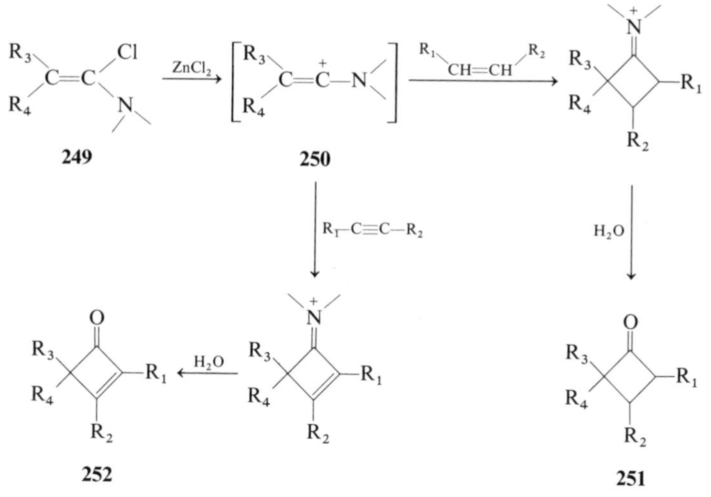

Scheme 3.36

alkynes, giving four-membered compounds [142]. On hydrolysis, the four-membered ketones **251** and **252** are formed in ca. 90% yield (Scheme 3.36).

A phosphorus-containing product (**255**) was isolated in 10 to 20% yield in addition to the expected propargylic and allenic bromides from the $PBr_3$ bromination of the alkynol **253** [143]. The authors proposed the intermediacy of the vinyl cation (**254**) formed by protonation under the strongly acidic conditions required for the reaction. This intermediate forms the product **255** by intramolecular trapping of the vinyl cation by the nucleophilic phosphorus (Scheme 3.37). Alternate reaction routes were ruled out [143].

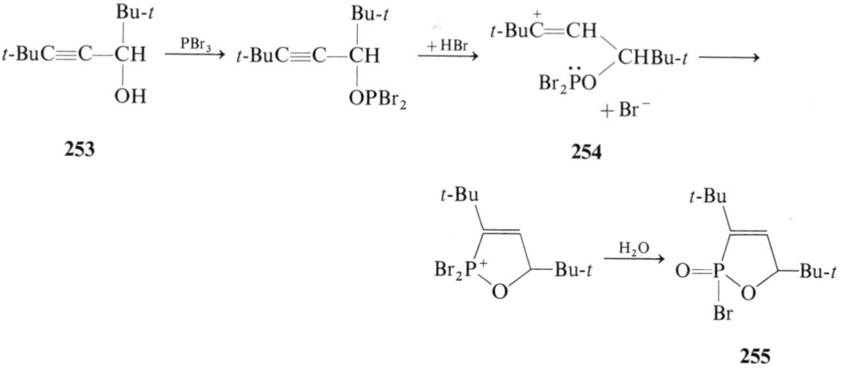

Scheme 3.37

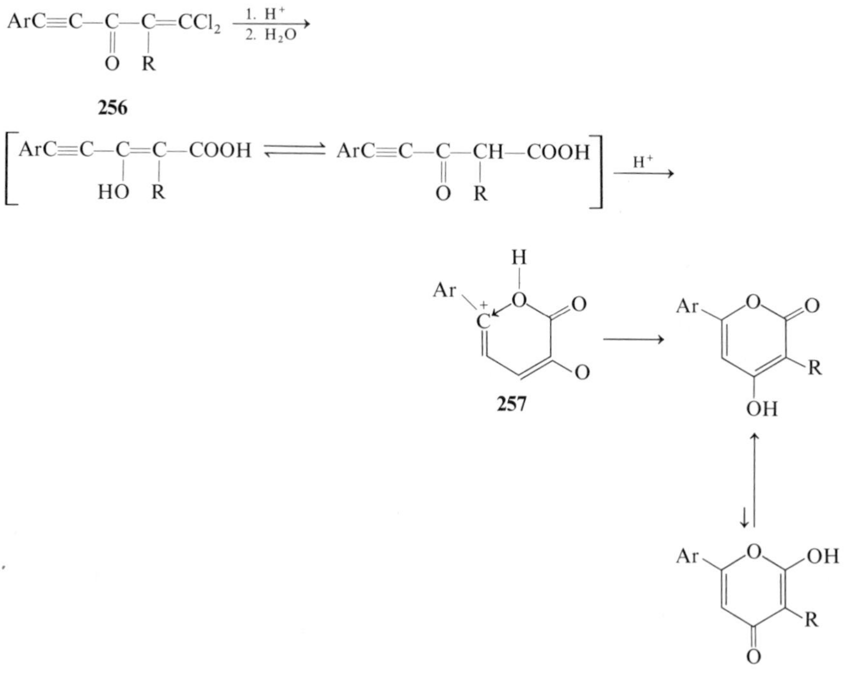

$$ArC{\equiv}C{-}\underset{\underset{O}{\|}}{C}{-}\underset{\underset{R}{|}}{C}{=}CCl_2 \xrightarrow[\text{2. H}_2\text{O}]{\text{1. H}^+}$$

**256**

$$\left[ ArC{\equiv}C{-}\underset{\underset{HO}{|}}{C}{=}\underset{\underset{R}{|}}{C}{-}COOH \rightleftharpoons ArC{\equiv}C{-}\underset{\underset{O}{\|}}{C}{-}\underset{\underset{R}{|}}{CH}{-}COOH \right] \xrightarrow{\text{H}^+}$$

**257**

Scheme 3.38

The acid-catalyzed cyclization of the $\beta,\beta$-dichlorovinylacetylenic ketones **256** to the substituted $\gamma$-pyrones presumably proceeds via the phenyl-stabilized vinyl cation **257** [144] (Scheme 3.38).

Chlorosulfonyl isocyanate reacts with phenyl-substituted alkynes to give addition products that are best explained by postulating vinyl cation intermediates [145]. For example, in the reaction of diphenylacetylene with chlorosulfonyl isocyanate, the vinyl cation **258** is intercepted by another molecule of chlorosulfonyl isocyanate (CSI) to give the six-membered heterocyclic adduct [145].

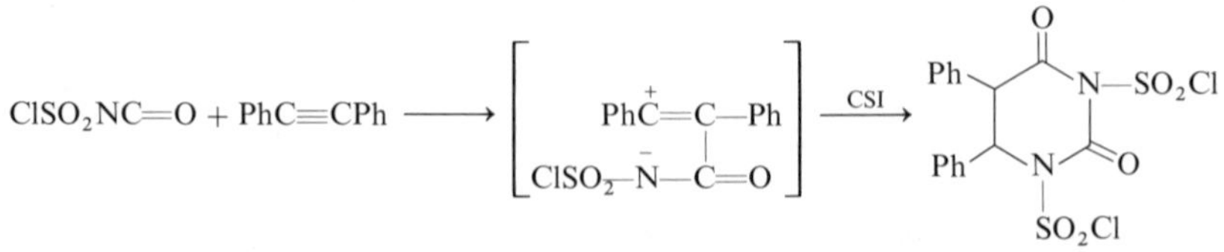

$$ClSO_2NC{=}O + PhC{\equiv}CPh \longrightarrow \left[ \begin{array}{c} Ph\overset{+}{C}{=}C{-}Ph \\ | \\ ClSO_2{-}\overset{-}{N}{-}C{=}O \end{array} \right] \xrightarrow{CSI}$$

**258**

Vinyl cations may also be involved in the cyclization of acetylenic acids to $\gamma$-lactone. No mechanistic studies on these reactions are available, but a

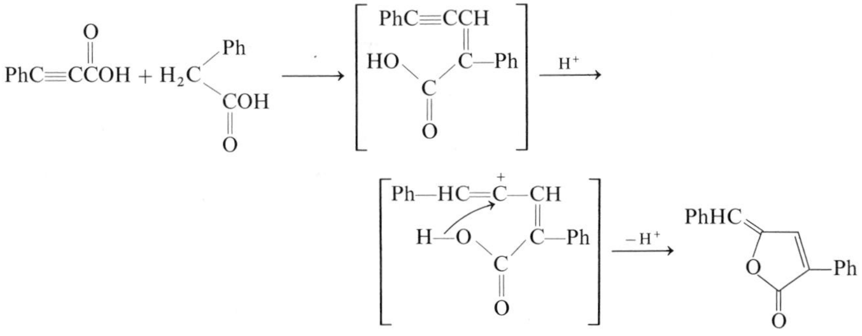

Scheme 3.39

recent example, in which the synthesis of cornicular lactone **260** is reported, depicts the vinyl cation intermediate **259** [146] (Scheme 3.39).

Vinyl cations **262** were suggested as intermediates in the formation of olefins by protonation of the ate complex **261** in the hydroboration reaction [147].

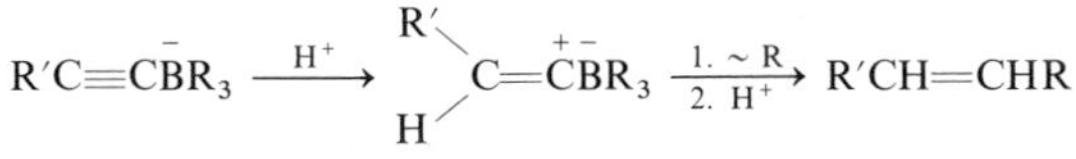

Electrophilic addition to alkynes in nitromethane as solvent gave solvent-incorporated products [148]. The reaction was rationalized as involving an initial electrophilic attack on the triple bond to give vinyl cation **263** followed by a subsequent reaction of the positive center with nitroalkane to form the

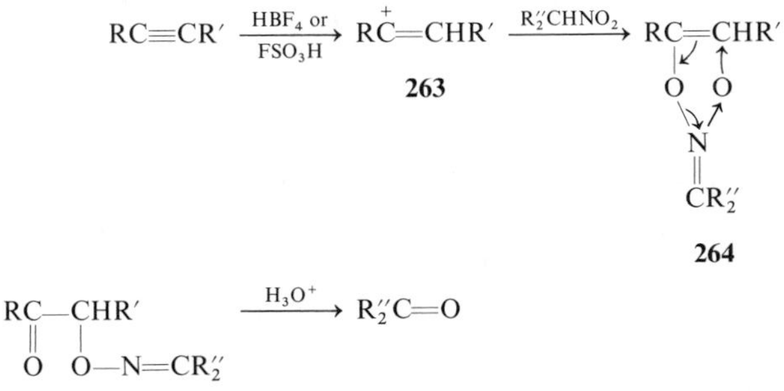

Scheme 3.40

vinylnitronic ester **264**. This rearranges to alkylideneiminoxy ketones **265**, which can be hydrolyzed to ketones (Scheme 3.40).

Interesting ferrocenyl-substituted cyclic compounds were prepared by protonation of ferrocenyl alkynes with trifluoroacetic acid [149]. Cyclization of 7-(ferrocenylethynyl)cycloheptatriene (**266**) with trifluoroacetic acid involves the initial formation of a mixture of the cations **269** and **270** in the ratio 2:1. These were observed by nmr (Chapter 8). The formation of these two ferrocenyl cations evidently proceeds through vinyl cation **267**, which then undergoes intramolecular cyclization to give the bicyclic allyl cation **268**. Quenching of the reaction mixture with excess aqueous sodium carbonate gave barbaralol (**271**), the triene **272**, and the trifluoroacetate **273** (Scheme 3.41).

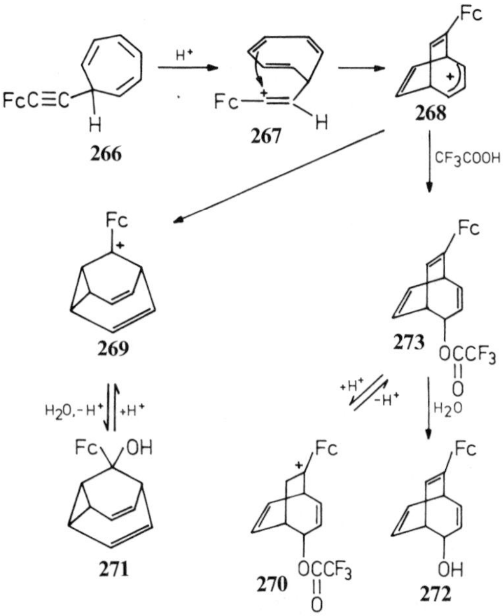

Scheme 3.41

Vinyl cation **274** is presumed to be an intermediate in Friedel–Crafts alkylations starting with either phenylacetylene or α-bromostyrene [150]. More of the alkylated product and less of acetophenone were found when α-bromostyrene was reacted with aromatic hydrocarbons in the presence of $Al_2Br_6$, whereas acetophenone is the major product in the reaction of phenylacetylene and aromatic hydrocarbons in the presence of 100% $H_3PO_6$ (Scheme 3.42).

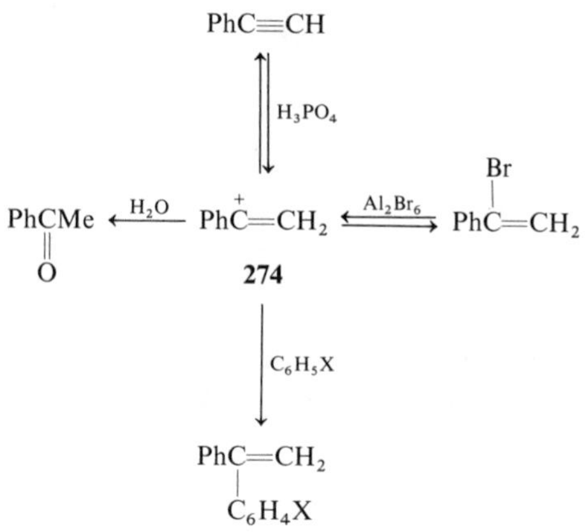

$$PhC\!\equiv\!CH$$

$$\Big\updownarrow H_3PO_4$$

$$PhCMe \xleftarrow{\ H_2O\ } Ph\overset{+}{C}\!\!=\!\!CH_2 \xrightleftharpoons{\ Al_2Br_6\ } Ph\overset{\overset{\displaystyle Br}{|}}{C}\!\!=\!\!CH_2$$

$$\underset{O}{\|} \qquad\qquad \mathbf{274}$$

$$\Big\downarrow C_6H_5X$$

$$Ph\underset{\underset{C_6H_4X}{|}}{C}\!\!=\!\!CH_2$$

X = *p*-Me, *o*-Me, *p*-OMe, *o*-OMe

Scheme 3.42

## III. TRIPLE-BOND PARTICIPATION IN SOLVOLYSIS

Participation of a double bond either from a homoallylic or from a more remote position in a solvolysis that involves interaction with the incipient positive charge has been known for over three decades. The many data that have accumulated show that double bonds indeed participate in reactions involving cations as intermediates, leading to rearranged products as well as to an increase in the solvolysis rate in comparison to suitable saturated reference compounds [151].

The triple bond was not used as a neighboring group in solvolysis reactions until 1965, when Hanack and co-workers [152] solvolyzed the homopropargyl derivative **275** and discovered a rearrangement reaction that was similar to a homoallylic rearrangement and was therefore called the "homopropargyl rearrangement."

The homopropargylic rearrangement has been systematically investigated by Hanack and co-workers from a mechanistic as well as a preparative viewpoint. The rearrangements of primary homopropargyl compounds of type **275** with various leaving groups (X) and different substituents (R) were studied in solvents of different nucleophilicities and ionizing power. The solvolyses were usually carried out in the presence of a buffer in order to neutralize the acid formed.

$$R\!-\!C\!\equiv\!CCH_2CH_2X$$

**275**

The solvolyses of 3-pentyn-1-yl derivatives **275** [R = Me; X = $p$-OSO$_2$-C$_6$H$_4$Me, $m$-OSO$_2$C$_6$H$_4$–NO$_2$, 3,5-OSO$_2$C$_6$H$_3$(NO$_2$)$_2$] in acetone/water, methanol, and acetic acid occurred predominantly by a direct S$_N$2 displacement ($k_s$ process) [153] to give the corresponding alcohol **275** (R = Me; X = OH), its methyl ether (**275**, R = Me; X = OMe), and the acetate **275** (R = Me; X = OAc), respectively [152]. When the solvent was changed from acetic acid to formic acid, 16% of the rearranged 2-methylcyclobutanone **282** (R = Me) was formed. In trifluoroacetic acid (TFA), a solvent of even lower nucleophilicity, the yield of **282** increased to 52% [154]. Consequently, an increase in the ionizing power of the solvent resulted in fewer of the unrearranged ester products. With better leaving groups such as $m$-nitro- or 3,5-dinitrobenzenesulfonate, the rearrangement to 2-methylcyclobutanone in TFA was almost quantitative [155]. These observations are in accord with a competition between the $k_\Delta$ and $k_s$ processes [153] in the homopropargylic rearrangement, the outcome of which depends on the solvent system.

Two mechanisms may explain the formation of the cyclized products: (1) addition of the solvent SOH to the triple bond followed by homoallyl rearrangement (Scheme 3.43); (2) participation of the triple bond ($k_\Delta$ process)

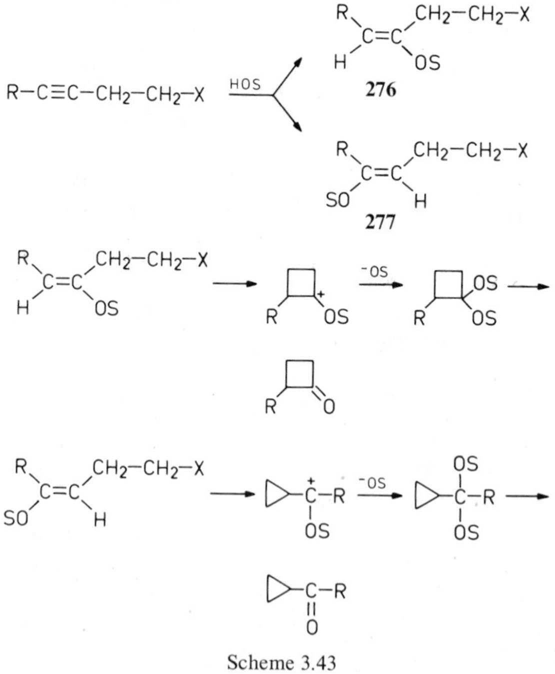

Scheme 3.43

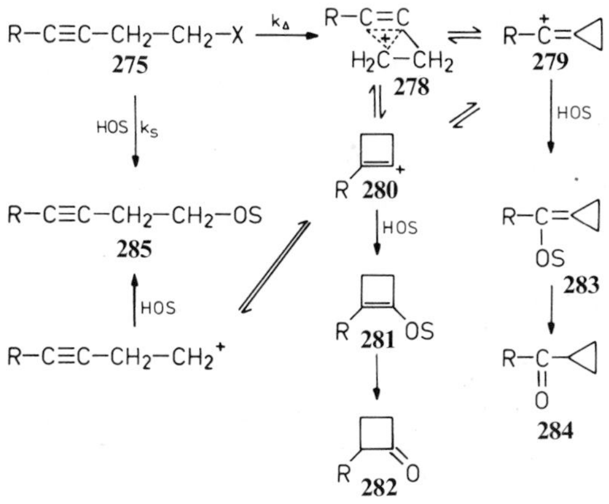

$$R-C\equiv C-CH_2-CH_2-X \xrightarrow{k_\Delta} \underset{\substack{H_2C-CH_2 \\ \textbf{278}}}{R-C\equiv C} \rightleftharpoons \overset{+}{R-C}=\triangleleft$$

$$\textbf{275} \qquad\qquad\qquad \textbf{279}$$

Scheme 3.44

via **278**, which is in equilibrium with vinyl cations **279** and **280** (Scheme 3.44). The experimental evidence discussed below strongly substantiates the involvement of vinyl cations **279** and **280** in the homopropargylic rearrangement.

Participation of the triple bond was supported by the solvolysis study of **275** ($R = Et; X = m\text{-}OSO_2C_6H_4NO_2$) in TFA as followed by nmr. This allowed the detection of the enol acetate **286**, indicating that the TFA does not add to the triple bond under the solvolytic conditions [156]. On the other hand, **275** underwent solvent addition at the triple bond in the presence of $Hg(OAc)_2$, which is known to catalyze additions to triple bonds. Under the latter conditions, mostly cyclopropyl derivatives were obtained via the intermediate addition product **277** (Scheme 3.43), which was unambiguously observed in the nmr spectrum [155].

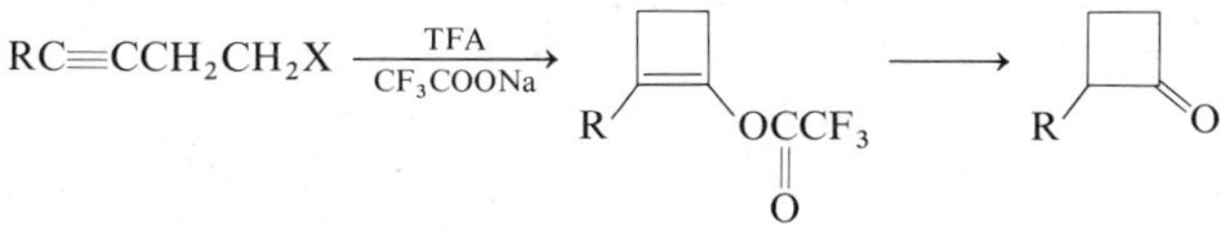

$$RC\equiv CCH_2CH_2X \xrightarrow[CF_3COONa]{TFA} \qquad\qquad \longrightarrow$$

$$\textbf{275} \qquad\qquad\qquad\qquad \textbf{286}$$

$$R = Et; X = m\text{-}OSO_2C_6H_4NO_2$$

Solvolysis of the deuterium-labeled precursor **287** in TFA gave 2-methyl-cyclobutanones **289** and **290**, with an equal distribution of deuterium in the 3 and 4 positions [156], pointing to the intermediate **288**, which can open up in two ways to give a mixture of **289** and **290**.

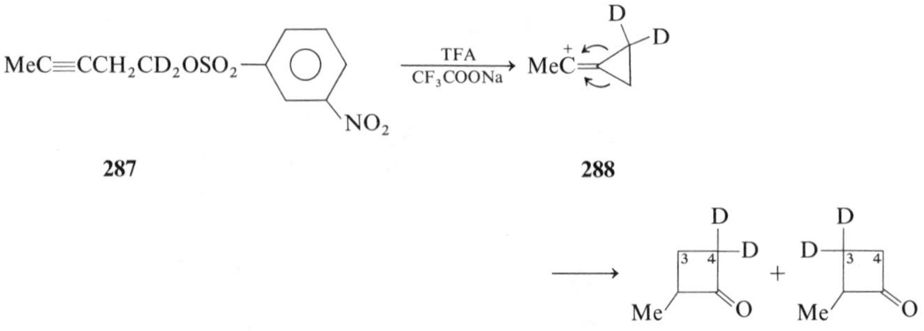

$$Mec{\equiv}CCH_2CD_2OSO_2-\text{[ring]}-NO_2 \xrightarrow[CF_3COONa]{TFA} MeC{\overset{+}{=}}\text{[cyclopropyl-D_3]}$$

287                                                        288

$$\longrightarrow \underset{Me}{\overset{D}{\text{[cyclobutanone 3,4-D]}}}O + \underset{Me}{\overset{D}{\text{[cyclobutanone 3,4-D]}}}O$$

289                    290

The substituent R in **275** affects the nature of the products. In the case of an electron-releasing ethoxy group as in **291**, only the cyclopropyl derivative, ethyl cyclopropylcarboxylate, was formed on solvolysis, and no cyclobutanone was detected [157]. On the other hand, only 2-trifluoromethylcyclobutanone and no cyclopropyl derivative was formed from **292** with the strongly electron-withdrawing $CF_3$ group as substituent [158].

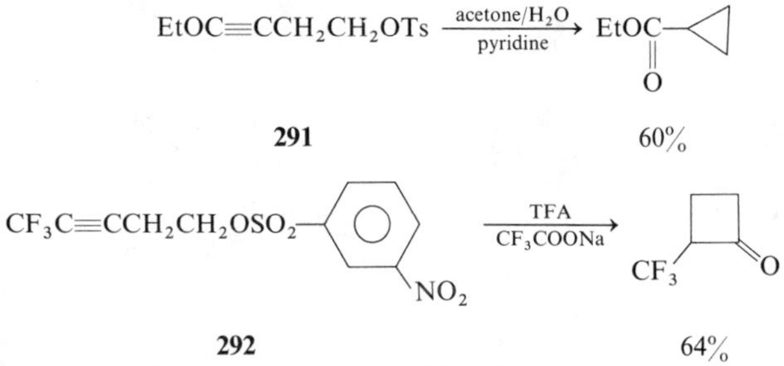

$$EtOC{\equiv}CCH_2CH_2OTs \xrightarrow[pyridine]{acetone/H_2O} EtOC{-}\text{[cyclopropyl]}$$
$$\underset{O}{\|}$$

291                                        60%

$$CF_3C{\equiv}CCH_2CH_2OSO_2-\text{[ring]}-NO_2 \xrightarrow{TFA} \underset{CF_3}{\text{[cyclobutanone]}}O$$

292                                        64%

The effect of a better leaving group on the solvolysis products of the 1-butyn-3-yl system further supports involvement of the $k_\Delta$ route in the homopropargylic rearrangement. Whereas the trifluoroacetolysis of 3-butyn-1-yl *m*-nitrobenzenesulfonate (**275**, R = H; X = *m*-$OSO_2C_6H_4NO_2$) gave only traces of cyclobutanone [158], the corresponding 3,5-dinitrobenzenesulfonate [**275**, R = H; X = 3,5-$OSO_2C_6H_3(NO_2)_2$] gave 20% cyclobutanone [158]. The yield of cyclobutanone could be increased to ca. 70% by the trifluoroacetolysis of 3-butyn-1-yl triflate (**275**, R = H; X = OTf) [159], indicating the preparative value of the homopropargyl rearrangement [1f, 160].

The above data clearly indicate that with an appropriate leaving group vinyl cations of type **279** and **280** are precursors of the products. The high stabilities of these ions are described in Chapter 5.

According to the mechanism of the homopropargyl rearrangement of Scheme 3.44, the enol esters or enol ethers **281** and **283** are formed as inter-

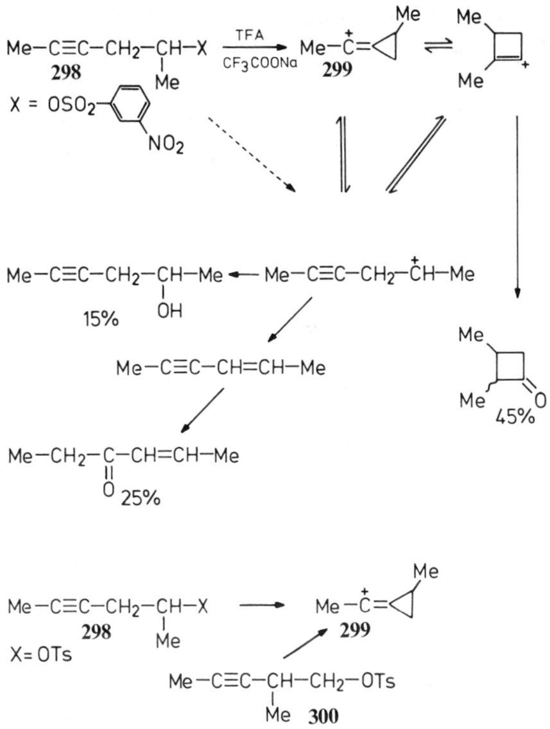

$$Me-C\equiv C-CH_2-\underset{\underset{\displaystyle Me}{|}}{C}H-X \xrightarrow[CF_3COONa]{TFA} Me-\overset{+}{C}= \langle \rightleftharpoons$$

**298**    **299**

$$X = OSO_2-\langle \rangle -NO_2$$

$$Me-C\equiv C-CH_2-\underset{\underset{\displaystyle OH}{|}}{C}H-Me \longleftarrow Me-C\equiv C-CH_2-\overset{+}{C}H-Me$$

15%

$$Me-C\equiv C-CH=CH-Me$$

$$Me-CH_2-\underset{\underset{\displaystyle O}{\|}}{C}-CH=CH-Me$$

25%

45%

$$Me-C\equiv C-CH_2-\underset{\underset{\displaystyle Me}{|}}{C}H-X \longrightarrow Me-\overset{+}{C}=\langle$$

**298**    **299**

X = OTs

$$Me-C\equiv C-\underset{\underset{\displaystyle Me}{|}}{C}H-CH_2-OTs$$

**300**

Scheme 3.45

mediates in the reaction. Hence, the intermediate cyclopropylidene (**283**) and/or the cyclobutenyl (**281**) products may be isolated. All attempts to isolate the enol esters **281** and **283** in formic acid or TFA were unsuccessful, since they were not stable under the reaction conditions but reacted further with formation of only the cyclopropyl alkyl ketone **284** and the cyclobutanone **282**. However, when anhydrous trifluoroethanol was used as a solvent, it was possible to isolate the trifluoroethyl ether **294** in the solvolysis of 3-penten-1-yl triflate **293** in about 86% yield [161].

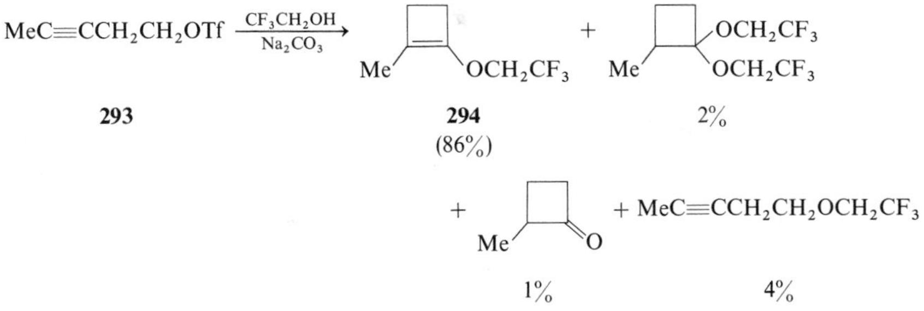

$$MeC\equiv CCH_2CH_2OTf \xrightarrow[Na_2CO_3]{CF_3CH_2OH}$$

**293**

294
(86%)

2%

1%           + $MeC\equiv CCH_2CH_2OCH_2CF_3$

4%

As discussed in detail in Chapter 5, the homopropargylic triflate **293** and compounds **295** to **297** (isomeric except for the leaving group) all solvolyze under comparable conditions to yield similar product mixtures, indicating that in all cases the reaction proceeds via the same intermediates, a result that is also found in the homoallyl-cyclobutyl–cyclopropylmethyl rearrangement.

Mec≡CCH$_2$CH$_2$OTf

Me $\diagdown$ OSO$_2$C$_4$F$_9$

Me $\diagdown$ Br Me

Me $\diagdown$ Br

**293**            **295** 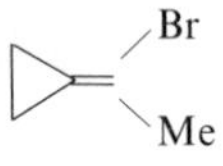            **296** 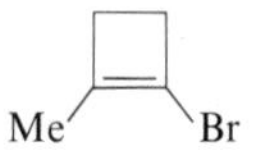            **297**

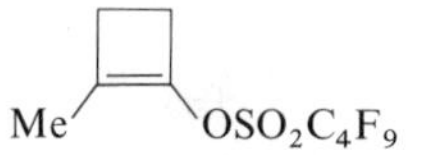

Secondary homopropargyl derivatives react by the same mechanism as the primary ones [157], with cyclization occurring even in a relatively nucleophilic solvent such as aqueous acetone. The product distributions from **298** and **300** in CF$_3$COOH and HCOOH were the same [158], suggesting the same vinyl cation (299) as a common intermediate [162] (Scheme 3.45).

In contrast to the secondary homopropargyl sulfonates, which also gave cyclobutanones, formolysis of **301**, in which the β-carbon carries a *gem-*

**TABLE 3.24**

**Rate of Formolysis of Saturated and Alkynyl Tosylates at 75° [162]**

| Tosylate | $10^4 k_1$ (sec$^{-1}$) |
|---|---|
| Mec≡CC—CH$_2$OTs, with Me above and Me below | 3.0 |
| Mec≡CC—CHOTs, with Me above and Me Me below | 53.0 |
| Mec≡CCH—CH$_2$OTs, with Me below | 1.14 |
| MeCH$_2$CH$_2$C—CH$_2$OTs, with Me above and Me below | 0.5 |
| MeCH$_2$CH$_2$CH—CH$_2$OTs, with Me below | 0.36 |

dimethyl group, gave mostly the open-chain ketone **305**, presumably via **304**. This reaction apparently occurs via a vinyl cation (**302**), which unlike the cation **299** undergoes exclusive ring opening to form the more stable tertiary cation **303**. The solvolysis rates of compounds **300** and **301** in formic acid are three and six times faster, respectively, than those of the saturated analogues, indicating anchimeric assistance by the triple bond during the solvolysis [163] (Table 3.24).

$$\text{MeC}\equiv\text{C}-\underset{\underset{\text{Me}}{|}}{\overset{\overset{\text{Me}}{|}}{\text{C}}}-\text{CH}_2\text{OTs} \xrightarrow[\text{HCOONa}]{\text{HCOOH}} \text{Me}-\overset{+}{\text{C}}=\!\!<\!\!\!\overset{\text{Me}}{\underset{\text{Me}}{}} \longrightarrow \text{MeC}\equiv\text{C}-\text{CH}_2\overset{+}{\text{C}}-\text{Me}_2$$

$$\qquad\quad \textbf{301} \qquad\qquad\qquad\qquad \textbf{302} \qquad\qquad\qquad\qquad \textbf{303}$$

$$\longrightarrow \text{MeC}\equiv\text{CCH}=\text{CMe}_2 \longrightarrow \text{MeCH}_2\underset{\underset{\text{O}}{\|}}{\text{C}}-\text{CH}=\text{CMe}_2$$

$$\qquad\qquad\qquad \textbf{304} \qquad\qquad\qquad\qquad \textbf{305}$$

Similar results were obtained in the formolysis of the $\beta,\beta$-disubstituted secondary homopropargyl compounds **306** and **307** [162]. No cyclic products could be identified due to the greater stability of the corresponding open tertiary cations compared with the intermediate vinyl cations.

$$\text{MeC}\equiv\text{C}-\underset{\underset{\text{Me}}{|}}{\overset{\overset{\text{Me}}{|}}{\text{C}}}-\underset{\underset{\text{H}}{|}}{\overset{\overset{\text{Et}}{|}}{\text{C}}}-\text{OTs} \qquad\qquad \text{MeC}\equiv\text{C}-\underset{\underset{\text{Me}}{|}}{\overset{\overset{\text{Me}}{|}}{\text{C}}}-\underset{\underset{\text{H}}{|}}{\overset{\overset{\text{Me}}{|}}{\text{C}}}-\text{OTs}$$

$$\qquad\qquad \textbf{306} \qquad\qquad\qquad\qquad\qquad \textbf{307}$$

Aryl-substituted homopropargyl derivatives were also investigated. The solvolysis of the 4-phenyl-3-butyn-1-yl ester **308** (X = OTs, $\beta$-naphthalenesulfonate) yielded 61% of the cyclopropylphenyl ketone **311** along with the formate **310** instead of the expected 2-phenylcyclobutanone [155]. Although cyclopropyl phenyl ketone (**311**) was obtained exclusively also from the solvolysis of 2-phenylcyclobutenyl bromide (**309**) [164], it is

$$\text{PhC}\equiv\text{CCH}_2\text{CH}_2\text{X} \xrightarrow[\text{HCOONa}]{\text{HCOOH}} \triangleright\!\!-\underset{\underset{\text{O}}{\|}}{\text{C}}\text{Ph} \xleftarrow[\text{TEA}]{80\%\ \text{CH}_3\text{CN}}$$

$$\qquad \textbf{308}\ \ \text{X}=\text{OTs} \qquad\qquad\qquad \textbf{311} \qquad\qquad\qquad \textbf{309}$$

$$\text{Ph}\underset{\underset{\text{O}}{\|}}{\text{C}}\text{CH}_2\text{CH}_2\text{OCHO}$$

$$\qquad\qquad\qquad \textbf{310}$$

probable that the rearrangement of **308** occurred via addition of the solvent to the triple bond and a subsequent homoallylic rearrangement. In strongly acidic medium, such as formic acid containing 1 equivalent of benzenesulfonic acid, or in TFA, addition to the triple bond of **308** (X = OBs) was observed [165]. The homoallylic rearrangement of the addition product **312** (X = OBs) was followed by nmr and was found to occur rapidly. As a result, it was argued [165] that there is no participation of the triple bond in the solvolysis of **308**. However, more recent studies indicated that under buffered conditions no addition to the triple bond occurs during solvolysis of **308**, as discussed below.

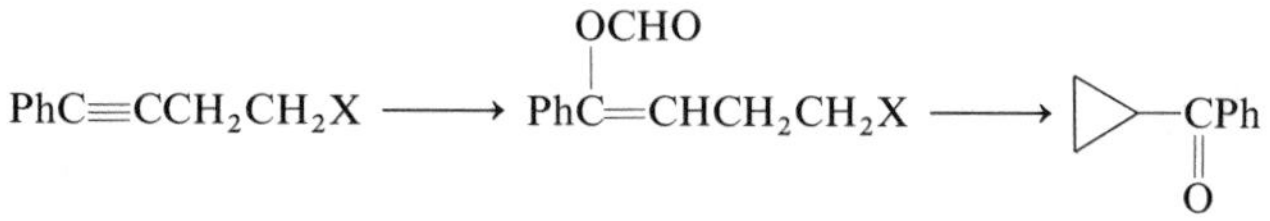

$$\underset{\textbf{308}}{PhC\equiv CCH_2CH_2X} \longrightarrow \underset{\textbf{312}}{Ph\overset{\overset{\displaystyle OCHO}{|}}{C}=CHCH_2CH_2X} \longrightarrow \underset{\textbf{311}}{\triangleright\!\!-\!\!\underset{\displaystyle O}{\overset{\displaystyle \|}{C}}Ph}$$

X = OBs

Deamination of homopropargylamines with nitrous acid, e.g., of **313** and **314**, was also investigated under different conditions. Very few or none of the cyclization products were observed. This was ascribed to the fact that formation of the carbenium ions from either **313** or **314** by the deamination reaction does not require anchimeric assistance of the triple bond, in contrast to the solvolysis reaction [166].

$$\underset{\textbf{313}}{PhC\equiv CCH_2CH_2NH_2} \qquad\qquad \underset{\textbf{314}}{MeC\equiv CCH_2CH_2NH_2}$$

As already mentioned, reaction in 2,2,2-trifluoroethanol (TFE), a solvent of low nucleophilicity and high ionizing power, made it possible to capture the intermediate vinyl cation as the trifluoromethyl ether **294** in the solvolysis

**TABLE 3.25**

Product Distribution from the Solvolysis of $RC\equiv CCH_2CH_2OTs$ in 100% Trifluoroethanol Buffered with $Na_2CO_3$ at 80° [167]

| R | Cyclobutanone[a] 282 (%) | Cyclopropyl ketone[a] 284 (%) | Enol ether 281 (%) | Hydrocarbon 317 (%) | Ether 318 (%) | Ketal 319 (%) |
|---|---|---|---|---|---|---|
| Ph | 6 | 9 | 5 | 12 | 42 | 21 |
| Tol | 3 | 31 | 2 | 10 | 28 | 23 |
| An | 2 | 37 | — | 8 | 18 | 30 |

[a] Formed from the corresponding enol ethers during work-up.

of 3-pentyn-1-yl triflate (293). When 4-phenyl-3-butyn-1-yl triflate (308, X = OTf) was solvolyzed in buffered absolute TFE, no addition product to the triple bond was observed [166, 167]. The product composition was found to depend on the solvent system and on the buffer employed. Besides cyclopropyl phenyl ketone and small amounts of 2-phenylcyclobutanone, the 2,2,2-trifluoroethyl ether 308 (X = $OCH_2CF_3$) and elimination products were identified (Table 3.25). The 4-tolyl-3-butyn-1-yl triflate 315 behaved similarly to 308 (X = OTf), but all attempts to isolate and study the solvolysis behavior of the highly reactive 4-anisyl-3-butyn-1-yl triflate 316 (X = OTf) were unsuccessful [167].

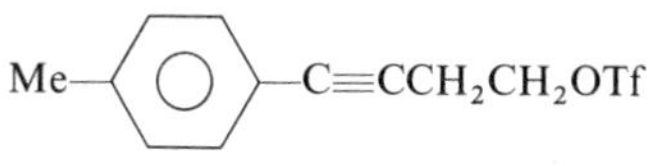

Me—⟨O⟩—C≡CCH₂CH₂OTf          MeO—⟨O⟩—C≡CCH₂CH₂X

              315 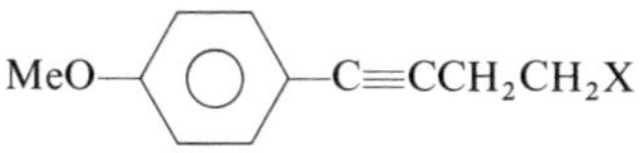                               316

However, the 4-anisyl-3-butyn-1-yl tosylate 316 (X = OTS) could be synthesized, and its solvolytic behavior in 100% TFE was compared with that of the 4-phenyl and 4-*p*-tolyl compounds 308 and 315. As shown in Table 3.25, the cyclopropyl ketone 284 and its ketal 319 were accompanied by minor quantities of the cyclobutanone 282 and its enol ether 281 (Scheme 3.46). The tendency to rearrange decreased in the series Ph < Tol < An, as expected [167] on the basis of the increasing stability of the appropriately substituted intermediate vinyl cation (279).

Solvolysis of 4-cyclopropyl-3-butyn-1-yl tosylate and triflate (320) was carried out in buffered TFE and in aqueous ethanol and gave products

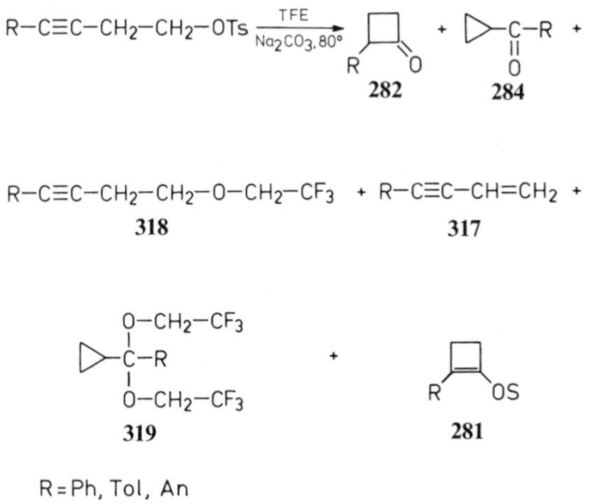

$$R-C≡C-CH_2-CH_2-OTs \xrightarrow[Na_2CO_3, 80°]{TFE} \quad \underset{R \quad O}{\square} \quad + \quad \triangleright\!\!-\!\!\underset{O}{\overset{C-R}{\|}} \quad +$$

                                          282              284

$$R-C≡C-CH_2-CH_2-O-CH_2-CF_3 \quad + \quad R-C≡C-CH=CH_2 \quad +$$

              318                               317

$$\triangleright\!\!-\!\!\underset{O-CH_2-CF_3}{\overset{O-CH_2-CF_3}{\underset{|}{\overset{|}{C}}-R}} \quad + \quad \underset{R \quad OS}{\square}$$

              319                               281

R = Ph, Tol, An

Scheme 3.46

derived from both the $k_\Delta$ and the $k_s$ routes [168]. In aqueous ethanol at 60°, the main products were enyne **321** together with the $k_s$ product **327** (86%) and traces of **325**. When the solvolysis was carried out at a lower temperature (30°), a higher (61%) yield of cyclized products was observed. In general, the yields of cyclized products were higher (81%) in TFE as compared with the solvolysis in aqueous ethanol [168].

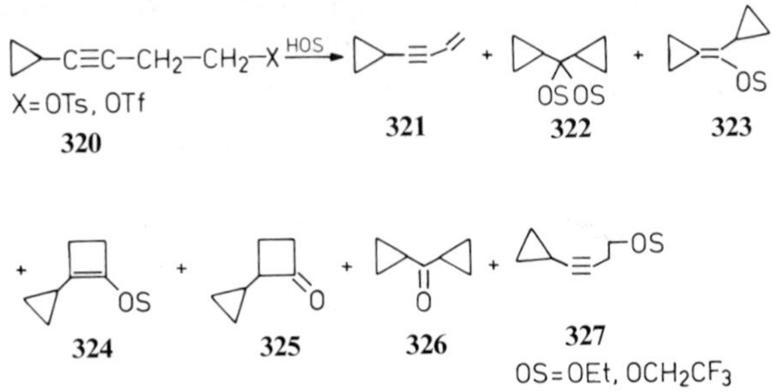

A direct solvent addition to the triple bond was ruled out by reacting the 4-cyclopropyl-3-butyn-1-ol **328** in the same buffered solvents as were employed for reaction of **320**. None of the enol ether or keto compounds were detected. In contrast, in unbuffered solvents, **328** gave 1,6-dioxaspiro-nonane (**330**) by the acid-catalyzed cyclization of the intermediate alcohol **329** [168]. Alcohol **329** was formed by the acid-catalyzed addition of $H_2O$ across the triple bond in the unbuffered medium.

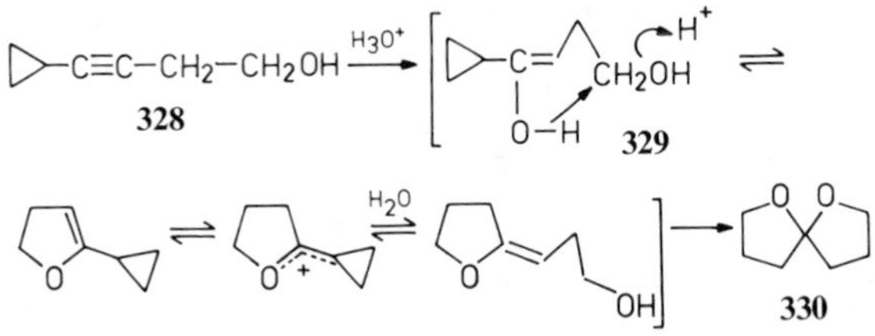

Carbon-14 and deuterium isotope effects were studied in the homo-propargyl rearrangements in order to ascertain the degree of rate acceleration due to neighboring-group participation in the reaction [169]. Triflates ($[1-^{14}C]$**293**, $[3-^{14}C]$**293**, and $[3-^{14}C]$**293**-$1,1$-$d_2$) were solvolyzed in TFE, and the observed isotope effects were compared with the results obtained for the neophyl system **331** [164]. The observed $k_H/k_D$ of 1.098 is consistent with the values found for neophyl derivatives, where anchimeric assistance by the phenyl group is known.

The primary heavy-atom isotope effect ($k/k^*$) of 1.048 for $[1-^{14}C]$**293** is smaller than that reported for the neophyl system. Therefore, an unsymmetric transition state is assumed for the reaction. The effect is large enough

$MeC{\equiv}CCH_2\overset{*}{C}H_2OTf$     $MeC{\equiv}\overset{*}{C}CH_2CH_2OTf$     $MeC{\equiv}\overset{*}{C}CH_2CD_2OTf$

[1-$^{14}$C]**293**          [3-$^{14}$C]**293**          [3-$^{14}$C]**293**-1,1-$d_2$

$(k/k^* = 1.048 \pm 0.003)$    $(k/k^* = 0.990 \pm 0.005)$    $[k_{H_2O}/k_{D_2O} = 1.098 \pm 0.004$ (per D)

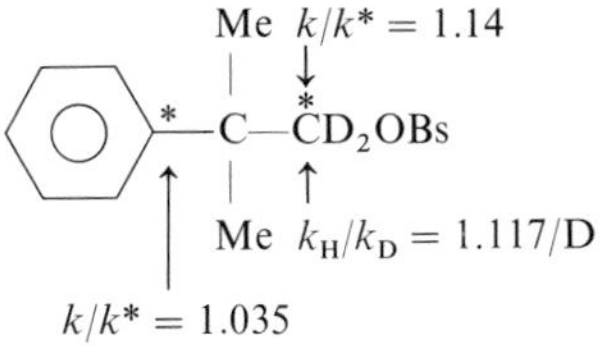

Me $k/k^* = 1.14$

$\langle\!\langle O \rangle\!\rangle$$\overset{*}{-}$$C$—$\overset{*}{C}D_2OBs$

Me $k_H/k_D = 1.117$/D

$k/k^* = 1.035$

**331**

to imply an $S_N2$-like transition state for the homopropargyl rearrangement, with the triple bond acting as an intramolecular nucleophile. The $k/k^*$ value of 0.990 for [3-$^{14}$C]**293**-1,1-$d_2$ is probably caused by the change of hybridization from sp to sp$^2$ at C-3. The foregoing results imply a lower activation energy for the solvolysis, an unsymmetric transition state (**332**), and considerable anchimeric assistance by the triple bond [169].

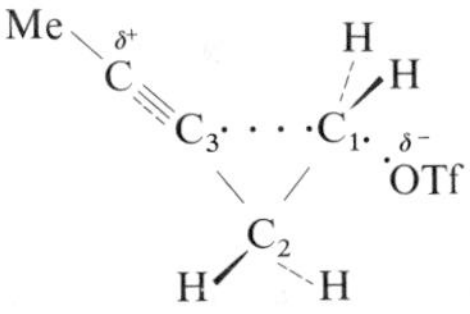

Me $\overset{\delta^+}{\underset{}{C}}$ H
$C_3 \cdots C_1 \underset{\delta^-}{}$ OTf
$C_2$
H H

**332**

The solvolyses of homopropargylic derivatives **333** and **334** have been studied [170]. In the trans configurations of **333** and **334**, the triple bond is favorably situated for interaction with the developing empty p orbital at the cationic center. However, when *trans*-2-ethynylcyclohexyl triflate (**333**, R = H; X = OTf) was solvolyzed in solvents of varyious ionizing powers, only unrearranged and elimination products were obtained. The activation of the triple bond by alkyl substituents reduces its retarding inductive effect and promotes participation of the triple bond, leading to cyclized products [171]. This was achieved for *trans*-2-propynylcyclohexyl nonaflate (**333**, R = Me; X = ONf) in 97% TFE to form 45% of the rearranged product **335** together with unrearranged and elimination products [170].

The cycloheptyl system **334** gave more rearrangement, probably due to its higher conformational flexibility, so that even the parent ethynyl-substituted homopropargyl derivative **334** (R = H; X = OTs) underwent 29% rearrangement to give **336** (R = H) [172]. *trans*-2-Propynylcycloheptyl tosylate (**334**, R = Me; X = OTs) gave 52% of the rearrangement product

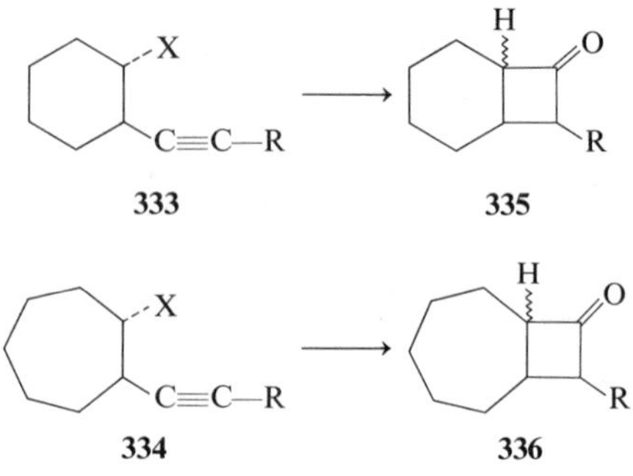

333                    335

334                    336

R = H, Me

336 (R = Me) in 80% aqueous TFE. For both 333 and 334, no cyclopropyl derivatives were formed under any of the reaction conditions investigated.

From the above data, it is clear that the ionizing power and nucleophilicity of the solvent play a major role in determining the extent of the homopropargyl rearrangement. Added buffer sometimes also has an effect, albeit less marked than that of the solvent. The effect of these parameters is illustrated in Table 3.26 for the homopropargyl derivatives 275. For example, 3-pentyn-1-yl triflate in 100% EtOH with 2,6-lutidine as buffer gave only 0.2% rearranged product, whereas use of 100% TFE with $Na_2CO_3$ as buffer gave 89% rearranged product [167].

**TABLE 3.26**

**Solvent Effects on the Homopropargyl Rearrangement of $RC\equiv CCH_2CH_2X$ [167]**

| R | X | Solvent/buffer | Temp (°C) | Extent of rearrangement (%) | Sum of cyclopropyl products | Sum of cyclobutyl products |
|---|---|---|---|---|---|---|
| Me | OTs | 100% TFE/Na$_2$CO$_3$ | 80 | 37 | 1 | 36 |
| | OTf | 100% EtOH/2,6-lutidine | 25 | 0.2 | | 0.2 |
| | OTf | 80% EtOH/2,6-lutidine | | 1.3 | | 1.3 |
| | OTf | 50% EtOH/2,6-lutidine | | 2.5 | | 2.5 |
| | OTf | 100% TFE/Na$_2$CO$_3$ | 25 | 89 | 1 | 88 |
| | OTf | 80% TFE/Na$_2$CO$_3$ | 25 | 47 | 1 | 46 |
| Ph | OTs | 100% TFE/Na$_2$CO$_3$ | 80 | 41 | 30 | 11 |
| | OTf | 100% TFE/Na$_2$CO$_3$ | 25 | 69 | 55 | 14 |
| | OTf | 80% TFE/Na$_2$CO$_3$ | 25 | 23 | 22 | 1 |
| Tol | OTs | 100% TFE/Na$_2$CO$_3$ | 80 | 59 | 54 | 5 |
| | OTf | 100% TFE/Na$_2$CO$_3$ | 25 | 80 | 77 | 3 |
| | OTf | 80% TFE/Na$_2$CO$_3$ | 25 | 31 | 28 | 3 |
| An | OTs | 100% TFE/Na$_2$CO$_3$ | 80 | 69 | 67 | 2 |
| Ph | OTf | 80% EtOH/Na$_2$CO$_3$ | 25 | 6 | 5 | 0.4 |
| Tol | OTf | 80% EtOH/Na$_2$CO$_3$ | 25 | 14 | 13 | 1 |

Solvolytic participation of a remote triple bond has also been reported [173]. Acetolysis of the tosylate **337** gave the cyclized product **339** in 36% yield, probably via the intermediate phenyl-stabilized vinyl cation **338**.

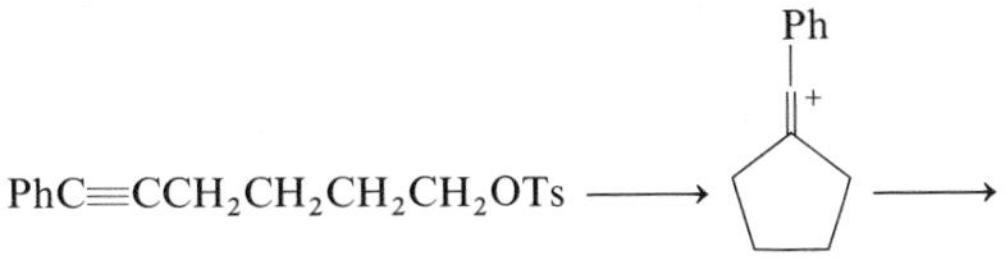

PhC≡CCH₂CH₂CH₂CH₂OTs  ⟶

**337**                    **338**

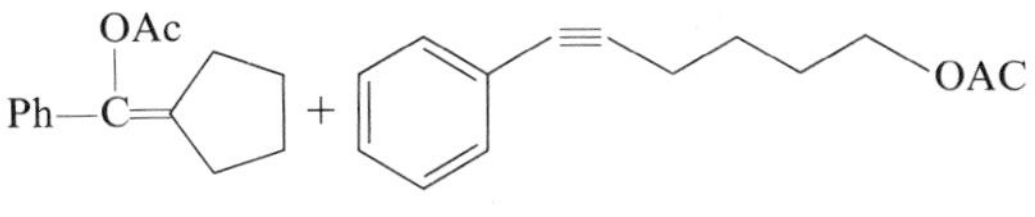

**339**

Participation of the remote triple bond also occurred in the solvolysis of the secondary tosylate **340**, as indicated by the isolation of the cyclic product **342** besides the open-chain products [171]. The extent of cyclization depended on the nature of the solvent, and rate studies and the product analysis indicated the intermediacy of the vinyl cation **341**. The reaction rate was similar to that for the alkene analogue **343**, and this was considered to

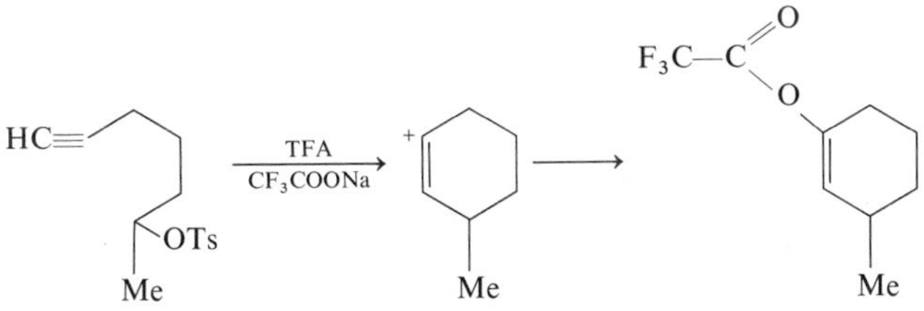

**340**          **341**          **342**

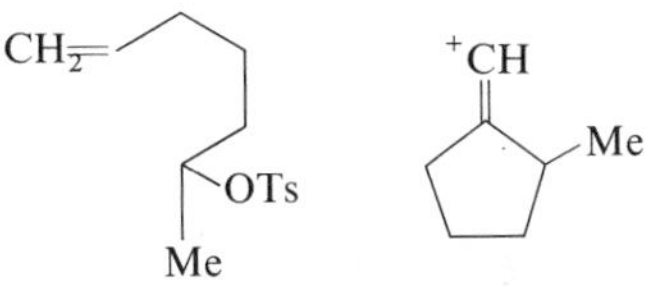

**343**          **344**

indicate a substantial anchimeric assistance, since it was estimated that the triple bond would have a 17.5-fold rate-reducing inductive effect. No cyclopentyl derivative was formed because the required intermediate **344** is a primary vinyl cation and therefore unfavorable.

When the solvolysis reaction was extended to 6-octyn-2-yl tosylate (**345**), participation by the triple bond was found to be most effective in TFA. Six-membered ring, five-membered ring, and open-chain products were formed [171]. The vinyl tosylates **346** and **347**, obtained in a 9:1 ratio in 44% yield, were probably formed by internal return. The trifluoroacetates **348** and **349** were obtained in 51% yield, and in both cases the five-membered ring products predominated. No elimination or acyclic products were observed in TFA.

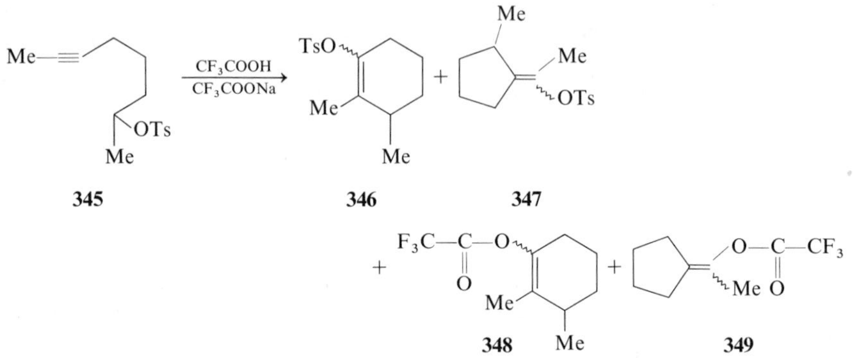

**345**                        **346**          **347**

**348**    Me                              **349**

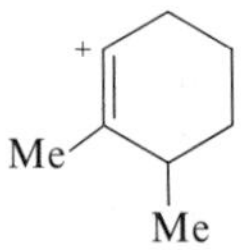

**350** 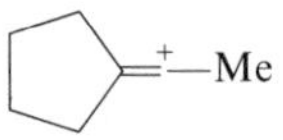                **351** 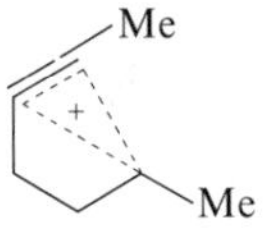                **352**

The formation of more of the cyclopentylidene derivatives was explained by the high energy of the bent vinyl cation **350** versus that of the linear vinyl cation **351**. Accordingly, the bridged ion **352** was suggested as an intermediate leading to products.

Participation of a remote triple bond in the solvolysis of the primary triflates **353** to **355** was observed in absolute TFE buffered by $Na_2CO_3$. 5-Hexyn-1-yl triflate (**353**) gave 36% cyclohexenyl derivatives and 58% 5-hexyn-1-yl trifluoromethyl ether, probably formed by a $k_s$ process [174].

$$HC{\equiv}C(CH_2)_4OTf \xrightarrow{\text{TFE}} HC{\equiv}C(CH_2)_4OCH_2CF_3 + $$

**353**                    58%              (12%)              24%

                                          **356**

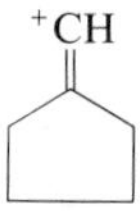

**357**

An ion-pair return is indicated by the fact that cyclohexenyl triflate (**356**) constitutes one-third of the rearrangement products. The yield of the cyclization products increased to 97% in the solvolysis of triflate **354**, in which a mixture of five- and six-membered ring cyclization products were obtained. In contrast, no five-membered ring product was formed in the rearrangement of the triflate **353**, since the precursor vinyl cation **357** is a primary cation of high energy.

The preferential formation of cyclopentylidene derivatives is reminiscent of Peterson and Kamat's work [171] on the solvolysis of **345** discussed above. It is interesting that 75% of the reaction products of **354** in TFE are the triflates **358** and **359**. These products, which are probably formed by internal return, are stable under the reaction conditions at 0° and do not solvolyze further. An ion-pair return with rearrangement was not detected in the solvolysis of **275** because of the instability and high reactivity of the expected cyclobuten-1-yl nonaflates (Chapter 5) under the solvolytic conditions employed.

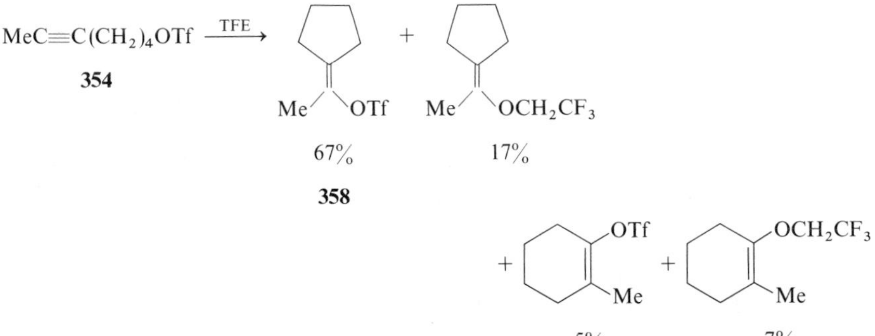
MeC≡C(CH2)4OTf
TFE
Me
OTf
Me
OCH2CF3
OTf
Me
OCH2CF3
Me

**354**; 67% **358**; 17%; 5% **359**; 7%

The tendency to form cyclic products decreases when the carbon chain becomes longer. Only 12% of the cyclic products were observed in the solvolysis of triflate **355** in TFE [174]. The cycloheptenyl triflate, which was detected to an extent of 3%, might have formed by ion-pair return.

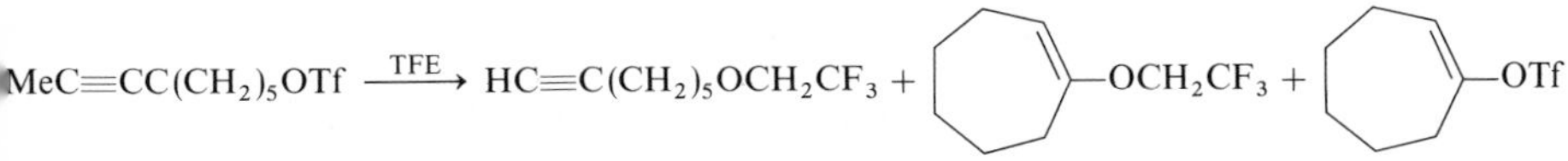
MeC≡CC(CH2)5OTf
TFE
HC≡C(CH2)5OCH2CF3 +
OCH2CF3 +
OTf

**355**; 74%; 9%; 3%

The primary triflate **360** (R = H, Me) gave no cyclization product on solvolysis in TFE [174], substantiating the early similar observation for the

**TABLE 3.27**

**Percentage of Cyclic Products Obtained in the Solvolysis of Alkynyl Triflates RC≡C(CH₂) CH₂OSO₂CF₃ in Anhydrous Trifluoroethanol [1f]**

| R | $n$ | Cyclic products (%) |
|---|---|---|
| H | 2 | 0 |
| Me | 2 | 0 |
| H | 3 | 50 (six-membered ring) |
| Me | 3 | 15 (six-membered ring) |
|  |  | 72 (five-membered ring) |
| H | 4 | 22 (seven-membered ring) |

corresponding tosylate [167]. The open-chain ether was the main product, suggesting that, due to the high energy of the strained cyclopentenyl cation [175], the expected intermediate diverted the reaction to the $k_s$ route. Table 3.27 gives the yield of the cyclized products obtained by the solvolysis of some acylic alkynyl triflates, showing how they depend on the position of the triple bond relative to the functional carbon atom [1f].

$$RC≡CCH_2CH_2CH_2OTf \xrightarrow{TFE} RC≡CCH_2CH_2CH_2OCH_2CF_3$$

**360**

An attempt to carry out stepwise cyclization via the ditosylate shown in Scheme 3.47 by participation of a triple bond with the developing cationic centers did not yield the expected products [176]. Instead, only mono-cyclization was observed even when TFE was used as a solvent (Scheme 3.47).

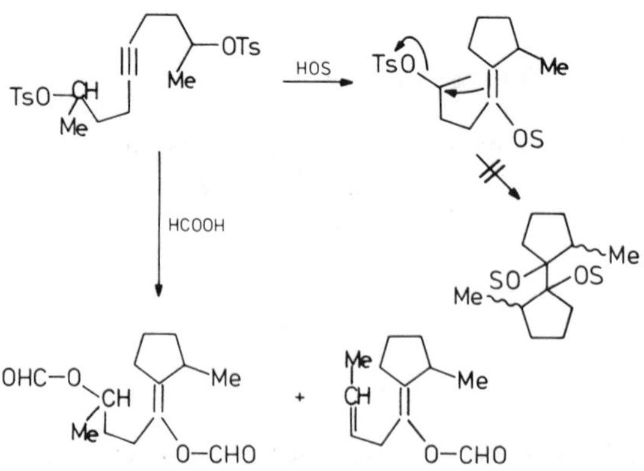

Scheme 3.47

An interesting extension of the participation of triple bonds is in its application in cyclic alkynyl systems. Stable cyclic alkynyl systems exist from eight-membered and higher rings [177]; hence, transannular participation can be studied in these rings. The first example of such participation by a triple bond was observed by Hanack and Heumann [178] in the solvolysis of 5-cyclodecyn-1-yl tosylate (361, X = OTs) in aqueous acetone. The main product was 1-decalone (362) along with a small amount of bicyclo-[5.3.0]decan-2-one (363), with no $S_N2$ product being detected [178].

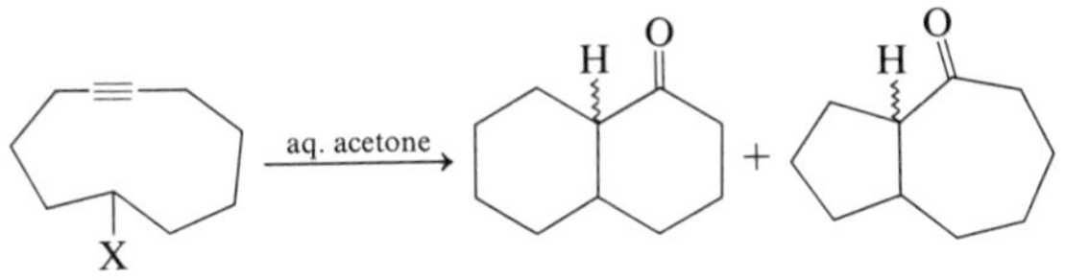

361                              362              363

X = OTs

The transannular rearrangement of 5-cyclodecyn-1-yl *p*-nitrobenzoate $(361, X = OCOC_6H_4NO_2\text{-}p)$ was studied in more detail [179]. The solvolysis of 361 $(X = OCOC_6H_4NO_2\text{-}p)$ in aqueous ethanol was about 20 times faster than that of the corresponding olefin isomer 364, and *cis*- and *trans*-1-decalone (362) were the only products observed. Another model compound, the open-chain derivative 365 with three $CH_2$ units between the triple bond and the leaving group, also solvolyzed more slowly in aqueous ethanol and gave no cyclization products. A change to the less nucleophilic acetic acid did not yield cyclized product either. Apparently, the spatial fixation of the triple bond in 361 is an important factor in its transannular interaction with the developing cationic center. In the cyclic alkyne series, the alcohol 361 (X = OH) underwent rearrangement to 1-decalone (362) with ethanolic HCl, and the corresponding amine 361, (X = $NH_2$) yielded at least 8% of the rearranged product on deamination [179]. In the open-chain alkyne derivatives 365, only the unrearranged alcohol 365 (X = OH) was obtained under the same reaction conditions.

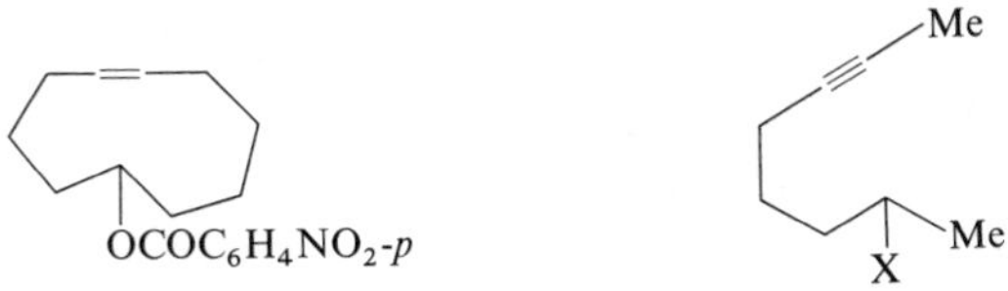

364                              365, X = $OCOC_6H_4NO_2\text{-}p$

The observed ratio of products 362 and 363 indicates that vinyl cation 366 must be more stable than ion 367. The greater stability of 366 compared to 367 was confirmed by the solvolysis studies of vinyl triflates 368 and 369.

In 50% aqueous ethanol as well as in 70% aqueous TFE, both triflates gave 1-decalone (362) as the major product [180].

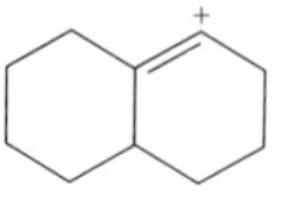

366 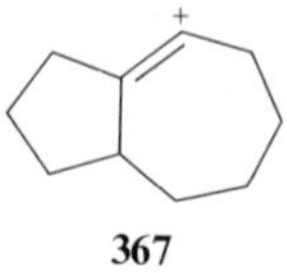    367

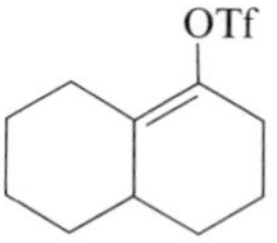

OTf    OTf

368 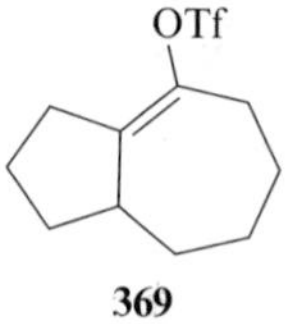    369

The above findings stimulated an investigation of the solvolysis of 361 in absolute TFE and an attempt to isolate any enol ethers formed. In fact, both enol ethers 370 and 371 were isolated and identified [181].

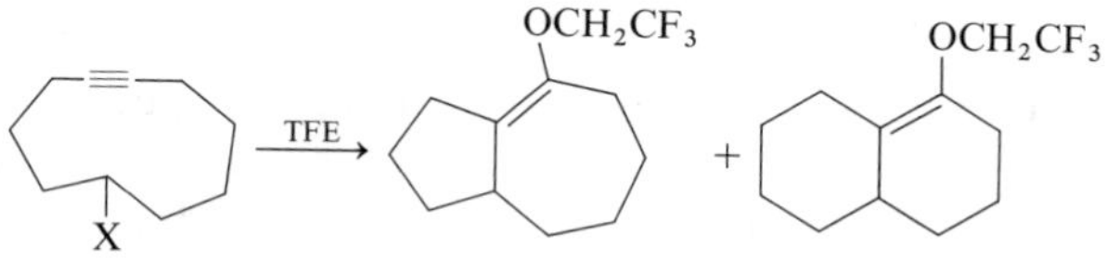

$OCH_2CF_3$    $OCH_2CF_3$

TFE →    +

X

$X = OCOC_6H_4NO_2\text{-}p,\ OTs$

361    370    371

Although vinyl cation 367 contains a seven-membered ring, which can better accommodate a bent vinyl cation than can 366 (Chapter 5), a careful examination reveals that the overall strain energy is lower for 366 than for 367. This is because an exocyclic double bond is present on the cyclopentane ring in 367, whereas this bond is exocyclic to a cyclohexane ring in 366. Indeed, methylenecyclopentane (a model for 367) is 5 kcal/mole less stable than 1-methylcyclohexane (a model for 366) owing to the ring strain [182]. Apparently, this effect more than compensates for the different stabilities of the bent vinyl cation moieties.

Acid-catalyzed rearrangement of the alkynol 361 (X = OH) with $BF_3 \cdot Et_2O$ gave the vinyl fluoride 372 as the single product [183]. Similarly, the secondary alkynol 373 yielded the vinyl fluoride 374 by treatment with the same reagent. However, the reaction of methanolic HCl with 373 gave a mixture of 10-methyl-1-decalone (375) and the bicyclo [5.3.0] compound 376 [183] (Scheme 3.48). A transannular participation of the triple bond leading to vinyl cations accounts for these results.

5-Cyclononyn-1-yl p-nitrobenzoate (377), in which the leaving group is symmetrically situated with respect to the triple bond, gave on solvolysis in absolute TFE exclusively the bicyclic enol ether 378 [184], as shown in

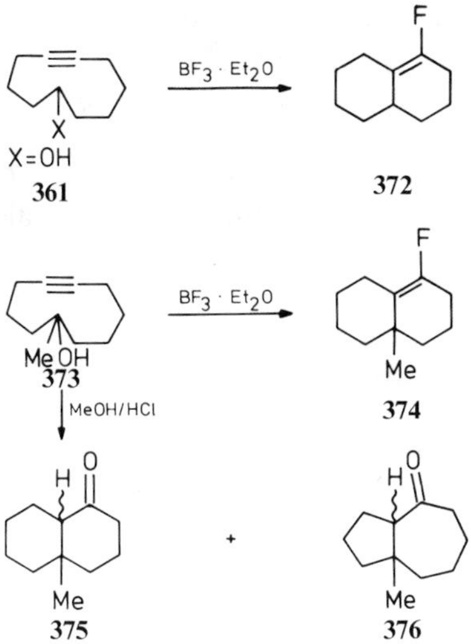

Scheme 3.48

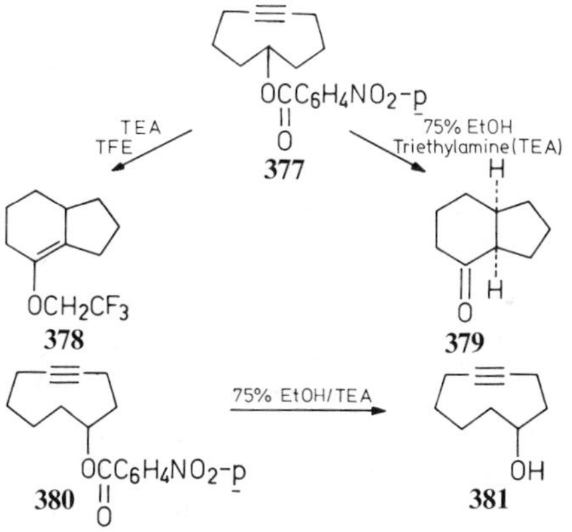

Scheme 3.49

Scheme 3.49. In 75% aqueous ethanol, **377** reacts ∼100 times faster than the corresponding *p*-nitrobenzoate **361** (X = OCOC$_6$H$_4$NO$_2$-*p*) with the ten-membered ring. The higher reactivity of **377** can be attributed to a more favorable overlap of its triple-bond orbitals with the developing carbenium ion compared to **361** (X = OCOC$_6$H$_4$NO$_2$-*p*).

4-Cyclononyn-1-yl *p*-nitrobenzoate (**380**) on solvolysis in aqueous ethanol, on the other hand, gave only the direct displacement product **381**, as shown in Scheme 3.49. Compound **380** does not rearrange with formation of bicyclic products because participation of the triple bond in **380** would lead either to a cyclopentenyl-like vinyl cation or to a vinyl cation with an exocyclic double bond to a four-membered ring, both of which are highly unlikely, as discussed in Chapter 5.

The solvolysis of 4-cyclooctyn-1-yl *p*-nitrobenzoate (**382**) in 75% aqueous ethanol gave an approximately 1:1 mixture of the alcohols **383** and **384**, without the formation of any rearranged product [184]. Alkynol **384** is probably formed by a simple $S_N2$ displacement and **383** is formed by a reduction process. As in the case of **380**, participation of the triple bond in **382** should again lead either to a cyclopentenyl-like vinyl cation or to a vinyl cation containing a four-membered ring. Due to this and the fact that it is itself an eight-membered cyclic acetylene and therefore highly strained, **382** does not rearrange with formation of bicyclic products upon solvolysis.

Intramolecular cyclization of alkynyl compounds is not restricted to cases in which the triple bond participates in a solvolysis reaction. Prelim-

$$OCC_6H_4NO_2\text{-}p$$

**382**                                          **383**        **384**

inary studies indicated that the alkynyl ketones 5-cyclodecyn-1-one (**385**) and 6-octyn-2-one (**387**) underwent intramolecular rearrangement on treatment with mineral acids [179, 185]; **385** gave rearranged ketone **386** as the only product, and **387** rearranged to 2,3-dimethyl-1-cyclohexen-2-one (**388**), as shown in Scheme 3.50. However, these reactions occurred slowly and in only low yields. The other possible product, 1-acetyl-2-methyl-1-cyclo-pentene (**389**), was unstable under the reaction conditions.

The effect of the strength of Lewis acids on the rate and the product composition in a similar reaction was recently studied [186]. The reaction of **387** with sublimed $AlCl_3$ in absolute ether was rapid and gave a mixture of **388** (60%) and **389** (40%). Both $BF_3 \cdot Et_2O$ and $SnCl_4$ also catalyzed the rearrangement, but reaction with $SnCl_4$ was very slow. The ratios of the products **388** and **389** were 45:55 with $BF_3$ and 60:40 with $SnCl_4$. The suggested mechanism involves addition of the Lewis acid ($AlCl_3$) to the carbonyl group, forming intermediate **392**. Cyclization of this intermediate gives vinyl cations **390** and **391**, which after capture by the nucleophile and hydrolysis give the final products **388** and **389** [185], as shown in Scheme 3.51.

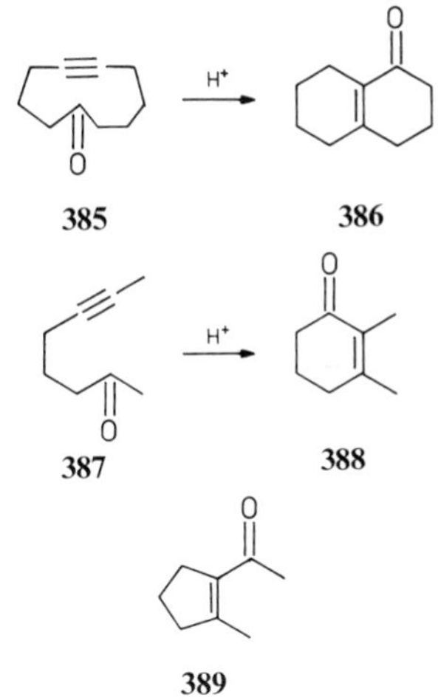

385                    386

387                    388

389
Scheme 3.50

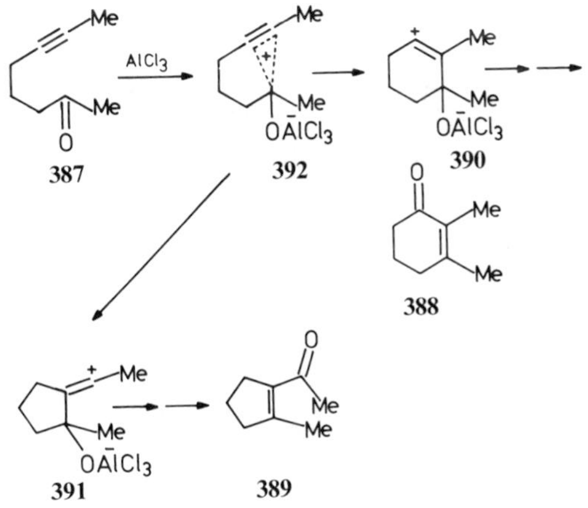

387        392        390

388

391        389
Scheme 3.51

The reaction of 5-cyclodecyn-1-one (385) with Lewis acids (BF$_3$·Et$_2$O, AlCl$_3$) was very fast, giving after addition of the nucleophile exclusively 386. The isomeric bicyclo[5.3.0]decen-2-one was not detected. Rearrangement apparently took place through intermediate 393, which rearranged to cation 394, which is apparently more stable than the bicyclic vinyl cation 395 [186].

The analogous rearrangement of 5-cyclononyn-1-one (396) to bicyclo [4.3.0]-1(6)-nonen-2-one (397) occurred with acid catalysts [187].

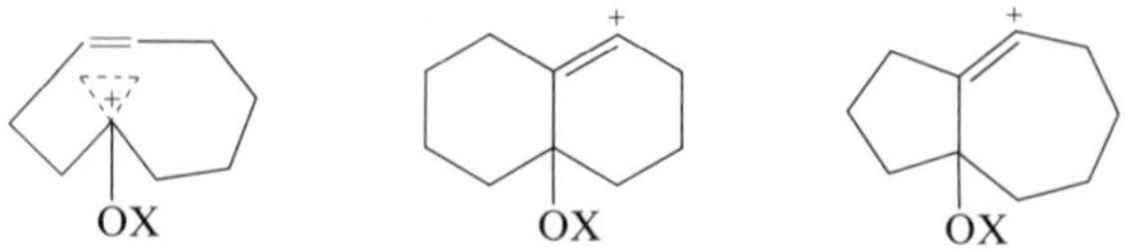

393                          394                          395

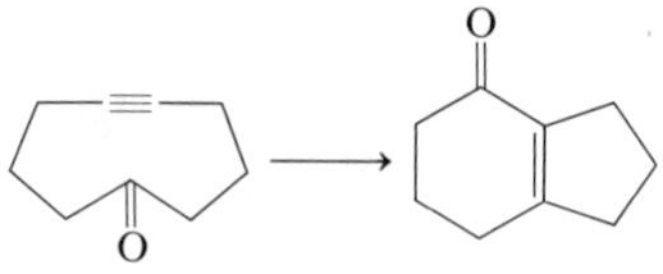

396                          397

Triple-bond participation also occurs in the solvolysis of a vinyl triflate. In the solvolysis of 6-octyn-2-en-2-yl triflate (**398**), rearranged products were observed only in solvents of low nucleophilicity and high ionizing power. In aqueous ethanol and aqueous TFE, the acyclic ketone **405** was the main product. About 35% of cyclization products were formed in 99.5% TFE [188]. The initially formed $\beta,\gamma$-unsaturated ketones **401** and **402** rearranged under the solvolytic conditions to the $\alpha,\beta$-unsaturated ketones **403** and **404**. The yield of the five-membered ketone **403** was three to four times higher than that of the six-membered ketone **404**. This indicates that the linear vinyl cation **399** is of a lower energy than the bent ion **400** (Scheme 3.52).

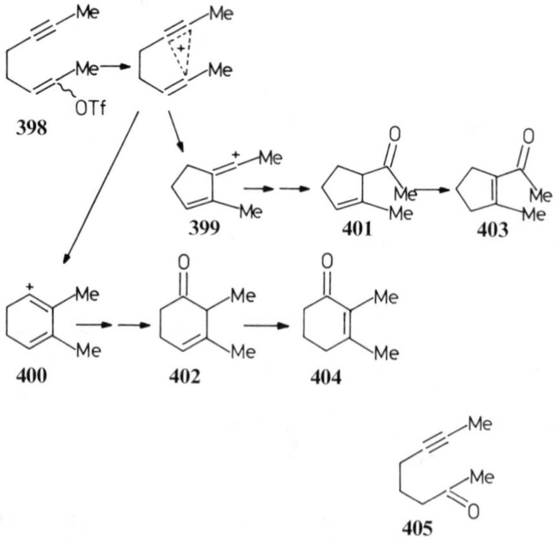

Scheme 3.52

Reaction of **398** is slow compared with that of the olefin analogue [189], and this was attributed to the inductive effect of the triple bond.

The participation of triple bonds leading to vinyl cations as intermediates has synthetic applications, as already mentioned for the synthesis of cyclobutanones. Syntheses of natural products have also been carried out by using the same principle.

Johnson and co-workers synthesized several steroids and triterpenes by biomimetic cyclizations with systems involving triple-bond participation [190]. The synthesis was achieved in one step, starting from suitable alkyne derivatives. The intermediate vinyl cations were trapped by several nucleophiles, and a preference for the formation of linear vinyl cation **407** was noted in the cyclization reaction of the trienyol **406** in the presence of TFA (Scheme 3.53) (however see ref. 1906 for more recent results).

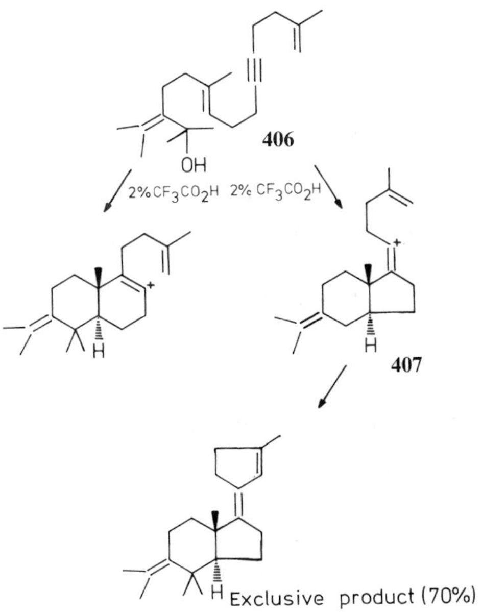

Scheme 3.53

The preferential formation of a five-membered product has been applied to the synthesis of dl-progesterone (**411**). The trienynol **408** cyclized in the presence of TFA in 1,2-dichloroethane via the intermediate vinyl cation **409** to the ketone **410**, from which **411** was obtained by subsequent steps (Scheme 3.54). A similar approach has been used for the synthesis of testosterone benzoate [191], longifolene [192], and other products [193]. These reactions are discussed at considerable length in an excellent review [190a].

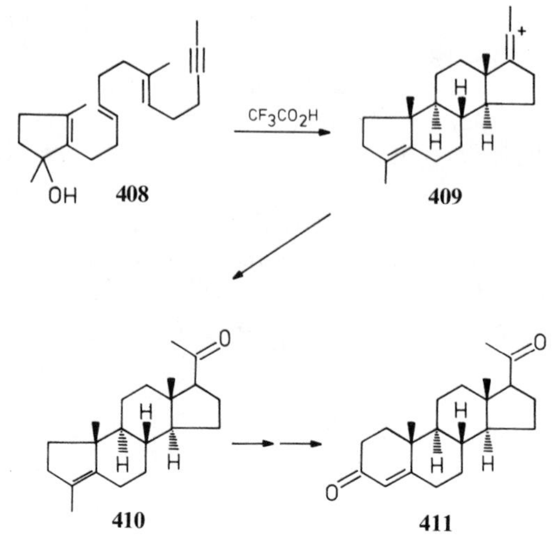


408 409 410 411

Scheme 3.54

Another example of triple-bond participation in the synthesis of natural products is the synthesis of the sesquiterpene cyclosativene, in which the key step is the cyclization reaction in TFE involving the triple bond in **412** to give the enol ether **414** via the vinyl cation **413** [194], as shown in Scheme 3.55.

It is clear from the extensive data compiled in this chapter that vinyl cations are definite intermediates in many electrophilic additions to alkynes. The intermediate vinyl cations involved are identical in their behavior to vinyl cations generated by solvolysis reactions. Reasonable explanations have been offered for the relative reactivity of alkynes and alkenes toward

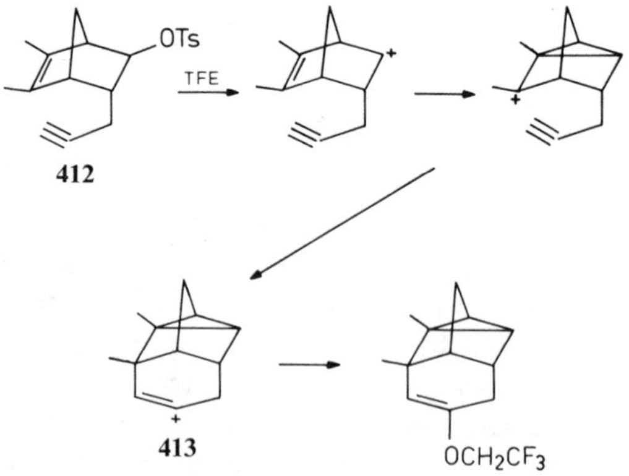


412 413 414

Scheme 3.55

electrophilic reagents that involve the nature of the electrophile, the solvent, and the substrate structure. Participation of triple bonds has been extensively employed in synthesis.

## REFERENCES

1.  (a) P. B. D. de la Mare and R. Bolton, "Electrophilic Addition to Unsaturated Systems." Elsevier, Amsterdam, 1966; (b) R. C. Fahey, *Top. Stereochem.* **3**, 237 (1968); (c) G. Modena and U. Tonellato, *Adv. Phys. Org. Chem.* **9**, 185 (1971); (d) P. J. Stang, *Prog. Phys. Org. Chem.* **10**, 205 (1973); (e) G. H. Schmid, *in* "The Carbon-Carbon Triple Bond" (S. Patai, ed.), Chap. 8. Wiley, New York, 1978; (f) M. Hanack, *Angew. Chem.* **90**, 346 (1978); *Angew. Chem., Int. Ed. Engl.* **17**, 333 (1978).
2.  R. Fittig and A. Schrohe, *Chem. Ber.* **8**, 367 (1875).
3.  M. G. Kutscheroff, *Chem. Ber.* **14**, 1532, 1540 (1881); **17**, 13 (1884).
4.  V. Jäger, M. Murray, U. Niedballa, and H. G. Viehe, "Methoden der Organischen Chemie" (Houben-Weyl), Vol. 5/2a. Thieme, Stuttgart, 1977.
5.  F. A. Long and M. A. Paul, *Chem. Rev.* **57**, 935 (1957).
6.  T. L. Jacobs and S. Searles, *J. Am. Chem. Soc.* **66**, 686 (1944).
7.  W. Drenth and H. Hogeveen, *Recl. Trav. Chim. Pays-Bas* **79**, 1002 (1960).
8.  H. Hogeveen and W. Drenth, *Recl. Trav. Chim. Pays-Bas* **82**, 375 (1963).
9.  H. Hogeveen and W. Drenth, *Recl. Trav. Chim. Pays-Bas* **82**, 410 (1963).
10. G. L. Hekkert and W. Drenth, *Recl. Trav. Chim. Pays-Bas* **80**, 1285 (1961).
11. E. J. Stamhuis and W. Drenth, *Recl. Trav. Chim. Pays-Bas* **80**, 797 (1961).
12. E. J. Stamhuis and W. Drenth, *Recl. Trav. Chim. Pays-Bas* **82**, 394 (1963).
13. E. J. Stamhuis and W. Drenth, *Recl. Trav. Chim. Pays-Bas* **82**, 385 (1963).
14. G. L. Hekkert and W. Drenth, *Recl. Trav. Chim. Pays-Bas* **82**, 405 (1963).
15. L. L. Schaleger and F. A. Long, *Adv. Phys. Org. Chem.* **1**, 1 (1963).
16. J. M. Williams, Jr. and M. M. Kreevoy, *Adv. Phys. Org. Chem.* **6**, 63 (1968).
17. B. Gutbezahl and E. Grunwald, *J. Am. Chem. Soc.* **75**, 559, 565 (1953).
18. L. Zucker and L. P. Hammett, *J. Am. Chem. Soc.* **61**, 2791 (1939).
19. J. F. Bunnett, *J. Am. Chem. Soc.* **83**, 4956, 4968, 4973, 4978 (1961).
20. W. F. Verhelst and W. Drenth, *J. Org. Chem.* **40**, 130 (1975).
21. W. F. Verhelst and W. Drenth, *J. Am. Chem. Soc.* **96**, 6692 (1974).
22. C. Friedel and M. Balsohn, *Bull. Soc. Chim. Fr.* p. 55 (1881).
23. A. Baeyer, *Chem. Ber.* **15**, 2705 (1882).
24. R. W. Bott, C. Eaborn, and D. R. M. Walton, *J. Organomet. Chem.* **1**, 420 (1964).
25. R. W. Bott, C. Eaborn, and D. R. M. Walton, *J. Chem. Soc.* p. 384 (1965).
26. J. Yukawa and Y. Tsuno, *Bull. Chem. Soc. Jpn.* **32**, 965, 971 (1959).
27. D. S. Noyce and M. D. Schiavelli, *J. Am. Chem. Soc.* **90**, 1020 (1968).
28. D. S. Noyce and M. D. Schiavelli, *J. Am. Chem. Soc.* **90**, 1023 (1968).
29. D. S. Noyce, M. A. Matesich, and P. E. Peterson, *J. Am. Chem. Soc.* **89**, 6225 (1967).
30. D. S. Noyce and K. E. DeBruin, *J. Am. Chem. Soc.* **90**, 372 (1968).
31. A. Streitwieser, Jr., R. H. Jagow, R. C. Fahey, and S. Suzuki, *J. Am. Chem. Soc.* **80**, 2326 (1958).
32. A. J. Kresge, *Chem. Soc. Rev.* **2**, 475 (1973).
33. S. Swaminathan and K. V. Narayanan, *Chem. Rev.* **71**, 429 (1971).
34. L. I. Olsson, A. Claesson, and C. Bogentoft, *Acta Chem. Scand.* **27**, 1629 (1973).
35. M. Edens, D. Boerner, C. R. Chase, D. Nass, and M. D. Schiavelli, *J. Org. Chem.* **42**, 3403 (1977).

36. S. I. Miller, *Adv. Phys. Org. Chem.* **6**, 185 (1968).
37. N. D. Epiotis, *J. Am. Chem. Soc.* **95**, 1191 (1973); N. D. Epiotis, "Theory of Organic Reactions." Springer-Verlag, Berlin and New York, 1978.
38. T. G. Traylor, *Acc. Chem. Res.* **2**, 152 (1969).
39. B. Zwanenburg and W. Drenth, *Recl. Trav. Chim. Pays-Bas* **82**, 879 (1963).
40. P. E. Peterson and J. E. Duddey, *J. Am. Chem. Soc.* **88**, 4990 (1966); P. E. Peterson and R. J. Bopp, *J. Am. Chem. Soc.* **89**, 1283 (1967); P. E. Peterson, *Acc. Chem. Res.* **4**, 407 (1971).
41. R. H. Summerville and P. von R. Schleyer, *J. Am. Chem. Soc.* **96**, 1110 (1974).
42. R. C. Fahey and D. J. Lee, *J. Am. Chem. Soc.* **88**, 5555 (1966).
43. R. C. Fahey, M. J. Payne, and D. J. Lee, *J. Org. Chem.*, **39**, 1124 (1974).
44. R. C. Fahey and D. J. Lee, *J. Am. Chem. Soc.* **90**, 2124 (1968).
45. R. Maroni, G. Melloni, and G. Modena, *Chem. Commun.* p. 857 (1972).
46. F. Mareuzzi, G. Melloni, and G. Modena, *Tetrahedron Lett.* p. 413 (1974).
47. Z. Rappoport and Y. Apeloig, *J. Am. Chem. Soc.* **96**, 6428 (1974).
48. Z. Rappoport and M. Atidia, *J. Chem. Soc., Perkins Trans. II* p. 2316 (1972).
49. A. Pross and Z. Rappoport, unpublished results.
50. K. Bowden and M. J. Price, *J. Chem. Soc. B.* p. 1466, 1472 (1970).
51. J. Tendil, M. Verny, and R. Vessière, *Bull. Chim. Soc. Fr.* p. 273 (1976).
52. J. Cousseau and L. Gouin, *Tetrahedron Lett.* p. 2889 (1974).
53. G. Borkent and W. Drenth, *Recl. Trav. Chim. Pays-Bas* **89**, 1057 (1970).
54. H. W. Whitlock, Jr, P. E. Sandvick, L. E. Overman, and P. B. Reichhardt, *J. Org. Chem.* **34**, 879 (1969).
55. H. A. Staab, H. Mack, and E. Wehinger, *Tetrahedron Lett.* p. 1465 (1968).
56. B. Bossenbroek, D. C. Sanders, H. M. Curry, and H. Shechter, *J. Am. Chem. Soc.* **91**, 371 (1969).
57. R. D. Howells and J. D. McCown, *Chem. Rev.* **77**, 69 (1977).
58. G. A. Olah and R. J. Spear, *J. Am. Chem. Soc.* **97**, 1845 (1975).
59. P. J. Stang and R. H. Summerville, *J. Am. Chem. Soc.* **91**, 4600 (1969).
60. D. D. Maness and L. D. Turrentine, *Tetrahedron Lett.* p. 755 (1973).
61. A. E. van der Hout-Lodder, J. W. de Haan, L. J. M. van de Ven, and H. M. Buck, *Recl. Trav. Chim. Pays-Bas* **92**, 1040 (1973); G. A. Olah, J. S. Staral, R. J. Spear, and G. Liang, *J. Am. Chem. Soc.* **97**, 5489 (1975), and references cited therein.
62. G. A. Olah and H. Mayr, *J. Am. Chem. Soc.* **98**, 7333 (1976).
63. G. A. Olah, R. J. Spear, P. W. Westerman, and J. M. Denis, *J. Am. Chem. Soc.* **96**, 5855 (1974).
64. T. Sasaki, S. Eguchi, and T. Toru, *Chem. Commun.* p. 780 (1968).
65. K. Bott, *Tetrahedron Lett.* p. 1747 (1969); *Chem. Commun.* p. 1349 (1969); *Justus Liebigs Ann. Chem.* **766**, 51 (1972).
66. D. R. Kell and F. J. McQuillin, *J. Chem. Soc., Perkin Trans. I* p. 2100 (1972).
67. J. K. Chakrabarti and A. Todd, *Chem. Commun.* p. 556 (1971).
68. F. Marcuzzi and G. Melloni, *J. Am. Chem. Soc.* **98**, 3295 (1976).
69. F. Marcuzzi and G. Melloni, *Gazz. Chim. Ital.* **105**, 495 (1975).
70. R. Maroni, G. Melloni, and G. Modena, *J. Chem. Soc., Perkin Trans. I* p. 2491 (1973).
71. R. Maroni, G. Melloni, and G. Modena, *J. Chem. Soc., Perkin Trans. I* p. 353 (1974).
72. F. Marcuzzi and G. Melloni, *J. Chem. Soc., Perkin Trans. II* p. 1517 (1976).
73. H. Hogeveen and C. F. Roobeek, *Tetrahedron Lett.* p. 3343 (1971).
74. G. Capozzi, V. Lucchini, F. Marcuzzi, and G. Melloni, *Tetrahedron Lett.* p. 717 (1976).
75. M. I. Rybinskaya, A. N. Nesmeyanov, and N. K. Kochetkov, *Russ. Chem. Rev.*, 433 (1969); A. E. Pohland and W. R. Benson, *Chem. Rev.* **66**, 161 (1966).

76. G. J. Martin, C. Rabiller and G. Mabon, *Tetrahedron* **28**, 4027 (1972); G. J. Martin and W. Kirschleger, *C. R. Acad. Sci., Ser.* **C 279**, 363 (1974).
77. H. Martens, F. Janssens, and G. Hoornaert, *Tetrahedron* **31**, 177 (1975).
78. A. A. Schegolev, W. A. Smit, G. V. Roitburd, and V. F. Kucherov, *Tetrahedron Lett.* p. 3373 (1974).
79. A. A. Schegolev, W. A. Smit, V. F. Kucherov, and R. Caple, *J. Am. Chem. Soc.* **97**, 6604 (1975).
80. M. I. Kanischev, W. A. Smit, A. A. Schegolev, and R. Caple, *Tetrahedron Lett.* p. 1421 (1978).
81. Cited in M. A. Wilson, *J. Chem. Educ.* **52**, 495 (1975).
82. Z. Rappoport, *Tetrahedron Lett.* p. 1073 (1978).
83. G. H. Schmid and D. G. Garratt, *in* "The Chemistry of Double-Bonded Functional Groups" (S. Patai, ed.), Part 2, Chap. 9. Wiley, New York, 1977.
84. J. J. Verbane and G. F. Hennion, *J. Am. Chem. Soc.* **60**, 1711 (1938); R. O. Norris, R. R. Vogt, and G. F. Hennion, *J. Am. Chem. Soc.* **61**, 1460 (1939); R. O. Norris and G. F. Hennion, *J. Am. Chem. Soc.* **62**, 449 (1940); G. F. Hennion and C. E. Welsh, *J. Am. Chem. Soc.* **62**, 1367 (1940).
85. M. L. Poutsma and J. L. Kartch, *Tetrahedron* **22**, 2167 (1966).
86. D. F. Shellhamer and M. L. Oakes, *J. Org. Chem.* **43**, 1316 (1978).
87. S. A. Sherrod and R. G. Bergman, *J. Am. Chem. Soc.* **93**, 1925 (1971).
88. K. Yates and A. Go, unpublished results, cited in reference 1e.
89. P. W. Robertson, W. E. Dasent, R. M. Milburn, and W. H. Oliver, *J. Chem. Soc.* p. 1628 (1950).
90. H. Sinn, S. Hopperdietzel, and D. Sauermann, *Monatsh. Chem.* **96**, 1036 (1965).
91. J. A. Pincock and K. Yates, *Can. J. Chem.* **48**, 3332 (1970).
92. J. A. Pincock and C. Somawardhana, *Can. J. Chem.* **56**, 1164 (1978).
93. R. Gelin and D. Pigasse, *Bull. Soc. Chim. Fr.* p. 2186 (1971).
94. G. Wittig and H. L. Dorsch, *Justus Liebigs Ann. Chem.* **711**, 46 (1968).
95. C. B. Reese and A. Shaw, *Tetrahedron Lett.* p. 4641 (1971).
96. M. H. Wilson and E. Berliner, *J. Am. Chem. Soc.* **93**, 208 (1971); V. L. Cunningham and E. Berliner, *J. Org. Chem.* **39**, 3731 (1974).
97. E. Mauger and E. Berliner, *J. Am. Chem. Soc.* **94**, 194 (1972).
98. N. Kharasch and S. J. Assony, *J. Am. Chem. Soc.* **75**, 1081 (1953).
99. V. Calo, G. Melloni, G. Modena, and G. Scorrano, *Tetrahedron Lett.* p. 4399 (1965); G. Modena and G. Scorrano, *Mech. React. Sulfur Compd.* **3**, 115 (1968).
100. G. Capozzi, V. Lucchini, G. Modena, and P. Scrimin, *Nouv. J. Chem.* **2**, 95 (1978).
101. G. H. Schmid and M. Heinola, *J. Am. Chem. Soc.* **90**, 3466 (1968).
102. T. Okuyama, K. Izawa, and T. Fueno, *J. Org. Chem.* **39**, 351 (1974).
103. G. Modena and U. Tonellato, *J. Chem. Soc. B* p. 374 (1971).
104. G. H. Schmid, A. Modro, D. G. Garrat, and K. Yates, *Can. J. Chem.* **54**, 3045 (1976).
105. G. H. Schmid, A. Modro, F. Lenz, D. G. Garrat, and K. Yates, *J. Org. Chem.* **41**, 2331 (1976).
106. W. Ried and W. Ochs, *Chem. Ber.* **107**, 1334 (1974).
107. J. Tendil, M. Verney, and R. Vessière, *Tetrahedron* **30**, 579 (1974).
108. G. Capozzi, O. De Lucchi, V. Lucchini, and G. Modena, *Chem. Commun.* p. 248 (1975).
109. G. Capozzi, V. Lucchini, G. Modena, and P. Scrimin, *Tetrahedron Lett.* p. 911 (1977).
110. R. Destro, T. Pilati, and M. Simonetta, *Chem. Commun.* p. 576 (1977).
111. G. H. Schmid and D. G. Garratt, *Chim. Scr.* **10**, 76 (1977).
112. T. Woolridge and T. D. Roberts, *Tetrahedron Lett.* p. 4007 (1973).
113. Mortreux and M. Blanchard, *Bull. Soc. Chim. Fr.* p. 4035 (1970).

114. B. J. Barry, W. J. Beale, M. D. Carr, S.-K. Hei, and I. Reid, *Chem. Commun.* p. 177 (1973).
115. R. O. C. Norman, W. J. E. Parr, and C. Barry Thomas, *J. Chem. Soc., Perkin Trans. I* p. 1983 (1976).
116. R. J. Spear and W. A. Jensen, *Tetrahedron Lett.* p. 4535 (1977).
117. R. A. Bell, M. H. Chisholm, D. A. Couch, and L. A. Rankel, *Inorg. Chem.* **16**, 677 (1977); R. A. Bell and M. H. Chisholm, *Inorg. Chem.* **16**, 687 (1977).
118. S. Uemura, A. Onoe, and M. Okano, *Chem. Commun.* p. 925 (1975).
119. P. Bassi and U. Tonellato, *J. Chem. Soc., Perkin Trans. II* p. 1283 (1974).
120. S. Uemura, A. Onoe, and M. Okano, *Chem. Commun.* p. 145 (1976).
121. K. Yates, G. H. Schmid, T. W. Regulski, D. G. Garrat, H. W. Leung, and R. McDonald, *J. Am. Chem. Soc.* **95**, 160 (1973).
122. (a) G. Modena, F. Rivetti, G. Scorrano, and U. Tonellato, *J. Am. Chem. Soc.* **99**, 3392 (1977); (b) G. Modena, F. Rivetti, and U. Tonellato, *J. Org. Chem.* **43**, 1521 (1978).
123. M.-F. Ruasse and J. E. Dubois, *J. Org. Chem.* **42**, 2689 (1977).
124. D. Kaufman and R. Kupper, *J. Org. Chem.* **39**, 1438 (1974).
125. F. Marcuzzi and G. Melloni, *Tetrahedron Lett.* p. 2771 (1975).
126. G. H. Schmid. A. Modro, and K. Yates, *J. Org. Chem.* **42**, 2021 (1977).
127. J. M. Kornprobst and J. E. Dubois, *Tetrahedron Lett.* p. 2203 (1974).
128. F. Garnier, R. H. Donnay, and J. E. Dubois, *Chem. Commun.* p. 829 (1971).
129. L. Radom, P. C. Hariharan, J. A. Pople, and P. von R. Schleyer, *J. Am. Chem. Soc.* **95**, 6531 (1973); Z. Rappoport and Y. Apeloig, *J. Am. Chem. Soc.* **91**, 6734 (1969).
130. G. Olah and T. R. Hokswender, Jr., *J. Am. Chem. Soc.* **96**, 3574 (1974).
131. P. Bassi and U. Tonellato, *J. Chem. Soc., Perkin Trans. I* p. 669 (1973).
132. G. Capozzi, G. Modena, and U. Tonellato, *J. Chem. Soc. B* p. 1700 (1971).
133. J. F. Bunnett and F. Olsen, *Can. J. Chem.* **44**, 1899, 1917 (1966).
134. K. Griesbaum, *Angew Chem.* **76**, 782 (1964).
135. K. Griesbaum and M. El-Abed, *Chem. Ber.* **106**, 2001 (1973).
136. K. Griesbaum and W. Seiter, *J. Org. Chem.* **41**, 937 (1976).
137. K. Griesbaum and W. Seiter, *Angew. Chem.* **88**, 59 (1976); *Angew. Chem., Int. Ed. Engl.* **15**, 55 (1976).
138. K. Griesbaum, A. Singh, and M. El-Abed, *Tetrahedron Lett.* p. 1159 (1978).
139. H. Schneider, Ph.D. Thesis, Univ. of Karlsruhe, 1977.
140. G. Stammann, Ph.D. Thesis, Univ. of Karlsruhe, 1977.
141. J. H. Lukas, F. Baardman, and A. P. Kouwenhoven, *Angew. Chem.* **88**, 412 (1976); *Angew. Chem., Int. Ed. Engl.* **15**, 369 (1976).
142. L. Ghosez and J. Marchand-Brynaert, in "Imminium Salts in Organic Chemistry" (H. Böhme and H. G. Viehe, eds.), p. 421. Wiley (Interscience), New York, 1976.
143. R. C. Elder, L. R. Florian, E. R. Kennedy, and R. S. Macomber, *J. Org. Chem.* **38**, 4177 (1973).
144. M. Julia and C. B. du Jassoneix, *Bull. Chim. Soc. Fr.* p. 751 (1975).
145. E. J. Moriconi and Y. Shimakawa, *J. Org. Chem.* **37**, 196 (1972).
146. Y. S. Rao and R. Filler, *Tetrahedron Lett.* p. 1457 (1975).
147. M. Miyaura, T. Yoshinari, M. Itoh, and A. Suzuki, *Tetrahedron Lett.* p. 2961 (1974).
148. G. V. Roitburd, W. A. Smit, A. V. Semenovsky, A. A. Shchegolev, V. F. Kucherov, O. S. Chizhov, and V. I. Kadentsev, *Tetrahedron Lett.* p. 4935 (1972).
149. T. S. Abram and W. E. Watts, *J. Chem. Soc., Perkin Trans. I* p. 1522, 1527 (1977).
150. R. M. Roberts and M. B. A-Baset, *J. Org. Chem.* **41**, 1698 (1976).
151. M. Hanack and H. J. Schneider, *Fortschr. Chem. Forch.* **8**, 554 (1967).
152. M. Hanack, J. Häffner, and I. Herterich, *Tetrahedron Lett.*, 875 (1965).
153. S. Winstein, E. Allred, R. Heck, and R. Glick, *Tetrahedron* **3**, 1 (1958); F. L. Schadt, T. W. Bently, and P. von R. Schleyer, *J. Am. Chem. Soc.* **98**, 7658, 7667 (1976).

154. M. Hanack and I. Herterich, *Tetrahedron Lett.* p. 3847 (1966).

155. M. Hanack, I. Herterich, and V. Vött, *Tetrahedron Lett.* p. 3871 (1967).

156. M. Hanack, V. Vött, and H. Ehrhardt, *Tetrahedron Lett.* p. 4617 (1968).

157. M. Hanack, S. Bocher, K. Hummel, and V. Vött, *Tetrahedron Lett.* p. 4613 (1968).

158. M. Hanack, S. Bocher, I. Herterich, K. Hummel, and V. Vött, *Justus Liebigs Ann. Chem.* **733**, 5 (1970).

159. K. Hummel and M. Hanack, *Justus Liebigs Ann. Chem.* **746**, 211 (1971).

160. M. Hanack, T. Dehesch, K. Hummel, and A. Nierth, *Org. Synth.* **54**, 84 (1974).

161. H. Stutz and M. Hanack, *Tetrahedron Lett.* p. 2457 (1974).

162. R. Gary and R. Vessière, *Tetrahedron Lett.* p. 2983 (1972).

163. J. W. Wilson, *J. Am. Chem. Soc.* **91**, 3238 (1969).

164. J. L. Deroque, F. B. Sundermann, N. Youssif, and M. Hanack, *Justus Liebigs Ann. Chem.* 419 (1973).

165. H. R. Ward and P. D. Sherman, Jr., *J. Am. Chem. Soc.* **89**, 1962 (1967).

166. D. Kammacher, Ph.D. Thesis, Univ. of Saarbrücken, 1972.

167. H. Stutz, Ph.D. Thesis, Univ. of Saarbrücken, 1978.

168. W. Schoberth, Ph.D. Thesis, Univ. of Saarbrücken, 1977.

169. C. J. Collins, B. M. Benjamin, M. Hanack, and H. Stutz, *J. Am. Chem. Soc.* **99**, 1669 (1977).

170. M. Hanack, W. Schumacher, and E. Kunzmann, *Tetrahedron Lett.* 239 (1979).

171. P. E. Peterson and R. J. Kamat, *J. Am. Chem. Soc.* **88**, 3152 (1966); **91**, 4521 (1969).

172. W. Schumacher and M. Hanack, unpublished results.

173. W. D. Closson and S. A. Roman, *Tetrahedron Lett.* p. 6015 (1966).

174. K. A. Fuchs, Ph.D. Thesis, Univ. of Saarbrücken, 1978.

175. M. Hanack, H. Bentz, R. Märkl, and L. R. Subramanian, *Justus Liebigs Ann. Chem.* 1894 (1978).

176. J. B. Lambert, J. Papay, and H. W. Mark, *J. Org. Chem.* **40**, 633 (1975).

177. M. Hanack, L. R. Subramanian, and W. Eymann, *Naturwissenschaften* **64**, 397 (1977).

178. M. Hanack and A. Heumann, *Tetrahedron Lett.* p. 5117 (1969).

179. M. Hanack, C. E. Harding, and J. L. Deroque, *Chem. Ber.* **105**, 421 (1972).

180. M. J. Chandy, L. R. Subramanian, and M. Hanack, *Chem. Ber.* **108**, 2212 (1975).

181. M. J. Chandy and M. Hanack, *Tetrahedron Lett.* p. 4377 (1977).

182. P. von R. Schleyer, J. E. Williams, and K. R. Blanchard, *J. Am. Chem. Soc.* **92**, 2377 (1970).

183. R. J. Balf, B. Rao, and L. Weiler, *Can. J. Chem.* **49**, 3135 (1971).

184. W. Spang and M. Hanack, *Chem. Ber.* in press.

185. C. E. Harding and M. Hanack, *Tetrahedron Lett.* p. 1253 (1971).

186. M. J. Chandy and M. Hanack, *Arch. Pharm. (Weinheim, Ger.)* **380**, 578 (1975).

187. G. S. Lange and T. W. Hall, *J. Org. Chem.* **39**, 3819 (1974).

188. M. J. Chandy and M. Hanack, *Tetrahedron Lett.* p. 4515 (1975).

189. T. C. Clarke and R. G. Bergman, *J. Am. Chem. Soc.* **96**, 7934 (1974).

190a. W. S. Johnson, *Angew. Chem., Int. Ed. Engl.* **15**, 9 (1976)[+] and references cited therein
    b. W. S. Johnson, T. M. Yarwell, R. F. Myers, and D. R. Morton, *Tetrahedron Lett.*, p. 2549 (1978).

191. D. R. Morton and W. S. Johnson, *J. Am. Chem. Soc.* **95**, 4419 (1973).

192. R. A. Volkmann, G. C. Andrews, and W. S. Johnson, *J. Am. Chem. Soc.* **97**, 4777 (1975).

193. P. T. Lansbury, T. R. Demmin, G. E. DuBois, and V. R. Haddon, *J. Am. Chem. Soc.* **97**, 394 (1975).

194. S. E. Baldwin and J. C. Tomesch, *Tetrahedron Lett.* p. 1055 (1975).

# 4

# ELECTROPHILIC ADDITIONS TO ALLENES AND PARTICIPATION OF THE ALLENYL BOND IN SOLVOLYSIS

The mechanisms of electrophilic additions to allenes have been investigated to a lesser extent than those of additions to alkynes [1]. The acid-catalyzed hydration of allene itself was reported as early as 1888 [2], but systematic investigations were undertaken much later. Allene (**1**) is a symmetric, relatively nonpolar molecule with sp hybridization at the central carbon atom and sp$^2$ hybridization at the two terminal carbon atoms; the three carbon atoms are linear, and the two double bonds are in perpendicular planes (**2**) [1b]. The central carbon is positively polarized with respect to the terminal carbons, due to the inductive effect of the C—H bonds (**3**).

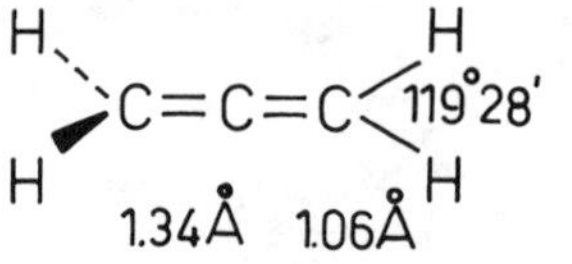

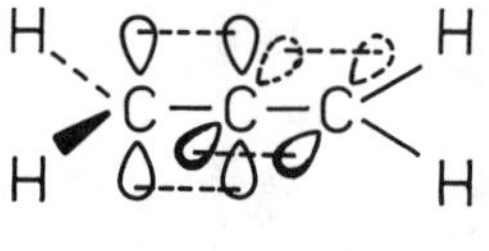

The fact that the double bonds in allene are shorter than that of ethylene is ascribed to hyperconjugation, whereby the double bonds acquire partial triple-bond character:

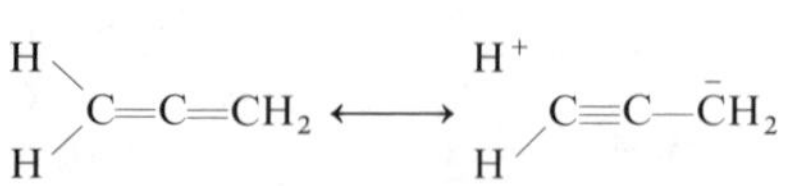

Such $\sigma$–$\pi$ overlap is particularly favored in allene, because the C—H $\sigma$ bonds at one end lie in the same plane as the C—C $\pi$ bond at the other end of the molecule. The problems involved in the preparation of allenes and their instability, which is reflected in their tendency to undergo alkyne–allene–diene conversions, polymerizations, and various rearrangements, may be one reason why few systematic studies on their electrophilic addition reactions have been undertaken so far. Another reason, discussed below, is the variety of products feasible due to attack on different carbon atoms of the allene, to rearrangements and eliminations of the intermediate cations, and, in some cases, to the instability of the products.

## I. ADDITIONS TO ALLENES

### A. Mechanisms of Addition

In general, the mechanisms of addition applicable to allenes present an additional dimension. Attack by the electrophile E at a terminal carbon atom gives the vinyl cation **4**, whereas allyl cation **5** results from attack at the central carbon atom.

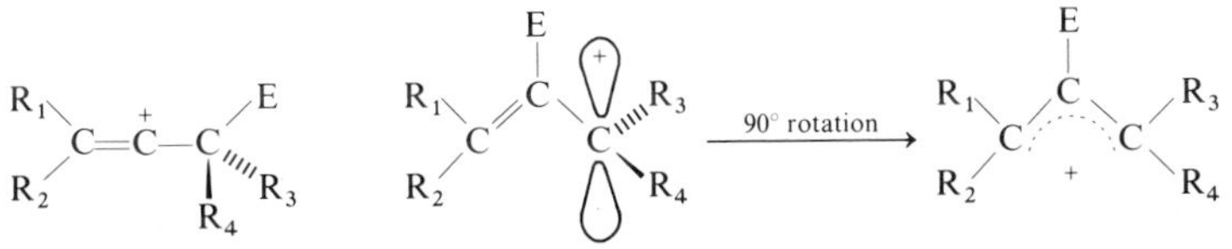

of the inductive effect of the C=C double bond.

A question of interest is, why are products from vinyl cations observed at all, when the alternative intermediate, the allyl cation, can be stabilized by interaction of the empty p orbital with the $\pi$ orbital of the C—C bond? Since the $\pi$ orbitals of the allene are perpendicular to one another in the initially formed allyl cation **5**, the p $(C^+)$ and $\pi$ (C=C) orbitals are also perpendicular, and a 90° rotation around the C—C bond is necessary for achieving the allylic stabilization of the conjugated cation **6** (Chapter 2). Consequently, the activation energy for the conversion of an allene to an allyl cation is raised compared with that for addition to an alkene, because of the inductive effect of the C=C double bond.

In practice, the preferred intermediate ions in the addition to allenes are tertiary allyl cation > vinyl cation $\simeq$ secondary allyl cation > primary allyl cation. When a secondary allyl cation is a possible intermediate, such factors as the substituents on the allene, the electrophile, and the solvent play an important role in determining whether an allyl or vinyl cation will be formed.

The rate-determining step in the electrophilic addition reaction to allenes is the formation of either the vinyl cation **4** or the twisted allyl cation **5**. The rotation of the allyl cation **5** to give the conjugated species **6** is presumably fast. The different sites of attack, especially at unsymmetrically substituted allenes, can result in the formation of several products. These possibilities for the $Ad_E2$ mechanism via a noncyclic carbenium ion [1c] are given in Scheme 4.1, where the delocalized allyl cations are obtained from the initial perpendicular ions by 90° rotation. Scheme 4.1 shows that with an unsymmetrically substituted allene, eight different isomeric products can be formed, since each isomer can exist in two geometric forms. Rearrangement and isomerization can increase the number of products. However, as shown

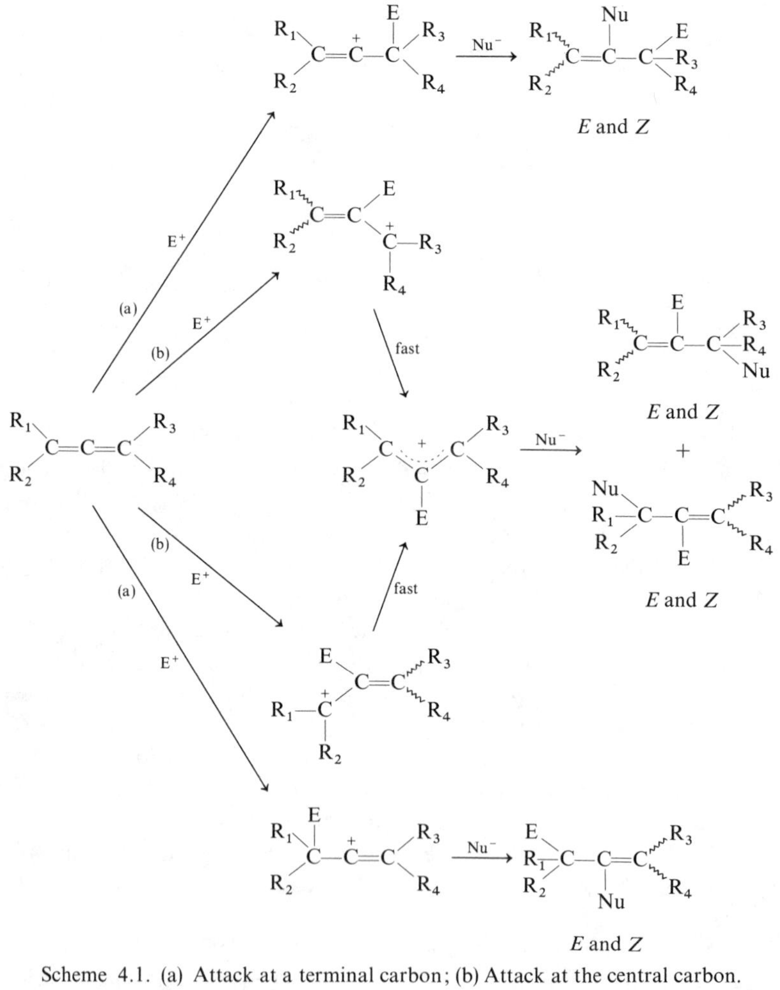

Scheme 4.1. (a) Attack at a terminal carbon; (b) Attack at the central carbon.

below, the number of products in many cases is lower, since the direction of the addition is strongly influenced by the nature of the substituents.

In addition, one should bear in mind the possibility that cyclic cations (7) may occur as intermediates. Various formalisms of bridged ions are employed [1c], and the symmetric $\sigma$-complex structure 7 will be used throughout this chapter to depict a bridged ion. Bridging probably occurs more readily in some cases because of the high activation energy necessary to generate the "twisted" allyl cation. This route is especially important in the addition of halogens and sulfenyl halides, as would be expected on the basis of analogous reactions of alkenes.

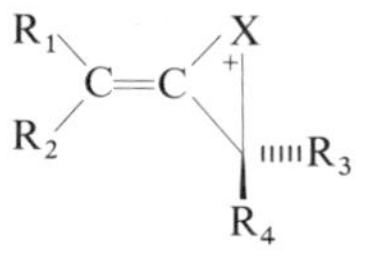

$$\begin{array}{c} R_1 \\ \diagdown \\ \phantom{x} \phantom{x} C{=}C \\ R_2 \diagup \end{array} \begin{array}{c} X \\ \diagup \phantom{} {+} \\ \phantom{} \\ \Big| \phantom{x}{}^{\prime\prime\prime\prime}R_3 \\ R_4 \end{array}$$

7  X = halogen, ArS

## B. Addition to Allene

The addition of hydrogen halides to allene is interesting, since it illustrates the behavior of the intermediate vinyl cation 8. Terminal addition takes place exclusively, and the products are identical with those formed by protonation reaction of hydrogen halides with propyne [3, 4].

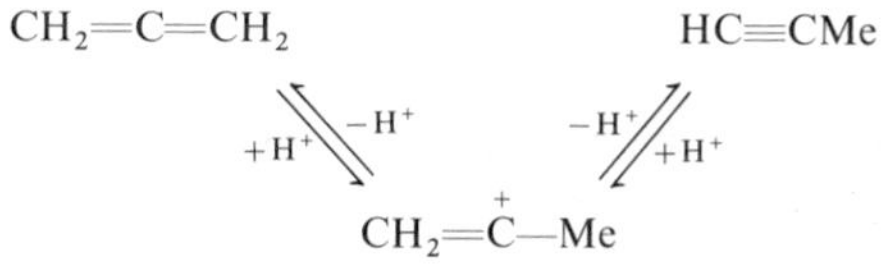

$$CH_2{=}C{=}CH_2 \qquad\qquad HC{\equiv}CMe$$

$$_{+H^+}\diagdown\phantom{x}\diagup^{-H^+} \qquad {}^{-H^+}\diagup\phantom{x}\diagdown_{+H^+}$$

$$CH_2{=}\overset{+}{C}{-}Me$$

8

Besides the expected mono- and diadducts 9 and 10, a considerable amount of 1,3-dihalo-1,3-dimethylcyclobutane (11) was also formed. The products are rationalized by initial formation of the vinyl cation 8, which can either react with X⁻ to give the addition products 9 and 10 (route A) or undergo cycloaddition with a second molecule of allene (route B) or propyne (route C), which is formed by isomerization of the starting allene (Scheme 4.2).

No cyclic products were found in the addition of hydrogen iodide, probably because capture of 8 by the highly nucleophilic iodide ion competes effectively with its reaction with the starting allene or propyne. Hydrogen fluoride was reported to add to allene to give 2,2-difluoropropane (10, X = F), but no reference was made to the formation of cyclic products [5].

The addition of DCl to allene with $BiCl_3$ as a catalyst was also reported [6]. The products and deuterium incorporation found were in accordance

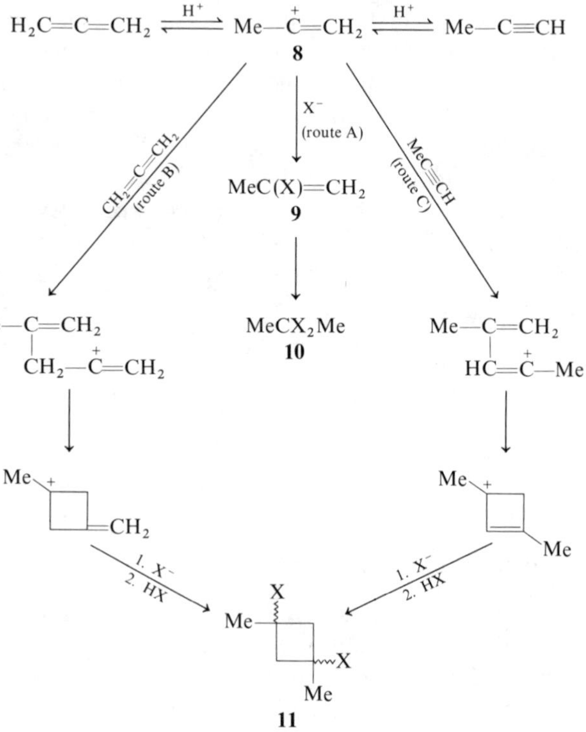

$$H_2C=C=CH_2 \overset{H^+}{\rightleftharpoons} Me-\overset{+}{C}=CH_2 \overset{H^+}{\rightleftharpoons} Me-C\equiv CH$$

Scheme 4.2

with routes A and B of Scheme 4.2, but no deuterium incorporation was found in the recovered allene. This seems to rule out an equilibrium between allene and propyne in the addition of HX to allenes. However, it should be borne in mind that the yield of chlorinated reaction products was very low in this case ($<1\%$), and the reaction conditions were different from those used by Griesbaum *et al.* [3, 4].

The uncatalyzed gas-phase hydrochlorination of allene was reported to give 2-chloropropene (**9**), 2,2-dichloropropane (**10**, X = Cl), and propyne as products [7]. This points to a vinyl cation-like intermediate for the addition, even in the gas phase. It seems, then, that hydrogen halides add to allene with the formation of a vinyl cation to give the Markownikoff-oriented reaction products.

Other electrophilic additions to allene either have not been thoroughly investigated from a mechanistic viewpoint or proceed via a cyclic cation (**12**), as suggested for the addition of HOBr [8], sulfenyl halides [9], and halogens [10, 11]. The cyclic halonium ions **12** (X = Cl, Br) were observed by nmr by treating 2,3-dihalopropenes with $SbF_5$ in $SO_2$ at $-70°$ [12]. The

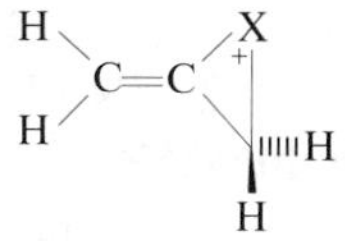

$$\begin{array}{c} H \\ \phantom{}\diagdown \\ \phantom{H}C{=}C \\ H\diagup \quad \overset{X}{\underset{H}{\diagup}}{}_{\text{\tiny III}H} \\ \phantom{}\mid \\ H \end{array}$$

**12**

formation of adducts **16** (4.3%) and **17** (1.5%) during the chlorination of neat allene at $-30°$ [11], in addition to the expected 2,3-dichloropropene **13** (23%) and propargyl chloride **14** (20%) [10], was suggested to proceed by an ionic mechanism involving the vinyl cation **15** (Scheme 4.3).

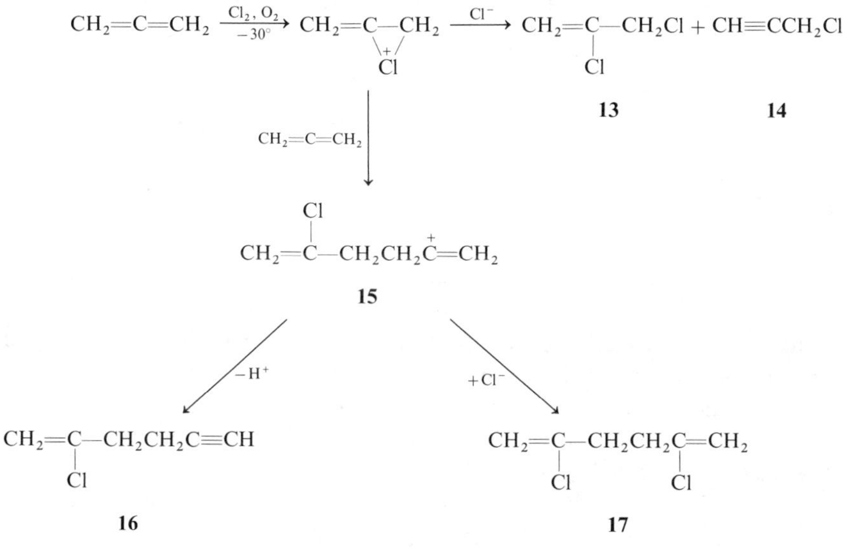

$$CH_2{=}C{=}CH_2 \xrightarrow[-30°]{Cl_2,\,O_2} CH_2{=}\underset{\underset{Cl}{+}}{C}{-}CH_2 \xrightarrow{Cl^-} CH_2{=}\underset{\underset{Cl}{|}}{C}{-}CH_2Cl + CH{\equiv}CCH_2Cl$$

**13** **14**

$$\downarrow CH_2{=}C{=}CH_2$$

$$CH_2{=}\underset{\underset{Cl}{|}}{C}{-}CH_2CH_2\overset{+}{C}{=}CH_2$$

**15**

$\overset{-H^+}{\swarrow}$ $\overset{+Cl^-}{\searrow}$

$$CH_2{=}\underset{\underset{Cl}{|}}{C}{-}CH_2CH_2C{\equiv}CH \qquad CH_2{=}\underset{\underset{Cl}{|}}{C}{-}CH_2CH_2\underset{\underset{Cl}{|}}{C}{=}CH_2$$

**16** **17**

Scheme 4.3

## C. Addition to Substituted Allenes

The picture presented by the electrophilic additions to substituted allenes is rather unclear since, in many cases, the exact mechanism is open to speculation. The main problem is usually in deciding whether the reaction occurs via an open vinyl and/or allyl cation or a cyclic cation. An equilibrium between the various cations is also possible.

The addition of sulfenyl chlorides to allenes presumably occurs via the cyclic thiiranium ion **7** (X = S) [13]. For the addition of halogens, both open and cyclic cations should be considered. The acid-catalyzed hydrations and the addition of acids take place via vinyl and/or allyl cations, and the exact nature of the intermediate depends on the substitution of the allene.

Generally, the product analysis indicates the exact site of proton attack, assuming that no isomerization takes place under the reaction conditions. A striking example of the influence of the substituents in directing the course of reaction is the addition of hydrogen halides to perfluoroallene (18) [14]. The product indicates proton attack on the central carbon, and, if an intermediate bridged cation is ruled out, the reaction must take place via the allyl cation 19, which is presumably stabilized by resonance. This is in contrast to the behavior of allene itself, in which a vinyl cation is the intermediate.

$$F_2C{=}C{=}CF_2 \xrightleftharpoons{\text{HX}} \underset{F}{\overset{F}{\diagup}}\!\!\overset{+}{C}{-}CH{=}CF_2 \longleftrightarrow \underset{|\underline{F}}{\overset{|\overline{F}^+}{\diagup}}C{-}CH{=}CF_2 \xrightarrow{X^-} F_2XC{-}CH{=}CF_2$$

   18                    19

Okuyama *et al.* [15] carried out a thorough investigation of the mechanism of hydrochlorination of several aryl allenes (20), which yields exclusively the cinnamyl chlorides 23. The high negative $\rho^+$ value of $-4.2$ for the reaction indicates that the positive center in the transition state is directly conjugated with the aryl group.

The most plausible mechanism is a rate-determining formation of the benzyl cation 21, in which the alkene $\pi$ bond is orthogonal to the empty p orbital. This unstable, perpendicularly twisted cation rapidly rotates to give the conjugated allyl cation 22, which subsequently reacts with the chloride ion. No evidence for vinyl cations 24 and 25, which would have been formed by terminal protonation, was obtained. If a cyclic transition state such as 26 were formed, the $\rho^+$ value would be smaller.

$$XC_6H_4CH{=}C{=}CH_2 \xrightleftharpoons{\text{H}^+} Ar\overset{+}{C}H{-}CH{=}CH_2 \longrightarrow ArC\overset{+}{\overbrace{H{\cdots}CH{\cdots}CH}}_2 \xrightarrow{+Cl^-}$$

   20                    21                    22

$$ArCH{=}CHCH_2Cl$$

   23

X = H, *p*-Me, *m*-Me, *p*-Cl, *m*-Cl
Ar = XC$_6$H$_4$

$$ArCH{=}\overset{+}{C}Me \qquad ArCH_2\overset{+}{C}{=}CH_2 \qquad \underset{H}{\overset{Ar}{\diagup}}\!\!\overset{H}{\overset{\|}{\overset{+}{C}}}{-}C{=}CH_2$$

   24                    25                    26

Further support for the proposed reaction mechanism comes from a comparison of the relative reaction rates of phenyl allene with its α-methyl- and γ-methyl-substituted analogues 27 and 28. If the conjugated allyl cation

$$\text{PhCH}=\text{C}=\text{CH}_2 \qquad \text{PhC(Me)}=\text{C}=\text{CH}_2 \qquad \text{PhCH}=\text{C}=\text{CHMe}$$

$$k_{rel} \quad 1 \qquad\qquad\qquad 27 \qquad\qquad\qquad 28$$

$$4 \times 10^3 \qquad\qquad 2 \times 10^2$$

**22** were formed in the rate-determining step, the rates of reaction of **27** and **28** would be of the same order of magnitude. The fact that the reactivity of α-methylphenyl allene (**27**) is 20 times greater than that of the γ-methyl isomer rules out the alternate structure **29** for the intermediate allyl cation [15].

$$\text{PhCH}=\overset{+}{\text{CH}}\text{CH}_2$$

**29**

Further proof for the existence of open cations as intermediates in the protonation of allenes comes from the direct observation of allyl cations **32** and **33** by nmr spectroscopy [16].

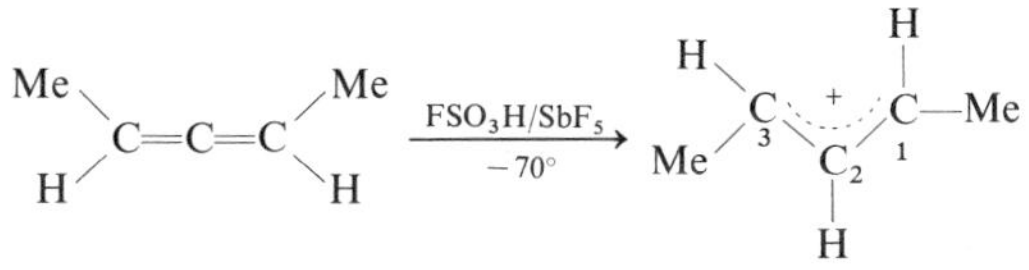

**30**                                    **32**

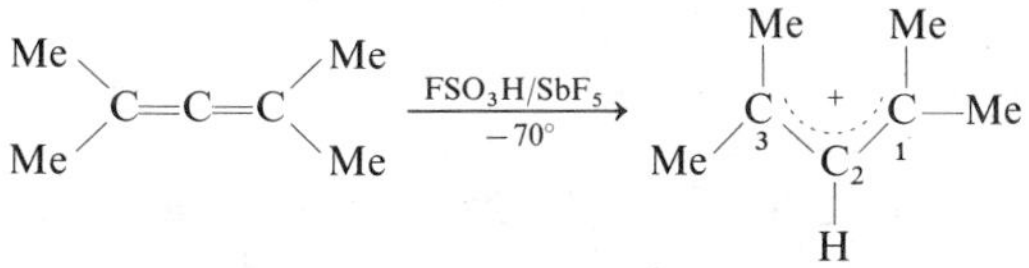

**31**                                    **33**

Exclusive protonation on the central carbon atom of 1,3-dimethyl allene (**30**) and tetramethyl allene (**31**) was confirmed by the use of $FSO_3D$. In this case, the signal of the proton on C-2 was absent, and no further deuterium incorporation took place.

However, the direction of attack varies with the substitution pattern of the allene. For example, 1,2-butadiene (**34**) reacted with hydrogen chloride to give the vinyl chlorides **36** and 2-butyne (**37**), presumably via the vinyl cation **35** [9a]. In contrast, the addition of HCl to 3-methyl-1,2-butadiene (**38**) yielded only a mixture of the allylic chlorides **39** and **40**, besides the

$$\text{MeCH}=\text{C}=\text{CH}_2 \underset{-78°}{\overset{H^+}{\rightleftharpoons}} \text{MeCH}=\overset{+}{\text{C}}-\text{Me} \xrightarrow{Cl^-} \text{MeCH}=\text{C(Cl)Me} + \text{MeC}\equiv\text{CMe}$$

**34**                          **35**                          *E*- and *Z*-**36**          **37**

(60%)                (40%)

isomerization product isoprene **41**, which in turn was hydrochlorinated under the conditions used. For this reason little can be said about the actual course of the reaction, but the delocalized allyl cation **42** can account for the observed products [9a] (Scheme 4.4).

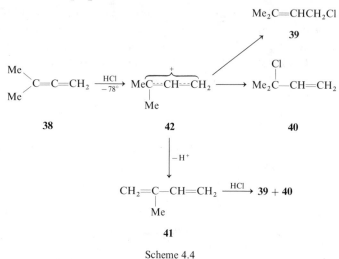

Scheme 4.4

Hydrochlorination of 1,2-hexadiene (**43**) with $BiCl_3$ as a catalyst yielded product **45**, which was formed by protonation at the terminal carbon atom [17]. Protonation occurred at the unsubstituted allenic carbon atom and not at the substituted allenic carbon, since the resulting vinyl cation **44** can be more effectively stabilized by hyperconjugation with the methyl group than the isomeric vinyl cation.

$$CH_2=C=CH-(CH_2)_2Me \xrightarrow[BiCl_3]{H^+} Me\overset{+}{C}=CH(CH_2)_2Me \xrightarrow{Cl^-} MeC(Cl)=CH(CH_2)_2Me$$

**43**                                 **44**                                 **45**

The direction of attack also varies with the nature of the electrophile. Whereas fluorosulfonic acid gave the allyl cation **32** by attack on the central carbon of 1,3-dimethyl allene and hydration of 1,3-dialkyl allenes **46** in sulfuric acid yielded the corresponding ketones [18], the addition of HBr to **46** at $-40°$ gave products of mixed orientation [19]. Thus, a mixture of four products (**47–50**) is formed in the hydrobromination of 2,3-nonadiene

$$MeCH=C=CHR \xrightarrow[-40°]{HBr} MeCH_2C(Br)=CHR + MeCH(Br)CH=CHR +$$

**46**                                 **47**                                 **48**

$R = n\text{-}C_5H_{11}, i\text{-}C_4H_9$                          $MeCH=C(Br)CH_2R + MeCH=CHCH(Br)R$

**49**                                 **50**

and 6-methyl-2,3-heptadiene, with the bromides **47** and **48** being formed predominantly. It seems that proton addition takes place preferably at the C-2–C-3 double bond, in contrast to bromination, in which the C-3–C-4 bond is attacked (see below). Consequently, **47** and **48** are presumably formed via the intermediate vinyl **51** or allyl cations **52**.

$$\text{MeCH}_2\overset{+}{-}\text{C}\!=\!\text{CHR} \qquad \text{Me}\overset{+}{\text{C}}\text{HCH}\!=\!\text{CHR}$$

**51**                    **52**

The 1,3-disubstituted allenes 2,3-pentadiene, 2,3-hexadiene, and 2,3-heptadiene (**53**) yielded a mixture of allyl bromides (**54** and **55**) (80%) and vinyl bromides (**56** and **57**) (20%) on reaction with HBr [20], indicating the intermediacy of both vinyl and allyl cations (Scheme 4.5). In contrast, the 1,1-disubstituted allenes 3-methyl-1,2-butadiene (**38**) and 3-methyl-1,2-pentadiene (**58**) gave a mixture of the allylic bromides on reaction with HBr [20], in line with the results of the hydrochlorination of **38** described previously. The addition of HCl to tri- and tetrasubstituted allenes gives a complex mixture of products due to isomerization. The authors postulated [20] the formation of a cyclic cation, which subsequently opens up to yield either the vinyl or allyl cation; however, the direct formation of a vinyl and/or allyl cation seems to be at least an equally reasonable assumption.

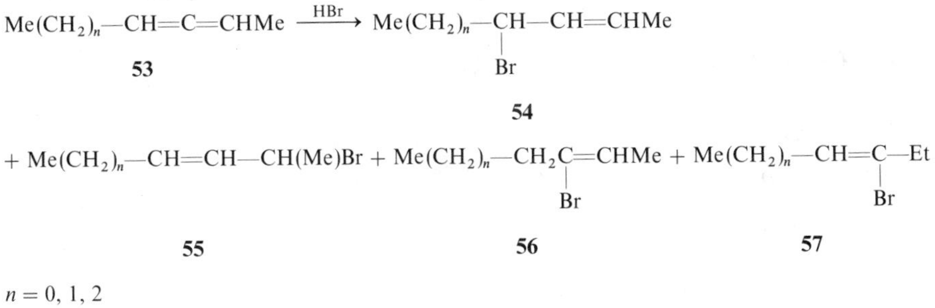

$$\text{Me(CH}_2)_n\!-\!\text{CH}\!=\!\text{C}\!=\!\text{CHMe} \xrightarrow{\ \text{HBr}\ } \text{Me(CH}_2)_n\!-\!\underset{\underset{\text{Br}}{|}}{\text{CH}}\!-\!\text{CH}\!=\!\text{CHMe}$$

**53**                                                    **54**

$$+\ \text{Me(CH}_2)_n\!-\!\text{CH}\!=\!\text{CH}\!-\!\text{CH(Me)Br} + \text{Me(CH}_2)_n\!-\!\underset{\underset{\text{Br}}{|}}{\text{CH}_2\text{C}}\!=\!\text{CHMe} + \text{Me(CH}_2)_n\!-\!\text{CH}\!=\!\underset{\underset{\text{Br}}{|}}{\overset{|}{\text{C}}}\!-\!\text{Et}$$

**55**                            **56**                        **57**

$n = 0, 1, 2$

Scheme 4.5

$$\text{CH}_2\!=\!\text{C}\!=\!\text{C}\!\!\begin{array}{c}\nearrow\text{Me}\\[4pt]\searrow\text{Et}\end{array}$$

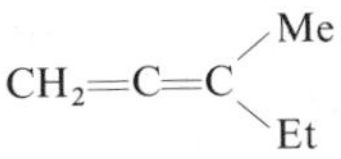

**58**

The addition of HCl to 1,2,3-pentatriene (**59**) in aqueous ethanol gave a complex mixture of mono- and diadducts [21]. The dependence of the first-order rate constant on the Hammett acidity function $H_0$ and the kinetic isotope effect $k_H/k_D$ of 1.7 point to a rate-determining proton addition. Apparently, addition takes place via protonation at the terminal carbon atom to give cation **60**.

$$\text{MeCH}{=}\text{C}{=}\text{C}{=}\text{CH}_2 \xrightarrow{\text{H}^+} \text{MeCH}{=}\text{C}{=}\overset{+}{\text{C}}{-}\text{Me} \longleftrightarrow \text{Me}\overset{+}{\text{C}}\text{H}{-}\text{C}{\equiv}\text{C}{-}\text{Me}$$

$$\qquad\quad \textbf{59} \qquad\qquad\qquad\quad \textbf{60}$$

1,2,4-Pentatriene and 1,2,4-hexatriene (61) gave complex mixtures of mono- and diadducts on addition of HCl and HBr [22]. The proton seems to attack the vinyl function, giving cation 62, which is stabilized by conjugation with the neighboring double bond (63).

$$\text{CH}_2{=}\text{C}{=}\text{CH}{-}\text{CH}{=}\text{CHR} \xrightarrow{\text{H}^+} \text{CH}_2{=}\text{C}{=}\text{CH}\overset{+}{\text{C}}\text{HCH}_2\text{R} \longleftrightarrow \text{CH}_2{=}\overset{+}{\text{C}}{-}\text{CH}{=}\text{CHCH}_2\text{R}$$

$$\textbf{61} \; \text{R} = \text{H, Me} \qquad\qquad\qquad \textbf{62} \qquad\qquad\qquad\qquad \textbf{63}$$

Allenes substituted by electron-withdrawing substituents also give products arising from vinyl cation intermediates. 3-Chloro-2-butenonitrile (65) was obtained as the exclusive product from the reaction of HCl with cyanoallene (64) [23].

$$\text{CH}_2{=}\text{C}{=}\text{CHC}{\equiv}\text{N} \xrightarrow{\text{HCl}} \text{CH}_2{=}\text{C(Cl)CH}_2\text{C}{\equiv}\text{N}$$

$$\qquad\quad \textbf{64} \qquad\qquad\qquad\quad \textbf{65}$$

The formation of ester 68 in the addition of HCl and HBr to allenyl ester 66 can be accounted for by protonation on oxygen, leading to the vinyl cation 67 [24]. A similar mechanism may also apply for addition to 64.

$$\text{CHI}{=}\text{C}{=}\text{CH}{-}\text{COOEt} \xrightarrow{\text{H}^+} \text{CHI}{=}\overset{+}{\text{C}}{-}\text{CH}{=}\text{C}\overset{\displaystyle \nearrow \text{OH}}{\searrow \text{OEt}} \xrightarrow{\text{X}^-}$$

$$\qquad\quad \textbf{66} \qquad\qquad\qquad\qquad\qquad \textbf{67}$$

$$\text{CHI}{=}\text{CX}{-}\text{CH}{=}\text{C}\overset{\displaystyle \nearrow \text{OH}}{\searrow \text{OEt}} \longrightarrow \text{CHI}{=}\text{CX}{-}\text{CH}_2\text{COOEt}$$

$$\qquad\qquad\qquad\qquad\qquad\qquad\qquad \textbf{68} \;\; \text{X} = \text{Br, Cl}$$

Allenes substituted by groups capable of stabilizing a positive charge at the terminal position add the proton to the central carbon, as shown above for the aryl allenes. Thus, acid-catalyzed hydrolysis of allenyl ethers 69 [25] proceeded via the carbenium ion 70, which is stabilized by an ethoxy group.

$$\text{RCH}{=}\text{C}{=}\text{CHOEt} \xrightarrow{\text{H}^+} \text{RCH}{=}\text{CH}\overset{+}{\text{C}}\text{HOEt} \xrightarrow{\text{OH}^-} \text{RCH}{=}\text{CH}\overset{\text{OH}}{\overset{|}{\text{C}}}\text{HOEt} \longrightarrow \text{RCH}{=}\text{CH}\overset{\text{O}}{\overset{\|}{\text{C}}}\text{H}$$

$$\textbf{69} \qquad\qquad\qquad \textbf{70}$$

In contrast to the usual central protonation of tetrasubstituted allenes, the cyclopropylidene allene 71 undergoes electrophilic attack at the terminal carbon atom of the allene function. The resulting cyclopropylidenemethyl cation 72 is efficiently stabilized by overlap of the empty p orbital with the

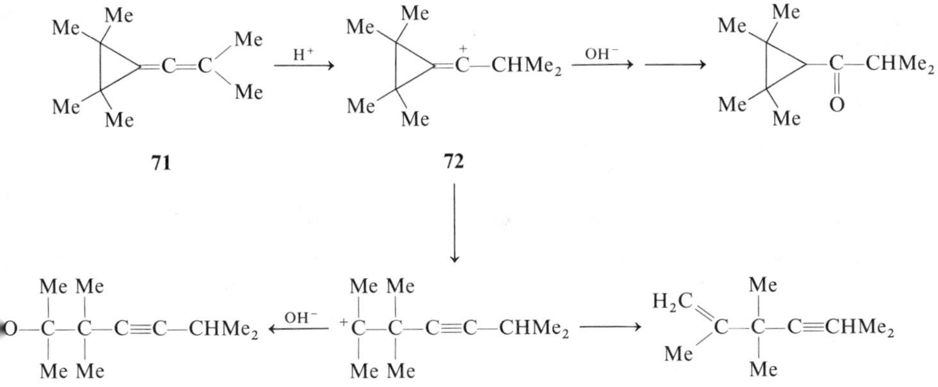

Scheme 4.6

Walsh orbitals of the cyclopropane ring, as discussed in Chapter 5 [26] (Scheme 4.6).

An entirely different reaction course was observed in the reaction of trichloroacetic acid with the phenyl-substituted derivatives 73 to 75 [27]. The proton attacks the central carbon atom with formation of a cyclopropyl

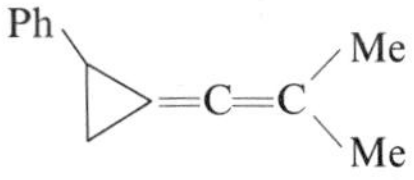

cation, which opens to the allyl cation intermediate 76 (Scheme 4.7). The reason for the difference in the direction of proton attack between the cyclopropylidene allenes 71 and 73 to 75 is not yet clear.

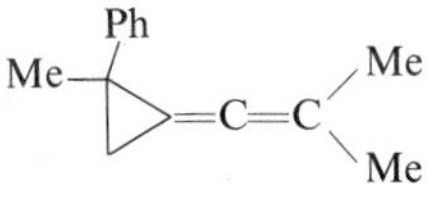

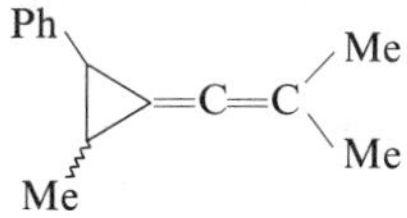

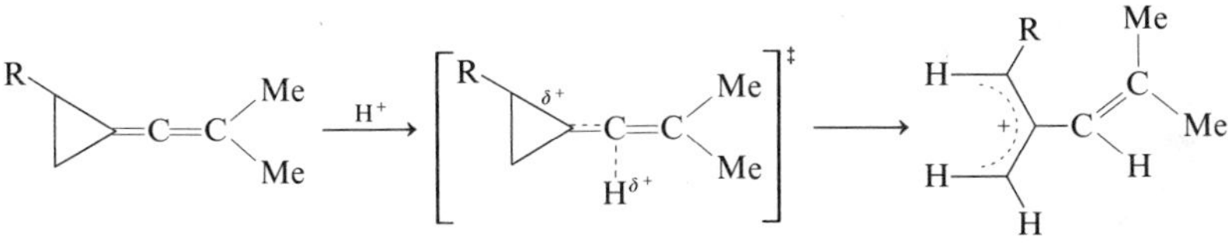

Scheme 4.7

Lewis acids catalyze the 2 + 2 cycloaddition of allenes to alkenes [28]. An initial polarization of the allenyl system to give intermediate 77, which is analogous to a vinyl cation, is a likely possibility for these cyclizations.

In the ionic halogenation reactions of most simple allenes, a cyclic halonium ion is considered to be the intermediate [29]. An example is the bromination of 2,3-pentadiene (30) in methanol and in $CCl_4$ [30]. Starting with optically active 2,3-pentadiene, the reaction yielded optically active products formed exclusively by an anti addition, strongly suggesting the bromonium ion 78 as the dissymetric reaction intermediate.

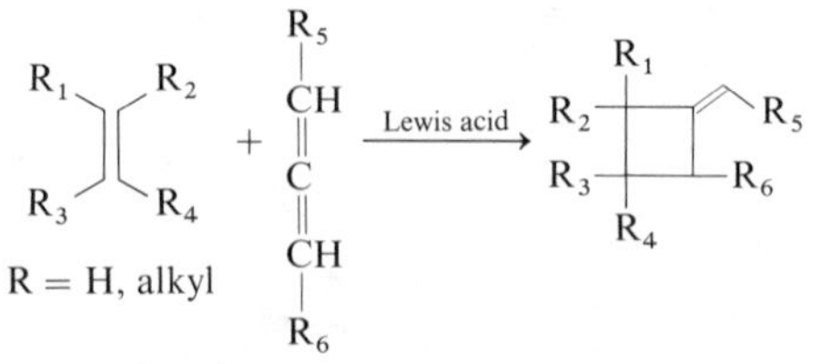

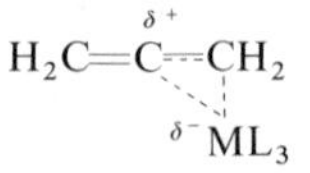

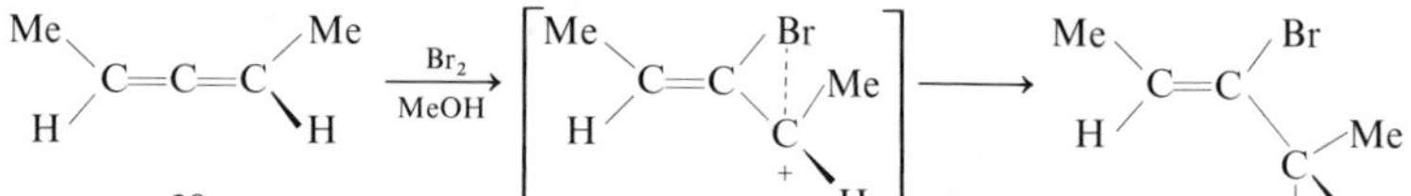

However, the iodination of 1,3-diarylallenes 79 was found to take place via the allyl cation 80 [31]. The rates of iodination and racemization of optically active 1,3-diphenylallene were equal, pointing to a planar allyl cation (80) as the common intermediate. The linear Hammett plot (based on $\sigma^+$ values) for 3-aryl-1-phenyl allenes was also taken by the authors as evidence for the formation of an allyl cation in the rate-determining step. However, it must be noted that the $\rho^+$ value of $-3.2$ is lower than the value of $-4.2$ found for the hydrochlorination of aryl allenes by Okuyama et al. [15]. Moreover, although the authors assume that the cation 80 is formed in the rate-determining step, it is more likely that the first-formed intermediate is the perpendicular ion 81, which then rapidly rotates to the planar form, provided that the addition does not give a bridged intermediate.

$$ArCH{=}C{=}CHAr \xrightleftharpoons{I_2} ArCH{\cdots}\overset{+}{CI}{\cdots}CHAr \longrightarrow products$$

79                                              80

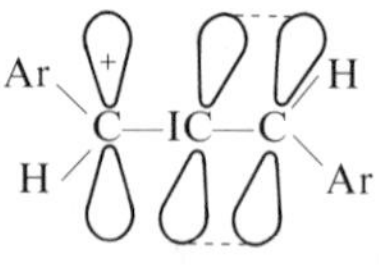

81

Studies of the products formed by halogenation of aryl allenes led Okuyama et al. [32] to propose a different mechanism. Bromination of phenyl allene in methanol at $0°$ led to $84\%$ of the 1,2 adduct 82 and to $16\%$ of the 2,3 adduct 83 (Nu = OMe, X = Br). The use of a solvent of lower nucleophilicity such as $CS_2$ enhanced the proportion of 2,3 adduct **83** (Nu = Br), whereas it decreased in the series Cl > Br > I, analogous to the tendency of forming cyclic halonium ions. Consequently, an attack of the halogen on the C-1–C-2 double bond to give a cyclic halonium ion (**84**) that can subsequently either open up to the resonance-stabilized allyl cation **85** or react directly with the nucleophile in two competing reactions has been suggested (Scheme 4.8) [32].

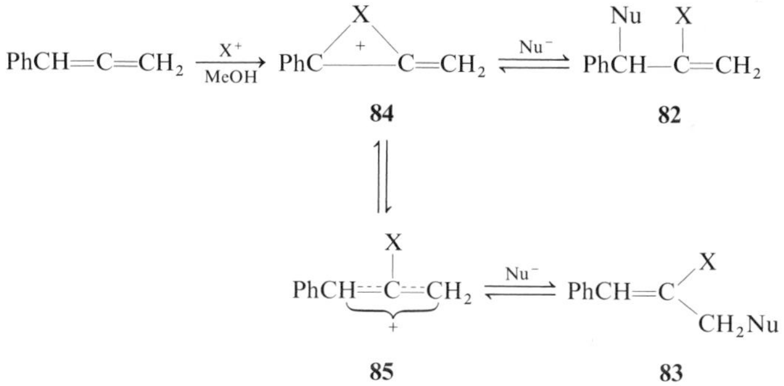

Scheme 4.8

The above-mentioned mechanism corresponds well with the observation that a greater amount of the 2,3 adduct is formed when the phenyl ring is substituted by electron-donating substituents or in the case of 3-phenyl-1-butadiene. Unfortunately, no kinetic data have been reported for this reaction. However, it is not unreasonable to postulate the same mechanism for the iodination of aryl allenes, in which the kinetics are complicated by the fact that the iodine could add to either allene $\pi$ bond with a similar probability.

In contrast, the chlorination of the allene **71** probably involves vinyl cation **86a**, originating from attack on the terminal dimethyl-substituted bond [26]. The propargyl chloride **87** (X = Cl), which is the sole product (Scheme 4.9), apparently arises by ring opening of the cyclopropyl-stabilized vinyl cation **86a**, analogous to the hydrohalogenation of **71** discussed above. Bromination of **71** yields a complex mixture of products, but in all cases the cyclopropyl group opens up, and **86b** is the most likely intermediate.

The addition of bromine to 1,2-cyclononadiene (**88**) gave the two isomeric dibromides **89** and **90** [33] in a 40 : 60 ratio [34], the latter being the trans-annular adduct (Scheme 4.10). In contrast, 1,2-cyclooctadiene (**91**) and 1,2-cyclodecadiene (**92**) gave normal addition products (**93** and **94**, respectively), without formation of any transannular product (Scheme 4.10). The general behavior of medium-ring allenes toward the addition of bromine, therefore,

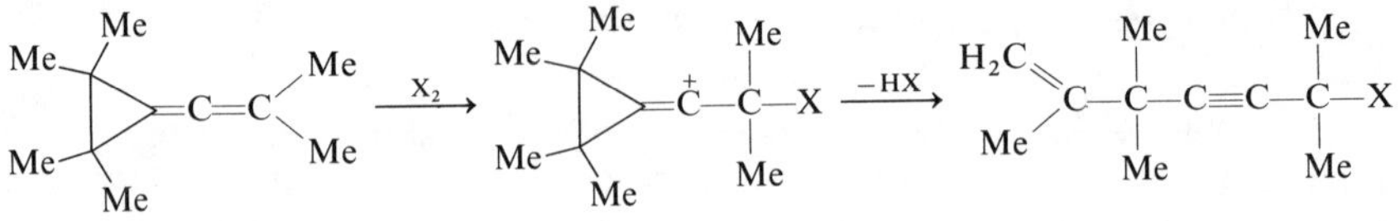

**71**                    **86a**  X = Cl                    **87**
                          **86b**  X = Br

Scheme 4.9

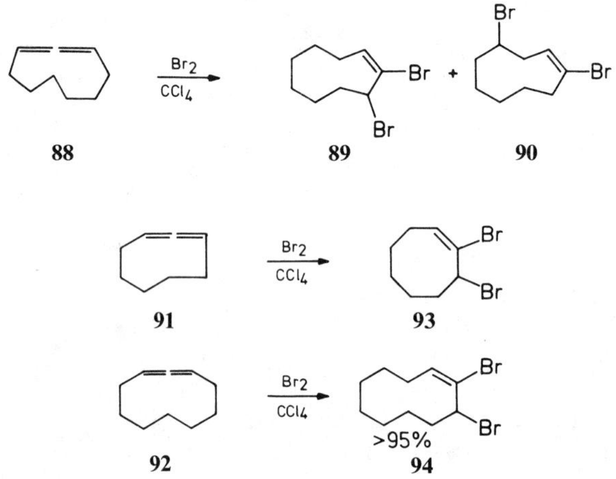

**88**            **89**            **90**

**91**                    **93**

                          >95%
**92**                    **94**

Scheme 4.10

**95**

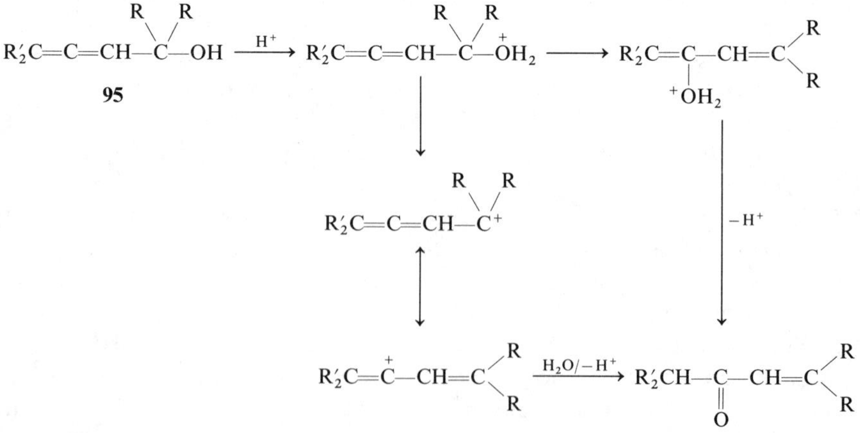

Scheme 4.11

does not correspond to that of Z-cycloalkenes of the same size [34], in which both 1,2 adducts and transannular products were obtained.

Substituted α-allenyl alcohols (95) were found to isomerize in acidic media to α,β-unsaturated aldehydes and ketones [35]. Vinyl cations were proposed as intermediates in this isomerization reaction, as shown in Scheme 4.11.

The acid-catalyzed reaction of α-allenic tertiary alcohols 96 in $H_2SO_4/$ AcOH gave vinyl ethers 98 as products [36]. The delocalized ion 97, shown in Scheme 4.12, is a probable intermediate. Ions similar to 97 are involved in the solvolyses of some 2-bromo-4-methyl-1,3-pentadienes, as discussed in detail in Chapter 5.

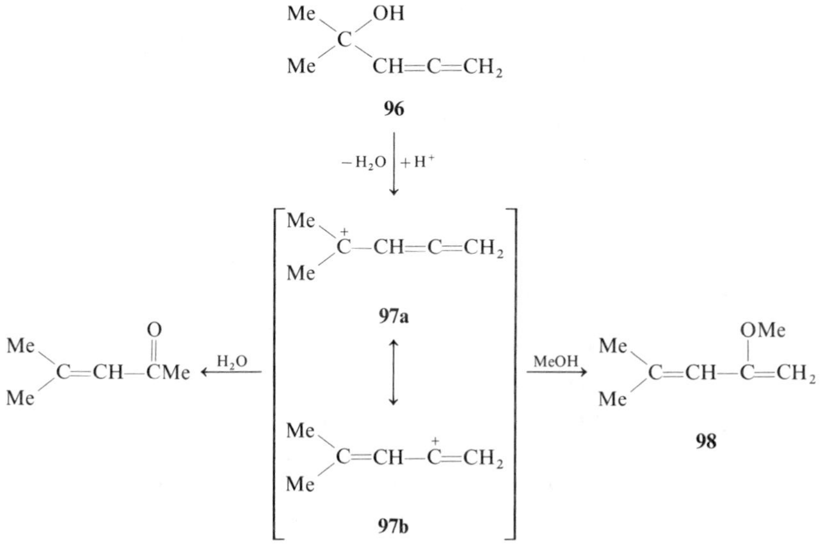

$$
\begin{array}{c}
\underset{Me}{\overset{Me}{>}}C\overset{OH}{\underset{CH=C=CH_2}{<}} \\
\mathbf{96}
\end{array}
$$

$$-H_2O \;\Big|\; +H^+$$

$$
\underset{Me}{\overset{O}{\underset{Me}{>}}}C=CH-\overset{O}{\overset{\|}{C}}Me \;\xleftarrow{\;H_2O\;}\;
\left[
\begin{array}{c}
\underset{Me}{\overset{Me}{>}}\overset{+}{C}-CH=C=CH_2 \\
\mathbf{97a} \\
\updownarrow \\
\underset{Me}{\overset{Me}{>}}C=CH-\overset{+}{C}=CH_2 \\
\mathbf{97b}
\end{array}
\right]
\;\xrightarrow{\;MeOH\;}\;
\underset{Me}{\overset{Me}{>}}C=CH-\overset{OMe}{\underset{|}{C}}=CH_2
$$

$$\mathbf{98}$$

Scheme 4.12

## II. PARTICIPATION OF ALLENYL BONDS IN SOLVOLYSES

A study similar to that on the participation of triple bonds in solvolysis reactions was carried out with allenyl groups. This reaction was investigated simultaneously but independently by Hanack and Häffner [37] and Bertrand and Santelli [38] starting in 1964 and is known as "the homoallenyl rearrangement." These investigators showed that, in analogy to double bonds in homoallyl compounds, the cumulative double bonds in allenes can participate in solvolysis reactions, leading to rearranged cyclic products ($k_\Delta$ process) besides the products from direct displacement by the nucleophile.

If a homoallenic compound of structure **99** pursues the same reaction course as the homoallylic rearrangement [39], it gives a vinyl cation (**100**) (Scheme 4.13), which is stabilized by delocalization of the positive charge by the neighboring cyclopropane ring (Chapter 5). The factors that influence the formation of vinyl cation **100** and its rearrangement will be discussed below.

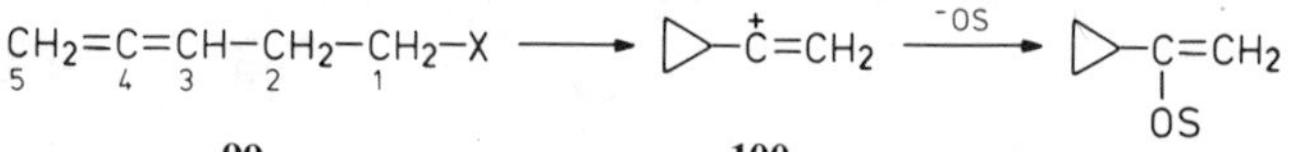

$$CH_2=\underset{5}{C}=\underset{4}{CH}-\underset{3}{CH_2}-\underset{2}{CH_2}-\underset{1}{CH_2}-X \longrightarrow \triangleright-\overset{+}{C}=CH_2 \overset{^-OS}{\longrightarrow} \triangleright-\underset{\underset{OS}{|}}{C}=CH_2$$

**99**     **100**    **101**

Scheme 4.13

In their early work, Hanack and Häffner investigated the solvolyses of 3,4-pentadienyl and 3,4-hexadienyl derivatives **102** [37]. Depending on the nucleophilicity and ionizing power of the solvents, primary substrates gave cyclized ketone **103**, as well as unrearranged products. Acetolysis of **102** (R = H; X = ONpS) gave 20% of the cyclopropyl methyl ketone **103** (R = H); use of formic acid increased the yield of **103** (R = H) to 80% (Scheme 4.14). Hence, the percentage of rearranged product **103** increased with increasing ionizing power and decreasing nucleophilicity of the solvent. Deamination of **102** (R = H; X = NH$_2$) gave only 9% of the cyclopropyl methyl ketone, besides 43% of the unrearranged alcohol, 20% of an unsaturated hydrocarbon, and 28% of an unidentified compound. Strong evidence for the intermediacy of the vinyl cation **100** will be the isolation of the corresponding enol derivatives **101** (Scheme 4.13) from the solvolysis reactions, as shown in the case of the homopropargyl rearrangement (Chapter 3). However, no enol esters could be isolated in the above reaction because they were easily hydrolyzed to the ketone during the work-up.

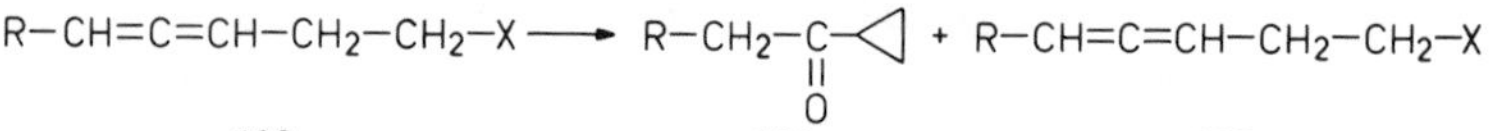

$$R-CH=C=CH-CH_2-CH_2-X \longrightarrow R-CH_2-\underset{\underset{O}{\|}}{C}\mathord{-}\triangleleft + R-CH=C=CH-CH_2-CH_2-X$$

**102**    **103**    **102**

R= H, Me         R=H, Me

X= Br, OSO$_2$-β-naphthyl(ONpS), NH$_2$    X=OH, OMe, OAc

Scheme 4.14

The rate of acetolysis of **102** (R = H) at 60° was 3.5 times faster than that of the *n*-pentyl model compound, and **102** (R = Me) reacted 9.5 times faster than the *n*-hexyl sulfonate [40]. These kinetic data, especially when allowance is made for inductive retardation for the allenyl group, clearly suggest homoallenic double-bond participation in the rate-determining step. The products are consistent with competition of reactions via the vinyl cation intermediate **100** and a direct solvolytic displacement.

Bertrand and Santelli [38] investigated the solvolysis of secondary homoallenyl tosylates **104** and **105** in aqueous acetic acid, in which 4,5-hexadien-2-yl tosylate (**104**) gave 67% of the cyclopropyl ketone **106** (cis and trans) and 30% of the unrearranged 4,5-hexadien-2-ol as the main products. For 4-methyl-4,5-hexadien-2-yl tosylate (**105**), having a substituent at C-3 with respect to the leaving group, solvolysis produced 84% of cyclobutanols **107** (cis and trans) and **108** (cis and trans). The formation of four-membered rather than three-membered ring products from **105** is due to the ring expansion of the intermediate vinyl cation **109** to give the cyclobutyl cations **110** and **111** (Scheme 4.15).

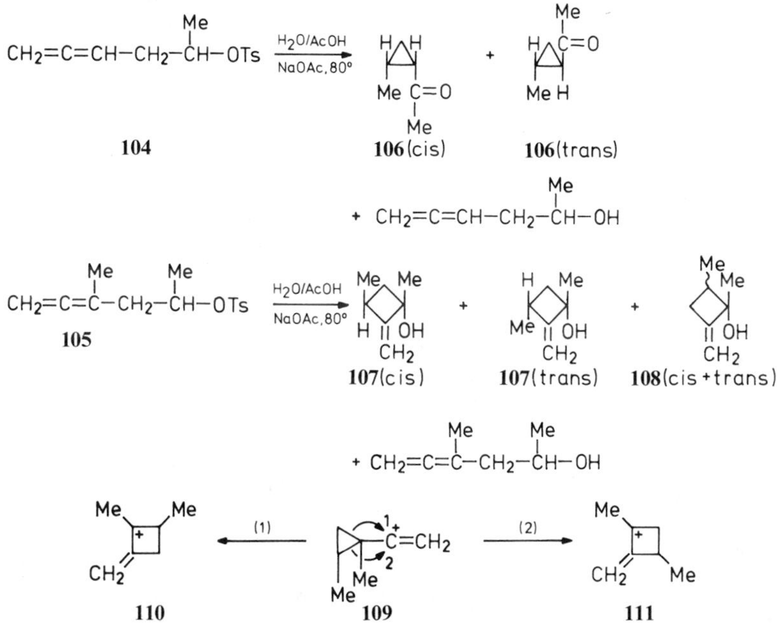

Scheme 4.15

Solvolysis of optically active homoallenyl tosylate **112** gave *cis*- and *trans*-2-methyl cyclopropyl methyl ketones **113** and **114** without loss of optical purity and with inversion of configuration at the carbon bearing the leaving group [41, 42]. This result strongly suggests allenyl bond participation in the formation of the rearranged product via a diastereoisomeric transition state. The allenic alcohol **115** was found to be racemic, indicating that it was formed via a symmetric carbenium ion.

Several additional homoallenyl systems substituted at different positions were investigated in order to ascertain the true extent of homoallenyl participation.

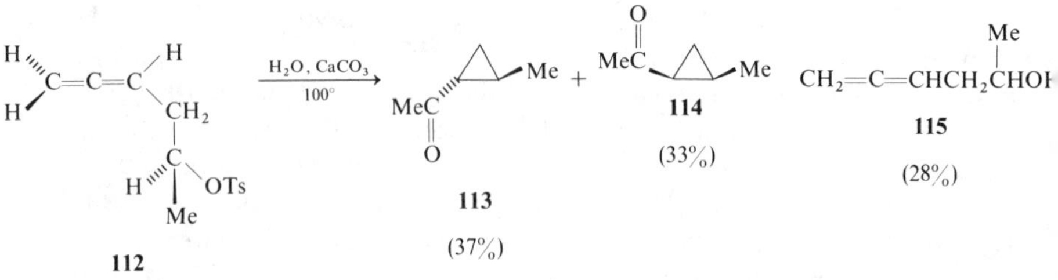

112 → (H₂O, CaCO₃, 100°) → 113 (37%) + 114 (33%) + 115 (28%)

Jacobs and Macomber investigated the rates and products in the solvolyses of eight homoallenyl compounds (**104**, **116–121**, and **125**) [43]. By comparing these rates with the rates of the saturated analogues, the authors [43] were able to estimate the ratio of the assisted solvolytic rate ($k_A$) for allenyl bond participation relative to the unassisted rate ($k_s$) in acetic acid at 60°. Only for the C-1 gem-substituted compound **120** was participation ruled out on the basis of both rate data and product distribution. The *gem*-dimethyl substitution prevented the formation of any cyclized products, and only normal substitution or elimination products were obtained. All others solvolyzed with anchimeric assistance from the allenyl double bonds and gave rearranged products, as shown in Scheme 4.16. The unsubstituted

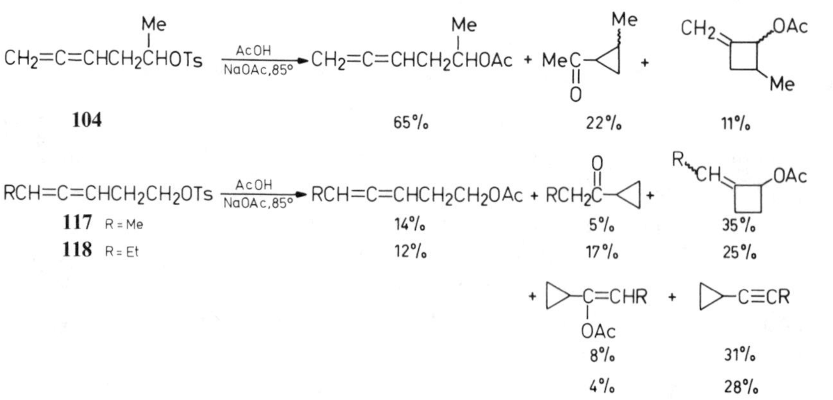

Scheme 4.16

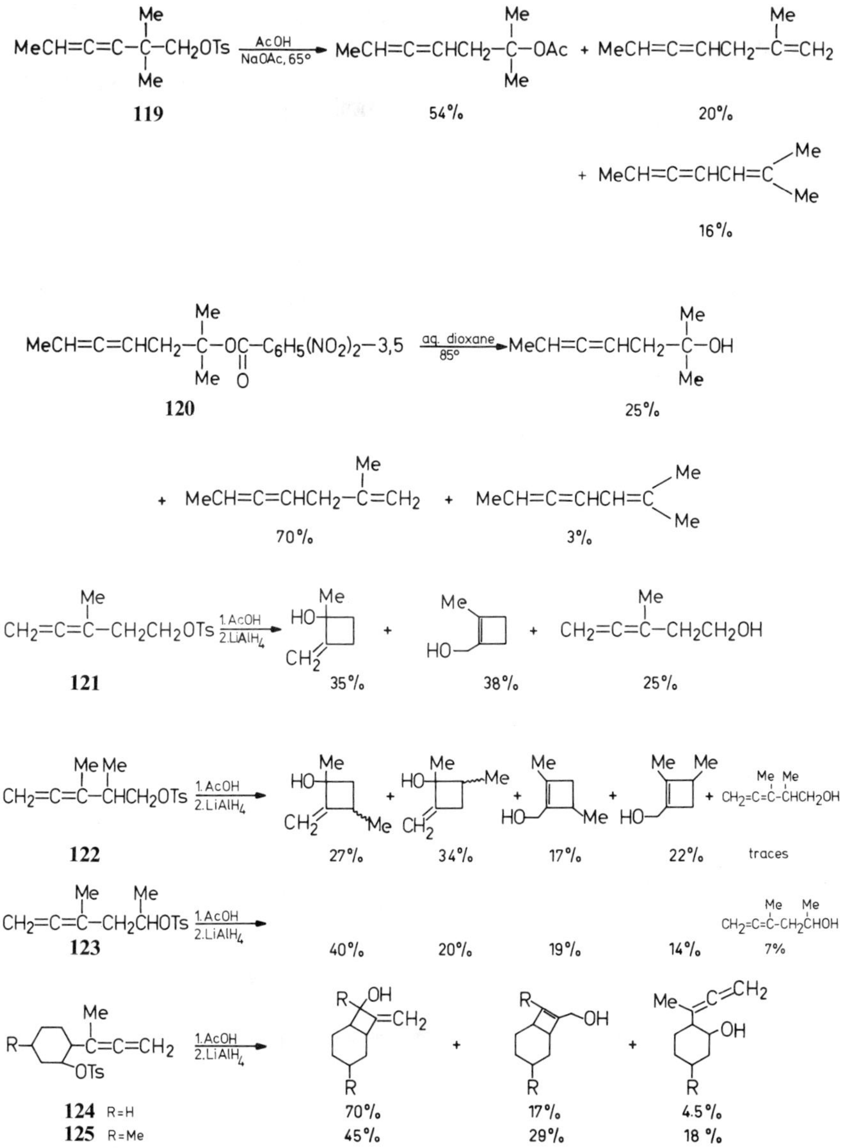

$$\underset{\textbf{119}}{\overset{\text{Me}}{\underset{\text{Me}}{MeCH=C=C-\overset{|}{\underset{|}{C}}-CH_2OTs}}} \xrightarrow[\text{NaOAc, 65°}]{\text{AcOH}} \underset{54\%}{\overset{\text{Me}}{\underset{\text{Me}}{MeCH=C=CHCH_2-\overset{|}{\underset{|}{C}}-OAc}}} + \underset{20\%}{\overset{\text{Me}}{\underset{}{MeCH=C=CHCH_2-\overset{|}{\underset{}{C}}=CH_2}}}$$

$$+ \underset{16\%}{MeCH=C=CHCH=C\overset{\diagup Me}{\diagdown Me}}$$

$$\underset{\textbf{120}}{\overset{\text{Me}}{\underset{\text{Me}\ O}{MeCH=C=CHCH_2-\overset{|}{\underset{|}{C}}-O\overset{|}{\underset{||}{C}}-C_6H_5(NO_2)_2-3,5}}} \xrightarrow[85°]{\text{aq. dioxane}} \underset{25\%}{\overset{\text{Me}}{\underset{\text{Me}}{MeCH=C=CHCH_2-\overset{|}{\underset{|}{C}}-OH}}}$$

$$+ \underset{70\%}{\overset{\text{Me}}{\underset{}{MeCH=C=CHCH_2-\overset{|}{\underset{}{C}}=CH_2}}} + \underset{3\%}{MeCH=C=CHCH=C\overset{\diagup Me}{\diagdown Me}}$$

$$\underset{\textbf{121}}{\overset{\text{Me}}{\underset{}{CH_2=C=C-CH_2CH_2OTs}}} \xrightarrow[\text{2.LiAlH}_4]{\text{1.AcOH}} \quad \underset{35\%}{\text{HO}} \quad + \quad \underset{38\%}{\text{Me}} \quad + \quad \underset{25\%}{\overset{\text{Me}}{\underset{}{CH_2=C=C-CH_2CH_2OH}}}$$

$$\underset{\textbf{122}}{\overset{\text{Me Me}}{\underset{}{CH_2=C=C-CHCH_2OTs}}} \xrightarrow[\text{2.LiAlH}_4]{\text{1.AcOH}} \quad \underset{27\%}{} \quad + \quad \underset{34\%}{} \quad + \quad \underset{17\%}{} \quad + \quad \underset{22\%}{} \quad + \quad \underset{\text{traces}}{\overset{\text{Me Me}}{\underset{}{CH_2=C=C-CHCH_2OH}}}$$

$$\underset{\textbf{123}}{\overset{\text{Me \quad Me}}{\underset{}{CH_2=C=C-CH_2CHOTs}}} \xrightarrow[\text{2.LiAlH}_4]{\text{1.AcOH}} \quad \underset{40\%}{} \quad \underset{20\%}{} \quad \underset{19\%}{} \quad \underset{14\%}{} \quad \underset{7\%}{\overset{\text{Me \quad Me}}{\underset{}{CH_2=C=C-CH_2CHOH}}}$$

$$\underset{\substack{\textbf{124} \ R=H \\ \textbf{125} \ R=Me}}{R-\text{cyclohexane-}\overset{\text{Me}}{\underset{\text{OTs}}{C=C=CH_2}}} \xrightarrow[\text{2.LiAlH}_4]{\text{1.AcOH}} \quad \underset{\substack{70\% \\ 45\%}}{} \quad + \quad \underset{\substack{17\% \\ 29\%}}{} \quad + \quad \underset{\substack{4.5\% \\ 18\%}}{}$$

Scheme 4.16 (*continued*)

homoallenyl compound **116** gave a $k_\Delta/k_s$ value of 0.9, and the C-2 *gem*-dimethyl-substituted compound **119** gave the highest participation ratio $k_\Delta/k_s$ of $\sim 5 \times 10^3$. The effect of methyl substitution in acetolysis on $k_\Delta$ values relative to the unsubstituted compound in acetic acid at 65° was summarized by the authors as follows [43]:

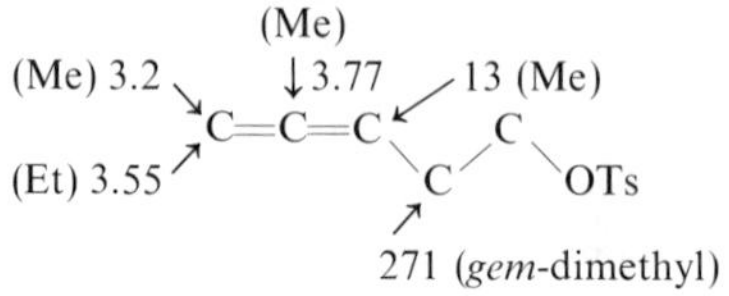

**TABLE 4.1**

**Correlation of $k_\Delta/k_s$ Ratio with Fraction of Cyclized or Rearranged Products**[a]

| Compound | $k_\Delta/k_s$ (rel) at 65° | Cyclized products, % (temp, °C) |
|---|---|---|
| **120** | 0.00 | 0 (65) |
| **125** | 0.44 | 50 (85) |
| **104** | 0.55 | 35 (85) |
| **116** | 1.00 | 24 (100) |
| **117** | 2.36 | 86 (85) |
| **118** | 3.44 | 88 (85) |
| **121** | 3.78 | 89 (85) |
| **119** | 5720.00 | 100 (65) |

[a] From Ref. 43.

A good correlation was found between the fraction of cyclized or rearranged products and the extent of participation $k_\Delta/k_s$, as shown in Table 4.1. Santelli and Bertrand [44] studied the systems **104** and **121** to **125**, and their results for the homoallenyl tosylates **121** to **125** are given in Scheme 4.16. From the kinetic data at 70° in acetic acid, they found that a methyl substitution at C-1, C-2, and C-3 in the parent homoallenyl tosylate **116** enhances the rate 42, 14, and 5 times, respectively, relative to the parent system [44]:

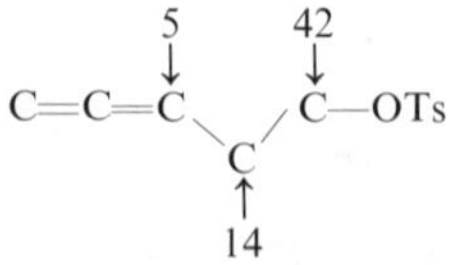

A bicyclobutonium ion (**126**) was suggested [43, 44] as the first intermediate (with greater charge accumulation at C-3 and C-4) in cases in which homoallenyl participation is significant. However, their results do not rule out an α-cyclopropylvinyl cation or other ions [42] as intermediates. Further work, as discussed below, has shown that the nature of the intermediate depends strongly on the substitution pattern.

$$H_2C = \overset{3}{\underset{1}{\underbrace{\phantom{xx}}}} \overset{4}{+} \rangle 2$$

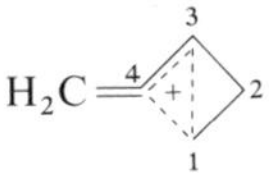

**126**

Bly *et al.* [45–47] supplied further evidence for homoallenyl participation by studying the comparative rate of solvolysis of the neopentyl brosylates **127** and **128**. The homoallenyl brosylate **127** solvolzyed $1.2 \times 10^3$ times

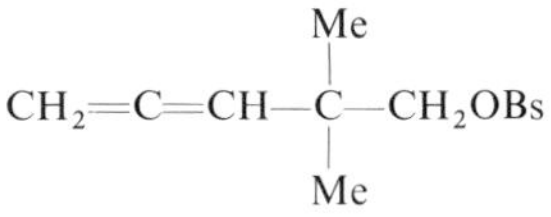

$$\underset{\underset{Me}{|}}{\overset{\overset{Me}{|}}{CH_2=C=CH-C-CH_2OBs}} \qquad \underset{\underset{Me}{|}}{\overset{\overset{Me}{|}}{CH_2=CH-C-CH_2OBs}}$$

**127** 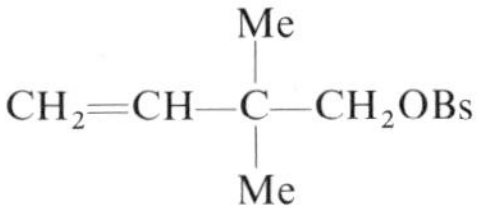 **128**

faster than the homoallyl analogue **128** in acetic acid at 75°. Even though part of the rate enhancement is due [45] to the higher ground state energy of the allenic compound compared to that of the alkene system, the anchimeric assistance in the solvolysis of **127** is significant.

Once again, the products obtained reflect the substitution pattern. *gem-*Dimethyl substitution at C-2 prevented the formation of cyclic products, but all of the linear compounds obtained were rearranged with *gem*-dimethyl groups at the C-1 position. Thus, the tertiary acetate **129** was the major product upon acetolysis of **127** and was accompanied by lesser amounts of the rearranged trienes **130** and **131**. Similarly, the open-chain ether **132**

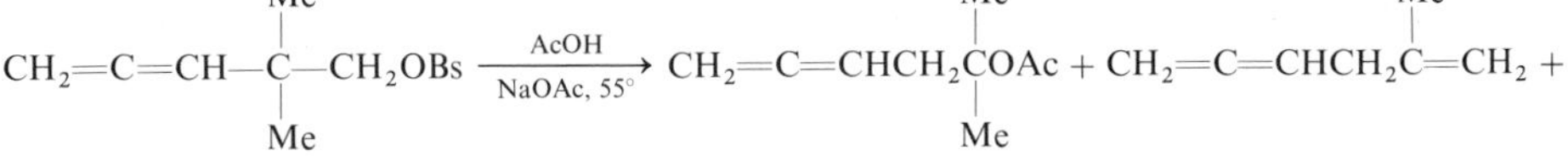

$$\underset{\underset{Me}{|}}{\overset{\overset{Me}{|}}{CH_2=C=CH-C-CH_2OBs}} \xrightarrow[\text{NaOAc, 55°}]{\text{AcOH}} \underset{\underset{Me}{|}}{\overset{\overset{Me}{|}}{CH_2=C=CHCH_2COAc}} + \underset{\underset{Me}{|}}{\overset{\overset{Me}{|}}{CH_2=C=CHCH_2C=CH_2}} +$$

**127** **129** **130**

(77%) (~12%)

$$CH_2=C=CHCH=C\overset{\diagup Me}{\diagdown Me}$$

**131**

(~11%)

(95%) and the alcohol **133** (85%) were the major products upon ethanolysis and hydrolysis of brosylate **127**. Bly *et al.* also found that the *gem*-dimethyl group present in **127** greatly increases the rate of the $k_\Delta$ process.

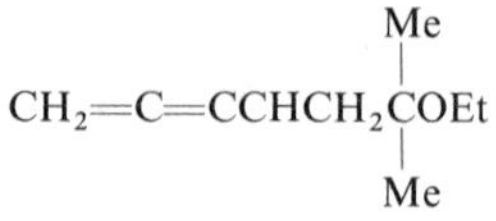

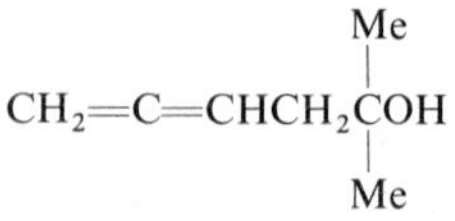

$$\underset{\underset{\underset{\textbf{132}}{|}}{\text{Me}}}{\overset{\overset{\text{Me}}{|}}{CH_2=C=CCHCH_2COEt}} \qquad \underset{\underset{\underset{\textbf{133}}{|}}{\text{Me}}}{\overset{\overset{\text{Me}}{|}}{CH_2=C=CHCH_2COH}}$$

Apparently, the initially formed cyclopropylvinyl cation **134** opens to the more stable tertiary cation **135**, from which the products are derived [45]. The higher stability of the tertiary cation **135** with respect to the vinyl

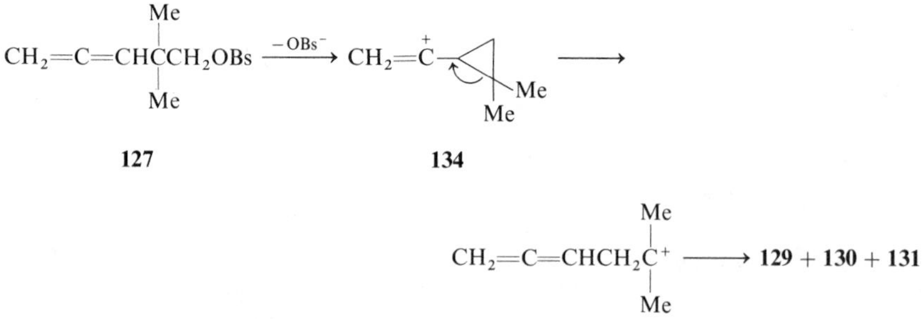

$$\underset{\textbf{127}}{\overset{\overset{\text{Me}}{|}}{\underset{\underset{\text{Me}}{|}}{CH_2=C=CHCCH_2OBs}}} \xrightarrow{-OBs^-} \underset{\textbf{134}}{CH_2=\overset{+}{C}} \qquad \longrightarrow$$

$$\underset{\textbf{135}}{\overset{\overset{\text{Me}}{|}}{\underset{\underset{\text{Me}}{|}}{CH_2=C=CHCH_2\overset{+}{C}}}} \longrightarrow \textbf{129} + \textbf{130} + \textbf{131}$$

cation **134** is similar to the result of solvolysis of the vinyl chloride **136**, in which the intermediate vinyl cation **137** opened up to give the more stable tertiary cation **138**, and the homopropargyl alcohol **139** was therefore the only product observed [48].

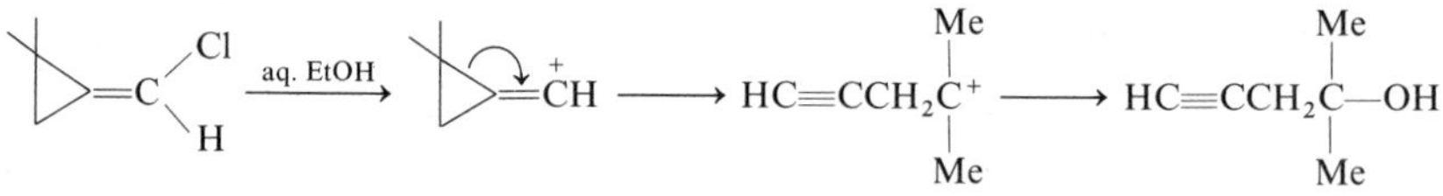

$$\underset{\textbf{136}}{\overset{Cl}{\underset{H}{>\!\!=\!\!C}}} \xrightarrow{\text{aq. EtOH}} \underset{\textbf{137}}{>\!\!\!=\!\!\overset{+}{CH}} \longrightarrow \underset{\textbf{138}}{\overset{\overset{\text{Me}}{|}}{\underset{\underset{\text{Me}}{|}}{HC\equiv CCH_2\overset{+}{C}}}} \longrightarrow \underset{\textbf{139}}{\overset{\overset{\text{Me}}{|}}{\underset{\underset{\text{Me}}{|}}{HC\equiv CCH_2C\!-\!OH}}}$$

To obtain more insight into the charge delocalization in the rate-determining step, the effects of methyl substitution at C-3 and C-5 on the acetolysis rates and product compositions of several homoallenic neopentyl-type brosylates (**140–145**) were investigated [46, 47]. As seen from the data in Table 4.2, each methyl substitution causes an increase in the overall acetolysis rate of a homoallenic brosylate. The 3-methyl substitution causes a rate enhancement of about 2.6, which is similar to that in the case of a homoallyl system (factor of 3.2) [49], as well as that of a saturated neopentyl-

TABLE 4.2

Kinetic Data on the Acetoylsis of Neopentyl-Type
Homoallenic Brosylates (151) at 75°[a]

| $R'_5$ | $R_5$ | $R_3$ | $10^4 k$ (sec$^{-1}$) | $k_{rel}$ | Estimated rate enhancement $\times 10^3$ |
|--------|-------|-------|----------------------|-----------|------------------------------------------|
| H  | H  | H  | 8.15  | 1.00 | 8.3–58  |
| H  | H  | Me | 21.5  | 2.64 | 6.2–44  |
| H  | Me | H  | 29.0  | 3.56 | 24–140  |
| H  | Me | Me | 68.9  | 8.46 | 15–70   |
| Me | Me | H  | 107.0 | 13.1 | 42–170  |
| Me | Me | Me | 221.0 | 27.1 | —       |

[a] From Refs. 46 and 47.

type brosylate (factor of 1.9 to 4.9) [50]. According to the authors [46], the similarity of these effects indicates that the origin of the rate enhancements is steric and that most of the positive charge is developing at the vinylic position (C-4) in the transition state. The increase in the rate by a factor of 3.6 for 5-methyl substitution is attributed to an electronic effect.

$CH_2{=}C{=}CHC(Me)_2CH_2OBs$      $CH_2{=}C{=}CC(Me)_2CH_2OBs$      $MeCH{=}C{=}CHC(Me)_2CH_2OBs$

                                        |
                                        Me

**140**                       **141**                       **142**

$MeCH{=}C{=}CC(Me)_2CH_2OBs$      $\begin{array}{c} Me{\searrow} \\ Me{\nearrow} \end{array} C{=}C{=}CHC(Me)_2CH_2OBs$      $\begin{array}{c} Me{\searrow} \\ Me{\nearrow} \end{array} C{=}C{=}CC(Me)_2CH_2OBs$

|                                                                                                            |
Me                                                                                                          Me

**143**                       **144**                       **145**

The Taft relationship was used to predict the ratio of the acetolysis rates of the homoallenyl and the corresponding neopentyl-type brosylates, $k_u/k_s$, in the absence of $\pi$-electron participation. From this and the measured $k_\Delta/k_s$ ratio, the rate enhancement due to $\pi$-electron delocalization was estimated to be $6.2 \times 10^3$ to $1.7 \times 10^5$ for compounds **140** to **145**, as summarized in Table 4.2 [46]. It is significant that the rate enhancements that result from homoallenic participation are about a factor of 10 larger than the homoallylic participation in related compounds.

Depending on the substituents, a bewildering array of products was obtained in the acetolysis of **140** to **145** [47]. In the absence of a methyl substituent at C-3 in the starting brosylates as in **142** and **144**, only rearranged allenic products were observed, as shown in Scheme 4.17. Generally, the major product, besides trienes, was the rearranged teritary acetate **146** [47].

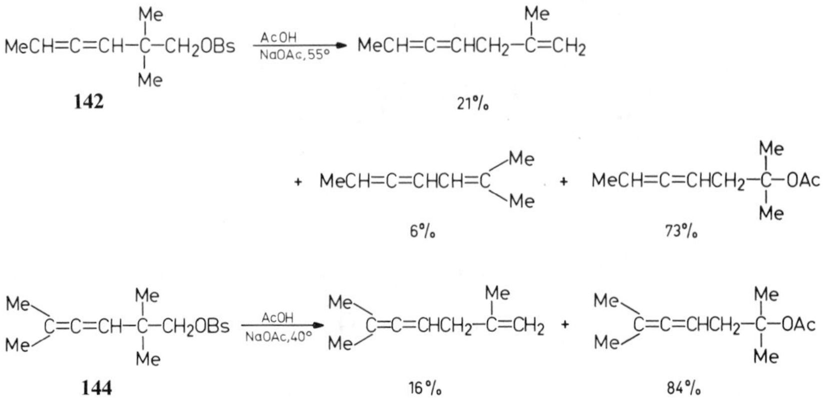

$$MeCH=C=CH-\underset{\underset{Me}{|}}{\overset{\overset{Me}{|}}{C}}-CH_2OBs \xrightarrow[NaOAc,55°]{AcOH} MeCH=C=CHCH_2-\underset{}{\overset{\overset{Me}{|}}{C}}=CH_2$$

**142**                                   21%

$$+ \; MeCH=C=CHCH=C\underset{Me}{\overset{Me}{<}} \;\; + \;\; MeCH=C=CHCH_2-\underset{\underset{Me}{|}}{\overset{\overset{Me}{|}}{C}}-OAc$$

                                6%                             73%

$$\underset{Me}{\overset{Me}{>}}C=C=CH-\underset{\underset{Me}{|}}{\overset{\overset{Me}{|}}{C}}-CH_2OBs \xrightarrow[NaOAc,40°]{AcOH} \underset{Me}{\overset{Me}{>}}C=C=CHCH_2-\overset{\overset{Me}{|}}{C}=CH_2 \; + \; \underset{Me}{\overset{Me}{>}}C=C=CHCH_2-\underset{\underset{Me}{|}}{\overset{\overset{Me}{|}}{C}}-OAc$$

**144**                                   16%                             84%

Scheme 4.17

$$\underset{R_2}{\overset{R_1}{>}}C=C=C\underset{\underset{R_3}{|}}{}CH_2\underset{\underset{Me}{|}}{\overset{\overset{Me}{|}}{C}}OAc$$

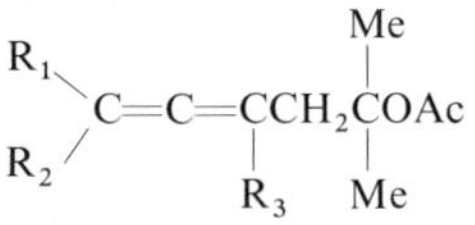

**146**

Products observed in the solvolysis of **141**, **143**, and **145** are illustrated in Scheme 4.18. The absence of unrearranged primary acetates indicates that the reaction proceeds entirely in an $S_N1$ manner. Moreover, despite the tendency of saturated neopentyl-type systems to give products resulting from methyl migration, no products from methyl migration were observed in the solvolysis of the homoallenic neopentyl systems investigated [51].

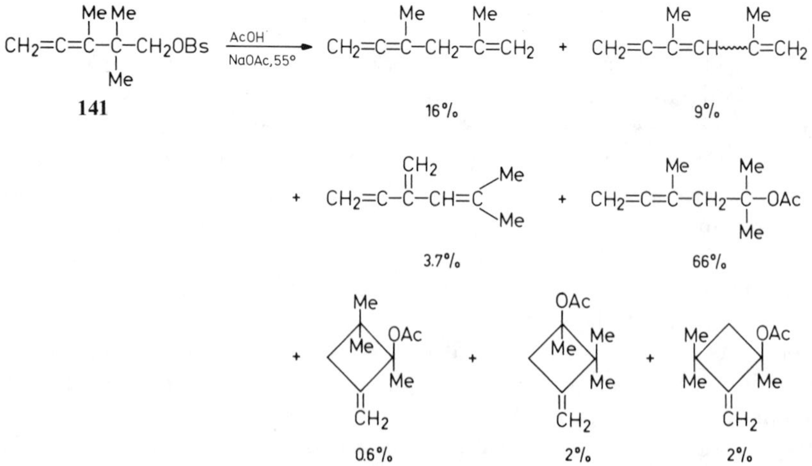

$$CH_2=C=\underset{\underset{Me}{|}}{\overset{\overset{Me}{|}}{C}}-\overset{\overset{Me}{|}}{C}-CH_2OBs \xrightarrow[NaOAc,55°]{AcOH} CH_2=C=\overset{\overset{Me}{|}}{C}-CH_2-\overset{\overset{Me}{|}}{C}=CH_2 \; + \; CH_2=C-\overset{\overset{Me}{|}}{C}=CH\text{---}\overset{\overset{Me}{|}}{C}=CH_2$$

**141**                                   16%                             9%

$$+ \; CH_2=C-\overset{\overset{CH_2}{||}}{C}-CH=C\underset{Me}{\overset{Me}{<}} \;\; + \;\; CH_2=C=\overset{\overset{Me}{|}}{C}-CH_2-\underset{\underset{Me}{|}}{\overset{\overset{Me}{|}}{C}}-OAc$$

                                3.7%                             66%

            0.6%                     2%                     2%

Scheme 4.18

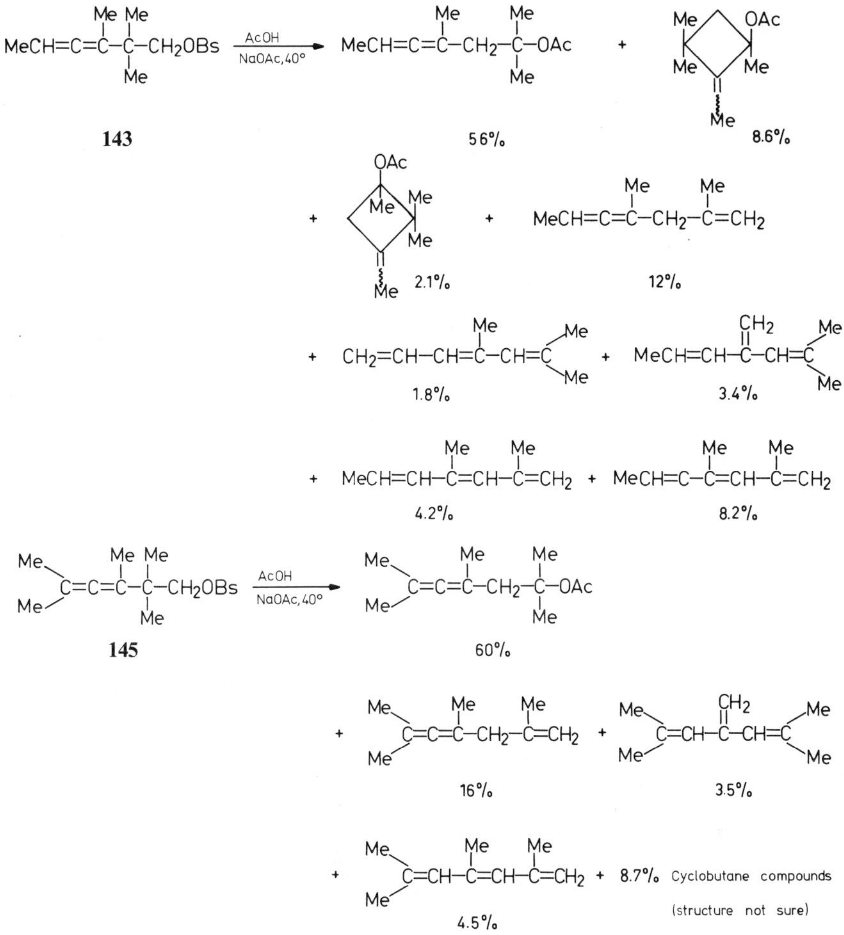

$$\underset{\textbf{143}}{\text{MeCH}=\text{C}=\overset{\overset{\text{Me}}{|}}{\text{C}}-\overset{\overset{\text{Me}}{|}}{\underset{\underset{\text{Me}}{|}}{\text{C}}}-\text{CH}_2\text{OBs}} \xrightarrow[\text{NaOAc,40°}]{\text{AcOH}} \underset{56\%}{\text{MeCH}=\text{C}=\overset{\overset{\text{Me}}{|}}{\text{C}}-\text{CH}_2-\overset{\overset{\text{Me}}{|}}{\underset{\underset{\text{Me}}{|}}{\text{C}}}-\text{OAc}} \quad + \quad \underset{8.6\%}{\text{[cyclobutane: Me, OAc, Me, Me, =Me]}}$$

$$+ \quad \underset{2.1\%}{\text{[cyclobutane: OAc, Me, Me, Me, =Me]}} \quad + \quad \underset{12\%}{\text{MeCH}=\text{C}=\overset{\overset{\text{Me}}{|}}{\text{C}}-\text{CH}_2-\overset{\overset{\text{Me}}{|}}{\text{C}}=\text{CH}_2}$$

$$+ \quad \underset{1.8\%}{\text{CH}_2=\text{CH}-\text{CH}=\overset{\overset{\text{Me}}{|}}{\text{C}}-\text{CH}=\text{C}\overset{\text{Me}}{\underset{\text{Me}}{}}} \quad + \quad \underset{3.4\%}{\text{MeCH}=\text{CH}-\overset{\overset{\overset{\text{CH}_2}{\|}}{}}{\text{C}}-\text{CH}=\text{C}\overset{\text{Me}}{\underset{\text{Me}}{}}}$$

$$+ \quad \underset{4.2\%}{\text{MeCH}=\text{CH}-\overset{\overset{\text{Me}}{|}}{\text{C}}=\text{CH}-\overset{\overset{\text{Me}}{|}}{\text{C}}=\text{CH}_2} \quad + \quad \underset{8.2\%}{\text{MeCH}=\text{C}-\overset{\overset{\text{Me}}{|}}{\underset{\underset{\text{Me}}{|}}{\text{C}}}-\text{CH}=\text{C}-\text{CH}_2}$$

$$\underset{\textbf{145}}{\overset{\text{Me}}{\underset{\text{Me}}{}}\text{C}=\text{C}=\overset{\overset{\text{Me}}{|}}{\text{C}}-\overset{\overset{\text{Me}}{|}}{\underset{\underset{\text{Me}}{|}}{\text{C}}}-\text{CH}_2\text{OBs}} \xrightarrow[\text{NaOAc,40°}]{\text{AcOH}} \underset{60\%}{\overset{\text{Me}}{\underset{\text{Me}}{}}\text{C}=\text{C}=\overset{\overset{\text{Me}}{|}}{\text{C}}-\text{CH}_2-\overset{\overset{\text{Me}}{|}}{\underset{\underset{\text{Me}}{|}}{\text{C}}}-\text{OAc}}$$

$$+ \quad \underset{16\%}{\overset{\text{Me}}{\underset{\text{Me}}{}}\text{C}=\text{C}=\overset{\overset{\text{Me}}{|}}{\text{C}}-\text{CH}_2-\overset{\overset{\text{Me}}{|}}{\text{C}}=\text{CH}_2} \quad + \quad \underset{3.5\%}{\overset{\text{Me}}{\underset{\text{Me}}{}}\text{C}=\text{CH}-\overset{\overset{\overset{\text{CH}_2}{\|}}{}}{\text{C}}-\text{CH}=\text{C}\overset{\text{Me}}{\underset{\text{Me}}{}}}$$

$$+ \quad \underset{4.5\%}{\overset{\text{Me}}{\underset{\text{Me}}{}}\text{C}=\text{CH}-\overset{\overset{\text{Me}}{|}}{\text{C}}=\text{CH}-\overset{\overset{\text{Me}}{|}}{\text{C}}=\text{CH}_2} \quad + \quad 8.7\% \text{ Cyclobutane compounds (structure not sure)}$$

Scheme 4.16 (*continued*)

Garry and Vessière [52], who studied the solvolysis of homoallenyl-neopentyl tosylate **147**, obtained rearranged products **148** and **149** and the unrearranged acyclic derivative **150** without formation of any cyclic products, as shown in Scheme 4.19.

In the mechanism shown in Scheme 4.20, suggested by Bly and Koock [47] to account for the observed products, the α-cyclopropylvinyl cation **152** is the initially formed intermediate in the reactions of all homoallenic derivatives (**151**) which solvolyze with π-electron participation. When the cyclopropyl ring in **152** is unsubstituted, it is relatively stable and reacts to yield predominantly cyclopropyl products (Schemes 4.14, 4.15, and 4.16). When **152** has a single substituent at C-1, route A still dominates, although

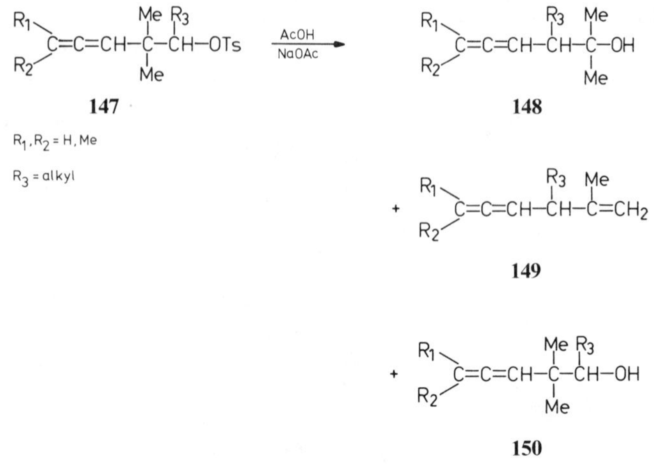

**147**

R$_1$,R$_2$ = H, Me

R$_3$ = alkyl

**148**

**149**

**150**

Scheme 4.19

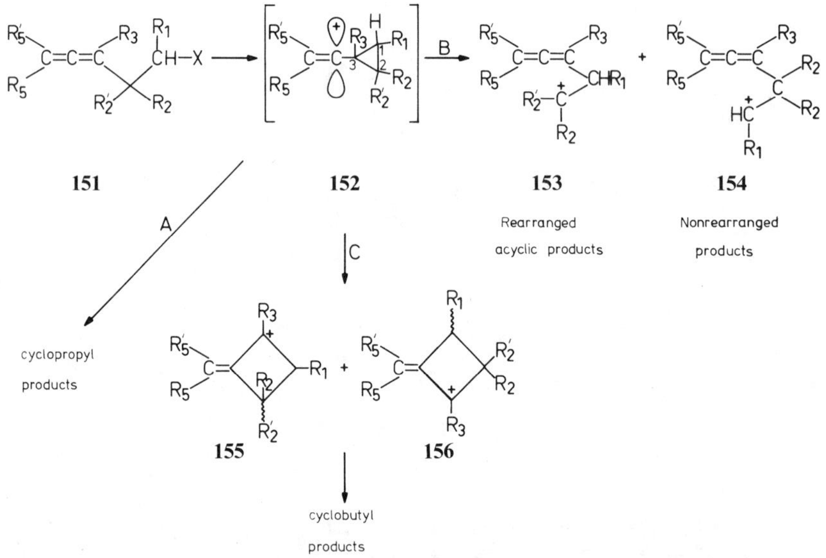

| **151** | | **152** | **153** | **154** |

Rearranged            Nonrearranged

acyclic products      products

cyclopropyl
products

**155**          **156**

cyclobutyl
products

Scheme 4.20

some ring opening to a secondary acyclic cation (**154**) and ring enlargement to a cyclobutyl cation may also occur, giving rise to acyclic and cyclobutyl compounds (Schemes 4.15 and 4.16). In the neopentyl derivatives, where R$_2$ and R$_2'$ are methyl, the C-3-unsubstituted (R$_3$ = H) derivative generates an α-cyclopropylvinyl cation that is less stable than the ring-opened tertiary cation **153** and the secondary cation **154** (R$_1$ = alkyl). Therefore, rearranged acyclic derivatives are formed exclusively (route B), as exemplified in Scheme 4.17.

In the secondary tosylate systems studied by Garry and Vessière [52], the formation of unrearranged product 150 (Scheme 4.19) may be accounted for via the vinyl cation 152 (route A) rather than a $k_s$ process, because the reactivity of homoallenic neopentyl-type derivatives is relatively insensitive to the nucleophilicity of the medium [47]. In C-3-substituted ($R_3$ = Me) cases, ring enlargement (route C) to the tertiary cyclobutyl-type cations 155 and 156 is important (Scheme 4.18).

Investigation of the C-2-monosubstituted system 157 throws additional light on the mechanism of homoallenic participation in solvolysis [53]. In this case, the value of the integrated rate constant increased significantly throughout acetolysis. This was interpreted in terms of rearrangement of the initial tosylate (157) to a comparably reactive tosylate (158) through the same ion pair that led to its formation (Scheme 4.21). This result agreed well with that for 104, the integrated rate constant of which decreased during the solvolysis, and therefore argues for rearrangement to a less reactive tosylate. Both 104 and 157 gave essentially identical product mixtures, as shown in Scheme 4.21. The rearranged tosylate 158 could be isolated when the acetolysis of 104 was interrupted before completion of the reaction.

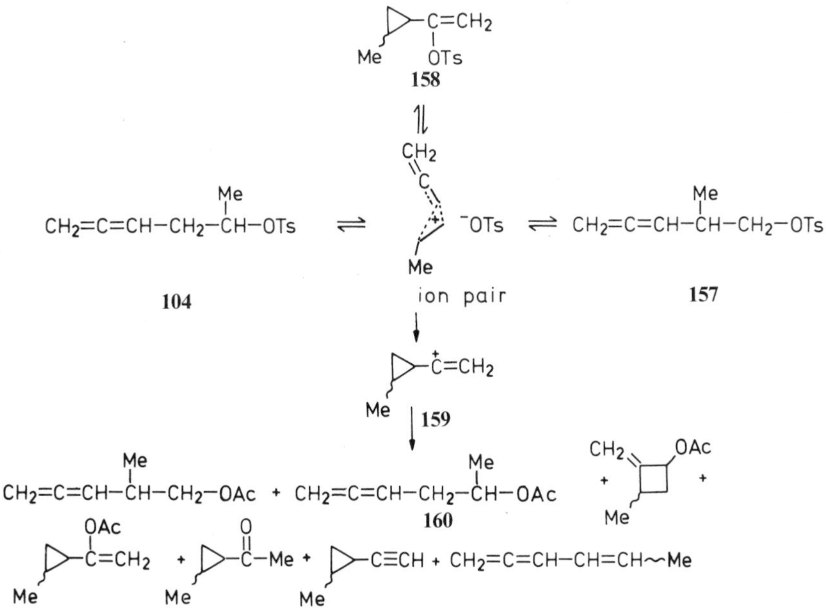

Scheme 4.21

Furthermore, **157** was shown to rearrange to both **104** and **158** during acetolysis [53].

The cyclopropylvinyl tosylate **158** (86% trans) was independently synthesized [54] and its acetolysis investigated. Unlike **104** and **157**, the cyclopropylvinyl tosylate **158** was found to solvolyze in a strictly first-order fashion, with no rearrangement to **104** and **157** during the solvolysis. The products obtained from **158** closely matched those from **104** and **157**, but the percentage of products containing a cyclopropyl group was greater (~62% versus ~35%) in the case of **158**. The allenyl acetate **160** was a major product in all three cases, suggesting that the product-determining intermediate is common to all three systems. The α-cyclopropylvinyl cation **159** satisfactorily accounts for the products formed in all cases (Scheme 4.21).

The exact position of the tosylate counterion was suggested to influence the reactivity of the ion-pair intermediates **161** and **162**, depending on whether the ion pairs are formed from **104**, **157**, or **158**. This may also account for differences in product distribution and to the ease of return versus product formation [54].

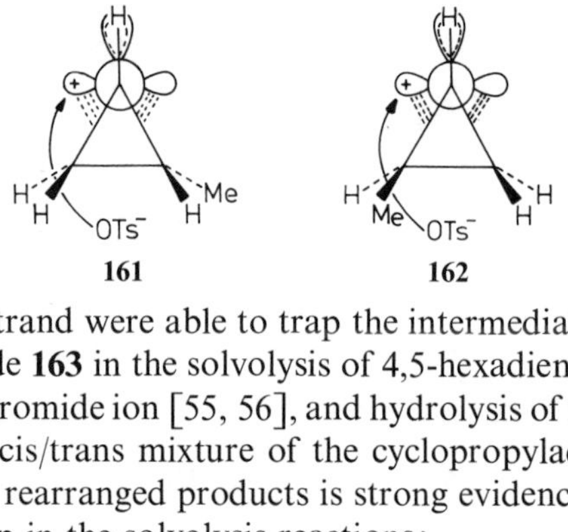

**161**                                              **162**

Santelli and Bertrand were able to trap the intermediate vinyl cation **159** as the vinyl bromide **163** in the solvolysis of 4,5-hexadien-2-yl tosylate (**104**) in the presence of bromide ion [55, 56], and hydrolysis of **104** in the presence of NaBH$_4$ gave a cis/trans mixture of the cyclopropylacetylenes **164**. The formation of these rearranged products is strong evidence for participation of the allenyl group in the solvolysis reactions:

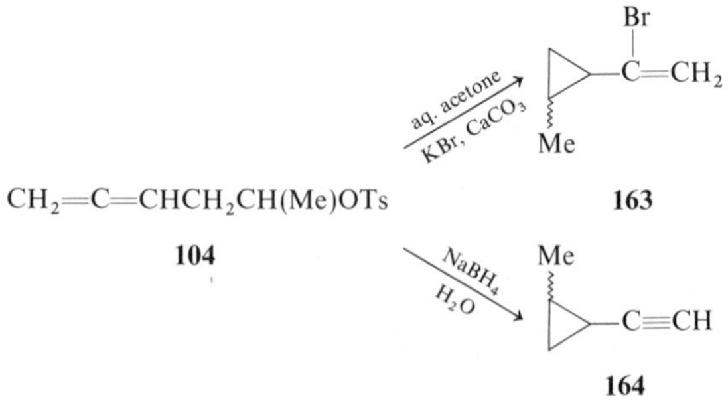

$CH_2=C=CHCH_2CH(Me)OTs$

**104**

**163**

**164**

Hydrolysis of C-1-deuterated tosylate **165** yielded the cyclobutyl derivatives **168** and **169** (75%) in a 40 : 60 ratio [56, 57], along with the unrearranged

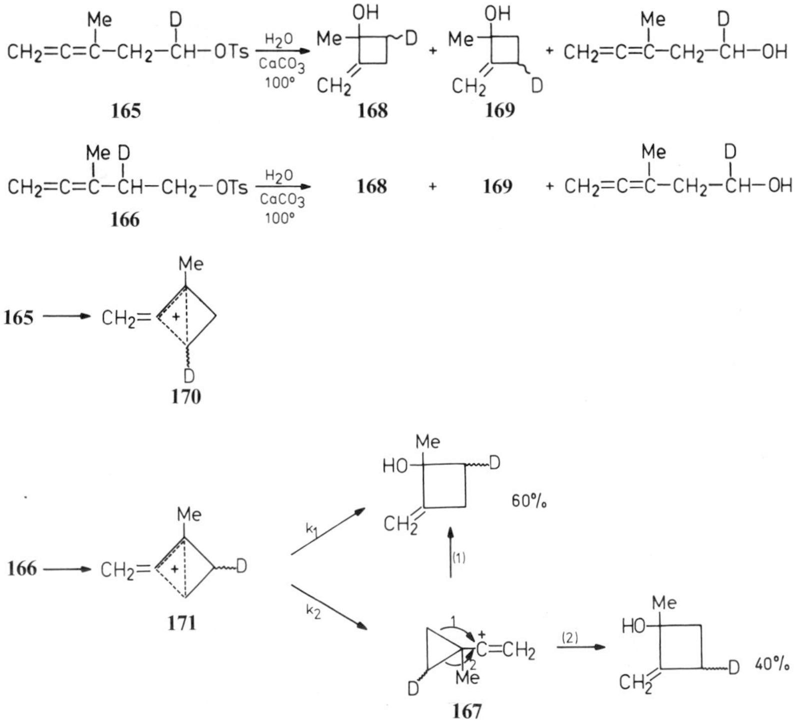

$$CH_2=C=\overset{\overset{\displaystyle Me}{|}}{C}-CH_2-\overset{\overset{\displaystyle D}{|}}{CH}-OTs \xrightarrow[\substack{CaCO_3 \\ 100°}]{H_2O}$$

**165**

Me⊢◻⊣~D + Me⊢◻ + $CH_2=C=\overset{\overset{\displaystyle Me}{|}}{C}-CH_2-\overset{\overset{\displaystyle D}{|}}{CH}-OH$

**168**     **169**

$$CH_2=C=\overset{\overset{\displaystyle Me\ D}{|\ |}}{C}-CH-CH_2-OTs \xrightarrow[\substack{CaCO_3 \\ 100°}]{H_2O} \quad \textbf{168} \quad + \quad \textbf{169} \quad + \quad CH_2=C=\overset{\overset{\displaystyle Me}{|}}{C}-CH_2-\overset{\overset{\displaystyle D}{|}}{CH}-OH$$

**166**

**165** ⟶ CH₂=

**170**

**166** ⟶ CH₂=

**171**

**167**

Scheme 4.22

alcohol (25%) (Scheme 4.22). The ratio was inverted to 60:40 in the case of the C-2-deuterated tosylate **166**. The fact that the methylene cyclobutanols **168** and **169** are not formed in a 50:50 ratio is taken as evidence that the bicyclobutonium ions **170** and **171** are the intermediates [58]. The α-cyclopropylvinyl cation **167** would have given an equal amount of **168** and **169**. Based on the 60:40 ratio of **168** to **169**, the authors calculated that $k_2 = 4k_1$. Part of the major product is therefore thought to arise from direct attack on the cyclobutonium ion, which is illustrated for **166** in Scheme 4.22. However, the unequal distribution can also be interpreted according to the mechanism of Lehman and Macomber discussed above [54], invoking the unsymmetric location of the tosylate counterion associated with the vinyl cation, as depicted in **161** and **162**.

Hydrolysis of tosylate **172** gave only 27.5% of cyclized products compared to 92% from tosylate **104** [56]. The cyclized products formed are mainly four-membered derivatives, and no trace of cyclopropyl compounds was detected. The lower amount of cyclization from **172** was explained as being due to the steric influence of the isopropyl substituent at C-1 hindering the homoallenyl participation. Therefore, hydride migration to form a more stable tertiary cation (from which **173** is derived) competes with the homo-allenyl participation [58].

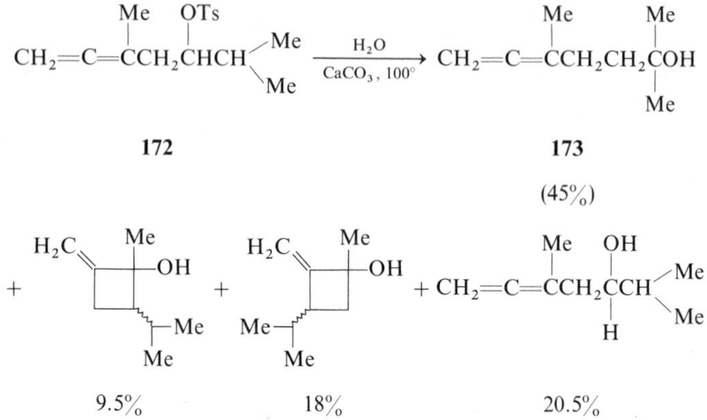

                172                                              173

                                                              (45%)

         9.5%                    18%                      20.5%

Hydride migration is apparently involved in the formation of the common product (**175**) from the hydrolysis of the *threo*- and *erythro*-vinylidene-cyclohexylcarbinyl tosylates **174t** and **174e** [59]. The other cyclized products arise from homoallenic participation with inversion of configuration at the carbon bearing the leaving group, except for the allenic alcohol **182**, which is formed with retention (Scheme 4.23). A set of cations (**183** and **184**) including

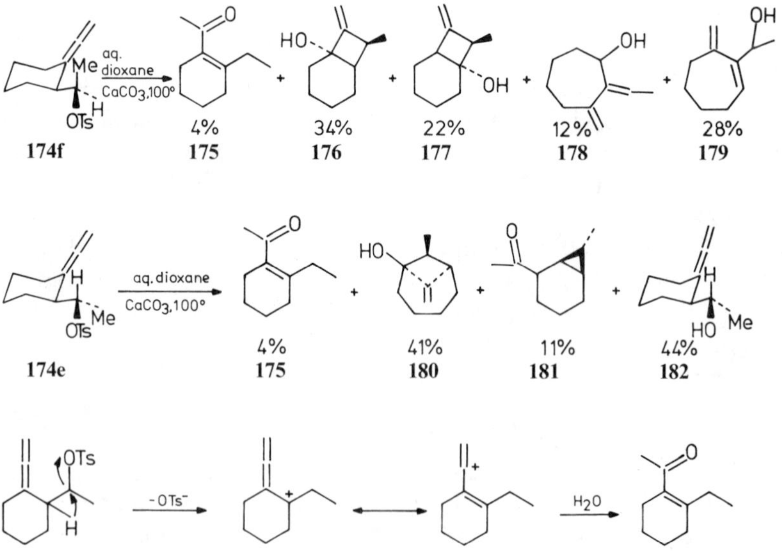

         174f        4%        34%        22%        12%        28%
                     175       176        177        178        179

         174e        4%        41%        11%        44%
                     175       180        181        182

                          Scheme 4.23

the bicyclobutonium ions **185** and **186** are postulated to explain the products of the reaction [59]. The set of cations **183** and **184**, **185** and **186** have different reactivities, as shown by the fact that they give rise to different products. The cations are indistinguishable except for the methyl group, and their effect can only be of steric origin. Thus, the cation **183** is more sterically hindered due to the three alkyl substituents of the cyclopropane ring being in a cis orientation (Scheme 4.23). The condition for the migration of bonds in **183** is therefore justified by the resulting steric compression. The cation has a short life span, as shown by the absence of any cyclopropane products [59]. The formation of the bicyclic compound from the cation **184** more probably occurs by a concerted process, as depicted in Scheme 4.24.

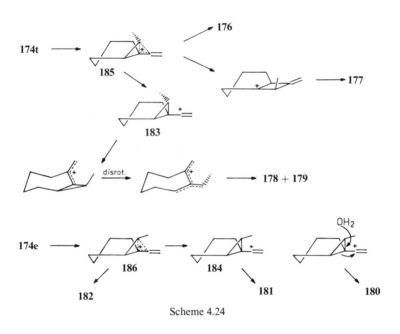

Scheme 4.24

Bicyclic products were also obtained from the solvolysis of allenyl tosylates **124** and **125** (see Scheme 4.16) [60]. This is most likely due to the high energy of the ground state allene bonds compensating for the highly strained transition states [60].

Further evidence for the intermediacy of stabilized α-cyclopropylvinyl cations as intermediates in the homoallenyl rearrangement was provided by Bergman et al. [61, 62]. The kinetics and products of solvolysis of cyclopropylvinyl iodide (**187**) were compared with those of the homoallenyl isomers (**188**). Cyclopropylvinyl iodide (**187**) and the isomeric homoallenyl iodide (**188**) gave in AcOH/AgOAc identical products in comparable amounts [61, 62] (see Schemes 4.16 and 4.21). Hence, the intermediacy of the same vinyl cation (**189**) was indicated.

In another control experiment, the cyclobutyl derivative **190** was prepared and solvolyzed and found to give only cyclobutyl compounds without any rearrangement [61]. Hence, the cyclobutyl cation **191** once formed does not rearrange, ruling out a direct formation of allyl cation **191** from the homoallenyl iodide **188** and confirming the formation of the α-cyclopropylvinyl cation **189**.

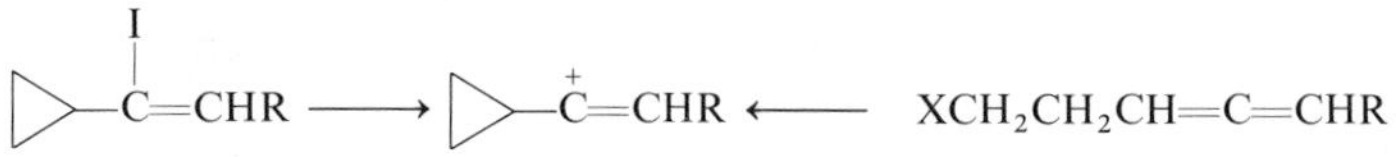


**187** **189** **188** R = H, Me; X = I, OTs

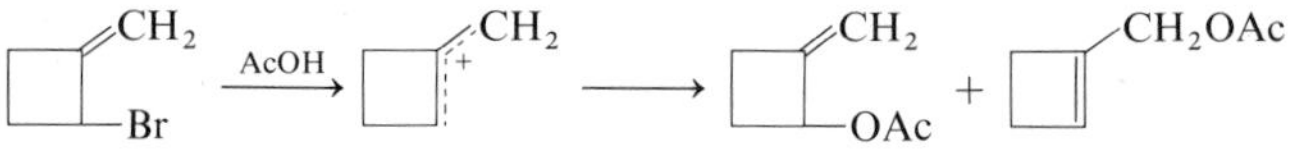


**190** **191**

A study of the hydrolysis of isomeric halides **192** and **193** by Santelli and Bertrand [58] gave different results, as illustrated in Scheme 4.25. Both **192**

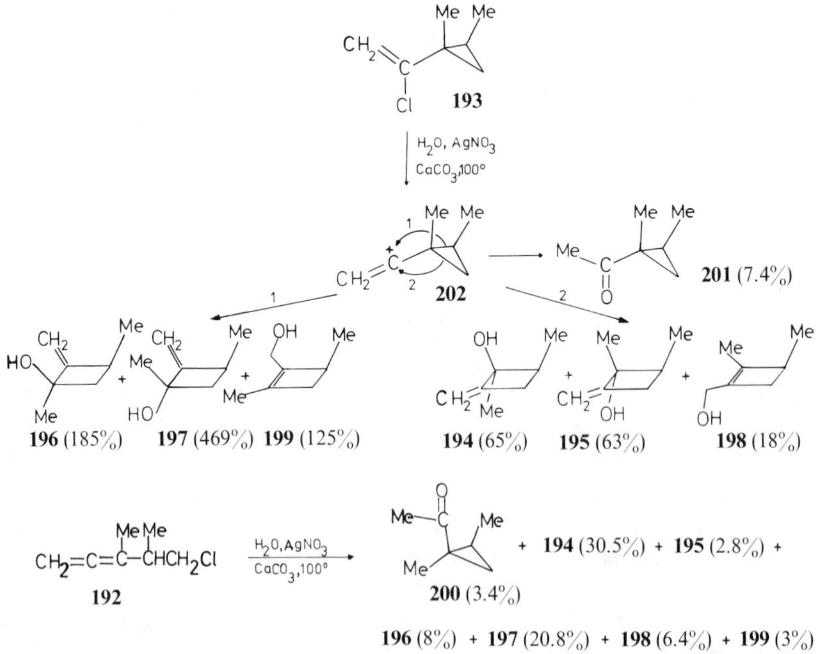


Scheme 4.25

and **193** gave mainly cyclized products consisting of more than 92% cyclobutane compounds. However, the product composition was considerably different, and therefore it was argued that no common intermediate was operative. The cyclopropylvinyl chloride **193** reacts via the vinyl cation **202**, which undergoes Wagner–Meerwein rearrangement with ring opening in two different ways. The cation **202** has retained its cis geometry with respect to the methyl groups, as seen from the 48% yield of cyclobutane compounds **195** and **197** with *cis*-methyl groups. No ketone corresponding to **200** was detected. If one postulates the same intermediates (**202** and **203**—both are possible from **192**) for the solvolysis of **192**, one would expect major amounts

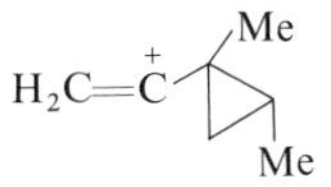

$$H_2C=\overset{+}{C}\overset{Me}{\underset{Me}{\diagup}}$$

**203**

of products **196**, **197**, and **199** resulting from the migration of the most substituted bond (path 1, Scheme 4.25), as was the case for the cyclopropylvinyl chloride **193**. Although this was not found to be the case, the authors [58] assumed a bicyclobutonium ion as an intermediate. However, the explanation given by Lehman and Macomber [54] (p. 180) can be applied here as well to satisfactorily account for the observed results.

Homoallenyl participation was found to be the major event in the hydrolysis of tosylate **204**, in which both homoallenyl and homoallyl rearrangement can occur [63]:

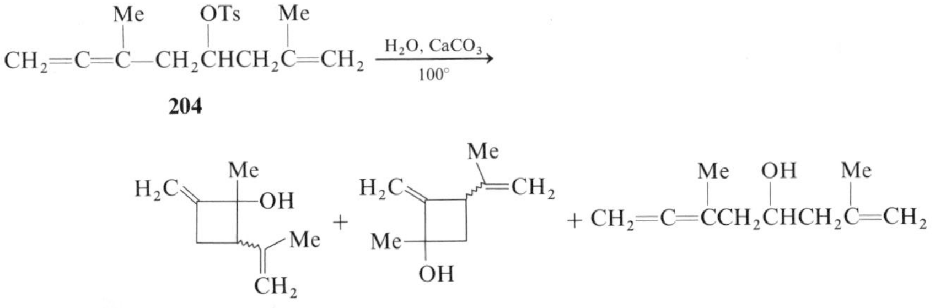

$$CH_2=C=\overset{\overset{\displaystyle Me}{|}}{C}-CH_2\overset{\overset{\displaystyle OTs}{|}}{C}HCH_2\overset{\overset{\displaystyle Me}{|}}{C}=CH_2 \xrightarrow[100°]{H_2O,\ CaCO_3}$$

**204**

    14%             31%             27%

Although no participation was found in the deamination of 2-amino-4,5-hexadiene (**102**, R = H; X = NH$_2$), the substituted derivative **205** gave cyclized products on deamination [64]. From the threo derivative **205**, the cyclic products all have inverted configuration at the functional carbon atom. This result shows that the homoallenyl rearrangement observed in the deamination reaction corresponds to an intramolecular nucleophilic substitution reaction without intervention of a planar carbocation.

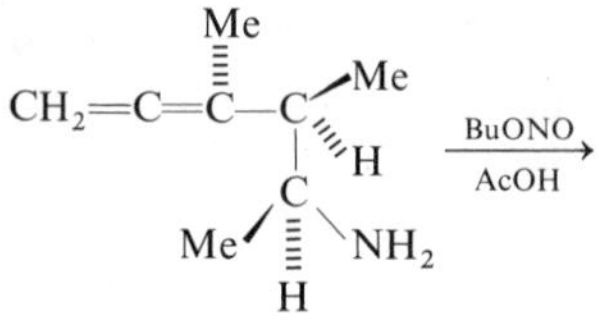

**205**

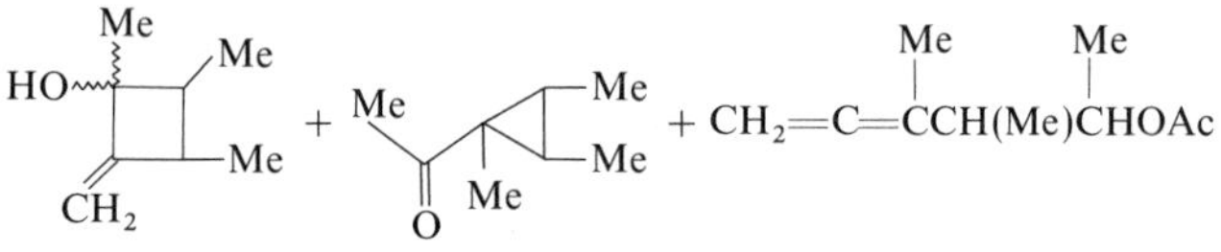

54%            27%            18%

Homoallenyl participation in cyclic allenyl tosylate **206** to give the bicyclic ketone **207** has also been reported [65]. On the other hand, compound **208** solvolyzed without participation of the allenyl system due to its conformationally unfavorable arrangement [65].

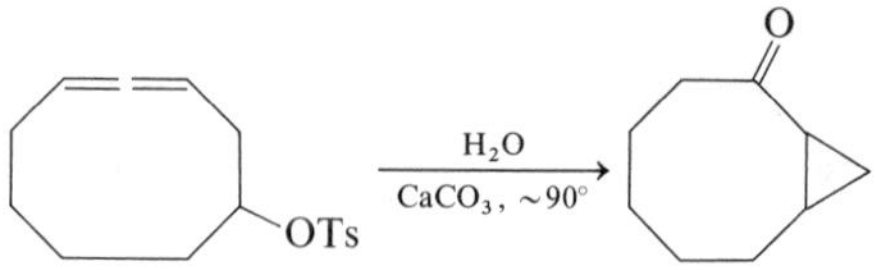

**206**                                    **207**

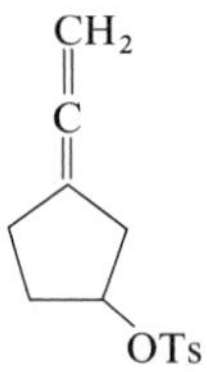

**208**

Solvolysis of γ-allenic tosylates was found to involve participation of the allenyl double bond. Thus, acetolysis of (S)-3-methyl-5,6-heptadien-2-yl tosylate (**209**) gave optically active cyclic products with inversion of configuration at the functional carbon [66].

Participation of even more remote allenic groups in solvolysis reactions also leads to cyclic products, but they do not involve vinyl cations as intermediates [67]:

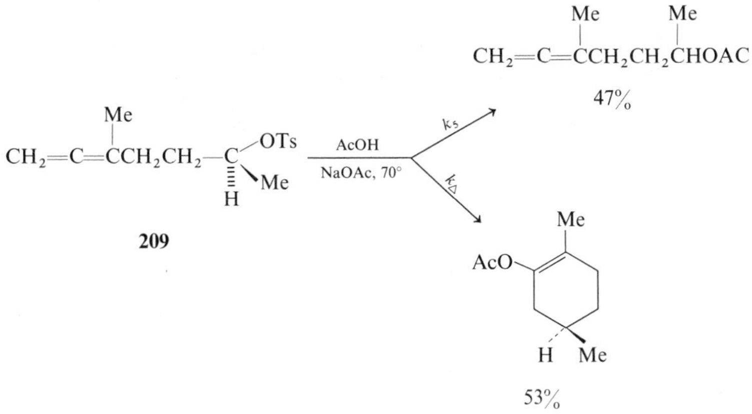

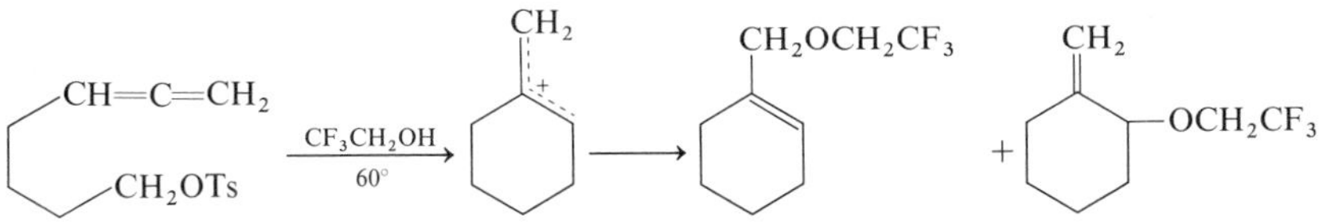

$$CH_2=C=CCH_2CH_2CHOAC$$
(with Me groups)

47%

$$CH_2=C=CCH_2CH_2-C\begin{smallmatrix}OTs\\Me\\H\end{smallmatrix}$$

(Me group on third carbon)

**209**

AcOH
NaOAc, 70°

$k_s$
$k_\Delta$

AcO— (cyclohexene ring with Me and H Me)

53%

(Structure: CH=C=CH₂ and CH₂OTs on ring)
CF₃CH₂OH
60°

$\overset{+}{CH_2}$ (ring)

$CH_2OCH_2CF_3$ (ring)

$CH_2$ (ring) + —OCH₂CF₃

A recent article [68] reports the intermediacy of vinyl cation **212** in the cyclization of the alcohol **210** in the presence of anhydrous formic acid. The high yield (70%) of the *cis*-octalone **213** makes this process attractive for preparative purposes. Cyclization of the corresponding cyclohexenone **211** gave the diketone **214** in 74% yield.

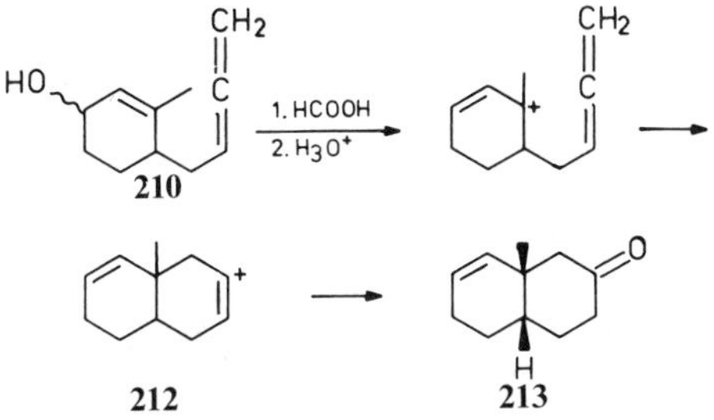

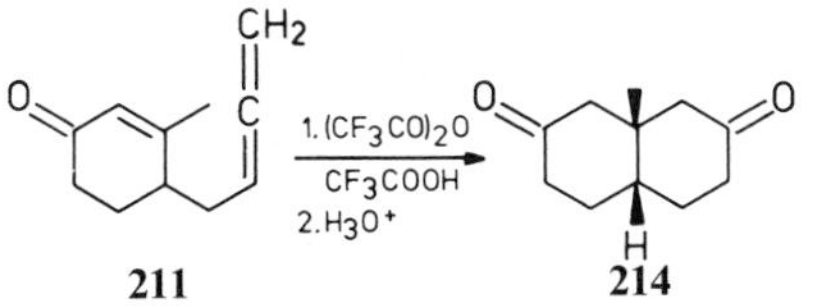

(Structure **210** with CH₂, C, HO)
1. HCOOH
2. H₃O⁺

(Structure with CH₂, C, +)

**210**

**212**

**213**

1. (CF₃CO)₂O
CF₃COOH
2. H₃O⁺

**211**

**214**

A summary of electrophilic additions to allenes and allenic participation involving vinyl cations appears in Tables 4.3 and 4.4, respectively.

**TABLE 4.3**

**Summary of Electrophilic Additions to Allenes Involving Vinyl Cations**

| Substrate | Reaction conditions | Temp (°C) | Major products | Reference |
|---|---|---|---|---|
| $CH_2=C=CH_2$ | 1:1 HCl | −70 | $MeC(Cl)=CH_2 + MeCCl_2Me +$  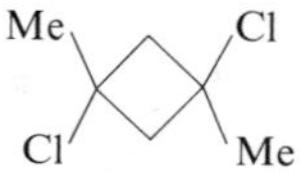 <br> 35%     54% <br> (cyclobutane products) 3% + 8% | [3, 4] |
| | 1:1 HBr | −70 | $MeC(Br)=CH_2 + CH_2CBr_2Me +$ 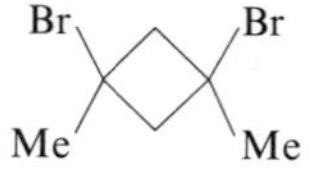 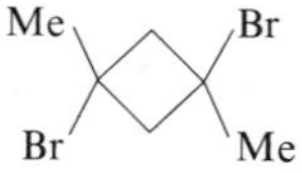 <br> 13%     35% <br> (cyclobutane products) 8% + 44% | [3, 4] |
| | 1:1 HI | −70 | $MeC(I)=CH_2 + MeCl_2Me$ <br> 6%     94% | [3, 4] |
| | HF | −76 | $MeCF_2Me$ | [5] |
| $MeCH=C=CH_2$ | HCl | −78 | $MeCH=C(Cl)Me + MeC≡CMe$ <br> $E$ and $Z$, 60%     40% | |
| $Me_2C=C=CH_2$ | HCl | −78 | $Me_2C(Cl)CH=CH_2 + Me_2C=CHCH_2Cl$ <br> 64%     36% | [9a] |
| $Me(CH_2)_2C=C=CH_2$ | HCl, BiCl_3 | ∼65 | $Me(CH_2)_2CH=C(Cl)Me + Me(CH_2)_2CH_2CCl_2Me$ | [17] |
| $RCH=C=CHMe$ | HBr | −40 | $RCH=C(Br)CH_2Me + RCH=CHCH(Br)Me$ | [19] |

| Substrate | Reagent | Temp | Product | Ref. |
|---|---|---|---|---|
| Me(CH₂)ₙCH=C=CHMe <br> n = 0, 1, 2 | HBr | 0 to −20 | $Me(CH_2)_nCH=C(Br)Me + Me(CH_2)_nCH_2C(Br)=CH_2$ <br> E and Z | [20] |
| $CH_2=C=CHCN$ | HCl | ~0 | $CH_2=C(Cl)CH_2CN$ <br> 75% | [23] |
| $CH(I)=C=CHCO_2Et$ | HX | r.t | $CH(I)=CXCH_2CO_2Et$ <br> X = Br, Cl | [24] |
| 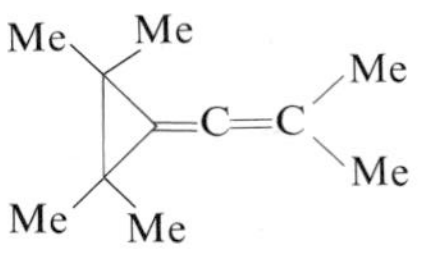 Me₂C(cyclopropane)C=C=C(Me)Me structure | $H_2O-THF-$ <br> $H_2SO_4$ <br> $AcOH-H_2SO_4$ <br> $CF_3CO_2H-C_6H_6$ <br> $MeOH-H_2SO_4$ | Reflux <br> 80 <br> 0–25 <br> Reflux | 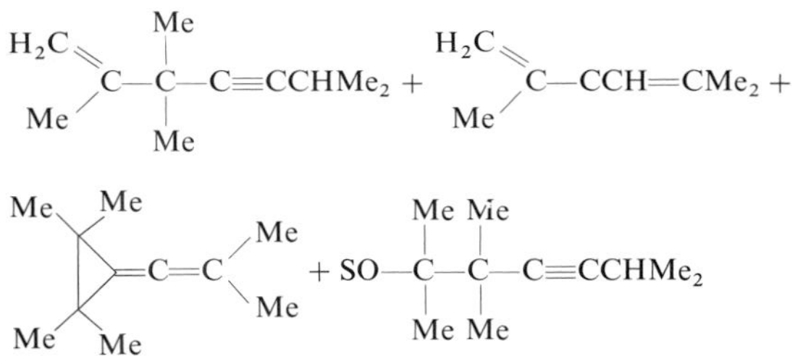 $H_2C=C(Me)_2C-C\equiv CCHMe_2 +$ <br> $H_2C=C-CCH=CMe_2 +$ <br> $Me_2C=C(Me)_2C-C\equiv CCHMe_2 + SO-C-C-C\equiv CCHMe_2$ <br> OS = OH, OAc, OMe | [26] |
| $CH_2=C=CH_2$ | $Cl_2, O_2$ | −30 | $CH_2=C(Cl)CH_2CH_2C\equiv CH + CH_2=C(Cl)CH_2CH_2C(Cl)=CH_2$ <br> 4.3%      1.5% | [11] |
| 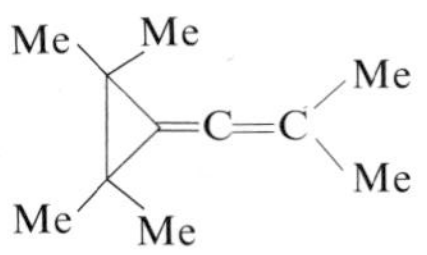 Me₂C(cyclopropane)C=C=C(Me)Me structure | $Cl_2,$ | −78 | 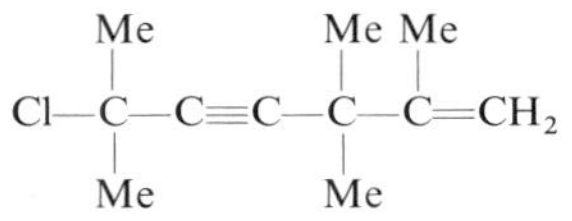 $Cl-C(Me)_2-C\equiv C-C(Me)_2-C=CH_2$ <br> 87% | [26] |

*(continued)*

**189**

**Table 4.3** (*Continued*)

| Substrate | Reaction conditions | Temp (°C) | Major products | Reference |
|---|---|---|---|---|
| | Br$_2$, MeOH | 0 to −15 | $\begin{array}{c}\text{Me}\quad\text{Me Me}\\ \text{Br--C--C}\equiv\text{C--C--C--OMe}\\ \text{Me}\quad\text{Me Me}\\ \text{Major}\end{array}$ + $\begin{array}{c}\text{Me}\quad\text{Me Me}\\ \text{Br--C--C}\equiv\text{C--C--C}=\text{CH}_2\\ \text{Me}\qquad\text{Me}\\ \text{Minor}\end{array}$ | [18] |
| | | | $\begin{array}{c}\text{Me}\quad\text{Me Me}\\ \text{Br--C--C}\equiv\text{C--C--C--Br}\\ \text{Me}\quad\text{Me Me}\\ \text{Trace}\end{array}$ | |
| MeCH=C=CHR<br>　　　　R | H$_3$O$^+$ | | MeCH$_2$CCH$_2$R <br> $\overset{\displaystyle O}{\Vert}$ | [18] |
| R$_2'$C=C=CHCOH<br>　　　　　R | H$_3$O$^+$ | Reflux | R$_2'$CHCH=CR$_2$ <br> $\overset{\displaystyle O}{\Vert}$ | [35] |
| Me$_2$CCH=C=CH$_2$<br>　OH | 1% H$_2$SO$_4$–MeOH | 20 | Me$_2$C=CHC(OMe)=CH$_2$ + Me$_2$C=CHCMe <br> 75%　　　　　　14%　$\overset{\displaystyle O}{\underset{\displaystyle \phantom{O}}{}}$=CHCMe | [36] |

**TABLE 4.4**

## Summary of Allenyl Participation Involving Vinyl Cations

| Substrate | Reaction Conditions | Temp (°C) | Major Products | Reference |
|---|---|---|---|---|
| $CH_2\!=\!C\!=\!CHCH_2CH_2OTs$ <br> **116** | AcOH–NaOAc | 100 | 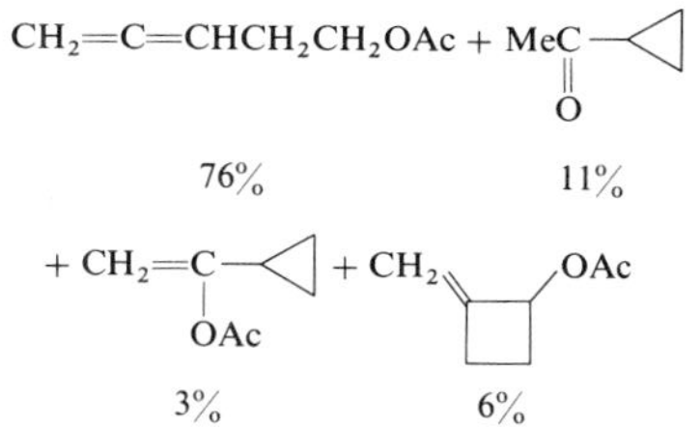 $CH_2\!=\!C\!=\!CHCH_2CH_2OAc$ + MeC(=O)△ <br> 76%  11% <br> + CH$_2$=C(OAc)△ + CH$_2$=◻(OAc) <br> 3%  6% | [37, 43] |
| $CH_2\!=\!C\!=\!CHCH_2CHOTs$ <br> Me <br> **104** | AcOH–NaOAc | 85 | 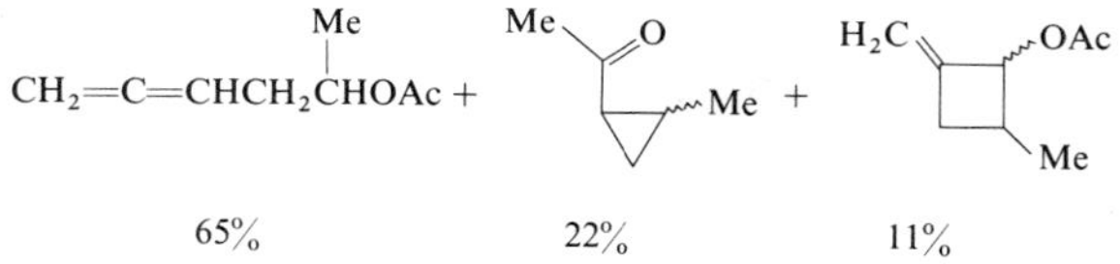 $CH_2\!=\!C\!=\!CHCH_2CHOAc$ + <br> Me <br> 65% <br> MeC(=O)△Me + H$_2$C=◻(OAc)(Me) <br> 22%  11% | [38] |
| Aqueous acetone– KBr–CaCO$_3$ | | 100 | 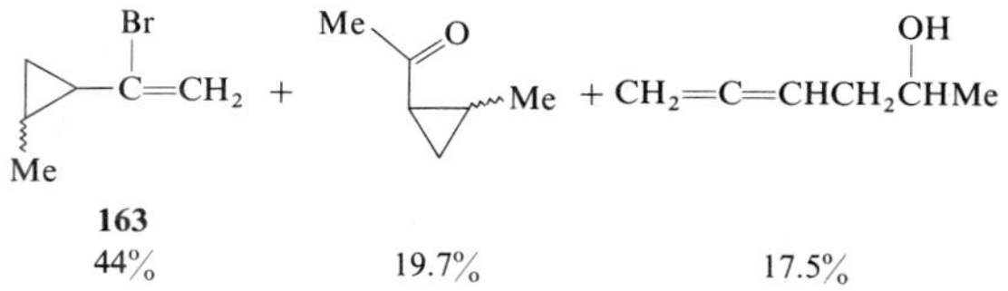 Br–C(=CH$_2$)△Me <br> **163** <br> 44% <br> MeC(=O)△Me + MeC(=O)△Me + CH$_2$=C=CHCH$_2$CHMe(OH) <br> 19.7%  17.5% | [56] |

(continued)

191

**TABLE 4.4** (*Continued*)

| Substrate | Reaction Conditions | Temp (°C) | Major Products | Reference |
|---|---|---|---|---|
| $\overset{\text{Me}}{\underset{\textbf{157}}{\text{CH}_2\text{=C=CH–CH–CH}_2\text{–OTs}}}$ | H₂O–NaBH₄ | 100 | C≡CH + CH₂=C=CHCH₂Et   **164** | [56] |
|  | AcOH–NaOAc | 85 | CH₂=C=CH–CH–CH₂–OAc + CH₂=C=CH–CH₂–CH–OAc (Me)   **160** 40%    1.9%    9%    1.5%    16%    C≡CH + CH₂=C=CH–CH=CH–CH⁓Me    24.5%    Traces | [53] |

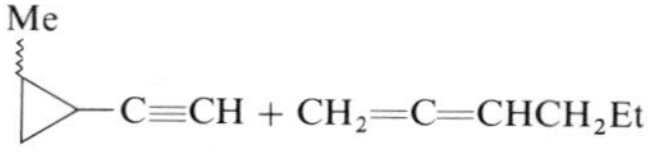

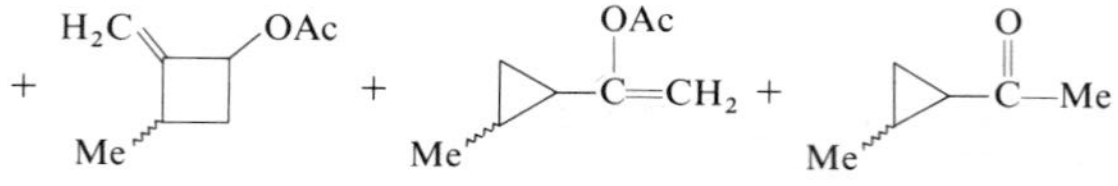

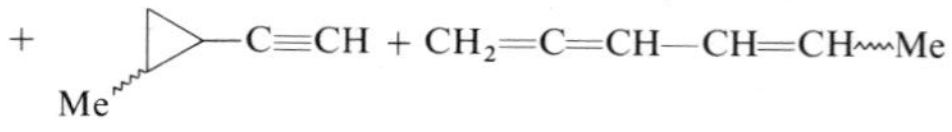

CH$_2$=C=C(Me)—CH$_2$CH$_2$OTs  **121**

1. AcOH
2. LiAlH$_4$

80

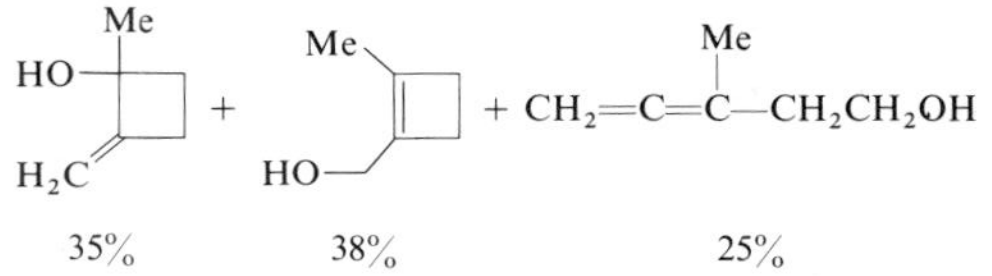

35%   38%   25%

[44]

RCH=C=CHCH$_2$CH$_2$OTs

**117** R = Me
**118** R = Et

AcOH–NaOAc

85

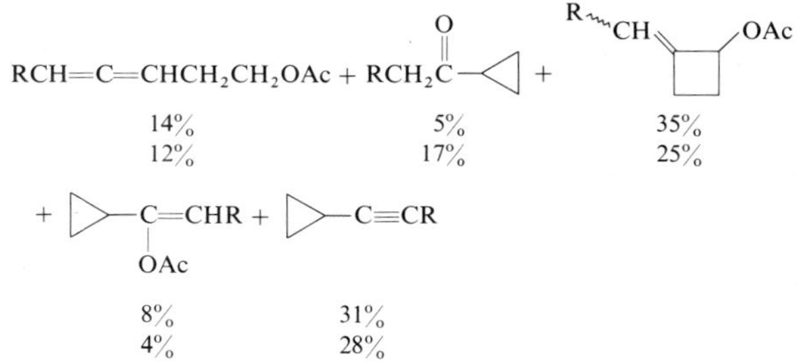

RCH=C=CHCH$_2$CH$_2$OAc + RCH$_2$C(=O)

14%   5%
12%   17%

C(OAc)=CHR + C≡CR

8%   31%
4%   28%

35%
25%

[43]

Me  Me
CH$_2$=C=C—CHCH$_2$OTs  **122**

1. AcOH
2. LiAlH$_4$

80

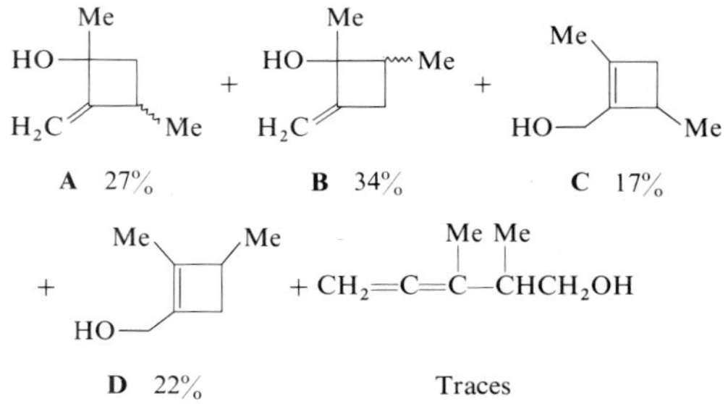

**A** 27%   **B** 34%   **C** 17%

**D** 22%

+ CH$_2$=C=C—CHCH$_2$OH   Traces

[44, 56]

*(continued)*

193

**TABLE 4.4** (*Continued*)

| Substrate | Reaction Conditions | Temp (°C) | Major Products | Reference |
|---|---|---|---|---|
| Me Me<br>CH₂=C—C—CH₂CHOTs<br>**123** | 1. AcOH<br>2. LiAlH₄ | 80 | **A** + **B** + **C** + **D** + CH₂=C=C—CH₂CHOH (Me Me)<br>40%  20%  19%  14%  7% | [44, 56] |
| Me<br>C=C=CH₂<br>(cyclohexyl ring with OTs)<br>**124** R = H<br>**125** R = Me | 1. AcOH<br>2. LiAlH₄ | 80 | 70% / 45%   17% / 29%   4.5% / 18% | [43, 44, 60] |
| Me  D<br>CH₂=C—C—CH₂—CH—OTs<br>**165** | H₂O–CaCO₃ | 100 | OH (···D)<br>Me—H₂C (C=CH₂)<br>**168** 30%<br>+<br>OH (D)<br>Me—H₂C (C=CH₂)<br>**169** 45% | [58] |
| Me D<br>CH₂=C—C—CH—CH₂—OTs<br>**166** | H₂O–CaCO₃ | 100 | **168** + **169** + CH₂=C=C—CH₂—CH—OH (Me, D)<br>45%   30%   25%<br>+ CH₂=C=C—CH₂—CH—OH (Me, D)<br>25% | [58] |

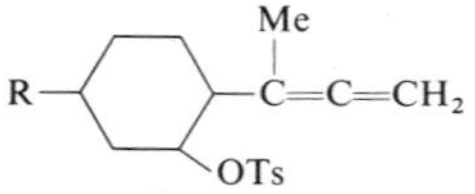

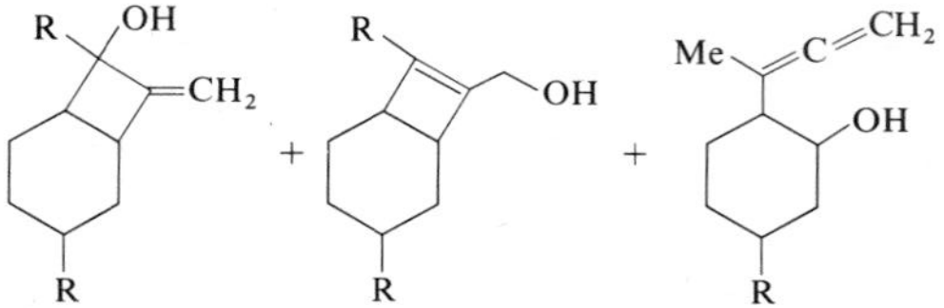

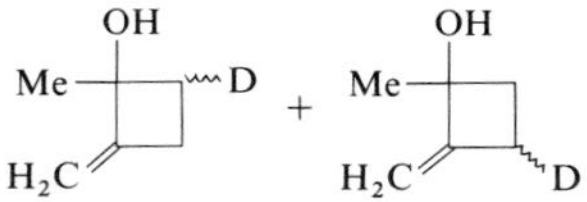

Me
|
CH₂=C=CH—C—CH₂OBs
|
Me

**127**

55    AcOH–NaOAc

$$\underset{\textbf{129}}{\underset{77\%}{CH_2=C=CHCH_2COAc}} \;\; \overset{Me}{|} \;\; + \;\; \underset{\textbf{130}}{\underset{\sim 12\%}{\overset{Me}{|}{CH_2=C=CHCH_2C=CH_2}}}$$

Me
|
CH₂=C=CHCH₂COAc

Me
|
+ CH₂=C=CHCH₂C=CH₂

**129**
77%

**130**
~12%

Me
/
+ CH₂=C=CHCH=C
\
Me

**131**
~11%

[45]

55    EtOH

Me
|
CH₂=C=CCHCH₂COEt
|
Me

**132**
95%

[45]

55    Aqueous acetone

Me
|
CH₂=C=CHCH₂COH
|
Me

**133**
85%

[45]

(continued)

# TABLE 4.4 (Continued)

| Substrate | Reaction Conditions | Temp (°C) | Major Products | Reference |
|---|---|---|---|---|
| 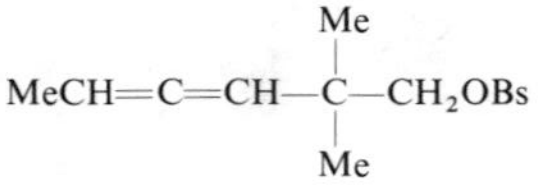 $\begin{array}{c}\text{Me}\\ \mid\\ \text{MeCH=C=CH-C-CH}_2\text{OBs}\\ \mid\\ \text{Me}\end{array}$  **142** | AcOH–NaOAc | 55 | 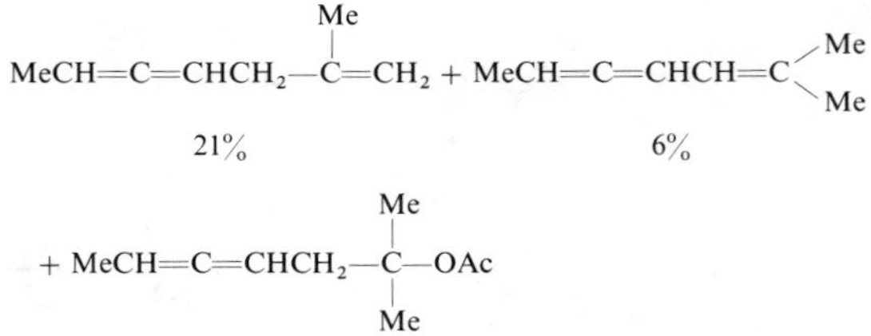 $\text{MeCH=C=CHCH}_2\text{-C(Me)=CH}_2 + \text{MeCH=C=CHCH=C(Me)}_2$  21%    6% $+ \text{MeCH=C=CHCH}_2\text{-C(Me)}_2\text{-OAc}$  73% | [47] |
| 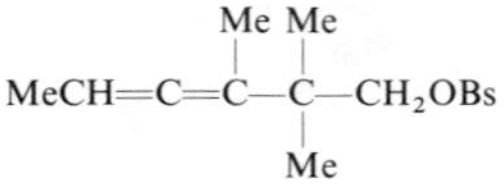 $\begin{array}{c}\text{Me Me}\\ \mid\;\;\mid\\ \text{MeCH=C=C-C-CH}_2\text{OBs}\\ \mid\\ \text{Me}\end{array}$  **143** | AcOH–NaOAc | 40 | 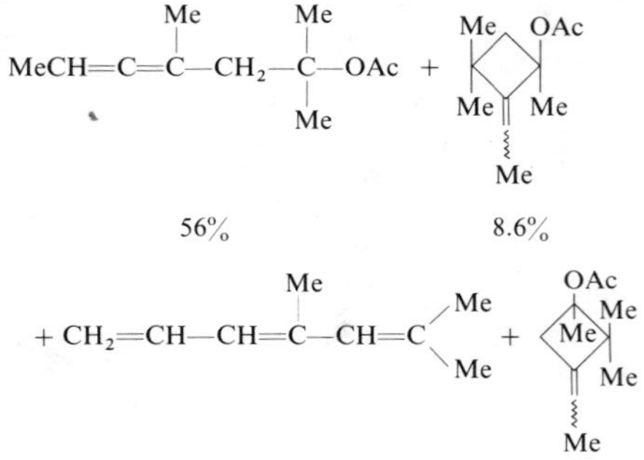 $\text{MeCH=C=C(Me)-C(Me)}_2\text{-CH}_2\text{-OAc}$ + (bicyclic OAc, Me structure)  56%    8.6% $+ \text{CH}_2\text{=CH-CH=C(Me)-CH=C(Me)}_2$ + (bicyclic OAc, Me structure)  1.8%    2.1% | [47] |

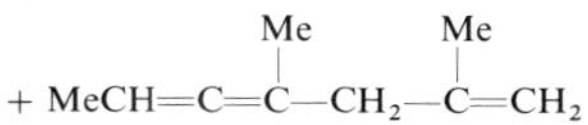

$\underset{\text{Me}}{\overset{\text{Me}}{\diagdown}}\text{C}=\text{CH}-\underset{\overset{|}{\text{Me}}}{\overset{\overset{\text{Me}}{|}}{\text{C}}}-\text{CH}_2\text{OBs}$

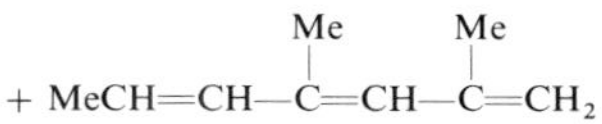

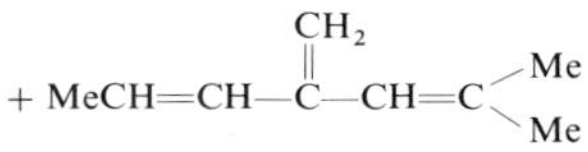

**144**

AcOH–NaOAc

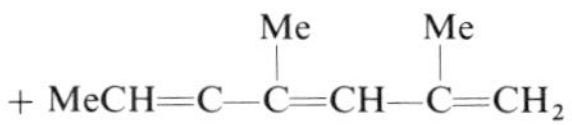

$+\ \text{MeCH}=\overset{\overset{\text{Me}}{|}}{\text{C}}-\text{CH}_2-\overset{\overset{\text{Me}}{|}}{\text{C}}=\text{CH}_2$

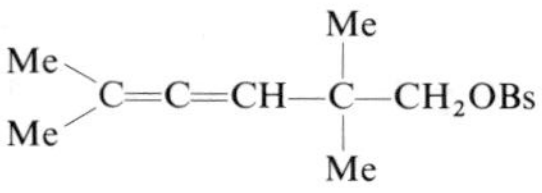

12%

$+\ \text{MeCH}=\text{CH}-\overset{\overset{\text{Me}}{|}}{\text{C}}=\text{CH}-\overset{\overset{\text{Me}}{|}}{\text{C}}=\text{CH}_2$

4.2%

$+\ \text{MeCH}=\text{CH}-\overset{\overset{\text{CH}_2}{\|}}{\text{C}}-\text{CH}=\text{C}\diagup^{\text{Me}}_{\diagdown\text{Me}}$

3.4%

$+\ \text{MeCH}=\text{CH}-\overset{\overset{\text{Me}}{|}}{\text{C}}=\text{CH}-\overset{\overset{\text{Me}}{|}}{\text{C}}=\text{CH}_2$

8.2%

**40**

$\underset{\text{Me}}{\overset{\text{Me}}{\diagdown}}\text{C}=\text{C}=\text{CHCH}_2-\overset{\overset{\text{Me}}{|}}{\text{C}}=\text{CH}_2\ +\ \underset{\text{Me}}{\overset{\text{Me}}{\diagdown}}\text{C}=\text{C}=\text{CHCH}_2-\overset{\overset{\text{Me}}{|}}{\underset{\underset{\text{Me}}{|}}{\text{C}}}-\text{OAc}$

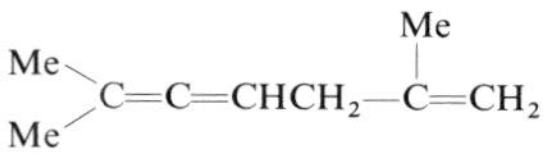

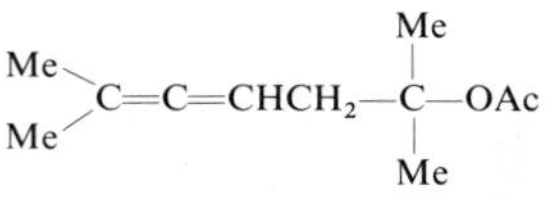

16%        84%

[47]

*(continued)*

**197**

**Table 4.4** (*Continued*)

| Substrate | Reaction Conditions | Temp (°C) | Major Products | Reference |
|---|---|---|---|---|
| 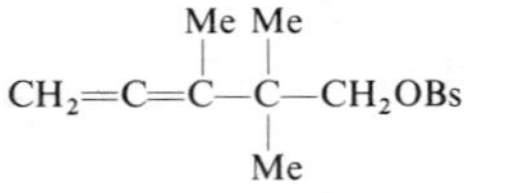 CH₂=C—C—CH₂OBs with Me Me and Me **141** | AcOH–NaOAc | 55 | 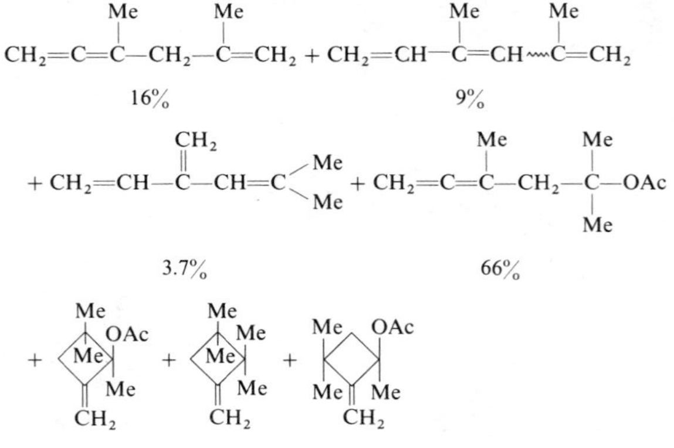 CH₂=C—C—CH₂—C=CH₂ + CH₂=CH—C=CH⋯C=CH₂ (Me, Me, Me, Me) 16%  9% <br> + CH₂=CH—CH—C—CH=C(Me)(Me) + CH₂=C—C—CH₂—C=CH₂—C—OAc (CH₂, Me) 66% <br> + (bicyclic structures: Me OAc Me / Me CH₂) 3.7%  2% <br> + (Me OAc Me Me / Me CH₂) 0.6%  2% | [47] |
| 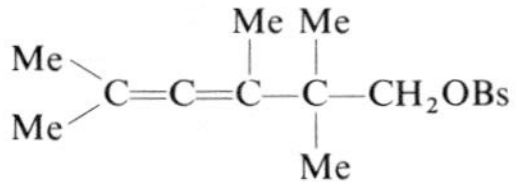 Me Me \ C=C—C—CH₂OBs with Me Me and Me **145** | AcOH–NaOAc | 40 | 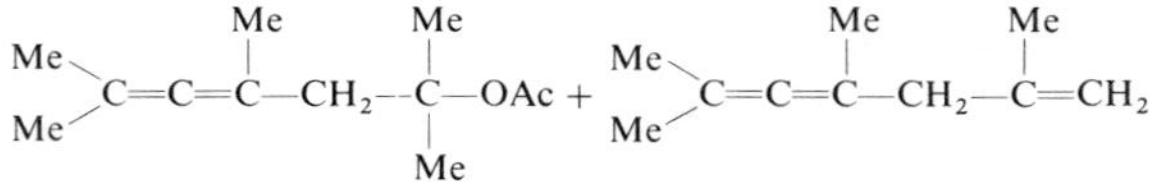 Me \ C=C—C—CH₂—C—OAc + Me \ C=C—C—CH₂—C=CH₂ (Me / Me, Me, Me) 60%  16% | [47] |

198

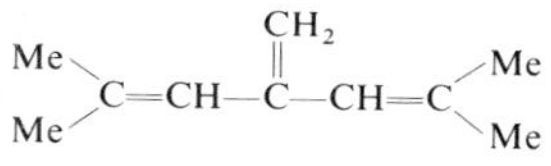

Me C=CH—C(=CH₂)—CH=C(Me)(Me)  +  Me(Me)C=

3.5%

Me(Me)C=CH—C(Me)=CH—C(Me)=CH₂  +  Me

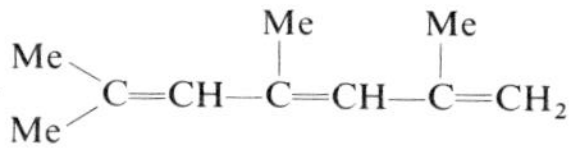

4.5%

+ Cyclobutane compounds (structure not certain)

8.7%

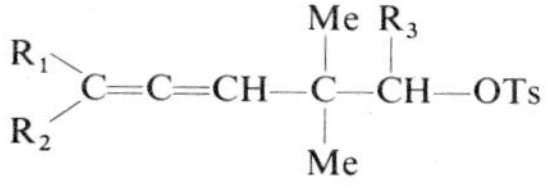

R₁(R₂)C=C=CH—CH(Me)—C(Me)—OH  +  R₁(R₂)C=C=CH—CH(R₃)—C(Me)—CH=CH₂

**148**                                    **149** $\qquad$ [52]

R₁(R₂)C=C=CH—CH(R₃)—C(Me)—CH—OH  +  R₁(R₂)C=CH—C(Me)=CH—CH—OH(Me)

**150**

Me(R₃)... Me

147  R₁, R₂ = H, Me; R₃ = alkyl

R₁(R₂)C=C=CH—CH(R₃)—CH(Me)—OTs

Aqueous AcOH–NaOAc      80

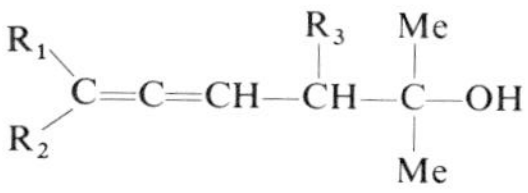

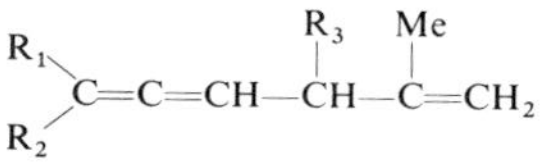

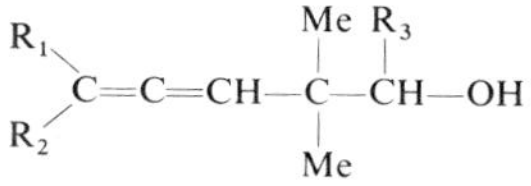

---

CH₂=C(Me)—CCH₂CH₂COH(Me)(Me)  +  (cyclobutane structure)

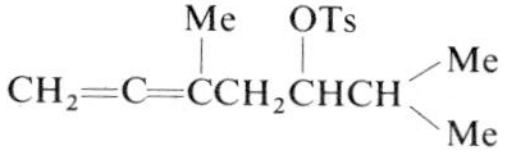

**173**
45%                                          9.5%

[56]

CH₂=C(Me)—CCH₂CH₂CHCH(Me)(Me), OTs

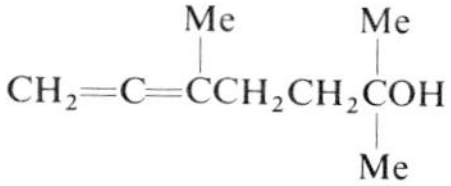

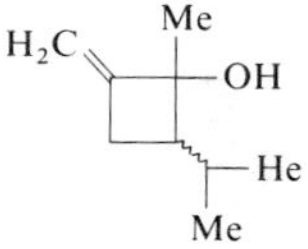

**172**

H₂O–CaCO₃      100

---

*(continued)*

**199**

**TABLE 4.4** (Continued)

| Substrate | Reaction Conditions | Temp (°C) | Major Products | Reference |
|---|---|---|---|---|
| 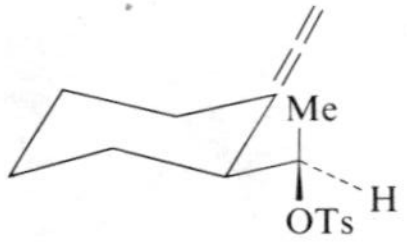 174t | Aqueous dioxane–CaCO$_3$ | 100 | 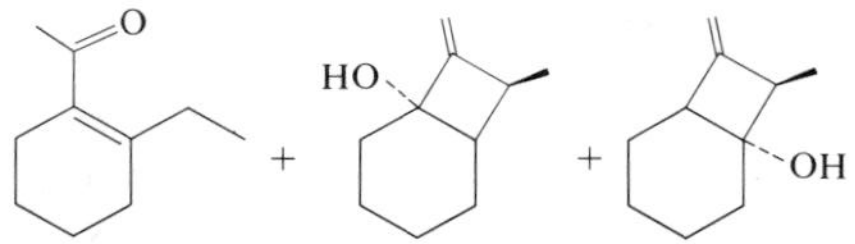 (see structures below) | [59] |

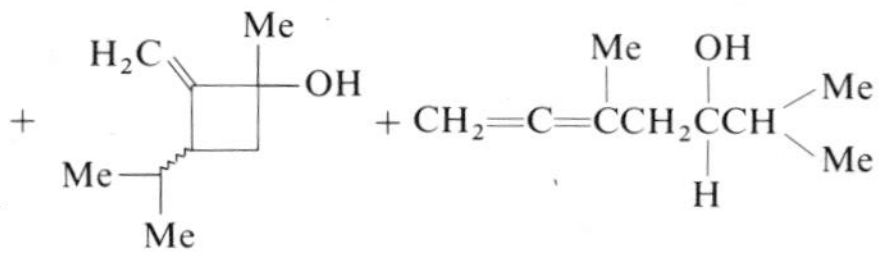

Substrate 174t:

Me / OTs / H

Major Products:

H$_2$C= (cyclobutane) Me, OH, Me, Me — 18%

+ CH$_2$=C=CCH$_2$CCH with Me, OH, Me, Me, H — 20.5%

+ (bicyclic) HO — **175** 4%

+ (bicyclic) — **176** 34%

+ (bicyclic) OH — **177** 22%

ethyl cyclohexenyl ketone (O)

+ OH (cycloheptenyl) — **178** 12%

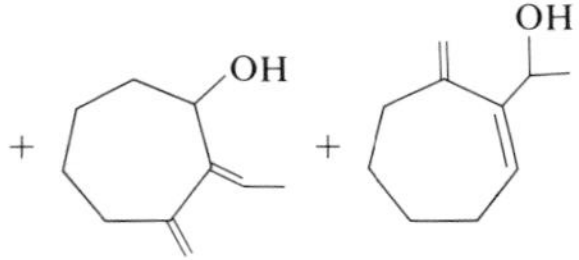

+ OH (cycloheptenyl) — **179** 28%

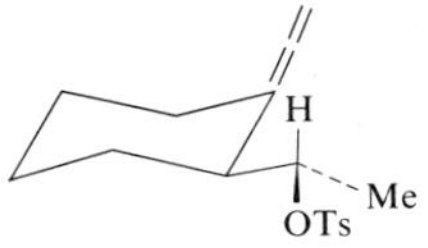

174e

Aqueous dioxane–CaCO$_3$

100

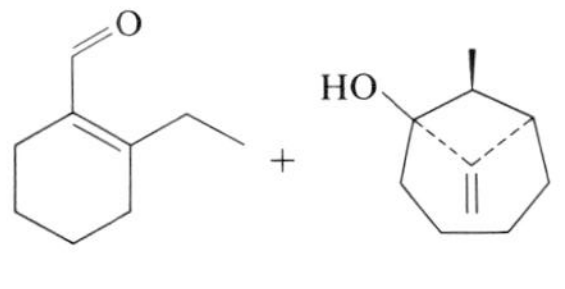

175
4%

180
41%

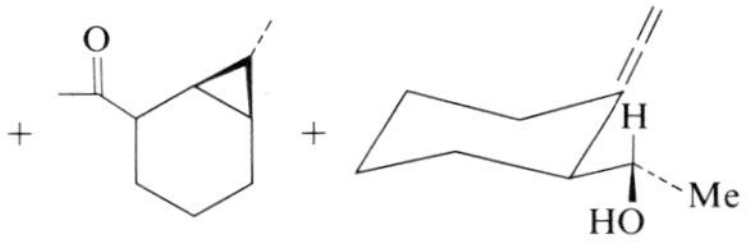

181
11%

182
44%

[59]

Me Me
CH$_2$=C=C—CHCH$_2$Cl

192

H$_2$O–AgNO$_3$–CaCO$_3$

100

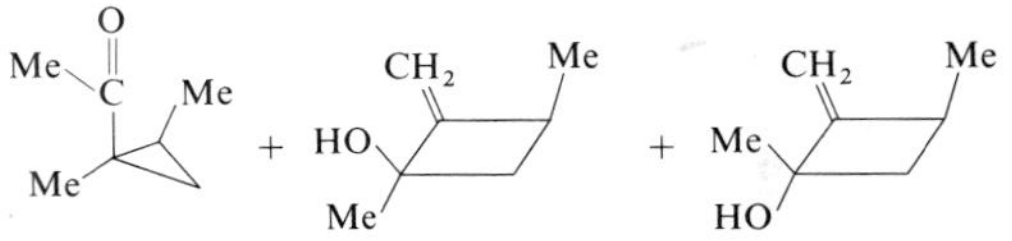

200
3.4%

196
8%

197
20.8%

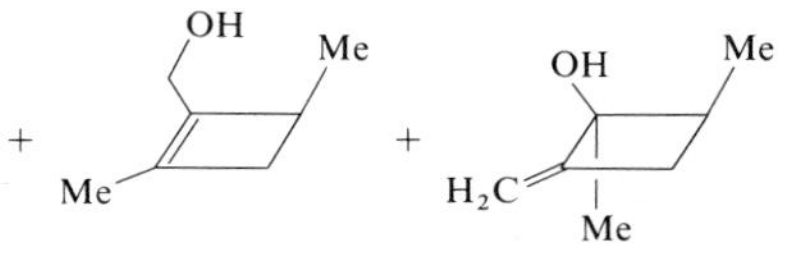

199
3%

194
30.5%

[58]

*(continued)*

**201**

**TABLE 4.4** (*Continued*)

| Substrate | Reaction Conditions | Temp (°C) | Major Products | Reference |
|---|---|---|---|---|
| $CH_2=C=C-CH_2CHCH_2C=CH_2$   (OTs, Me, Me)   **204** | $H_2O-CaCO_3$ | 100 | (195, 28%) + (198, 6.4%) + 14% + 31% + 27% | [63] |
| (Me, Me, Me, $CH_2=C=C-C-C$, H, $NH_2$, H)   **205** | BuONO–AcOH | 50 | 54% + 27% + $CH_2=C=CCH(Me)CHOAc$ 18% | [64] |

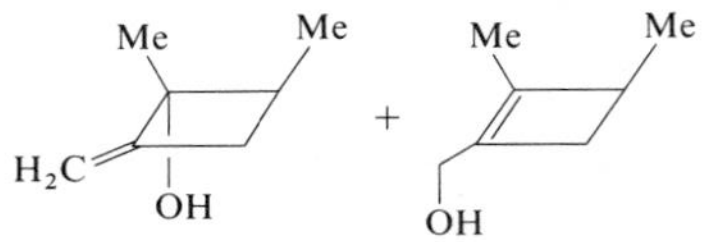

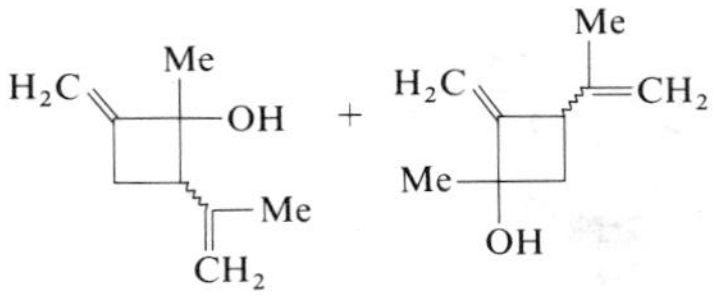

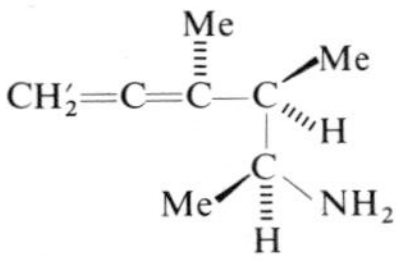

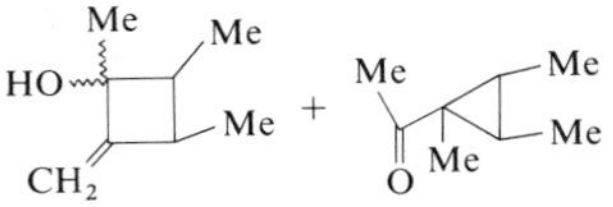

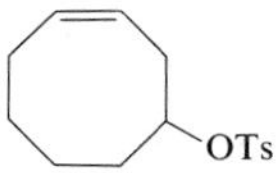

$$\text{206}$$

$$\text{OTs (cyclooctene tosylate)}$$

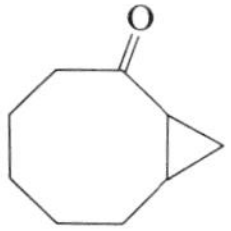

H$_2$O–CaCO$_3$    90

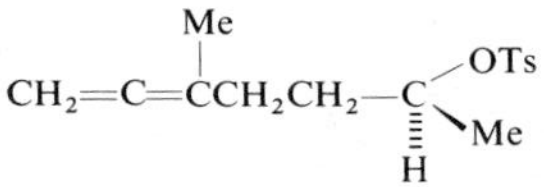

**207**

[65]

$$CH_2=C=CCH_2CH_2-C(OTs)(Me)(H)$$ with Me

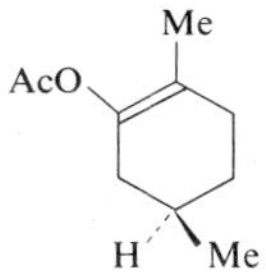

AcOH–NaOAc    70

Me, AcO (cyclohexene) Me, H   53%

$+ \; CH_2=C=CCH_2CH_2CHOAc$ with Me, Me   47%

[66]

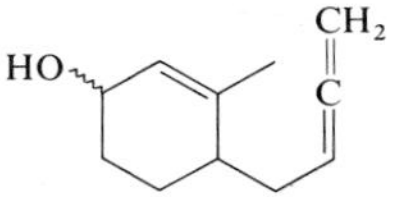

**210**

1. HCOOH
2. OH$^-$    r.t.

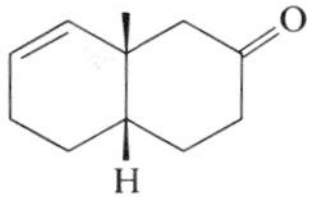

**213**
70%

[68]

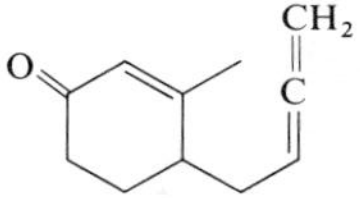

**211**

1. CF$_3$COOH
2. OH$^-$    r.t.

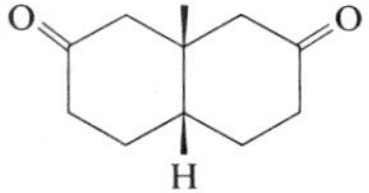

**214**
74%

[68]

## REFERENCES

1. (a) P. B. D. de la Mare and R. Bolton, "Electrophilic Addition to Unsaturated Systems." Elsevier, Amsterdam, 1966; (b) D. R. Taylor, *Chem. Rev.* **67**, 317 (1967); (c) M. C. Caserio, *in* "Selective Organic Transformations" (B. S. Thyagarajan, ed.), Vol. 1, p. 239. Wiley (Interscience), New York, 1970; (d) G. Modena and U. Tonellato, *Adv. Phys. Org. Chem.* **9**, 185 (1971); (e) P. J. Stang, *Prog. Phys. Org. Chem.* **10**, 205 (1973).
2. G. Gustavson and N. Demjanoff, *J. Prakt. Chem.* **38**, 201 (1888).
3. K. Griesbaum, *J. Am. Chem. Soc.* **86**, 2301 (1964).
4. K. Griesbaum, *Angew. Chem.* **76**, 782 (1964); *Angew. Chem., Int. Ed. Engl.* **3**, 697 (1964); K. Griesbaum, W. Naegele, and G. G. Wanless, *J. Am. Chem. Soc.* **87**, 3151 (1965); K. Griesbaum, *Angew. Chem.* **81**, 966 (1969); *Angew. Chem., Int. Ed. Engl.* **8**, 933 (1969).
5. P. R. Austin, U.S. Patent 2,585,529; *Chem. Abstr.* **46**, 3799 (1952).
6. B. S. Charleston, C. K. Dalton, S. S. Washburne, and D. R. Dalton, *Tetrahedron Lett.* p. 5147 (1969).
7. F. Amar, D. R. Dalton, G. Eisman, and M. J. Haugh, *Tetrahedron Lett.* p. 3037 (1974).
8. W. R. Dolbier, Jr. and B. H. Al-Sader, *Tetrahedron Lett.* p. 2159 (1975).
9. (a) T. L. Jacobs and R. N. Johnson, *J. Am. Chem. Soc.* **82**, 6397 (1960); (b) W. H. Mueller and P. E. Butler, *J. Org. Chem.* **33**, 1533 (1968); (c) L. Rasteikiene, D. Greiciute, M. G. Lin'kova, and I. L. Knunyants, *Russ. Chem. Rev.* **46**, 548 (1977).
10. H. G. Peer, *Recl. Trav. Chim. Pays-Bas* **81**, 113 (1962).
11. M. L. Poutsma, *J. Org. Chem.* **33**, 4080 (1968).
12. J. M. Bollinger, J. M. Brinich, and G. Olah, *J. Am. Chem. Soc.* **92**, 4025 (1970).
13. K. Izawa, T. Okuyama, and T. Fueno, *J. Am. Chem. Soc.* **95**, 4090 (1973); T. L. Jacobs and R. C. Kammerer, *J. Am. Chem. Soc.* **96**, 6213 (1974).
14. R. E. Banks, R. N. Haszeldine, and D. R. Taylor, *Proc. Chem. Soc. London* p. 121 (1964).
15. T. Okuyama, K. Izawa, and T. Fueno, *J. Am. Chem. Soc.* **95**, 6749 (1973); K. Izawa, T. Okuyama, T. Sakagami, and T. Fueno, *J. Am. Chem. Soc.* **95**, 6752 (1973).
16. C. U. Pittman, Jr., *Chem. Commun.* p. 122 (1969).
17. G. F. Hennion and J. J. Sheehan, *J. Am. Chem. Soc.* **71**, 1964 (1949).
18. A. V. Fedorova and A. A. Petrov, *J. Gen. Chem. USSR* **32**, 1740 (1962).
19. A. V. Fedorova, *J. Gen. Chem. USSR* **33**, 3508 (1963).
20. J. P. Bianchini and A. Guillemonat, *Bull. Soc. Chim. Fr.* p. 2120 (1968).
21. D. Mirejovsky, J. F. Arens, and W. Drenth, *Recl. Trav. Chim. Pays-Bas* **95**, 270 (1976).
22. J. Grimaldi, A. Cozzone, and M. Bertrand, *Bull. Soc. Chim. Fr.* p. 2723 (1967).
23. P. Kurtz, H. Gold, and H. Disselnkötter, *Justus Liebigs Ann. Chem.* **624**, 1 (1959).
24. J. Tendil, M. Verney, and R. Vessière, *Tetrahedron* **30**, 579 (1974).
25. J. H. van Boom, P. P. Montijn, L. Brandsma, and J. F. Arens, *Recl. Trav. Chim. Pays-Bas* **84**, 31 (1965).
26. M. L. Poutsma and P. A. Ibarbia, *J. Am. Chem. Soc.* **93**, 440 (1971).
27. D. J. Pasto, M. F. Miles, and S.-K. Chou, *J. Org. Chem.* **42**, 3098 (1977).
28. J. H. Lukas, A. P. Kouwenhoven, and F. Baardman, *Angew. Chem.* **87**, 740 (1975).
29. G. H. Schmid and D. G. Garratt, *in* "The Chemistry of the Double-Bonded Functional Groups" (S. Patai, ed.), Part 2, p. 725. Wiley (Interscience), New York, 1977.
30. W. L. Waters, W. S. Linn, and M. Caserio, *J. Am. Chem. Soc.* **90**, 6741 (1968).
31. A. J. G. van Rossum and R. J. F. Nivard, *J. Chem. Soc., Perkin Trans. II* p. 1322 (1976).
32. T. Okuyama, K. Ohashi, K. Izawa, and T. Fueno, *J. Org. Chem.* **39**, 2255 (1974).
33. D. K. Wedegaertner and M. J. Millam, *J. Org. Chem.* **33**, 3943 (1968).
34. C. B. Reese and A. Shaw, *Tetrahedron Lett.* p. 4641 (1971).
35. R. Gelin, S. Gelin, and M. Albrand, *Bull. Soc. Chim. Fr.* p. 720 (1972).

36. L.-I. Olsson, A. Claesson, and C. Bogentoft, *Acta Chem. Scand.* **27**, 1629 (1973).
37. M. Hanack and J. Häffner, *Tetrahedron Lett.* p. 2191 (1964).
38. M. Bertrand and M. Santelli, *C. R. Acad. Sci.* **259**, 2251 (1964).
39. M. Hanack and H. J. Schneider, *Fortschr. Chem. Forsch.* **8**, 554 (1967).
40. M. Hanack and J. Häffner, *Chem. Ber.* **99**, 1077 (1966).
41. M. Bertrand and M. Santelli, *Chem. Commun.* p. 718 (1968).
42. M. Santelli and M. Bertrand, *Tetrahedron* **30**, 235 (1974).
43. T. L. Jacobs and R. S. Macomber, *J. Am. Chem. Soc.* **91**, 4824 (1969).
44. M. Santelli and M. Bertrand, *Tetrahedron Lett.* p. 3699 (1969).
45. R. S. Bly, A. R. Ballantine, and S. U. Koock, *J. Am. Chem. Soc.* **89**, 6993 (1967).
46. R. S. Bly and S. U. Koock, *J. Am. Chem. Soc.* **91**, 3292 (1969).
47. R. S. Bly and S. U. Koock, *J. Am. Chem. Soc.* **91**, 3299 (1969).
48. A. Ghenciulescu and M. Hanack, *Tetrahedron Lett.* p. 2827 (1970).
49. K. L. Servis and R. D. Roberts, *J. Am. Chem. Soc.* **87**, 1331 (1965).
50. E. N. McElrath, R. M. Fritz, C. Brown, C. Y. LeGall, and R. B. Duke, *J. Org. Chem.* **25**, 2195 (1960).
51. J. E. Nordlander, S. P. Jindal, P. von R. Schleyer, R. C. Fort, Jr., J. J. Harper, and R. D. Nicholas, *J. Am. Chem. Soc.* **88**, 4475 (1966).
52. R. Garry and R. Vessière, *Bull. Soc. Chim. Fr.* p. 1542 (1968).
53. R. Macomber, *J. Am. Chem. Soc.* **92**, 7101 (1970).
54. T. von Lehman and R. S. Macomber, *J. Am. Chem. Soc.* **97**, 1531 (1975).
55. M. Santelli and M. Bertrand, *Tetrahedron Lett.* p. 2511 (1969).
56. M. Santelli and M. Bertrand, *Tetrahedron* **30**, 227 (1974).
57. M. Santelli and M. Bertrand, *Tetrahedron Lett.* p. 2515 (1969).
58. M. Santelli and M. Bertrand, *Tetrahedron* **30**, 243 (1974).
59. M. Santelli and M. Bertrand, *Tetrahedron* **30**, 257 (1974).
60. M. Santelli and M. Bertrand, *Tetrahedron* **30**, 251 (1974).
61. S. A. Sherrod and R. G. Bergman, *J. Am. Chem. Soc.* **93**, 1925 (1971).
62. D. R. Kelsey and R. G. Bergman, *J. Am. Chem. Soc.* **93**, 1941 (1971).
63. G. Markarian, B. Ragonnet, M. Santelli, and M. Bertrand, *Bull. Soc. Chim. Fr.* p. 1407 (1975).
64. J. P. Dulcère, M. Santelli, and M. Bertrand, *C. R. Acad. Sci., Ser. C* **274**, 1087, 1304 (1972).
65. M. Bertrand and C. S. Rouvier, *Bull. Soc. Chim. Fr.* p. 1800 (1973).
66. B. Ragonnet, M. Santelli, and M. Bertrand, *Helv. Chim. Acta* **57**, 557 (1974).
67. M. H. Sekera, B. A. Weissman, and R. G. Bergman, *Chem. Commun.* p. 679 (1973).
68. K. E. Harding, J. L. Cooper, and P. M. Puckett, *J. Am. Chem. Soc.* **100**, 993 (1978).

# 5

# BOND HETEROLYSIS

## I. GENERAL CONSIDERATIONS

The generation of vinyl cations by bond heterolysis of suitable vinyl derivatives according to Eq. (1) is one of the most exciting reactions in vinyl cation chemistry. Until the mid-1960's, vinyl cations in general were

$$\begin{array}{c} \diagdown \\ C = C \diagup X \\ \diagup \diagdown \end{array} \longrightarrow \begin{array}{c} \diagdown \\ C = C - \\ \diagup \end{array} \overset{\oplus}{+} X^{\ominus} \longrightarrow \longrightarrow \text{products} \qquad (1)$$

not considered to be very attractive intermediates because they were thought to be highly unstable and therefore difficult to generate by a simple bond heterolysis such as a solvolysis reaction. The breakthrough in the chemistry of vinyl cations came when it became clear that vinyl cations are not elusive intermediates but, as in the case of saturated carbenium ions, can be easily generated in simple solvolysis reactions (see Section III,A).

The reason that the solvolytic pathway for the generation of vinyl cations was only recently successfully applied can be found in an old prejudice of organic chemists against these intermediates. They were reluctant to formulate vinyl cations not only in solvolysis reactions, but also in additions to triple bonds or allenes (see Chapters 3 and 4), for two main reasons: first, all vinyl cations were considered to have high energy relative to trisubstituted carbenium ions. However, as discussed in Chapter 2, vinyl cations are not especially unstable. Second, the reactivities of simple vinyl halides were generally known to be low. This low reactivity and therefore inability of vinyl halides to undergo $S_N1$-type solvolysis with formation of vinyl cations, even in the presence of silver salts, was and still is a standard legend in some organic chemistry textbooks. The assumed unreactivity of vinyl halides was ascribed not only to the low stability of the intermediate vinyl cation (relative to saturated carbenium ions), but also to the ground state stabilization of the vinyl halides themselves, due to a strong carbon–

halogen bond. The strength of the carbon–halogen bond was attributed to the increased $\sigma$ character [1] of the carbon–halogen bond as a result of the change in hybridization of carbon from $sp^3$ in alkyl halides to $sp^2$ in vinyl halides, as well as to the partial double-bond character of this bond according to **1** [2]:

$$\underset{\textbf{1a}}{\overset{R}{\underset{X}{\diagup}}C{=}C\diagup} \longleftrightarrow \underset{\textbf{1b}}{\overset{\ominus}{C}{-}\underset{\oplus}{C}\overset{R}{\underset{X}{\diagup}}}$$

In fact, carbon–halogen bonds are shorter in vinyl halides than in alkyl halides. The C—X bond in vinyl bromide is shortened to 1.86 Å from 1.91 Å in $CH_3Br$, and in vinyl chloride it is 1.69 Å compared to 1.76 Å in $CH_3Cl$ [3]. Other reasons for the large differences in solvolytic reactivity between alkyl halides and vinyl halides are differences in solvation and in the extent of backside nucleophilic assistance by the solvent during the solvolytic process.

A bond heterolysis with formation of a vinyl cation can be performed by various methods. Before discussing the work carried out in the last few years on solvolysis reactions of vinyl compounds, we will first turn our attention to some older attempts to generate vinyl cations via diazonium ions. Although this method never attained real significance in vinyl cation chemistry, loss of nitrogen from vinyldiazonium ion precursors is a special case in which vinyl cations are generated.

## II. VINYL CATIONS GENERATED VIA DIAZONIUM IONS

Vinyl cations have been suggested as intermediates in the deamination of 2,2-diphenylvinylamine (**2**) and 9-(aminomethylene)fluorene (**9**) with nitrosyl chloride or with isoamyl nitrite. The reaction of 2,2-diphenylvinylamine (**2**) with nitrosyl chloride in dichloromethane leads to the rearranged compounds **3**, **4**, and **5** in addition to other products [4].

$$\underset{\textbf{2}}{\overset{H_5C_6}{\underset{H_5C_6}{\diagup}}C{=}C\overset{NH_2}{\underset{H}{\diagdown}}} \xrightarrow[CH_2Cl_2]{NOCl} \underset{\textbf{3}}{C_6H_5{-}C{\equiv}C{-}C_6H_5} + \underset{\textbf{4}}{\overset{H_5C_6}{\underset{Cl}{\diagup}}C{=}C\overset{C_6H_5}{\underset{H}{\diagdown}}} + \underset{\textbf{5}}{\overset{H_5C_6}{\underset{Cl}{\diagup}}C{=}C\overset{H}{\underset{C_6H_5}{\diagdown}}}$$

(13%)        (6%)        (5%)

The formation of the three rearranged products **3**, **4**, and **5** was explained in terms of the rearrangement of an intermediate vinyl cation. Loss of nitrogen from the intermediate diazonium ion **6** leads to the vinyl cation **7**, which rearranges to the more stable cation **8**, from which **3** is formed by elimination, and **4** and **5** by capture by $Cl^-$. The formation of **4** and **5** in almost equal amounts could be interpreted by the formation of a linear vinyl cation (**8**) as an intermediate (see, however, Chapter 6).

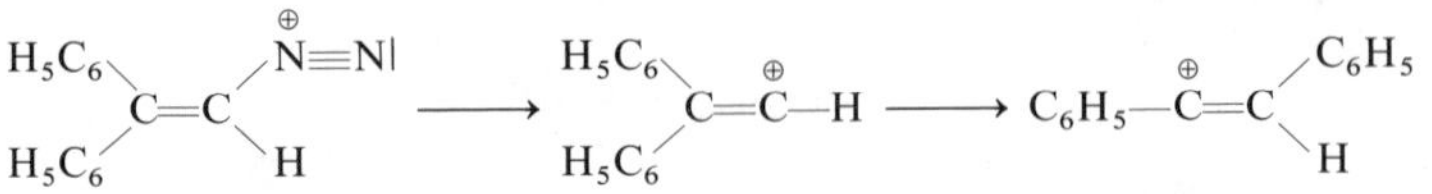

      **6**                     **7**                   **8**

An alternate mechanism for this reaction, however, has not been ruled out. The reaction of **9** with isoamyl nitrite in cyclohexene gave adduct **11** in yields of 55 to 80%, suggesting a carbene mechanism involving **10** for this deamination, rather than a vinyl cation. Therefore, the formation of vinyl cations by deamination of vinylamines under neutral conditions is not firmly established, and the alternative mechanism involving unsaturated carbenes must be considered [5].

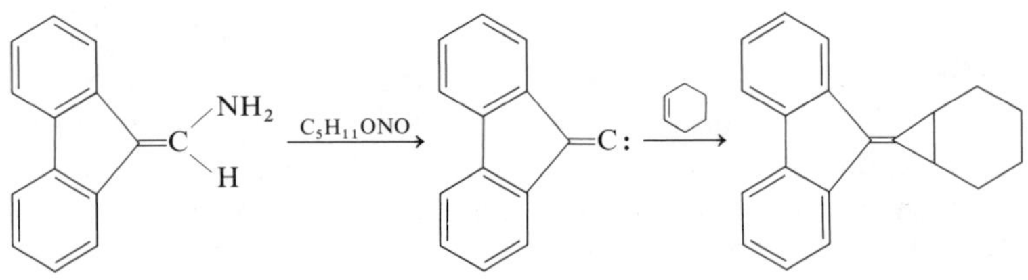

      **9**                     **10**                   **11**

Another example in which the generation of vinyl cations from diazonium ion precursors has been proposed is the alkaline decomposition of 3-nitroso-2-oxazolidones **12** [6]. Depending on the reaction conditions, however, other intermediates may also be involved. In aqueous suspension or alcoholic solution after treatment with potassium hydroxide, **12** rapidly decomposes with formation of acetylenes, ketones, aldehydes, and vinyl ethers. The relative ratio of the products is dependent on the substituents $R_1$, $R_2$, and

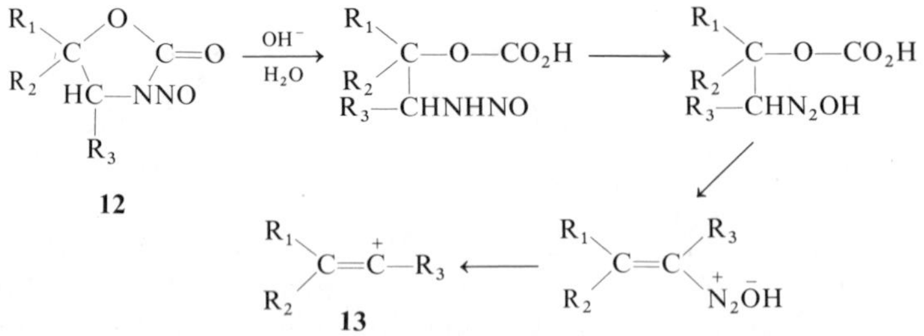

      **12**                              **13**

Scheme 5.1

$R_3$ in **12**. The observed products were explained as shown in Scheme 5.1, in which a vinyl cation intermediate (**13**) formed from a diazonium ion precursor [6, 7] is trapped by halide ions, yielding vinyl halides as products [7]. However, the formation of vinyl cations from 3-nitroso-2-oxazolidones **12** is suggested to occur only in protic solvents whereas, in aprotic media, unsaturated carbenes [5] are the suggested intermediates [7, 8].

Jones and Miller [9] found a potentially useful way to generate vinyl-diazonium ions by treating triazenes of structure **14** with acids. The diazonium ion **15** may then lose nitrogen, with formation of the corresponding vinyl cation **16**, as shown in Scheme 5.2. In all cases studied, the products were those that would be expected to arise from a vinyl cation intermediate. For example, treatment of **14a** with acetic acid gave triphenylvinyl acetate **18** (X = OAc) in 100% yield. The reactions in general were accompanied by the rearrangement **16** → **17**.

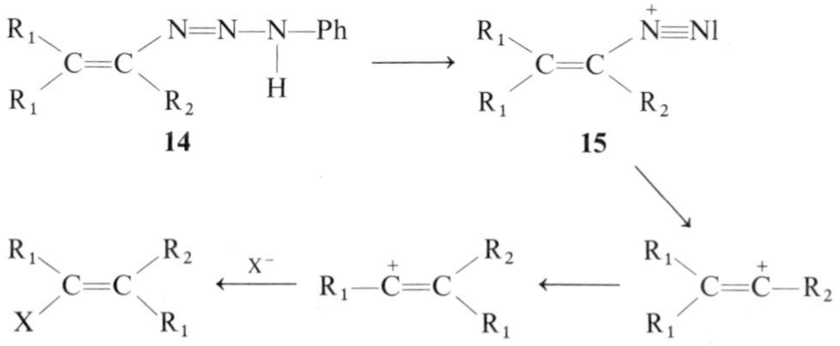

a $R_1 = R_2 = Ph$          c $R_1 = Ph; R_2 = CH_3$
b $R_1 = p\text{-tolyl}; R_2 = Ph$     d $R_1 = Ph; R_2 = H$

Scheme 5.2

The intermediacy of a free linear vinyl cation in the rearrangements of **14b** and **14d** is indicated by two observations. The 20% rearrangement of **14b** is completely suppressed by the addition of 10 molar equivalents of NaOAc, although KOAc does not affect the complete rearrangement in the reaction of **14d**. The *cis*-acetate predominates in the mixture of the rearranged isomers Z-**18a** and E-**18b** from the reaction of **14d**, as expected (see Chapter 6, Section IV). Other products from the reaction of **14d** via the less stable ion are diphenylacetylene and deoxybenzoin [9].

E-**18c**          Z-**18d**

Vinyl cation intermediates were also proposed in the photolysis of pyrazoline derivatives **19**, which decomposed with formation of the allene

**21** and the 1,3-diene derivative **22** [10]. The vinyl cation **20** was stabilized by the allylic double bond:

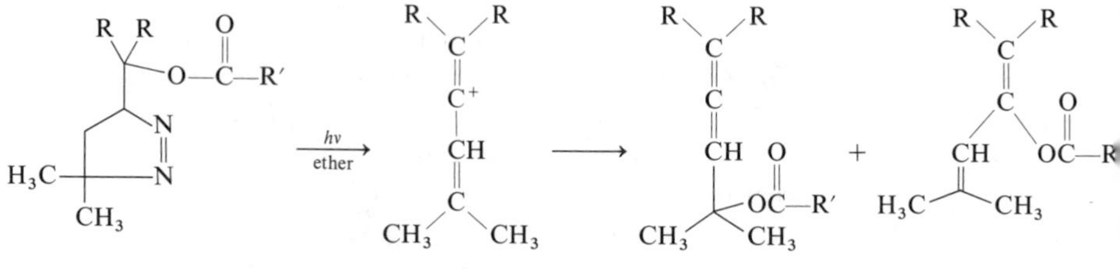

19                20                21                22

Except for the reactions of compounds **14**, the mechanisms of reactions described in this section leading to vinyl cations via diazonium ions were not studied in detail. Although the observed products are consistent with vinyl cation intermediates, alternate or competing mechanisms not involving vinyl cations cannot be ruled out as long as further studies of the mechanisms involved are lacking.

A reaction that indirectly generates a vinyl cation from a diazonium ion is the diazotation of amine **23**, as shown in Scheme 5.3. Treatment of **23** with sodium nitrite in aqueous acetic acid gave cyclopropyl methyl ketone **26**, alcohols **27** and **28**, as well as the acetate **29**. The product composition is somewhat comparable with the product composition in homoallenyl rearrangements (Chapter 4) and in the solvolysis of cyclopropylvinyl derivatives (Section IV,B). This points to the intermediacy of the stabilized cyclopropylvinyl cation **25** formed by a rearrangement reaction via the diazonium ion **24** [11].

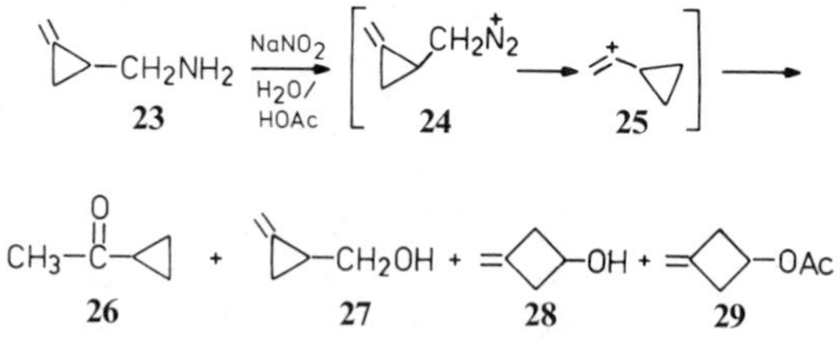

Scheme 5.3

Little is known about another possible indirect route to vinyl cations from diazonium ions. Whereas solvolysis reactions with participation of allenyl and triple bonds leading to cyclic products via vinyl cation intermediates have been extensively studied (Chapters 3 and 4), the participation of these bonds in deamination reactions of suitable amines has been observed only in very few cases. Cyclization occurred to some extent when the homoallenylamine **30** was treated with butyl nitrite in acetic acid; however, the cyclization products (methylenecyclobutanols) were not formed via vinyl cations [12].

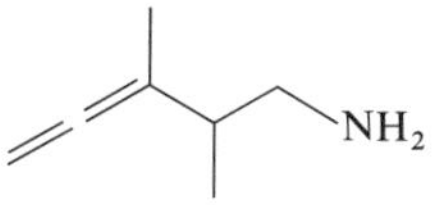

**30**

The participation of triple bonds in amine deamination has been studied in somewhat more detail [13]. As described in Chapter 3, solvolysis of acetylene sulfonates often leads to cyclic products due to participation of the triple bond in the carbenium ion-forming step. The participation of a triple bond can result in the formation of cyclic vinyl cation intermediates of different structure, from which the products are formed by solvent capture. An example is the solvolysis of homopropargyl sulfonates **31a**. Whereas the sulfonates **31a** solvolyze in various solvents with formation of cyclobutanones and cyclopropyl ketones via cyclic vinyl cations, the homopropargylamines **31b** and **31c** on deamination with nitrous acid in acetic acid formed only noncyclic alcohols or acetates [14]. Another example is the solvolysis of 6-octyn-2-yl sulfonates **32a**, in which six- and five-membered ring compounds were formed in high yields. On the other hand, the reaction of 6-octyn-2-ylamine **32b** with nitrous acid gave only unrearranged alcohol **32c** [13]. Also, the thermal decomposition of the nitrosoamide **32d** in acetic acid did not lead to cyclized compounds; the main product was the acetate **32e** [13].

$$R—C{\equiv}C—CH_2—CH_2—X$$

**31a**   R = alkyl, aryl; X = OTs, OTf
**31b**   R = CH$_3$; X = NH$_2$
**31c**   R = phenyl; X = NH$_2$

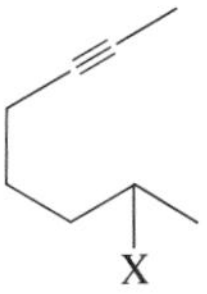

**32a**   X = OTs, OTf
**32b**   X = NH$_2$
**32c**   X = OH
**32d**   X = N(NO)COCH$_3$
**32e**   X = OAc

The only well-documented example of a cyclization of an acetylenic amine after treatment with nitrous acid is the cyclodecynylamine **33**. Whereas reactive 5-cyclodecynyl derivatives (tosylate, p-nitrobenzoate) solvolyze with

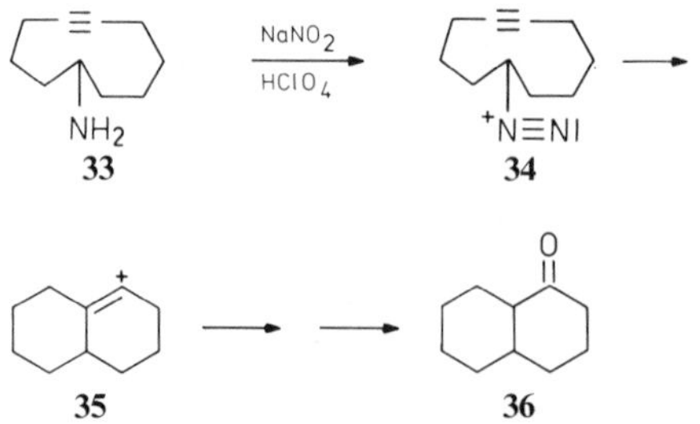

Scheme 5.4

quantitative rearrangement and formation of 1-decalone (**36**) and small amounts of bicyclo[5.3.0]decan-2-one, **33** reacted with sodium nitrite in perchloric acid with formation of only 8% of **36** [13]. This was explained by the intermediate formation of the bicyclic vinyl cation **35** from the diazonium ion **34** by transannular participation of the triple bond, as shown in Scheme 5.4. Ion **35** then reacts further with the solvent to give the observed decalone (**36**) [14].

## III. DIRECT SOLVOLYTIC GENERATION

### Mechanisms of Direct Solvolytic Generation

Contrary to earlier views, the generation of vinyl cations is easily accomplished by simple solvolysis reactions if at least two requirements are met. First, in place of leaving groups of low nucleofugacity such as the halides, leaving groups that are more prone to $S_N1$ reaction should be used. Particularly favorable for solvolysis reactions are the so-called super leaving groups: trifluoromethanesulfonate ($CF_3SO_2O-$) (triflate) [15] or the still faster reacting nonafluorobutanesulfonate ($C_4F_9SO_2O-$) (nonaflate) [16]. The second requirement is that the intermediate vinyl cations be stabilized by $\alpha$ substituents with electron-donating ability or $\beta$ substituents that are capable of positive charge dispersal. When leaving groups less inclined to undergo $S_N1$ reactions, e.g., the already mentioned halides or even tosylates, are used in solvolysis reactions, the vinyl compounds will undergo a heterolysis of the carbon–halogen or carbon–oxygen bond only if the resulting vinyl cations are stabilized by such groups.

The requirements for the occurrence of vinyl cation mechanisms will be discussed in detail in forthcoming sections in connection with the many attempts to generate stabilized and simple alkylvinyl cations by solvolysis reactions. Therefore, here we will only briefly delineate some of the major requirements for a vinyl cation mechanism in solvolysis reactions. Besides the formation of vinyl cations by bond heterolysis, a number of alternative mechanisms may lead to the same substitution or elimination products and must therefore be considered [17, 18].

## 1. The $S_N1$ Vinyl Cation Mechanism

The products formed from the vinyl cation **38**, which is formed by the heterolytic cleavage of the C—X bond in the vinyl compound **37**, are shown in Scheme 5.5.

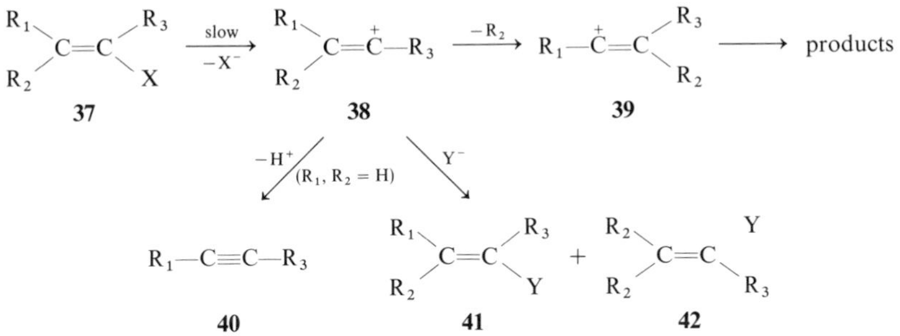

Scheme 5.5

Many effects must be taken into account in the evaluation of whether a vinyl cation mechanism is operative. These include kinetics, solvent effects, the effect of the leaving group, the special influence of the $\alpha$ and $\beta$ substituents, and the products formed.

A vinyl cation mechanism requires first-order kinetics, except when the ions are highly stable (see Chapter 6, Section V), and the reaction rate should be independent of both the solvent pH and the concentration of an added base. On the other hand, the solvolysis rate should be dependent on the ionizing power of the solvent; the Winstein–Grunwald $m$ value [19] should be between 0.5 and 1.0, except in special cases. If the solvolysis is carried out in deuterated solvents, the solvent isotope effect should be close to unity, and the products in deuterated solvents (e.g., $CH_3COOD$ or $CF_3CH_2OD$) must not show deuterium incorporation. Besides these solvent effects, there may be other effects characteristic of carbenium ion intermediates, such as salt effects (especially the common ion rate effect), as well as ion-pair formation and internal return. Capture of the intermediate vinyl cation by externally added nucleophiles is possible.

Furthermore, as in the case of saturated carbenium ions, the intermediacy of vinyl cations can be deduced from the formation of rearranged solvolysis products, as discussed in Chapter 7.

## 2. E2 Elimination from Vinylic Substrates

For vinyl compounds such as **43** with a trans hydrogen at the double bond with respect to the leaving group X, an additional mechanism must be considered in which, under neutral or especially basic conditions, a synchronous $\beta$ elimination and acetylene (**44**) formation can occur. For vinyl substrates **43**, in which $R_2 = CH_2R$, the concerted elimination can also lead to the allene **45**:

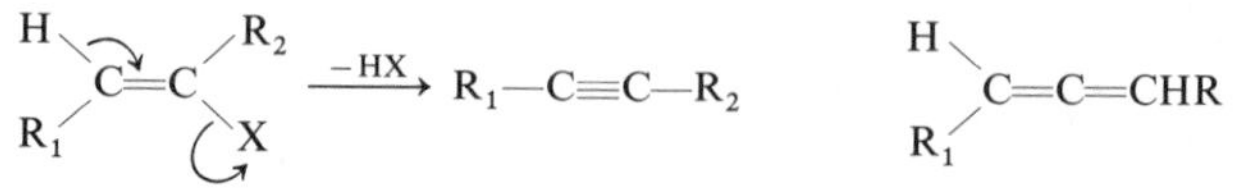

considered in which, under neutral or especially basic conditions, a synchronous $\beta$ elimination and acetylene (**44**) formation can occur.

Such a synchronous elimination mechanism is indicated by a considerable increase in the reaction rate when base is added and by a large primary deuterium isotope effect. The E2 products are identical with E1 products, which can be obtained from the vinyl cation **38** substituted by a $\beta$-hydrogen.

### 3. Addition–Elimination Pathways

Additional mechanisms that can explain the products in the reaction of vinyl derivatives are the addition–elimination mechanisms according to paths A and B in Scheme 5.6. In path A [electrophilic addition–elimination $(Ad_E$–E)] the products are formed by an electrophilic attack of a proton (from the solvent) on the double bond of the vinyl compound **46**, forming first an intermediate trisubstituted carbenium ion (**47**). This ion in turn reacts with the solvent nucleophile $Y^\ominus$, leading to the addition product **48**. Subsequent elimination of the leaving group as HX gives the stereoisomers **50** and **51**.

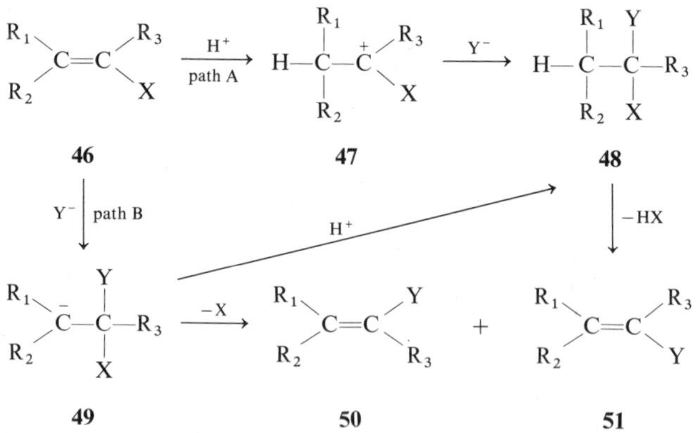

Scheme 5.6

In the case of $Ad_E$–E route (path A), the reaction rate will be strongly dependent on the $H^+$ concentration of the solvent medium. If the solvolysis is carried out in deuterated solvents such as $CH_3COOD$ or $CF_3CH_2OD$, contrary to the vinyl cation mechanism, deuterium incorporation into the products must be observed if $R_1$ or $R_2$ is hydrogen. Depending on the substituents in **46**, path A should also exhibit a solvent isotope effect. Special effects of different leaving groups X in **46** are also indicative of path A; e.g., a better electron-attracting leaving group, leading to a rate enhancement in the case of a vinyl cation mechanism, here results in rate retardation. Such leaving group effects will be discussed elsewhere. Acetylene formation is not possible via path A (see Chapter 6).

Path B in Scheme 5.6 shows an alternative addition–elimination mechanism, the nucleophilic addition–elimination ($Ad_N$–E) route. Here the first step consists of an attack of a nucleophile on the double bond, leading to **49**, which, after losing $X^-$, gives mainly or exclusively **51** and sometimes also its isomer **50**. Nucleophilic addition–elimination may dominate if especially strong nucleophiles, such as thiophenolate, are used [20]. The $Ad_N$–E route is second order, and its rate is strongly dependent on the added nucleophile. The reaction is accelerated by electron-attracting groups in the $\beta$ position of **46**, and this route leads to retention of configuration [21–23].

### 4. $S_N2$ Displacement on Vinylic Substrates

Yet another alternative to the two-step vinyl cation process is the one-step, direct $S_N2$ displacement by nucleophiles at the vinylic carbon atom:

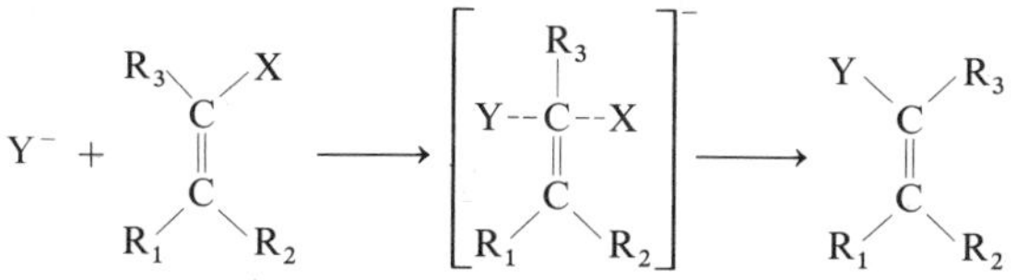

The $S_N2$ route, which predicts inversion of configuration of the substitution product **54**, has not yet been observed [21]. This route is energetically unlikely on steric grounds [24], since the $\alpha$ and $\beta$ substituents $R_1$ and $R_3$ prevent a backside approach of the nucleophile to the $\alpha$-carbon with the formation of transition state **53**. Extended Hückel calculations confirm the relative difficulty of $S_N2$ processes in vinyl systems [24].

### 5. Nucleophilic Attack on Sulfur

If the leaving group in a vinyl compound is a sulfonate, the products under certain conditions can be formed by nucleophilic attack of the solvent or its conjugate base on sulfur, as exemplified by $HO^-$ in Scheme 5.7. Such nucleophilic attack on sulfur is a well-known phenomenon [25]. Vinyl sulfonates **55** were reported to react with S–O cleavage only in a few cases in which special structural requirements were met. This was the case in the

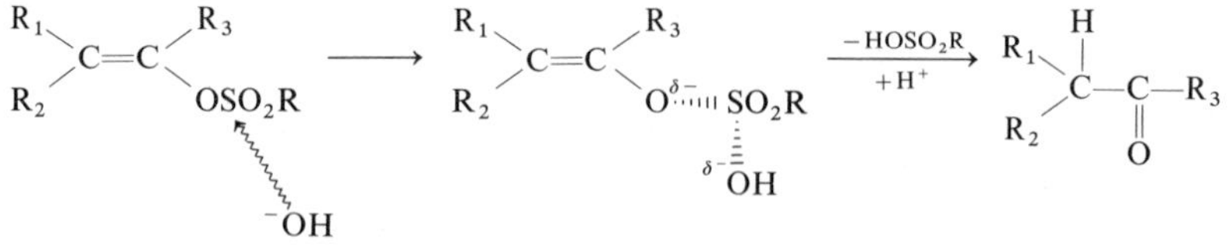

**55**                                    **56**                                    **57**

Scheme 5.7

solvolysis of certain cyclic vinyl tosylates [26] and cyclic vinyl triflates [e.g., cyclopentenyl triflate and cyclohexadienyl triflate in strongly nucleophilic solvents (see Section V,B)]. When the solvolysis of the vinyl sulfonate **55** is carried out in an [18]O-labeled solvent system such as $C_2H_5OH/H_2{}^{18}O$, no incorporation of [18]O in the product **57** points to a sulfur–oxygen cleavage. Furthermore, this mechanism, like the $\beta$-hydrogen elimination, the $Ad_N$–E route, and the direct $S_N2$ displacement, is bimolecular and hence there is a first-order dependence on added base or nucleophile, if the attack on sulfur is not by the neutral solvent.

## IV. STABILIZED VINYL CATIONS

As mentioned above, one of the conditions for easily generating vinyl cations by solvolysis reactions is their stabilization by substituents with electron-donating ability. Because vinyl cations are disubstituted carbenium ions, the stabilizing substituents can be linked to the positive carbon atom in two ways. A singly bonded substituent produces a secondary vinyl cation such as **58** whereas, when the $\beta$-carbon atom of the vinylic structure is part of the substituent, we obtain structures such as the allenyl cation **60** and cyclopropylidenemethyl cation **63**. For stabilizing vinyl cations, all electron-donating substituents that are effective in stabilizing trigonal carbenium ions are suitable. Aryl and vinyl groups have been used to stabilize a vinyl cation in the classic way. The extensive investigations of arylvinyl cations are discussed separately in Chapter 6. Stabilization by a vinyl group results in a dienyl ion (**59**). Vinyl cation **61**, in which the positive charge is interacting with a triple bond, has only been generated very recently by a solvolysis reaction [26a]. The few experimental results obtained so far agree very well with the predictions of calculations presented in Chapter 2.

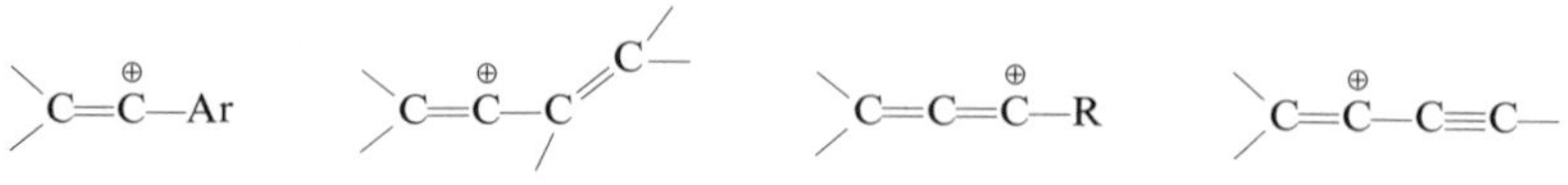

**58**                  **59**                  **60**                  **61**

Vinyl cations can be stabilized especially well in a nonclassical way by the overlap of the vacant p orbital of the ion with the bent bonds of a cyclopropyl group. Again, there are two possibilities for utilizing the powerful stabilizing effect of the cyclopropane ring on a positive charge. A cyclopropane ring bonded singly leads to the secondary vinyl cation **62** whereas, if the $\beta$-carbon atom of the vinyl structure is part of the cyclopropane ring, we obtain a cyclopropylidenemethyl cation (**63**). The special stability of **63** arises from its unique geometry, which allows a particularly suitable overlap of the participating cyclopropyl orbitals with the empty p orbital.

$$\text{>C=}\overset{\oplus}{\text{C}}\text{—}\triangleleft \qquad\qquad \triangleright\text{=}\overset{\oplus}{\text{C}}\text{—R}$$

**62**          **63**

## A. Dienyl Cations

The stabilizing effect of a vinylic double bond has been observed in solvolysis reactions of acyclic 1,3-bromodienes [27] and certain cyclic 1,3-dienyl triflates [28] and bromides [29]. Bromodienes **64** solvolyzed in 80% aqueous ethanol with first-order rates that were independent of added triethylamine, and therefore addition–elimination mechanisms were ruled out. Further support for a vinyl cation mechanism was given by the high sensitivity of the rates to the ionizing power of the solvent, with a Winstein–Grunwald $m$ value of 0.80. The observed products **66** to **68** were derived from the mesomeric vinyl cation **65**. The unsubstituted 1,3-bromodiene **64a** solvolyzed very slowly. However, methyl substitution greatly enhanced the solvolysis rate by stabilizing the mesomeric cation **65**, with the following relative rates for compounds **64a** through **64g**, respectively [27]: 0.01, 535, 1.0, 1.5, 716, 297, 765.

$$\begin{array}{c}
\text{R}_4 \qquad\;\; \text{Br} \\
\;\;\;\text{C=C} \qquad\quad \text{R}_1 \\
\text{R}_5 \qquad\quad \text{C=C} \\
\qquad\;\; \text{R}_3 \qquad\;\; \text{R}_2
\end{array}$$

**64a** $R_1 = R_2 = R_3 = R_4 = R_5 = H$
**64b** $R_3 = R_4 = R_5 = H; R_1 = R_2 = CH_3$
**64c** $R_1 = R_4 = R_5 = H; R_2 = R_3 = CH_3$
**64d** $R_2 = R_4 = R_5 = H; R_1 = R_3 = CH_3$
**64e** $R_3 = R_4 = H; R_1 = R_2 = R_5 = CH_3$
**64f** $R_3 = R_5 = H; R_1 = R_2 = R_4 = CH_3$
**64g** $R_3 = H; R_1 = R_2 = R_4 = R_5 = CH_3$

$$\text{>C=}\overset{+}{\text{C}}\text{—C}\diagdown_{\underset{4}{\text{C—}}} \quad\longleftrightarrow\quad \text{>C=C=C}\diagdown_{\overset{+}{\text{C}}}$$

**65**

$$H-C{\equiv}C-\underset{R_2}{\overset{R_3}{C}}{=}C\underset{R_2}{\overset{R_1}{\diagup}} \qquad CH_3-\overset{O}{\overset{\|}{C}}-\underset{}{\overset{R_3}{C}}{=}C\underset{R_2}{\overset{R_1}{\diagup}} \qquad CH_2{=}\underset{}{\overset{R_3}{C}}{=}\underset{R_2}{\overset{R_1}{C}}-COEt$$

**66**                          **67**                          **68**

A maximal charge delocalization in **65** is possible only when the planes of the double bonds in the bromodienes **65** are perpendicular to each other, thereby providing overlap of the developing empty p orbital on the functional carbon atom in **64** and the orbitals of the C-3–C-4 double bond. Hence, the bromodienes **64** should react most readily in a nonplanar conformation. This is the case if either $R_1$ or $R_3$ in **64** is methyl, thus leading to the required nonplanar conformation. This is due to the steric interaction of $R_1$ and $R_3$ with the bromine or the 1,2 double bond, respectively. The nonplanarity of these bromodienes was proved by uv measurements, in which the absorption maxima shifted to shorter wavelengths by 19 to 34 nm relative to the calculated values [27].

The requirement of noncoplanarity of the double bonds for the formation of 1,3-dienyl cations has also been demonstrated by solvolysis reactions of cyclic 1,3-dienyl triflates [28]. Cycloheptadienyl triflate **69a** and cyclo-octadienyl triflate **69b** both solvolyzed in 60% aqueous ethanol with formation of the corresponding vinyl cations. This was shown by the solvolysis products and the relative solvolysis rates. Cycloheptadienyl triflate **69a**, when solvolyzed in 60% aqueous ethanol (100°, 4 days) buffered with triethylamine, gave 70% 2-cycloheptenone **70a**, 27% 2-ethoxy-1,3-cycloheptadiene **71a**, and 3% 1-acetylcyclopentene. The latter was formed by a secondary reaction from the 2-cycloheptenone [30]. The products formed in the solvolysis of cyclooctadienyl triflate **69b** under the same conditions were 77% 3-cyclo-octenone, which was formed by a double-bond shift from 2-cyclooctenone **70b** [29], 16% 2-cyclooctenone **70b**, and 7% 2-ethoxy-1,3-cyclooctadiene **71b**. Table 5.1 shows the rates of **69a** and **69b**, which were measured at different temperatures and different $H^+$ concentrations. The rates were independent of pH, thus excluding an addition–elimination mechanism [28].

**TABLE 5.1**

**Rates of Reaction of Triflates 69, 72, and 73 in 50% Ethanol**

| Triflate | Temp (°C) | $k$ (sec$^{-1}$) |
|----------|-----------|------------------|
| **69a** | 76.2 | $4.03 \times 10^{-5}$ |
| **72** | 75.2 | $4.31 \times 10^{-5}$ |
| **69b** | 50.3 | $1.26 \times 10^{-2}$ |
| **73** | 49.4 | $1.87 \times 10^{-5}$ |

The cyclic dienyl triflates **69a** and **69b** show characteristic rate differences in comparison with the cyclic vinyl triflates of the same ring size (**72** and **73**, respectively) (Table 5.1). Cycloheptadienyl triflate **69a** solvolyzes at almost the same rate as 1-cycloheptenyl triflate **72**, but cyclooctadienyl triflate **69b** reacts about $10^3$ times faster than 1-cyclooctenyl triflate **73**. The introduction of a second double bond in **69a** and **69b** should lead to an increased strain in comparison with **72** and **73** if a linear vinyl cation is formed. Thus, for **69a** and **69b** a reaction rate lower than that for **72** and **73** should be expected. The

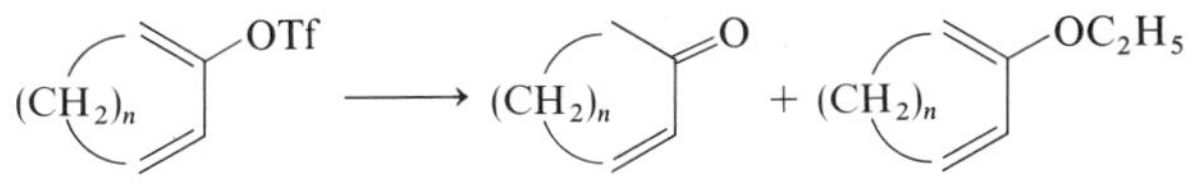

|            |           |           |           |           |           |
|------------|-----------|-----------|-----------|-----------|-----------|
| **69a**    | $(n = 3)$ | **70a**   | $(n = 3)$ | **71a**   | $(n = 3)$ |
| **69b**    | $(n = 4)$ | **70b**   | $(n = 4)$ | **71b**   | $(n = 4)$ |

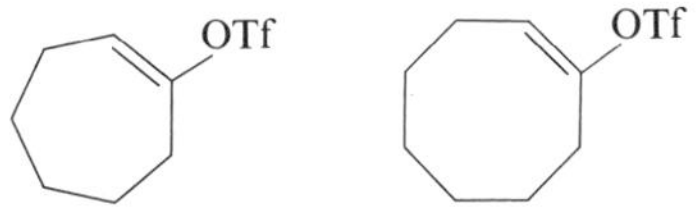

72                    73

comparatively high rates of **69a** and **69b** therefore indicate that the intermediate vinyl cation has been stabilized by the second double bond in the allylic position. The necessary condition of a stabilizing interaction of the participating orbitals in **69b** can be achieved, as shown in **75**, owing to the flexibility of the eight-membered ring. In **69a** the stabilizing effect between the interacting orbitals is less suitable, as shown in **74**. The planes of the double bonds in this energetically favorable conformation form an angle of 50° with each other, thereby allowing only a partial overlap of the participating orbitals.

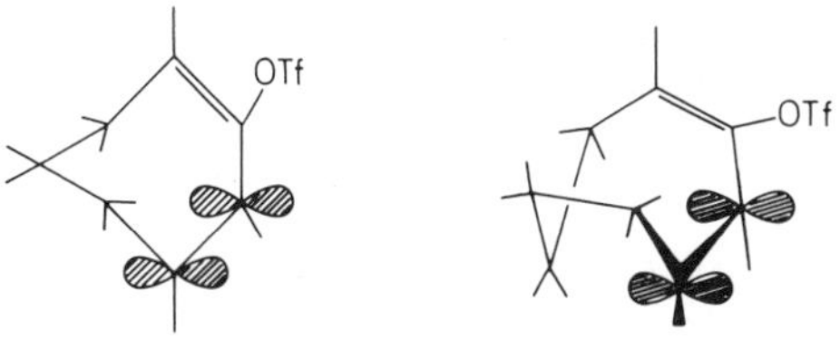

74                    75

When the geometry of the ring does not allow an overlap of the vacant p orbital of the vinyl cation and the neighboring double bond, as in case of 2-bromo-1,3-cyclohexadiene **76**, no reaction occurs. Compound **76** does not solvolyze in 80% aqueous ethanol, even in the presence of silver salts or at temperatures as high as 180° [29]. Even if a super leaving group is used, a cyclohexadienyl cation cannot be generated by a solvolysis reaction. Cyclohexadienyl triflate **77** does not react in solvents of low nucleophilicity;

for example in $CF_3CH_2OH$, even at 130°, it is recovered unchanged after 20 days [31]. In nucleophilic solvents such as 60% aqueous ethanol and in absolute methanol, 77 reacts with formation of 2-cyclohexenone. In both solvents the cyclohexenone is formed via a sulfur–oxygen cleavage (p. 215), without the intermediacy of a vinyl cation [31].

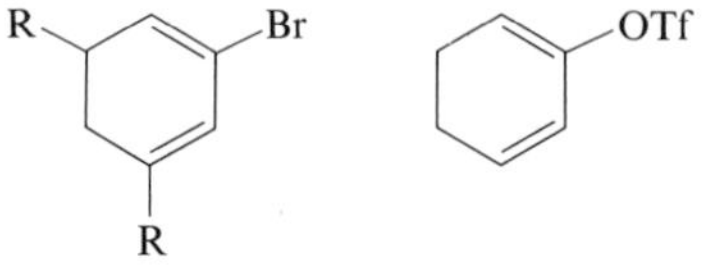

76   R = H, CH$_3$                    77

Cyclooctatetraenyl triflate 78 (COT-triflate) has also been synthesized [32]. Whether COT-triflate 78 would solvolyze with formation of the corresponding vinyl cation was difficult to predict, although the vinyl cation was expected to be stabilized by the neighboring double bond, since 78 could also react by valence isomerization via the intermediates 79, 80, and 81, as shown in Scheme 5.8. Similar isomerizations were shown to occur in

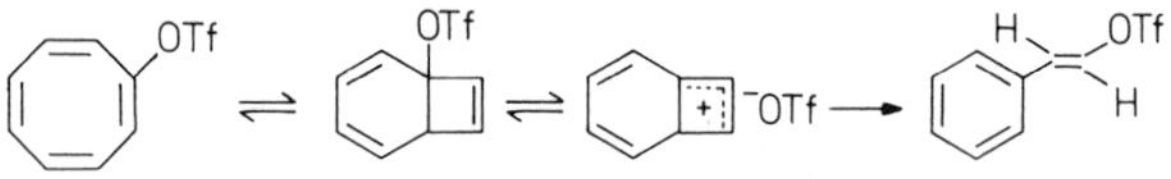

78              79              80              81

Scheme 5.8

the case of the analogous cyclooctatetraenyl bromide (COT-bromide) [33]. COT-Triflate 78 was solvolyzed in solvents of different ionizing power and nucleophilicity at room temperature, and the products are given in Table 5.2.

**TABLE 5.2**

**Solvolysis of COT-Triflate 78[a] in Various Solvents[b]**

|              | Products (%) |     |    |       |
| ------------ | ------------ | --- | -- | ----- |
| Solvent      | 82a          | 82b | 83 | 81[a] |
| Absolute TFE | 79           | —   | 5  | 14    |
| Absolute EtOH| —            | 77  | 10 | 11    |
| 80% TFE      | 35           | —   | 43 | 19    |
| 50% EtOH     | —            | 42  | 42 | 14    |

[a] COT-Triflate 78 was not obtained in a pure state. It always contained 14–19% 81 and 2–3% COT-bromide.

[b] Solvents were buffered with 1.1 mole equivalents of pyridine. The reaction was carried out at room temperature for 1 hr.

**TABLE 5.3**

**Solvolysis Rates of COT-Triflate 78 at 1°**

| Solvent | $k$ (sec$^{-1}$) | $t_{1/2}$ (sec) |
|---|---|---|
| Absolute TFE | $0.72 \pm 0.02 \times 10^{-2}$ | 96 |
| 80% EtOH | $0.3 \pm 0.1 \times 10^{-2}$ | 210 |

Due to its high reactivity, the solvolysis rates of **78** had to be measured at 1°. The results are shown in Table 5.3. The solvolysis rates of **78** were found to be 100 times faster than the rates of the cyclooctadienyl triflate **69b**. The solvolysis products **82a,b**, and **83**, which were formed exclusively with retention of the eight-membered ring, led to the conclusion that COT-triflate **78** solvolyzes with formation of the intermediate vinyl cation **84**. A reaction

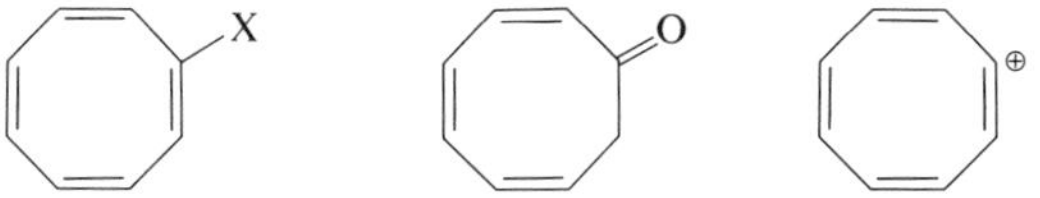

**82a**  X = OCH$_2$CF$_3$
**82b**  X = OC$_2$H$_5$             **83**            **84**

path via ion **80** is unlikely, since this would lead to products different from those actually observed. The 10$^3$ times faster solvolysis rate of COT-triflate **78** as compared to the cyclooctadienyl triflate **69b** can be explained as follows: the conjugation effect of the double bonds in **69b** is still present in the corresponding vinyl cation **85**. This conjugation obstructs the favorable overlap of the vacant $p$-orbital of the cation **85** with the neighboring double bond and must be overcome before its stabilization is possible. Dreiding

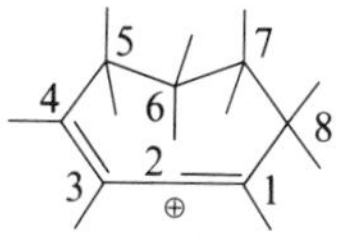

**85**

models of **85** show that one of the hydrogen atoms at C-6 in **85** lies very close to C-2. The strain of **85** is increased by twisting the ring, whereby the eclipsed conformations between the hydrogen atoms of C-5, C-6 and C-7 are selectively unfavorable. Their steric interactions will even be increased by twisting the transannular C-6 atom out of the ring. In all possible conformational maneuvers the overlap of the empty p orbital with the neighboring double bond was not optimal. In the most acceptable case an angle of about 30° was found between the π-orbitals of the neighboring double bond and the vacant p orbital, which shows only a modest stabilization of **85**.

On the other hand the hard tub form of **78** allows instant stabilization of the vinyl cation without any conformational change. Transannular inter-actions do not occur and the positions of the hydrogen atoms in **84** do not differ significantly from their positions in **78**.

To avoid the valence isomerization of Scheme 5.8, a process of low activation energy, an especially good leaving group such as triflate and high ionizing solvents had to be used in the solvolytic generation of **84**. Then the solvolysis could be carried out at a temperature (room temperature) low enough to favor an ionization process leading to **84** rather than the valence isomerization [32].

## B. Cyclopropylvinyl Cations

In 1969 two short articles appeared describing the use of the well-known stabilizing effect of a cyclopropane ring on a positive charge to stabilize a vinyl cation generated in a solvolysis reaction [34]. Both the cyclopropylvinyl chloride **86a** and the cyclopropylvinyl iodide **86b** reacted, even at room temperature, within minutes with silver salts in acetic acid. The products (Scheme 5.9) were cyclopropyl methyl ketone (**88**), cyclopropylvinyl acetate (**89**), and, depending on the reaction conditions, cyclopropylacetylene (**90**).

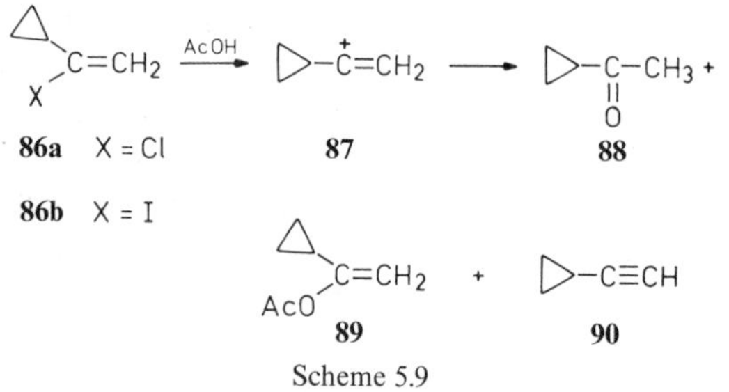

Scheme 5.9

The analogous iodide **91** did not react with silver salts in acetic acid at room temperature and underwent a silver salt-catalyzed conversion to isopropyl methyl ketone only at 150°. These data were consistent with the formation of a vinyl cation (**87**) stabilized by the neighboring cyclopropane ring.

91          92

This cyclopropylvinyl cation (**87**) is the unsaturated equivalent of the cyclopropylcarbinyl cation **92**, which is especially stable due to the well-documented nonclassical interaction of the positive charge and the cyclopropane ring [35]. Like the cyclopropylcarbinyl cation **92**, the cyclopropylvinyl cation **87** adopts a preferred conformation in which the highest stabilizing effect of the cyclopropane ring occurs. Molecular orbital calculations have shown that the linear cyclopropylvinyl cation **87** is more stable in the "bisected" conformation (**93**) than in the "perpendicular" conformation (**94**) [24], in analogy with the cyclopropylcarbinyl cation **92**, which also prefers the bisected form. In the bisected conformation (**93**), a maximal overlap of the vacant p orbital of the cation with the orbitals of the C–C–cyclopropane bonds is provided.

**93**   bisected

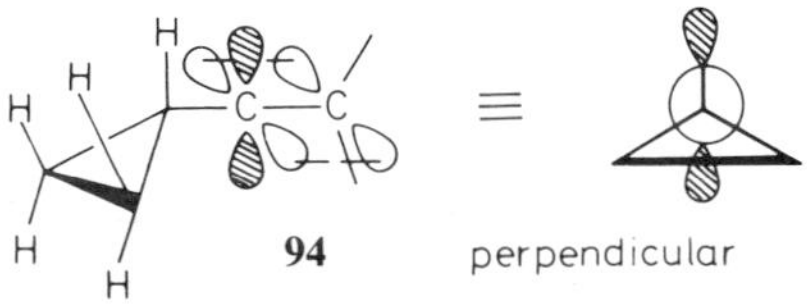

**94**   perpendicular

The intermediacy of a cyclopropylvinyl cation (**87a**) was first proposed by Hanack and Häffner [36] in 1964, when they studied the solvolysis reaction of homoallenyl derivatives **95**. As shown independently by Bertrand and Santelli [37], homoallenyl compounds **95** (X = OTs, halogens) solvolyzed in various aqueous solvents to form cyclopropyl alkyl ketones **96**, indicating the intermediate formation of the cyclopropylvinyl cation **87a**. In comparison with the homoallylic rearrangement, this reaction, which proceeds with participation of one of the allenyl double bonds, was called the "homoallenyl rearrangement" (see Chapter 4).

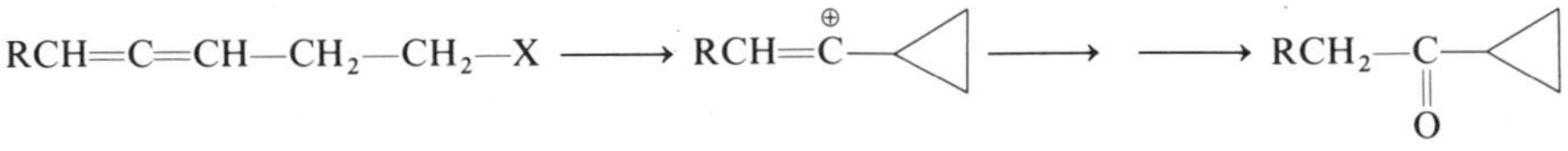

**95**                **87a**                **96**

That a cyclopropylvinyl cation (**87**) was formed in the solvolysis of the cyclopropylvinyl iodide **86b** as well as from the homoallenyl iodide **95** was shown by more detailed product studies [38]. The solvolysis of **86b** and **97a** with silver acetate in acetic acid at 25° yielded more than 95% of the above-mentioned products **88**, **89**, and **90** in similar relative amounts. In addition, small quantities of the acetates **97b**, **100**, and **101** were found, as shown in Scheme 5.10. The similarity of the solvolysis products starting

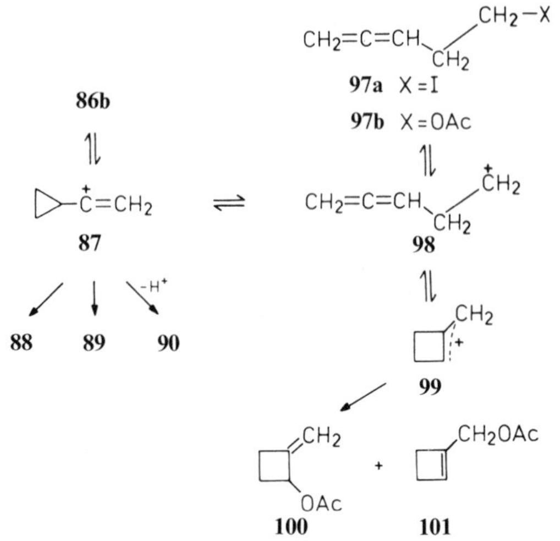

$$CH_2=C=CH\diagdown_{CH_2}^{CH_2-X}$$

**86b**

**97a**  X = I

**97b**  X = OAc

$$\Updownarrow$$

$$\triangleright-\overset{+}{C}=CH_2 \quad \rightleftharpoons \quad CH_2=C=CH\diagdown_{CH_2}^{\overset{+}{C}H_2}$$

**87**                        **98**

$$\diagup \quad \downarrow \quad \overset{-H^+}{\diagdown}$$

**88    89    90**

$$\Updownarrow$$

$$\square\overset{CH_2}{\vdots^+}$$

**99**

$$\square\overset{CH_2}{\underset{OAc}{}} \quad + \quad \square\overset{CH_2OAc}{}$$

**100**              **101**

Scheme 5.10

with either the cyclopropylvinyl or the homoallenyl compound, was explained by the similar or identical intermediate carbenium ions **87**, **98**, and **99**. The product composition, which showed that more than 95% of the compounds were formed from the cyclopropylvinyl cation **87**, indicated that **87** was especially stable and was therefore the main intermediate cation.

The solvolysis of the $Z$–$E$ isomers of 1-cyclopropyl-1-iodopropene ($Z$-**102** and $E$-**103**) was also examined with respect to the stereochemistry of the solvolysis products [39]. Both **102** and **103** reacted in acetic acid with an excess of silver acetate at room temperature via the intermediate cyclopropylvinyl cation **104** to form the products shown in Scheme 5.11. The product ratios given in Scheme 5.11 were identical within experimental error for the solvolyses of **102** and **103** and were proved to be stable under the solvolysis conditions. Of particular significance is the almost 1:1 ratio of the cyclopropylvinyl acetates **105** and **106** formed from both vinyl iodides **102** and **103** under kinetic control. The nonstereospecificity was evidence for a vinyl cation mechanism and pointed to a linear sp-hybridized vinyl cation (**104**) as the intermediate, with a small $CH_3/H$ steric control in the product-forming step.

The silver-catalyzed acetolysis of isomers **102** and **103**, as well as other 2-substituted 1-cyclopropylvinyl iodides such as $Z$-**107** and $E$-**107**, showed a higher rate of reaction for the $E$ isomers, which have the cyclopropyl and the $\beta$ substituent in a cis relationship.

The $k_E/k_Z$ ratio for **103/102** was found to be 10.3, and that for $E$-**107**/$Z$-**107** was 12.5. The higher rates of the $E$ isomers were explained by the relief of strain afforded by nonbonded interactions between the alkyl groups in the $\beta$

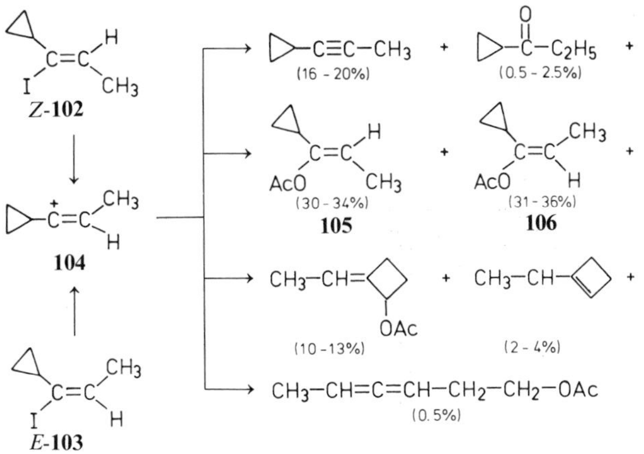

Scheme 5.11

position and the cyclopropyl group as C-1 a carbon that undergoes rehy-bridization during ionization [40]. The cyclopropyl to cyclobutyl rearrange-ment in Scheme 5.11 was also found to depend on the substituent in the $\beta$ position of the 1-cyclopropylvinyl iodides [40]. The cyclobutyl/cyclopropyl product ratio increased from 0.018 to ca. 0.2 on going from H in the $\beta$ position in **86b** (Scheme 5.10) to $CH_3$ in the $\beta$ position in **102** and **103** in Scheme 5.11. With a cyclopropyl in the $\beta$ position as in **107**, the percentage

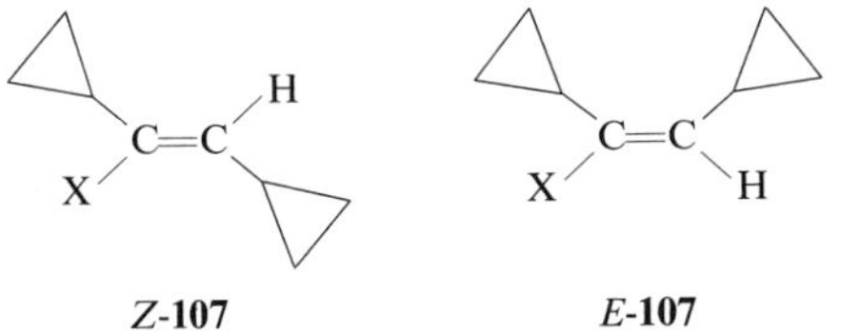

of rearranged cyclobutyl product jumped to ca. 45% of the product mixture. The influence of the $\beta$ substituents was explained by their ability to stabilize the positive charge in the rearrangement transition state (see Scheme 5.10, rearrangement **87 → 99**) [40].

In comparison with the cyclopropylvinyl compounds discussed so far, sub-stitution of one or two alkyl groups in the cyclopropane ring itself leads to more rearranged products. The acetolysis of *trans*-1-(2-methylcyclopropyl)-vinyl tosylate (**108**) led to a complex mixture of several products (Scheme 5.12) in which the *trans*-ketone **109** and the 4,5-hexadien-2-yl acetate **110** pre-dominated [41]. Careful kinetic and product studies showed that the prod-ucts and accelerated rates could be best explained by postulating a cyclopropylvinyl cation intermediate. The formation of only the *trans*-ketone **109** in the solvolysis of the vinyl tosylate **108** without any of the cis isomer

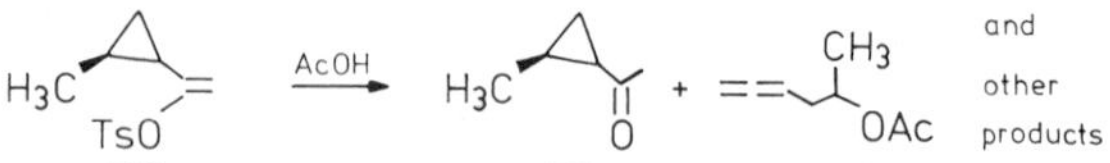

Scheme 5.12

led to the conclusion that the corresponding *trans*-cyclopropylvinyl cation **111** was the only intermediate and that there was no crossover to a *cis*-cyclopropylvinyl cation. The formation from **111** of the other solvolysis product, such as **110**, was also explained by solvent attack taking place at C-2 or C-3 of the cyclopropane ring [41].

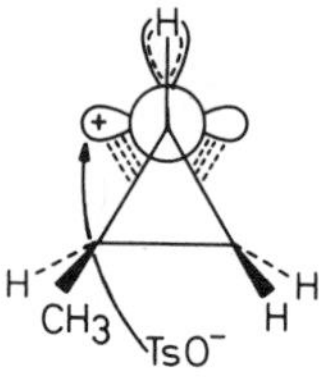

**111**

Solvolysis of the cyclopropylvinyl chloride **112**, which possesses two *cis*-methyl groups in the cyclopropane ring, was also studied [42]. The hydrolysis in water of **112** at 100° catalyzed with silver nitrate in the presence of calcium carbonate gave the products **113** to **119** as shown in Scheme 5.13.

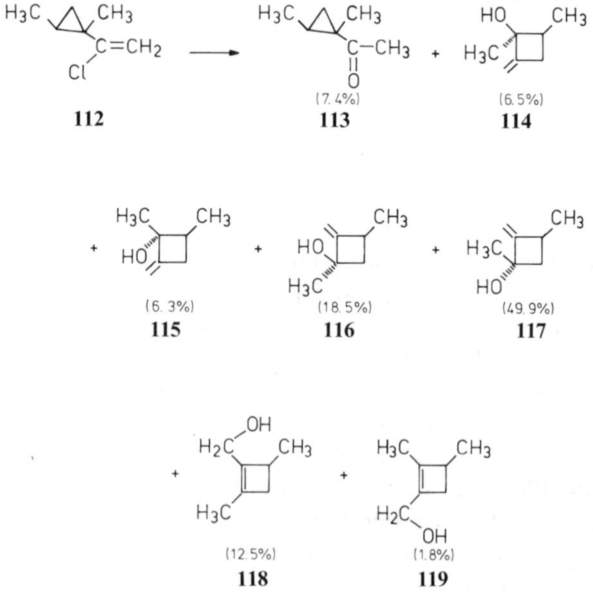

Scheme 5.13

More then 92% were rearranged products containing the cyclobutane skeleton. From these compounds it can be easily seen that in the ring enlargement the most substituted bond of the cyclopropane ring in **112** is preferentially involved. As in the case of the cyclopropylvinyl tosylate **108**, the formation of ketone **113** from the vinyl chloride **112** occurred stereospecifically, indicating that an interconversion of the two diastereoisomeric cyclopropylvinyl cations **120** and **121** was not possible.

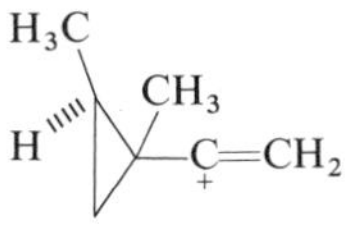

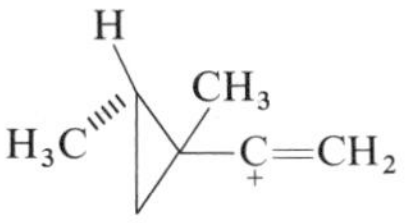

The powerful stabilizing effect of a cyclopropane ring on a vinyl cation was also demonstrated by its ability to provide the driving force for a hydride shift across the double bond, as discussed in Chapter 7, Section III,A.

The solvolysis reactions leading to cyclopropyl-stabilized vinyl cations discussed so far are all related to acyclic vinyl derivatives. Cyclic vinyl cations stabilized by a neighboring cyclopropane ring are unknown, although several attempts were made to synthesize precursors for their generation. The vinyl iodide **122** was prepared [38], but its solvolysis was not studied.

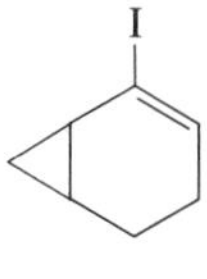

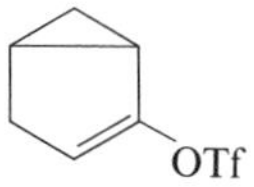

Cyclic vinyl triflates containing a cyclopropane ring in the β position could not be synthesized. The triflate **123**, for example, turned out to be highly unstable and could not be isolated [31]. The vinyl triflate **124**, with the cyclopropane ring in the position γ to the potential vinyl cation, was synthesized and its solvolysis studied in 50% aqueous ethanol and absolute trifluoroethanol (TFE) [43]. Although **124** solvolyzed in aqueous ethanol at 93° with formation of ketone **124a**, no reaction occurred in absolute TFE,

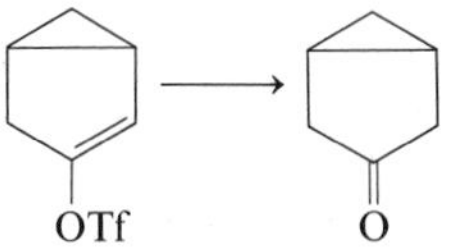

a solvent of low nucleophilicity, even at 155°. This observation, as well as the measured reaction rates, which were strongly dependent on added base, indicated that the triflate **124** solvolyzed not by a vinyl cation mechanism, but by S–O cleavage [43] (Section III,A,5).

## C. Cyclopropylidenemethyl and Related Cations

Cyclopropylidenemethyl cations **63**, in which one carbon of the cyclopropane ring is already a part of the vinyl system, are especially stable. The high stability of **63** is due to its special geometry [44]. The electron-donating cyclopropane ring, because of the short C–C distance of the double bond, is closer to the positive center, and the axis of the vacant orbital of the vinyl cation lies in the plane of the three-membered ring, as represented in **125**. In this way there is an especially favorable geometry for overlap of the participating orbitals. The stability of **125** was confirmed by MO, *ab initio*,

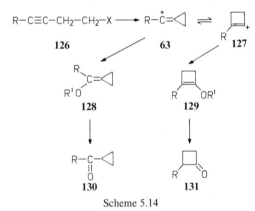

**125**

and MINDO/3 calculations, as discussed in Chapter 2. Cyclopropylidenemethyl cations **63** were first proposed by Hanack and co-workers [45, 46] as intermediates in the "homopropargyl rearrangement" in which, during solvolysis in aqueous alcohol, the reactive homopropargyl derivatives **126** can be rearranged, with the formation of the enol ethers **128** and **129** or cyclopropyl ketones **130** and cyclobutanones **131**, depending on the solvent.

Scheme 5.14

The mechanism of this reaction, which includes the vinyl cation **63** and the cyclobutenyl cation **127** of Scheme 5.14, is discussed in detail in Chapter 3.

The generation of a cyclopropylidenemethyl cation **63** by solvolysis of a cyclopropylidenemethyl derivative was first reported with the chloride **132**

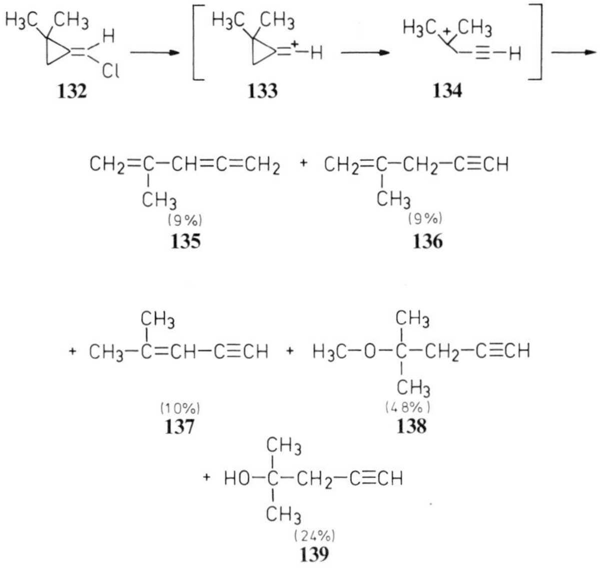

H3C  CH3                    ⎡ H3C  CH3              H3C  CH3      ⎤
  ⟩=⟨ H                     ⎢   ⟩=⁺−H          ⟩−≡−H  ⎥
       Cl                    ⎣                              +          ⎦
  **132**                        **133**                  **134**

$CH_2=C-CH=C=CH_2$  +  $CH_2=C-CH_2-C≡CH$
    |                              |
   $CH_3$                           $CH_3$
  (9%)                           (9%)
  **135**                           **136**

       $CH_3$                    $CH_3$
        |                              |
+ $CH_3-C=CH-C≡CH$  +  $H_3C-O-C-CH_2-C≡CH$
                                    $CH_3$
   (10%)                        (48%)
   **137**                        **138**
                 $CH_3$
                 |
       + $HO-C-CH_2-C≡CH$
                 |
                 $CH_3$
              (24%)
               **139**

Scheme 5.15

[47], as shown in Scheme 5.15. In aqueous methanol, **132** reacted in the presence of silver salts even at room temperature with formation of silver chloride. Without silver salt catalysis, **132** solvolyzed in the presence of triethylamine with formation of the rearranged compounds **135** to **139**. Compound **132** showed a first-order rate constant in the same solvent at 140° which was independent of added base, thus ruling out an addition–elimination mechanism. The comparatively high solvolysis rate as well as the product composition was explained by the formation of the intermediate vinyl cation **133**. Due to the geminal dimethyl groups in the 2 position of the cyclopropane ring, **133** rearranged with ring opening and formation of the more stable tertiary cation **134**. Elimination of a proton from **134** with and without rearrangement gave **135**, **136**, and **137**, respectively. Capture by the solvent led to the ether **138** and the alcohol **139** [47]. Because in the solvolysis of **132** products with retention of the cyclopropane ring were not found, the formation of the tertiary carbenium ion **134** by a concerted process from **132** cannot be ruled out with certainty. Ion **133** was the first primary vinyl cation to be proposed as an intermediate in a solvolysis reaction. Without the stabilizing effect of the cyclopropane ring, the formation of a primary vinyl cation by a solvolysis reaction would be energetically impossible, as discussed in Chapter 2.

The unsubstituted cyclopropylidenemethyl bromide **140** was also solvolyzed in solvents of different nucleophilicity [44, 48]. Solvolysis of **140**

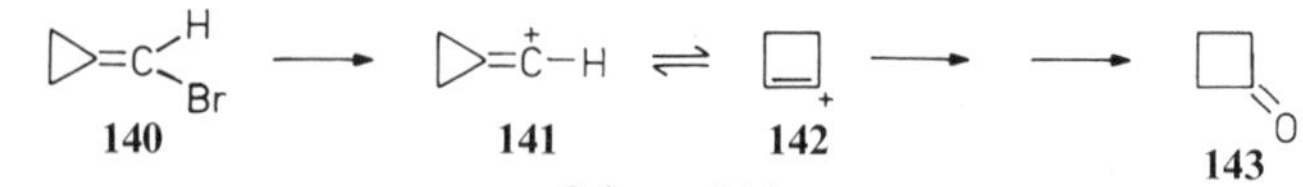

140                 141        142
143
Scheme 5.16

in 50% aqueous methanol in the presence of 1.2 mole equivalents of triethylamine gave, as shown in Scheme 5.16, cyclobutanone 143 as the only
solvolysis product detectable by gas chromatography [44]. The rates of
solvolysis of cyclopropylidenemethyl bromide 140 in 60 and 80% aqueous
ethanol at different temperatures are shown in Table 5.4. The rate constants
in Table 5.4 clearly show that, in contrast to other unstabilized vinyl halides,
cyclopropylidenemethyl bromide 140 is an appreciably reactive system
under solvolytic conditions. The rate constants were almost independent
of the concentration of the added base, and the possibility of an acid-
catalyzed addition–elimination mechanism was therefore ruled out. The $m$
value of the Winstein–Grunwald equation was determined to be 0.53, which
does not contradict a vinyl cation mechanism. The formation of the cyclobutanone 143 as the product of the solvolysis indicated that the intermediate
primary vinyl cation 141 rearranged to the cyclobutenyl cation 142. Cyclobutenyl cations such as 142, in contrast to other cyclic vinyl cations, are
stabilized through a nonclassic interaction between the positive charge
and the C-2–C-3 bond, so that the formation of 142 is relatively easy, as
discussed in Chapter 2.

**TABLE 5.4**

**Solvolysis Rates of Cyclopropylidenemethyl Bromide 140 in
Aqueous Ethanol with Different Base Concentrations[a]**

| Solvent (% EtOH) | Triethylamine (equivalents) | Temp (°C) | $k$ (sec$^{-1}$) |
|---|---|---|---|
| 80 | 3.0 | 100 | $7.1 \times 10^{-8}$ |
| 60 | 3.0 | 130 | $5.73 \times 10^{-6}$ |
| 80 | 3.0 | 130 | $1.44 \times 10^{-6}$ |
| 60 | 1.1 | 140 | $1.35 \times 10^{-5}$ |
| 60 | 3.0 | 140 | $1.29 \times 10^{-5}$ |
| 60 | 1.1 | 150 | $3.41 \times 10^{-5}$ |
| 60 | 3.0 | 150 | $2.97 \times 10^{-5}$ |

[a] From Ref. [48].

That a vinyl cation is involved in the solvolysis of such primary cyclopropylidenemethyl halides as 132 and 140 was definitely shown from the
solvolysis of another pair of primary vinyl bromides, the $E$- and $Z$-2-
methylcyclopropylidenemethyl bromides $E$-144 and $Z$-144 [49]. These

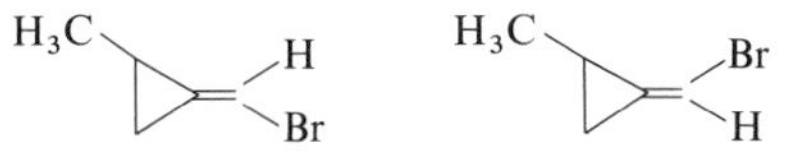

E-144                    Z-144

compounds were solvolyzed in ethanol–water mixtures of different ionizing power, the products were determined, and the rates of solvolysis were measured [49]. Table 5.5 shows the rates in 60 and 80% aqueous ethanol. By comparing the rates of E- and Z-144 with the rate of cyclopropylidenemethyl bromide 140 under the same conditions, it can be seen that a methyl group in the 2 position increases the rate by a factor of 10 for E-144 and by 27 for Z-144. Accordingly, both isomers solvolyze with rate enhancements typical for compounds that are methyl-substituted in the cyclopropane ring of a cyclopropylmethyl system, compared with the nonsubstituted analogue [50]. Compounds E- and Z-144 in 50% aqueous ethanol solvolyzed with

**TABLE 5.5**

**Solvolysis Rates of E- and Z-2-Methylcyclopropylidenemethyl Bromide (E- and Z-144)**[a]

| Vinyl bromide | Solvent (% EtOH) | $k \ (\text{sec}^{-1})$ |
|:---:|:---:|:---:|
| E-144 | 60 | $3.69 \times 10^{-4}$ |
|  | 80 | $4.97 \times 10^{-5}$ |
| Z-144 | 60 | $1.43 \times 10^{-4}$ |
|  | 80 | $2.74 \times 10^{-5}$ |

[a] Solvent was buffered with 1.1 mole equivalents triethylamine. The reaction was carried out at 140°C.

complete rearrangement and formation of the cyclobutanones 149 and 150 as well as the acetylenic alcohol 151, as shown in Scheme 5.17. The formation of the same cyclopropylidenemethyl cation (145) from both the E and the Z isomers was proven by the fact that E- and Z-144 formed the rearranged products in exactly the same ratio [49]. This ruled out a synchronous ring opening without the intermediacy of 145, for which the products, depending on the stereochemistry of the starting bromide 144, would have been very different. Although no rearrangement with formation of acetylenic derivatives was observed in the solvolysis of cyclopropylidenemethyl bromide 140 (vide supra), as in the case of the dimethyl compound 132, E- and Z-144, with a methyl substituent on the cyclopropane ring, yielded an acetylenic compound (151) as the main product. The preferred formation of the homopropargyl alcohol 151 was in accordance with the expectation that the vinyl cation 145 in Scheme 5.17 will partly rearrange into the

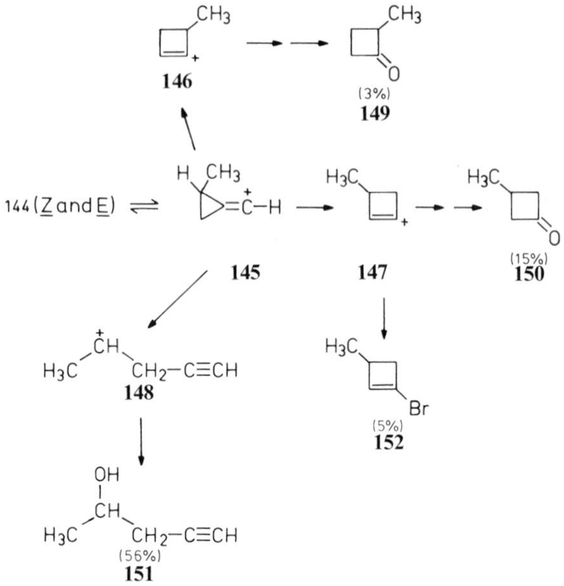


Scheme 5.17

homopropargyl cation **148**. The formation of 3-methylcyclobutenyl bromide **152** pointed to the fact that ion-pair return occurs during the solvolysis of **144**.

A comparison with the homologous cyclobutylidenemethyl bromide **154** is well suited to show how the reactivity of a vinyl bromide is increased through the stabilizing effect of the cyclopropane ring in **132**, **140**, and **144**. Even at temperatures up to 180° **154** was recovered unchanged after 21 days in methanol–water, showing that it exhibits the low reactivity of a "normal" primary vinyl bromide [44].

A series of cyclopropylidenemethyl bromides (**155**–**159**) leading to secondary cyclopropylidenemethyl cations were synthesized, and their

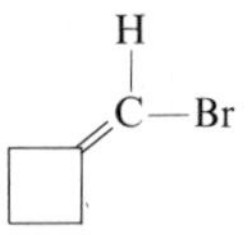


**154**

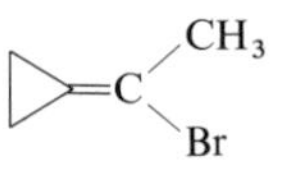


**155**

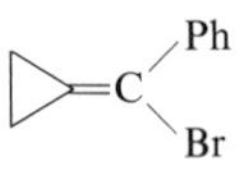


**156**

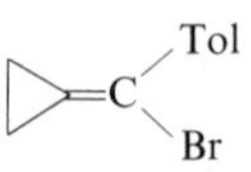


**157**

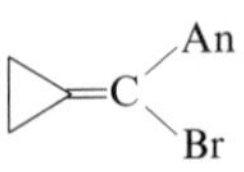


**158**

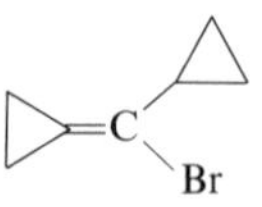


**159**

solvolytic behavior was studied [48, 51, 52]. If one adds a methyl group to cyclopropylidenemethyl bromide **140**, the resulting vinyl bromide (**155**) is distinguished by a noticeably increased reaction rate. The solvolysis rate, which is a factor of $10^3$ higher than that of cyclopropylidenemethyl bromide **140**, indicates a higher stabilization of the intermediate vinyl cation **160**, as shown in Scheme 5.18. Of special interest are the products of **155** solvolyzed

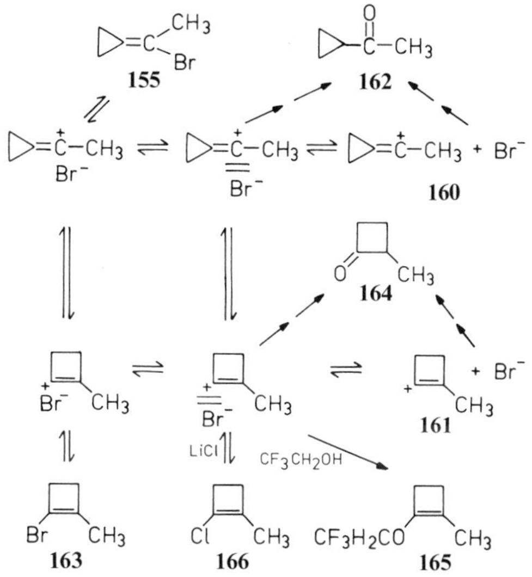

Scheme 5.18

in the presence of triethylamine in various solvent systems. Cyclopropyl methyl ketone (**162**), 2-methylcyclobutanone (**164**), and 1-bromo-2-methyl-cyclobutene (**163**) were produced. The exact product ratios are shown in Table 5.6.

In all solvolysis reactions of **155**, 1-bromo-2-methylcyclobutene (**163**) was the main product. A rearrangement in this manner, which can typically be described as an ion-pair return with simultaneous structural rearrange-ment, was observed with **155** for the first time in the case of a vinyl cation [48]. The rearrangement of **155** to **163** was direct evidence for a vinyl cation mechanism, since the formation of the cyclobutenyl bromide **163** by another mechanism, such as the addition–elimination mechanism, was clearly ruled out. As shown in Table 5.6, addition of chlorides in the solvolyses caused the formation of a considerable amount of 1-chloro-2-methylcyclobutene (**166**). The predominance of the rearranged bromide **163** under all solvolysis conditions suggested an ion-pair mechanism. The formation of rearranged chloride **166** in the trapping experiment with lithium chloride showed the presence of solvent-separated ion pairs, whereas the bromide **163** resulted

TABLE 5.6

Solvolysis Products of 155 in Different Solvent Systems[a]

| Solvent | Salt (added) | Products (%) | | | | | |
|---|---|---|---|---|---|---|---|
| | | 162 | 164 | 163 | 166 | 165 | 155 |
| 60% EtOH | — | <1 | 35 | 65 | — | — | 0 |
| 80% EtOH | — | <1 | 30 | 70 | — | — | 0 |
| 50% TFE | — | <1 | 27 | 65 | — | 8 | 0 |
| 80% TFE | — | <1 | 15 | 65 | — | 20 | 0 |
| SO$_2$ (20°, 45 days)[b] | — | — | — | 32 | — | — | 18 |
| 100% EtOH | ZnCl$_2$ (30 equivalents) | — | Trace | 50 | 50 | — | 0 |
| 100% DMF[c] | ZnCl$_2$ (30 equivalents) | — | Trace | 55 | 45 | — | 0 |
| 50% TFE | LiCl (20 equivalents) | <1 | 10 | 60 | 15 | 15 | 0 |
| 50% DMSO[d] | LiCl (20 equivalents) | <1 | 20 | 55 | 25 | — | 0 |

[a] The reaction was carried out at 90° for 4 days with 2.5 equivalents of Et$_3$N.
[b] Two unidentified components (12 and 38%).
[c] Dimethylformamide.
[d] Dimethylsulfoxide.

from tight ion pairs. Scheme 5.18 shows the formation of the individual reaction products with respect to ion-pair formation.

The substitution of a phenyl group at the α position of cyclopropylidene-methyl bromide 140 markedly increased the stability of the vinyl cation. Thus, cyclopropylidenebenzyl bromide 156 solvolyzed 2.3 × 10$^3$ times faster than the nonsubstituted vinyl bromide 140 in 80% ethanol at 100° [52]. The increased stability of intermediate vinyl cations such as 167 in Scheme 5.19 was also shown by the solvolysis products of 156. Cyclopropyl-idenebenzyl bromide 156 was solvolyzed in 80% ethanol buffered with triethylamine at 80°. The main product was the cyclopropyl phenyl ketone 168 besides the ene–yne 170, which probably was formed from the inter-mediate enol ether 169, depending on the amount of added base [52]. The high percentage of cyclopropyl phenyl ketone 168 formed with few re-arranged products points to the higher stability of the phenyl-substituted cyclopropylidenemethyl cation 167 in comparison with the parent cyclo-propylidenemethyl cation 141, which, after its formation, completely rearranged to form cyclobutanone 143 (vide supra). On the other hand, some rearrangement of 167 to form 2-phenylcyclobutanone 171 (Scheme 5.19) cannot be ruled out, because 171 was found to be unstable under the reac-tion conditions and therefore might have escaped isolation [52].

Cyclopropyl phenyl ketone 168 was also obtained as the main reaction product when 1-bromo-2-phenylcyclobutene (172) was solvolyzed in various solvent systems under different reaction conditions [52]. As discussed in

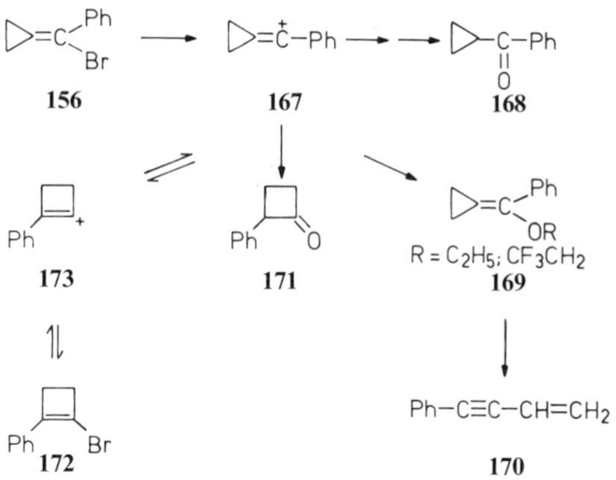

$$\triangleright = C \overset{Ph}{\underset{Br}{}} \longrightarrow \triangleright = \overset{+}{C} - Ph \longrightarrow \longrightarrow \triangleright - \overset{O}{\underset{\|}{C}} - Ph$$

**156**        **167**        **168**

$$\underset{Ph}{\square}^{+} \;\rightleftharpoons\; \underset{Ph}{\square}{O} \;\searrow\; \triangleright = C \overset{Ph}{\underset{OR}{}}$$

**173**        **171**        $R = C_2H_5; CF_3CH_2$
**169**

$\updownarrow$

$$\underset{Ph \quad Br}{\square}$$
**172**

$$Ph-C{\equiv}C-CH{=}CH_2$$
**170**

Scheme 5.19

detail below, cyclobutenyl derivatives, e.g., **172**, also solvolyze via a vinyl cation mechanism. The favored formation of cyclopropyl phenyl ketone **168** from **172** via the intermediate 2-phenylcyclobutenyl cation **173** shows the increased stability of the cyclopropylidenebenzyl cation **167**.

Cyclopropylidenemethyl bromides substituted with even better stabilizing substituents at the functional carbon atom are the tolyl and anisyl compounds **157** and **158**, respectively. Both vinyl bromides showed the same pattern of solvolysis products, pointing to a high stability of the corresponding cyclopropylidenemethyl cations. In 80% aqueous ethanol (80°, buffered with triethylamine), **157**, for example, solvolyzed as shown in Scheme 5.20 with formation of 31% ene–yne **179a**, 49% cyclopropyl tolyl

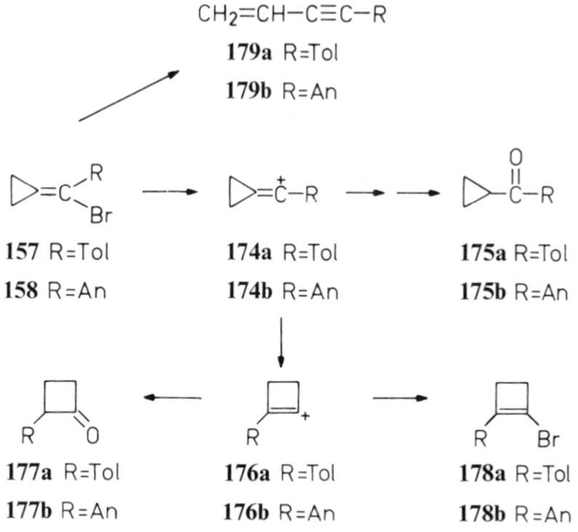

$$CH_2{=}CH-C{\equiv}C-R$$
**179a** R=Tol
**179b** R=An

$$\triangleright = C \overset{R}{\underset{Br}{}} \longrightarrow \triangleright = \overset{+}{C} - R \longrightarrow \longrightarrow \triangleright - \overset{O}{\underset{\|}{C}} - R$$

**157** R=Tol     **174a** R=Tol     **175a** R=Tol
**158** R=An     **174b** R=An     **175b** R=An

$$\underset{R \quad O}{\square} \longleftarrow \underset{R}{\square}^{+} \longrightarrow \underset{R \quad Br}{\square}$$

**177a** R=Tol     **176a** R=Tol     **178a** R=Tol
**177b** R=An     **176b** R=An     **178b** R=An

Scheme 5.20

ketone **175a**, and 8% 2-tolylcyclobutenyl bromide **178a**. Twelve percent of the products were not identified [51]. The anisylvinyl bromide **158** under the same conditions yielded 34% **178b**, 54% nonrearranged ketone **175b**, and 4% 2-anisylcyclobutenyl bromide **178b** [51]. The ene–ynes **179a** and **179b** were formed only under basic conditions. Without triethylamine the solvolysis yielded the cyclopropyl ketones **175a** and **175b** as the main products. The ene–yne formation was explained by a synchronous opening of the cyclopropane ring by base attacking at the $\beta$-hydrogens during the ionization process [51]. The aryl cyclopropyl ketones **175a** and **175b**, obtained from the stabilized vinyl cations **174a** and **174b**, respectively, were the main products in the solvolysis of **157** and **158** (Scheme 5.20). The cyclobutenyl bromides **178a** and **178b** were formed by ring enlargement and ion-pair return.

An especially thorough investigation of the products was made in the solvolysis of the cyclopropylcyclopropylidenemethyl bromide **159**. This vinyl bromide was expected to give a highly stabilized vinyl cation, due to the increased stabilizing effect of two cyclopropane rings on the positive charge. The solvolysis of **159** was also investigated with respect to internal return phenomenon [53]. The exact product composition from the solvolysis of **159** in ethanol–water mixtures under different conditions is given in Table 5.7. In all aqueous solvents, dicyclopropyl ketone **191** was the major product, indicating that vinyl cation **188** of Scheme 5.21 was especially stable.

**TABLE 5.7**

**Solvolysis Products of Cyclopropylcyclopropylidenemethyl Bromide 159 in Aqueous Ethanol**[a]

|  | Products (%) | | | | | | |
|---|---|---|---|---|---|---|---|
| Solvent | **182** | **184** | **185** | **190** | **191** | **192** | **193** |
| 80% EtOH | 7 | 11 | 7 | 4 | 65 | 1 | 5 |
| 50% EtOH | 6 | <1 | 9 | 7 | 73 | 2 | 3 |

[a] Solvent was buffered with 2 equivalents of triethylamine. The reaction was carried out at 60° for 24 hr.

Detailed studies on the influence of added salts on the product composition were also carried out. This and extensive kinetic studies, including the investigation of salt effects, led to the mechanism given in Scheme 5.21 [53]. Accordingly, the cyclopropylcyclopropylidenemethyl bromide **159** rearranged via the corresponding internal ion pairs **180**, **181**, and **183** with formation of the cyclobutenyl bromide **182** and the homopropargyl bromide

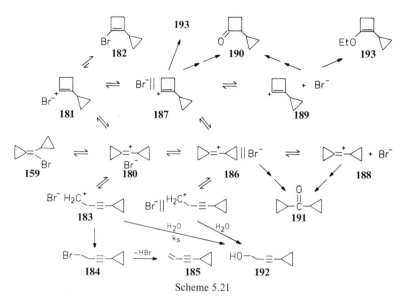

Scheme 5.21

**184**. The internal ion pair **181** dissociates further to the external ion pair **187**. Internal return in a thermodynamically controlled reaction yielded, besides bromide **159**, either of the two vinyl bromides **182** and **184**, both of which were stable under the solvolysis conditions. In the presence of base, **184** was partly dehydrobrominated with formation of the ene–yne **185**. The external ion pairs **186** and **187** preferentially reacted with solvent and formed ketones **191** and **190** and/or ether **193**, depending on the nucleophilicity of the solvent. The formation of external ion pairs **186** and **187** was proven by the solvolysis of **159** in 50% EtOH in the presence of lithium chloride, in which vinyl chlorides **194**, **195**, and **196** were formed in addition

| | | |
|---|---|---|
| **194** | **195** | **196** |

to the products given in Table 5.7 [53]. The special salt effect found in the kinetics of **159** in 50% EtOH in the presence of LiClO$_4$ also established the formation of external ion pairs. The formation of the products may also occur via relatively free ions, such as **188** in Scheme 5.21. The rates of reaction of vinyl bromides **140**, **155** to **159**, and for comparison **197** to **199** are summarized in Table 5.8.

As the data in Table 5.8 indicate, an increase in the solvolytic rate (implying an increase in the stabilization of the intermediate vinyl cation) was

TABLE 5.8

Rates of Reaction of Vinyl Bromides 140,
155–159, and 197–199 in 80% Ethanol at 100°

| Compound | $k$ (sec$^{-1}$) | $k_{rel}$ | $m$ |
|---|---|---|---|
| 140 | $7.1 \times 10^{-8a}$ | 1 | $0.53^b$ |
| 155 | $6.6 \times 10^{-5}$ | $1 \times 10^3$ | $0.64^c$ |
| 156 | $1.5 \times 10^{-4}$ | $2.2 \times 10^3$ | — |
| 157 | $7.6 \times 10^{-4a}$ | $1.1 \times 10^4$ | $0.85^d$ |
| 158 | $3.8 \times 10^{-2a}$ | $5.3 \times 10^5$ | — |
| 159 | $6.4 \times 10^{-3a}$ | $9.0 \times 10^4$ | $0.89^e$ |
| 197 | $4.2 \times 10^{-9a}$ | $6 \times 10^{-2}$ | $0.62^f$ |
| 198 | $3.6 \times 10^{-5}$ | $8.6 \times 10^2$ | $0.68^g$ |
| 199 | $9.0 \times 10^{-7a}$ | $1.3 \times 10^1$ | — |

$^a$ Extrapolated.
$^b$ At 130°.
$^c$ At 90°.
$^d$ At 74°.
$^e$ At 49°.
$^f$ At 170°.
$^g$ At 120°.

observed when the α-carbon of the vinyl bromide was successively substituted by more powerful electron-releasing groups. In going from methyl
(155) through phenyl (156), p-tolyl (157), and cyclopropyl (159) to p-anisyl
(158), a rate increase of $10^2$ was observed [48, 51, 52]. Furthermore, the
secondary methyl derivative 155 showed a noticeable increase in reaction
rate ($\sim 10^3$) compared to the parent compound 140, indicating the higher
stability of a secondary vinyl cation versus a primary vinyl cation in general.
The rate ratio between the phenyl (156) and the tolyl (157) derivatives
compared with the methyl derivative (155) was low, indicating considerable
ground state stabilization of 156 and 157 due to conjugation between the
aryl ring and the double bond (see Chapter 6). Within the range of aryl-
substituted cyclopropylidenemethyl compounds 156, 157, and 158, the
rates increased, as expected, with the increasing electron-releasing ability
of the substituent in the para position of the aromatic ring. The bromide
157 reacted five times faster than 156 owing to the inductive effect of the

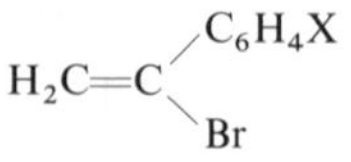

$$H_2C=C\begin{array}{c} /C_6H_4X \\ \backslash Br \end{array} \qquad (CH_3)_2C=C(C_6H_5)Br$$

197  X = H                          199
198  X = p-OCH$_3$

$p$-methyl group in the phenyl ring. The effect of a $p$-methoxy group was small, with **158** reacting 245 times faster than the parent compound **156** under these conditions. Such a substituent effect on the solvolysis rates suggested a unimolecular ionization process and the formation of a vinyl cation. The especially high stabilizing effect of a cyclopropyl group was shown in the solvolysis rate of **159**, which was of the same order ($10^5$ times faster than that of the parent compound **140**) as that obtained with the vinyl bromide **158** with a $p$-methoxy substituent. The contribution of the cyclopropylidene structure to the stabilization of the intermediate vinyl cations **174a** and **174b** in Scheme 5.20 was reflected, for example, by the higher reactivity of the cyclopropylidenebenzyl bromide **156** compared with α-bromostyrene (**197**) [54], the rate ratio being $10^4$. Comparison of the solvolysis rate of **156** with the rate of the bromide **199** also shows that **156** solvolyzed $\sim 10^4$ times faster than **199**. Another indication of the large stabilizing effect of the cyclopropylidene substituent is the rate enhancement due to the effect of a $p$-methoxy group in **158**, which is relatively small ($k_{158}/k_{156} = 245$), whereas $k_{198}/k_{197}$, without the influence of the cyclopropylidene substituent, reaches 8500 [51] (see also Chapter 6).

The formation of cyclopropylidenemethyl cations was further supported by the high sensitivity of the reaction rates to the ionizing power of the solvent; e.g., at 74.5°, the vinyl bromide **157** reacted 26 times faster in 50% aqueous ethanol than in 80% ethanol, corresponding to a Winstein–Grunwald $m$ value of 0.85, which, together with the value for **159** ($m = 0.89$), is one of the highest $m$ values observed for a reaction involving a vinyl cation.

For the vinyl bromide **156, 157,** and **158**, the plots of log $k_x/k_H$ correlated linearly with Brown $\sigma^+$ substituent constants, with a $\rho$ value of $-2.8$. Compared to the values obtained in the solvolysis of α-arylvinyl substrates such as **197** and **198** [55] listed in Table 6.7, this value seems small. However, it is reasonable when one considers the delocalization of the positive charge of the cyclopropylidenemethyl cations over the adjacent cyclopropane ring.

The high stability of a cyclopropylidenemethyl cation suggested the idea of generating this cation by a solvolysis reaction using an even poorer leaving group than bromine. Fluorine is probably the poorest leaving group among the sulfonates and halogens, and for this reason the solvolysis of the cyclopropylidenemethyl fluorides **200** to **202** and the anisyl-substituted vinyl fluoride **203** was investigated [56].

|     200     |     201     |     202     |     203     |

The tolylcyclopropylidenemethyl fluoride **200** was solvolyzed in 50 and 80% aqueous ethanol with and without triethylamine as a buffer. The reaction was complete after 30 hr at 140°. Whereas without triethylamine the main product was the cyclopropyl tolyl ketone **175a**, with triethylamine 90% polymeric material, in addition to small amounts of **175a** and 2-tolylcyclobutanone (**177a**), was formed. The solvolysis of isopropylcyclopropylidenemethyl fluoride **201** in 50% aqueous ethanol or trifluoroethanol was even more controlled by an added base. Without the addition of base, **201** reacted after 12 hr at 140° with quantitative formation of the cyclopropyl isopropyl ketone **205**. In the presence of triethylamine, the reaction was complete only after 14 days at 170°. The main products in this case were 2-isopropylcyclobutanone **204** (66%) and **205** (6%), as well as some **206** and **207** [56] (Scheme 5.22).

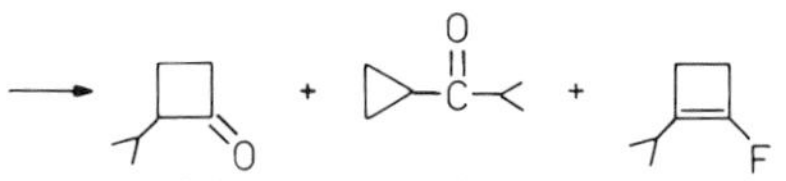

$$+ \ (CH_3)_2CH-C{\equiv}C-CH_2-CH_2-OH$$
**207**

Scheme 5.22

In 50% aqueous ethanol (with triethylamine as buffer), the cyclopropylcyclopropylidenemethyl fluoride **202** gave 4% homopropargyl alcohol **208**, 12% 2-cyclopropylcyclobutanone **190**, and, as the main product, 87% dicyclopropyl ketone **191**, as shown in Scheme 5.23.

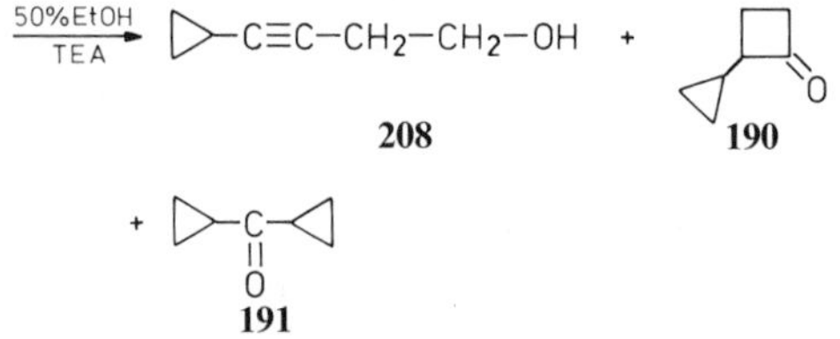

Scheme 5.23

The anisylvinyl fluoride **203**, which was solvolyzed for comparison in 50% aqueous ethanol (buffered with 1.2 mole equivalents of triethylamine) for 20 days at 160°, yielded the ketone **209** (33%) and the acetylene **210** (65%) [56]:

$$203 \ \xrightarrow[160°]{50\% \ EtOH} \ An-\underset{\underset{O}{\|}}{C}-C_2H_5 \ + \ An-C{\equiv}C-CH_3$$

**209**                      **210**

Since the $S_N1$ process seems inherently difficult, while the double bond is strongly polarized, vinyl fluorides may undergo electrophilic or nucleophilic addition reactions of the solvent. The solvolysis of the vinyl fluorides **200** to **203** must therefore be discussed especially critically with regard to the possibility of mechanisms other than vinyl cation intermediacy leading to the product formation. The solvents used for the solvolysis of the vinyl fluorides (ethanol/water and TFE) allow only the addition of a proton as an electrophile. In the case of **201** or **202**, the addition of a proton should lead to the intermediate cyclopropylmethyl cation **211**, as shown in Scheme 5.24. The other cation (**212**) that could be formed by addition of a proton to **201** or **202** is less stable and should react further with ring opening. The intermediate cation **211** could explain the formation of the products **204** to **206** as well as **190** and **191** in the solvolysis of **201** and **202**, respectively. However, electrophilic addition in the presence of base is very unlikely, because the intermediate cation **211** does not explain the homopropargyl alcohol **207**, but it may be the mechanism for the exclusive formation of the ketone **205** in the absence of base.

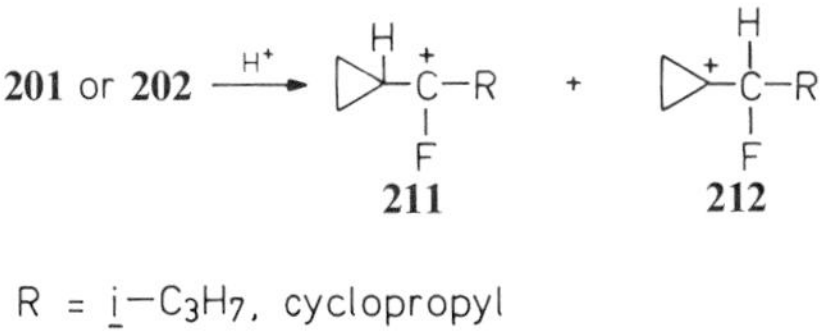

$$ 201 \text{ or } 202 \xrightarrow{\ H^+\ } \quad 211 \quad + \quad 212 $$

R = $\underline{i}-C_3H_7$, cyclopropyl

Scheme 5.24

An $Ad_N$–E mechanism as well as other routes, such as a synchronous rearrangement or a concerted fragmentation, were also ruled out for the solvolyses of vinyl fluorides **200** to **203** [56]. The products as well as the kinetic data were in agreement with an $S_N1$ solvolysis of these vinyl fluorides, again demonstrating the high stability of a cyclopropylidenemethyl cation and showing that even vinyl fluorides can be solvolyzed with formation of a vinyl cation if this is stable enough.

## D. Cyclobutenyl Cations

A great deal of work has been devoted to the question of whether the third "homopropargyl isomer," a cyclobutenyl derivative (**213**), could also undergo a solvolysis reaction. A cyclobutenyl cation generated from **213** should be highly strained since there is no possibility of achieving a linear structure that is energetically favorable for a vinyl cation. On the other hand, the experimental results obtained with cyclobutenyl derivatives **213** (R = H, alkyl, aryl) point to the fact that they solvolyze via a vinyl cation mechanism. The most convincing evidence for this mechanism was obtained with cyclobutenyl nonafluorobutanesulfonate (**213**, R = H) (nonaflate) itself [57]. Comparing, for example, the relative rates of a series of cyclic vinyl nonaflates in aqueous ethanol, it was found that with increasing ring size the reaction rate increased considerably [16]. The cyclooctenyl nonaflate **216** is solvolyzed $1.3 \times 10^4$ times faster than the cyclohexenyl nonaflate **215**, whereas the next lower homologue, the cyclopentenyl nonaflate **214**, reacted very slowly and not via a vinyl cation mechanism [57, 58]. The solvolysis rate of the cyclobutenyl nonaflate **213** (R = H) is $3.7 \times 10^3$ greater than that of the cyclohexenyl nonaflate **215**, as shown in Scheme 5.25. The high

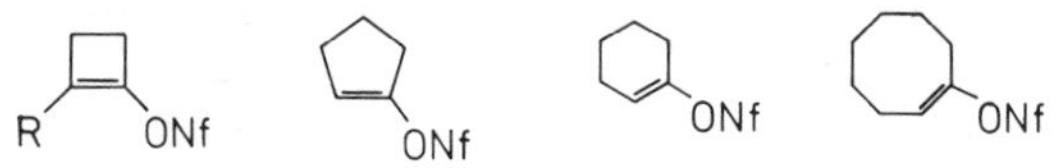

R = H ;  Nf = SO$_2$C$_4$F$_9$

| $k_{rel}$ | $3.7 \times 10^3$ | $5 \times 10^{-2}$ | 1 | $1.3 \times 10^4$ |

Scheme 5.25

solvolytic reactivity of the cyclobutenyl nonaflate **213** was explained by postulating a cationic intermediate (**217**) in which the positive charge is stabilized by nonclassical interaction [57]. As discussed in Chapter 2, calculations supported the view that the $\sigma$-bond orbital of the C-2–C-3 bond overlapped with the vacant p orbital of the cation at C-1 in **217**.

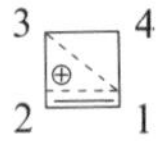

**217**

To obtain insight into the exact mechanism of solvolysis of the cyclobuten-1-yl nonaflate **218**, several experiments were carried out. In each case the results clearly pointed to a cyclic vinyl cation intermediate. Besides the vinyl cation mechanism, other possible pathways in the solvolysis reaction of **218** are the oxygen–sulfur cleavage and the Ad$_E$–E and Ad$_N$–E reactions (Section III,A). They were, however, ruled out by kinetic studies, which showed that the solvolysis rate of **218** was independent of the pH in the range of pH 3.2 to 9.2 [57]. In order to capture the intermediate cyclobuten-1-yl cation, **218** was solvolyzed in absolute trifluoroethanol (TFE) buffered with triethylamine at 75° for 10 days. Cyclobuten-1-yl trifluoroethyl ether (**223**) and the ketal **224**, as shown in Scheme 5.26, were formed in the ratio 10:1. The formation of the enol ether **223** can be explained by postulating an intermediate cyclobuten-1-yl cation. The ketal **224** was formed by the addition of TFE to the enol ether **223**. An addition–elimination mechanism for the solvolysis of cyclobutenyl nonaflate **218** in TFE was unequivocally

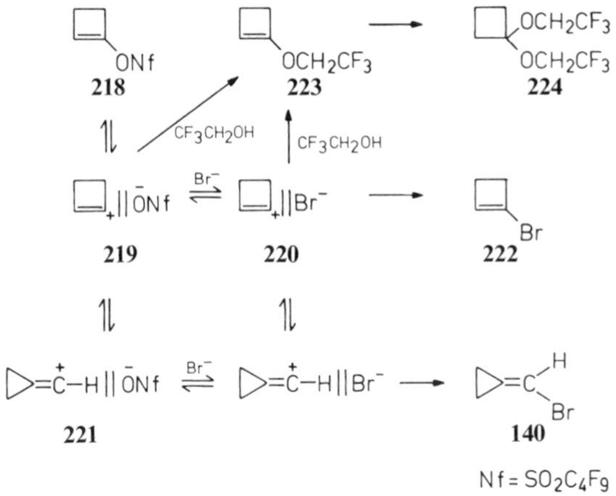

Scheme 5.26

ruled out by carrying out the solvolysis of **218** in absolute $CF_3CH_2OD$; the enol ether (**223**) obtained did not contain any deuterium.

An experiment to prove conclusively the formation of the four-membered cyclic vinyl cation was performed as follows. The solvolysis of **218** was carried out in absolute TFE buffered with triethylamine and containing a tenfold excess of tetraethylammonium bromide at 75° for 10 days. The product analysis showed that cyclobuten-1-yl bromide **222** and cyclo-propylidenemethyl bromide **140** were formed (53.3% in total) in the ratio 85:15, along with 34% of **223** and 0.9% of **224**. As shown in Scheme 5.26, the formation of these products was considered to occur via ion pairs [57]. Solvation of the leaving group leads to the solvent-separated ion pair **219**. From **219** the product **223** or ion pair **220** was formed, and the latter reacted either with the solvent or with the nucleophilic bromide, leading to **223** and **222**. The rearrangement to form the cyclopropylidenemethyl bromide **140** occurred in the solvent-separated ion pair **219**. On addition of excess tetraethylammonium bromide, the ion pair was preferentially captured by the more nucleophilic bromide ion. Both **140** and **222** were stable under the solvolysis conditions.

2-Methylcyclobutenyl nonaflate **225** was found to solvolyze $1.3 \times 10^2$ times faster than the parent compound **218**. The Winstein–Grunwald $m$ value for **225** was 0.67 [61]. The solvolysis products in aqueous EtOH, as shown in Scheme 5.27, were formed mostly without rearrangement and were identified as 2-methylcyclobutanone **226** (95%) and enol ether **227** (4%). Only a very small amount (1%) of the rearranged cyclopropyl methyl ketone **228** was found in the product mixture. Solvolysis of **225** in absolute TFE gave, as expected for a vinyl cation mechanism, the corresponding enol ether **229**

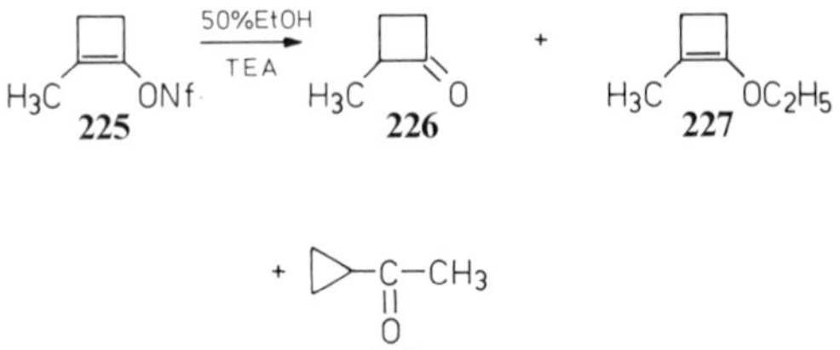

Scheme 5.27

$Nf = SO_2C_4F_9$

and the ketal **230**, as shown in Scheme 5.28. Compound **230** was formed by addition of TFE to the double bond of **229** [62]. The higher rate of reaction of 2-methylcyclobutenyl nonaflate **225** compared to that of the parent cyclobutenyl nonaflate **218** was another indication of the formation of a bridged vinyl cation (**217**) that is further stabilized by the methyl group in

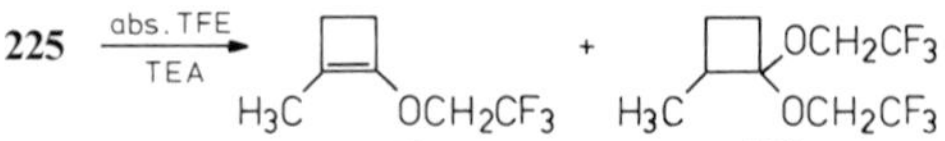

Scheme 5.28

the 2 position. Comparison of the rate of 2-methylcyclohexenyl triflate **231** with the rate of cyclohexenyl triflate **232** under the same conditions reveals that the former solvolyzes only ten times faster than the latter [63]. Relief of steric strain in the transition state can therefore be only partially responsible for the large increase in the rate of 2-methylcyclobutenyl nonaflate **225** over that of **218**.

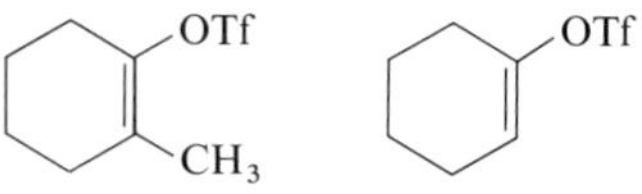

Other 2-substituted cyclobutenyl derivatives investigated were the 2-phenylcyclobutenyl nonaflate **233** [64], the 2-anisylcyclobutenyl nonaflate **236** [64], and the 2-cyclopropylcyclobutenyl triflate **237** [65]. Their rates of solvolysis are summarized in Table 5.9. As seen in Table 5.9, 2-phenylcyclobutenyl nonaflate **233** solvolyzed approximately at the same rate as the parent compound **218**, and 2-anisylcyclobutenyl nonaflate **236** reacted only 3.7 times faster than **218** [64]. The small influence of the phenyl and the anisyl group in the 2 position of the cyclobutenyl system on the solvolysis rate in contrast to the methyl group was explained by the negative inductive effect of both groups. A mesomeric stabilization of the intermediate bridged

TABLE 5.9

Solvolysis Rates of Cyclobutenyl Nonaflates in
50% Ethanol

| Nonaflate | Temp (°C) | $k$ (sec$^{-1}$) | $k_{rel}$ | Ref. |
|-----------|-----------|------------------|-----------|------|
| **218** | 51.2 | $2.57 \times 10^{-5}$ | 1 | [57] |
| **225** | 50.8 | $3.20 \times 10^{-3}$ | 130 | [61] |
| **233** | 51.8 | $2.61 \times 10^{-5}$ | 1 | [64] |
| **236** | 51.8 | $9.50 \times 10^{-5}$ | 3.7 | [64] |
| **240** | 50.5 | $5.92 \times 10^{-5}$ | 2.3 | [66] |

2-substituted vinyl cation **217** would require a rotation of the phenyl group, which is in the same plane as the cyclobutenyl ring in the starting nonaflates **233** and **236** [64]. The solvolysis products of 2-phenylcyclobutenyl nonaflate **233** were cyclopropyl phenyl ketone and 2-phenylcyclobutanone, along with the homopropargyl compounds **235a** and **235b**, as shown in Scheme 5.29 [64]. The solvolysis products of 2-anisylcyclobutenyl nonaflate **236** in 50%

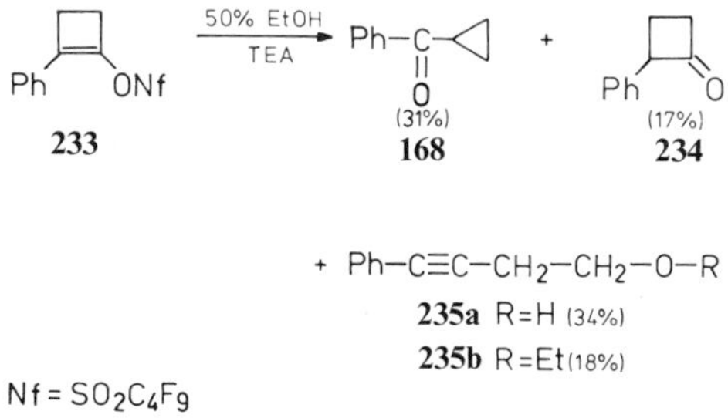

Nf = $SO_2C_4F_9$

Scheme 5.29

aqueous ethanol (buffered with triethylamine) were not fully analyzed [64]; however, cyclopropyl anisyl ketone **175b** and 2-anisylcyclobutanone **177b** were found in the ratio 1.3 : 1, as shown in Scheme 5.30. 2-Cyclopropyl-cyclobutenyl triflate **237** was solvolyzed in 80% aqueous ethanol. Dicyclo-propyl ketone **238** was found to be the main product (86%); there were also small amounts of the homopropargyl alcohol **239a** and ether **239b** and two

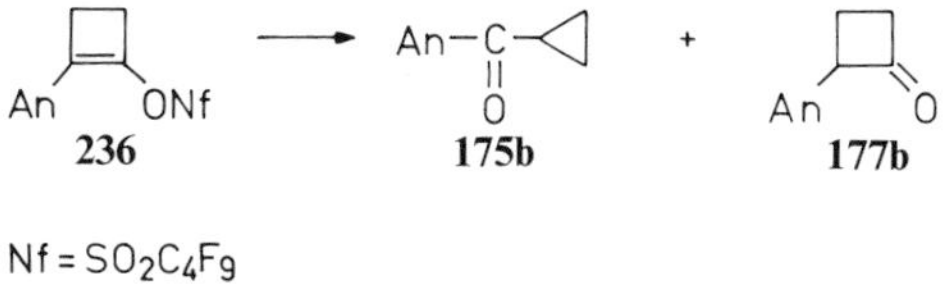

Nf = $SO_2C_4F_9$

Scheme 5.30

unidentified products [65], as shown in Scheme 5.31. As can be seen from the solvolysis products, 2-substituted cyclobutenyl nonaflates in general solvolyze mostly with rearrangement, leading qualitatively to the same product compositions as in the case of the cyclopropylidenemethyl derivatives **140** and **155** to **159** (*vide supra*).

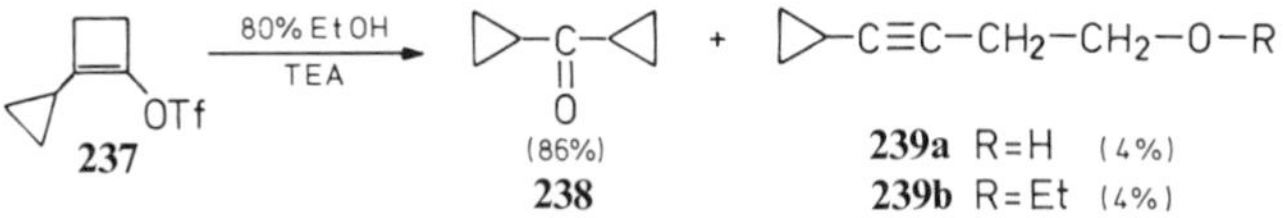

$$
\underset{\substack{\textbf{237}}}{\overset{\text{OTf}}{\Longleftarrow}} \xrightarrow[\text{TEA}]{80\% \text{ EtOH}} \underset{\substack{\text{(86\%)} \\ \textbf{238}}}{\triangleright\!-\!\overset{\text{O}}{\underset{\|}{\text{C}}}\!-\!\triangleleft} \; + \; \triangleright\!-\!\text{C}\!\equiv\!\text{C}\!-\!\text{CH}_2\!-\!\text{CH}_2\!-\!\text{O}\!-\!\text{R}
$$

**239a** R=H  (4%)
**239b** R=Et  (4%)

Scheme 5.31

One 3-substituted cyclobutenyl nonaflate, the 3-methylcyclobutenyl nonaflate **240**, was synthesized and its solvolysis studied. The methyl group in the 3 position also leads to a rate increase of 2.3 relative to the parent compound **218** (see Table 5.9), thus indicating a small stabilizing effect of the methyl group on the positive charge in the bridged vinyl cation intermediate. The solvolysis products of **240** were not all fully identified. However, besides 3-methylcyclobutanone **241**, the secondary homopropargyl alcohol **242** was formed [66], as shown in Scheme 5.32. The rearranged acetylenic alcohol **242** and the cyclobutanone **241** were also found in the solvolysis of

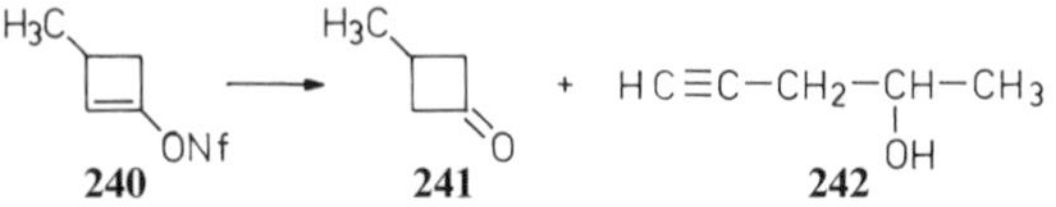

$$
\underset{\textbf{240}}{\overset{\text{H}_3\text{C}}{\Longleftarrow}\text{ONf}} \longrightarrow \underset{\textbf{241}}{\overset{\text{H}_3\text{C}}{\Longleftarrow}\text{O}} \; + \; \text{HC}\!\equiv\!\text{C}\!-\!\text{CH}_2\!-\!\underset{\underset{\textbf{242}}{\text{OH}}}{\overset{}{\text{CH}}}\!-\!\text{CH}_3
$$

Scheme 5.32

2-methylcyclopropylidenemethyl bromide **144**, which points to the formation of the same cationic intermediates (*vide supra*).

## E. Allenyl Cations

### 1. General Aspects

The activation of vinylic systems for solvolysis reaction by aryl and cyclopropyl groups was described in the preceding sections. Another potential activating group is a double bond, since allylic halides are much more reactive than their saturated analogues [67]. The enhanced solvolytic reactivity is due to overlap of the incipient vacant p orbital with the π system of the double bond in the allylic ion (see **243**) and in the transition state

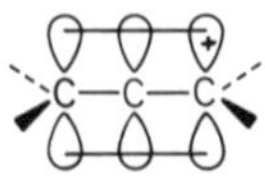

**243**

leading to it. An analogous stabilization in vinylic systems can be achieved in two different ways. The second double bond can be either conjugated to the first one or directly attached to it in an allenic system. Both types of stabilization were investigated. The former was discussed in Section IV,A.

The other unsaturated allylic analogue is the allenyl system. From the viewpoint of orbital overlap, it has an advantage over the butadiene-type systems in that the allylic $\pi$ orbital is geometrically constrained to the most favored geometry for overlap with the p orbital due to the orthogonality of the two double bonds. Formation of the allenyl ion **245** is not accompanied by any loss of ground state conjugation, whereas formation of **244** requires ground state deconjugation by distortion.

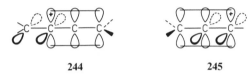

244                                         245

The solvolysis of allenic derivatives is related in different ways to several other reactions, which were discussed previously. The formation of vinyl cations by electrophilic addition to allenes and by participation of the allenyl double bond in solvolytic reactions was discussed in Chapter 4. The solvolysis of the cyclopropenylidenemethyl systems, which was discussed in Section IV,C, is analogous to the solvolysis of the allenic systems since a built-in orbital can overlap the developing vacant p orbital. The solvolysis of the propargylic derivatives **246** [Eq. (2)], which was extensively investigated [68–70], is also a closely related reaction since an extreme valence bond structure of the alkynylcarbenium ion **247** is the allenyl ion structure **248**. These hybrid ions were also prepared [71] under stable ion conditions from the propargyl alcohols, as discussed in Chapter 8.

$$R_1C\equiv C-CR_2R_3X \longrightarrow R_1C\equiv C-\overset{+}{C}R_2R_3 \longleftrightarrow R_1\overset{+}{C}=C=CR_2R_3 \longrightarrow products$$

246                            247                            248                   (2)

The first suggestion for the intermediacy of the allenyl cation in the solvolysis of allenic derivatives was that of Jacobs and Fenton [72]. However, the main work on the mechanism of reaction is the recent work of Schiavelli and co-workers [73–79], who showed with the aid of kinetics and substituent, solvent, salt, and isotope effects that the reactions proceed via an initial cleavage of the bond to the leaving group with the formation of the ion **245**.

Table 5.10 gives an extensive list of relative solvolysis rates of saturated and unsaturated chlorides and bromides and enables one to compare the reactivity of allenic compounds **249** with saturated and unsaturated

TABLE 5.10

Relative Solvolysis Rates of Saturated and Unsaturated
Halides in 80% EtOH at 25°

| Compound | Relative rate | Reference |
|---|---|---|
| $Ph_2CHCl$ | 192 | [19a] |
| $Ph_2C=C=C(Cl)Ph$ | $72^a$ | [74] |
| $t\text{-BuC(Me)}=C=C(Br)Me$ | 65 | [78] |
| $MeC\equiv CC(Cl)Me_2$ | 60 | [80] |
| $(CH_3)_3CBr$ | 40 | [19a] |
| $PhCH(Me)Br$ | 22 | [81] |
| $t\text{-BuC(Me)}=C=C(Br)Bu\text{-}t$ | 4.1 | [78] |
| $(CH_3)_3CCl$ | 1.0 | [82] |
| $HC\equiv CCMe_2Br$ | 0.5 | [82] |
| $E\text{-MeCH}=C(Br)-CH=CMe_2$ | 0.033 | [27] |
| $HC\equiv CCMe_2Cl$ | 0.025 | [68] |
| $CH_2=C(Br)-CH=CMe_2$ | 0.02 | [27] |
| $Me_2C=C=CHBr$ | 0.011 | [78] |
| $CH_2=C(OTf)Me$ | 0.01 | [85] |
| $Me_2CHBr$ | 0.008 | [83, 96, 106] |
| $E\text{-MeCH}=C(OTf)Me$ | 0.0038 | [85] |
| $CH\equiv CCH_2Br$ | $0.0017^b$ | [84] |
| $E\text{-CH}_2=C(Br)-CMe=CHMe$ | $1.5 \times 10^{-5}$ | [27] |
| 2-Adamantyl-Br | $1.2 \times 10^{-5}$ | [83, 96, 106] |
| 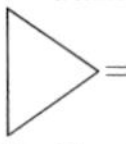 ▷=CHBr | $2.9 \times 10^{-6c}$ | [44] |
| $Ph_2C=C(Cl)Ph$ | $10^{-6d}$ | [20] |
| $CH_2=C=CHBr$ | $4 \times 10^{-7e}$ | [84] |
| $CH_2=C(Br)-CH=CH_2$ | $2 \times 10^{-7f}$ | [27] |

$^a$ Calculated by assuming $m = 0.69$ in aqueous EtOH.
$^b$ Based on the relative reactivity to allenyl bromide in 50% EtOH.
$^c$ Based on the relative reactivity to allenyl bromide in 60% EtOH at 160°.
$^d$ Based on data for $An_2C=C(Cl)Ph$, assuming a $\beta\text{-}An_2/\beta\text{-}Ph_2$ ratio of 5.45, as found in 60% EtOH [86].
$^e$ Calculated by assuming that $m = 0.44$ at 25°.
$^f$ Based on the relative reactivity to $E\text{-CH}_2=C(Br)-CMe=CHMe$ at 150°

analogues. The powerful activation by the allylic double bond is reflected by the fact that substituted allenyl systems, such as 3-*tert*-butyl-1,3-dimethyl bromoallene (**249**) (i.e., $R_3 = Me$; $R_2 = t\text{-Bu}$) [78] and triphenylallenyl chloride [74], are only three times less reactive than benzhydryl chloride [19a], whereas 3-*tert*-butyl-1,3-dimethyl chloroallene should have a reactivity similar to that of *tert*-butyl chloride [82]. The higher reactivity of the allenic compounds is even more pronounced when they are compared with their vinylic analogues. Triphenylallenyl chloride is nearly 8 orders of magnitude more reactive than triphenylvinyl chloride [20, 74]. 3-*tert*-Butyl-1,3-dimethyl bromoallene is $10^4$ times more reactive than *E*-2-butenyl triflate [85, 95] and, if a reasonable $k_{OTf}/k_{Br}$ reactivity ratio of $10^7$ is assumed, the allenic double bond enhances the reactivity by 11 orders of magnitude. This is also demonstrated by the fact that α-unsubstituted allenyl cations were formed from the precursor bromides under relatively mild conditions [75–79], whereas α-unsubstituted $\beta,\beta$-dialkylvinyl cations were never generated solvolytically. The activation to solvolysis by the allylic double bond in the vinylic system greatly exceeds the rate enhancement by the allylic bond in saturated systems. In 50% EtOH the unimolecular relative rate constants are $k(CH_2=CHCH_2Cl)/k(CH_3CH_2CH_2Cl) = 25$ [87], $k(CH_2=CHCHClCH_3)/k((CH_3)_2CHCl) = 675$, and $k(CH_2=CHC(CH_3)_2Cl)/k((CH_3)_3CCl) = 262$ [88]. This difference is due to the higher demand for positive charge stabilization in the vinylic system and to the favorable geometry for overlap in the allenic systems.

$$R_3R_2C=C=C(Br)R_1 \qquad R_1C\equiv CCR_2R_3Br$$

$$\textbf{249} \qquad\qquad\qquad \textbf{250}$$

The two modes of allylic stabilization in the vinylic system can be compared by using the data of Table 5.10. The reactivities of the unsubstituted bromoallene **249** ($R_1 = R_2 = R_3 = H$) and 2-bromo-1,3-butadiene are almost the same [27, 84], and similar reactivities are also shown by 3,3-dimethyl bromoallene (**249**, $R_1 = H$; $R_2 = R_3 = Me$) and 2-bromo-4-methyl-1,3-pentadiene, the corresponding dimethyl derivatives [27, 78]. Comparison of the reactivities of a double bond and a cyclopropyl group shows that cyclopropylidenemethyl bromide **140** is somewhat more reactive than bromoallene [22, 84], but it should be recognized that a $k_s$ route may intervene in the solvolysis of one or both of these unsubstituted compounds.

A comparison between an allenic halide and a propargylic halide that lead to the same ion shows that the relative reactivity depends on the system. The solvolysis of propargyl bromide **250** ($R_1 = R_2 = R_3 = H$) is 4230 times faster than that of bromoallene **249** ($R_1 = R_2 = R_3 = H$) [93], but the reactivity of the $\gamma$, $\gamma$-dimethyl derivative 3-bromo-3-methyl-1-butyne (**250**, $R_1 = H$; $R_2 = R_3 = Me$) is 45 times greater than that of 1-bromo-3-methyl-1,2-butadiene (**249**, $R_1 = H$; $R_2 = R_3 = Me$) [78, 82b]. It was suggested in the case of the unsubstituted compounds that this is due both to a ground state difference, which reflects the stronger $sp^2$ C–Br bond in **249** than the $sp^3$ C–Br bond in **250**, and to a difference in the transition state geometries [84]. For the $\gamma,\gamma$-dimethyl derivatives it was suggested that the ground state differences may be smaller than expected since it was assumed that the transition states are energetically similar, both being expected to rely heavily on the tertiary center for carbenium ion stabilization [78].

## 2. Exclusion of Non-$S_N1$ Substitution Routes

The possibility of competing substitution routes was considered and excluded for several of the allenyl systems. An electrophilic addition–elimination of proton by the solvent was avoided by conducting all the studies of solvolysis mechanisms in nonacidic media. However, since most of the reactions were conducted in unbuffered solvents, an electrophilic addition of the HX formed is still a possibility. This was ruled out for the reaction of triphenyl chloroallene in 90% EtOH since the reaction of $4 \times 10^{-5}$ $M$ substrate gives a pH of 4.4 at infinity, whereas the kinetic behavior and the rate constant were the same either at a tenfold higher substrate concentration, which gives a lower pH at infinity, or after the addition of 1250 times molar excess of triethylamine [74]. An $Ad_N$–E route was ruled out for the same reaction by the insensitivity to the added base [74], whereas the $k_{Br}/k_{Cl}$ ratio of 56 eliminated this route for 1,3-di-*tert*-butyl-2-phenyl haloallenes [75]. An $\alpha,\beta$-elimination–addition is impossible for the allenyl systems, but an $\alpha,\alpha$-elimination–addition via an intermediate unsaturated carbene $R_3R_2C=C=C:$ [5] should be considered. This mechanism was established for the solvolysis of $\alpha$-unsubstituted allenes in the presence of bases [5, 69, 82, 89, 90]. The $\alpha$- and $\delta$-isotope effects eliminate this possibility for several $\gamma$-substituted–$\alpha$-unsubstituted haloallenes [78]. The $\alpha$-isotope effects for the elimination should be primary and much higher than the observed secondary isotope effect (Section IV,E,5), whereas the $\delta$-isotope effect $k_{6\text{-}H}/k_{6\text{-}D}$ of 1.31 observed for the solvolysis of **249** ($R_1 = H$, $R_2 = R_3 = Me$) in 60% EtOH [78] is higher than the effect of 1.15 obtained for the base-catalyzed reaction of this substrate in 80% EtOH [89].

The first-order kinetics in all cases, the nature of the products (Section IV,E,3), the substituent effects (Section IV,E,4), the isotope effects (Section IV,E,5), the leaving group effects (Section IV,E,6), the solvent effects (Section IV,E,7), and the appearance of common ion rate depression (Section IV,E,8) are all in line with a $k_c$ variant of the solvolysis reaction [91], as discussed below. A possible exception is the unsubstituted bromo-allene, in which the $k_s$ route may be important.

## 3. Product Distribution

The solvolysis of the allenyl derivatives **249** can give two structurally isomeric products. Allenyl derivatives, or their tautomeric ketones when water is the nucleophile, could be visualized as arising by capture of the hybrid **248** at C-1, whereas propargylic derivatives should be formed by capture of **247** at C-3 [Eq. (3)]. The product distribution will be an important probe for the charge distribution in the hybrid ion **248–247**.

$$R_3R_2C{=}C{=}C(X)R_1 \xrightarrow{\;-X^-\;} R_3R_2C{=}C{=}\overset{+}{C}{-}R_1 \longleftrightarrow R_3R_2\overset{+}{C}{-}C{\equiv}C{-}R_1$$

$$\quad\quad\; \textbf{249} \quad\quad\quad\quad\quad\quad\quad \textbf{248} \quad\quad\quad\quad\quad\quad\quad \textbf{247}$$

$$\Big\downarrow {\scriptstyle H_2O} \quad\quad\quad\quad\quad \Big\downarrow {\scriptstyle H_2O} \quad\quad\quad (3)$$

$$\overset{\displaystyle O}{\overset{\|}{R_3R_2C{=}CHC{-}R_1}} \quad\quad\quad R_3R_2C(OH)C{\equiv}C{-}R_1$$

Both products were formed in the earlier work of Jacobs and Fenton, which suggested the intermediacy of allenyl cations [72]. 1-Bromo-4,4-dimethyl-1,3-diphenyl-1,2-pentadiene (**251**) gave, with sodium methoxide in methanol, the methyl ether **252** [Eq. (4)]. This was ascribed to an initial

$$PhC(t\text{-Bu}){=}C{=}C(Br)Ph \xrightarrow{\;-Br^-\;} PhC(t\text{-Bu}){=}C{=}\overset{+}{C}{-}Ph \longleftrightarrow Ph(t\text{-Bu})\overset{+}{C}{-}C{\equiv}C{-}Ph$$

$$\quad\quad \textbf{251} \quad\quad\quad\quad\quad\quad\quad \textbf{248a} \quad\quad\quad\quad\quad\quad \textbf{247a}$$

$$\Big\downarrow {\scriptstyle MeO^-} \quad\quad\quad\quad\quad\quad\quad (4)$$

$$PhC(t\text{-Bu}){=}C{=}C(OMe)Ph$$

$$\textbf{252}$$

ionization to the allenyl–propargyl cation **248a–247a**, which reacted preferentially with a nucleophile at its allenyl position, since attack at the propargyl position was sterically hindered by the *tert*-butyl group.

The hydrolysis of other allenyl halides (**253**) in aqueous solution, which gave the propargyl alcohols **254** rather than the ketones **255**, was ascribed to a similar ionization to the cation **247b–248b**, which was attacked more rapidly at the propargylic position [Eq. (5)]. The propargyl alcohol survived and was therefore the main product under the conditions of irreversibility.

$$PhC(R){=}C{=}C(X)R' \longrightarrow$$

$$\textbf{253}$$

$$[Ph{-}C(R){=}C{=}\overset{+}{C}{-}R' \longleftrightarrow Ph{-}\overset{+}{C}(R){-}C{\equiv}C{-}R'] \longrightarrow Ph{-}C(R)(\overset{+}{O}H_2)C{\equiv}C{-}R'$$

$$\quad\;\; \textbf{248b} \quad\quad\quad\quad\quad\quad \textbf{247b} \quad\quad\quad\quad\quad\quad\quad\quad H^+ {\Big\updownarrow} {-}H^+$$

$$\Big\downarrow {\scriptstyle H_2O} \quad\quad\quad\quad\quad\quad\quad\quad\quad\quad\quad\quad\quad\quad\quad Ph{-}C(R)(OH)C{\equiv}C{-}R'$$

$$\quad\quad\quad\quad\quad\quad\quad\quad\quad\quad\quad\quad\quad\quad\quad\quad\quad\quad\quad \textbf{254}$$

$$Ph{-}C(R){=}C{=}C(\overset{+}{O}H_2)R' \longrightarrow Ph{-}C(R){=}C{=}C(OH)R' \longrightarrow Ph{-}C(R){=}CHCOR'$$

$$R, R' = t\text{-Bu, or Ph} \quad\quad\quad\quad\quad\quad\quad\quad\quad\quad\quad\quad\quad\quad \textbf{255}$$

$$(5)$$

252                                                                                    5. Bond Heterolysis

Whereas the initial products from the reaction of 1,1,3-triphenyl-2-propyn-1-ol (**256**) with an excess of methyl or ethyl alcohol in the presence of sulfuric acid or from triphenylallenyl bromide (**257**) are only the propargyl ethers **258**, only the ketone **259** was obtained from **256** in alcohol/sulfuric acid after a long reaction time [Eq. (6)]. An acid-catalyzed hydrolysis of the propargyl methyl ether to the ketone via an intermediate formation of **247b**–**248b** and an acetal was suggested [72].

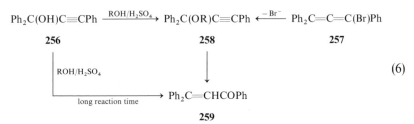

(6)

Table 5.11 gives the available product distributions for systems that were also studied kinetically. The propargylic product was either the exclusive or the main product of solvolysis and was even the exclusive product for the α-anisyl-substituted cation. Allenyl products were observed only when both γ substituents were *tert*-butyl groups. An exception was the formation of the ether **252** from **251** and methoxide ion, in contrast to the exclusive formation of the propargyl alcohol from the ion **248a**, which was formed from the chloroallene in aqueous acetone [75]. The different reaction conditions may account for this difference.

The conclusion from the product distribution is that most of the charge resides on C-3 and on its substituents and that the tertiary propargylic

**TABLE 5.11**

**Product Distributions in the Solvolysis of Chloroallenes**
$R_3R_2C=C=C(Cl)R_1$ **in Aqueous Acetone**

| $R_1$ | $R_2$ | $R_3$ | $R_3R_2C(OH)C{\equiv}CR_1$ $(\%)^a$ | $R_3R_2C=CHCOR_1$ $(\%)^a$ | Reference |
|-------|-------|-------|------|------|-----------|
| *t*-Bu | *t*-Bu | *t*-Bu | 80 | 20 | [75] |
| *t*-Bu | *t*-Bu | Ph | 100 | —[b] | [75, 93] |
| Ph | Ph | *t*-Bu | 100 | —[b] | [75] |
| Ph | *t*-Bu | *t*-Bu | 94 | 6 | [75] |
| Ph | Ph | Ph | 100 | —[b] | [73, 74] |
| An | Ph | Ph | 100 | —[b] | [74] |

[a] Percentage of the compound in the propargyl–allenyl products.
[b] No evidence for the formation of the ketone was found.

resonance structure **247** contributes more to the hybrid than the allenyl structure **248**, in accord with theoretical calculations of Drenth and co-workers [92] on the relative charge distributions in the ions **247** and **248**, as indicated in Chapter 2. Capture by the solvent takes place at C-3, except when the approach to this carbon is hindered by two *tert*-butyl groups. The importance of structure **247** is further shown in the following sections.

### 4. Effect of $\alpha$ and $\gamma$ Substituents

The high efficiency of the allylic double bond of the allenic system in charge delocalization is strongly reflected by the effects of the $\alpha$ and the $\gamma$ substituents on the solvolysis rates. The activation by aryl groups is shown by the fact that the reactivity of triphenylallenyl chloride **260a** is $10^{10}$ times higher than the extrapolated value for the unsubstituted chloroallene. The effect of the substituents in the aryl groups of aryl-substituted chloroallenes is evaluated with the aid of the $\rho$ values. The log $k$ values for the solvolysis of the triarylallenyl chlorides **260a** through **260d** [Eq. (7)] give a linear plot when Brown and Okamoto's $\sigma^+$ values [94] are used, with $\rho^+ = -2.0$ [74]. This value is significantly lower than the values of $< -3.6$ obtained for the

$$(YC_6H_4)_2C{=}C{=}C(Cl)C_6H_4X\text{-}p \longrightarrow (YC_6H_4)_2\overset{+}{C{\cdots}C{\cdots}}CC_6H_4X\text{-}p \longrightarrow$$

| | |
|---|---|
| **260a**  X = H; Y = H | **261** |
| **260b**  X = Cl; Y = H | |
| **260c**  X = Me; Y = H | $(YC_6H_4)_2C(OH){-}C{\equiv}CC_6H_4X\text{-}p$ |
| **260d**  X = MeO; Y = H | |
| **260e**  X = H; Y = $p$-Me | (7) |

solvolyses of $\alpha$-arylvinyl derivatives in the absence of anchimeric assistance (Table 6.7) and indicates that charge dispersal by the $\alpha$-aryl group is less extensive than in the vinylic systems. On the other hand, **260e**, which is substituted by two $\gamma$-$p$-tolyl groups, is 3.7 times more reactive than the $\gamma,\gamma$-diphenyl compound **260a** (Table 5.12) [74] and, assuming additivity of the effects and $\sigma^+$ dependency, the $\rho^+$ value is $-0.9$ for a change of the $\gamma$-aryl substituents. Although this value is not very negative, it is still more negative than the values for $\beta$-aryl groups in triarylvinyl systems in nucleophilic solvents (Chapter 6, Section III). Nevertheless, this value does not indicate that extensive charge delocalization to the $\gamma$ position takes place, as expected. The reason for this is probably the mutual steric interaction of the two $\gamma$-aryl groups in the ion **261**, the result of which is the inability of the two groups to achieve planarity at the same time. The importance of this effect is demonstrated when coplanarity is enforced on the $\gamma$-aryl group. The indanyl derivative **262** is 6800 times more reactive in 90% acetone at

**TABLE 5.12**

**Relative Solvolysis Rates of $R_3R_2C{=}C{=}C(Cl)R_1$ in Aqueous Acetone at 35°**

| $R_1$ | $R_2$ | $R_3$ | Solvent | $k_{rel}{}^a$ | $\Delta H^{\ddagger}$, kcal/mole | $\Delta S^{\ddagger}$ (25°), eu | Reference |
|---|---|---|---|---|---|---|---|
| $t$-Bu | $t$-Bu | $t$-Bu | 50% $Me_2CO$ | 1.00 | 23.4 | −5.0 | [75] |
| $t$-Bu | $t$-Bu | Ph | 50% $Me_2CO$ | 1.84 | — | — | [75] |
| Ph | Ph | $t$-Bu | 50% $Me_2CO$ | 10.6 | — | — | [75] |
| Ph | $t$-Bu | $t$-Bu | 50% $Me_2CO$ | 14.4 | 20.2 | −10.4 | [75] |
| $t$-Bu | Ph | Ph | 50% $Me_2CO$ | 87.6 | $19.4^b$ | $-11.7^b$ | [75] |
| Ph | Ph | Ph | 50% $Me_2CO$ | $3.45 \times 10^2$ | $20.2^c$ | $-10.7^c$ | [73, 74] |
| $t$-Bu | 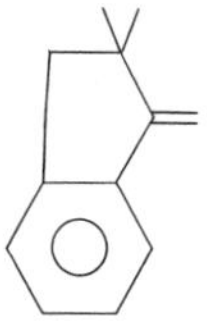 | | 90% $Me_2CO$ | $1.25 \times 10^4$ | $20.4^d$ | $-11.0^d$ | [76] |
| $p$-$ClC_6H_4$ | Ph | Ph | 80% $Me_2CO$ | $2.16 \times 10^2$ | $21.1^c$ | $-8.8^c$ | [74] |
| Ph | $Tol^e$ | $Tol^e$ | 80% $Me_2CO$ | $1.27 \times 10^3$ | — | — | [74] |

$^a$ At 44.8°, $k = 3.75 \times 10^{-5}$.
$^b$ In 60% acetone.
$^c$ In 80% acetone.
$^d$ In 70% acetone, $\Delta H^{\ddagger} = 19.5$ kcal/mole and $\Delta S^{\ddagger} = -10.3$ eu.
$^e$ Tol = $p$-tolyl.

35° [76] than the open-chain counterpart 1,3-di-*tert*-butyl-3-phenylallenyl chloride (**263**), as shown in Table 5.12 [75].

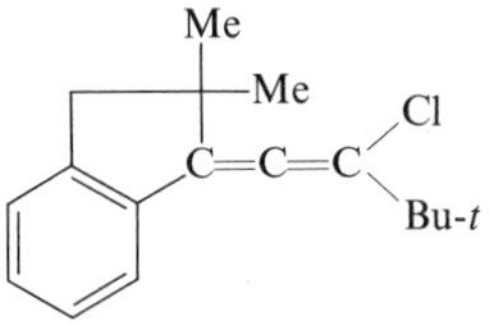

**262** 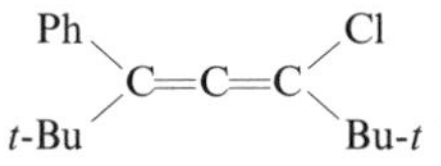                    **263**

In view of the effect of the $\gamma$-tolyl groups, only a small fraction of this difference is due to the presence of the *ortho*-alkyl group of **262**, which is absent in **263**. Furthermore, since C-3 is remote from the reaction center and does not undergo any hybridization change, the strain introduced by the five-membered ring is identical in the ground state and in the transition state and should not be reflected in the rate difference. A major portion of the rate difference was therefore ascribed to the constrained coplanarity of the aromatic ring and the $C_\beta$—$C_\gamma$ double bond and to the consequently higher stabilization of the transition state for the solvolysis of **262**. A similar rate difference was obtained for related saturated derivatives, since **264** solvolyzes $10^4$ times faster than **265** in 80% EtOH [95].

**264**                              **265**

The relative rates for various combinations of phenyl and *tert*-butyl substituents in chloroallene are given in Table 5.12. Whereas triphenyl-allenyl chloride is 345 times faster than tri-*tert*-butylallenyl chloride, the $k_{Ph}/k_{t\text{-}Bu}$ ratio, which reflects the effect of substitution of a *tert*-butyl group by a phenyl group, depends on the nature of the other substituents, and consecutive substitution does not result in an additive change. The ratios for a substitution at C-1 are 14.4, 5.8, and 3.9 when zero, one, or two phenyl groups are present, respectively, on C-3 [76]. This is consistent with an increased charge delocalization by the aromatic rings on C-3, with a resultant reduction of the charge on C-1. A similar explanation applies to the $k_{Ph}/k_{t\text{-}Bu}$ ratios at C-3, which are 1.8 and 0.74 for the 1-*tert*-butyl- and the 1-phenyl-substituted system, respectively, when a *tert*-butyl group is present on C-3. Substitution of the second *tert*-butyl by a phenyl at C-3 results in 47.6- and 32.5-fold rate enhancements for the 1-*tert*-butyl and the 1-phenyl derivatives. This change is more pronounced than the change of the first $\gamma$-*tert*-butyl group and is apparently due to a much larger steric interaction between a $\gamma$-phenyl and a $\gamma$-*tert*-butyl group than between two $\gamma$-phenyl groups. Consequently, a closer approach to planarity of both groups is achieved in the $\gamma,\gamma$-diphenyl derivatives and is reflected in the higher rate [76].

Table 5.13 summarizes the effect of $\alpha$- and $\gamma$-alkyl groups in aqueous ethanol. The $k_{\alpha\text{-}Me}/k_{\alpha\text{-}H}$ reactivity ratios are $2 \times 10^4$ when the two $\gamma$ substituents are methyl groups [79], $1.33 \times 10^4$ when they are *tert*-butyl and methyl, and $3.4 \times 10^3$ when they are *tert*-butyl [77, 78]. The $k_{\alpha\text{-}t\text{-}Bu}/k_{\alpha\text{-}H}$ ratio is $1.07 \times 10^3$ when the $\gamma$ substituents are methyl and *tert*-butyl [77, 78]. A somewhat surprising result is that the effect of the $\gamma$ substituents does not differ much when the $\alpha$-carbon is either unsubstituted or substituted by an alkyl group. A $\gamma$-methyl and a $\gamma$-*tert*-butyl group together increase the reactivity of the $\alpha$-unsubstituted compound by four orders of magnitude, or by three orders of magnitude if the $\alpha$-carbon is substituted by a *tert*-butyl group. The effect of two $\gamma$-methyl groups is nearly the same for the unsubstituted and for the $\alpha$-methylallenyl chloride. These and similar values in Table 5.13 indicate the importance of charge delocalization by the $\gamma$-alkyl substituents. The difference between the reactivities of different $\gamma$-alkyl

TABLE 5.13

Effects of $\alpha$ and $\gamma$ Substituents for $R_3R_2C=C=C(Br)R_1$

| $R_1$ | $R_2$ | $R_3$ | $k_{rel}^a$ | Solvent | Temp (°C) | $k_{rel}^b$ | $\Delta H^{\ddagger}$ (kcal/mole) | $\Delta S^{\ddagger}$ (eu) | Reference |
|---|---|---|---|---|---|---|---|---|---|
| t-Bu | t-Bu | Me | 1.0 | 60% EtOH | 64 | $1.25 \times 10^7$ | 21.1$^c$ | -3.3$^c$ | [77] |
| H | t-Bu | Me | $1.07 \times 10^{-3}$ | 60% EtOH | 64 | $1.34 \times 10^4$ | — | — | [78] |
| H | Me | Me | $3.94 \times 10^{-3}$ | 60% EtOH | 64 | $4.93 \times 10^4$ | — | — | [78] |
| H | Me | Et | $3.94 \times 10^{-3}$ | 60% EtOH | 54.5 | $4.93 \times 10^4$ | — | — | [79] |
| H | H | H | $2.94 \times 10^{-7}$ | 60% EtOH | 160 | — | — | — | [84] |
| t-Bu | t-Bu | Me | 1.0 | 80% EtOH | 24.6 | — | 22.3 | -3.5 | [77] |
| Me | t-Bu | Me | 14.3 | 80% EtOH | 24.6 | $1.79 \times 10^8$ | — | — | [78] |
| Me | t-Bu | t-Bu | 0.64 | 80% EtOH | 24.6 | $8.0 \times 10^6$ | 23.4 | -1.0 | [78] |
| t-Bu | t-Bu | Me | 1.0$^d$ | 50% EtOH | 74.4 | $1.25 \times 10^7$ | — | — | [77] |
| H | t-Bu | t-Bu | $1.87 \times 10^{-4}$ | 50% EtOH | 74.4 | 2338 | — | — | [78] |
| H | H | H | $8.0 \times 10^{-8}$ | 50% EtOH | 74.4 | 1.0 | 24.9 | -23.4 | [84] |

$^a$ Relative rate constant compared with the first compound in each solvent.

$^b$ Relative $k$ in aqueous EtOH. Based on all the data in the temperature range $25°-75°$.

$^c$ In 70% EtOH, $\Delta H^{\ddagger} = 22.1$ kcal/mole and $\Delta S^{\ddagger} = -2.3$ eu.

$^d$ Extrapolated by using an $m$ value of 0.81 and an Ea value similar to that in 60% EtOH.

groups is larger for the $\alpha$-unsubstituted haloallenes. For $\alpha$-unsubstituted $\alpha$-methylallenyl bromide the relative reactivities of the second $\gamma$-alkyl substituents are similar, being Me, 1; Et, 1; $t$-Bu, 0.17. For $\gamma$-$tert$-butyl-$\alpha$-methylallenyl chloride the effect of the second $\gamma$-alkyl group is Me, 1; $t$-Bu, 22. The effects of the $\gamma$-alkyl groups are approximately additive. For the $\alpha$-unsubstituted system two $\gamma$-$tert$-butyl groups increase the reactivity 2340-fold, one $\gamma$-methyl and one $\gamma$-$tert$-butyl increase it 13,400 times, and two $\gamma$-methyl groups increase it 49,300-fold. The combined effect of three alkyl groups is shown by the fact that the reactivity of 3-$tert$-butyl-1,3-dimethyl-allenyl chloride is $1.8 \times 10^8$ higher than that of the unsubstituted compound [78, 84].

As discussed in Section IV,E,7, the relative reactivities are also solvent dependent, and the $k(70\% \text{ TFE})/k(50\% \text{ EtOH})$ rate ratio decreases in the order $t$-Bu$_2$C=C=CHBr(8.2) > $t$-BuC(Me)=C=CHBr(3.7) > Me$_2$C=C=CHBr (1.8). These data raise the possibility of nucleophilic solvent assistance by solvent attack on the neutral substrate or the ion pair [77].

The $\alpha$-substituent effects $k_{\alpha\text{-Me}}/k_{\alpha\text{-H}}$ of $3.4 \times 10^3$ to $2 \times 10^4$ in aqueous EtOH are relatively low when compared with the limiting value of $10^8$ suggested by Schleyer and co-workers as indicative of a $k_c$ route [96]. Moreover, since $tert$-butyl has a greater electron-donating effect than methyl, a $k_{\alpha\text{-}t\text{-Bu}}/k_{\alpha\text{-H}}$ ratio higher than the $k_{\alpha\text{-Me}}/k_{\alpha\text{-H}}$ ratio is expected, but the $k_{\alpha\text{-}t\text{-Bu}}/k_{\alpha\text{-H}}$ of $1.3 \times 10^3$ in 80% EtOH is lower than the $k_{\alpha\text{-Me}}/k_{\alpha\text{-H}}$ ratios. On the other hand, in the nonnucleophilic 97% TFE, a $tert$-butyl group is much more efficient, and $k[t\text{-Bu}_2\text{C=C=C(Br)}t\text{-Bu}]/k(t\text{-Bu}_2\text{C=C=CHBr}) \geq 10^{5.1}$ [79]. Consequently, these ratios may substantiate some nucleophilic assistance in aqueous EtOH, probably at the ion-pair stage.

### 5. $\alpha$-Deuterium and $\delta$-Deuterium Isotope Effects

The high reactivity of allenyl systems makes it possible to measure the $\alpha$-deuterium secondary isotope effect in the solvolysis, whereas such values are unavailable for the solvolyses of the much less reactive vinyl systems (see Section V,A,3). In addition, isotope effects of substituents on the $\gamma$-carbon can also be measured, and both $\alpha$- and $\delta$-deuterium isotope effects were used as sensitive probes of the nature of the transition state in the solvolysis of the allenyl halides.

The many isotope effects measured for the allenyl systems are listed in Table 5.14, together with the differential free energy of activation per deuterium associated with the isotope effect. The $\alpha$-deuterium secondary isotope effects $k_H/k_D$ for three different systems in 50% EtOH, 70% TFE, or 97% TFE are all high, being 1.20 to 1.28, with associated $\Delta\Delta F^{\ddagger}$ values per deuterium of 118 to 168 cal/mole per deuterium [77, 78]. These values are the largest reported for solvolysis reactions. For "limiting" or $k_c$ solvolysis,

**TABLE 5.14**

**Isotope Effects in the Solvolysis of Haloallenes**

| Substrate | Isotope effect | Solvent | Temp (°C) | $k_H/k_D{}^a$ | $\Delta\Delta F^{\ddagger}$ per D (cal/mole) | Reference |
|---|---|---|---|---|---|---|
| t-Bu₂C=C=CHBr | α | 50% EtOH | 74.4 | 1.28 ± 0.03 | 168 | [78] |
| | α | 70% TFE | 74.4 | 1.20 ± 0.01 | 126 | [78] |
| t-BuC(Me)C=C=CHBr | α | 50% EtOH | 64.2 | 1.20 ± 0.01 | 121 | [78] |
| | α | 70% TFE | 60.2 | 1.218 ± 0.009 | 130 | [77, 78] |
| Me₂C=C=CHBr | α | 97% TFE | 64.2 | 1.28 ± 0.05 | 169 | [78] |
| | α | 70% TFE | 54.2 | 1.23 ± 0.02 | 136 | [78] |
| | α | 50% EtOH | 54.2 | 1.20 ± 0.01 | 118 | [78] |
| t-BuC(Me)=C=C(Br)Bu-t | δ | 60% EtOH | 24.6 | 1.21 ± 0.03 | 37 | [77, 78] |
| | δ | 70% EtOH | 24.6 | 1.19 ± 0.02 | 35 | [77] |
| | δ | 80% EtOH | 45.3 | 1.23 ± 0.03 | 44 | [77, 78] |
| | δ | 90% EtOH | 45.3 | 1.27 ± 0.04 | 50 | [77] |
| | δ | 70% TFE | 60.2 | 1.23 ± 0.01 | 44 | [77] |
| t-BuC(Me)=C=C=CHBr | δ | 97% TFE | 64.1 | 1.33 ± 0.02 | 64 | [78] |
| | δ | 70% TFE | 60.2 | 1.21 ± 0.02 | 42 | [78] |
| | δ | 60% EtOH | 64.1 | 1.17 ± 0.04 | 34 | [78] |
| | δ | 50% EtOH | 64.2 | 1.18 ± 0.01 | 38 | [78] |
| Me₂C=C=CHBr | δ | 97% TFE | 64.2 | 1.33 ± 0.02$^b$ | 32 | [78] |
| | δ | 70% TFE | 54.2 | 1.33 ± 0.01$^b$ | 31 | [78] |
| | δ | 60% EtOH | 64.2 | 1.31 ± 0.04$^b$ | 31 | [78] |
| | δ | 50% EtOH | 54.2 | 1.33 ± 0.04$^b$ | 31 | [78] |
| t-BuC(Me)=C=C(Cl)Bu-t | δ | 97% TFE | 15.3 | 1.121 ± 0.009 | 24 | [78] |
| | δ | 60% EtOH | 44.1 | 1.13 ± 0.01 | 27 | [78] |

$^a$ Corrected to 100% deuteration.
$^b$ Isotope effect for two δ-methyl groups.

Shiner and co-workers suggested maximal $k_H/k_D$ values which are dependent on the nature of the leaving group [81]. For example, for a saturated chloride, the $k_H/k_D$ value of 1.15 (83 cal/mole per deuterium) is considered to be a maximum [97], whereas the value for a bromide leaving group is even lower (1.125; 70 cal/mole per deuterium) [81]. The maximal $k_H/k_D$ values are higher for oxygen leaving groups [81].

The expected maximal $k_{\alpha\text{-H}}/k_{\alpha\text{-D}}$ values for an $sp^2 \to sp$ hybridization change was estimated [77] from two equilibrium constants: $K_1 = 1.075$ for the $sp^3 \to sp^2$ change of reaction (8) [98] and $K_2 = 1.257$ for the $sp^2 \to sp$ change of reaction (9) [98]. Assuming that the same differential effect applies

$$CH_3CH_3 + CH_2{=}CHD \xrightarrow{K_1} CH_3CH_2D + CH_2{=}CH_2 \qquad (8)$$

$$CH_2{=}CH_2 + HC{\equiv}CD \xrightarrow{K_2} CH_2{=}CHD + HC{\equiv}CH \qquad (9)$$

for the solvolysis data, the $k_{\alpha\text{-H}}/k_{\alpha\text{-D}}$ for the $sp^2 \to sp$ hybridization change is equal to $K_2/K_1$, i.e., 1.17 times that for the $sp^3 \to sp^2$ change. When a value of 1.125 for the latter change with a bromide leaving group is taken, the maximal $k_{\alpha\text{-H}}/k_{\alpha\text{-D}}$ value expected for the solvolysis of a vinyl or an allenyl bromide is 1.32 [77]. A more recent estimate of this value [78] is 1.28, which is the value obtained for the solvolysis of 249 ($R_1 = H$; $R_2 = R_3 = t\text{-Bu}$) or 249 ($R_1 = H$; $R_2 = R_3 = CH_3$) in 97% TFE (Table 5.14).

Although the $k_{\alpha\text{-H}}/k_{\alpha\text{-D}}$ values of Table 5.14 are high, they do not approach the maximal expected values. However, it was noted [78] that the use of these values to assess the contribution of the $k_s$ route is restricted in the allenyl system due to the presence of potentially two sites for solvent assistance. Whereas nucleophilic solvent assistance at $C_\alpha$ lowers the $k_{\alpha\text{-H}}/k_{\alpha\text{-D}}$ values, as is the case with saturated systems, solvent assistance at $C_\gamma$ results in an increased s character of the $C_\alpha{-}H$ bond, thus increasing $k_{\alpha\text{-H}}/k_{\alpha\text{-D}}$. It was further argued that, whereas attack at $C_\alpha$ is unlikely in the neutral allenyl bromide, it is a possibility in the tighter ion pair, in which the charge is delocalized on both $C_\alpha$ and $C_\gamma$. Keeping in mind this limitation, a constancy of the $k_{\alpha\text{-H}}/k_{\alpha\text{-D}}$ values in solvents of different nucleophilicities would argue for a $k_c$ route, whereas a lower value in a more nucleophilic solvent would indicate a $k_s$ contribution. The $[k_{\alpha\text{-H}}/k_{\alpha\text{-D}}$ (50% EtOH)]/$[k_{\alpha\text{-H}}/k_{\alpha\text{-D}}$ (70% TFE)] ratios are 1.067, 0.985, and 0.976, respectively, for the $\gamma,\gamma$-di-tert-butyl, $\gamma$-tert-butyl-$\gamma$-methyl, and the $\gamma,\gamma$-dimethyl $\alpha$-bromoallenes. The last two values are indistinguishable from unity within the combined experimental errors. Consequently, the $\alpha$-isotope effects do not support a $k_s$ contribution.

It was suggested that the $\delta$-isotope effects would be a better probe of the transition state structure [78]. The $\delta$-hydrogen isotope effects observed on replacing one or two $\gamma$-methyl groups by a $CD_3$ group are 1.17 to 1.33

$(\Delta\Delta F^{\ddagger} = 31$ to $64$ cal/mole per deuterium) for a bromide leaving group and $1.12$ to $1.13$ $(\Delta\Delta F^{\ddagger} = 24$ to $27$ cal/mole per deuterium) for a chloride leaving group [77, 78]. These values for the $\gamma$-substituted allenes could be compared with the $\beta$-isotope effects $k_{\beta\text{-CH}_3}/k_{\beta\text{-CD}_3}$ for the solvolysis of a propargyl system that also leads to a trialkyl-substituted propargyl-allenyl cationoid intermediate. The isotope effect for the solvolysis of **266** $[k(\textbf{266a})/k(\textbf{266b})]$ is $1.65$, and $\Delta\Delta F^{\ddagger}$ is $50$ cal/mole per deuterium [80]. The $k_{\beta\text{-CH}_3}/k_{\beta\text{-CD}_3}$ values

$$(\text{CH}_3)_2\text{C(Cl)}\text{—C}{\equiv}\text{CCH}_3 \qquad (\text{CD}_3)_2\text{C(Cl)}\text{—C}{\equiv}\text{CCH}_3$$

**266a**            **266b**

for the solvolysis of this system and of *tert*-butyl chloride in aqueous EtOH are $1.33$ and $1.31$, respectively. It was suggested that the rate-determining step in these cases is the conversion of a tight ion pair to a solvent-separated ion pair [97] and that this value is characteristic of such a process for other systems. The relatively high values for a remote substituent in the allenic system, either substituted or unsubstituted at the $\alpha$ position, strongly support a solvolysis mechanism via a positively charged intermediate, the charge of which is highly delocalized to the $\gamma$ position. Moreover, the $k_{\delta\text{-CD}_3}/k_{\delta\text{-CH}_3}$ of $1.33$ for 3-*tert*-butyl-3-methylallenyl bromide (**249**, $R_2 = \text{Me}$; $R_3 = t\text{-Bu}$) is consistent with a rate-determining conversion of a tight ion pair to a solvent-separated ion pair. Substantially lower $\delta\text{-CD}_3$ effects (Table 5.14) are ascribed to a nucleophilic attack in the rate-determining step at either $C_\alpha$ or $C_\gamma$. The simplified scheme shown below was suggested for such cases

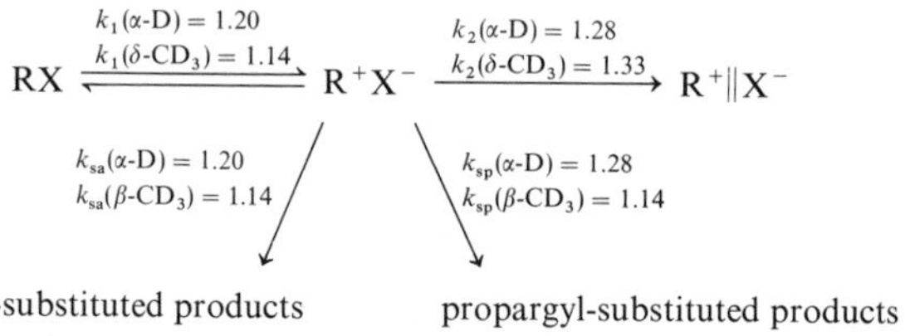

$k_1(\alpha\text{-D}) = 1.20$      $k_2(\alpha\text{-D}) = 1.28$

$\text{RX} \xrightleftharpoons{\dfrac{k_1(\delta\text{-CD}_3) = 1.14}{}} \text{R}^+\text{X}^- \xrightarrow{\dfrac{k_2(\delta\text{-CD}_3) = 1.33}{}} \text{R}^+\|\text{X}^-$

$k_{\text{sa}}(\alpha\text{-D}) = 1.20$      $k_{\text{sp}}(\alpha\text{-D}) = 1.28$
$k_{\text{sa}}(\beta\text{-CD}_3) = 1.14$     $k_{\text{sp}}(\beta\text{-CD}_3) = 1.14$

allenyl-substituted products        propargyl-substituted products

[99] by the values for maximal $\alpha$-isotope effect and for the interconversion of the ion pairs for the $k_2$ step. According to this scheme, the $\delta$-isotope effect should be reduced by half for a rate-determining heterolysis $(k_1)$ or attack on the propargyl $(k_{\text{sp}})$ or the allenyl $(k_{\text{sa}})$ center. The $\alpha$-isotope effect is reduced only for nucleophilic attack on the tight ion pair at $C_\alpha$, i.e., for cases when $k_1$ or $k_{\text{sa}}$ is rate limiting. When $k_{\text{sp}}$ is rate determining, the value of the $\alpha$-isotope effect is the maximal value.

Analysis of the isotope effects suggests that $k_{\text{sp}}$ is rate determining for 3,3-dimethylallenyl bromide in $97\%$ TFE, whereas in aqueous EtOH $k_1$ is rate determining. For hindered substrates in aqueous TFE, $k_2$ is rate determining, since $k_{\text{sp}} \ll k_2$. This is the case for the 3-*tert*-butyl-3-methylallenyl bromide and possibly for the 3,3-di-*tert*-butylallenyl bromide.

Support for the involvement of ion pairs in the solvolysis comes from the polarimetric–titrimetric rate ratio discussed in Section IV,E,8. Moreover, the Meyer–Schuster rearrangement has been shown to involve a rate-determining nucleophilic attack at the allenyl center of an ion–dipole pair.

$$Ar_2C(OH)C\equiv CH \xrightarrow{\ H^+\ } Ar_2C=CHCHO$$

The $\delta$-isotope effects for most of the substrates are remarkably constant from 50% EtOH to 97% TFE, suggesting that for these substrates the rate-determining step is the same in the different solvents. An exception is 1,3-di-*tert*-butyl-1,3-methylallenyl bromide for which the value increases with the decrease in the water content of the solvent [77] (Table 5.14), but this may be due to an experimental error, as well as to a change in the rate-determining step. The constancy of the isotope effects argues strongly for a $k_c$ process for all the substrates, including the less reactive ones. Variations are ascribed to a change in the rate-determining step, as discussed above.

The larger $\delta$-isotope effect for 3-*tert*-butyl-3-methylallenyl bromide in 97% TFE as compared with aqueous EtOH is reminiscent of the behavior for *tert*-butyl chloride [82, 100]. This was ascribed for *tert*-butyl chloride to the increased importance of a rate-determining elimination from an ion pair in 97% TFE and was supported by the higher fraction of olefin in this solvent.

## 6. Effect of the Leaving Group

Bromide and chloride were the only leaving groups investigated in the solvolysis of allenyl systems. The $k_{Br}/k_{Cl}$ reactivity ratios were reported to be 20 to 58 [78]. Examples are the value of 56 for 1,3-di-*tert*-butyl-3-phenyl 1-haloallenes in 50% acetone [75], 31 for 1-*p*-anisyl-4,4-diphenyl-1-halobutatrienes in 70% EtOH [101], and 52 for the solvolysis–cyclization of 1-[2-(3-*o*-anisyl)benzo[*b*]furanyl]-4,4-di-*o*-anisyl-1-halobutatrienes (**267**) in 70% EtOH [101]. The values for the 1,3-di-*tert*-butyl-3-methyl haloallenes, which are calculated by using temperature extrapolations [77, 78], are 28 and 29 in 60% EtOH and 70% EtOH, respectively.

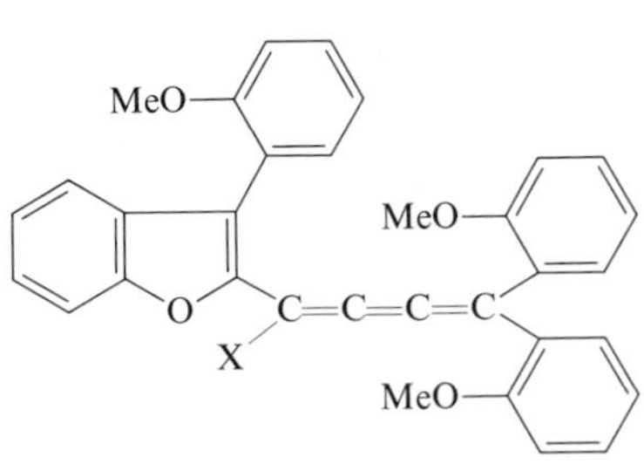

**267**

## 7. Solvent Effects

Due to the sensitivity of the allenyl systems to electrophilic additions, solvent effects were measured only in nucleophilic solvents, such as aqueous ethanol, aqueous trifluoroethanol, and aqueous acetone. Although only two or three solvent compositions were investigated at the same temperature, the data in Table 5.15 are presented for convenience as Grunwald–Winstein $m$ values. Despite the limited data, it is significant that almost all the $m$ values are high, being 0.7 to 1.2, i.e., in the region assigned for the nucleophilically unassisted $k_c$ route. The three exceptions are the lowest value of 0.46 for the unsubstituted bromoallene [84], the value of 0.69 for triphenyl chloroallene [80, 81], and the value of 0.46 for the butatrienyl bromide **267** in 60–80% EtOH, although in the last case the log $k$ versus $Y$ plot is curved, and $m = 0.94$ at 80–90% aqueous EtOH [101]. Since the latter two systems are α-aryl-substituted and reactive, a $k_s$ contribution is highly unlikely, and the low $m$ values can be ascribed to a relatively low demand for charge dispersal by the solvent owing to an extensive charge delocalization on $C_\alpha$, $C_\gamma$, and the aryl substituents. The low $m$ value for bromoallene, which is similar to the $m$ value of 0.455 for the allyl system [67, 88], together with the low reactivity, high activation energy, and highly negative activation entropy, suggests nucleophilic solvent assistance for the solvolysis of this compound.

The $m$ values are higher than those for similarly substituted vinylic systems (Table 6.17), and this difference supports the suggestion that the low $m$ values for heavily substituted vinyl halides are due to steric hindrance to solvation. Two factors enhance the solvation of the allenyl systems compared with that of the analogous vinylic systems. First, the α and γ substituents in the allenyl system are one additional double bond away from one another compared with the α and β substituents on a vinylic carbon. Second, whereas the substituents in the vinylic systems are in the same plane, they are in orthogonal planes in the allenyl systems and steric hindrance to solvation is therefore minimal.

The $(k_{aq.\ TFE}/k_{aq.\ EtOH})_Y$ values also rule out the solvent-assisted $k_s$ route. When nucleophilic solvent assistance is important, the reaction rate is much higher in a solvent of high nucleophilicity than in a solvent with low nucleophilicity, but with the same ionizing power $Y$. In this way, it was shown that the $k_s$ route is important for 2-propyl tosylate but is unimportant for *tert*-alkyl halides [83, 102]. An exception is the reaction of 1-adamantyl bromide, in which the solvolysis rates in aqueous TFE mixtures are higher than the values in other solvents with the same $Y$ value [102]. In spite of this, a comparison of the solvolysis rates of haloallenes in 97% TFE ($Y = 1.15$; $N = -2.59$) with those in 60% EtOH ($Y = 1.12$; $N = -0.1$) or of those in 70% TFE ($Y = 1.66$; $N = -1.09$) with those in 50% EtOH ($Y = 1.66$; $N =$

**TABLE 5.15**

**Values of $m$ for the Solvolysis of $\alpha$-Haloallenes**

| Compound | Solvent[a] | Temp (°C) | $n^b$ | $m$ | Reference |
|---|---|---|---|---|---|
| $CH_2\!=\!C\!=\!CHBr$ | 60–50% E–W | 160 | 2 | 0.46 | [84] |
| $Me_2C\!=\!C\!=\!CHBr$ | E–W[c] | —[c] | —[c] | 0.88 | [78] |
| $t\text{-BuC(Me)}\!=\!C\!=\!CHBr$ | 60–50% E–W | 64.2 | 2 | $0.93^d$ | [78] |
| $t\text{-BuC(CD}_3)\!=\!C\!=\!CHBr$ | 60–50% E–W | 64.2 | 2 | 0.92 | [78] |
| $t\text{-Bu}_2C\!=\!C\!=\!C(Me)Br$ | 80–60% E–W | 24.6 | 3 | $1.00 \pm 0.06$ | [78] |
| $t\text{-BuC(Me)}\!=\!C\!=\!C(t\text{-Bu})Cl$ | 70–50% E–W | 44.1 | 3 | $0.90 \pm 0.05$ | [78] |
| $t\text{-BuC(Me)}\!=\!C\!=\!C(t\text{-Bu})Br$ | 70–60% E–W | 24.6 | 2 | 0.95 | [77] |
| | 80–60% E–W | 34.7 | 3 | $0.81 \pm 0.04$ | [77] |
| | 90–80% E–W | 45.3 | 2 | 0.72 | [77] |
| $t\text{-BuC(CD}_3)\!=\!C\!=\!C(t\text{-Bu})Br$ | 70–60% E–W | 24.6 | 2 | 0.95 | [77] |
| | 90–80% E–W | 45.3 | 2 | 0.73 | [77] |
| $t\text{-Bu}_2C\!=\!C\!=\!C(t\text{-Bu})Cl$ | 50–40% E–W | 44.8 | 2 | 1.21 | [75] |
| | 60–40% E–W | 55.2 | 3 | 1.22 | [75] |
| $t\text{-BuC(Ph)}\!=\!C\!=\!C(t\text{-Bu})Cl$ | 50–40% A–W | 35.2 | 2 | 0.90 | [75] |
| $Ph_2C\!=\!C\!=\!C(t\text{-Bu})Cl$ | 60–50% A–W | 35.2 | 2 | 0.87 | [75] |
| | 70–60% A–W | 44.8 | 2 | 0.75 | [75] |
| 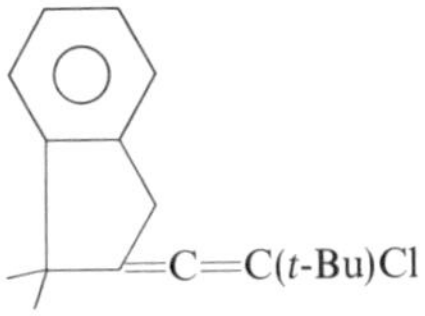 ![structure] =C=C(t-Bu)Cl | 90–70% A–W | 24.6 | 3 | $0.73 \pm 0.04$ | [76] |
| $t\text{-Bu}_2C\!=\!C\!=\!C(Ph)Cl$ | 60–40% A–W | 35.2 | 3 | $1.13 \pm 0.001$ | [75] |
| | 50–40% A–W | 25.0 | 2 | 1.08 | [75] |
| $t\text{-BuC(Ph)}\!=\!C\!=\!C(Ph)Cl$ | 50–40% A–W | 35.2 | 2 | 0.95 | [75] |
| $t\text{-BuC(Ph)}\!=\!C\!=\!C(Ph)Br$ | 60–50% A–W | 35.2 | 2 | $1.04^e$ | [75] |
| $Ph_2C\!=\!C\!=\!C(Ph)Cl$ | 80–60% A–W | 26.1 | 3 | $0.69 \pm 0.06$ | [73, 74] |
| | 80–70% A–W | 45.0 | 2 | 0.63 | [74] |
| $Ph_2C\!=\!C\!=\!C(An)Cl$ | 90–80% A–W | 26.1 | 2 | 0.77 | [74] |
| $Ph_2C\!=\!C\!=\!C(An)Br$ | 80–70% E–W[f] | 25.0 | 2 | 0.79 | [101] |
| Butatriene **267** | 80–60% E–W[f] | 50.0 | 3 | 0.46 | [101] |
| | 90–80% E–W[f] | 50.0 | 2 | 0.94 | [101] |

[a] Abbreviations: E–W, ethanol–water (v/v); A–W, acetone–water (v/v).
[b] Number of solvent compositions studied.
[c] Reaction conditions were not reported.
[d] Reported $m = 0.90$ [78].
[e] Reported $m = 0.96$ [75].
[f] Nonlinear $\log k$ versus $Y$ plot.

**TABLE 5.16**

$(k_{aq,TFE}/k_{aq,EtOH})_Y$ Ratios for $R_3R_2C{=}C{=}C(X)R_1$[a]

| $R_1$ | $R_2$ | $R_3$ | X | Solvent | Temp (°C) | $(k_{aq.TFE}/k_{aq.EtOH})_\alpha$ |
|-------|-------|-------|---|---------|-----------|--------------------------------|
| t-Bu | t-Bu | t-Bu | Cl | 97% TFE/60% EtOH | —[b] | 900 |
| t-Bu | t-Bu | Ph | Cl | 97% TFE/60% EtOH | —[b] | 280 |
| t-Bu | t-Bu | Me | Cl | 97% TFE/60% EtOH | —[b] | 12 |
| H | t-Bu | t-Bu | Br | 70% TFE/50% EtOH | 75 | 8.2 |
| H | t-Bu | Me | Br | 97% TFE/60% EtOH | 64 | 2.4 |
|   |   |   |   | 70% TFE/50% EtOH | 64 | 3.7 |
| H | Me | Me | Br | 97% TFE/60% EtOH | 54 | 0.63 |
|   |   |   |   | 70% TFE/50% EtOH | 54 | 1.8 |

[a] From Ref. [78].
[b] Temperature was not reported.

$-0.20)$ [103] (Table 5.16) reveals an interesting trend. The $(k_{aq.\ TFE}/k_{aq.\ EtOH})_Y$ ratios increase with increasing steric bulk of the substituents, especially in the $\gamma$ positions. The effect is significant, as shown by the decrease in the ratio from 900 for tri-*tert*-butylallenyl chloride to 0.63 for 3,3-dimethyl-allenyl bromide [78]. Consequently, for the trisubstituted chloroallenes the response to the change in the size of the substituent is higher in 60% EtOH than in 97% TFE. Replacing a $\gamma$-*tert*-butyl group by a $\gamma$-methyl group in tri-*tert*-butylallenyl chloride reduces the rate by 1.7 times in 97% TFE and by 7 times in 60% EtOH. The trend in Table 5.16, but not the ratios themselves, is that expected for a $k_s$ contribution since this should be more important for the less hindered compounds. It was noted that, if a $k_s$ route contributes to the overall reaction, the solvent probably attacks at $C_\gamma$ rather than at $C_\alpha$ [78]. However, in view of the other mechanistic criteria and the $(k_{aq.\ TFE}/k_{aq.\ EtOH})_Y$ ratios, which are close to unity even for the least hindered compounds of Table 5.16, a $k_s$ route seems unlikely. Moreover, for the two less hindered substrates, two $(k_{aq.\ TFE}/k_{aq.\ EtOH})_Y$ values are available, and by using the complete Grunwald–Winstein Eq. (10) [104] and assuming that it applies for both aqueous EtOH and aqueous TFE, one obtains two

$$\log(k/k_0) = mY + lN \tag{10}$$

values for $l$ for each of these substrates. For example, for 3,3-dimethyl-allenyl bromide the $l$ values obtained, $-0.29$ and $0.82$, are very different. It seems, therefore, that the $k_s$ route is unimportant and that the high $(k_{aq.\ TFE}/k_{aq.\ EtOH})_Y$ ratios require a special interpretation. It is known that the $Y$ values for aqueous TFE mixtures, which are based on *tert*-butyl chloride, give unusual $\log k$ versus $Y$ plots for both saturated [105] and vinylic [106] solvolyses. These unusual solvent effects, which may reflect a balance between decreasing solvent electrophilicity and increasing dielectric constant in more aqueous mixtures or the intervention of ion pairs [106], may operate also for the solvolysis of the allenyl systems. It was suggested that a shift to a rate-limiting elimination, as proposed for *tert*-butyl chloride in TFE, may be a possible explanation [78].

Another discrepancy is that the solvolysis rate of triphenylallenyl chloride in 90% EtOH is 13.9 times higher than in aqueous acetone of the same $Y$ value [74]. Since the nucleophilicity of the aqueous acetone mixture is lower, as judged by the $N$ values [103], this may again reflect a nucleophilic-assisted route. Although an explanation for this behavior was not given, it was mentioned that similar deviations were observed for several $S_N1$ reactions [107].

## 8. Nature of the Cationic Intermediates

Two probes were applied in the detailed investigation of the solvolysis reactions of haloallenes. These were the common ion rate depression and the disparity between the titrimetric and the polarimetric rate constants.

A single case of common ion rate depression was recorded for the solvolysis of haloallenes. The spectrophotometric or the conductometric rate constants $(k_t)$ for $4 \times 10^{-4}$ to $4 \times 10^{-5}$ $M$ triphenylallenyl chloride in 90% EtOH are constant during the kinetic run. However, addition of 0.010, 0.038, 0.075, and 0.100 $M$ LiCl reduces $k_t$ to 89, 78, 72, and 70.6% of its value in the absence of added salt $(k_t^0)$. At the same concentrations of LiBr, the $k_t$ values are enhanced to 100.4, 101.4, 101.4, and 106% of $k_t^0$ [74]. The $\alpha'$ $(= k_{-1} \ k_{H_2O})$ values (see Chapter 6, Section V,A) calculated from these data by assuming that the normal salt effects of LiCl and LiBr are identical decrease from 12.8 to 5, indicating that there may be a limit to the rate depression and that part of the products may be formed from ion pairs. By applying Eq. (39) of Chapter 6, it was found that $\geq 33.5\%$ of the products are derived from the free ion. The $\alpha'$ values resemble the value of 12 for benzhydryl chloride in 80% acetone (Table 6.26) and are similar to or higher than most of the values obtained for $\alpha$-phenyl systems in nucleophilic solvents (Table 6.25). Consequently, the triphenylallenyl cation **261a** shows an appreciable selectivity and stability in a nucleophilic solvent.

The dissymmetry of the allenic system provides a rare opportunity to evaluate the extent of ion-pair return by investigating the gap between the polarimetric and the titrimetric rates in a vinylic system. Schiavelli and co-workers resolved 1-bromo-3-methyl-1,2-pentadiene (**268**), the simplest

$$\begin{array}{c} CH_3 \diagdown \qquad\qquad \diagup Br \\ \quad\quad C{=}C{=}C \\ CH_3CH_2 \diagup \qquad\qquad \diagdown H \end{array}$$

**268**

3,3-dialkyl bromoallene capable of optical resolution, to its optical antipodes and compared the polarimetric rate constant $k_\alpha$ for the $R$ isomer with the titrimetric rate constant $k_t$ for the $d,l$ mixture [79]. The extent of racemization of the unreacted **268** was not measured directly, but its rate constant $k_{rac}$ was calculated from the difference between $k_\alpha$ and $k_t$. The data for solvolysis in 60% EtOH and 97% TFE in the presence of various salts are given in Table 5.17, which shows the following. (1) Common ion rate depression by LiBr was not observed. (2) There is a gap between $k_\alpha$ and $k_t$, both in the absence and in the presence of added salts, but the $k_\alpha/k_t$ ratio is nearly the same in 60% EtOH (1.13 ± 0.04) and in 97% TFE (1.07 ± 0.04) in the absence of added salt. Consequently, the $k_{rac}$ values in the two solvents are similar. (3) No special salt effect was observed with $LiClO_4$ in 60% EtOH, and $k_t$ exhibits the shallow linear dependence on the salt concentration characteristic of a normal salt effect [108] [Eq. (11)] with LiCl, $LiClO_4$, and LiBr in both solvents. A similar dependence on [LiCl] and [LiBr] in

$$k = k_0(1 + b[\text{salt}]) \tag{11}$$

60% EtOH was shown by $k_\alpha$, but a larger effect was found with LiBr in 97% TFE. The $b$ values of Eq. (11), which are given in Table 5.17, are small for these cases, except the last. (4) Both $k_t$ and $k_\alpha$ are markedly enhanced by

**TABLE 5.17**

**Polarimetric and Titrimetric Data for the Solvolysis of 1-Bromo-3-methyl-1,2-pentadiene at 54.5°[a]**

| Solvent | Salt | $10^5 k_\alpha$ (sec$^{-1}$) | $10^5 k_t$ (sec$^{-1}$) | $10^5 k_{rac}$ (sec$^{-1}$)[b] | $b$ for $k_\alpha$ | $b$ for $k_t$ |
|---|---|---|---|---|---|---|
| 60% EtOH | — | 4.65 | 4.10 | 0.55[c] 0.31 | — | — |
| | 0.15 $M$ LiCl | — | 4.24 | — | — | 0 |
| | 0.05 $M$ LiBr | 4.83 | 4.28 | 0.47 | 1.4 | 0.7 |
| | 0.10 $M$ LiBr | 4.90 | 4.42 | 0.64 | — | — |
| | 0.15 $M$ LiBr | 5.50 | 4.55 | — | — | — |
| | 0.05 $M$ LiClO$_4$ | 4.59 | 4.38 | — | 0 | 0.8 |
| | 0.10 $M$ LiClO$_4$ | 4.96 | 4.52 | — | — | — |
| | 0.05 $M$ NaN$_3$ | 7.49 | 6.41 | 1.09 | 9.5 | 7.4 |
| | 0.10 $M$ NaN$_3$ | 10.8 | 8.29 | 2.37 | — | — |
| 97% TFE | — | 6.18 | 5.78 | 0.43[d] | — | — |
| | 0.05 $M$ LiBr | 8.46 | 6.43 | 1.96 | 8.1 | 0.8 |
| | 0.10 $M$ LiBr | 10.9 | 6.69 | 4.10 | — | — |
| | 0.10 $M$ LiCl | — | 6.87 | — | — | 0 |

[a] From Ref. [79].
[b] From the difference between the $k_\alpha$ versus [salt] and the $k_t$ versus [salt] lines.
[c] From the extrapolated values for the [NaN$_3$] lines.
[d] From the extrapolated values for the [NaBr] lines.

added $NaN_3$ in 60% EtOH. When corrected for the normal salt effect, $k_\alpha$ and $k_t$ are proportional to $[N_3{}^-]^{0.31}$ and $[N_3{}^-]^{0.28}$, respectively. The difference in the response of the two terms to the salt concentration results in a fourfold increase of $k_{rac}$ in the presence of 0.1 $M$ $NaN_3$. (5) Whereas $k_t$ shows a normal salt effect with LiBr in 97% TFE, $k_\alpha$ increases significantly ($b$ = 8.1), with a consequent 9.5-fold increase of $k_{rac}$ at 0.1 $M$ LiBr.

These results were interpreted as indicating a solvolysis mechanism that involves the initial formation of the configurationally stable ion pair **269** [Eq. (12)] [79]. In 97% TFE only $k_\alpha$ and $k_{rac}$ increase beyond the normal salt effect, and this is understood if the external bromide ion captures the ion pair from the rear, giving the $S$ isomer of **268** in a process that is degenerate with respect to $k_t$ but not to $k_\alpha$. This effect can be significant only if the reaction of the ion pair with bromide ion competes favorably with the capture by the TFE solvent (i.e., $k_{SOH} < k_{Br}$). A return to the racemized RBr **270** occurs at the expense of ion-pair return to covalent RBr of retained configuration without enhancing $k_t$. This competition becomes inefficient in the much more nucleophilic 60% EtOH (i.e., $k_{SOH} > k_{Br}$), and $k_{rac}$ increases only slightly on addition of 0.1 $M$ LiBr.

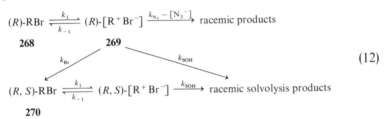

The addition of $NaN_3$ increases $k_\alpha$ and $k_t$ beyond the expectation that is based on the normal salt effect, but the higher response of $k_\alpha$ results in a consequent increase in $k_{rac}$. Since $N_3{}^-$ is more nucleophilic than $Br^-$, capture of the ion pair **269** is preferred over either capture by the solvent or the internal return in 60% EtOH. The reduced internal return results in a larger $k_\alpha$ and $k_t$ if the inverted $RN_3$ formed initially decomposes faster to racemic products than it returns to **269**. It is known that allenyl and propargyl azides are unstable under the reaction conditions [89]. A complete capture of the ion pair is suggested by the fact that the extrapolated $k_t$ and $k_\alpha$ values from the $k$ versus [salt] linear plots are identical.

An alternative explanation for the increase in $k_t$ is a special salt effect by $NaN_3$. However, the absence of an appreciable increase by $LiClO_4$ suggests that this possibility is unlikely and that **269** is an intimate rather than a solvent-separated ion pair.

The availability of two potential positively charged sites in the cationoid intermediate has a bearing on the use of the common ion rate depression and the $k_\alpha$–$k_t$ criteria as probes for the nature of the intermediates.

Product formation by nucleophilic attack on the ion pair at either C-1 or C-3 should lead to the formation of a racemic product. Attack at the allenyl terminus gives the $\alpha,\beta$-unsaturated ketone, and attack at the propargyl center gives the propargyl alcohol, and the dissymetry due to the allenyl moiety is lost in both cases. When C-3 is attacked, it has to be assumed that the counterion of the tight ion pair is close to C-1 and sufficiently remote from C-3, so that it has a negligible effect on the stereochemistry of the reaction at C-3. It is important to realize that an internal return from **269** to the propargyl bromide **271** is a hidden process since the expected faster reaction of **271** than of **268** (Table 5.10) makes **271** kinetically undetectable.

$$\begin{array}{c} CH_3 \\ \diagdown \\ \qquad C(Br)\text{---}C\equiv C\text{---}H \\ \diagup \\ CH_3CH_2 \end{array}$$

**271**

However, return to the propargylic center has a bearing on conclusions concerning the selectivity of the intermediate. It is very likely that the primary bromoallene **268** gives products via ion pairs and that the absence of common ion rate depression is due to the fact that the free ion is not formed at all. In contrast, the solvolysis of triphenylallenyl chloride proceeds, at least partially, via the free cation. However, since triphenylpropargyl chloride would solvolyze much faster than triphenylallenyl chloride, the external ion return to the propargyl center would not be reflected in the common ion rate depression and in the $\alpha'$ values. As discussed in Chapter 6, Section V,A, the observed selectivity should then be divided by the fraction of the capture rate at C-1 in order to obtain the "real" selectivity of the ion at the two positions together. Since capture of the triphenylallenyl cation by the solvent occurs almost exclusively at C-3 (Table 5.11) (73, 74), a similar behavior may also be shown by chloride ion and $\alpha'$ should be much higher than the values of 5 to 13 calculated by neglecting return to C-3. If this is indeed the case, triphenylallenyl chloride is much more reactive than the corresponding vinyl chloride, and the triphenylallenyl cation is much more selective than the triphenylvinyl cation.

This problem should also be considered when one is comparing the extent of ion-pair return in the allenyl and the vinyl systems. If $k_\alpha$ of Table 5.17 is identical to the ionization rate constant $k_{ion}$, then $k_{rac}/k_\alpha$ is the fraction of ion pairs that gives internal return, and the fraction of ion-pair return, $1 - F$, is 0.10 to 0.20. These values are lower than those for the ion pair formed in the solvolysis of 1,2-dianisyl-2-phenylvinyl bromide in AcOH [109] (Chapter 6, Section V,B) but, in view of the different solvents and the possibility of return to the propargylic carbon, such comparison has a very limited value in the study of the reaction mechanisms involved.

## 9. Reactions of Allenyl Cations

The formation of $\alpha,\beta$-unsaturated ketones and propargyl alcohols was discussed in Section IV,E,2. Reactions with a neighboring oxygen involving cyclization were also observed. 4,4-Bis(o-substituted)aryl-1-[2-(3-arylbenzo[b]furanyl)]butatrienyl halides give reactions analogous to those described for 2-(o-substituted)arylvinyl bromides [110, 111], with the following modification: cyclization, which is sterically impossible at the allenyl position, takes place at the propargylic position [Eq. (13)]. The tri-o-anisylbutatrienyl bromide **272** cyclizes in 80% EtOH to the bis(benzo-[b]furanyl)acetylene **274**, probably via the cation **273**. The intermediacy of **273** is deduced from the solvent and leaving group effects discussed above.

The diphenyl analogue of **272**, i.e., **275**, gives on alumina or on silica gel in cyclohexane the corresponding bis(benzo[b]furanyl)acetylene **276** [101], probably via a similar intermediate.

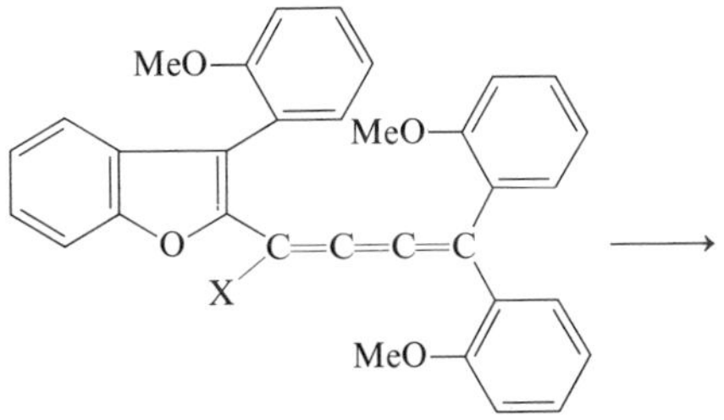

**272**

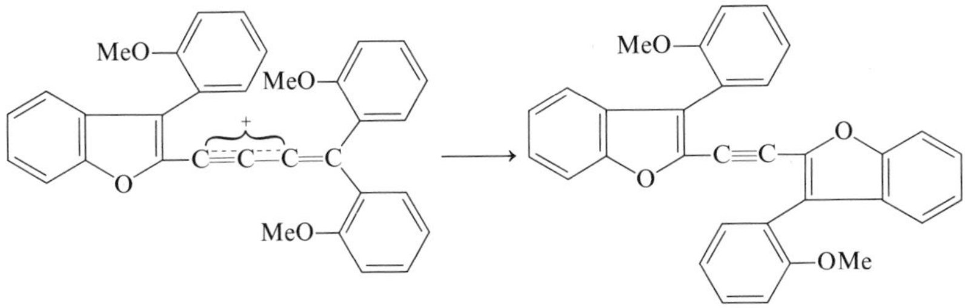

(13)

**273**    **274**

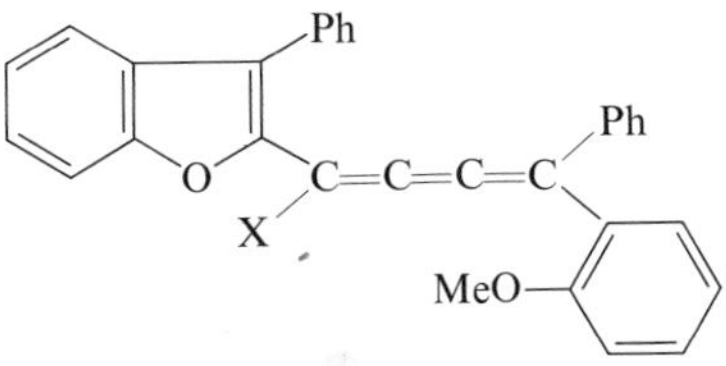

**275** 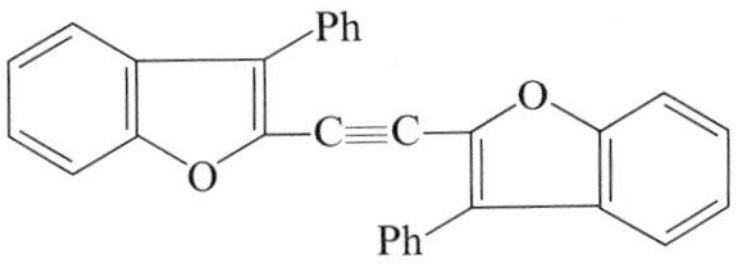    **276**

Photochemical cyclization of **272** or its chloro analogue also gives the alkyne **274**, and a C—X bond fission to the butatrienyl radical, followed by an electron transfer to form **273**, which then cyclizes, seems plausible [101]. This route is analogous to the photochemical solvolysis–cyclization of 2,2-di-*o*-anisyl- or 2,2-di-*o*-thioanisyl-1-bromoethylenes [112].

## V. ALKYLVINYL SUBSTRATES

### A. Acyclic Systems

#### 1. Medium and Structural Effects

Among the most investigated and best understood processes in organic chemistry are solvolytic displacement reactions at saturated carbon [113]. The availability of the fluorosulfonate super leaving groups has made possible the investigation of such processes with simple alkylvinyl substrates.

Vinylic solvolyses are somewhat anomalous when compared with the behavior of saturated analogues. On the one hand, there is evidence of nucleophilic solvent involvement in the reaction of alkylvinyl substrates; on the other hand, there is a high degree of carbenium ion character, and added external nucleophiles have no significant effect on rates or products [85]. The solvolyses of a series of alkylvinyl triflates (**277–282**) in various solvents have been reported [85]. The solvent $m$ values as well as the values of

$$CH_2{=}C\diagup\diagdown{\overset{OSO_2CF_3}{\underset{R}{}}}$$

| | |
|---|---|
| **277a** | R = CH$_3$ |
| **277b** | R = *n*-C$_4$H$_9$ |
| **277c** | R = *i*-C$_3$H$_7$ |
| **277d** | R = *t*-C$_4$H$_9$ |

$$\overset{H_3C}{\underset{H}{}}C{=}C\diagup\diagdown{\overset{OSO_2CF_3}{\underset{R}{}}}$$

| | |
|---|---|
| **278a** | R = CH$_3$ |
| **278b** | R = C$_2$H$_5$ |
| **278c** | R = *i*-C$_3$H$_7$ |

$$\overset{H}{\underset{H_3C}{}}C{=}C\diagup\diagdown{\overset{OSO_2CF_3}{\underset{R}{}}}$$

| | |
|---|---|
| **279a** | R = CH$_3$ |
| **279b** | R = C$_2$H$_5$ |
| **279c** | R = *i*-C$_3$H$_7$ |

$$(CH_3)_2C{=}C\diagup\diagdown{\overset{OSO_2CF_3}{\underset{R}{}}}$$

| | |
|---|---|
| **280a** | R = CH$_3$ |
| **280b** | R = C$_2$H$_5$ |
| **280c** | R = *i*-C$_3$H$_7$ |

$$\overset{R}{\underset{H}{}}C{=}C\diagup\diagdown{\overset{OSO_2CF_3}{\underset{CH_3}{}}}$$

| | |
|---|---|
| **281a** | R = CH$_3$ |
| **281b** | R = C$_2$H$_5$ |
| **281c** | R = *i*-C$_3$H$_7$ |

$$\overset{H}{\underset{R}{}}C{=}C\diagup\diagdown{\overset{OSO_2CF_3}{\underset{CH_3}{}}}$$

| | |
|---|---|
| **282a** | R = CH$_3$ |
| **282b** | R = C$_2$H$_5$ |
| **282c** | R = *i*-C$_3$H$_7$ |

$k_{ROH}/k_{AcOH}$ and $k_{ROH}/k_{TFE}$ are given in Table 5.18 together with those of appropriate model compounds. The solvent $m$ and $l$ values [104] [see Eq. (10) in Section IV,E,7] of the vinylic substrates **277** to **280** are in between those of 2-adamantyl tosylate and 2-propyl tosylate. It is well known [83, 96] that 2-propyl tosylate reacts with considerable solvent assistance, whereas 2-adamantyl tosylate, a secondary substrate blocked for backside solvent

TABLE 5.18

Solvent $m$ Values for Some Alkylvinyl Triflates and Alkyl Tosylates at $75°$ [a]

| Compound | $m$ | $l$ | $k_{ROH}/k_{AcOH}$ ($Y = -1.64$) | $k_{ROH}/k_{TFE}$ ($Y = 0.93$) |
|---|---|---|---|---|
| 277a | 0.70 | 0.41 | — | 30 |
| 277b | 0.71 | 0.32 | 7.5 | 11.9 |
| 278a | 0.71 | 0.40 | 15.9 | 15.9 |
| 279a | 0.76 | 0.15 | 2.3 | 3.0 |
| 280a | 0.85 | 0.12 | 2.5 | 2.2 |
| $CH_3OSO_2C_6H_4CH_3$-$p$ | 0.30 | 1.00 | 97 | — |
| $i$-$C_3H_7OSO_2C_6H_4CH_3$-$p$ | 0.61 | 0.35 | 4.2 | 12.3 |
| 2-Adamantyl-$OSO_2C_6H_4CH_3$-$p$ | 1.0 | 0.0 | 0.19 | 0.21 |

[a] From Ref. [85].

attack, solvolyzes by a limiting mechanism. Hence, the vinyl triflate values imply at least some solvent assistance in the solvolyses of these substrates. This is confirmed by the respective $l$ values and the relative rates in solvents of different nucleophilicity but similar ionizing power.

The effects of added nucleophiles on the rates and products of solvolysis are summarized in Table 5.19. As these data indicate, trifluoroethanolyses of vinyl triflates 277a and 280a are as fast or faster in the presence of $LiClO_4$ than in the presence of LiBr, indicating little sensitivity to a strong nucleophile. Similarly, in contrast to $i$-PrOTs, which gives about 50% $i$-PrBr under these conditions, the added bromide ion has little effect on the product composition from reaction of vinyl triflates 277a and 280a.

In order to rule out E2 elimination with solvent acting as base, the effect of added $HO^-$ on the solvolysis of alkylvinyl triflates was determined [85]. The results show that triflates 279a and 279c, as well as cyclohexenyl triflate

TABLE 5.19

Rates and Products of Solvolysis of Vinyl Triflates in the Presence of Added Salts in $CF_3CH_2OH$ at $90°$ [a]

| Triflate | Salt (conc) | $10^4 k$ (sec$^{-1}$) | $k_{rel}$ | Allene | Alkyne | Ether | Bromide |
|---|---|---|---|---|---|---|---|
| | | | | | Product (%) | | |
| 277a | — | 1.45 | 1 | $9.8 \pm 1.2$ | $72.1 \pm 0.4$ | $18.1 \pm 0.8$ | — |
| | $LiClO_4$ (0.137 $M$) | 3.30 | 2.27 | $11.0 \pm 0.6$ | $77.1 \pm 1.4$ | $11.8 \pm 1.9$ | — |
| | LiBr (0.137 $M$) | 3.22 | 2.22 | $8.7 \pm 0.7$ | $68.2 \pm 2.6$ | $14.6 \pm 2.4$ | $8.5 \pm 0.8$ |
| 280a | — | 15.65 | 1 | $38.7 \pm 0.7$ | — | $61.3 \pm 0.7$ | — |
| | $LiClO_4$ (0.137 $M$) | 39.8 | 2.55 | $36.2 \pm 0.5$ | — | $63.8 \pm 0.5$ | — |
| | LiBr (0.137 $M$) | 33.1 | 2.12 | $31.1 \pm 0.5$ | — | $64.6 \pm 0.5$ | $4.3 \pm 0.1$ |

[a] From Ref. [85].

(a substrate that cannot eliminate due to the high strain involved), which have rates differing by $10^6$ in the absence of base, all react within a factor of 4 in the presence of $HO^-$. This strongly suggests [85] a change of mechanism to S–O cleavage in the solvolyses of vinyl triflates in the presence of $HO^-$ and at the same time rules out E2 elimination. Such nucleophilic attacks on sulfur are well-known processes [25, 26] (see Section III,A,5).

**TABLE 5.20**

**Effect of Alkyl Substitution on Solvolysis of Vinyl Triflates in 50% EtOH at 75°[a]**

|            |             | $k_{rel}$          |                    |
| :--------: | :---------: | :----------------: | :----------------: |
| Compound   | R = $CH_3$  | R = $C_2H_5$       | R = $i$-$C_3H_7$   |
| 277        | 1           | 2.49 ($n$-$C_4H_9$)| 2.51               |
| 278        | 1           | 1.3                | 2.46               |
| 279        | 1           | 5.59 $(2.61)^b$    | 44.5 $(6.25)^a$    |
| 280        | 1           | 7.58 $(3.54)^b$    | 83.3 $(11.7)^a$    |
| 281        | 1           | 0.97               | 2.17               |
| 282        | 1           | 2.97               | 7.71               |

[a] From Ref. [85].
[b] Corrected for ground state steric effects. See Summerville et al. [85].

The effect of substituents on the solvolysis of simple vinyl systems is summarized in Table 5.20. These substituent effects are particularly interesting. Even the limited data in Table 5.20 allow a Taft correlation [114] and hence a measure of the charge development at the incipient electron-deficient carbon during the solvolysis [113]. The $\rho^*$ values for compounds **279, 280,** and **282** are $-5.6$, $-4.7$, and $-4.2$, respectively. The correlations for compounds **277, 278,** and **281** are poor, and the respective $\rho^*$ values are smaller, being $-1.4$, $-2.0$, and $-1.7$, respectively. These values of $\rho^*$ for the vinylic systems can be compared with those of simple secondarly aliphatic substrates, which vary from 0 in ethanol to $-7$ in trifluoroacetic acid and $-1.6$ in 50% EtOH at 75°. The $\rho^*$ for saturated tertiary chlorides in 80% EtOH is $-3.3$. Hence, the average $\rho^*$ value of $-4.8$ for substrates **279, 280,** and **282** indicates that the response to electronic effects in the case of vinyl cations is greater than in the case of tertiary and secondary carbenium ions. This is in accord with the greater need for stabilization of the less stable vinyl cations in comparison with their saturated analogues. Of particular interest is the magnitude of the effect of substituent changes trans to the leaving group in vinyl systems **282**. These effects are almost as large

($\rho^* = -4.2$ for **282**) as those at the $\alpha$-carbon ($\rho^* = -5.6$ for **279**), despite the extra bond between the substituent group and the reaction center. The large stabilizing effect of $\beta$ substituents on vinyl cations is also present in the gas phase, as evidenced by the data in Table 2.2. Furthermore, this greater demand on the positions trans to the leaving group is also reflected in the magnitude of isotope effects (*vide infra*). These results suggest that the transition state for ionization of vinyl substrates is bent, with the electron-deficient orbital interacting with the trans substituent [85]. A similar trans effect has been observed in certain saturated systems [115].

The medium and structural effects, together with the results of stereo-chemical data discussed in the following section, indicate that solvolysis of alkylvinyl triflates proceeds through ion-pair pathways. There is no evidence for direct $S_N2$ or E2 reactions of vinyl substrates with strong base, $OH^-$, or nucleophile $Br^-$. However, solvent $m$ and $l$ values and dependence on solvent nucleophilicity at constant ionizing power suggest that solvent nucleophilicity is important in these reactions, particularly in systems with $\beta$-hydrogen trans to the leaving group. Since only partial inversion is observed (*vide infra*) in the solvolyses of these alkylvinyl substrates, the nucleophilic sensitivity of these systems is likely to be due to specific solvation of a trans $\beta$-hydrogen or other site not directly at the rear of the incipient electron-deficient vinylic carbon.

In contrast to the behavior of simple alkylvinyl triflates, the corresponding tosylates react via an addition–elimination process in formic acid, as evidenced [116] by the three times greater reactivity of the tosylates compared to that of the brosylates at 60°. However, the vinyl cation mechanism prevails in aqueous methanol, as shown by the product and rate data [117].

## 2. Stereochemistry of Solvolyses

The stereochemistry of solvolytic displacement at a chiral tetrahedral carbon has been the subject of intensive investigations. The demonstration of extensive racemization in numerous solvolytic displacements on optically active substrates played an important role in establishing the $S_N1$ mechanism. Correlation of the absolute configuration of secondary and even primary optically active substrates with the configuration of the solvolysis products helped to establish the importance of ion pairs and solvent participation in displacement reactions.

The planarity of trigonal carbons in vinylic systems rules out chirality as a means of investigating the stereochemistry of solvolytic displacement at a vinyl center except in special cases (see Section IV,E,8 and Chapter 6, Section V,B). However, the relationship of the incoming nucleophile and the leaving group with the $\beta$ substituents on the double bond may be used to

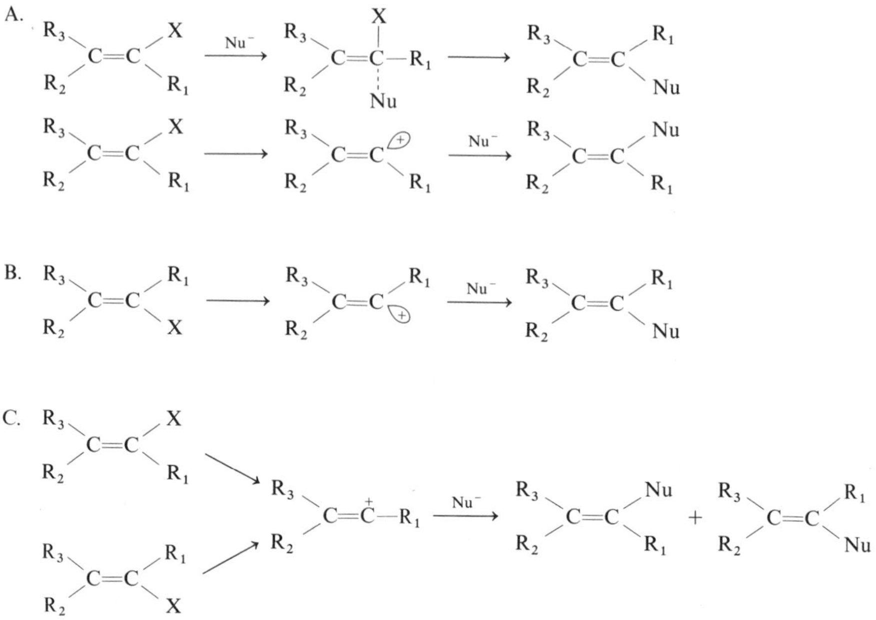

A.

B.

C.

Scheme 5.33

gain insight into the stereochemistry of vinylic substitutions. Analogous to displacement on a chiral center, three extreme possibilities exist for the stereochemical outcome of vinylic displacements, as shown in Scheme 5.33.

Inverted stereochemistry (path A) would be the result of a direct backside displacement analogous to $S_N2$ processes or of backside attack by solvent or nucleophile on an ion pair. Retained stereochemistry, as depicted in path B, would imply formation and trapping of configurationally stable bent (trigonal) vinyl cations. Alternatively, participation of $R_2$ during solvolysis and formation of a bridged ion could give retained stereochemistry (see Chapter 7). Finally, complete stereochemical randomization could arise by formation and capture of free linear vinyl cations, as shown in route C of Scheme 5.33, or by capture of a rapidly equilibrating pair of bent vinyl cations. Hence, the same ratio of products being formed from either pure starting isomer is the vinylic equivalent of racemization at the chiral tetrahedral carbon.

Extensive investigations by Rappoport and co-workers of the stereochemistry of solvolysis of α-arylvinyl substrates established (see Chapter 6,IV) that they proceed by complete stereochemical randomization. Initial results by Sherrod and Bergman [38] and Kelsey and Bergman [39] indicated that the cyclopropyl-stablized systems *E*-**283** and *Z*-**283** also lead to the same product distribution upon solvolysis. Hence, in analogy with tetrahedral saturated systems, highly stabilized vinyl substrates undergo extensive or complete "racemization" upon solvolysis. However, careful reexamination

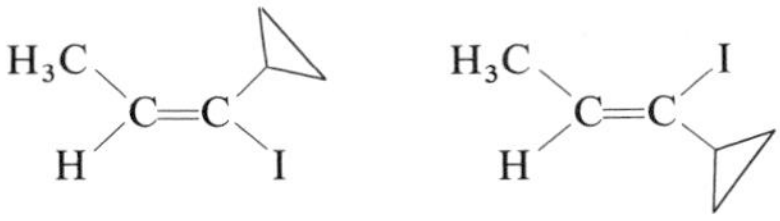

*E*-283        *Z*-283

[118] with more sophisticated analytical techniques revealed an inversion component in the reaction of *E*-**283** and *Z*-**283** in AcOH–AgOAc at 25°, and a small inversion component was detected [118] as well in the reaction of the dicyclopropylvinyl iodides *E*-**284** and *Z*-**284**.

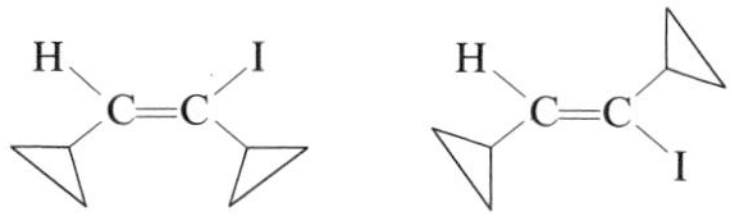

*E*-284        *Z*-284

A detailed investigation of the stereochemistry of simple alkylvinyl cations was reported by Summerville and Schleyer [119]. The 2-buten-2-yl cation **285** produced by protonation of 1,2-butadiene and 2-butyne was observed to capture solvent to give *Z* in preference to *E* product in the ratios 3.3:1 at 75° and 5.1:1 at 35°:

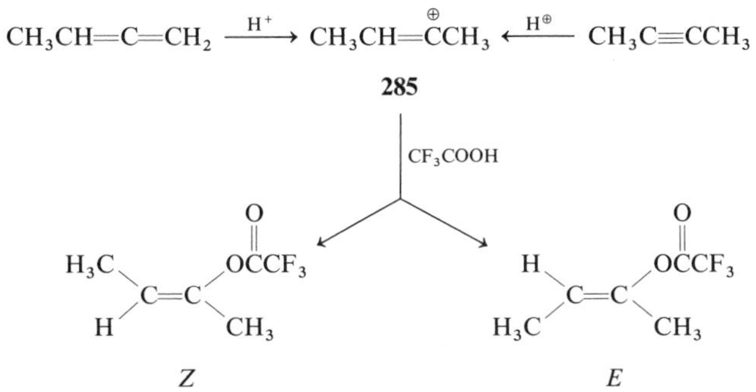

$$CH_3CH{=}C{=}CH_2 \xrightarrow{\ H^+\ } CH_3CH{=}\overset{\oplus}{C}CH_3 \xleftarrow{\ H^{\oplus}\ } CH_3C{\equiv}CCH_3$$

**285**

CF₃COOH

*Z*               *E*

An attractive effect between the β-methyl group, rendered electron-deficient by hyperconjugation, and the attacking nucleophile was postulated to account for the preference of *Z* product [119]. Trifluoroacetolyses of *Z*- and *E*-2-buten-2-yl triflates *Z*-**286** and *E*-**286** gave more than 90% 2-butyne along with trifluoroacetates, the *Z/E* ratios being 4.4 and 8.0, respectively [119]. Acetolysis products of these vinyl triflates were formed with a larger fraction of inversion.

Similar results were obtained by Clarke and Bergman [120] in the reaction

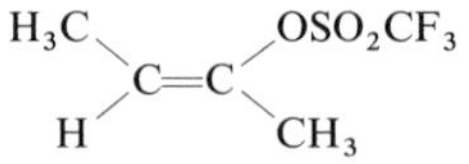

Z-286 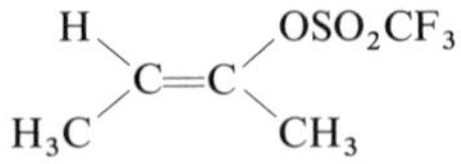                          E-286

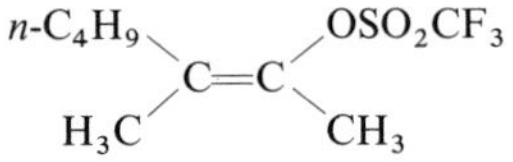

Z-287 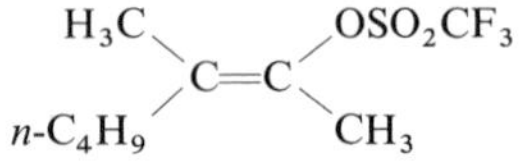                          E-287 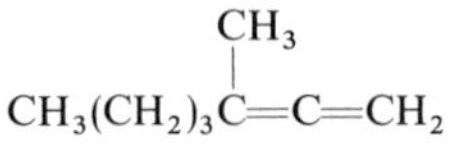                          288

of Z- and E-3-methyl-2-hepten-2-yl triflates Z-287 and E-287. Trifluoroethanolysis of Z-287 and E-287 at 60° gave, besides 15% allene 288, trifluoroethyl ethers with a Z/E ratio of 4.5 and 2.5, respectively.

The observation of partial inversion in the solvolysis of these systems
rules out the exclusive involvement of either free vinyl cations or direct $S_N2$
displacement by solvent or nucleophile. The amount of inversion necessary
to explain the observed product ratios resembles that observed in tertiary
rather than secondary aliphatic substitutions, further suggesting an $S_N1$
rather than $S_N2$ process, and these product ratios are best explained by an
ion-pair mechanism, as shown in Scheme 5.34.

In analogy with the results of displacements at saturated tetrahedral
carbon centers, the initial reaction at vinyl systems results in isomeric ion

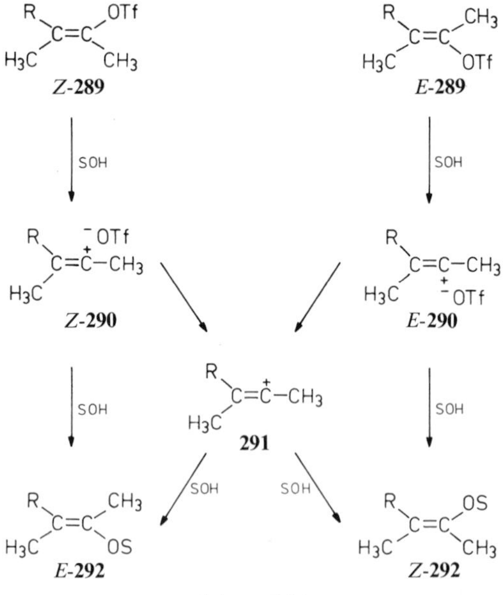

Scheme 5.34

pairs $Z$-290 and $E$-290 from vinyl triflates $Z$-289 and $E$-289, respectively. These ion pairs are protected from solvent attack on the side of the departing nucleofugic group. Solvent capture of these ion pairs from the backside results in inverted products. Competition between solvent trapping of ion pairs $Z$-290 and $E$-290 and the free vinyl cation 291 leads to the net inverted products $E$-292 and $Z$-292 observed in the solvolysis of isomeric vinyl triflates $Z$-289 and $E$-289, respectively.

Inversion was also observed [120] in the solvolysis of triflates $Z$-293 and $E$-293, as shown in Scheme 5.35. Besides the elimination product allene 294 and acyclic ethers $E$-295 and $Z$-295, cyclic derivatives 296 to 299 were observed. The fact that 42% of cyclic products were observed to be formed from $E$-293, compared to only 27% from the $Z$ isomer $Z$-293, further supports the idea of ion pairs in the solvolysis of simple alkylvinyl triflates. The ion pair resulting from $Z$-293 is less favorably disposed for internal nucleophilic attack by the $\pi$ system of the double bond than is the ion pair resulting from triflate $E$-293 with its gegenion on the opposite side of the remote double bond. Direct double-bond participation in the solvolysis of 293 was ruled out [120] by the fact that the reactivity of triflates $Z$-293 and $E$-293, is slower than that of their saturated analogues $Z$-287 and $E$-287 by a factor of 3.

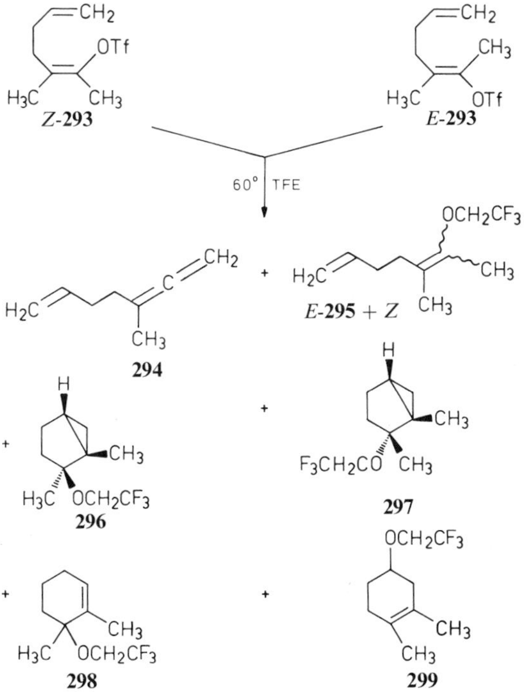

Scheme 5.35

3. Kinetic Deuterium Isotope Effects

Kinetic deuterium isotope effects have proved to be a valuable tool in studies of reaction mechanisms in general and in solvolytic displacement reactions in particular [97, 121]. As with deuterium isotope effects observed in normal carbenium ions, there may be several kinds of isotope effects, such as $\alpha$, $\beta$, and $\delta$, in the solvolytic generation of vinyl cations. Since an $\alpha$-isotope effect in a vinyl cation would require the formation of highly energetic "primary" vinyl cations (see Chapter 2) hitherto not possible, such effects have not been observed in simple vinyl cations, except for allenyl systems (see Section IV,E,5).

In vinyl substrates there may be two different kinds of $\beta$-deuterium isotope effects, one in which the isotopic substitution is $\beta$ to the leaving group on an adjacent saturated carbon, as in 300a, and one in which it is on the neighboring unsaturated carbon, as in 300b, a distinction that can be extended to $\gamma$- as well as $\delta$-deuterium isotope effects. In the latter case (300b), a further

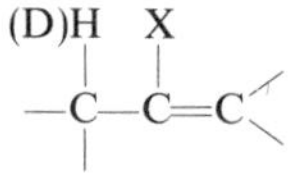

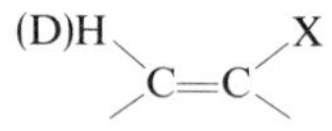

distinction can be made on the basis of stereochemistry, depending on whether the deuterium is cis or trans to the leaving group. $\beta$-Deuterium isotope effects of both types, along with the effect of stereochemistry on the isotope effects, have been reported in the solvolytic generation of vinyl cations and are summarized in Table 5.21.

A comparison of the deuterium isotope effects in the solvolysis of saturated and vinylic systems is most revealing. For the simple alkyl systems, comparisons are best made between isopropyl substrate 311 and vinylic systems 301, 303, and 304. For solvolysis of isopropyl substrates, the maximal $\beta$-

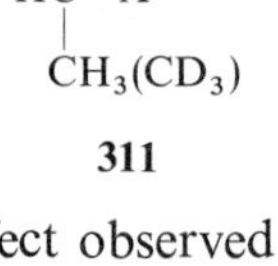

deuterium isotope effect observed [127] was $k_H/k_D = 2.12$ for the tosylate at 25° in $CF_3COOH$, corresponding to a $\Delta\Delta F^{\ddagger}$ of 75 cal/mole per deuterium, with lower values of $k_H/k_D = 1.2$ in aqueous ethanol. Assuming that the solvolyses of 311 as well as 301 to 303 are limiting $S_N1$ processes, which seems likely under the given reaction conditions, and neglecting differences due to leaving group, solvent, and temperature on the isotope effect, which is not unreasonable in limiting cases [121], the isotope effect for 311 is some

**TABLE 5.21**

**Kinetic Deuterium Isotope Effects in the Solvolytic Generation of Vinyl Cations**

| Substrate | Solvent (Temp, °C) | $k_H/k_D$ | Ref. |
|---|---|---|---|
| $(CH_3)_2C{=}C(OTf)CD_3$ <br> **301** | 60% EtOH (75) | 1.54 | [122] |
| $(CD_3)_2C{=}C(OTf)CH_3$ <br> **302** | 60% EtOH (25) | 0.86 | [122] |
| 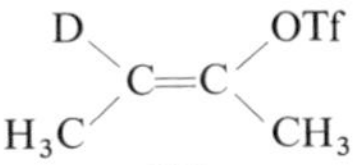 D, H₃C / C=C \ OTf, CH₃ <br> **303** | 60% EtOH (75) | 1.25 | [122] |
| 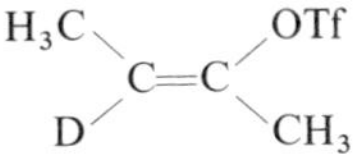 H₃C, D / C=C \ OTf, CH₃ <br> **304** | 60% EtOH (75) | 2.01 | [122] |
| 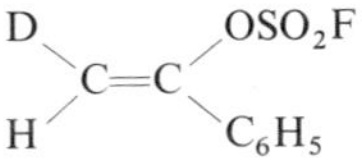 D, H / C=C \ OSO₂F, C₆H₅ <br> **305** | AcOH (52.9) | 1.01 | [123] |
| 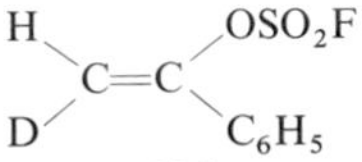 H, D / C=C \ OSO₂F, C₆H₅ <br> **306** | AcOH (52.9) | 1.57 | [123] |
| 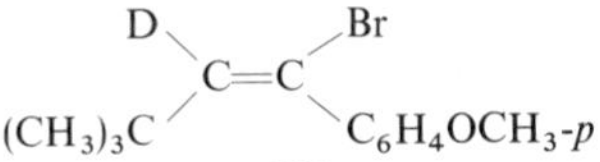 D, (CH₃)₃C / C=C \ Br, C₆H₄OCH₃-$p$ <br> **307** | 50% EtOH (80) | 1.10 | [124] |
| 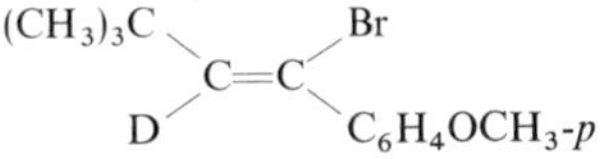 (CH₃)₃C, D / C=C \ Br, C₆H₄OCH₃-$p$ <br> **308** | 50% EtOH (80) | 1.26 | [124] |
| 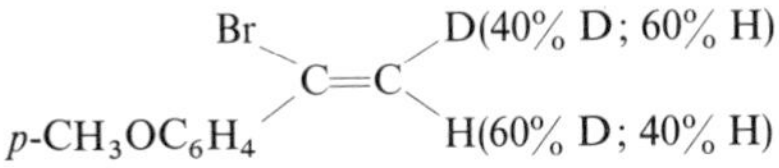 Br, $p$-CH₃OC₆H₄ / C=C \ D(40% D; 60% H), H(60% D; 40% H) <br> **309** | AcOH (120) | 1.21 | [125] |
| 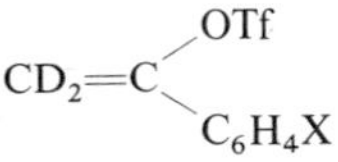 $CD_2{=}C$ \ OTf, C₆H₄X <br> **310** | | | |
| **310a** X = H | 80% EtOH (75) | 1.45 | [126] |
| **310b** X = $m$-Cl | 80% EtOH (75) | 1.61 | [126] |
| **310c** X = $p$-CF₃ | 80% EtOH (75) | 1.64 | [126] |
| **310d** X = $p$-NO₂ | 80% EtOH (75) | 1.71 | [126] |

25% lower per deuterium than the corresponding $\beta$-isotope effect of **301**, with a $\Delta\Delta F^{\ddagger}$ of 100 cal/mole per deuterium. The difference is even larger when the deuterium in the vinylic system is of type **300b**, with $\Delta\Delta F^{\ddagger}$ of 154 cal/mole per deuterium for **303** versus 75 cal/mole per deuterium for **311**. A higher $\beta$-deuterium isotope effect per deuterium in vinyl systems as compared with the analogous saturated systems is in line with the greater need for stabilization by hyperconjugation of a vinyl cation as compared with that of a saturated carbenium ion and parallels the observed substituent effects, as discussed in Section V,A,1.

The $\gamma$-deuterium isotope effects seem to be similarly larger for a vinyl system, with $k_H/k_{D_6} = 0.86$ for **302** at 25° in 60% EtOH compared with $k_H/k_{D_6} = 0.975$ for **312** also at 25° but in 80% EtOH. The direction of this effect (i.e., inverse) is in agreement with the normal inductive electron-donating ability of deuterium as determined [128] by p$K$ measurements for $CD_3COOH$ relative to $CH_3COOH$.

Comparison of the isotope effects within the different vinylic substrates is instructive as well. Although the solvolyses of triflates **302** to **304** proceed through similar linear vinyl cations, the average reduction in free energy of activation per deuterium ($\Delta\Delta F^{\ddagger}$) is some 55% (154 versus 100 cal/mole) greater for **303** with $\beta$-isotopic substitution on the double bond, as in **300b**, than for **301** with $\beta$-isotopic substitution on an adjacent saturated $\alpha$-carbon, as in **300a**. This confirms the strong dependence of $\beta$-deuterium isotope effects on the dihedral angle between the isotopically substituted C—H bond and the developing empty p orbital of the carbenium ion [121]. Indeed, Shiner and co-workers and others have shown [100, 129] that the $\beta$-deuterium isotope effect is at a maximum when the dihedral angle is 0° and becomes inverse with a dihedral angle of 90°. In the cation resulting from **303** and **304**, the $\beta$-hydrogen on the double bond is rigidly held in the same plane as the empty p orbital, whereas in the ion resulting from **301** free rotation is still possible around the single bond $CD_3\overset{+}{-}C{=}C$, resulting in an average C–H(D) dihedral angle of 30° with the p orbital and hence less hyperconjugative overlap and a lower isotope effect. Furthermore, overlap is clearly more effective across the shorter C=C double bond in **300b** than across the single bond in **300a**.

Another manifestation of the strong hyperconjugative interaction and its dependence on the dihedral angle as it affects $\beta$-deuterium isotope effects in solvolyses is exhibited by the stereochemical effects summarized in Table 5.21. The isotope effect $k_H/k_D$ of 2.01 for a trans deuterium in **304** is four times as large as the effect of a cis deuterium in **303** with a $k_H/k_D$ of 1.25. A similarly large difference in cis ($k_H/k_D = 1.10$) and trans ($k_H/k_D = 1.57$) isotope effects was observed [123] in the acetolysis of **305** and **306**, respectively. A somewhat smaller, 2.5-fold, difference with $k_H/k_D = 1.10$ for **307** and $k_H/k_D = 1.26$ for **308**, was observed [124] in the aqueous ethanolysis of these isomeric vinyl bromides. This spread in $\beta$-deuterium isotope effects as a function of stereochemistry is mirrored by a similar spread in $\beta$-alkyl substituent effects (*vide supra*) in the solvolysis as a function of stereochemistry, but it may also be due to intrusion of minor amounts of E2 elimination in the geometrically favorable trans isomers. These effects clearly show the importance of a trans periplanar arrangement and further suggest that the transition state for ionization of vinyl substrates may be bent, with the nascent electron-deficient orbital more strongly interacting with the trans substituents. Moreover, these large kinetic $\beta$-deuterium isotope effects provide further proof that even the solvolysis of simple alkylvinyl substrates proceeds by a unimolecular $S_N1$-type mechanism.

The importance of hyperconjugation in vinyl cations is further confirmed by the deuterium isotope effects for solvolysis of ring-substituted vinyl triflates. The parent compound **310a** shows a lower isotope effect ($\Delta\Delta F^{\ddagger} = $ 128 cal/mole per deuterium) than the analogous alkyl systems **303** and **304**, with 154 and 483 cal/mole per deuterium, respectively. This is in accord with the stabilization of the resultant vinyl cation by the adjacent phenyl ring in **310**. Therefore, there is a lower charge density on the cationic center and consequently less need for hyperconjugative stabilization by the $\beta$ substituents in the case of **310** than in the case of the simple alkylvinyl cation resulting from **303** and **304**.

Once again, the magnitude of the $\beta$-deuterium isotope effects is considerably larger in the case of vinyl substrates **310** than in the analogous saturated systems **313**, in which the $k_H/k_{D_3}$ effects varied [81] from 1.113 for the $p$-CH$_3$O isomer to 1.224 for the parent compound. The effect of

$$\begin{array}{c} Y \\ | \\ XC_6H_4CHCD_3 \end{array}$$

**313**  Y = Cl, Br

substituents on the isotope effect in **310** is interesting. The more destabilizing substituents, such as $p$-NO$_2$ and $p$-CF$_3$, result in *increased* isotope effects (Table 5.21), indicating a greater need for hyperconjugative stabilization of the developing empty p orbital with increasing electron withdrawal by the substituents. Extrapolation [126] of the data in Table 5.21 for **310** predicts a $k_H/k_D$ of 1.36 and 1.21 for the $p$-CH$_3$ and $p$-CH$_3$O isomers, respectively. The $k_H/k_D$ value of 1.21 for the $p$-methoxystyryl system is in reasonable agreement with the value predicted by Noyce and Schiavelli [130] from the hydration of $p$-CH$_3$OC$_6$H$_4$C$\equiv$CD and the $k_H/k_D$ value of 1.21 for **309** observed by Rappoport and Apeloig [125]. Finally, the fact that the kinetic $\beta$-deuterium isotope effects for **310** correlate [126] with $\sigma^*$ is in accord with the Hammond [131] and Swain–Thornton postulates [132].

## B. Cyclic Vinyl Systems

The solvolytic behavior of cyclic vinyl systems is of interest because of the question of the effect that ring strain has on the rate of reaction. The first attempt to generate cyclic vinyl cations by solvolysis was that of Peterson and Indelicato [116] via reaction of 1-cyclohexenyl tosylates and brosylates. However, only addition–elimination was observed in formic acid and only recovered starting material was observed in 50% aqueous methanol at 130° after 18 days [116, 117]. This early work indicated the difficulty of forming cyclic vinyl cations with relatively small rings.

### 1. Small Rings (C$_4$ and C$_5$)

The solvolytic behavior of cyclobutenyl fluorosulfonates was discussed in Section IV,D. Although at first glance it would seem that the cyclobutenyl system would be much too strained to react via an S$_N$1 process and the intermediacy of a vinyl cation, abundant experimental evidence in fact indicates that it is one of the most reactive vinylic substrates [57, 61, 64, 66]. The high solvolytic reactivity of the cyclobutenyl substrates was accounted for (Section IV,D) by the involvement of the nonclassical ion **217** as the reactive intermediate.

Cyclopentenyl triflate **315** was originally reported [63] to react via a vinyl cation mechanism. This was based on the exclusive formation of cyclopentanone as product in 50% aqueous ethanol buffered with TEA at 100°, as well as the 10$^5$ *slower* reactivity of **315** compared to that of the acyclic model compound **314**, that was explained by the high energy of the intermediate bent vinyl cation resulting from **315**. Recent evidence [57, 58], however, indicates that **315** does not react via a vinyl cation intermediate. In particular, it was observed [57, 58] that **315** did not react and was recovered practically unchanged in a nonnucleophilic solvent such as TFE buffered with TEA after 10 days at 100°. Furthermore, solvolysis in $^{18}$O-labeled aqueous ethanol did not result in $^{18}$O incorporation in the product cyclopentanone. These data indicate that **315** reacted via nucleophilic attack on the sulfur with concomitant S–O cleavage (see Section III,A,5) rather than via unimolecular ionization and vinyl cation formation. Apparently, the five-membered ring cannot accommodate the preferred linear sp-hybridized geometry (see Chapter 2) of a vinyl cation due to the very high strain energy of such an arrangement. Furthermore, the unfavorable bent sp$^2$-hybridized vinyl cation must be of sufficiently high energy to be unattainable under these conditions, and hence the competing S–O cleavage via attack on sulfur takes over. However, the cyclopentenyl cation has been generated [133] by irradiation of 1-iodocyclopentene. Photolysis in CH$_2$Cl$_2$ at −25° gave, besides reduced products, 18% 1-chlorocyclopentene and irradiation in $CH_3OH$, also at $-25°$, gave 13% 1,1-dimethoxycyclopentane. These products were explained [133] by initial formation of radicals followed by a rapid electron transfer to give the cyclopentenyl cation and nucleophilic capture of the ion.

## 2. Medium Rings ($C_6-C_{12}$)

The solvolytic behavior of the medium-ring cycloalkenyl triflates **316** to **322** has been extensively examined [63, 134, 135]. In the reaction of all cycloalkenyl triflates **316** to **322**, first-order kinetics were observed. Moreover, treatment of cyclohexenyl triflate **316** with buffered $CH_3COOD$ (4 weeks at 135°) gave a mixture of cyclohexanone (85%) and 1-cyclohexenyl acetate (15%). Cyclohexanone was treated with $CH_3COOD$ under the same conditions. Mass spectral analysis showed that the amount of deuterium incorporated was the same in both samples of cyclohexanone. This rules out reaction of **316** via an addition–elimination mechanism, in contrast to the behavior [116] of the corresponding cyclohexenyl tosylates and brosylates, which were shown to react via such a mechanism in formic acid.

$E\text{-}CH_3CH{=}C(CH_3)OSO_2CF_3$

**314**

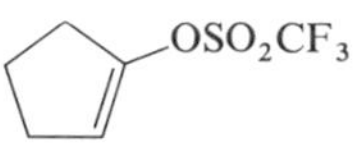

**315**

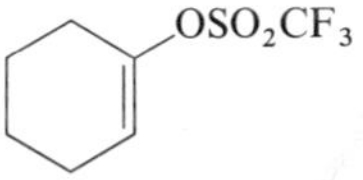

**316**

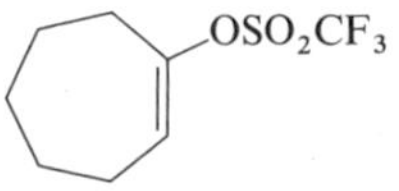

**317**

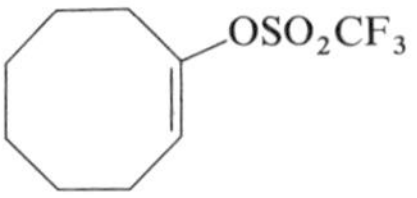

**318**

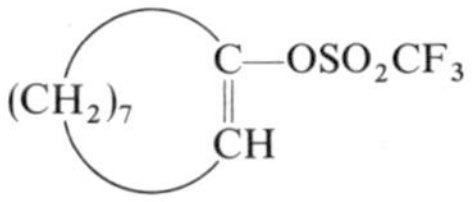

**319**

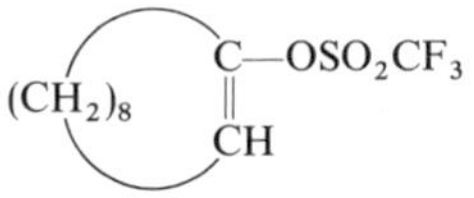

**320**

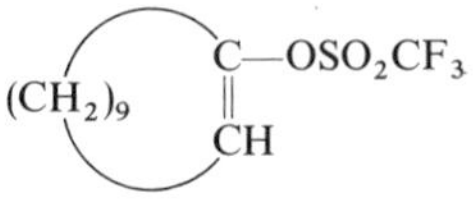

**321**

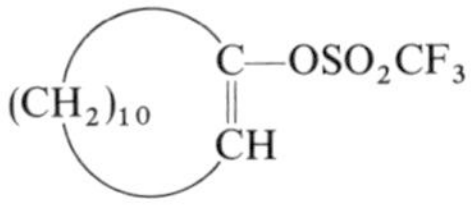

**322**

A summary of the rates of reaction of cycloalkenyl systems **315** to **322** is given in Table 5.22. For comparison, the rate of the acyclic model compound **314** is included as well. A plot of the logarithm of the relative rates in Table 5.22 versus the ring size is shown in Fig. 5.1. A number of interesting facts emerge from the data in Table 5.22 and Fig. 5.1. The effect of solvent on the relative reactivities of triflates **314** to **322** is minor, at least with 50% aqueous ethanol and 70% aqueous TFE. Cyclohexenyl triflate **316** reacts some $10^4$ times slower than the acyclic model compound **314**. Since a six-membered ring is unable to accommodate the linear geometry of a vinyl cation due to Baeyer and Pitzer strain [137], this slower reactivity reflects the higher energy

**TABLE 5.22**

**Summary of Solvolysis Rates of Cycloalkenyl Triflates 314-322 at 70°[a]**

| Vinyl triflate | 70% TFE | | 50% EtOH | | Reference |
|---|---|---|---|---|---|
| | $k$ (sec$^{-1}$) | $k_{rel}$ | $k$ (sec$^{-1}$) | $k_{rel}$ | |
| 314 | $1.8 \times 10^{-4}$ | 1 | $1.7 \times 10^{-4}$ | 1 | [136] |
| 315 | $1.0 \times 10^{-9}$ | $5.6 \times 10^{-6}$ | $1.0 \times 10^{-9}$ | $5.9 \times 10^{-6}$ | [63] |
| 316 | $1.1 \times 10^{-8}$ | $6.1 \times 10^{-5}$ | $1.2 \times 10^{-8}$ | $7.1 \times 10^{-5}$ | [63] |
| 317 | $3.3 \times 10^{-5}$ | $1.8 \times 10^{-1}$ | $4.2 \times 10^{-5}$ | $2.5 \times 10^{-1}$ | [63] |
| 318 | $5.5 \times 10^{-4}$ | 3.1 | $4.5 \times 10^{-4}$ | 2.6 | [63] |
| 319 | $2.2 \times 10^{-2}$ | 120 | $2.3 \times 10^{-2}$ | 135 | [134, 135] |
| 320 | $3.3 \times 10^{-3}$ | 18 | $3.3 \times 10^{-3}$ | 19 | [134] |
| 321 | $1.1 \times 10^{-2}$ | 61 | $1.1 \times 10^{-2}$ | 65 | [135] |
| 322 | $6.5 \times 10^{-3}$ | 36 | $6.8 \times 10^{-3}$ | 40 | [135] |

[a] Extrapolated; see Hargrove and Stang [135].

of the bent vinyl cation. Models indicate that seven- and eight-membered rings can accommodate the linear geometry of an sp-hybridized carbon coupled to a trigonal carbon, and hence the reactivities of cycloalkenyl triflates 317 and 318 are comparable to that of the geometrically unrestricted acyclic analogue 314.

The kinetic data and the relative reactivities are in accord with a vinyl cation mechanism for the reaction of cycloalkenyl triflates 316 to 322. Furthermore, at least a qualitative statement can be made about the relative strain energies of the medium-ring ($C_8$–$C_{12}$) cycloalkenyl cations. A "sawtoothed" plot analogous to that of Fig. 5.1 was obtained [138] in the solvolysis of 1-methylcycloalkyl chlorides relative to *tert*-butyl chloride. This phenomenon has been attributed to the higher strain energy of the odd member relative to the next-nearest even-membered ring in medium-sized

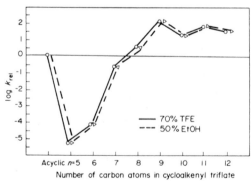

**Fig. 5.1.** Plot of logarithm of relative rates of solvolysis versus number of carbon atoms in cycloalkenyl triflates.

rings. This conclusion is supported by heats of hydrogenation of cycloalkenes in acetic acid [137].

The products of solvolyses of cycloalkenyl triflates **316** to **318** at 100° in 50% aqueous ethanol buffered with $Et_3N$ were exclusively the corresponding cycloalkanones. Triflates **319** and **320** gave, under the same conditions, 35% cyclononyne, 24% cyclononanone, 18% 1,2-cyclononadiene, 12% unidentified product, and 11% 1-ethoxycylononene from **319**, and 60% cyclodecyne, 31% 1,2-cyclodecadiene, and 9% cyclodecanone from **320**. Triflate **319** in 70% TFE buffered with pyridine at 75° gave 47% cyclononyne, 22% 1,2-cyclononadiene, 3.5% cyclononanone, and 28% unidentified products. Under similar conditions, triflates **321** and **322** gave, respectively, besides 80 and 87% of cycloallene, 17 and 8% of cycloalkynes, along with 3 and 5% of the corresponding cycloalkanones, respectively.

These products provide additional support for a unimolecular ionization and a vinyl cation intermediate in the solvolyses of cycloalkenyl triflates **316** to **322**. The cycloalkanones result from solvent capture of the intermediate cycloalkenyl cation, whereas the cycloalkynes and cycloallenes come from loss of a proton from the same ion. Of course, cycloallenes and cycloalkynes were observed only in the rings ($C_9$–$C_{12}$) that are capable of accommodating the required ring strain for their formation.

Interestingly, contrary to the behavior of the corresponding cycloalkyl cations [139], no transannular hydride migration was observed in the solvolyses of any of the cycloalkenyl systems **319** to **322**. This lack of transannular hydride migration in the medium-ring cyclic vinyl cations may be due to unfavorable conformations of the intermediate ions or to the higher energy, and hence shorter lifetime, of the vinyl cation compared to that of the corresponding cycloalkyl cations. Therefore the vinyl cation does not have a sufficient lifetime to rearrange (i.e., undergo transannular hydride migration) before it is captured by external nucleophile or loses a $\beta$ proton. Furthermore, as discussed in Chapter 7, even 1,2-hydride migrations, which are highly favored in carbenium ion chemistry, are scarce in vinyl cation reactions and occur only in special systems.

## REFERENCES

1. W. Moffit, *Proc. R. Soc. London, Ser. A* **202**, 548 (1950); E. D. Hughes, *Trans. Faraday Soc.* **34**, 185 (1938); **37**, 603 (1941).
2. J. D. Roberts and V. C. Chambers, *J. Am. Chem. Soc.* **73**, 5034 (1951).
3. L. O. Brockway, J. Y. Beach, and L. Pauling, *J. Am. Chem. Soc.* **57**, 2693 (1935); J. A. C. Hugill, I. E. Coop, and L. E. Sutton, *Trans. Faraday Soc.* **34**, 1518 (1938).
4. D. Y. Curtin, J. A. Kampmeier, and R. O'Connor, *J. Am. Chem. Soc.* **87**, 863 (1965).
5. P. J. Stang, *Chem. Rev.* **78**, 383 (1978).
6. M. S. Newman and A. Kutner, *J. Am. Chem. Soc.* **73**, 4199 (1951); M. S. Newman and A. E. Weinberg, *J. Am. Chem. Soc.* **78**, 4654 (1956).

7. M. S. Newman and C. D. Beard, *J. Am. Chem. Soc.* **91**, 5677 (1969); M. S. Newman and C. D. Beard, *J. Am. Chem. Soc.* **92**, 7564 (1970).
8. M. S. Newman and A. O. M. Okorodudu, *J. Org. Chem.* **34**, 1220 (1969), and references therein.
9. W. M. Jones and F. W. Miller, *J. Am. Chem. Soc.* **89**, 1960 (1967).
10. A. C. Day and M. C. Whiting, *J. Chem. Soc. B* p. 991 (1967).
11. A. Nishimura, H. Kato, and M. Ohta, *J. Am. Chem. Soc.* **89**, 5083 (1967); *Bull. Chem. Soc. Jpn.* **43**, 1530 (1970).
12. J. P. Dulcere, M. Santelli, and M. Bertrand, *C. R. Acad. Sci., Ser. C* **274**, 1087 (1972).
13. M. Hanack, C. E. Harding, and J. L. Derocque, *Chem. Ber.* **105**, 421 (1972).
14. D. Kammacher and M. Hanack, unpublished results.
15. M. Hanack, *Acc. Chem. Res.* **3**, 209 (1970); G. Modena and U. Tonellato, *Adv. Phys. Org. Chem.* **9**, 185 (1971); P. J. Stang, *Prog. Phys. Org. Chem.* **10**, 205 (1973); L. R. Subramanian and M. Hanack, *J. Chem. Educ.* **52**, 80 (1975); M. Hanack, *Acc. Chem. Res.* **9**, 364 (1976); P. J. Stang, *Acc. Chem. Res.* **11**, 107 (1978); M. Hanack, *Angew. Chem., Int. Ed. Engl.* **17**, 333 (1978).
16. L. R. Subramanian and M. Hanack, *Chem. Ber.* **105**, 1465 (1972).
17. Z. Rappoport, T. Bässler, and M. Hanack, *J. Am. Chem. Soc.* **92**, 4985 (1970).
18. Z. Rappoport and J. Kaspi, *J. Chem. Soc., Perkin Trans. II* p. 1102 (1972).
19. (a) E. Grunwald and S. Winstein, *J. Am. Chem. Soc.* **70**, 846 (1948); (b) A. H. Fainberg and S. Winstein, *J. Am. Chem. Soc.* **78**, 2770 (1956).
20. Z. Rappoport and A. Gal, *J. Org. Chem.* **37**, 1174 (1972).
21. Z. Rappoport, *Adv. Phys. Org. Chem.* **7**, 1 (1969); see also *Tetrahedron Lett.* p. 1073 (1978).
22. S. Patai and Z. Rappoport, *in* "The Chemistry of Alkenes" (S. Patai, ed.), Vol. 1, p. 469. Wiley (Interscience). New York, 1964.
23. G. Modena, *Acc. Chem. Res.* **4**, 73 (1971).
24. D. R. Kelsey and R. G. Bergman, *J. Am. Chem. Soc.* **93**, 1953 (1971).
25. L. R. Subramanian, M. Hanack, L. W. K. Chang, M. A. Imhoff, P. von R. Schleyer, F. Effenberger, W. Kurtz, P. J. Stang, and T. E. Dueber, *J. Org. Chem.* **41**, 4099 (1976).
26. N. Frydman, R. Bixon, M. Sprecher, and Y. Mazur, *Chem. Commun.* p. 1044 (1969).
26a. F. Hassdenteufel and M. Hanack, *Tetrahedron Lett.*, in press.
27. C. A. Grob and R. Spaar, *Helv. Chim. Acta* **53**, 2119 (1970); *Tetrahedron Lett.* p. 1439 (1969).
28. E. Lamparter and M. Hanack, *Chem. Ber.* **106**, 3216 (1973).
29. C. A. Grob and H. R. Pfaendler, *Helv. Chim. Acta* **53**, 2130 (1970).
30. R. Helwig and M. Hanack, *Justus Liebigs Ann. Chem.* p. 614 (1977).
31. M. Hanack, H. Bentz, R. Markl, and L. R. Subramanian, *Justus Liebigs Ann. Chem.* 1894 (1978).
32. M. Hanack and L. Sproesser, *J. Am. Chem. Soc.* **100**, 7066 (1978).
33. R. Huisgen and W. E. Konz, *J. Am. Chem. Soc.* **92**, 4102 (1970).
34. S. A. Sherrod and R. G. Bergman, *J. Am. Chem. Soc.* **91**, 2115 (1969); T. Bässler and M. Hanack, *J. Am. Chem. Soc.* **91**, 2117 (1969).
35. M. Hanack and H. J. Schneider, *Angew. Chem., Int. Ed. Engl.* **6**, 666 (1967); K. B. Wiberg, B. A. Hess, Jr., and A. J. Ashe, III, *in* "Carbonium Ions" (G. A. Olah and P. von R. Schleyer, eds.), Vol. 3, p. 1295. Wiley (Interscience), New York, 1972.
36. M. Hanack and J. Häffner, *Tetrahedron Lett.* p. 2191 (1964); M. Hanack and J. Häffner, *Chem. Ber.* **99**, 1077 (1966).
37. M. Bertrand and M. Santelli, *C. R. Acad Sci., Ser. C* **259**, 2251 (1964); M. Bertrand and M. Santelli, *Tetrahedron Lett.* p. 2511, 2515 (1969).
38. S. A. Sherrod and R. G. Bergman, *J. Am. Chem. Soc.* **93**, 1925 (1971).
39. D. R. Kelsey and R. G. Bergman, *J. Am. Chem. Soc.* **93**, 1941 (1971).

40. D. R. Kelsey and R. G. Bergman, *J. Chem. Soc., Chem. Commun.* p. 589 (1973).
41. T. V. Lehman and R. S. Macomber, *J. Am. Chem. Soc.* **97**, 1531 (1975).
42. M. Santelli and M. Bertrand, *Tetrahedron* **30**, 243 (1974).
43. W. Eymann and M. Hanack, *Tetrahedron Lett.* p. 3507 (1976).
44. T. Bässler and M. Hanack, *Tetrahedron Lett.* p. 2171 (1971).
45. M. Hanack, J. Häffner, and H. Herterich, *Tetrahedron Lett.* p. 875 (1965).
46. M. Hanack, S. Bocher, J. Herterich, K. Hummel, and V. Vott, *Justus Liebigs Ann. Chem.* **733**, 5 (1970).
47. A. Ghenciulescu and M. Hanack, *Tetrahedron Lett.* p. 2870 (1970).
48. M. Hanack, T. Bässler, W. Eymann, W. E. Heyd, and R. Kopp, *J. Am. Chem. Soc.* **96**, 6686 (1974).
49. G. Hammen, T. Bässler, and M. Hanack, *Chem. Ber.* **107**, 1676 (1974).
50. P. von R. Schleyer and G. W. van Dine, *J. Am. Chem. Soc.* **88**, 2321 (1966).
51. J. Salaun and M. Hanack, *J. Org. Chem.* **40**, 1994 (1975).
52. W. E. Heyd and M. Hanack, *Angew. Chem., Int. Ed. Engl.* **12**, 318 (1973); J. L. Derocque, F. B. Sundermann, N. Youssif, and M. Hanack, *Justus Liebigs Ann. Chem.* p. 419 (1973).
53. R. Kopp and M. Hanack, *Chem. Ber.* **112**, 2453 (1979).
54. C. A. Grob and G. Cseh, *Helv. Chim. Acta* **47**, 194 (1964).
55. C. A. Grob and H. R. Pfaendler, *Helv. Chim. Acta* **54**, 2060 (1971).
56. L. Eckes and M. Hanack, *Chem. Ber.* **111**, 1253 (1978).
57. L. R. Subramanian and M. Hanack, *Angew. Chem.* **84**, 714 (1972); *J. Org. Chem.* **42**, 174 (1977).
58. R. Märkl and M. Hanack, unpublished results.
59. H. Fischer, K. Hummel, and M. Hanack, *Tetrahedron Lett.* p. 2169 (1969).
60. Y. Apeloig, P. von R. Schleyer, J. D. Dill, J. B. Collins, and J. A. Pople, *Natl. Meet. Am. Chem. Soc., 170th, Chicago, Ill., 1975*; J. B. Collins and P. von R. Schleyer, personal communication.
61. K. Subramanian and M. Hanack, *Tetrahedron Lett.* p. 3365 (1973).
62. L. R. Subramanian and M. Hanack, unpublished results.
63. W. D. Pfeifer, C. A. Bahn, P. von R. Schleyer, S. Bocher, C. E. Harding, K. Hummel, M. Hanack, and P. J. Stang, *J. Am. Chem. Soc.* **91**, 1513 (1971).
64. M. Hanack, E. J. Carnahan, A. M. Krowczinski, W. Schoberth, L. R. Subramanian, and K. Subramanian, *J. Am. Chem. Soc.* **106**, 100 (1979).
65. W. Schoberth and M. Hanack, *Chem. Ber.* (in press).
66. K. Subramanian and M. Hanack, unpublished results.
67. R. H. DeWolfe and W. G. Young, *Chem. Rev.* **56**, 753 (1956).
68. A. Buroway and E. Spinner, *J. Chem. Soc.* p. 3752 (1964).
69. G. F. Hennion and D. E. Maloney, *J. Am. Chem. Soc.* **73**, 4735 (1951); R. S. Macomber, *Tetrahedron Lett.* p. 4639 (1970).
70. J. Salaun, *J. Org. Chem.* **41**, 1237 (1976); **42**, 28 (1977).
71. H. G. Richey, Jr. *et al., J. Am. Chem. Soc.* **87**, 1381, 4017 (1965); G. A. Olah *et al., J. Am. Chem. Soc.* **87**, 5632 (1965); **96**, 5855 (1974).
72. T. L. Jacobs and D. M. Fenton, *J. Org. Chem.* **30**, 1808 (1965).
73. M. D. Schiavelli, S. C. Hixon, and H. W. Moran, *J. Am. Chem. Soc.* **92**, 1082 (1970).
74. M. D. Schiavelli, S. C. Hixon, H. W. Moran, and C. J. Boswell, *J. Am. Chem. Soc.* **93**, 6989 (1971).
75. M. D. Schiavelli, R. P. Gilbert, W. A. Boynton, and C. J. Bosewell, *J. Am. Chem. Soc.* **94**, 5061 (1972).
76. M. D. Schiavelli, P. L. Timpanaro, and R. Brewer, *J. Org. Chem.* **38**, 3054 (1973).
77. M. D. Schiavelli and D. E. Ellis, *J. Am. Chem. Soc.* **95**, 7916 (1973).
78. M. D. Schiavelli, T. C. Germroth, and J. W. Stubbs, *J. Org. Chem.* **41**, 681 (1976).

79. D. Scheffel, P. J. Abbott, G. J. Fitzpatrick, and M. D. Schiavelli, *J. Am. Chem. Soc.* **99**, 3769 (1977).
80. V. J. Shiner, Jr. and G. S. Kriz, Jr., *J. Am. Chem. Soc.* **86**, 2643 (1964).
81. V. J. Shiner, Jr., W. E. Buddenbaum, B. L. Murr, and G. Lamaty, *J. Am. Chem. Soc.* **90**, 418 (1968).
82. (a) V. J. Shiner, Jr., W. Dowd, R. D. Fisher, S. R. Hartshorn, M. A. Kessick, L. Milakofsky, and M. W. Rapp, *J. Am. Chem. Soc.* **91**, 4838 (1969); (b) V. J. Shiner, Jr. and J. W. Wilson, *J. Am. Chem. Soc.* **84**, 2402 (1962).
83. J. L. Fry, C. J. Lancelot, L. K. M. Lam, J. M. Harris, R. C. Bingham, D. J. Raber, R. E. Hall, and P. von R. Schleyer, *J. Am. Chem. Soc.* **92**, 2538 (1970); S. H. Liggero, J. O. Harper, P. von R. Schleyer, A. P. Krapcho, and D. E. Horn, *J. Am. Chem. Soc.* **92**, 3789 (1970).
84. C. V. Lee, R. J. Hargrove, T. E. Dueber, and P. J. Stang, *Tetrahedron Lett.* p. 2519 (1971).
85. R. H. Summerville, C. A. Senkler, P. von R. Schleyer, T. E. Dueber, and P. J. Stang, *J. Am. Chem. Soc.* **96**, 1100 (1974).
86. Z. Rappoport and Y. Houminer, *J. Chem. Soc., Perkin Trans. II* p. 1506 (1973).
87. C. A. Vernon, *J. Chem. Soc.* p. 423 (1954).
88. A. Streitwieser, Jr., *Chem. Rev.* **56**, 571 (1965).
89. V. J. Shiner, Jr. and J. S. Humphrey, Jr., *J. Am. Chem. Soc.* **89**, 622 (1967).
90. H. D. Hartzler, *J. Am. Chem. Soc.* **83**, 4990, 4997 (1961); *J. Org. Chem.* **29**, 1311 (1964); W. J. Le Noble, Y. Tatsukami, and H. F. Morris, *J. Am. Chem. Soc.* **92**, 5681 (1970).
91. P. von R. Schleyer, J. L. Fry, L. K. M. Lam, and C. J. Lancelot, *J. Am. Chem. Soc.* **92**, 2542 (1970).
92. D. Mirejowsky, W. Drenth, and F. B. von Duijneveldt, *J. Org. Chem.* **43**, 763 (1978).
93. J. H. Ford, C. D. Thompson, and C. S. Marvel, *J. Am. Chem. Soc.* **57**, 2619 (1935).
94. H. C. Brown and Y. Okamoto, *J. Am. Chem. Soc.* **80**, 4979 (1958).
95. G. Baddeley and J. Chadwick, *J. Chem. Soc.* p. 368 (1951); G. Baddeley, J. W. Rasburn, and R. Rose, *J. Chem. Soc.* p. 3168 (1958).
96. J. L. Fry, J. M. Harris, R. C. Bingham, and P. von R. Schleyer, *J. Am. Chem. Soc.* **92**, 2540 (1970).
97. V. J. Shiner, Jr., *in* "Isotope Effects in Chemical Reactions" (C. J. Collins and N. S. Bowman, eds.), ACS Monograph No. 167, pp. 105–120. Van Nostrand-Reinhold, New York, 1970.
98. S. R. Hartshorn and V. J. Shiner, Jr., *J. Am. Chem. Soc.* **94**, 9002 (1972).
99. V. J. Shiner, Jr., *J. Am. Chem. Soc.* **75**, 2925 (1953); V. J. Shiner, Jr., *J. Am. Chem. Soc.* **83**, 240 (1961).
100. V. J. Shiner, Jr., B. L. Murr, and G. Heinemann, *J. Am. Chem. Soc.* **85**, 2413 (1963).
101. H. Taniguchi, unpublished results.
102. D. J. Raber, R. C. Bingham, J. M. Harris, J. L. Fry, and P. von R. Schleyer, *J. Am. Chem. Soc.* **92**, 5977 (1970).
103. F. L. Schadt, T. W. Bentley, and P. von R. Schleyer, *J. Am. Chem. Soc.* **98**, 7667 (1976).
104. S. Winstein, E. Grunwald, and H. W. Jones, *J. Am. Chem. Soc.* **73**, 2700 (1951); S. Winstein, A. H. Fainberg, and E. Grunwald, *J. Am. Chem. Soc.* **79**, 4146 (1957).
105. D. E. Sunko, I. Szele, and M. Tomič, *Tetrahedron Lett.* p. 1827 (1972).
106. Z. Rappoport and J. Kaspi, *J. Am. Chem. Soc.* **96**, 4518 (1974).
107. V. J. Shiner, Jr. and R. D. Fisher, *J. Am. Chem. Soc.* **93**, 2553 (1971).
108. (a) A. H. Fainberg and S. Winstein, *J. Am. Chem. Soc.* **78**, 2763 (1956); (b) D. J. Raber, J. M. Harris, and P. von R. Schleyer, *in* "Ions and Ion Pairs in Organic Reactions" (M. Szwarc, ed.), Vol. 2, 247. Wiley (Interscience), New York, 1973.
109. Z. Rappoport and Y. Apeloig, *J. Am. Chem. Soc.* **97**, 821 (1975).

110. T. Sonoda, M. Kawakami, T. Ikeda, S. Kobayashi, and H. Taniguchi, *Chem. Commun.* p. 612 (1976).
111. T. Sonoda, S. Kobayashi, and H. Taniguchi, *Bull. Chem. Soc. Jpn.* **49**, 2560 (1976)
112. T. Suzuki, T. Sonoda, S. Kobayashi, and H. Taniguchi, *Chem. Commun.* p. 180 (1976).
113. E. R. Thornton, "Solvolysis Mechanisms." Ronald Press, New York, 1964; D. J. Raber and J. M. Harris, *J. Chem. Educ.* **49**, 60 (1972); J. M. Harris, *Prog. Phys. Org. Chem.* **11**, 89 (1974).
114. R. W. Taft, Jr., *J. Am. Chem. Soc.* **75**, 4231 (1953).
115. R. C. Bingham and R. von R. Schleyer, *Tetrahedron Lett.* p. 23 (1971); *J. Am. Chem. Soc.* **93**, 3189 (1971).
116. P. E. Peterson and J. M. Indelicato, *J. Am. Chem. Soc.* **90**, 6516 (1968).
117. P. E. Peterson and J. M. Indelicato, *J. Am. Chem. Soc.* **91**, 6194 (1969).
118. T. C. Clarke, D. R. Kelsey, and R. G. Bergman, *J. Am. Chem. Soc.* **94**, 3626 (1972).
119. R. H. Summerville and P. von R. Schleyer, *J. Am. Chem. Soc.* **96**, 1110 (1974); **94**, 3629 (1972).
120. T. C. Clarke and R. G. Bergman, *J. Am. Chem. Soc.* **96**, 7934 (1974).
121. For reviews and leading references, see E. A. Halevi, *Prog. Phys. Org. Chem.* **1**, 109 (1963); L. Melander, "Isotope Effects on Reaction Rates." Ronald Press, New York, 1960; see also reference 107.
122. P. J. Stang, R. J. Hargrove, and T. E. Dueber, *J. Chem. Soc., Perkin Trans. II* p. 843 (1974).
123. D. D. Maness and L. D. Turrentine, *Tetrahedron Lett.* p. 755 (1973).
124. Z. Rappoport, A. Pross, and Y. Apeloig, *Tetrahedron Lett.* p. 2015 (1973).
125. Z. Rappoport and Y. Apeloig, *J. Am. Chem. Soc.* **96**, 6428 (1974).
126. P. J. Stang, R. J. Hargrove, and T. E. Dueber, *J. Chem. Soc., Perkin Trans. II* p. 1486 (1977).
127. A. Streitwieser, Jr., and A. Dafforn, *Tetrahedron Lett.* p. 1263 (1969).
128. A. Streitwieser, Jr. and H. S. Klein, *J. Am. Chem. Soc.* **85**, 2759 (1963).
129. V. J. Shiner, Jr. and J. S. Humphrey, Jr., *J. Am. Chem. Soc.* **85**, 2416 (1963); V. J. Shiner, Jr. and J. G. Jewett, *J. Am. Chem. Soc.* **86**, 945 (1964); G. J. Karabatsos, G. C. Sonnichsen, C. G. Papaionnon, S. E. Scheppele, and R. L. Shone, *J. Am. Chem. Soc.* **89**, 463 (1967); B. L. Murr and J. A. Conklin, *J. Am. Chem. Soc.* **92**, 3464 (1970).
130. D. S. Noyce and M. D. Schiavelli, *J. Am. Chem. Soc.* **90**, 1023 (1968).
131. G. S. Hammond, *J. Am. Chem. Soc.* **77**, 334 (1955).
132. C. G. Swain and E. R. Thornton, *J. Am. Chem. Soc.* **84**, 822 (1962); E. R. Thornton, *J. Am. Chem. Soc.* **89**, 2915 (1967).
133. S. A. McNeely and P. J. Kropp, *J. Am. Chem. Soc.* **98**, 4319 (1976).
134. E. Lamparter and M. Hanack, *Chem. Ber.* **105**, 3789 (1972).
135. R. J. Hargrove and P. J. Stang, *Tetrahedron* **32**, 37 (1976).
136. P. J. Stang and R. H. Summerville, *J. Am. Chem. Soc.* **91**, 4600 (1969).
137. J. Sicher, *Prog. Stereochem.* **3**, 202 (1962).
138. H. C. Brown and M. Borkowski, *J. Am. Chem. Soc.* **74**, 1894 (1952).
139. J. L. Fry and G. L. Karabatsos, "Carbonium Ions," Vol. II. Wiley (Interscience), New York, 1970; A. C. Cope, M. M. Martin, and M. A. McKervey, *Q. Rev. Chem. Soc.* **20**, 119 (1969); V. Prelog, *Pure Appl. Chem.* **6**, 545 (1963); V. Prelog and J. G. Traynham, *in* "Molecular Rearrangements" (P. de Mayo, ed.), Vol. 1, p. 593. Wiley (Interscience), New York, 1963.

# 6

# ARYLVINYL CATIONS
# VIA SOLVOLYSIS

## I. INTRODUCTION

The intermediacy of $\alpha$-arylvinyl cations in heterolysis reactions was first suggested by Grob in the solvolytic fragmentation of the $Z$-$\beta$-bromo-cinnamate ion [Eq. (1)] [1]. Nevertheless, Grob and Cseh's solvolytic generation of $\alpha$-arylvinyl cations from para-substituted $\alpha$-bromostyrenes in

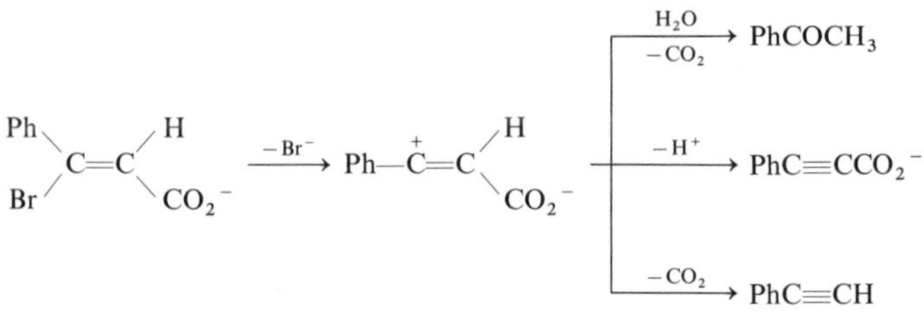

$$
\underset{Br}{\overset{Ph}{>}}C{=}C\underset{CO_2^-}{\overset{H}{<}} \xrightarrow{-Br^-} Ph{-}\overset{+}{C}{=}C\underset{CO_2^-}{\overset{H}{<}}
\begin{cases}
\xrightarrow[-CO_2]{H_2O} PhCOCH_3 \\
\xrightarrow{-H^+} PhC{\equiv}CCO_2^- \quad (1) \\
\xrightarrow{-CO_2} PhC{\equiv}CH
\end{cases}
$$

buffered aqueous ethanol [Eq. (2)] [2] is usually considered the first example

$$
ArC(Br){=}CH_2 \xrightarrow[Et_3N]{80\% \ EtOH} Ar\overset{+}{C}{=}CH_2 \xrightarrow{(H_2O)} ArCOCH_3 \quad (2)
$$

of the intermediacy of vinyl cations in solvolysis reactions. The highly activating $\alpha$ substituent that was required for the generation of these allegedly highly unstable intermediates was obtained by replacing an $\alpha$-aryl group with a strong electron-donating para substituent, such as an amino or methoxy group. The demonstration of first-order kinetics and high sensitivity to substituent and solvent effects under conditions that were not too drastic [2] paved the way for further studies, among which those of the solvolysis of $\alpha$-arylvinyl systems have been the most extensive [3].

The qualitative similarities between the solvolyses of vinylic and saturated compounds were demonstrated in Chapter 5. However, important quantitative differences between the solvolyses of these different systems and in the nature of the derived intermediates have been observed mainly with aryl-substituted vinylic systems. Some differences, such as the presence of alternative reaction centers or different stereochemistry of the ions, are also common to α-alkylvinyl systems, but they have been studied in much more detail with the α-aryl-substituted systems, in which the high activation makes it possible to observe phenomena not usually observed with α-alkylvinyl systems. Other phenomena, especially those connected with conjugation effects, are either very significant or unique to the aryl-substituted systems [3].

The double bond of a vinylic substrate conceivably may conjugate with any or all of the four vinylic substituents. Conjugation with an aryl group gives a styrene moiety with several kilocalories per mole of delocalization energy [4]. If more than one aryl substituent is attached to the double bond, a $\pi(Ar)-\pi(C=C)$ conjugation will take place with all the aryl substituents. Obviously, the extent of this conjugation is highly dependent on the planarity of the system and hence on steric interaction between the substituents. Consequently, when the steric–conjugation interactions become important, mechanistic criteria that are applicable for assigning the detailed mechanism and for estimating the polarity of the transition state in saturated systems may fail in vinylic systems. The ambiguity of several such criteria when applied to the solvolysis of vinylic systems is discussed in detail elsewhere in this chapter, but here the solvolysis mechanism of trianisylvinyl derivaties $An_2C=C(X)An$ will serve as an example. The low sensitivity to solvent and leaving-group effects, as reflected by the Winstein–Grunwald $m$ value of 0.34 in aqueous ethanol when $X = Br$ [5] and the $k_{OTs}/k_{Br}$ ratio of 32 [6], suggests a solvent-assisted $(k_s)$ [7] or a neighboring-group-assisted $(k_\Delta)$ [7] solvolysis, in analogy with saturated compounds. However, the high sensitivity to substituent effects, as reflected by the $\rho^+$ value of $-4.5$ for a change in the α-aryl group [5, 8], and the high sensitivity to the arylsulfonate leaving group, as reflected by a $\rho$ value of 1.91 [6b], indicate a highly polar transition state and an unassisted $k_c$ route [7]. This route is substantiated by an extensive common ion rate depression and an extensive degenerate β-anisyl rearrangement in 2,2,2-trifluoroethanol (TFE) [9, 10] and in AcOH [9, 11]. Most of these contradictory results could be explained by the accumulation of the three aryl groups, all of which are conjugated with the double bond, in the plane of this bond [3].

The ground state of an unhindered α-arylvinyl system is stabilized by a $\pi(\alpha\text{-}Ar)-\pi(C=C)$ conjugation (1). The incipient cationic orbital is developed in a perpendicular plane. Stabilization of the ion and of the transition state leading to it is achieved by overlap of the vacant $p$ orbital and the $\pi$ system of the α-aryl group (2). However, this stabilizing $\pi(\alpha\text{-Ar})\text{-}p(C^+)$ conjugation can be achieved only at the expense of a simultaneous, complete $\pi(\alpha\text{-Ar})\text{-}$ $\pi(C{=}C)$ deconjugation (1 → 2, Scheme 6.1).

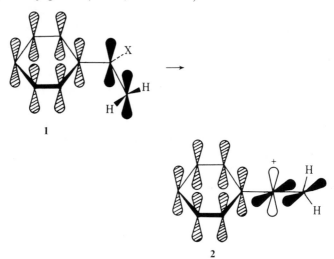

Scheme 6.1. From Rappoport [3].

The presence of an additional β-aryl substituent or substituents complicates this simple picture. The situation is not much different if the α-aryl and β-aryl groups are in trans positions and the other two substituents are small. However, if the two aryl groups are cis to one another as in 3 (Scheme 6.2), both groups are already twisted from the plane of the double bond in the ground state. The loss of $\pi(\alpha\text{-Ar})\text{-}\pi(C{=}C)$ conjugation during the ionization to 4 will be lower than for 1, whereas increased $\pi(\beta\text{-Ar})\text{-}\pi(C{=}C)$ conjugation in the ion 4 and the transition state for its formation (Scheme 6.2) will enhance the reactivity of the compound. The situation will become even more complex for an α,β,β-triarylvinyl system, in which several conjugation terms and steric interactions between the various substituents should be taken into account.

The extent of the ground state $\pi(\alpha\text{-Ar})\text{-}\pi(C{=}C)$ conjugation is equivalent to the delocalization energy of styrene (4.4 kcal/mole) [4]. Solvolytic data can also give an estimate of this value. Table 6.1 gives the appropriate data. The value of $1.1 \times 10^6$ obtained by Peterson and Indelicato for the relative reactivity $k(2\text{-butyl-OTs})/k(E\text{-2-buten-2-yl-OTs})$ [12] can be taken as a reference value, notwithstanding the fact that it involves $\pi(\alpha\text{-Me})\text{-}\pi(C{=}C)$ and $\pi(\beta\text{-Me})\text{-}\pi(C{=}C)$ conjugation terms and that 2-butyl tosylate may react via a $k_s$ route [13]. The much higher reactivity ratio of $2 \times 10^{10}$ for α-phenyl-

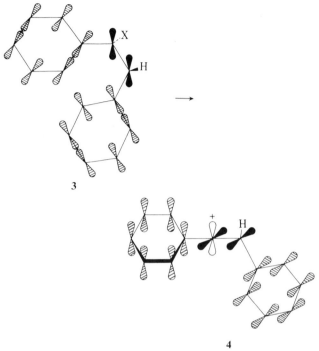

**3**

**4**

Scheme 6.2. From Rappoport [3].

ethyl bromide versus α-bromostyrene [2, 5, 14] shows clearly the rate-retarding effect of a phenyl group on this ratio. For these two compounds, almost all the rate difference is due to the 14 kcal/mole higher activation energy for the vinylic compound.

A distortion of planarity caused by ortho substituents, such as the 2,6-dimethyl groups of the mesityl system [15] or the substituted C-1 and C-8 of the 9-anthryl system [17], results in an effective ground state deconjugation, and consequently the reactivity ratios of Table 6.1 decrease again to values similar to or lower than those obtained in the absence of an α-aryl group. Estimation suggests that the difference between the saturated and the unsaturated mesityl systems is also mainly due to activation enthalpy differences [15]. However, it should be realized that the comparisons of Table 6.1 can give only an approximate value for the $\pi(\alpha\text{-Ar})-\pi(C{=}C)$ conjugation since they involve four ground states and four transition states and steric enhancement of the rates may play a role in the solvolysis of the crowded saturated compounds [16]. Another estimate of the $k(\text{saturated})/k(\text{vinylic})$ ratio of ca. $10^6$ is based on a less appropriate model and involves severe extrapolations [19].

**TABLE 6.1**

**Relative Reactivities of Saturated and Vinylic Compounds**

| Saturated RX | Vinylic RX | Solvent | Temp (°C) | $\pi(Ar)-\pi(C=C)$ conjugation | $k(\text{saturated})/k(\text{vinylic})$ | Reference |
|---|---|---|---|---|---|---|
| 2-Butyl tosylate | E-2-Buten-2-yl tosylate | 50% MeOH | 130 | None | $1.1 \times 10^6$ | [12] |
| $C_6H_5CH(Br)Me$ | $C_6H_5C(Br)=CH_2$ | 80% EtOH | 25 | Complete | $2 \times 10^{10}$ | [2, 5, 14] |
| α-Mesitylethyl chloride | α-Mesitylvinyl chloride | EtOH | 36.5 | Small | $1.4 \times 10^6$ | [15, 16] |
| α-(9-Anthryl)ethyl chloride | α-(9-Anthryl)vinyl chloride | 90% $Me_2CO$ | 140 | Very small | $6.9 \times 10^4$ | [17, 18] |

A similar problem is involved in the relative reactivities for the generation of vinyl and trigonal cations by electrophilic additions to phenylalkynes and phenylalkenes, respectively. These additions, which are discussed in Chapter 3, show high sensitivity to the nature of the electrophile and to the polarity of the medium. The relative reactivities $k$(alkene)/$k$(alkyne) are especially high for bromine addition in low-dielectric media and follow the order $Br_2 > Cl_2 > ArSCl$. However, addition of $Br_2$ and of $H_3O^+$ in water gave a $k$(alkene)/$k$(alkyne) ratio of ca. 1, and it was therefore suggested that solvation effects are very important in these reactions [20]. This conclusion is challenged by the $k$(arylalkene)/$k$(arylalkyne) ratios of ca. 1 for the addition of benzhydryl cation in the low-dielectric dichloromethane [21] and for the addition of the acetic acid solvent to $ArCH=CH_2$ and $ArC≡CH$, when $Ar$ = anisyl [22] or 9-anthryl [17]. Moreover, recent work by Modena and co-workers [23] on the addition of $H_3O^+$ in aqueous sulfuric acid suggests that the solvation requirements of both types of additions are modest and closely similar for analogous pairs of olefins and acetylenes.

When the activating effect of an unsubstituted phenyl group is compared with those of other substituents, the relative reactivities are found to be dependent on the system investigated. A $k(E\text{-}PhC(OTs)=CHMe)/k(E\text{-}MeC(OTs)=CHMe)$ ratio of ca. 1600 can be estimated from related data [12, 24], giving an $\alpha$-Ph/$\alpha$-Me ratio of the same magnitude. When appropriate extrapolations for the different temperatures, leaving groups, and solvents were used, the solvolytic data for $\alpha$-bromostyrene in aqueous ethanol [2] and for cyclopropylidenemethyl chloride in methanol [25] gave an $\alpha$-cyclopropyl/$\alpha$-phenyl ratio of ca. 6000 for the $\beta,\beta$-unsubstituted system.

On the other hand, when charge dispersal to the $\beta$ position is extensive, as in the substituted cyclopropylidenemethyl bromides 5 to 8, the range of

| 5 | 6 | 7 | 8 |

reactivity is much narrower, as shown by the relative reactivities of 1, $10^3$, $2.5 \times 10^3$, and $10^5$ for 5, 6, 7, and 8, respectively, in 80% EtOH at 100° [26]. A $\pi(\alpha\text{-}Ph)-\pi(C=C)$ conjugation probably accounts for the relatively low $\alpha$-Ph/$\alpha$-Me and $\alpha$-cyclopropyl/$\alpha$-Ph ratios.

An unsubstituted $\alpha$-vinyl group has a similar activating effect to an $\alpha$-phenyl: 2-bromobutadiene is 1.6 times more reactive than $\alpha$-bromostyrene in 80% EtOH [2, 27].

The solvolysis reactions leading to $\alpha$-arylvinyl cations and sometimes to $\beta$-arylvinyl cations will be presented in the following order. (1) Competing reactions that lead to the same solvolysis or elimination products as formed via the vinyl cations will be discussed in terms of the criteria that distinguish them from the $S_N1$ heterolysis. (2) Substituent effects, solvent effects, and leaving-group effects that are related to the structure and the polarity of the transition state of the solvolysis will be analyzed. (3) The stereochemistry of the solvolysis and the structure of the α-arylvinyl cations will be discussed. (4) The details of the solvolysis scheme, including the intervention of free ions and ion pairs and their reactions, such as external ion return, internal return, and capture by the solvent, will be outlined. (5) Neighboring-group participation and rearrangement across the double bond by aryl groups will be discussed. These will also include the closely related ions formed by the deamination reaction. (6) Some related nonsolvolytic methods for the generation of α-arylvinyl cations will be discussed.

Certain related topics, such as deuterium isotope effects in the solvolysis of arylvinyl derivatives (Chapter 5, Section V,A), nmr observation of arylvinyl cations (Chapter 8), rearrangements of β-heteroatoms and β-hydrogens (Chapter 7), and synthetic uses of arylvinyl cations (Chapter 9), are discussed in the indicated chapters for the sake of unity and completeness of the topics under consideration.

## II. COMPETING SUBSTITUTION AND
## ELIMINATION ROUTES

The $S_N1$ route is only one of a diversity of substitution routes at a vinylic carbon [28, 29]; similarly, the E1 route may compete with other elimination routes. Competing substitution routes may involve an initial attack, either electrophilic, nucleophilic, or radical, on the double bond or on the leaving group, whereas competing elimination may involve attack on the vinylic hydrogen.

Discussions of these competing reactions were obviously more extensive in the early days of vinyl cation chemistry, when methods of generating vinyl cations solvolytically had not been established and these alternatives had to be ruled out. However, it is still important to recognize these possibilities and to delineate the structural and medium effects that direct the reaction toward one of these substitution–elimination routes (see also Chapter 5, Section III,A).

Not all the substitution routes are of interest in relation to their possible competition with the solvolytic routes of α-arylvinyl systems. Nevertheless, the variety of such routes that have actually been observed to compete with the $S_N1$ route for α-anisylvinyl systems is demonstrated in Table 6.2. The table is divided into families of closely related compounds in which two

**TABLE 6.2**

Competing Substitution and Elimination Routes for α-Anisylvinyl Compounds

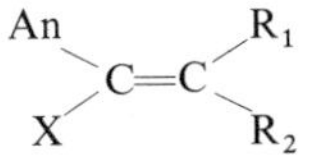

$$\begin{array}{c} An \\ \diagdown \\ X \diagup \end{array} C{=}C \begin{array}{c} R_1 \\ \diagup \\ \diagdown R_2 \end{array}$$

| X | $R_1$ | $R_2$ | Solvent/base | Mechanism | Reference |
|---|---|---|---|---|---|
| Br | Ph | Ph | 80% EtOH/RS⁻ | $S_N 1$ | [5, 30] |
| Br | Fluorene[a] | | 80% EtOH/RS⁻ | $Ad_N$–E | [30] |
| Br | An | H | 80% EtOH/OH⁻ | $S_N 1$ + E1 | [31] |
| Br | H | An | 80% EtOH/OH⁻ | E2 | [31] |
| OCOC₆H₃(NO₂)₂-3,5 | Me | Me | TFE/CF₃CH₂O⁻ | $B_{Ac}2$ | [32] |
| OCOC₆H₃(NO₂)₂-3,5 | Me | Me | AcOH/AcO⁻ | $A_{Ac}2(+ S_N 1?)$ | [32] |
| OCOCF₃ | Me | Me | AcOH/AcO⁻ | $Ad_E$–E | [32] |
| OTs[b] | Me | Me | AcOH/AcO⁻ | $S_N 1$ | [33] |
| OTs[b] | Me | Me | CF₃COOH/CF₃COO⁻ | $S_N 1$ | [34] |
| Br | Tol[c,d] | Ph[c] | CF₃COOH/CF₃COO⁻ | $Ad_E$–E | [35] |
| Cl | Ph | An | AcOH/AcO⁻ | $S_N 1$ | [36] |
| Cl | H | H | AcOH/AcO⁻ | $S_N 1$ + $Ad_E$–E | [22] |
| Br | Ph | An | AcOH/AcO⁻ | $S_N 1$ | [36] |
| Br | H | H | AcOH/AcO⁻ | $S_N 1(+ Ad_E$–E?) | [22] |

[a] $R_1 R_2 C{=}$ = fluorenylidene.
[b] OTs, tosylate (OSO₂C₆H₄Me-$p$).
[c] The α-aryl group is $p$-tolyl.
[d] Tol, $p$-tolyl.

substitution or elimination routes compete under similar conditions for different members of the family.

## A. Nucleophilic Addition–Elimination Route

The nucleophilic addition–elimination route (Ad$_N$–E) [Eq. (3)] [29] is the most important route for vinylic substitution and is the closest analogue of

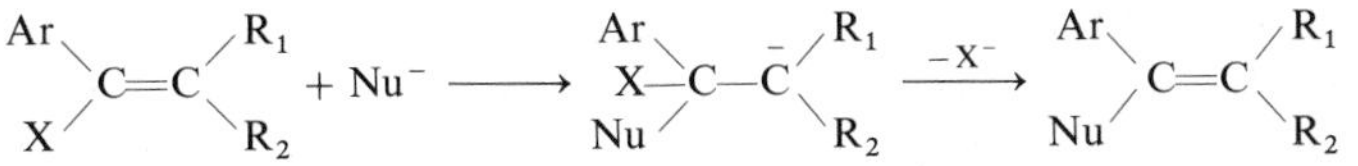

$$\begin{array}{ccc} Ar \diagdown \quad \diagup R_1 \\ C{=}C \\ X \diagup \quad \diagdown R_2 \end{array} + Nu^- \longrightarrow \begin{array}{c} Ar \diagdown \quad _- \diagup R_1 \\ X{-}C{-}C \\ Nu \diagup \quad \diagdown R_2 \end{array} \xrightarrow{-X^-} \begin{array}{c} Ar \diagdown \quad \diagup R_1 \\ C{=}C \\ Nu \diagup \quad \diagdown R_2 \end{array} \quad (3)$$

the S$_N$2 route for saturated systems. The rate-determining nucleophilic addition leads to a carbanion, or to a zwitterion if Nu is neutral (e.g., the solvent SOH). Since the S$_N$1 route leads to a carbonium ion, the structural requirements of both routes are sufficiently different that competition between them for a single substrate is rarely a mechanistic problem, and there are many mechanistic criteria to distinguish between them.

The $S_N1$ reaction is first order with respect to the substrate, and its rate is independent of the nature or the concentration of added nucleophiles. It is strongly activated by $\alpha$ electron-donating substituents, whereas the effect of $\beta$ substituents is mainly steric. It shows an "element effect" of the leaving group, it is accelerated in polar solvents, and it gives identical product mixtures starting from either an $E$ or a $Z$ precursor. On the other hand, the $Ad_N–E$ route is second order with respect to the nucleophile, and its rate is strongly dependent on the nucleophile. It is mainly activated by $\beta$ electron-withdrawing substituents, it shows a small element effect (e.g., $k_{Br}/k_{Cl} \sim 1$), and it gives retention of the geometric configuration starting from either isomer [29].

A search for a system capable of reacting via either route under different conditions showed that, when activating groups of opposite types are the substituents at the $\alpha$ and the $\beta$ positions, the $Ad_N–E$ route takes over. For example, competition between the powerfully $S_N1$-activating $\alpha$-$p$-dimethyl-aminophenyl group and the two $Ad_N–E$-activating $\beta$-cyano groups of $p$-$Me_2NC_6H_4C(Cl){=}C(CN)_2$ resulted in reaction via the latter route alone [37].

The only system shown to react by both routes is **10**, which is obtained by replacing the two $\beta$-aryl groups of the typical $S_N1$ substrate **9** by a fluorenyl

moiety. The triarylvinyl skeleton, which favors the $S_N1$ route, is retained in **10**, but the fluorenyl group is capable of stabilizing the partial negative charge formed in the transition state of the $Ad_N–E$ route [30].

The behavior of systems **9** and **10** in 80% EtOH buffered by NaOAc and in the presence of $PhCH_2S^-$ and $p$-$MeC_6H_4S^-$ ions is compared in Table 6.3 in terms of several mechanistic criteria [30]. It is clear from the solvent effect, the $\alpha$-substituent effect, the leaving-group effect, the zero order with respect to the nucleophile $Nu^-$, and the formation of the vinyl acetates in AcOH and the vinyl ethers and triarylethanones in 80% EtOH that both systems react via C—X bond heterolysis in the presence of NaOAc and that the quantitative difference between **9** and **10** is very small. Even a change to thio nucleophiles, such as benzylthiolate or $p$-toluenethiolate ions, has no effect on the rate-determining step in the reaction of **9**. The $S_N1$ nature of the process is shown by the rate–product disparity criterion, since the products in these cases are the corresponding triarylvinyl thiolates.

**TABLE 6.3**

**Mechanistic Criteria for the Reactions of Systems 9 and 10 in 80% EtOH[a]**

| System | Order in Nu$^-$ | Product | $k_{Br}/k_{Cl}$ | $k_{\alpha\text{-An}}/k_{\alpha\text{-Ph}}$ | $m$ (aq. EtOH) | $k_{PhCH_2S^-}/k_{p\text{-MeC}_6H_4S^-}$ | $\Delta H^{\ddagger}$ (kcal/mole) | $\Delta S^{\ddagger}$ (eu) |
|---|---|---|---|---|---|---|---|---|
| **9** + NaOAc | 0[b] | **9**, X = OAc | 53 | 3670 | 0.45 | 1 | 22 | −22 |
| **10** + NaOAc | 0 | **10**, X = OAc | 75 | 3100 | 0.57 | | 24 | −17 |
| **10** + PhCH$_2$SNa | 1[c] | **10**, X = SCH$_2$Ph | 1.8 | 0.93 | ~0 | 12.6 | 16 | −21 |

[a] From Rappoport and Gal [30].
[b] Also with NaOH, when the products are the ketone and the ether, and with PhCH$_2$SNa and p-MeC$_6$H$_4$SNa, when **9**, X = SCH$_2$Ph, and **9**, X = p-MeC$_6$H$_4$S, are formed.
[c] Also with p-MeC$_6$H$_4$SNa, when the product is **10**, X = p-MeC$_6$H$_4$S.

In contrast, the much faster reaction of **10** (Ar = An) with thiolate ions is of the $Ad_N$–E type, since comparison of lines 2 and 3 of Table 6.3 shows that the highly nucleophilic thiolate ions are involved in the rate-determining step. The evidence is the kinetic order, the element effect, and the faster reaction with the more reactive $PhCH_2S^-$ as compared with the $p$-$MeC_6H_4S^-$ nucleophile. The Grunwald–Winstein $m$ value [38] of ca. 0 reflects the expected rate decrease of the ion–molecule reaction superimposed on increased dissociation of the thiol in the more aqueous solvent. The most sensitive tool for distinguishing between the two routes and for detecting a small contribution of the $S_N1$ route in a predominantly $Ad_N$–E process is the α-An/α-Ph ratio. Since Hammett's $\rho$ values [39] for nucleophilic attack on carbon–carbon double bonds are always positive (mostly $\geq 1$) [40], $k_{\alpha\text{-An}}/k_{\alpha\text{-Ph}}$ ratios should be below unity. On the other hand, $\rho$ values for the anchimerically unassisted $S_N1$ processes are $\leq -3.5$ (see Section III,A). The ratio of 0.93 ($\rho = 0.06$) obtained for **10** with $PhCH_2S^-$ should be contrasted with the ratio of 3100 ($\rho = -4.5$) for the $S_N1$ route with $AcO^-$. Since the ratios are somewhat higher than those expected for a pure $Ad_N$–E route, an intervention of 0.5 to 1.0% of the $S_N1$ route in the overall process was suggested [30].

Hence, the expected change in mechanism from a C—X bond cleavage ($S_N1$ via **11**) to a rate-determining C—Nu bond formation ($Ad_N$–E via **12**)

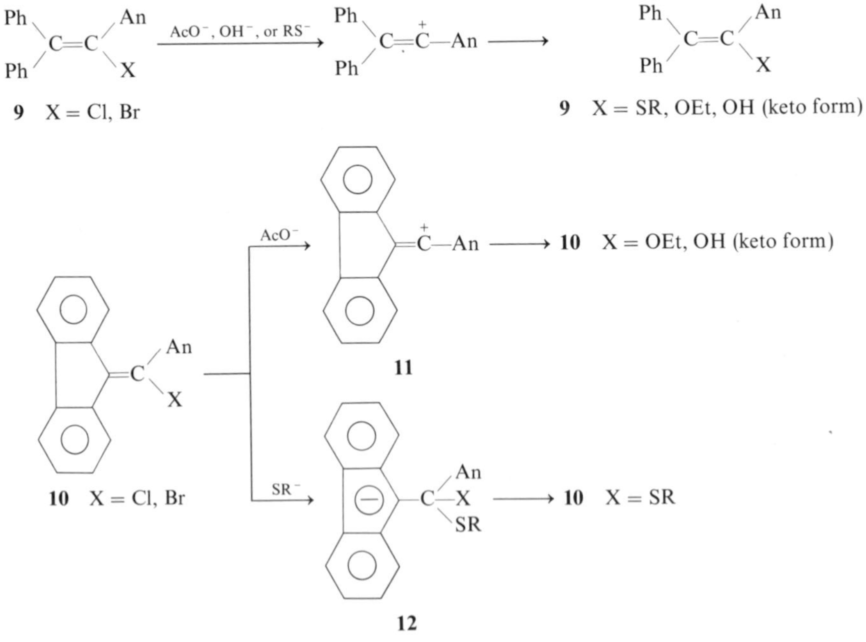
Ph
An
Ph
X
9 X = Cl, Br
AcO⁻, OH⁻, or RS⁻
Ph
C—An
Ph
Ph
An
Ph
X
9 X = SR, OEt, OH (keto form)
AcO⁻
C—An
10 X = OEt, OH (keto form)
An
X
10 X = Cl, Br
11
SR⁻
An
X
SR
10 X = SR
12

Scheme 6.3

was achieved for system **10** by changing a moderate nucleophile (AcO$^-$) to a powerful one (RS$^-$), as shown in Scheme 6.3.

The main structural change involved an increased resonative negative charge dispersal ability of the $\beta$ substituent, since fluorene is 10 p$K_a$ units more acidic than diphenylmethane [41]. A more modest structural change in **9** is apparently not sufficient to cause a shift to the Ad$_N$–E route. Preliminary experiments show that compound **13**, in which one of the $\beta$-phenyl groups is substituted by a $p$-nitro group, reacts both in aqueous ethanol and in 2,2,2-trifluoroethanol in the presence of OAc$^-$ or RS$^-$ ions via S$_N$1 [42].

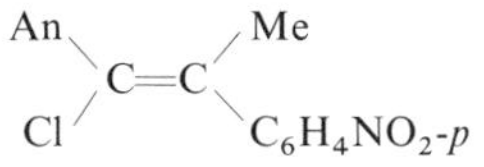

$$\text{An}\diagdown \quad \diagup \text{Me}$$
$$\text{C}{=}\text{C}$$
$$\text{Cl}\diagup \quad \diagdown \text{C}_6\text{H}_4\text{NO}_2\text{-}p$$

**13**

In some cases deuterium incorporation via this route should be observed. The absence of deuterium in the cyclopropyl phenyl ketone formed from 1-bromo-2-phenylcyclobutene (**14**) rules out this possibility [Eq. (4)] [43].

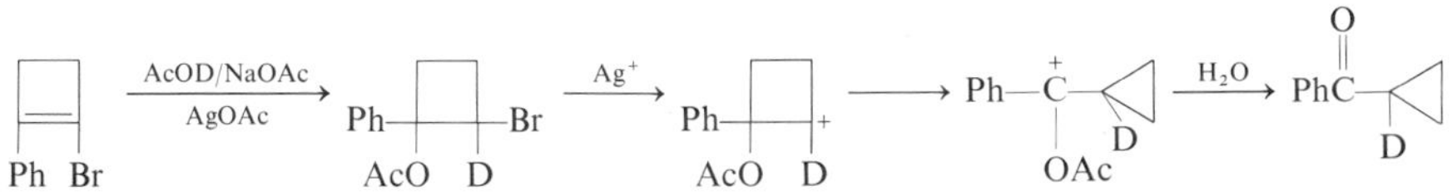

$$\tag{4}$$

## B. Electrophilic Addition–Elimination Route

The route most likely to compete with the S$_N$1 route in carboxylic acid solvents is the electrophilic addition–elimination route (Ad$_E$–E), in which rate-determining proton addition from the solvent to the vinylic carbon is followed by a rapid addition of a nucleophile and elimination of HX from the saturated adduct [Eq. (5)]. This route was first invoked for the solvolyses

$$\text{R}_1\text{C(X)}{=}\text{CR}_2\text{R}_3 \xrightarrow{\text{H}^+} \text{R}_1\overset{+}{\text{C}}\text{(X)}{-}\text{CHR}_2\text{R}_3 \xrightarrow{\text{Nu}^-} \text{R}_1\text{C(X)(Nu)}{-}\text{CHR}_2\text{R}_3 \xrightarrow{-\text{HX}} \text{R}_1\text{C(Nu)}{=}\text{CR}_2\text{R}_3$$
$$\tag{5}$$

of 1-cyclohexenyl and $cis$-2-buten-2-yl tosylate and brosylate in HCOOH/HCOONa on the basis of the $k_{\text{OTs}}/k_{\text{OBs}}$ reactivity ratios of 3.3 to 3.4 [44].

Schubert and Barfknecht [45] studied in 1970 the behavior of a compound they believed to be the hydrobromide salt of $p$-amino-$\alpha$-bromostyrene (**15**), which was studied earlier by Grob and Cseh in 80% EtOH [2], in aqueous buffer, and in perchloric acid solutions. They reported a linear increase of

$$p\text{-H}_2\text{NC}_6\text{H}_4\text{C(Br)}{=}\text{CH}_2$$

**15**

$k_{obs}$ with $H_3O^+$ in media of low acidity, and pH independence at high acidity, and claimed that they observed the spectrum of a reaction intermediate at intermediate pH. They suggested a rate-determining protonation of the double bond at low acidity and protonation of the free amine at high acidity, and they concluded that the "vinyl carbonium ion mechanism for this compound is definitely incorrect" [45].

A reinvestigation by Grob and Pfaendler [46] showed a large rate discrepancy between their results and those of Schubert. A reexamination of the hydrolysis of **15** as a function of pH in phosphate, borate, and citrate buffers [Eq. (6)] revealed an opposite profile: the rate constant was practically unchanged at pH 3 to 13, whereas it decreased sharply at pH 2 to 0. These

$$p\text{-}H_2NC_6H_4C(Br){=}CH_2 \xrightarrow{\text{aq. buffers}} p\text{-}H_2NC_6H_4COCH_3 + p\text{-}H_2NC_6H_4C{\equiv}CH \quad (6)$$

$$\textbf{15} \qquad\qquad\qquad\qquad\qquad \textbf{16} \qquad\qquad\qquad \textbf{17}$$

are the expected results for an $S_N1$–E1 route which is deactivated at high acidities by protonation on the amino group.

It was suggested that Schubert and Barfknecht probably followed the protonation of *p*-aminophenylacetylene (**17**), a solvolysis product of **15**. However, Schubert [47a] reported that in acidic solutions the $k_{obs}$ values and uv spectrum of **17** did not correspond to those reported in Barfknecht's thesis [47b] for the hydrobromide of **15**. Moreover, Barfknecht's salt was shown not to be a salt of **15**, although it had been prepared by the method described by Grob and Cseh [2]. Difficulties with the preparation were later described [46]. Schubert and Green [47a,c] found the behavior of **15** and its *N,N*-dimethyl derivative to be extremely complex. For example, depending on the method of sample introduction into aqueous solutions for uv, dimerization occurred either locally (before the compound was dispersed) or within a few seconds at $10^{-5}$ *M* concentration. Nevertheless, it eventually was established that the first event was probably a very fast $S_N1$ ionization of the bromostyrene.

Insensitivity to pH in the pH range of 6 to 9 rules out the $Ad_E$–E route for the solvolytic rearrangement of 1-bromo-2-phenylcyclobutene (**14**) [43]. Nevertheless, the problem of the competing $Ad_E$–E route cannot be ignored, especially in acidic media. Since the two routes lead to an α-aryl-substituted cation, some mechanistic criteria, such as the effect of the α-aryl substituent, the kinetic order, the stereochemistry and cis–trans isomerization during the reaction, cannot be used to distinguish between them. Rappoport, Bässler, and Hanack summarized in 1970 the predictions for reactions via the $Ad_E$–E versus the $S_N1$ route for 20 different phenomena [48], and all the data (except for one system [44]) were compatible with only the $S_N1$ route. Of these, only the criteria for distinguishing between the routes for $\alpha$-arylvinyl systems will be discussed.

*A priori*, with a sterically unhindered system in the absence of stabilizing electronic effects, the $Ad_E$–E route will be more facile than the $S_N1$ route. Substitution by an $\alpha$-aryl group introduces the stabilizing effect of conjugation and modifies the relationship between the two routes. The steric requirements of the transition states of the two routes are opposite. In the planar ground state of the vinylic system, steric interaction between the ortho substituents on the aryl group and a cis $\beta$ substituent is at a maximum when the $\pi(Ar)$–$\pi(C{=}C)$ overlap is at a maximum. In the transition state, the $\pi(Ar)$ and $\pi(C{=}C)$ systems are perpendicular and the steric interaction is at a minimum. On the other hand, the ion formed by the $Ad_E$–E route gains maximal electronic stabilization when the $\pi(Ar)$ and $\pi(C{=}C)$ orbitals are in the same plane, i.e., when the steric interactions of the ortho and the $\alpha$ substituents are at a maximum. Since the ground states for the two reactions are identical, it is clear that increased bulk of the $\alpha$-aryl group will make the $S_N1$ more favored and the $Ad_E$–E less favored.

The strongest tool for distinguishing between the two routes [48] and the one most frequently applied is the solvent isotope effect (SIE). For an $S_N1$ route, only a small SIE is expected, and the $k_{AcOH}/k_{AcOD}$ values of 1.04 to 1.12 for the acetolyses of several alkyl tosylates, such as 1-adamantyl- or *p*-methoxyneophyl tosylates, are typical [49a]. On the other hand, appreciable isotope effects are expected for rate-determining proton transfer on the basis of the extensive data for such reactions in water [49b]. Calibration of the expected value for the $Ad_E$–E route for systems structurally related to those studied solvolytically is shown in Table 6.4 for $\alpha$-anisyl-$\beta,\beta$-unsubstituted and $\beta,\beta$-diaryl-substituted systems. The isomerization of the ethylene **18** to its trans isomer **19** [Eq. (7)] and the electrophilic addition

$$
\underset{\textbf{18}}{\overset{An}{\underset{H}{\Large\diagup}}C{=}C\overset{An}{\underset{Ph}{\Large\diagdown}}}
\quad\overset{H^+}{\rightleftharpoons}\quad
\underset{\textbf{19}}{\overset{An}{\underset{H}{\Large\diagup}}C{=}C\overset{Ph}{\underset{An}{\Large\diagdown}}}
\tag{7}
$$

of AcOH to the ethylene **20** [Eq. (8)] [22] proceed via an initial proton transfer from the acid to the double bond and show solvent isotope effects of 2.55 [36] and 3.40 [22], respectively.

$$
\underset{\textbf{20}}{AnCH{=}CH_2} + AcOH \longrightarrow AnCH(OAc)CH_3
\tag{8}
$$

Table 6.4 gives an extensive list of $k_{RCOOH}/k_{RCOOD}$ values for many systems solvolyzed in RCOOH media. The table is arranged according to the nature

## TABLE 6.4

### Solvent Isotope Effect for the Solvolysis of $\alpha$-Arylvinyl Systems $ArC(X)\!=\!CR_1R_2$ in Carboxylic Acids

| Ar | X | $R_1$ | $R_2$ | RCOOH | $k_{\text{RCOOH}}/k_{\text{RCOOD}}$[a] | Mechanism | Reference |
|---|---|---|---|---|---|---|---|
| Ph | OSO$_2$F | Ph | Ph | AcOH | 1.04 | $S_N1$ | [19b] |
| An | OBs | An | An | AcOH | 1.05 | $S_N1$ | [6] |
| An | OTs | An | An | AcOH | 0.85 | $S_N1$ | [6] |
| Ph | OTs | Ph | Ph | AcOH | 0.93 | $S_N1$ | [19b] |
| An | OTs | Me | Me | CF$_3$COOH | 1.07 | $S_N1$ | [34] |
| An | Br | Ph | An | AcOH | 1.11[b,c] | $S_N1$ | [36] |
| An | Br | Fluorenylidene[d] | | AcOH | 1.20 | $S_N1$ | [50] |
| An | Br | An | An | Me$_3$CCOOH | 1.05 | $S_N1$ | [51] |
| Tol[e] | Br | Tol[e] | Tol[e] | AcOH | 0.92 | $S_N1$ | [52] |
| Tol[e] | Br | Tol[e] | Tol[e] | CF$_3$COOH | 2.6 (1.6)[f] | $Ad_E$–E ($Ad_E$–E + $S_N1$)[f] | [52] |
| Tol[e] | Br | Ph | Tol[e] | AcOH | 0.96[b] | $S_N1$ | [35] |
| Tol[e] | Br | Ph | Tol[e] | CF$_3$COOH | 3.4[b], 3.9[g] | $Ad_E$–E | [35] |
| An | Br | H | H | AcOH | 1.45 | $S_N1$ (+ $Ad_E$–E?) | [22] |
| An | Br | Ph | An | AcOH | 1.10[b,c] | $S_N1$ | [36] |
| An | Cl | H | H | AcOH | 1.94 | $S_N1$ + $Ad_E$–E | [22] |
| 9-Anthryl | Cl | H | H | AcOH | 0.91 | $S_N1$ | [17] |
| An | OCOC$_6$H$_3$(NO$_2$)$_2$-3,5 | Me | Me | AcOH | 1.03 | $A_{Ac}2$ (+ $S_N1$?) | [32] |
| An | OCOCF$_3$ | Me | Me | AcOH | 5.9 | $Ad_E$–E | [32] |
| An | OAc | H | H | AcOH | 3.45 | $Ad_E$–E | [22] |
| An | H | An | Ph | AcOH | 2.55[b,h] | $Ad_E$[h] | [36] |
| An | H | H | H | AcOH | 3.40[i] | $Ad_E$[i] | [22] |

[a] Ratios of the titrimetric rate constants for reaction in the presence or absence of RCOONa.
[b] Data for the cis isomer.
[c] Ratio of the ionization rate constants.
[d] $R_1R_2C\!=\!$ fluorenylidene.
[e] Tol, p-tolyl.
[f] In the presence of AgOCOCF$_3$.
[g] Data for a 1:1 cis–trans mixture.
[h] Data for the cis–trans isomerization reaction.

of the leaving group from fluorosulfonate [19], which is one of the best leaving groups, to acetate [22], which is a poor leaving group. With good leaving groups, such as fluorosulfonate [19], brosylate [6], and tosylate [6, 19b], the SIE is close to unity. Apparently, the increased leaving ability, coupled with the reduced nucleophilicity of the double bond due to electron withdrawal by these substituents, results in an exclusive $S_N1$ reaction even in an acidic solvent such as trifluoroacetic acid. With poor leaving groups, such as acetate [22] and even trifluoroacetate [32], the slow $S_N1$ reaction, coupled with resonative electron donation by the leaving group, which increases the stability of the $sp^2$-hybridized ion, leads to an exclusive $Ad_E$–E route. With the moderate leaving groups, the halide ions, resonative electron donation is counterbalanced by inductive electron withdrawal, and a delicate balance exists between the two routes. For the sterically crowded triarylvinyl bromides and chlorides, the steric hindrance to electrophilic addition and to planarity in the $sp^2$-hybridized ion leads to an exclusive $S_N1$ reaction in AcOH [36]. However, in the more acidic trifluoroacetic acid, the reaction is predominantly (or exclusively) via $Ad_E$–E [35, 52]. It is noteworthy that the SIE indicates a predominant $S_N1$ reaction in $CF_3COOH$ in the presence of the $Ag^+$ ion, as expected [52].

For the $\beta,\beta$-unsubstituted system, the SIE indicates a possible small contribution of the $Ad_E$–E route to a predominant $S_N1$ reaction for the bromo compound **21** but a significant contribution of both routes for the

$$AnC(Br)\!=\!CH_2 \qquad AnC(Cl)\!=\!CH_2$$

**21**  **22**

chloro compound **22** [22]. This system was thoroughly investigated, and other mechanistic criteria lead to the same conclusion. That reduced steric interactions for the $Ad_E$–E route are responsible for this is shown by the SIE for the α-(9-anthryl)vinyl chloride **23**, which again reacts via $S_N1$ due to the larger steric effects of the α-aryl group [17].

C(Cl)=CH$_2$

**23**  **24**

Solvent isotope effects in noncarboxylic solvents were also used as a criterion for ruling out an $Ad_E$–E route. Examples are the $k_{SOH}/k_{SOD}$ value of 1.04 found by Stang and co-workers for the solvolysis of $\beta$-styryl triflates in 80% EtOH [53] and the $k_{SOH}/k_{SOD}$ value of 1.22 obtained by Rappoport and Kaspi for the solvolysis of trianisylvinyl tosylate in 70% acetone [6].

An SIE of 1.33 was found for $p$-amino-$\alpha$-bromostyrene (15) when the solvolysis rates in 50% dioxane–water and in 50% dioxane–$D_2O$ were compared [46]. A case in which an SIE of ca. 1 does not necessarily point to the $S_N1$ route is discussed in Section II,D.

Other mechanistic criteria were applied mainly to the solvolyses of compounds 21 to 24, and some of them are demonstrated in Table 6.5. For 21 and 24, the cis–trans isomerization that accompanies the solvolysis was used as a tool for investigating the nature of the solvolysis intermediate [36, 55, 56] (Section V,B), and a possible isomerization via an $Ad_E$ route had to be ruled out. These criteria are the leaving-group effect, the solvent effect, the nature of the products and the effect of a catalyst.

**TABLE 6.5**

**Criteria for Distinguishing between the $S_N1$ and $Ad_E$–E Routes in AcOH**

| Compound | $k_{RCOOH}/k_{RCOOD}$ | $k_{80\%EtOH}/k_{AcOH}$ | $(k_{aq.\ EtOH}/k_{RCOOH})_Y$ | $k_{Br}/k_{Cl}(AcOH)$ | Reference |
|---|---|---|---|---|---|
| 21 | 1.45 | 15 | — | — | [22] |
| 22 | 1.94 | 0.13 | — | 0.46 | [22] |
| 23 | 0.91 | 11 | 0.76 | 21.5 | [17] |
| 24, X = Cl | 1.10 | 1.3 | — | 24 | [36] |
| 24, X = Br | 1.11 | 4.5 | 0.46 | — | [36] |
| $(o\text{-An})_2C{=}C(Br)An^a$ | — | 4.4 | — | 27.8 | [54] |

$^a$ $o$-An, $ortho$-anisyl.

The leaving-group effect was the first that indicated the operation of the $Ad_E$–E route. The expected order of nucleofugacity of leaving groups in $S_N1$ reactions is $OBs > OSO_2Ph \geq OTs \geq OMs > Br > Cl > OCOCF_3 > OCOC_6H_3(NO_2)_2$-3,5 $> OAc$ [32, 36]. For electrophilic addition to vinylic systems substituted by these groups, the relative reactivity is predicted to be $OAc \geq OCOC_6H_3(NO_2)_2$-3,5 $> OCOCF_3 > Cl \sim Br > OTs > OBs$. For compounds 24 in AcOH, the solvolysis rate constants are in the order $OTs > OMs > Br > Cl \geq OAc$ [36], and for the accompanying isomerization the order is $OMs > Br > Cl > OAc$ [36], indicating that both the solvolysis and the isomerization proceed via an initial heterolysis of the C–X bond, in line with the SIE. Likewise, $k_{OBs}/k_{OTs}$ and $k_{Br}/k_{Cl}$ reactivity ratios, which are higher than unity and which were observed for several vinylic solvolyses (Section III,D), indicate an $S_N1$ route: $k_{OBs}/k_{OTs}$ of ca. 0.3 [44] and $k_{Br}/k_{Cl}$ of $<1$ [57] were demonstrated for electrophilic addition. They fit the greater electron-withdrawing ability of brosylate compared with that of tosylate and the similar $\sigma_I$ and $\sigma_R$ values of the two halogens [58].

The $k_{21}/k_{22}$ ratios of 0.46 to 0.56 (Table 6.5) [22] are reminiscent of the $k_{Br}/k_{Cl}$ ratio of 0.23 for the addition of trifluoroacetic acid to 2-X-propenes

[57]. They support an $Ad_E$ component for the solvolysis of **22**. However, this criterion is not unequivocal. For example, the $k_{Cl}/k_{OCOCF_3}$ value of 0.5 for the acetolysis of **25** (X = Cl, $OCOCF_3$) is consistent with both an $S_N1$ and an $Ad_E$–E route [32], but the SIE (Table 6.4) clearly indicates that the acetolysis occurs via the former mechanism. However, the $k_{Cl}/k_{OCOC_6H_3(NO_2)_2-3,5}$ value of **22** for compounds **25** (X = Cl, $OCOC_6H_3(NO_2)_2$–3,5) is two to three

$$\underset{X}{\overset{An}{>}}C=C\underset{Me}{\overset{Me}{<}}$$

**25**

orders of magnitude lower than similar ratios for the solvolyses of saturated compounds [32]. The SIE of 1.03 is consistent with an $S_N1$ mechanism, but a normal ester hydrolysis mechanism ($A_{Ac}2$), with $k_{AcOH}/k_{AcOD} < 1$ [59], together with an $Ad_E$–E component, cannot be ruled out (Section II, D).

A similar criterion is the $k_X/k_H$ ratios. Halogen is deactivating compared with hydrogen in $Ad_E$, and $k_{Br}/k_H = 0.08$ and $k_{Cl}/k_H = 0.35$ for electrophilic addition of $CF_3COOH$ to 2-X-propenes [57]. Comparison of the addition of AcOH to **20** with the solvolyses of **21** and **22** gives $k_{Br}/k_H = 0.09$ and $k_{Cl}/k_H = 0.19$, indicating the possibility of an appreciable $Ad_E$–E contribution for both **21** and **22** [22]. In contrast, $k_{Br}/k_H = 4$ and $k_{Cl}/k_H = 0.15$ for the isomerization of **18** indicating an $S_N1$ process for **24** (X = Br) [36].

Since the $Ad_E$–E route will be unimportant in the nucleophilic aqueous EtOH, the reactivity ratio $(k_{aq.\ EtOH}/k_{COOH})_Y$, where $Y$ is the ionizing power [38], should be $\sim 1$ for an $S_N1$ process but much lower than unity when the $Ad_E$–E route contributes. The $k_{80\%EtOH}/k_{AcOH}$ ratios are more readily determined, and since $Y(80\%\ EtOH)$ is greater than $Y(EtOH)$ [38], a ratio $\ll 1$ will clearly indicate a contributing $Ad_E$–E route. These predictions are borne out for compounds **21** through **24**, as shown in Table 6.5.

Silver salt catalysis is important for the $S_N1$ process but is relatively less important, if at all, for the $Ad_E$–E process. The much higher rate acceleration for the solvolysis of **24** (X = Br) than for the isomerization of **18** was used as evidence for the $S_N1$ route for **24** (X = Br) [36]. That a predominant $Ad_E$–E route can be converted to a mixture of both routes in the presence of a silver salt is shown by the SIE of Table 6.4 for tri-p-tolylvinyl bromide [52].

When both the $\alpha$ and the $\beta$ substituents can stabilize a positive charge, an electrophilic addition may take place on both carbons, leading to a saturated 1,2 diadduct (**26**), as demonstrated in Scheme 6.4 [48]. However, no adduct of type **26** has been observed so far in the solvolysis of $\alpha$-arylvinyl systems.

Electrophilic addition of the solvent can still be a part of the overall substitution process even if it is not rate determining. The intermediate vinyl cation formed from $\beta,\beta$-unsubstituted compounds partitions between

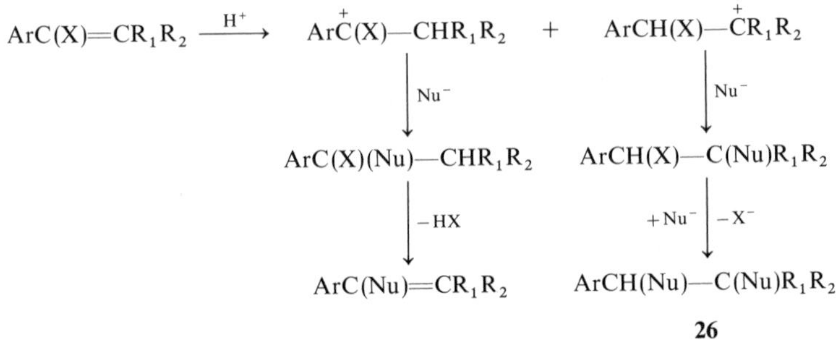
ArC(X)=CR₁R₂
H⁺
ArC⁺(X)—CHR₁R₂
+
ArCH(X)—C⁺R₁R₂
Nu⁻
Nu⁻
ArC(X)(Nu)—CHR₁R₂
ArCH(X)—C(Nu)R₁R₂
−HX
+Nu⁻
−X⁻
ArC(Nu)=CR₁R₂
ArCH(Nu)—C(Nu)R₁R₂
26

Scheme 6.4

substitution ($S_N1$) and elimination (El) routes and, in an acidic solvent such as AcOH, electrophilic addition to the acetylene will give the same substitution product as in the $S_N1$ route [22]. An example is the acetolysis of α-(9-anthryl)vinyl chloride (**23**) in AcOH/$Bu_4NOAc$, which gave initially a 1 : 1 distribution of the acetylene **27** and the vinyl acetate **28** but which slowly yielded more of **28** [eq. (9)] [17]. On the other hand, in the slower reaction

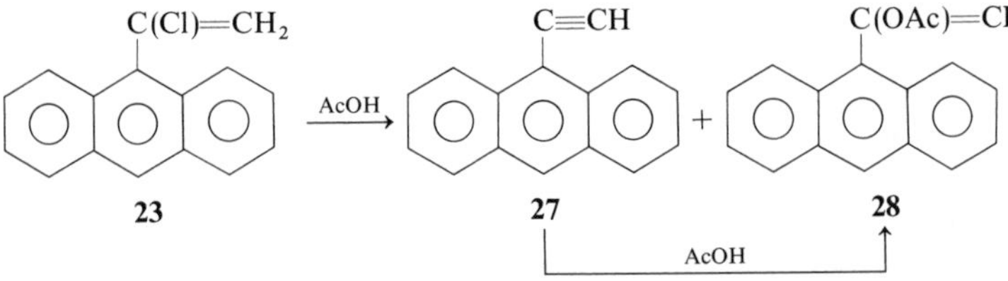
C(Cl)=CH₂
C≡CH
C(OAc)=CH₂
AcOH
+
23
27
28
AcOH

(9)

in the presence of NaOAc, the exclusive product was initially **28**, which was then converted slowly during the reaction to the ketone **29** [Eq. (10)]. These results are consistent with the rates of the various processes. The rate of addition of AcOH in the presence of $Bu_4NOAc$ to **27** and the rate of solvolysis of **23** in this medium are similar, allowing detection of the acetylene, but in AcOH/NaOAc the solvolysis of **23** is much slower, and **27** is therefore not detected [17].

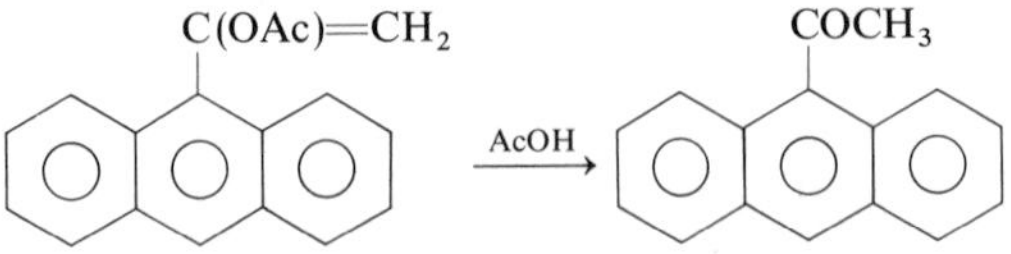
C(OAc)=CH₂
COCH₃
AcOH

(10)

**28** **29**

In contrast, no *p*-methoxyphenylacetylene (**31**) was detected in the acetolyses of **21** or **22**, which led exclusively to *p*-methoxyacetophenone (**30**) [Eq. (11)] [22]. However, the addition of AcOH to **31** [Eq. (12)] is 17 times faster than the solvolysis of **21**, and formation of **31** will therefore be hidden in this system. *A priori*, such an El–Ad$_E$ sequence could be distinguished

$$AnC(X) = CH_2 \xrightarrow{AcOH} AnCOCH_3 \tag{11}$$

$$\text{21 or 22} \qquad\qquad \text{30}$$

$$AnC \equiv CH \xrightarrow{AcOH} AnC(OAc) = CH_2 \tag{12}$$

$$\text{31} \qquad\qquad\qquad \text{32}$$

from the $S_N1$ route by the phenomenon of deuterium incorporation from a deuterated solvent during the addition step [48]. Incorporation is expected for neither the $S_N1$ route nor the $Ad_E$–E route, in which the proton addition to the haloalkene is irreversible. Unfortunately, this tool cannot be used for the reactions of **21** and **22**, since the rate of hydrolysis of the vinyl acetate **32** to the ketone **30** [Eq. (13)] is 314 times faster than the acetolysis rate of **21** and the ketone **30** exchanges its three hydrogens at a rate 51 times faster per

$$AnC(OAc) = CH_2 \xrightarrow{AcOH} AnCOCH_3$$

$$\text{32} \qquad\qquad\qquad \text{30}$$

hydrogen than the hydrolysis [22]. Consequently, the reaction product by either mechanism is $CD_3$-labeled **30**. Hydrolysis of vinyl acetates to the corresponding ketones in dry AcOH was observed for other α-arylvinyl acetates [see Eq. (10)], and a possible mechanism was suggested [60].

These aspects of the solvolysis of β,β-unsubstituted α-arylvinyl halides are summarized in Scheme 6.5. Routes (i) and (iii) were observed for **23** [17], routes (i) and (ii) were suggested for **22** [22], route (i) and possibly some route (ii) were suggested for **21**, whereas if route (iii) is operative it is hidden for both **21** and **22** [22]. The last step (iv) is common to compounds **21** to **23**. Ion pairing, which probably plays an essential role in these reactions, is not included in Scheme 6.5.

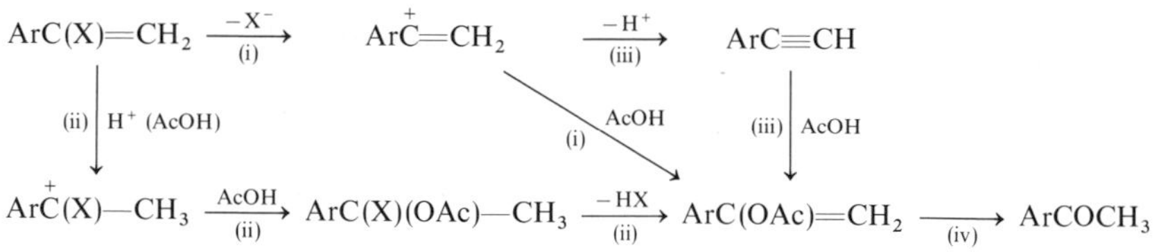

$$ArC(X) = CH_2 \xrightarrow[\text{(i)}]{-X^-} Ar\overset{+}{C} = CH_2 \xrightarrow[\text{(iii)}]{-H^+} ArC \equiv CH$$

(ii) ↓ H$^+$ (AcOH)          (i) ↘ AcOH          (iii) ↓ AcOH

$$Ar\overset{+}{C}(X) - CH_3 \xrightarrow[\text{(ii)}]{AcOH} ArC(X)(OAc) - CH_3 \xrightarrow[\text{(ii)}]{-HX} ArC(OAc) = CH_2 \xrightarrow[\text{(iv)}]{} ArCOCH_3$$

Scheme 6.5

The rate of the $Ad_E$–E route increases with increasing acidity of the solvent, but since increased acidity of a carboxylic acid is associated with increased ionizing power $Y$, the rate of the $S_N1$ process should increase as well. If the rate of the carbenium ion-forming $Ad_E$–E process follows the Grunwald–Winstein equation [38], the relative $\log k_0$ and $m$ values for both processes determine the mechanistic outcome. If both $\log k_0$ and $m$ are higher for one process than for the other, this process will predominate or will be exclusive in all the solvents for which $Y > 0$. However, if log $k_0$ is higher for the process with the lower $m$, an inversion of the relative importance of the two routes may occur at the intersection of the two log $k$ versus $Y$ plots. Examples of both types of behavior are known. The solvolysis of **25** (X = OTs, OBs) in AcOH [33] and in $CF_3COOH$ [34] is $S_N1$, as judged by the SIE values of Table 6.4 and the $k_{OBs}/k_{OTs}$ ratios of 3.42 (AcOH) [33], 3.56 (1:1 AcOH–HCOOH) [33, 34], and 2.27 ($CF_3COOH$) [34]. The lower $k_{OBs}/k_{OTs}$ ratio in the last solvent may indicate either some intervention of the $Ad_E$–E route or an increased electrophilic assistance in the expulsion of the slower leaving group in the more acidic solvent, in line with the Hammond postulate [61]. On the other hand, with 1,2-ditolyl-2-arylvinyl bromides **33**, Table 6.4 shows that the $S_N1$ route in AcOH is changed to an $Ad_E$–E route in $CF_3COOH$ [35, 52], and the color change accompanying the solvolysis of trianisylvinyl bromide in $CF_3COOH$ [62] also indicates the formation of the $sp^2$-hybridized ion. Even in 1:1 AcOH–HCOOH, the $k_{Br}/k_H$ ratio of 0.6 and the $k_{Cl}/k_H$ ratio of 0.13 for the isomerizations of **18** and **24** [36] may indicate a substantial contribution of the $Ad_E$–E route in the isomerization, although a differential solvation of the incipient chloride and bromide ions is an alternative explanation.

$$\underset{Br}{\overset{Tol}{\diagdown}}C{=}C\underset{Ar}{\overset{Tol}{\diagup}}$$

**33**   Ar = Tol, Ph

Not enough data exist to discuss the difference between the $\beta,\beta$-dimethyl system **25** and the $\beta,\beta$-diaryl systems **24** and **33** in acetic and trifluoroacetic acids. However, the regioselectivity of the protonation could differ in the two systems. The most stable ion is obtained by protonation at the $\beta$ position of **25**, but protonation of **33** at the $\alpha$ position may lead to a more stable ion (see Scheme 6.4), which on rearrangement [Eq. (14)] would lead to the substitution product.

$$Ar''Ar'C{=}C(X)Ar \xrightarrow{H^+} Ar''Ar'\overset{+}{C}{-}CH(X)Ar \xrightarrow{\sim H}$$
$$Ar''Ar'CH{-}\overset{+}{C}(X)Ar \xrightarrow[2.\ -HX]{1.\ SOH} Ar''Ar'C{=}C(OS)Ar$$

(14)

Support for protonation on the $\alpha$-carbon is provided by an interesting variant of the $Ad_E$–E route. In the reactions of triphenylvinylmercuric bromide and acetate **34** (X = Br, OAc) with $HClO_4$ in AcOH or MeOH, the exclusive product is the electrophilic substitution product, triphenylethylene (**36**) [Eq. (15)] [63]. Apparently, protonation at $C_\alpha$ is faster under

$$Ph_2C\!\!=\!\!C(Ph)HgX \xrightarrow{\text{HClO}_4} Ph_2\overset{+}{C}\!\!-\!\!CH(Ph)HgX \longrightarrow Ph_2C\!\!=\!\!CHPh \;(+\;HgX^+) \qquad (15)$$

$$\qquad\quad 34 \qquad\qquad\qquad\quad 35 \qquad\qquad\qquad\qquad 36$$

the reaction conditions than protonation at $C_\beta$ or cleavage of the C—Hg bond, and $HgX^+$ is expelled as an electrofugal rather than as a nucleofugal leaving group. Although the intermediacy of the ion **35** seems highly likely, the reaction is more complicated, since a second-order dependency on the acid was found. The absence of an $S_N1$ component can be also due to the very high $HClO_4$ concentrations ($\geq 1\ M$) used, since the $Ad_E$–E route is acid dependent, whereas the expected acid dependency of the $S_N1$ route is lower, if it exists at all.

The above examples require that the $Ad_E$–E route always be considered for solvolyses in RCOOH media. Moreover, buffering of any solvent with a base is necessary in order to avoid the $Ad_E$–E route with the strong acid (HX) formed during the solvolysis. Indeed, solvolysis of **21** in unbuffered AcOH showed an autocatalysis with a 28-fold increase in the rate constant during the reaction, and this was interpreted as being due to an $Ad_E$–E route [22]. The irregular kinetics of the solvolyses of several triarylvinyl halides in unbuffered AcOH [36] may be due to a similar reason.

## C. Radical Addition–Elimination Route

Silver salts are frequently used as catalysts for the solvolysis of vinyl halides [36, 62, 64, 65], but the rapid precipitation of silver halide can sometimes arise from a non-$S_N1$ reaction. Kaufman and Miller found that, in 80% aqueous acetonitrile, $\beta$-bromostyrene (**37**) and silver nitrate gave quantitatively AgBr within 30 min at 100° and that the analogous reactions of 2-bromo-1,1-diphenylethylene and 1-bromo-2-phenylpropene were complete after 2 hr at 130° [66]. Formal substitution products, i.e., the corresponding nitrovinyl compounds, were formed [Eq. (16)].

$$PhCH\!\!=\!\!CHBr + AgNO_3 \xrightarrow{\text{MeCN–H}_2\text{O}} \left\{ \begin{array}{l} \longrightarrow PhCH\!\!=\!\!CHNO_2 \\[2ex] \xrightarrow{\;\;/\!/\;\;} PhCH_2CH\!\!=\!\!O \end{array} \right. \qquad (16)$$

$$\qquad 37$$

That a silver-assisted heterolysis of the C–Br bond is not involved in this reaction is indicated by several facts. (1) The reaction rates are too fast for the generation of the highly unstable $\alpha$-unsubstituted vinyl cations. (2) The reaction with triphenylvinyl bromide is much slower than those of the three $\beta$-phenylvinyl bromides mentioned above, although an $\alpha$-phenyl group should stabilize the vinyl cation by many orders of magnitude compared with a $\beta$-phenyl group. (3) 2-Bromo-1,1-diphenylethylene gave only the nitro derivative **38** with the unrearranged skeleton, and the expected rearranged products with the 1,2-diphenyl skeleton (e.g., **39**) were not formed [Eq. (17)]. (4) Aldehydes, which are the expected capture products of the primary vinyl cations, were not found among the products.

$$Ph_2C{=}CHBr \xrightarrow[MeCN-H_2O]{AgNO_3} \begin{cases} \longrightarrow Ph_2C{=}CHNO_2 \\ \qquad\qquad \textbf{38} \\ \\ \longrightarrow Ph_2C{=}\overset{+}{C}H \longrightarrow Ph\overset{+}{C}{=}CHPh \longrightarrow \end{cases} \tag{17}$$

$$PhC(NO_2){=}CHPh + PhC{\equiv}CPh + PhCOCH_2Ph$$

**39**

A plausible reaction mechanism involves a thermal decomposition of the silver nitrate to form $NO_2$ and substitution by a radical addition–elimination mechanism, as demonstrated in Scheme 6.6 for the reaction of $\beta$-

$$2AgNO_3 \xrightarrow{\Delta} 2Ag + 2NO_2 + O_2 \tag{a}$$

$$NO_2 + PhCH{=}CHBr \longrightarrow Ph\overset{\cdot}{C}H{-}CH(Br)NO_2 \xrightarrow{-Br\cdot} PhCH{=}CHNO_2 \tag{b}$$

$$2Br\cdot + 2NO_3^- \longrightarrow 2Br^- + 2NO_2 + O_2 \tag{c}$$

Scheme 6.6

bromostyrene. There are precedents for both steps (a) and (b) of the reaction [67]. Since silver nitrate was found to react catalytically in the presence of sodium nitrate, the reaction may have a radical chain mechanism [steps (b) and (c)] initiated by step (a). The decisive role of the nitrate ion is shown by the inertness of **37** to silver tetrafluoroborate for 6 hr at 130°.

## D. Attack on the Leaving Group

Attack on the leaving group should be considered for multiatom leaving groups, such as carboxylate or sulfonate esters, which have an electrophilic center. Such a route may gain importance with a poor leaving group for an $S_N1$ reaction such as a carboxylate ester and, indeed, no clear-cut case of an $S_N1$ solvolysis of a substituted vinyl carboxylate has been recorded. Even the reaction of **25** [X = OCOCF$_3$, 3,5-dinitrobenzoate (ODNB)], in which the ester group is substituted by electron-withdrawing substituents, still proceeds in basic solution in a good ionizing solvent by the $B_{Ac}2$ mechanism [32]. In the presence of strongly basic amines, such as $Et_3N$, $Et_2NH$, or 1,8-bis(dimethylamino)naphthalene, or with $CF_3CH_2ONa$ in trifluoroethanol, **25** (X = ODNB) gives both trifluoroethyl 3,5-dinitrobenzoate and the ketone **40** in a second-order reaction. The rates are rather insensitive to the nature of the amine, and the active species is probably the trifluoroethoxide ion formed in an acid–base equilibrium with the solvent [Eq. (18)].

$$R_3N + CF_3CH_2OH \rightleftharpoons R_3NH^+ + CF_3CH_2O^- \quad (18)$$

Attack on the carbonyl group is a slow step [Eq. (19)], and decomposition of the tetrahedral intermediate gives the observed products. A direct attack

$$
\begin{array}{c}
\underset{R'OCO}{\overset{An}{\diagdown}}C=C\underset{Me}{\overset{Me}{\diagup}} + CF_3CH_2O^- \longrightarrow An—C=C\underset{Me}{\overset{Me}{\diagup}} \longrightarrow \\
\end{array}
$$

$$R' = CF_3, C_6H_3(NO_2)_2\text{-}3,5$$

$$
\begin{array}{c}
O \\
| \\
C—OCH_2CF_3 \\
R' \overset{|}{\underset{O}{\phantom{.}}}
\end{array}
\quad (19)
$$

$$R'COOCH_2CF_3 + AnCOCHMe_2$$
$$\mathbf{40}$$

of the amine on the carbonyl group is ruled out by the high reactivity of 1,8-bis(dimethylamino)naphthalene, which is a good proton base but a very reluctant carbon nucleophile [32].

The overall reaction order for **25** (X = ODNB) with much weaker amines, such as pyridine and 2,6-lutidine, is 2 and attack by the amine on the carbonyl group is plausible. The same is probably true for the reactions of **25**, when X = $OCOCF_3$. The absence of an $S_N1$ contribution to this reaction is shown by the zero intercept of the $k_2$ versus [amine] plot for all these amines [32].

An example of attack on the sulfur atom of a vinyl sulfonate is the reaction of the rather inert α-alkyl-activated 17β-acetoxy-3-tosyloxyandrosten-2-ene in the presence of labeled NaOMe in MeOH [68]. Modena and co-workers raised the possibility of the intervention of this route in the reaction of strong nucleophiles with 2,4,6-trinitrobenzenesulfonate esters, but evidence for this was not presented [69]. This route was ruled out for triarylvinyl arenesulfonates in basic aqueous acetone since it requires a first-order rate in the base and a slower reaction in the more polar media. Water was ruled out as the nucleophile since the reaction calls for α-An/α-Ph and β-An/β-Ph ratios of lower than unity, contrary to what was observed [6]. This route was also ruled out for the acetolysis of triphenylvinyl fluorosulfonate, since only triphenylvinyl acetate was formed and no crossover product, resulting from acylation of a ketone or an enolate intermediate, was observed when the acetolysis was conducted in the presence of α,α-ditolylacetophenone [19b].

As mentioned earlier, the SIE is clear evidence for the $Ad_E$–E mechanism in the hydrolysis of α-acetoxy-p-methoxystyrene [22] and of **25** (X = $OCOCF_3$) [32]. However, the $k_{AcOH}/k_{AcOD}$ ratio of 1.03 ± 0.05 for the first-order hydrolysis of **25** (X = ODNB) may indicate an $S_N1$ route, although 3,5-dinitrobenzoate is a poorer leaving group than trifluoroacetate [32]. The products of this reaction are 2:1 **25** (X = OAc) and **40**. The formation of **40** and the isotope effect can be accommodated by the usual acid-catalyzed ester hydrolysis $A_{Ac}2$, in which a preequilibrium protonation of the oxygen of the carbonyl group is followed by a rate-determining attack of the nucleophile on the positive carbon atom. However, the vinylic acetate **25** (X = OAc), cannot be formed via this route, and an $S_N1$ route is inconsistent with the $k_{Cl}/k_{ODNB}$ ratio of 22 for system **25**. The hydrolysis mechanism may therefore be a combination of the $Ad_E$–E and the $A_{Ac}2$ routes, and we note that, whereas α-acetoxy-*p*-methoxystyrene hydrolyzes in aqueous acid via an $Ad_E$–E route, other α-acetoxystyrenes hydrolyze by the "normal" $A_{Ac}2$ route for ester hydrolysis [70].

Even with a highly reactive leaving group, attack at a nonvinylic center may take place. Modena and co-workers [69, 71] found that the reaction of *E*-2-(*p*-chlorophenylthio)-1,2-diphenylvinyl 2,4,6-trinitrobenzenesulfonate (**41**) with *p*-chlorobenzenethiolate ion and with piperidine gave, respectively, *p*-chlorophenyl 2,4,6-trinitrophenyl sulfide (**42**) and *N*-(2,4,6-trinitrophenyl)-piperidine as the major products [Eq. (20)]. Nucleophilic aromatic substitution by these strong nucleophiles is apparently faster than the formation

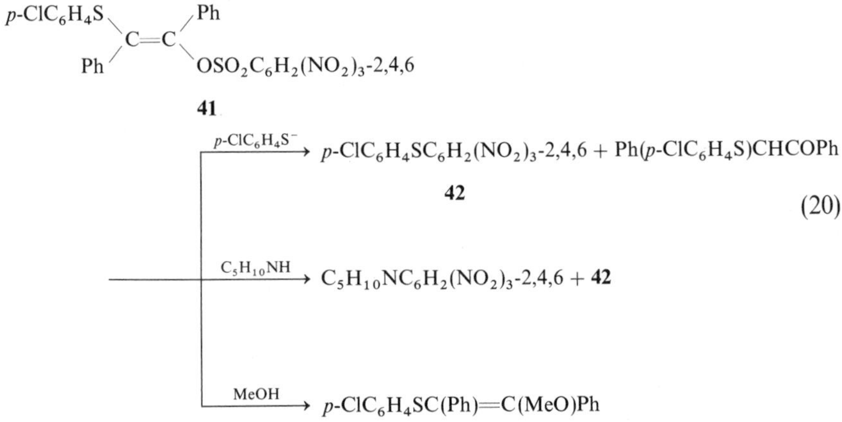

$$p\text{-}ClC_6H_4S \diagdown \quad \diagup Ph$$
$$C=C$$
$$Ph \diagup \quad \diagdown OSO_2C_6H_2(NO_2)_3\text{-}2,4,6$$

**41**

$$\xrightarrow{p\text{-}ClC_6H_4S^-} p\text{-}ClC_6H_4SC_6H_2(NO_2)_3\text{-}2,4,6 + Ph(p\text{-}ClC_6H_4S)CHCOPh$$

**42**

(20)

$$\xrightarrow{C_5H_{10}NH} C_5H_{10}NC_6H_2(NO_2)_3\text{-}2,4,6 + \textbf{42}$$

$$\xrightarrow{MeOH} p\text{-}ClC_6H_4SC(Ph)=C(MeO)Ph$$

of the vinyl cation by a slow $C$—$OSO_2Ar$ bond heterolysis. With a weaker nucleophile, such as methanol, the formation of the vinyl cation is faster than attack on the aromatic carbon, and the $S_N1$ product is obtained [Eq. (20)].

## E. E2 Elimination

Systems carrying β-hydrogens frequently give acetylenes via the E1 route and, when the leaving group and the hydrogen are in a trans relationship, an E2 reaction may compete with the E1 route. β-Deuterium isotope effects indicate that the E2 route has some importance in reactions of alkylvinyl triflates [72], but it has little or no importance in reactions of $RCH=C(Br)An$ ($R = Me$, $t$-Bu) with NaOAc in 50% EtOH [73].

A favorable geometry and a strong base may direct the reaction to the E2 route, as shown by comparing the reactions of $E$- and $Z$-$\alpha$-bromo-4,4'-dimethoxystilbenes **43** and **44** in the presence of a strong (NaOH) and a moderate (NaOAc) base (Table 6.6) [31]. The elimination of the $E$ isomer is

**TABLE 6.6**

**Solvolysis of $\alpha$-Bromo-4,4'-dimethoxystilbenes in 80% EtOH at 120°** [a]

| Compound | Base | $10^5 k_1$ | $k_E/k_Z$ |
|---|---|---|---|
| $E$-AnCH=CAn(Br)[b] | 0.5 $M$ NaOH | 113 | $\leq 0.06$ |
| **43** | 0.5 $M$ NaOAc | 132 | 44 |
| $Z$-AnCH=CAn(Br) | 0.5 $M$ NaOH | 1880[c] | |
| **44** | 0.5 $M$ NaOAc | 3 | |

[a] From Rappoport and Atidia [31].
[b] An, $p$-$CH_3OC_6H_4$.
[c] Estimated value since the reaction is very fast.

accompanied by some solvolysis, and its rate is practically independent of the nature of the base. The $Z$ isomer, in which the hydrogen and the bromide are trans to one another, reacts 627 times faster in the presence of sodium hydroxide. The $k_E/k_Z$ ratio increases 730-fold from 0.06 in the presence of NaOH to 44 in the presence of NaOAc. Since the $E$ isomer is more reactive in the $S_N1$ reaction than the $Z$ isomer, an almost exclusive E2 mechanism is indicated for the reaction of **44** in the presence of NaOH (Scheme 6.7).

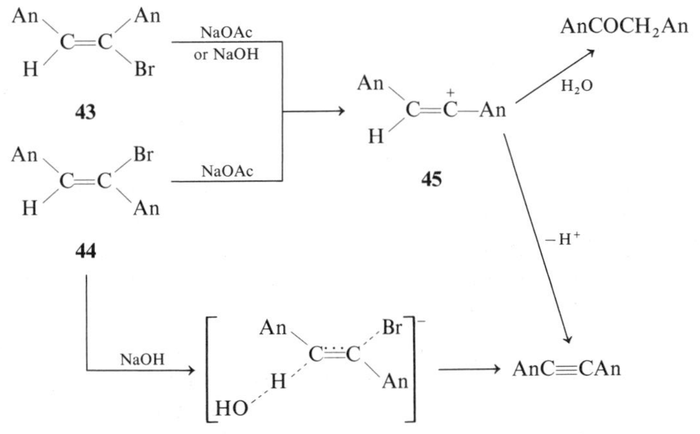

Scheme 6.7

This is supported by the high $k_Z/k_E$ ratio for the NaOH-promoted elimination of α-chlorostilbenes in EtOH [74]. The other three reactions proceed via the carbenium ion 45.

An increase in the electron-withdrawing ability of the α-aryl substituent reduces the $S_N1$ reactivity and simultaneously increases the E2 reactivity by increasing the acidity of the β-hydrogen. This was observed with the first system studied solvolytically by Grob and Cseh [2]. α-Bromostyrenes substituted by electron-donating para substituents solvolyze in 80% EtOH via the $S_N1$ route with first-order kinetics and form mainly substituted acetophenones. However, the p-nitro derivative 46 reacted immeasurably slowly at 190° in the absence of base, although it gave p-nitrophenylacetylene (47) in a second-order reaction in the presence of triethylamine at 170°

$$p\text{-}O_2NC_6H_4C(Br)\text{=}CH_2 \xrightarrow{\text{Et}_3N} p\text{-}O_2NC_6H_4C\text{≡}CH \qquad (21)$$

        46                                    47

[Eq. (21)]. A change from an $S_N1$ reaction for the electron-donating sub-substituents to an E2 reaction for the p-nitro compound is therefore apparent [2].


## III. STRUCTURAL AND MEDIUM EFFECTS ON THE SOLVOLYSIS

### A. Effect of α-Aryl Substituents

As discussed in Section I, the effect of an α-aryl substituent on the rate of solvolysis is strongly dependent on the extent of the π(Ar)–π(C=C) conjugation in the ground state and the π(Ar)–p(C$^+$) conjugation in the transition state. Consequently, the electronic effect of the α-aryl substituent can be isolated only when the substituent is at the meta or the para position.

The number of substituents studied is usually small due to the low reactivity of the vinylic systems; a large number of substituents is studied only in cases in which the leaving group is highly reactive. Nevertheless, in all the systems for which more than two substituents have been investigated, a better correlation with Brown's $\sigma^+$ values [75] than with Hammett's $\sigma$ values [39] has been obtained ($\rho^+$ values are presented in Table 6.7), even when based on only two substituents. The $\rho^+$ values are highly negative, being around $-4$ in the absence of anchimeric assistance except for the original system of Grob [2], in which the more negative $\rho^+$ value may reflect a large error in $\sigma^+$ for the p-amino substituent. The sign and the magnitude of $\rho^+$ indicate a highly polar transition state with high cationic character at $C_\alpha$. Lower $\rho^+$ values were obtained when charge dispersal in

TABLE 6.7

Values of $\rho^+$ for the Solvolysis of $\alpha$-Arylvinyl Derivatives

| System[a] | Solvent | Temp (°C) | $n$[b] | $\rho^+$ | Reference |
|---|---|---|---|---|---|
| $H_2C$=C(Br)Ar | 80% EtOH | 100 | 4 | $-6.6$[c] | [2, 46] |
| $H_2C$=C($OSO_2CF_3$)Ar | 80% EtOH | 75 | 5 | $-4.1$ | [76] |
| $Ph_2C$=C(OTs)Ar | 70% $Me_2CO$ | 120 | 2 | $-3.6$ | [6] |
| $Ph_2C$=C(Br)Ar | 80% EtOH | 120 | 2 | $-4.6$ | [30] |
| $Ph_2C$=C(Br)Ar | AcOH | 160 | 2 | $-3.4$ | [50] |
| $Ph_2C$=C(I)Ar | 70% DMF | 189.5 | 3 | $-3.6$ | [77] |
| Fl=C(Br)Ar | 80% EtOH | 155.5 | 2 | $-4.5$ | [30] |
| $(o\text{-}MeOC_6H_4)_2C$=C(Br)Ar | 80% EtOH | 160 | 3 | $-4.1$ | [54] |
| An($o\text{-}MeSC_6H_4$)C=C(Br)Ar | 50% EtOH | 160 | 2 | $-3.9$ | [78] |
| $E$-MeC(Ar)=C($OSO_2CF_3$)Me | 97% TFE | 100 | 7 | $-3.76$ | [79] |
| $Z$-MeC(Ar)=C($OSO_2CF_3$)Me | 97% TFE | 100 | 7 | $-1.96$ | [79] |
| $E$-PhSCH=C(OTNB)Ar | $MeNO_2$ | 25 | 4 | $-4.8$ | [80] |
| $E$-$p$-ClC$_6$H$_4$SCH=C(OTNB)Ar | $MeNO_2$ | 25 | 4 | $-4.7$ | [80] |
| $E$-PhSC(Me)=C(OTNB)Ar | $MeNO_2$ | 25 | 2 | $\sim-3.0$ | [80] |
| $E$-PhC(PhS)=C(OTNB)Ar | AcOH | 110 | 3 | $-2.3$ | [81b] |
| $E$-Ar'C(SPh)=C(OTNB)Ar | $MeNO_2$ | 25 | 5 | $-2.85$[d] | [82] |
| ▷=C(Br)Ar | 80% EtOH | 100 | 3 | $-2.8$ | [83] |
| $E$-Ar'C(I)=C(OTNB)Ar | 19:1 $MeNO_2$–MeOH | 25 | 3[e] | $-4.1$[d] | [84] |
| $E$-Ar'C(I)=C(OTNB)Ar | AcOH | 25 | 3[e] | $-4.1$[d] | [84] |
| $E$-Ar'C(Br)=C(OTNB)Ar | 19:1 $MeNO_2$–MeOH | 25 | 3[e] | $-4.8$[d] | [84] |
| $E$-Ar'C(Br)=C(OTNB)Ar | AcOH | 25 | 2[e] | $-4.9$[d] | [84] |
| $Z$-Ar'C(Cl)=C(OTNB)Ar | 19:1 $MeNO_2$–MeOH | 25 | 2[e] | $-5.1$[d] | [84] |
| $Z$-Ar'C(Cl)=C(OTNB)Ar | AcOH | 25 | 2[e] | $-5.0$[d] | [84] |
| $Ph_2C$=C=C(Cl)Ar | 80% $Me_2CO$ | 25 | 4 | $-2.0$ | [85] |
| ArC≡CH/$H_2SO_4$ | $H_2SO_4$/$H_2O$[f] | 25 | 4 | $-3.8$ | [86] |

[a] Abbreviations: Fl, fluorenylidene; OTNB, 2,4,6-trinitrobenzenesulfonate.
[b] Number of $\alpha$-aryl substituents used for the correlation.
[c] From Miller and Kaufman [77]: $<-4.5$.
[d] Based on the Yukawa–Tsuno equation for substituents on both $C_\alpha$ and $C_\beta$.
[e] Number of different $\alpha$ substituents.
[f] Data for the addition of a proton to the arylacetylenes.

the transition state was extensive and occurred either directly to the $\beta$-carbon [83] or via the substituted sulfur for the systems that show $\beta$-sulfur participation ($\rho = -2.3$ to $-3.0$) [80–82] or to the $\gamma$-carbon in the solvolysis of the allenyl halides ($\rho = -2.0$) [85].

The increased reactivity associated with $\pi(Ar)$–$\pi(C$=C) deconjugation is discussed in Section I. The effect is important for $\beta$-substituted systems (Section III,B), but it was also observed for the $\beta$-unsubstituted compounds

**48a** through **48d**. The p-methyl substituent of **48b** has a rate enhancement effect on the solvolysis of **48a** which is similar to that of the 1-phenylethyl analogues [16], but substitution by o-methyl groups gives more than an additivity effect in **48c** and **48d** and less than a cumulative effect in the o-methyl-substituted saturated systems. The relative rates for **48a**, **48b**, **48c**, and **48d** when X = OTs are 1, 23, 7300, and 420,000, respectively, in 50%

$$C(X){=}CH_2$$

R$_2$ —⟨ ⟩— R$_1$

R$_3$

| | |
|---|---|
| **48a** | R$_1$ = R$_2$ = R$_3$ = H |
| **48b** | R$_1$ = R$_2$ = H; R$_3$ = Me |
| **48c** | R$_1$ = R$_3$ = Me; R$_2$ = H |
| **48d** | R$_1$ = R$_2$ = R$_3$ = Me |

MeOH and 1, 40, 3000, and 180,000, respectively, when X = Br in 80% EtOH at 88° [15]. These differences are mainly due to differences in the activation enthalpies. Models show that the presence of the two o-methyl groups results in an almost complete deconjugation of the $\pi(Ar)-\pi(C{=}C)$ systems, which become nearly orthogonal when X = OTs [15].

## B. Effect of β-Aryl and Other β Substituents

A change from an α-aryl β-unsubstituted compound to a β-mono- or a β,β-disubstituted compound, especially when the substituents are β-aryl groups, involves both steric and electronic effects of the β substituents. For example, in the ground state of a triarylvinyl halide, the cis-stilbene interaction at the rear and the interaction of the leaving group and the β-aryl group at the front distort the three aryl groups from planarity and raise the ground state energy over that of the hypothetical planar structure. In the transition state of the ionization, the α-aryl group approaches linearity with the $C_\alpha-C_\beta$ bond (Section IV). The interaction at the rear is then partially relieved, and an increased steric interaction at the front, due to the movement of X to a position perpendicular to the $C_\alpha-C_\beta$ bond, is counterbalanced by lengthening of the C—X bond. These changes also affect the conjugation between all the substituents and the double bond and the extent of solvation of the ground and the transition states.

Gas-phase calculations suggest that charge dispersal into the β-carbon is appreciable for several β-substituted α-unsubstituted vinyl cations, including the β-phenylvinyl cation [87]. A smaller but still appreciable charge dispersal is predicted for the β-unsubstituted α-arylvinyl cation [88]. The main effect is due not to the carbene–carbenium ion structure (**49**) [Eq. (22)], but to a polarization of the $\pi$ electrons, which introduces a charge density

$$R-\overset{+}{C}=CH_2 \longleftrightarrow R-\overset{\cdot\cdot}{C}-\overset{+}{C}H_2 \qquad (22)$$

**49**

into an orbital on the $\beta$-carbon that is perpendicular to the vacant $p$ orbital. The effect should be strongly attenuated by solvation.

Steric, resonance, inductive, and hyperconjugative effects of the $\beta$ substituents on the reaction rates, as well as their involvement in an anchimeric assistance during the ionization, can sometimes be separated and identified.

**TABLE 6.8**

Solvolysis[a] of

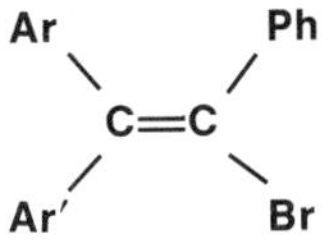

| | | Relative $k_1$ in TFE[b] | | Relative $k_1$ in 60% EtOH[b] |
|---|---|---|---|---|
| Ar | Ar' | At 120° | At 140° | |
| Ph | Ph | 1 | 1 | 1 |
| An | Ph | 3.4 | 2.9 | 2.14 |
| Ph | An | 3.7 | 3.4 | 2.44 |
| An | An | 9.5 | 12.5 | 5.45 |

[a] From Rappoport and Houminer [8].
[b] Buffered by 2,6-lutidine.

Table 6.8 gives the solvolysis data for a limited group of triarylvinyl bromides in which the geometry of the system remains constant. The data show that the effects of these groups are small and additive. Replacing a $\beta$-phenyl by a $\beta$-anisyl group increases the rate approximately threefold, regardless of the geometry, and a similar effect is brought about by a change of the second $\beta$-phenyl group [8]. A small effect of $k(\mathrm{Ar} = \mathrm{An})/k(\mathrm{Ar} = \mathrm{Ph}) = 2.95$ in 70% acetone was also found for the tosylates $\mathrm{AnC(OTs)}{=}\mathrm{CAr}_2$ [6], and similar data were obtained for analogous $\alpha$-anisyl-substituted systems (see Table 6.11). This observation is important for two reasons. First, since the steric effects are identical within the series, it is clear that the inductively electron-withdrawing aryl groups increase the solvolysis rate. Second, isomeric pairs in which an anisyl group is trans to the leaving group or a phenyl group occupies the same position give almost the same rate. Since a neighboring $\beta$-aryl participation involves the $\beta$-aryl group trans to bromine, reaction via transition state **50** should be faster than that via transition state **51**. The similar rates, therefore, rule out the $k_\Delta$ route for the α-phenyl-substituted triarylvinyl bromides of Table 6.8.

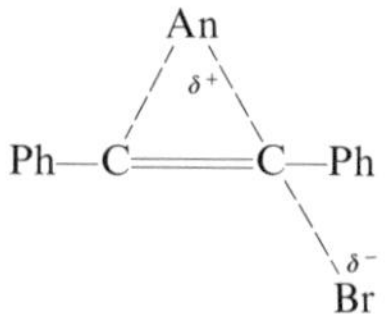

|  50 | 51 |

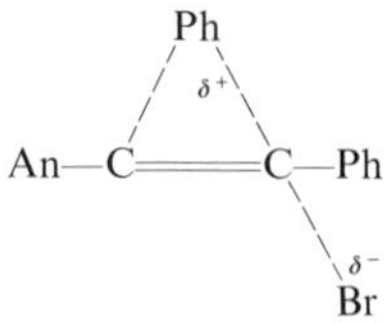

The $\rho^+$ values calculated from the data of Table 6.8 are low, $-0.6$ to $-0.7$. Table 6.9 gives data for other $\beta,\beta$-disubstituted systems in which anchimeric assistance by $\beta$-sulfur or $\beta$-halogen is possible. The $\rho$ values are

**TABLE 6.9**

Values of $\rho$ for the Solvolysis of

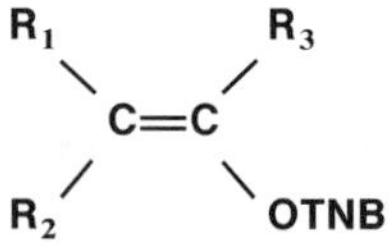

| $R_1$ | $R_2$ | $R_3$ | Solvent | Temp (°C) | $\rho(R_1)$ | $\rho(R_2)^a$ | Reference |
|-------|-------|-------|---------|-----------|-------------|---------------|-----------|
| ArS | Ar | Ar | MeNO$_2$ | 25 | $-1.45^a$ | $-1.25$ | [82] |
| ArS | H | Ph | MeNO$_2$ | 25 | $-0.88$ | | [80] |
| ArS | Me | Ph | MeNO$_2$ | 25 | $-1.55$ | | [80] |
| ArS | Pr | Pr | MeNO$_2$ | 25 | $-1.7$ | | [89] |
| I | Ar | Ar | 19:1 MeNO$_2$–MeOH | 25 | | $-0.8^b$ | [84] |
| | | | AcOH | 25 | | $-0.8^b$ | [84] |
| Br | Ar | Ar | 19:1 MeNO$_2$–MeOH | 25 | | $-0.5^b$ | [84] |
| | | | AcOH | 25 | | $-1.0^b$ | [84] |
| Cl | Ar | Ar | 19:1 MeNO$_2$–MeOH | 80 | | $-0.5^b$ | [84] |
| | | | AcOH | 80 | | $-0.5^b$ | [84] |

$^a$ Calculated from a Yukawa–Tsuno type of equation for the $R_1$, $R_2$, and $R_3$ substituents.
$^b$ Approximate value.

similarly low for the nonparticipating $\beta$-aryl group in the absence of $\beta$-halogen participation when $R_1 = $ Cl or Br and even in the case of iodine participation [84]. On the other hand, for systems carrying arylthio groups at a position trans to the leaving group, the $\rho$ value for a change in the aryl group is more appreciable [80, 82], especially since the substituents are more remote from the $\beta$-carbon than are those of the compounds of Table 6.8. These values strongly support a neighboring-group participation by the sulfur via a thiirenium-like transition state (Chapter 7, Section III,D). In the single case in which an aryl group is present in a nonparticipating position, the $\rho$ value of $-1.25$ indicates that the charge at the $\beta$ position of the thiirenium-like transition state is higher than that in nonparticipating systems [82].

Comparison of the $\rho^+$ values for a change at the $\beta$ position with those of Table 6.7 for a change at the $\alpha$ position can give only a very rough estimate of the extent of charge transfer to the $\beta$-carbon. In the triarylvinyl systems the two $\beta$-aryl groups are severely distorted from the plane of the double bond, and maximal overlap with the polarized $\pi$ orbital is not achieved. This distortion may affect the $\rho^+$ values for participating systems due to a possible loss of $\pi(C{=}C)-\pi(\beta$-Ar) conjugation, as discussed in Section VI.

TABLE 6.10

Reactivity Difference in the $ArC(X){=}CR_2$ Series

| Ar | X | Solvent | Temp (°C) | $k_{\beta\text{-Ph}}/k_{\beta\text{-H}}$ | Reference |
|----|---|---------|-----------|------------------------------------------|-----------|
| Ph | I | 70.4% DMF | 189.5 | 0.28 | [77] |
| An | Br | AcOH/NaOAc | 120.3 | 1.11 | [11, 56] |
| | | 80% EtOH | 120 | 0.80 | [5] |
| Ph | $OSO_2F$ | AcOH | 25 | 4.0 | [19b] |

Comparison of the effects of two $\beta$-phenyl groups with those of two $\beta$-hydrogens is of some interest, since conjugation and steric effects are absent in the unsubstituted system. Table 6.10 shows that the reactivity of the two systems is rather similar, although the $k_{\beta\text{-Ph}}/k_{\beta\text{-H}}$ ratios decrease with increasing bulk of the leaving group. However, the conclusion that a single $\beta$ substitutent will be relatively uneffective or that the similar rates are evidence for the absence of $\beta$-aryl participation is unfounded, as shown in Table 6.11. The similar reactivities in Table 6.10 are therefore due to fortuitous cancellation of some of the effects discussed above and below.

A large body of data is available for the solvolysis of $\beta$-substituted $\alpha$-anisylvinyl bromides under identical conditions (Table 6.11). It is clear that the relative reactivities are highly dependent on the geometry of the systems, leading to a four order of magnitude difference in reactivity. A dramatic contrast exists between the $E$ and $Z$ families of the mono-$\beta$-substituted systems. For the $Z$ series **52**, the rates are almost the same when R = H, Me, or $t$-Bu [73, 90], and the rate decreases seven to eightfold when R = An [31]. On the other hand, there is a telescopic increase in the rate of the $E$ series **53** with increased bulk of R in the order H < Me < An $\ll$ $t$-Bu. The

**TABLE 6.11**

Relative Solvolysis Rates of

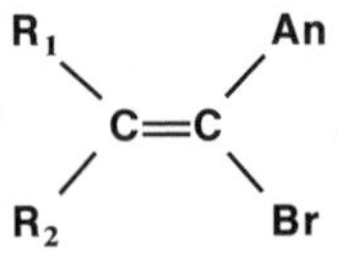

$$R_1\diagdown \diagup An$$
$$C{=}C \quad \text{at } 120°$$
$$R_2\diagup \diagdown Br$$

| $R_1$ | $R_2$ | Relative $k_1$ in AcOH[a] | Relative $k_1$ in 80% EtOH | $\Delta H^{\ddagger}$ (80% EtOH), kcal/mole | $\Delta S^{\ddagger}$ (80% EtOH), eu | Reference |
|---|---|---|---|---|---|---|
| H | H | 1.0 | 1.0[b] | 27.0 | −7 | [2] |
| Me | H | | 6.90[b,c] | 25.1 | −10.2 | [90] |
| H | Me | | 0.83[b,c] | 22.3 | −21.5 | [90] |
| Me | Me | 1.44 | 3.70[a] | 20.8 | −19 | [22, 50] |
| t-Bu | H | | 1362[d] | | | [73] |
| H | t-Bu | | 0.83[d] | | | [73] |
| An | H | 5.06 | 5.45[a] | | | [31] |
| H | An | 0.26 | 0.12[a] | | | [31] |
| An | Br | 0.013[e] | | | | [31] |
| Br | An | 0.036[e] | | | | [31] |
| OAc | An | 0.086[f] | | | | [31] |
| Ph | Ph | 1.23 | 0.80[g] (0.65[a]) | 22.3 | −19 | [11, 30] |
| An | Ph | 2.45 | 1.02[g] | 21.3 | −20 | [36, 55] |
| Ph | An | 2.50 | 1.18[g] | 20.0 | −20.3 | [36, 55] |
| An | An | 4.11 | 1.70[g] (1.48[a]) | 21.2 | −20 | [5, 11] |
| o-An | o-An | 3.74 | 1.00[g] | 26.8 | −7 | [54] |
| o-An | An | | 1.18[g] | 25.3 | −11 | [78] |
| An | o-An | | 1.59[g] | 25.2 | −11 | [78] |
| o-EtOPh | o-EtOPh | | 1.03[g] | | | [91] |
| An | o-MeSPh | | 0.25[g] | | | [91] |
| Fluorenylidene | | 0.15 | 0.26[a] | 24.0 | −17 | [30, 50] |
| Anthronylidene | | 7.36 | 5.75[h] | | | [92] |
| Xanthenylidene | | ca. 9 | 2.6[h] | | | [93] |

[a] Base: NaOAc.
[b] Base: Et$_3$N.
[c] Relative $k_1$ at 100°.
[d] Based on data in 50% EtOH at 80°.
[e] At 140°. Corrected for the presence of two bromines.
[f] At 141.2°.
[g] Base: NaOH.
[h] At 105°.

corresponding $k_E/k_Z$ ratios of 8.3 (Me), 49 (An), and 1640 (t-Bu) follow the same order [31, 73, 90].

**52**                    **53**

Analysis of these data requires breaking down the effect of the $\beta$ substituent into its inductive and steric components. From a study of the effect of methyl groups on the solvolysis of methyl-substituted 2-bromo-1,3-butadienes, Grob and Spaar concluded [27b] that the electron-releasing effect of a methyl group compared with that of hydrogen may be negligible or even reversed when the methyl group is located on the $\beta$-carbon atom of a vinyl cation. On the other hand, Schleyer and co-workers [94b] and Stang and Dueber [79] found very high $\rho^*$ values for the effect of $\beta$-alkyl substituents in the solvolysis of $\alpha,\beta$-dialkylvinyl and $\alpha,\beta,\beta$-trialkylvinyl triflates. These differences can be reconciled if the effect of the $\alpha$ activating group is considered. For the $\alpha$-alkylvinyl cations, charge dispersal to the $\beta$-carbon is higher than for the more stabilized $\alpha$-vinyl-substituted vinyl cations.

Further extrapolation to the more stable $\alpha$-anisylvinyl cations suggests that the inductive effect of a $\beta$-alkyl substituent is of little importance in determining the solvolysis rate of these compounds. The relative rates of the $Z$-alkyl isomers therefore reflect the insensitivity of the rates to the inductive effect.

On the other hand, the high sensitivity to R in the $E$ series, the order of which is based on the bulk of R, probably reflects a composite steric–conjugation effect. Models show that the steric interaction between R = H and the $\alpha$-anisyl group of **53** is negligible, but it increases rapidly with the bulk of R. Relief of this interaction is achieved by rotation around the $C_\alpha$—Ar bond and results in deconjugation of the $\pi(\alpha\text{-Ar})$–$\pi(C{=}C)$ system, which achieves a maximum when R = t-Bu. Consequently, the ground state energy increases accordingly, and the $\Delta\Delta F^{\ddagger}$ value of 5.2 kcal/mole for the difference between the $E$- and $Z$-$\beta$-tert-butyl systems [73] suggests a complete ground state deconjugation when R = t-Bu. This increase in the ground state energy is similar to that observed for a bulky $\alpha$-aryl substituent and similarly enhances the solvolysis rate.

The importance of the steric interaction between the cis R and $\alpha$-anisyl groups calls for a similar effect, although of a different magnitude, for the $E$ series, since the $\alpha$-bromine is also bulky. Here, relief of ground state strain cannot be achieved by rotation around the $C_\alpha$–leaving-group bond, and the ground state energy and the rate are expected to increase with increased bulk of R, although probably less than for the $E$ isomer. The almost similar rates for **52** when R = Me, $t$-Bu [73] indicate that the relative transition state energies are similar to the relative ground state energies. On approaching the transition state, the angle between R and the bromine, which leaves perpendicularly to the $\pi$ system, decreases, but the C—Br bond is simultaneously elongated. The two effects increase and decrease, respectively, the steric interactions between R and Br in the transition state. A delicate balance between them and with the ground state effects gives an apparent insensitivity to R in the $Z$ series. An alternative explanation is that the solvolytic transition state is bent, so that the steric interactions in the ground state and in the transition state are nearly identical.

An interesting case is that of compounds **52** and **53** when R = An. Superimposed on the effects described above and on the inductive electron withdrawal by the $\beta$-anisyl are conjugation effects of the $\beta$-anisyl with the double bond. In **53** the nonplanarity of its $cis$-stilbene moiety results in a ground state $\pi(\alpha$-An)$-\pi$(C=C) deconjugation. The linearization at the $\alpha$-carbon in the transition state is accompanied by an increased $\pi(\beta$-An)$-\pi$(C=C) conjugation which follows the reduced $\alpha$-An$-\beta$-An interaction. The effect is smaller than for $\beta$-methyl, reflecting again a mixture of an inductive and a conjugation effect. The $\pi(\beta$-An)$-\pi$(C=C) conjugation in the ground state of the Z-$\beta$-anisyl compound should reduce the solvolysis rate compared with $\beta$-H analogue, but both the electronic effect and the steric interaction with the bromine in the transition state should enhance the solvolysis rate. That the outcome is a rate reduction [31] indicates that the steric effect on the side of the leaving group in the transition state is either smaller than or similar to that in the ground state. A bent transition state or a highly elongated C—Br bond in the transition state accounts equally well for this result. The fact that the free energy of activation for **52** when R = An is 1.1 kcal/mole (in AcOH) to 1.6 kcal/mole (in 80% EtOH) higher than that for **52** when R = H [31] nicely fits an enhanced $\pi(\beta$-An)$-\pi$(C=C) conjugation in an unhindered transition state. The $k_E/k_Z$ reactivity ratio of 49 therefore reflects a complex combination of three $\pi$(An)$-\pi$(C=C) conjugation terms.

The effect of a cyclopropylidene group at the $\beta,\beta$ positions is of interest. The solvolysis of compound **7** is ca. $10^4$ times faster than that of the noncyclic analogue **54**, in spite of the extensive charge dispersal ability of the $\alpha$-phenyl group [43]. It is clear that a conjugation rather than inductive effect is involved, and this is discussed in more detail in Chapter 5, Section IV,C.

TABLE 6.12

Relative Solvolysis Rates of Substituted $\beta$-Halovinyl Sulfonates 55[a]

| | | | Relative $k_1$ | | |
|---|---|---|---|---|---|
| $R_1$ | $R_2$ | $R_3$ | In 9:1 $MeNO_2$–MeOH at 100° | In 19:1 $MeNO_2$–MeOH at 25° | In AcOH at 80° |
| Me | Me | Me | 1.0 | | |
| | | I | 93 | | |
| | | Br | 0.0027 | | |
| | | Cl | [b] | | |
| Ph | Ph | Ph | | 1.0 | 1.0 |
| | | I | | 9.6 | 6.0 |
| | | Br | | 0.0076 | 0.012 |
| | | Cl | | $0.00032^c$ | $0.00028^c$ |
| Ph | Br | Ph | | $0.00064^c$ | $0.001^c$ |
| | Cl | Ph | | $0.00028^c$ | $0.00024^c$ |

[a] From Bassi and Tonellato [84a,b].
[b] Too slow to measure.
[c] Estimated value.

With $\beta$-halogen substituents, both their strongly rate-reducing inductive effect and a possible anchimeric assistance have to be considered. Table 6.12 gives the available data for compounds 55 ($R_2$ or $R_3$ = halogen) together with data for the $\alpha$-methyl-substituted systems for comparison. The relative reactivity is always I ≫ Br > Cl. Since $\beta$-iodine is a rate-enhancing substituent compared with a $\beta$-methyl or a $\beta$-phenyl, reaction via an iodine-bridged species, e.g., 56, is suggested, and it is supported by the stereochemistry of the reaction [84] (Chapter 7).

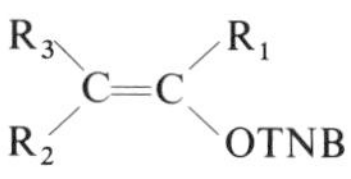

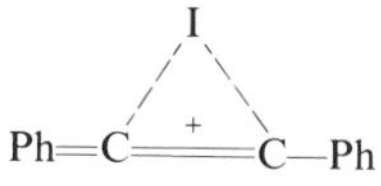

The interpretation with the other two halogens is not as clear-cut. Both are strongly rate retarding, but the inductive effect may mask some participation. Comparison of the $k_E/k_Z$ reactivity ratios for 55 ($R_1$ = Ph; $R_2$ or $R_3$ = Ph) gives values of 12 and 1.1 for $\beta$-Br and $\beta$-Cl, respectively. Notwithstanding the different $\pi(Ar)–\pi(C{=}C)$ interactions in the $E$ and $Z$ isomers, it was suggested that no participation occurs for $\beta$-chlorine and that the $\beta$-bromine participation via 57 occurs to a smaller extent than $\beta$-iodine participation via 56 [84]. However, since the steric effects in both isomers are unknown, it is difficult to assess the contribution of the inductive effect.

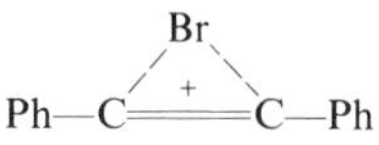

The situation is somewhat more complicated for compounds **58** and **59**, in which both bromines can solvolyze [31]. The $k_{58}/k_{59}$ ratio in AcOH/NaOAc is 0.4 under conditions in which $k_{43}/k_{44}$ is 19.7. In both cases, $\beta$-bromine reduces the rate, but to a different extent: $k_{43}/k_{58} = 375$, and $k_{44}/k_{59} = 6.6$ (Table 6.11). The former ratio for the cis isomers is reminiscent

$$
\begin{array}{cccc}
\underset{Br}{\overset{An}{>}}C{=}C\underset{Br}{\overset{An}{<}} & \underset{Br}{\overset{An}{>}}C{=}C\underset{An}{\overset{Br}{<}} & \underset{H}{\overset{An}{>}}C{=}C\underset{Br}{\overset{An}{<}} & \underset{H}{\overset{An}{>}}C{=}C\underset{An}{\overset{Br}{<}} \\
\mathbf{58} & \mathbf{59} & \mathbf{43} & \mathbf{44}
\end{array}
$$

of the values found for a nonparticipating $\beta$-bromine in saturated systems or for the $Z$ isomer of **55** ($R_1 = R_3 = Ph$; $R_2 = Br$) (Table 6.12). The low value for the trans isomers together with the favorable geometry is consistent with anchimeric assistance by the bromine. However, the higher rate reduction for the $E$ isomer of **55** ($R_1 = R_2 = Ph$; $R_3 = Br$) and the presence of the $\alpha$-anisyl group, which reduces participation, suggests a different origin for the $k_{44}/k_{59}$ ratio.

The departure from planarity of the substituents is higher for **59** than for **44**. If the consequent loss of delocalization energy is assumed to be between that for *cis*- and *trans*-stilbene, the ground state effect will account for the reactivity difference between the two systems. The similar reactivities of **58** and **59** suggest a similar ground state energy for both, in line with other evidence [31].

The rigid geometry suggests that the effect of two $\beta$-methyl groups will be approximately the product of the individual contributions of the *cis*- and the *trans*-methyl groups. The reactivity of 1-anisyl-2-methylpropen-1-yl bromide (**25**, X = Br) relative to that of $\alpha$-bromo-$p$-methoxystyrene (**21**) is 3.7 [22], whereas additivity predicts a higher ratio of 5.7. The situation is reversed with the $\beta$-anisyl groups. Additivity suggests a relative reactivity of 1.36 (AcOH) and 0.65 (80% EtOH) for trianisylvinyl bromide **60** versus $\alpha$-bromo-$p$-methoxystyrene, whereas the actual numbers are higher: 4.11 [11, 22] and 1.48 [2, 5], respectively.

All the effects discussed above are encountered with triarylvinyl halides. Steric interactions in the ground state cause distortion of all three aryl groups from planarity, and the ionization results in a relief of the back strain and probably in a minor increase in the front strain (see above). Steric hindrance to solvation should also contribute to the relative reactivity (Section III,E). Hence, only in the case of a constant geometry is normal behavior anticipated, as shown in Table 6.8. When geometric constraints are introduced, the range of reactivity increase as shown by the data of Table 6.11.

Although the presence of an $o$-methoxy substituent on the $\beta$-aryl group is manifested in steric effects in the reaction of the ions (Section V,A), the effect on the solvolysis rate is minor, and $k_{60}/k_{61}$ is 1.7 in 80% EtOH and

1.1 in AcOH [54]. However, when the two $\beta$-aryl groups are tied up in a ring, the reactivity difference between various systems can be appreciable. For example, $k_{62}/k_{(10,X=Br)} = 49$ in AcOH and 22 in 80% EtOH [30, 50, 92], whereas $k_{63}/k_{(10,X=Br)} = 60$ in AcOH [50, 93] (Table 6.11). Although the data in AcOH are less accurate since they are based on extrapolated $k$ values, it is clear that compound **10** is less reactive and compounds **62** and **63** are

| 60 | Ar = p-An | 62 | X = C=O | 10 | X = Br |
| 61 | Ar = o-An | 63 | X = O | | |

more reactive than the open-chain analogue. The common features of the ring systems are the inductive electron withdrawal by the ring and the rigidity of the systems. Although the inductive effect reduces the reactivity compared with two $\beta$-phenyl groups, the rigidity of the system, which enforces the two $\beta$-aryl groups in the plane of the double bond both in the ground state and in the transition state is the major factor. Because of the low flexibility of these groups, steric interaction between the $\alpha$ and the $\beta$ substituents is reduced mainly by rotation around the $C_\alpha$—$\alpha$-Ar bond. The five-membered ring of **10** forces the aryl groups of the fluorenylidene moiety away from the $\alpha$-aryl group, as compared with the free $\beta$-aryl groups in the open-chain system **9**. Consequently, the $\alpha$-aryl group is less buttressed out of the plane of the double bond in **9** than in **10**, and the corresponding stabilization of **10** is reflected in its reduced relative reactivity [30]. In contrast, space-filling models show that, in the six-membered anthronylidene and xanthenylidene rings of **62** and **63**, the $\beta$-aryl groups are closer to the $\alpha$-aryl group than in **9** [92]. The increased buttressing of the $\alpha$-aryl group out of the plane increases the ground state energy, and the increased buttressing of the leaving group at the front results in a steric acceleration (see Section III,D). Consequently, these combined effects cause the reactivity to increase, in spite of the electron withdrawal by the carbonyl group of **62** or the oxygen of **63**. The role of deuterium as a substituent is discussed in Chapter 5 (Section V,A,3) and in Chapter 7 (Section III,C).

## C. Effects of the Leaving Group

Interest in the effect of the leaving group is based on the information it reveals about the intervention of competing substitution routes, the position of the transition state, and special effects in the ground and transition states of the vinylic systems.

The first is discussed in Sections II,A and B. With slow leaving groups such as trifluoroacetate, the reaction takes a non-$S_N1$ course. With moderate or good leaving groups, $k_{Br}/k_{Cl}$ and $k_{OBs}/k_{OTs}$ ratios lower than unity [44] are among the best indicators of the $Ad_E$–E route, whereas $k_{Br}/k_{Cl}$ ratio of ca. 1 was used as an evidence for the $Ad_N$–E route for system 10 [30]. Three ratios $k_{Br}/k_{Cl}$, $k_{OBs}/k_{OTs}$, and $k_{OTs}/k_{Br}$, will be discussed in relation to the other two aspects.

**TABLE 6.13**

The $k_{Br}/k_{Cl}$ Ratios for α-Anisylvinyl Systems $AnC(X)=CR_1R_2$

| $R_1$ | $R_2$ | Solvent | Temp (°C) | $k_{Br}/k_1$ | Reference |
|---|---|---|---|---|---|
| An | An | 80% EtOH | 120 | 57.6 | [5] |
|  |  | AcOH | 120.3 | 19 | [50] |
|  |  |  | 141.5 | 11 | [11] |
| Ph | Ph | 80% EtOH | 140.2 | 53 | [30] |
|  |  | AcOH | 140 | 17 | [50] |
| Fluorenylidene |  | 80% EtOH | 140.2 | 75 | [30] |
|  |  | AcOH | 140 | 21 | [50] |
| H | H | 80% EtOH | 120 | 54 | [22] |
| Me | Me | 80% EtOH | 120 | 20 | [22] |
|  |  | TFE | 35.2 | 21 | [95b] |
|  |  | 60% TFE | 35.2 | 73 | [95b] |
|  |  | AcOH | 140 | 5 | [50] |
| Ph | An[a] | AcOH | 120.3 | 23.9 (24.6)[b] | [36] |
|  |  | 80% EtOH | 120 | 83 (93)[b] | [55] |
|  |  | 1:1 AcOH–HCOOH | 99.7 | 15.1 | [55] |
| o-An[c] | o-An[c] | 80% EtOH | 140 | 85 | [54] |
|  |  | AcOH | 140 | 27.8 | [54] |

[a] Cis isomer.
[b] Based on $k_{ion}$ (see Section V,B).
[c] o-An, ortho-anisyl.

Table 6.13 summarizes the available $k_{Br}/k_{Cl}$ ratios for α-arylvinyl systems. The ratios are similar to those found in saturated systems solvolyzing via the $k_c$ route [96, 97]. The ratios of 20 to 85 in 80% EtOH are higher than those in more acidic solvents, such as AcOH, 1:1 AcOH–HCOOH, and TFE. This is most easily explained in terms of solvent-assisted explusion of the leaving group, since hydrogen bonding would be stronger to the smaller incipient chloride ion. The absence of any special "element effect" with halide leaving groups is also shown by the "normal" $k_I/k_{Br}$ ratios of 3.6 and 2.5 for the 1-anisyl-2,2-diphenylvinyl and the triphenylvinyl systems, respectively, in 70.4% dimethylformanide [77]. The $k_{Br}/k_{Cl}$ ratios, which are based on the ionization rate constants (Section V,B), do not differ greatly from those based on the titrimetric constants [36, 55].

A $k_{OBs}/k_{OTs}$ ratio lower than unity demonstrates the operation of the $Ad_E$–E route in the *cis*-2-buten-2-yl system [44], but the ratios are usually higher than unity for α-arylvinyl systems. For 1-anisyl-2-methylpropen-l-yl sulfonates, the ratios are between 2.27 and 5.70 in various solvents (Table 6.14).

**TABLE 6.14**

The $k_{OBs}/k_{OTs}$ Ratios for the Solvolysis of An(X)=CMe$_2$

| Solvent | Temp (°C) | $k_{OBs}/k_{OTs}$ | Reference |
|---|---|---|---|
| 70% Me$_2$CO | 60 | 5.70 | [6, 33] |
| AcOH | 35 | 3.42 | [33] |
| 1:1-AcOH–HCOOH | 35 | 3.56 | [34] |
| 1:3 AcOH–HCOOH | 15 | 3.84 | [33] |
| CF$_3$COOH | 15 | 2.27 | [34] |
| CF$_3$CH$_2$OH | 15 | 2.82 | [95] |
| 60% CF$_3$CH$_2$OH | 15 | 3.14 | [95] |

A more complete analysis of five trianisylvinyl arenesulfonates in 70% acetone gave $\rho$ values of 1.91 and 1.67 at 50° and 75° ($k_{OBs}/k_{OTs}$ of 5.54 and 5.02), respectively [6]. The $\rho$ value at 75° is higher than those of 1.17 to 1.57 for the $k_s$ hydrolyses of ethyl, methyl, and isopropyl arenesulfonates in aqueous ethanol and methanol [97]. The values are similar to the $\rho$ values of 1.76 and 1.86 for 1-adamantyl and 2-adamantyl arenesulfonates, respectively, [98], in which solvent participation is not possible, indicating a highly polar transition state for the ionization. Data for more reactive sulfonate leaving groups comes from the work of Jones and Maness [19] on the acetolysis of triphenylvinyl triflate, fluorosulfonate, and tosylate. A larger range of values is obtained when data for the 2,4,6-trinitrobenzenesulfonate group from the work of Modena and Tonellato [81] and for the relative reactivities of trianisylvinyl sulfonates from the work of Rappoport and Kaspi [6] are included (Table 6.15).

The difference between the very reactive triflate and fluorosulfonate leaving groups and the much less reactive tosylate is due mainly to differences in the activation entropy. The large range of reactivity between the strong electron-withdrawing leaving groups and the more moderate ones is shown by the $\rho$ value of 11.0 for a plot of log $k$ versus $\sigma_m$ for the acetolysis of Ph$_2$C=C(Ph)OSO$_2$R (R = F, CF$_3$, Me, Tol) [6b]. Since a similar plot for saturated compounds reacting via the $k_c$ route yields a $\rho$ value of 10.3 [100],

**TABLE 6.15**

**Leaving-Group Ability of Sulfonates in the Solvolysis of Triarylvinyl Systems $Ar_2C=C(Ar)OSO_2R$ at 25°**

| Leaving group | Relative rate[a] | $\Delta H^{\ddagger}$ (kcal/mole) | $\Delta S^{\ddagger}$ eu | Reference |
|---|---|---|---|---|
| $n\text{-}C_4H_9SO_2O^b$ | 150,000 | | | [99] |
| $CF_3SO_2O^c$ | 81,000 | $23.6^d$ | $-2.0^d$ | [19b] |
| $FSO_2O$ | 42,000 | $22.6^d$ | $-6.5^d$ | [19b] |
| $2,4,6\text{-}(O_2N)_3C_6H_2SO_2O$ | 26,000 | | | [81b] |
| $p\text{-}BrC_6H_4SO_2O$ | 3.5 | | | [6] |
| $p\text{-}MeC_6H_4SO_2O$ | 1.0 | $23.8^d$ | $-20.5^d$ | [6, 19b] |
| $MeOSO_2$ | 0.7 | | | [6] |

[a] Relative to tosylate.
[b] Nonaflate. Data extrapolated from nonaflate/triflate ratios for aliphatic systems.
[c] Triflate.
[d] Values in AcOH for Ar = Ph.

a substantial negative charge is developed on the sulfonate leaving group in the transition state, the magnitude of which is similar to that for the solvolysis of saturated compounds.

The reactivity order of Table 6.15 changes only slightly in the case of neighboring-group participation. Extrapolated data for $E\text{-}Ph(PhS)C=C(Ph)OSO_2R$, in which $\beta$-sulfur participation is evident, gives the following order of reactivity in AcOH for R: $2,4,6\text{-}(O_2N)_3C_6H_2$ (55,000), $p\text{-}BrC_6H_4$ (3.4), $p\text{-}MeC_6H_4$ (1.0); $k_{OBs}/k_{OTs}$ is 4.5 in 80% EtOH [81].

Jones and Maness [19] argued that the similarity of the $k_{OTf}/k_{OTs}$ ratio of $10^{4.9}$ to those of $10^{4.3}$ to $10^{5.3}$ found for several saturated systems [101] points to the lack of importance of the resonance form **64** in the ground state.

$$\overset{\diagdown}{\underset{\diagup}{C}}-C\overset{\diagup}{\underset{\diagdown}{\underset{+}{OSO_2R}}}$$

**64**

A significant lone-pair–$\pi(C=C)$ interaction is more important for the vinyl tosylate, and consequently the $k_{OTf}/k_{OTs}$ ratio in the vinylic system should be higher than the ratios for the saturated systems.

The fact that the reactivity of trianisylvinyl *o*-nitrobenzenesulfonate is 2.04-fold higher than that of the *p*-nitrobenzenesulfonate [6] as well as the similar activation parameters suggest a relatively minor steric effect of the arenesulfonate group. The small reactivity difference may be due to some stabilization of the transition state for the *o*-nitro compound by sulfonate–*o*-nitro interactions. The insensitivity to steric effects in the sulfonate group is supported by the $k_{OMs}/k_{OTs}$ ratio of 0.58 for the trianisylvinyl system [6] and by the ratios of 0.73 to 0.90 for the 1,2-dianisyl-2-phenylvinyl system in 80% EtOH [55] and in AcOH [36], respectively, which are similar to those found for saturated systems.

The most interesting ratios are the $k_{OTs}/k_{Br}$ ratios. Hoffman suggested that the magnitude of these ratios in aliphatic series could be used as a probe for the extent of bond breaking in the solvolysis transition state [102]. According to this, large ratios ($> 10^3$) would indicate a very ionic transition state and vice versa, and it was generalized that the faster an $S_N1$ reaction, the more ionic would be its transition state [102]. On the other hand, nucleophilic participation [103] and relief of ground state strain [104] were suggested as factors contributing to the low ratios obtained for several aliphatic systems. The characteristic feature of these values in vinylic solvolysis is that, whereas some of the ratios (Table 6.16) approach and even exceed the value of 231 for the 2-adamantyl system, which was suggested as typical for secondary substrates reacting via the $k_c$ route [104], most of the ratios are lower than this value. Moreover, the ratios are solvent and structure dependent. An extreme case are the anthronylidene derivatives 62 and 65, the values for which in AcOH, TFE, and 90% TFE are $<1$ [24], and the value of 0.157 in TFE is the lowest known for solvolysis either via

62   X = Br
65   X = OTs

the $k_c$ or the $k_s$ route [24]. This is especially remarkable since temperature extrapolations were not required for obtaining the ratios for this system and since 62 shows the highest extent of common ion rate depression among the α-anisylvinyl systems, so that its solvolysis mechanism is unquestionably $k_c$ [92]. Comparison of the reactivities of aliphatic and vinylic compounds with their selectivities as measured by the $k_{OTs}/k_{Br}$ ratios shows that the reactivity–selectivity principle does not hold for the two families of compounds together: contrary to Hoffmann's generalization [102], the $k_{OTs}/k_{Br}$ ratio is not necessarily higher for the faster reaction. For example, $k$(2-adamantyl-Br)/$k$(62) = 0.0033 in AcOH, whereas $(k_{OTs}/k_{Br})$(2-adamantyl)/$(k_{OTs}/k_{Br})$(62 and 65) = 358. Moreover, the same is true even for the α-anisylvinyl systems themselves: $k$[AnC(Br)=CMe$_2$]/$k$(62) = 0.2, whereas $(k_{OTs}/k_{Br})$ [AnC(X)=CMe$_2$]/$(k_{OTs}/k_{Br})$(62 and 65) = 149 in AcOH [24]. Since the

**TABLE 6.16**

**The $k_{OTs}/k_{Br}$ Reactivity Ratios in the Solvolysis of $\alpha$-Arylvinyl Systems**

| System | Solvent | Temp (°C) | $10^4 k_{Br}$ (sec$^{-1}$) | $10^4 k_{OTs}$ (sec$^{-1}$) | $k_{OTs}/k_{Br}$ | Reference |
|---|---|---|---|---|---|---|
| PhC(X)=CH$_2$ | 40% Me$_2$CO | 140 | 0.0414 | 2.65 | 64 | [33] |
| 2,4-Me$_2$C$_6$H$_3$C(X)=CH$_2$ | 80% EtOH | 88 | 0.026 | 10.5 | 404 | [15] |
| 2,4,6-Me$_3$C$_6$H$_2$C(X)=CH$_2$ | 80% EtOH | 67.8 | 0.18 | 60.3 | 335 | [15] |
| AnC(X)=CMe$_2$ | 70% Me$_2$CO | 120 | 5.37 | 217 | 40.4 | [33] |
| | 80% EtOH | 120 | 8.78 | 590.4 | 67.2 | [33] |
| | AcOH | 120 | 0.235[a] | 180 | 765 | [24] |
| | TFE | 35 | 0.461[a] | 18.0 | 38.9 | [95b] |
| | TFE | 35 | 0.42[a,b] | 13.5 | 32.5 | [95b] |
| | 97% TFE | 35 | 0.37[a] | 9.33 | 25.2 | [95b] |
| | 90% TFE | 35 | 0.33[a] | 7.72 | 23.4 | [95b] |
| | 80% TFE | 35 | 0.34[a] | 7.97 | 23.4 | [95b] |
| | 70% TFE | 35 | 0.35[a] | 7.90 | 22.5 | [95b] |
| | 60% TFE | 35 | 0.36[a] | 7.16 | 19.7 | [95b] |
| | 50% TFE | 35 | 0.37[a] | 7.13 | 19.5 | [95b] |
| AnC(X)=CPh$_2$ | 70% Me$_2$CO | 120 | 0.75 | 18.8 | 25.1 | [24] |

| Compound | Solvent | Temp | | | | Ref. |
|---|---|---|---|---|---|---|
| AnC(X)=CAn$_2$ | 70% Me$_2$CO | 120 | 2.14 | 68.0 | 31.8 | [24] |
| | 90% EtOH | 75 | 0.043 | 1.18 | 27.4 | [6] |
| | AcOH | 120 | 0.67$^a$ | 58.2 | 86.8 | [24] |
| | TFE | 90 | 33$^a$ | 430 | 130.3 | [24] |
| cis-AnC(Ph)=C(X)An | 80% EtOH | 120 | 2.92 | 116 | 39.7 | [24, 55] |
| | AcOH | 120 | 0.4$^a$ | 113 | 282.5 | [24, 36] |
| Fl=C(X)An$^c$ | 70% Me$_2$CO | 120 | 0.322 | 6.15 | 19.1 | [6] |
| Anth=C(X)An$^d$ | 80% EtOH | 105 | 3.5 | 2.63 | 0.75 | [24] |
| | AcOH | 120 | 1.2$^a$ | 6.18 | 5.15 | [24] |
| | TFE | 35 | 0.108$^a$ | 0.017 | 0.157 | [24] |
| | 90% TFE | 35 | 0.068$^a$ | 0.0158 | 0.23 | [24] |
| Xant=C(X)An$^e$ | 80% EtOH | 105 | 1.68 | 6.4 | 3.8 | [93] |
| | AcOH | 120 | 1.98$^a$ | 4.27 | 2.16 | [93] |
| | TFE | 35 | 0.132$^a$ | 0.077 | 0.58 | [93] |

$^a$ Extrapolated value due to common ion rate depression.
$^b$ In the presence of Et$_3$N.
$^c$ Fl, fluorenylidene; compounds 10.
$^d$ Anth, 9-anthronylidene; compounds 62 and 65.
$^e$ Xant, xanthenylidene; compound 63 and the corresponding tosylate.

ion derived from **62** is more stable than the $AnC=CMe_2$ ion, it is clear (Section V,A) that, if the stability of the ions is related to the structure of the transition states leading to them, the $k_{OTs}/k_{Br}$ ratios are poor guides to the polarity of the latter [24].

Several explanations could be offered for the low $k_{OTs}/k_{Br}$ ratios. (1) the extrapolation of some of the $k_{Br}$ values due to common ion rate depression may distort the values. However, since the extrapolations give minimal $k_{Br}$ values, the $k_{OTs}/k_{Br}$ ratios are maximal values. (2) A higher extent of ion-pair return for a vinylic tosylate than for the corresponding bromide may make the titrimetric $k_{OTs}/k_{Br}$ ratios lower than the mechanistically more meaningful values, which are based on the ionization rate constants. This possibility can be ruled out since the extent of ion-pair return is lower for *cis*-1,2-dianisyl-2-phenylvinyl mesylate than for the bromide. The $k_{OMs}/k_{Br}$ ratio in AcOH is 492 when based on $k_{ion}$ but only 200 when based on $k_t$ [36], indicating that ion-pair return may further decrease, rather than increase, the $k_{OTs}/k_{Br}$ ratios. (3) The low ratios may be due to $k_s$ (or $k_\Delta$) solvolysis, but the strong arguments for the $k_c$ route, especially for the systems that show the lowest ratios, ruled out these routes. (4) A $k_c$ process with a transition state of low polarity is eliminated by the data of Tables 6.13 to 6.15, which indicate an extensive charge on the leaving group in the transition state. (5) An n–π conjugation, which was the first explanation for the inertness of vinylic systems [105], may be more important for the tosylate (**64, R =** Tol) than for the bromide (**66**), making the latter relatively more reactive. The evidence for and against this explanation is indirect. The $k_{OTf}/k_{OTs}$ values

$$\diagup\overset{-}{C}-C\diagup_{\diagdown Br^+}$$

**66**

discussed above [19] and the similarity of the $k_{OTs}/k_{Br}$ values for the fluorenylidene and for the $\beta,\beta$-diphenyl compounds [6] argue against it. On the other hand, Pople, Schleyer, and co-workers used thermodynamic data to show that attachment of ethoxy (a model for tosylate) to vinyl is 5.6 kcal/mole more favored than attachment of ethyl, whereas bromine attachment to vinyl is 0.8 kcal/mole less favored than attachment of ethyl [106]. However, it is not clear to what extent an ethoxy group with a high $\sigma_R^+$ value [107a] may serve as a model for the electron-withdrawing tosylate. (6) The solvolysis of the tosylates may be sterically retarded or the solvolysis of the bromides may be sterically accelerated, and space-filling models corroborate a steric explanation. The bulk of the tosylate at the reaction site is smaller than that of the bromine, since the covalent radii of oxygen and bromine are 0.66 and 1.14 Å, respectively. Moreover, if the transition state involves some steric compression due to interaction of the leaving group with the $\beta$ substituent cis to it, this will be larger for the bromide. Steric interaction of the tosylate oxygen with remote hydrogens, as suggested for the solvolysis of saturated bicyclic systems [104], is not revealed by the models of the vinylic systems. The models show that the steric interaction for the bromides increases markedly with increased bulk of the $\beta$ substituents, and a model of **62** falls apart; the corresponding models for the tosylates are less hindered, and it is possible to build them. The corresponding decrease in the $k_{OTs}/k_{Br}$ ratios points to an increased steric acceleration for the solvolysis of the bromides, with the increased bulk of the $\beta$ substituent as the main contributor to the low ratios [24]. The low ratios for the $\beta$-anthronylidene and $\beta$-xanthenylidene systems reflect the rigidity that forces planarity on the systems and brings the 1,8-hydrogens of the cyclic moiety into the plane of the leaving group, thus increasing the steric effects. Indeed, it was recently found that the C–Br bond in **62** is 0.065 Å longer than the C–Br bond in vinyl bromide [107b].

Inspection of Table 6.16 shows that this effect accounts for only part of the low ratios, since the $k_{OTs}$ values decrease with increased bulk of the $\beta$ substituents in all the solvents studied. The reason for the decrease in $k_{OTs}$ is not clear, although some n(Ö)–$\pi$(C=C) conjugation may be involved in the cyclic systems. Steric effects cannot be the sole effects involved, as shown by the high ratios of 404 and 335 found for the $\alpha$-2,4-dimethylphenyl and the $\alpha$-2,4,6-trimethylphenyl systems, respectively [15]. The ratios are also higher in AcOH than in more nucleophilic solvents, as found for sautrated systems [108]. Since the $k_s$ route is not involved, solvation by hydrogen bonding of the acidic proton with the incipient tosylate ion is greater than that of the incipient bromide ion.

## D. Solvent Effects

The nature of the solvent affects the solvolysis of $\alpha$-arylvinyl compounds in many ways. It can change the rate and products of the reaction, the extent of external ion return and ion-pair return (Section V), the extent of rearrangement across the double bond (Section VI), and sometimes the solvolysis mechanism. In this section we will discuss only the effect on the reaction rate.

In a $k_c$ solvolysis with a highly polar transition state, the sensitivity to changes in solvent is relatively high, since charge dispersal by the entering solvent molecule (as in the $k_s$ process) or by a neighboring group (as in the $k_\Delta$ process) is not possible. The quantitative measure of this solvent sensitivity is the slope $m$ of the Winstein–Grunwald equation [Eq. (23)] [38], where $Y$ is the ionizing power of the medium. High $m$ values, around unity,

$$\log(k/k_0) = mY \tag{23}$$

are typical for sp$^3$-hybridized substrates solvolyzing via $k_c$, whereas low values, 0.25 to 0.35, are typical for compounds solvolyzing via $k_s$ [38]. The extent of rearside nucleophilic assistance by the solvent in the transition state is estimated from the complete Winstein–Grunwald equation [Eq. (24)] [38b,c], where $N$ is the nucleophilicity of the solvent and $l$ is the

$$\log(k/k_0) = mY + lN \qquad (24)$$

response to this parameter [38, 109]. It was previously determined from the ratio of rates in a nucleophilic solvent (e.g., aqueous EtOH) to rates in a less nucleophilic one (e.g., AcOH) at the same $Y$ [38] or from Schleyer's $Q$ value, which is based on the comparison of the solvent effects on two extreme models: methyl tosylate, for which solvent assistance is extensive ($Q = 0$), and 2-adamantyl tosylate, for which there is no rearside solvent assistance ($Q = 1$) [110]. The extensive set of $N$ values determined recently [109] makes it possible to determine the $l$ values by application of Eq. (24).

All the $m$ values for the solvolysis of α-arylvinyl compounds in binary systems for which $Y$ values are available are listed in Table 6.17. In most of the cases, only two solvent compositions are available, and the $m$ values should be regarded as approximate. In several cases in which three solvent compositions are employed, the log $k$ versus $Y$ plots are nonlinear [5, 22]. Nevertheless, there are enough compounds of different structures in Table 6.17 for which sufficient data are available to discern structure–solvent sensitivity relationships. The most characteristic feature of Table 6.17 is that most of the $m$ values are lower than expected for the $k_c$ route [38], and the approximate trend is that the $m$ values decrease with increased bulk of the $\beta$ substituent. The comparison suffers from the fact that the $m$ values, which are temperature dependent, were determined at different temperatures. The values were therefore extrapolated to 25° by assuming that $m_1/m_2 = T_2/T_1$ as found for α-phenylethyl bromide [96], although it is not known whether this extrapolation is applicable for other substrates. The extrapolated $m$ values are in the $k_c$ region (0.8–1.0) [38] for $\beta,\beta$-unsubstituted systems in which the α-aryl group is not sterically hindered. With increased bulk of the α-aryl substituent by ortho substitution or of the $\beta$ substituent, the $m$ values become smaller. They are 0.6 to 0.5 for the α-mesityl-substituted systems [15] and for the triarylvinyl systems for which sufficient data are available, and the $m$ value for the strongly hindered **65** is 0.28 at 105° [3, 33], a value that should be regarded as low even for a compound that solvolyzes via the $k_s$ route. These trends are similar when the leaving group is either bromide or tosylate.

The few data of Table 6.17 suggest that $m$ values in nonnucleophilic solvents, such as AcOH–HCOOH mixtures, are higher than those in nucleophilic solvents. This trend was clearly verified in an extensive study of the

**TABLE 6.17**

**Values of $m$ for the Solvolysis of $\alpha$-Arylvinyl Compounds**

| Compound | Solvent[a] | Temp (°C) | $n$[b] | $m$ | $m$ (25°)[c] | Reference |
|---|---|---|---|---|---|---|
| $CH_2=C(Br)Ph$ | 80–50% E–W | 170 | 2 | 0.62 | 0.92 | [2] |
| $CD_2=C(OTf)Ph$ | 80–70% E–W | 50 | 2 | 0.57 | 0.62 | [53] |
| $CH_2=C(Br)An$[d] | 80–70% E–W | 120 | 2 | 0.77 | 1.01 | [22] |
|  | 90–80% E–W | 120 | 2 | 0.58 | 0.76 | [22] |
| $CH_2=C(Br)C_6H_4NH_2$-$p$ | 80–0% E–W | 0 | 2 | 0.58 | 0.53 | [46] |
| $CH_2=C(OTs)Ph$ | 80–20% A–W | 120 | 7 | $0.63 \pm 0.02$ | 0.83 | [33] |
| $CH_2=C(OTs)C_6H_3Me_2$-2,4 | 98–50% E–W | 67.6 | 6 | $0.5 \pm 0.02$ | 0.65 | [15] |
| $CH_2=C(OTs)C_6H_2Me_3$-2,4,6 | 98–50% E–W | 36.5 | 6 | $0.60 \pm 0.02$ | 0.62 | [15] |
| $Me_2C=C(Br)An$ | 80–70% E–W | 110 | 2 | 0.73 | 0.93 | [22] |
|  | 90–80% E–W | 110 | 2 | 0.70 | 0.90 | [22] |
| $CH_2=C(Cl)Anthryl$-9 | 90–70% E–W | 120 | 3 | 0.77 | 1.01 | [17] |
|  △—C(Br)Tol | 80–50% E–W | 74.5 | 2 | 0.85 | 0.99 | [83] |
| $E$-AnCH$=C(Br)An$ | 90–80% E–W | 120 | 2 | 0.64 | 0.84 | [31] |
| $Z$-AnCH$=C(Br)An$ | 90–80% E–W | 120 | 2 | 0.59 | 0.78 | [31] |
| $E$-AnC(Br)$=C(Br)An$ | 100–95% Ac–W | 140 | 2 | 0.61 | 0.85 | [31] |
| $Z$-AnC(Br)$=C(Br)An$ | 100–95% Ac–W | 140 | 2 | 0.64 | 0.89 | [31] |
| $Ph_2C=C(OTf)Ph$ | 100–90% E–W | 50 | 4 | $0.57 \pm 0.04$ | 0.62 | [19b] |
| $E$-AnC(Ph)$=C(Br)An$ | 100–50% Ac–F | 100 | 2 | 0.72 | 0.78 | [55] |

*(Continued)*

**TABLE 6.17** (*Continued*)

| Compound | Solvent[a] | Temp (°C) | n[b] | m | m (25°)[c] | Reference |
|---|---|---|---|---|---|---|
| Z-AnC(Ph)=C(Br)An | 100–50% Ac–F | 100 | 2 | 0.75 | 0.80 | [55] |
| Ph$_2$C=C(Br)An | 90–80% E–W | 140 | 2 | 0.49 | 0.68 | [30] |
| An$_2$C=C(Cl)An[d] | 90–80% E–W | 120 | 2 | 0.42 | 0.55 | [5] |
| | 80–65% E–W | 120 | 2 | 0.53 | 0.70 | [5] |
| An$_2$C=C(Br)An | 90–80% E–W | 120 | 2 | 0.34 | 0.49 | [5] |
| Fl=C(Br)An[e] | 90–70% E–W | 140 | 3 | 0.56 | 0.77 | [30] |
| An$_2$C=C(OTs)An | 90–60% A–W | 75 | 4 | 0.42 ± 0.04 | 0.49 | [6] |
| (o-An)$_2$C=C(Br)An | 90–70% E–W | 120 | 3 | 0.53 | 0.70 | [54] |
| E-(o-An)C(An)=C(Br)An | E–W | 120 | | 0.49 | 0.65 | [78] |
| Z-(o-An)C(An)=C(Br)An | E–W | 120 | | 0.48 | 0.63 | [78] |
| E-p-O$_2$NC$_6$H$_4$C(Me)=C(Cl)An | 80–50% E–W | 120 | 4 | 0.62 | 0.82 | [42] |
| Z-(o-MeSC$_6$H$_4$)C(An)=C(Br)An | 80–50% E–W | 120 | 2 | 0.53 | 0.70 | [91] |
| Anth=C(OTs)An[f] | 100–60% E–W | 105 | 7 | 0.28 ± 0.01 | 0.36 | [33] |

[a] Abbreviations: Ac–W, aqueous acetone; E–W, aqueous ethanol; Ac–W, aqueous acetic acid; Ac–F, acetic acid–formic acid.
[b] Number of solvent compositions.
[c] Extrapolated value.
[d] The $mY$ plot is nonlinear.
[e] Fl, fluorenylidene; compound **10**.
[f] Anth, anthronylidene; compound **65**.

TABLE 6.18

Values of $m$ for AnC(OTs)=CMe$_2$ in Various Solvents$^a$

| Solvent$^b$ | Temp (°C) | $n^c$ | $m$ | $m$ (25°)$^d$ |
|---|---|---|---|---|
| 90–40% Me$_2$CO–H$_2$O | 60 | 6 | 0.51 ± 0.01 | 0.57 |
| 90–50% EtOH–H$_2$O | 40 | 5 | 0.48 ± 0.01 | 0.50 |
| 30–10% EtOH–H$_2$O | 40 | 3 | 1.35 | 1.42 |
| 100–50% AcOH–H$_2$O | 60 | 8 | 0.49 ± 0.03 | 0.55 |
| 100–60% AcOH–H$_2$O | 60 | 7 | 0.54 ± 0.02 | 0.60 |
| 100–25% AcOH–HCOOH | 35 | 5 | 0.79 ± 0.07 | 0.82 |
| 90–25% AcOH–HCOOH | 35 | 4 | 0.68 ± 0.02 | 0.70 |
| 100–0% TFE–EtOH | 35 | 9 | 0.76 | 0.79 |
| 100–50% TFE–H$_2$O | 35 | 7 | −0.14 | −0.14 |

$^a$ From Kaspi [33] and Rappoport and Kaspi [95].
$^b$ Numbers are volume of the first solvent to volume of the second solvent.
$^c$ Number of solvent compositions.
$^d$ Extrapolated value.

effect of solvent on the solvolysis of 1-anisyl-2-methylpropen-1-yl tosylate (67) (Table 6.18) [33]. Most of the log $Y$ plots for 67 are linear, and the $m$

$$AnC(OTs)=CMe_2$$

**67**

values or the $m$ values extrapolated to 25° in nucleophilic solvents are low for the $k_c$ route. In contrast, the values at 35° in AcOH–HCOOH ($m = 0.79$) or in TFE–EtOH ($m = 0.76$) are in the $k_c$ region, although they are lower than most of the values observed for the 1- and 2-adamantyl systems [111]. An extreme case is the reaction in TFE–H$_2$O mixtures, the $m$ value being negative in the high TFE region [95], as discussed below.

Since there is clear evidence from substituent effects and common ion rate depression that the solvolysis of almost all the compounds of Tables 6.17 and 6.18 are of the $k_c$ type, the low $m$ values can be due to the use of highly activating aryl (especially anisyl) groups, to the use of good leaving groups, or to steric hindrance to solvation.

The most plausible explanation for the decrease in the $m$ values with the increase in the bulk of the $\beta$ substituent is steric hindrance to solvation [3, 6, 77]. As already mentioned, space-filling models, the $\alpha$- and $\beta$-substituent effects, and the $k_{OTs}/k_{Br}$ ratios suggest that steric interactions between neighboring substituents in the planar vinyl systems are highly important and that these interactions increase strongly with increased bulk of the $\beta$ substituent. Approach of the solvent to either the ground state or to the polar transition state from the rear is highly hindered, and the extent of hindrance to nucleophilic solvation from this side is proportional to the bulk of the $\beta$ substituent. However, approach of the solvent from the side of the leaving group is less hindered, especially if C—X bond cleavage is extensive in the transition state. Consequently, since the solvation of the leaving group is electrophilic, a higher $m$ value is expected in the more electrophilic solvent, and this was indded found to be the case. Bartlett and Tidwell found a similar rate retardation due to steric inhibition of solvation in the solvolysis of highly substituted tertiary systems such as tri-*tert*-butylcarbinyl *p*-nitrobenzoate [112].

The importance of electrophilic solvation is shown by the fact that the reactivity of **67** in AcOH–H$_2$O is one order of magnitude higher than in aqueous acetone of the same $Y$ value [33]. The linear log $k$ versus $Y$ plot for AcOH–H$_2$O covers the $Y$ region of $-1.68$ to $+1.6$, but the $k$ values in 60% and 50% AcOH are nearly identical and are very similar to the value for pure water extrapolated from the aqueous acetone line, in spite of the difference in the $Y$ values. Saturation of the electrophilic properties of the solvent seems to be an adequate explanation [33].

The reaction of **67** in trifluoroacetic acid is much enhanced compared with the reaction in the other solvents [34]. By using the AcOH–HCOOH line, a $Y$ value of 3.8 is calculated for CF$_3$COOH, and this value should be compared with $Y = 4.57$, which is based on 2-adamantyl tosylate [109].

A specific solvation of the transition state of the most hindered substrates by water, the less bulky component of the solvent mixture, seems plausible. Hence, the microscopic environment around the transition state does not change much when the macroscopic composition of the solvent is changed.

Although Eq. (24) has not yet been applied to vinylic solvolysis, solvent effects were used to rule out the $k_s$ route. The $(k_{aq.\ EtOH}/k_{RCOOH})_Y$ values should be appreciably higher than unity when solvent participation is important and around unity when it is unimportant. The $(k_{aq.\ EtOH}/k_{RCOOH})_Y$ values of 0.46 and 0.76 (Table 6.5) for compounds **23** and **24** (X $=$ Br), respectively [36, 55], as well as the values of 0.80 and 0.88 for $\alpha$-(2,4-dimethyl phenyl)vinyl tosylate and $\alpha$-(2,4,6-trimethylphenyl)vinyl tosylate, respectively [15], show that a significant change in the nucleophilicity of the solvent produces no significant change in the rate. The data for the last two compounds gave Schleyer's $Q$ values of 0.76 for the 2,4-dimethyl compound and 0.67 for the 2,4,6-trimethyl compound, and comparison with the $Q$ values for saturated systems points to a weak or negligible solvent participation in the solvolysis.

The solvent effect in an apparently $k_\Delta$ process was also measured. The solvolysis rate of $E$-1,2-ditolyl-2-(phenylthio)vinyl 2,4,6-trinitrobenzene-

**TABLE 6.19**

**Solvent Effect in the Solvolysis of TolC (PhS)=C(Tol)OTNB[a]**

| Solvent | $10^4 k_1$ (sec$^{-1}$) | Solvent | $10^4 k_1$ (sec$^{-1}$) |
|---------|------------|---------|------------|
| MeNO$_2$ | 42.3 | AcOH | 4.72 |
| MeOH | 14.5 | Dioxane | 0.34 |
| EtNO$_2$ | 17.5 | 4:1 MeOH–dioxane | 10.5 |
| EtOH | 7.06 | 13:1 Me$_2$CO–MeOH | 3.56 |
| Me$_2$CO | 3.17 | 9:1 Me$_2$CO–MeOH | 4.03 |

[a] From Modena and Tonellato [82] and Modena *et al.* [113].

sulfonate (**68**) increases with increased polarity of the solvent [82, 113]. The rate increases 124-fold between the relatively nonpolar dioxane and the highly polar nitromethane (Table 6.19). No attempt was made to discern a linear free-energy relationship in solvent properties for these data.

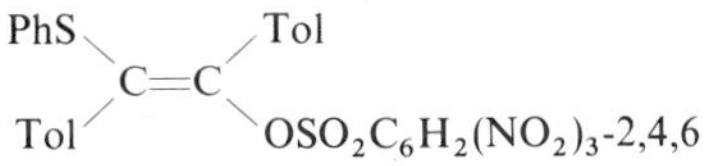

$$
\begin{array}{c}
\text{PhS} \diagdown \qquad \diagup \text{Tol} \\
\text{C}{=}\text{C} \\
\text{Tol} \diagup \qquad \diagdown \text{OSO}_2\text{C}_6\text{H}_2(\text{NO}_2)_3\text{-2,4,6}
\end{array}
$$

**68**

An unusual solvent effect is found in the solvolysis of 1-(o-methoxyphenyl)-2-methylpropen-1-yl tosylate (**69**) [Eq. (25)] and of **67** or its bromo analogue **25** (X = Br) in aqueous TFE mixtures [95]. On addition of water to pure

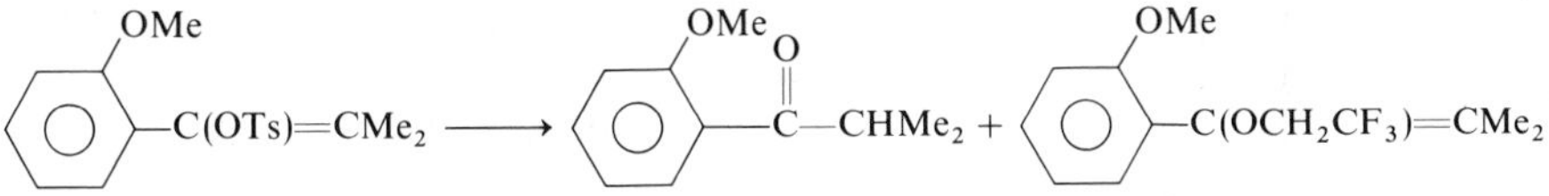

$$
\underset{\mathbf{69}}{\text{C(OTs)=CMe}_2} \longrightarrow \underset{\mathbf{70}}{\overset{\text{O}}{\overset{\|}{\text{C}}}-\text{CHMe}_2} + \underset{\mathbf{71}}{\text{C(OCH}_2\text{CF}_3)\text{=CMe}_2}
$$

$$
(25)
$$

TFE, the rate constant first shows a slight decrease for all three compounds. The rate constants then start to increase at a high mole fraction of water ($X_{\text{H}_2\text{O}}$). The reaction was followed up to high $X_{\text{H}_2\text{O}}$ values only with **69**, and the increase in $k_1$ with $X_{\text{H}_2\text{O}}$ was very sharp in this region. This is shown by the data of Table 6.20 and Fig. 6.1, which gives log $k_1$ versus $X_{\text{H}_2\text{O}}$ and versus $Y_{t\text{-BuCl}}$ plots. Common ion rate depression was observed with **25** (X = Br) (Section V,A), but the extrapolated $k_1^0$ values increased only slightly at the $X_{\text{H}_2\text{O}}$ region of 0.38 to 0.84 (see Table 6.29), whereas a rate increase for **67** was not achieved in this region. The position of the minimum was different for all three substrates [95].

**TABLE 6.20**

**Solvolysis of $o$-MeOC$_6$H$_4$C(OTs)=CMe$_2$ in Aqueous TFE at 35°**

| % TFE[a] | $X_{H_2O}$ | $E_T^b$ | $10^4 k_1$ (sec$^{-1}$) | Products 71 (%) | Products 70 (%) | $k_{H_2O}/k_{TFE}^c$ |
|---|---|---|---|---|---|---|
| 100 | 0.00 | 73.2 | 1.70 | 100 | 0 | |
| 97 | 0.15 | 73.0 | 1.56 | 86 | 14 | 0.93 |
| 94 | 0.26 | 72.6 | 1.41 | 77 | 23 | 0.85 |
| 90 | 0.38 | 72.1 | 1.31 | 65.5 | 34.5 | 0.96 |
| 80 | 0.58 | 71.3 | 1.07 | 51.5 | 48.5 | 0.76 |
| 70 | 0.70 | 70.8 | 0.99 | 43 | 57 | 0.56 |
| 60 | 0.79 | | 1.16 | | | |
| 50 | 0.84 | 70.2 | 1.15 | 30 | 70 | 0.18 |
| 40 | 0.89 | | 1.34 | | | |
| 30 | 0.93 | 69.4 | 1.73 | | | |
| 20 | 0.96 | 69.4 | 3.09 | | | |
| 15 | 0.97 | | 4.19 | | | |
| 10 | 0.98 | 69.5 | 5.29 | | | |
| 5 | 0.99 | | 6.11 | | | |
| 2.5 | 0.995 | | 7.84 | | | |

[a] Percent TFE (w/w) in TFE–H$_2$O mixtures.
[b] Transition energy for the charge transfer band of **72**.
[c] Relative rate of capture of R$^+$ by the solvent components.

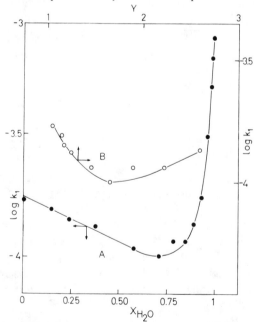

**Fig. 6.1.** Plots of log $k_1$ for **69**; (A) versus $X_{H_2O}$; (B) versus $Y_{t\text{-BuCl}}$. [Reprinted by permission of the publisher from *J. Am. Chem. Soc.* **96**, 4518 (1974).]

Since the ionizing power parameter $Y$, which is based on *tert*-butyl chloride, increases with an increase in $X_{H_2O}$ [114], the rate decrease associated with the increase in the $X_{H_2O}$ values from 0.00 to 0.70 for **69** amounts to a negative Winstein–Grunwald $m$ value of $-0.14$. The appearance of such a minimum in a $\log k_1$ versus $Y$ (or versus $X_{H_2O}$) plot has precedents only in the solvolyses of *tert*-butyl chloride and of 7-methyl-7-norbornyl tosylate in aqueous 1,1,1,3,3,3-hexafluoro-2-propanol (HFIP) [115] and probably in the reaction of the latter compound in aqueous TFE [116]. The minimum for these compounds was suggested to be due to the formation of the hydrate $HFIP \cdot 2H_2O$ [115]. A similar possibility was rejected for the reactions in aqueous TFE due to the different positions of the minima for the three vinylic substrates and the absence of independent evidence for hydrates of TFE. Two alternative explanations for the unusual solvent effect are opposing effects on $k_1$ and the intervention of ion pairs during the solvolysis [95].

Although an increased ionization rate is expected in solvent mixtures with higher dielectric constants, electrophilic assistance by hydrogen bonding in the departure of the leaving group is also of great importance. Since there are conflicting data on the problem of whether water or TFE is a better solvent for electrophilically assisted processes, this was probed by measuring the solvatochromic shifts of the internal charge-transfer band of 1-(*p*-hydroxyphenyl)-2,4,6-triphenylpyridinium betaine (**72**) [117] in aqueous TFE. The values of the corresponding transition energies $E_T$, which are given in Table 6.20, decrease linearly with the increase in $X_{H_2O}$ in the region

**72**

from 0.0 to 1.0, substantiating the conclusion that solvation is actually reduced when the water content is increased. Although this effect is apparently more important than the effect of the dielectric constant at low and medium $X_{H_2O}$ values, the large increase in $k_1$ at high $X_{H_2O}$ values may reflect a saturation of the electrophilic assistance effect at high $\varepsilon$ values. Superimposed on these two opposing effects on $k_1$ is the probable occurrence of ion pairs, which contributes to the observed shape of the $k_1$ versus $X_{H_2O}$ plot for **69** [95].

As discussed in Section V,A, the titrimetric rate constants $k_t^0$ are lower than the ionization rate constants $k_{ion}$, and $k_t^0 = k_{ion}F$, where $F$ is the fraction of ion pairs that dissociate to the product-forming free ions. The

$F$ and $k_{ion}$ might show opposite responses to a change in the blend of dissociating and ionizing power with $X_{H_2O}$. Ion-pair dissociation constants increase with $\varepsilon$, and $F$ should increase with $X_{H_2O}$ in a concave curve, which reaches a plateau when $F = 1$. Concurrently, $k_{ion}$ may increase with $X_{H_2O}$ due to electrophilic solvation as discussed above, at least when $X_{H_2O} \ll 1$. The product of $k_{ion}$ and $F$ may give, under certain conditions, a $k_t^0$ versus $X_{H_2O}$ plot with a minimum. Analysis of the possible shapes of such curves suggests that ion pairs are indeed involved in the solvolysis and that the difference in the $F$ and the $k_{ion}$ terms is reflected in the shapes of the curves for the different substrates [95b].

Consequently, the $Y$ values, which are based on *tert*-butyl chloride solvolysis, may be inadequate for the solvolysis of vinylic systems in aqueous TFE. The spectroscopic $E_T$ values or the rate constants for the solvolysis of **69** may be better models for the effect of solvent on the solvolysis of vinylic compounds in these media [95].

## IV. STEREOCHEMISTRY OF VINYLIC SOLVOLYSIS AND THE STRUCTURE OF α-ARYLVINYL CATIONS

The possible structures of vinyl cations are discussed in Chapters 2 and 5. The bent structure is ruled out for noncyclic α-alkylvinyl cations on the basis of the solvolysis rates of cyclic vinyl triflates.

The formation of the cyclopropyl phenyl ketone **75** as the main solvolysis product of the 1-bromo-2-phenylcyclobutene **14** in various solvents [43]

$$
\begin{array}{cccc}
\text{14} & \text{73} & \text{74} & \text{75}
\end{array}
$$
(26)

[Eq. (26)] is evidence for the preference of the linear α-phenylcyclopropylidene cation **74** over the β-phenylcyclobutenyl cation **73** with its forced, "bent" geometry. However, the driving force for the **73** → **74** rearrangement is probably the formation of an α-phenyl-stabilized cation from the resonatively less stabilized β-phenyl-substituted cation [26, 43].

The question of a bridged versus linear vinyl cation is connected with the migrations across the double bond and is discussed in Section VI. This section discusses the stereochemistry of solvolysis and its relation to the structure of α-arylvinyl cations.

The rate-determining step differs from the product-determining step in vinylic solvolysis. The structure of the product(s) is determined solely by the stereochemistry of the intermediate vinyl cation and by the direction of

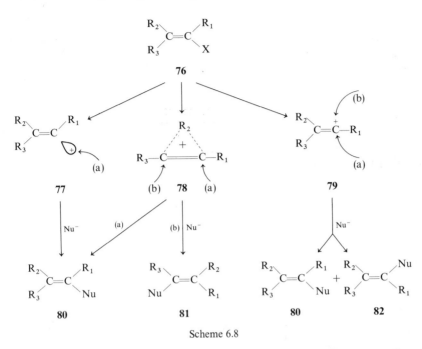

Scheme 6.8

approach of the capturing nucleophile to it. Scheme 6.8 shows the predicted reaction course for the solvolysis of one geometric isomer (76) via the possible intermediates. Reaction of the bent ion 77 with the nucleophile Nu⁻ gives the retained product 80. The same product is also formed from the bridged ion 78, together with the rearrangement product 81, if attack on C-2 is energetically feasible. Attack on the linear ion 79 would lead to both isomers 80 and 82. If 77 does not rearrange to its geometric isomer before capture by Nu⁻, the isomer of 76 will give retained products via the bent and the bridged ions but the same products as from 76 via the same free ion (79) [118]. Hence, formation of the same product mixture starting from either a pure E or a pure Z isomer (so-called "racemization") [118] strongly indicates a free linear sp-hybridized intermediate vinyl cation or a pair of rapidly interconverting bent stereoisomers. As discussed earlier, distinguishing between these two alternatives is very difficult.

Unlike the planar $sp^2$-hybridized ion that is derived from saturated precursors and presents equal faces to an approaching nucleophile, a linear free vinyl cation carrying two different $\beta$ substituents has two diastereotopic sides that must react with the nucleophile with different rates. The product distribution under kinetic control is determined by these relative rates. A steric argument suggests that the side of the smaller $\beta$ substituent is more accessible to the nucleophile and predicts that capture from this side will predominate. The more crowded product will then be obtained under kinetic control, and one isomer will give excess retention whereas the other isomer will give excess inversion.

It was shown in Chapter 5 that, in $\alpha$-alkyl-substituted systems, the $E$ and $Z$ isomers give different product distributions and excess inversion [94, 119], and this was ascribed to ion pairing of $R^+$ and $X^-$, which shields the face of the ion that is closer to the leaving group. Moreover, steric considerations fail to predict the geometry of the product from the 1,2-dimethylvinyl cation, and this was ascribed to electronic effects [94c]. Obviously, it is difficult to assess the expected product distribution in the capture of a free ion when ion-pair formation intervenes.

The stereochemistry of the solvolysis of several pairs of $\alpha$-arylvinyl halides, especially of the 1,2-dianisyl-2-phenylvinyl systems [36, 51, 55, 118], was investigated. The solvolysis of the isomers **83** and **84**, which carry different leaving groups X, in the presence of different nucleophiles in different solvents (Scheme 6.9) gave the same product distribution from

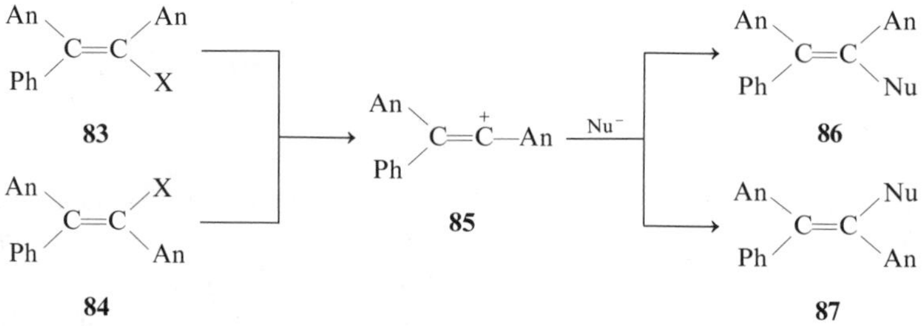

Scheme 6.9

either isomer (Table 6.21). These data strongly indicate a linear cation as the product-forming intermediate. The product distribution is very close to a 50:50 $E–Z$ mixture for all the systems as expected, since the steric bulk of the two $\beta$-aryl groups at the vicinity of the reaction center is identical [36]. Additional stereochemical evidence that the free cation **85** and not an ion pair is the product-forming intermediate comes from common ion rate depression data (Section V,A) [36].

A linear $\alpha$-phenylvinyl cation is also indicated in the solvolysis of the closely related 2-anisyl-1,2-diphenylvinyl bromides **88** and **89**. Although the main solvolysis product in TFE buffered by 2,6-lutidine is the $\beta$-phenyl rearranged product **90** (Scheme 6.10), racemization was observed in the 15% unrearranged product, which is a 1:1 mixture of **91** and **92**. Likewise, the reaction of **89** with $p$-toluenethiolate ion in TFE gives a 1:1 mixture of the thiolates **93** and **94** [8].

# TABLE 6.21

## Products from the Solvolysis of the 1,2-Dianisyl-2-phenylvinyl System

| Isomer | X | Nu | Solvent | 86 (%) | 87 (%) | Reference |
|---|---|---|---|---|---|---|
| 83 | Br | AgOAc or NaOAc | AcOH | 54 | 46 | [36] |
| | | NaOAc | AcOH–HCOOH | 54 | 46 | [55] |
| | | NaOAc | AcOH–Ac$_2$O | 54 | 46 | [55] |
| 84 | Br | AgOAc or NaOAc | AcOH | 54 | 46 | [36] |
| | | NaOAc | AcOH–HCOOH | 54 | 46 | [55] |
| | | NaOAc | AcOH–Ac$_2$O | 54 | 46 | [55] |
| 83 | Br | PhCH$_2$S$^-$ | 80% EtOH | 50 | 50 | [118] |
| | | p-MeC$_6$H$_4$S$^-$ | 80% EtOH | 50 | 50 | [118] |
| 84 | Br | PhCH$_2$S$^-$ | 80% EtOH | 50 | 50 | [118] |
| | | p-MeC$_6$H$_4$S$^-$ | 80% EtOH | 50 | 50 | [118] |
| 1:1 **83–84** | Br | AgOMs | MeCN | 54 | 46 | [36] |
| 83 | Br | LiCl | DMF | 50 | 50 | [118] |
| 84 | Br | LiCl | DMF | 50 | 50 | [118] |
| | | LiCl | AcOH | 50 | 50 | [36, 118] |
| 83 | Cl | NaOAc | AcOH | 54 | 46 | [36] |
| 56% **83** + 44% **84** | OTs | NaOAc | AcOH | 54 | 46 | [36] |
| 83 | OMs | NaOAc | AcOH | 54 | 46 | [36] |
| | | EtOH | 80% EtOH | 50[a] | 50[a] | [55] |
| 84 | OMs | NaOAc | AcOH | 54 | 46 | [36] |
| | | EtOH | 80% EtOH | 50[a] | 50[a] | [55] |
| | | Et$_4$NBr | AcOH | 52[a] | 48[a] | [36] |

[a] Approximate value.

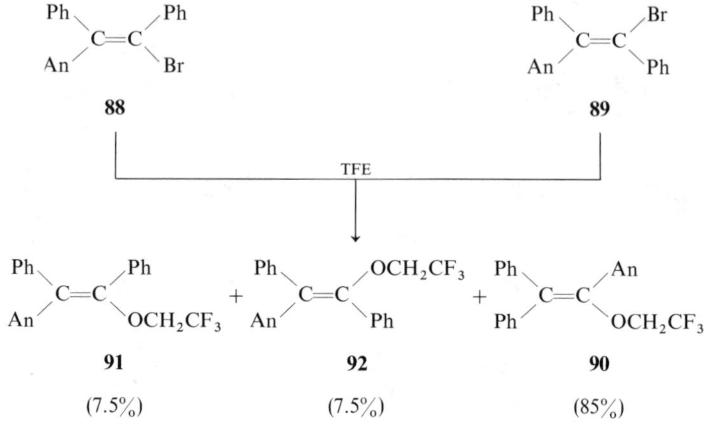

$$
\underset{\substack{\text{An} \\ }}{\overset{\text{Ph}}{>}} C = C \underset{\substack{\text{Br} \\ }}{\overset{\text{Ph}}{<}}
$$

**88**

$$
\underset{\substack{\text{An} \\ }}{\overset{\text{Ph}}{>}} C = C \underset{\substack{\text{Ph} \\ }}{\overset{\text{Br}}{<}}
$$

**89**

TFE

$$
\underset{\substack{\text{An} \\ }}{\overset{\text{Ph}}{>}} C = C \underset{\substack{\text{OCH}_2\text{CF}_3 \\ }}{\overset{\text{Ph}}{<}} + \underset{\substack{\text{An} \\ }}{\overset{\text{Ph}}{>}} C = C \underset{\substack{\text{Ph} \\ }}{\overset{\text{OCH}_2\text{CF}_3}{<}} + \underset{\substack{\text{Ph} \\ }}{\overset{\text{Ph}}{>}} C = C \underset{\substack{\text{OCH}_2\text{CF}_3 \\ }}{\overset{\text{An}}{<}}
$$

**91**          **92**          **90**

(7.5%)        (7.5%)        (85%)

Scheme 6.10

When the bulk of the two $\beta$ substituents is different, both isomers still give the same distribution. Acetolysis of the 1,2-dianisylvinyl bromides **43** and **44** gives 92% dianisylacetylene, but the 8% substitution products consists of two acetates in a 89:11 ratio [31]. Tentatively, the acetate **95** is the major component of the mixture [Eq. (27)], in line with the capture of ion **45** from its least hindered side [31].

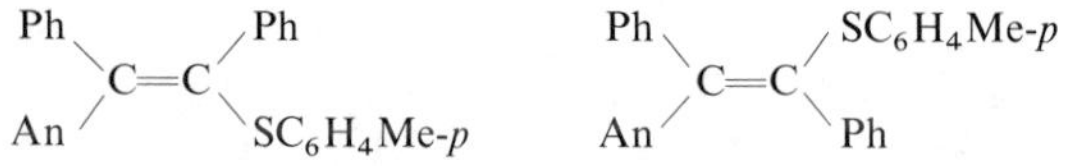

$$
\underset{\substack{\text{An} \\ }}{\overset{\text{Ph}}{>}} C = C \underset{\substack{\text{SC}_6\text{H}_4\text{Me-}p \\ }}{\overset{\text{Ph}}{<}}
$$

**93**

$$
\underset{\substack{\text{An} \\ }}{\overset{\text{Ph}}{>}} C = C \underset{\substack{\text{Ph} \\ }}{\overset{\text{SC}_6\text{H}_4\text{Me-}p}{<}}
$$

**94**

$$
\left. \begin{array}{l} \underset{\substack{\text{H} \\ }}{\overset{\text{An}}{>}} C = C \underset{\substack{\text{Br} \\ }}{\overset{\text{An}}{<}} \\ \textbf{43} \\[6pt] \underset{\substack{\text{H} \\ }}{\overset{\text{An}}{>}} C = C \underset{\substack{\text{An} \\ }}{\overset{\text{Br}}{<}} \\ \textbf{44} \end{array} \right\} \xrightarrow[\text{NaOAc}]{\text{AcOH}} \text{An} - \overset{+}{C} = C \underset{\substack{\text{An} \\ }}{\overset{\text{H}}{<}} \longrightarrow
$$

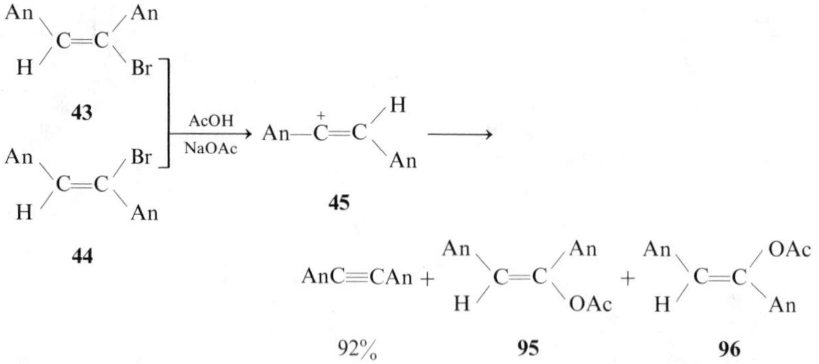

**45**

$$
\text{AnC} \equiv \text{CAn} + \underset{\substack{\text{H} \\ }}{\overset{\text{An}}{>}} C = C \underset{\substack{\text{OAc} \\ }}{\overset{\text{An}}{<}} + \underset{\substack{\text{H} \\ }}{\overset{\text{An}}{>}} C = C \underset{\substack{\text{An} \\ }}{\overset{\text{OAc}}{<}} \qquad (27)
$$

92%        **95**        **96**

Other evidence for the linearity of $\alpha$-phenylvinyl cations and for capture from the least hindered side comes from the extensive work, especially by Melloni and co-workers [120–124], on the electrophilic addition to phenyl-

**TABLE 6.22**

**Product Distribution from the Ion 97 Formed by Electrophilic Addition to Acetylenes**[a]

| Reaction | R | $E$ Isomer, % (e.g., **98**) | $Z$ Isomer, % (e.g., **99**) |
|---|---|---|---|
| PhC≡CH + RCl | $t$-Bu | 100 | 0 |
| | Ph₂CH | 90 | 10 |
| | PhCH₂ | 80 | 20 |
| PhC≡CR + HCl | Me | 70 | 30 |
| | Et | 80 | 20 |
| | $i$-Pr | 95 | 5 |
| | $t$-Bu | 100 | 0 |
| | Ph₂CH | 95 | 5 |
| | PhCH₂ | 85 | 15 |
| PhC≡CH + RBr | $t$-Bu | 100 | 0 |
| | Ph₂CH | 90 | 10 |
| | PhCH₂ | 80 | 20 |
| PhC≡CR + $t$-BuCl | Me | 100 | 0 |
| | Et | 95 | 5 |
| | Ph | 95 | 5 |
| PhC≡CR + $t$-BuBr | Me | 100 | 0 |
| | Et | 95 | 5 |
| PhC≡CR + HBr | Me | 75 | 25 |
| | Et | 80 | 20 |
| | $i$-Pr | ∼100 | ∼0 |
| | $t$-Bu | 100 | 0 |
| | PhCH₂ | 90 | 10 |
| | Ph₂CH | 98 | 2 |

[a] From Maroni et al. [120], Marcuzzi and Melloni [122], and Marcuzzi and Melloni [124].

substituted acetylenes (Chapter 3). Addition of alkyl chlorides (RCl) to phenylacetylene in the presence of a Friedel–Crafts catalyst, such as zinc chloride in methylene chloride, or addition of hydrogen chloride to alkyl-phenylacetylenes gave mixtures of $E$ and $Z$ adducts, with almost identical product distributions for both reactions [120]. Similar results were obtained in the addition of the corresponding alkyl bromides to phenylacetylene or of HBr to alkylphenylacetylenes in the presence of ZnBr₂ [122] (Table 6.22). This similarity argues strongly for the intermediacy of a common intermediate, the vinyl cation **97** [Eq. (28)]. Some deviation from the nearly identical product distribution is due either to a minor molecular syn addition of HCl to the acetylene or to a $Z \rightleftarrows E$ isomerization of the vinyl bromides under the reaction conditions, but this can be avoided by using a low [RBr]/[PhC≡CH] ratio.

$$RX + PhC{\equiv}CH \left.\begin{array}{c} \\ \\ \\ \\ \\ \end{array}\right] \xrightarrow[CH_2Cl_2]{ZnCl_2/} \underset{97}{\overset{R}{\underset{H}{>}}C{=}\overset{+}{C}{-}Ph + X^-} \longrightarrow \begin{array}{c} \underset{H}{\overset{R}{>}}C{=}C\overset{Ph}{\underset{X}{<}} \\ \mathbf{98} \\ + \underset{H}{\overset{R}{>}}C{=}C\overset{X}{\underset{Ph}{<}} \\ \mathbf{99} \end{array} \qquad (28)$$

$$HX + PhC{\equiv}CR$$

R = t-Bu, Ph$_2$CH, PhCH$_2$
X = Cl, Br

The product distribution clearly indicates the preference for capture of **97** from its least hindered side, since capture from the side of the hydrogen is always dominant. Moreover, the $E$ stereoselectivity decreases with the bulk of R from an exclusive formation of the $E$ isomer when R = $t$-Bu to only 80% of the $E$ isomer when R = PhCH$_2$, as expected.

An exclusive formation of the thermodynamically less stable $E$ isomer was also observed in the analogous addition of HBr to 1-anisyl-2-*tert*-butylacetylene [Eq. (29)] [73].

$$An{-}C{\equiv}C{-}Bu\text{-}t + HBr \longrightarrow \underset{Br}{\overset{An}{>}}C{=}C\overset{Bu\text{-}t}{\underset{H}{<}} \qquad (29)$$

Capture of a symmetric ion should lead to a 1:1 mixture of the $E$ and $Z$ products. Indeed, the addition of HBr to 1-anisyl-2-deuterioacetylene gave a 1:1 distribution of the two isomeric bromides **100** and **101** in AcOH, or in chloroform in the presence of HgBr$_2$ at 0° [Eq. (30)]. In many aprotic solvents the $E$ isomer **100** is formed in excess, suggesting the intervention of a molecular syn addition of HBr [56]. Addition of HCl to 2-deuterio-1-phenylacetylene gave preferential syn addition, except in sulfolane in the presence of ZnCl$_2$, when a 1:1 mixture of the $E$- and $Z$-α-chlorostyrenes was formed [121].

$$An{-}C{\equiv}C{-}D + HBr \longrightarrow \underset{Br}{\overset{An}{>}}C{=}C\overset{D}{\underset{H}{<}} + \underset{Br}{\overset{An}{>}}C{=}C\overset{H}{\underset{D}{<}} \qquad (30)$$

$$\qquad\qquad\qquad\qquad\qquad \mathbf{100} \qquad\qquad \mathbf{101}$$

Additions of alkyl chlorides to diphenylacetylene [123] enabled Maroni and co-workers to test further the steric effects in the addition to the vinyl cation since the ions formed carry two bulky $\beta$ substituents. Diphenylmethyl chloride gave substituted indenes (Chapter 3), but ZnCl$_2$-catalyzed addition of *tert*-butyl chloride gave mainly the adducts **102** and **103** in a 95:5 ratio [Eq. (31)]. In contrast, addition of benzyl chloride gave the corresponding vinyl chlorides **104** and **105** in a 15:85 ratio, in addition to the indene derivatives. Consequently, the addition of *tert*-butyl chloride may be

$$Ph{-}C{\equiv}C{-}Ph + RCl \xrightarrow{\text{ZnCl}_2} \underset{R}{\overset{Ph}{\diagdown}}C{=}C\underset{Ph}{\overset{Cl}{\diagup}} + \underset{R}{\overset{Ph}{\diagdown}}C{=}C\underset{Cl}{\overset{Ph}{\diagup}} \qquad (31)$$

| | | | |
|---|---|---|---|
| **102** | R = *t*-Bu | **103** | R = *t*-Bu |
| **104** | R = PhCH$_2$ | **105** | R = PhCH$_2$ |

regarded as predominantly anti, whereas that of benzyl chloride is predominantly cis, as predicted for the capture of a linear vinyl cation from its least hindered side. The steric requirements of the $\beta$ groups are *t*-Bu > Ph > PhCH$_2$, making the phenyl side more accessible in the ion **106** but less accessible in the ion **107**.

$$\underset{t\text{-Bu}}{\overset{Ph}{\diagdown}}C{=}\overset{+}{C}{-}Ph \qquad \underset{PhCH_2}{\overset{Ph}{\diagdown}}C{=}\overset{+}{C}{-}Ph$$

$$\textbf{106} \qquad\qquad\qquad \textbf{107}$$

Two identical bulky groups at the $\beta$ position of a vinyl cation have the same effect on the direction of attack by a nucleophile. This is shown by the formation of a 1 : 1 mixture of the isomeric chlorides **108** and **109** in the addition of benzyl-$\alpha,\alpha$-$d_2$ chloride to 1,3-diphenylpropyne in 1,2-dichloroethane [Eq. (32)] [124].

$$PhCD_2Cl + PhCH_2C{\equiv}CPh \longrightarrow \underset{PhCD_2}{\overset{PhCH_2}{\diagdown}}C{=}C\underset{Ph}{\overset{Cl}{\diagup}} + \underset{PhCH_2}{\overset{PhCD_2}{\diagdown}}C{=}C\underset{Ph}{\overset{Cl}{\diagup}} \qquad (32)$$

$$\textbf{108} \qquad\qquad\qquad \textbf{109}$$

These results, together with the data of Table 6.22, also indicate that the $\alpha$-phenylvinyl cationic species formed in the low-dielectric media are free ions, since the intervention of oriented ion pairs is expected to form preferentially the syn adducts. This may be the mechanism of addition of HCl to phenylacetylene-$d$ in the presence of ZnCl$_2$ in dichloromethane, which gives a 2.3 : 1 ratio of the *E*- to the *Z*-$\alpha$-chloro-$\beta$-deuteriostyrene [121, 124], or the mechanism of addition of HBr to 1-anisyl-2-deuterioacetylene in CHCl$_3$ [56].

In view of the extensive evidence for the dominant role of steric effects on the stereochemistry of capture of the linear $\alpha$-arylvinyl cations by nucleophiles, it is somewhat surprising that two studies apparently suggest that the above generalizations do not always hold.

Kernaghan and Hoffman [125] studied the reactions of *E*- and *Z*-1-phenylpropen-1-yl bromides (**110** and **111** respectively) with silver trifluoroacetate in isopentane under heterogeneous conditions and the reaction of **110** in diethyl ether under homogeneous conditions (Scheme 6.11). The products were the vinyl trifluoroacetates **112** and **113** and the acetylene **114**.

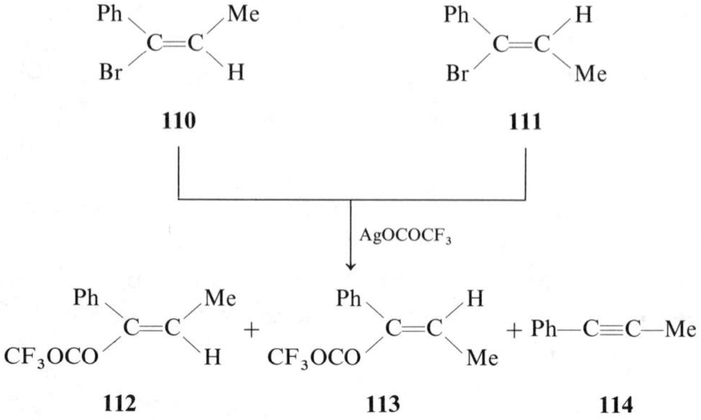

$$\underset{Br}{\overset{Ph}{\diagdown}}C=C\underset{H}{\overset{Me}{\diagup}} \qquad \underset{Br}{\overset{Ph}{\diagdown}}C=C\underset{Me}{\overset{H}{\diagup}}$$

**110**                                    **111**

AgOCOCF$_3$

$$\underset{CF_3OCO}{\overset{Ph}{\diagdown}}C=C\underset{H}{\overset{Me}{\diagup}} \; + \; \underset{CF_3OCO}{\overset{Ph}{\diagdown}}C=C\underset{Me}{\overset{H}{\diagup}} \; + \; Ph-C\equiv C-Me$$

**112**                       **113**                    **114**

Scheme 6.11

The product distribution (Table 6.23) shows preferential retention of configuration of the vinyl trifluoroacetates starting from either isomer. The extent of net retention in the heterogeneous reaction was the same, 13%, for both isomers [125].

**TABLE 6.23**

Products from the Reaction of 110 and 111 with AgOCOCF$_3$ at 25°[a]

| Isomer | Reaction conditions | 114 (%) | 112/113 |
|---|---|---|---|
| **110** (E) | Heterogeneous, isopentane, 1 day | 57 | 1.31 ±0.02 |
| **111** (Z) | Heterogeneous, isopentane, 3 days | 53 | 0.77 ± 0.02 |
| **110** (E) | Homogeneous, Et$_2$O, 3 days | 57 | 1.9 ± 0.09 |

[a] From Kernaghan and Hoffmann [125].

The excess retention in the heterogeneous reaction was ascribed to the generation of the vinyl cation on the surface of the silver salt. The inter-action with trifluoroacetate ion from the crystal lattice results in a preferential frontside attack by CF$_3$COO$^-$ ion from the surface over attack by a CF$_3$COO$^-$ ion from the solution, and the outcome is a preferential retention. This explanation does not contradict the assumption made above concerning the capture of a free cation by a nucleophile, and this reaction is the counter-part of the mechanism for excess inversion by ion pairing with leaving group.

However, the greater than 30% net retention in the homogeneous re-action in diethyl ether was explained in terms of Sneen's mechanism [126] of a double inversion process involving an intermediate oxonium ion. This mechanism was not presented explicitly, but the sequence is presumably that of Eq. (33), in which the first inversion gives the oxonium ion **115**, which gives the inverted product **112** by a second inversion by the trifluoro-acetate ion. Addition of trifluoroacetic acid in isopentane or in dichloro-methane to the acetylene **114** gave preferentially the *E*-trifluoroacetate **112**.

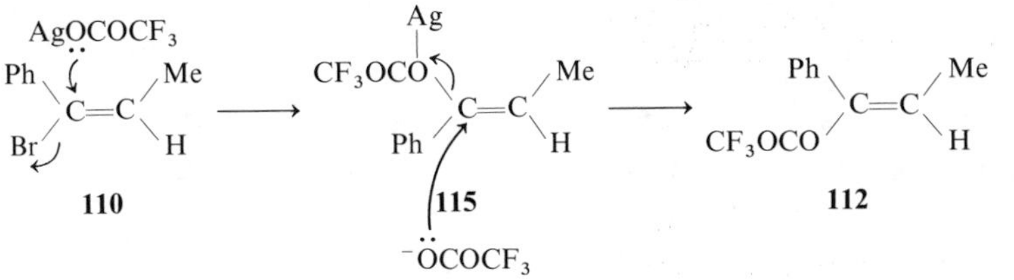

$$\tag{33}$$

This was suggested to be reasonable if a vinylic cation–trifluoroacetate pair of similar configuration is involved in both reactions [125].

Kernaghan and Hoffman compared their results with those of Rappoport and Apeloig on the α-anisyl-substituted systems **83** and **84** [118] and with those of Kelsey and Bergman on the 1-cyclopropyl-1-iodopropenes [127]. The difference in stereochemistry was ascribed to the fact that the stability of the α-phenylvinyl cation **116** is lower than of the ions **85** and **117**. They concluded, therefore, that their example is the first of a stereoselective heterolysis of vinyl halides.

This mechanism is in conflict with several features of the vinylic substitution reactions. First, the stability of cation **116** should be similar to that of other α-phenylvinyl cations that are intermediates in solvolysis and addition reactions and that give sterically controlled products, as discussed above. An example is the cation **118**, which is the intermediate in Scheme 6.10 [8]. Second, the mechanism suggests that, if a symmetric solvation shell is

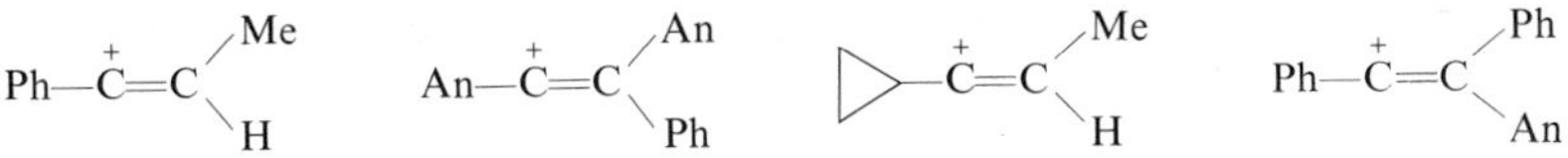

not attained before covalent collapse takes place, retention rather than inversion will be observed. Finally, although the mechanism involves an in-plane double inversion via a highly implausible oxonium ion intermediate, not even a single inversion has yet been observed in the substitution at a vinylic carbon of an acyclic system [29].

However, Hoffmann's results for the homogeneous reaction do fit into the general framework of the stereochemistry suggested above. The linear free vinyl cation **116** has two β substituents with different bulk, and capture from the side of the smaller hydrogen would lead to excess retention, as observed starting from **110**. Indeed, the 66 : 34 distribution of **112** and **113** is very similar to the ratio obtained in the solvolysis of the two isomeric *p*-methoxy analogues [128]. The same steric argument predicts that the same products will be formed via **116**, starting from the isomer **111**; i.e., excess inversion will be observed for this isomer. Unfortunately, the stereochemistry of the solvolysis of **111** was not studied since this compound was much less reactive than **110** [125], in analogy with their *p*-methoxy derivatives (Table 6.11). Hence, the double inversion mechanism should be rejected.

An attempt to repeat the reaction of **110** with $AgOCOCF_3$ in ether gave mainly propiophenone and only a small percentage of **112** and **113** [128]. If the decomposition of **112** and **113** to propiophenone occurs at different rates, the ratio of the two products is of little use as a stereochemical tool.

A discrepancy of a different type was found by Reich and Reich [129]. The silver salt-assisted solvolysis of the isomeric *E,Z*- and *Z,Z*-1-bromo-4-chloro-1,4-diphenylbutadienes (**119** and **120**, respectively) in acetic acid or in acetic anhydride gave the *E,Z*-monoacetate **121** in 21 to 27% yield and the *Z,Z*-monoacetate **122** in 51 to 53% yield, together with 20 to 27% acetylene **123**. When **119**-2-*d* and **120**-2-*d* were used, the acetylene **123** still contained 56 and 50% of the deuterium, respectively, whereas **119**-3-*d* and **120**-3-*d* gave **123**, which retained 79 and 83% of the deuterium, respectively. From these experiments it was concluded that the vinyl acetates are formed mainly from the open ion **124**, whereas most of the acetylene is derived

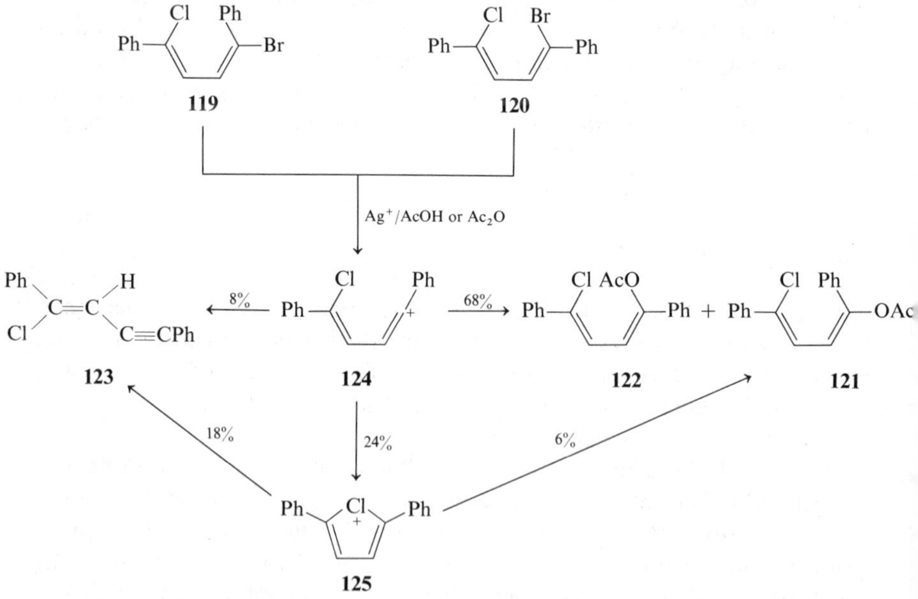

Scheme 6.12

from the chlorolium ion **125**, which is formed via $\beta$-chlorine participation (Scheme 6.12).

The important stereochemical result is that nearly identical product compositions were obtained from either **119** or **120**, indicating again that an $\alpha$-phenyl-stabilized ion is stable enough to give products from the free linear ion **124**. However, since there is no doubt that the $\beta$-($\beta$-chloro-$\beta$-phenylvinyl) substituent of the vinyl cation **124** is bulkier than the $\beta$-hydrogen, the product distribution is the opposite of that expected by assuming that steric effects govern the capture ratio at the two faces of the vinyl cation. The product of capture from the side of the substituted vinyl group is **122**, which is formed in 2.0- to 2.5-fold excess over the isomer **121**, which is formed by capture from the side of the hydrogen. This is even more surprising since the cisoid form of the vinyl cation, which is not easily converted to the transoid form [129], makes the hindrance by the $\beta$ substituent more severe than that by a simple alkyl group. It must be concluded that the reason for the product distribution should be other than steric hindrance, but this reason is not yet clear.

The stability and the selectivity of $\alpha$-arylvinyl cations (Section V,A) suggest that they can be generated and observed by nmr under stable ion conditions. Indeed, as discussed in Chapter 8, this has been the case.

# V. NATURE OF THE CATIONIC INTERMEDIATES

## A. Selective Free Ions as the Product-Forming Intermediates

The nature of the intermediates in several vinylic solvolyses was discussed previously, and several lines of evidence, especially the partial inversion of configuration in the solvolysis products of $\alpha$-alkyl-substituted systems [94, 119], indicated the importance of ion pairs as product-forming intermediates in the solvolysis. The kinetics were usually first order and without complications.

On the other hand, the kinetics of the solvolysis and of an accompanying isomerization are an important tool for determining the nature of the intermediates in the solvolysis of $\alpha$-arylvinyl systems [36, 51, 55, 56, 130]. Consequently, our knowledge of the details of the solvolysis of $\alpha$-arylvinyl systems is much more complete than that of any other class of vinylic substrates.

A most unusual and surprising feature of the solvolysis of many $\alpha$-arylvinyl systems is the widespread appearance of common ion rate depression (mass law effect) during the solvolysis. This was observed in the early days of vinylic solvolysis in several vinyl systems. In 1968 Miller and Kaufman found the effect in the solvolysis of 1-anisyl-2,2-diphenylvinyl iodide in 70% dimethyl-formamide [77]. Modena and Tonellato reported it in 1968 during the cyclization of $E$-1,2-ditolyl-2-phenylthiovinyl 2,4,6-trinitrobenzenesulfonate in acetone [131], and Rappoport and Apeloig observed it in 1969 in the acetolysis of 1,2-dianisyl-2-phenylvinyl bromides **83** and **84** [118].

Common ion rate depression is the reduction in the rate of a solvolysis reaction caused by an anionic leaving group. In most of the solvolysis reactions, addition of a salt of the leaving group $M^+X^-$ results in a small rate acceleration due to a "normal" salt effect [132], and only in a few cases does such addition result in a rate reduction. In even a fewer number of cases, the low concentrations of $X^-$ formed during the solvolysis are sufficient to cause a severe rate reduction. All these types of behavior were observed for the solvolysis of $\alpha$-arylvinyl systems [36].

Following Ingold [133, 134] and Winstein [135], common ion rate depression is ascribed to the reaction of a *free* carbonium ion with $X^-$, which leads to the reformation of the covalent RX [134, 135]. It indicates a partial or a complete formation of the products from the free ion. If we neglect for the time being the possibility that ion pairs play a role in the reaction, Scheme 6.13 is the most simplified scheme that accounts for this phenomenon.

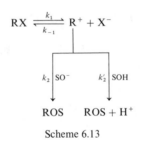

Scheme 6.13

The ion $R^+$ formed in the $k_1$ step is usually highly reactive and reacts rapidly with excess solvent SOH in a $k_2'$ step. The reaction is then first order with a rate constant $k_1$. However, if $R^+$ is sufficiently long-lived and selective, the more nucleophilic $X^-$ may compete with the less nucleophilic solvent, which is present in much higher concentration. In vinylic solvolysis, in which buffering is necessary to avoid an $Ad_E$–E route with the formed HX, competition between capture of $R^+$ by $X^-$ $(k_{-1})$ and by the lyate conjugate base $SO^-$ $(k_2)$ should also be considered. Although all three nucleophiles compete for $R^+$, it is more convenient (although not always correct) to deal with competition of $X^-$ with either SOH or $SO^-$ for $R^+$. A steady-state treatment leads to Eqs. (34) and (35) for competition of $X^-$ and $SO^-$ and of $X^-$ with SOH, respectively. The $k_{obs}$ is the first-order rate constant, and

$$k_{obs} = \frac{k_1}{1 + (k_{-1}[X^-]/k_2[SO^-])} = \frac{k_1}{1 + (\alpha[X^-]/[SO^-])} \quad (34)$$

$$k_{obs} = \frac{k_1}{1 + (k_{-1}[X^-]/k_2')} = \frac{k_1}{1 + \alpha'[X^-]} \quad (35)$$

$\alpha = k_{-1}/k_2$ and $\alpha' = k_{-1}/k_2'$ are the mass law constants, which measure the selectivities of the cation in its reactions with the two nucleophiles [36].

However, as seen from Scheme 6.13 and Eq. (35), the $\alpha'$ value includes the solvent concentration. For most solvolysis reactions, the $\alpha$ or the $\alpha'$ values are 0, indicating that the ion is sufficiently reactive that capture by the solvent is the exclusive process. However, with the increased stability of $R^+$, its reactivity decreases and its selectivity increases, and $\alpha$ and $\alpha'$ values also increase [133, 136]. The phenomenon of external ion return [135a] is manifested by common ion rate depression, since the $[X^-]$ term appears in the denominators of Eqs. (34) and (35) and formed or added $X^-$ will reduce the rate. However, this is observed only for the most stable ions.

The first-order titrimetric rate constants for the solvolyses of several $\alpha$-anisylvinyl bromides decrease strongly during the kinetic run especially in AcOH [31, 36, 50, 51, 55, 118a, 130] and in trifluoroethanol [8, 33, 95], since the $[X^-]$ term increases with the progress of the reaction. For example, the extrapolated $k_t^0$ ($= k_1$) value at zero reaction time for **83** in AcOH is four times higher than the value at 50% reaction [36].

This large rate decrease introduces several problems and complications in the determination of both the extrapolated $k_t^0$ values and the $\alpha$ and $\alpha'$ values. (1) As seen from Eqs. (34) and (35), $\alpha$ and $\alpha'$ (selectivities) and $k_1$ (reactivity) values are derived from the same equation, and the errors in both parameters are correlated. (2) The magnitude of the "normal" salt effect that increases the value of $k_1$ should be known for evaluating $\alpha$, but this effect is masked by the rate decrease due to common ion rate depression. Moreover, when $\alpha$ is determined within a kinetic run, the buffering salt of the lyate ion is progressively replaced by the salt of the leaving group, resulting in an unknown differential salt effect. (3) Ion pairing in low-dielectric media will reduce the concentration of free $X^-$, which is the most active nucleophile in the capture reaction. (4) The presence of both SOH and $SO^-$ raises the question as to which of them is the capturing nucleophile, i.e., whether $\alpha$ or $\alpha'$ is being measured. (5) When the $\beta$ substituents on the vinylic system are different, the external ion return gives the two geometric isomers. This is of no concern if the $\beta$ substituents are of a similar size and therefore both isomers solvolyze with similar rates [36, 55] but, since return occurs preferentially from the less hindered side of the cation (Section IV), the thermodynamically less stable and usually the more reactive isomer is formed when the two $\beta$ substituents differ in size [31, 42]. Starting from the "fast" isomer, the common ion rate depression will give some of the "slow" isomer, the rate decrease will be superimposed on that due to external ion return, and the $\alpha$ values will be lower than the real values. Starting from the "slow" isomer, return to the "fast" isomer will be only slightly registered in the $\alpha$ values, which will be lower than the real values.

A promising means of resolving the complication arising from the correlation between $k_t{}^0$ and $\alpha$ was applied by Lodder and co-workers [137]. They studied both the vinyl acetate formation and the exchange of radioactive bromide ion in the reactions of several triarylvinyl bromides in AcOH/NaOAc under the solvolysis conditions used by Rappoport and co-workers. According to Scheme 6.13, the incorporation of the labeled bromide ion into the unreacted vinyl bromide should be a first-order reaction, and the sum of the corresponding rate constant ($k_{exch}$) and that for the concurrent formation of the solvolysis product ($k_t$) should be equal to $k_1$, since the two processes account for both the reverse and the forward reactions of the intermediate $R^+$. Indeed, a first-order bromide exchange was observed in all the cases investigated and, from the directly measured $k_{exch}$ and $k_t$ values, the $\alpha$ values were calculated. The $k_{exch}$ values were found to be higher in most cases than the $k_t{}^0$ values, and the derived $\alpha$ values also differed from those calculated from the extent of common ion rate depression. Whereas the $\alpha$ values based on $k_{exch}$ are more reliable if Scheme 6.13 is applicable, capture of an intermediate ion pair (Section V,B) will increase $k_{exch}$ and will explain the discrepancy between the $k_{exch}$ and the $\alpha$ values so calculated and the $k_t{}^0$ and the $\alpha$ values based on the common ion rate depression. Since this problem is not yet solved, the more extensive data from the common ion rate depression serve as the basis for the discussion below.

As a consequence of the above-mentioned complications, the accuracy of the $\alpha$ values is lower than desired. Nevertheless, in spite of the enormous amount of work on the solvolysis of saturated compounds, the number of $\alpha$ and $\alpha'$ values measured for vinylic systems is similar to or even exceeds the number in saturated systems. Values of $\alpha$ in RCOOH/RCOONa (or $RCOO^-Bu_4N^+$) for competition between the leaving group and the $RCOO^-$ anion are given in Table 6.24. The added salt is only slightly dissociated to free ions, and both free carboxylate ions and the less reactive carboxylate

**TABLE 6.24**

**The $\alpha$ Values for $ArC(X)\!=\!CR_1R_2$[a]**

| Ar | $R_1$ | $R_2$ | X | Solvent | Base | Temp (°C) | $\alpha$ | Reference |
|---|---|---|---|---|---|---|---|---|
| An | H | H | Br | AcOH | NaOAc | 120 | 0 | [11, 22, 56] |
| 9-Anthryl | H | H | Cl | AcOH | NaOAc | 120 | 9 | [17] |
|  |  |  |  | AcOH | Bu$_4$NOAc | 120 | 11 | [17] |
| An | Me | Me | Br | AcOH | NaOAc | 120 | 4.3 | [50] |
| An[b] | Br | An | Br | AcOH | NaOAc | 141 | 0.2 | [31] |
| An[c] | An | Br | Br | AcOH | NaOAc | 141 | 0.8 | [31] |
| An[c] | An | OAc | Br | AcOH | NaOAc | 141 | 0.8 | [31] |
| An[b] | An | Ph | OMs | AcOH | NaOAc | 75 | 0.1 | [36] |
| An | Ph | Ph | Br | AcOH | NaOAc | 120 | 15 | [50] |
| An | Ph | Ph | Cl | AcOH | NaOAc | 120 | 6 | [50] |
| An[b] | An | Ph | Br | AcOH | NaOAc | 120 | 21 | [36] |
| An[c] | Ph | Ph | Br | AcOH | NaOAc | 120 | 21 | [36] |
| An[b] | An | Ph | Cl | AcOH | NaOAc | 120 | 5 | [36] |
| An | An | Ph | Br | AcOH | NaOAc | 120 | 18 | [50] |
| An | An | An | Cl | AcOH | NaOAc | 120 | 7 | [50] |
| An |  | Fl[d] | Br | AcOH | NaOAc | 120 | 9.5 | [50] |
| An |  | Fl[d] | Cl | AcOH | NaOAc | 120 | 3.3 | [50] |
| An |  | Anth[e] | Br | AcOH | NaOAc | 120 | 32 (~63)[f] | [92, 137] |
| Tol |  | Anth[e] | Br | AcOH | NaOAc | 140 | 9 | [138] |
| An | Ph | An | Br | Me$_3$CCOOH | Me$_3$CCOONa | 185 | 0 | [51] |
| An | An | Anth[e] | Br | Me$_3$CCOOH | Me$_3$CCOONa | 185 | 2.4 | [51] |

[a] $[RX] = 0.04\ M$.
[b] Cis isomer.
[c] Trans isomer.
[d] Fl, fluorenylidene.
[e] Anth, anthronylidene.
[f] Calculated by assuming that $k_{exch}$ obtained from the bromide exchange experiment is equal to $k_t^0$.

ion–cation ion pairs are involved in the capture process. This is demonstrated by the higher $\alpha$ value obtained for the more dissociated $Bu_4NOAc$ as compared with that obtained for NaOAc [17]. Moreover, the $X^-$ formed will also be distributed between the dissociated and the ion-pair species, as shown by the increase in the $\alpha$ values with the decrease in the substrate concentration (which is related to the concentration of free $X^-$), as expected [36]. Hence, the $\alpha$ values are not an absolute measure of the competition between the free ions, and only their relative values are significant. Comparison of the values should be made for identical substrate and salt concentrations; the data of Table 6.24 are for 0.04 $M$ substrate.

The data in $RCOOH/RCOO^-$ obey Eq. (34) somewhat better than they obey Eq. (35). The rate of product formation is also higher at higher $OAc^-$ concentrations in AcOH, indicating that capture by $OAc^-$ is an important source of the product [36, 50]. Nevertheless, a significant and even dominant capture by the carboxylic acid in these media cannot be unequivocally ruled out. The $[RCOOH]/[RCOO^-]$ ratios are usually greater than 200, so that $k_2/k_2' \geq 1000$ is required for a very dominant capture of the ion by $RCOO^-$. It is difficult to evaluate the $k_2/k_2'$ ratios under solvolytic conditions due to the mixture of a normal salt effect, common ion rate depression, and ion-pairing effects. The values estimated from the study of rearrangements in vinylic systems (Section VI) are lower and, if they are also applicable for the systems of Table 6.24, the $\alpha$ values of the table should be recalculated according to Eq. (35) to give $\alpha'$ values that are two orders of magnitude larger.

The reactions in an aqueous organic medium, in which the buffer is usually an organic base, are free from such complications, and Eq. (35) is obeyed. However, comparison under identical conditions is still advantageous. Table 6.25 gives the $\alpha'$ values for competition between the leaving group and the solvent. In this medium, elimination [17] and $\beta$-aryl rearrangement [8] gave apparent $\alpha$ values ($\alpha_{obs}$) that are lower than the "real" $\alpha$ values (see below); both $\alpha_{obs}$ and $\alpha'$ values are given in Table 6.25.

Tables 6.24 and 6.25 demonstrate convincingly that many $\alpha$-arylvinyl cations are selective. If the selectivity is directly related to the lifetime and the stability of the ions and inversely related to their reactivity, then the $\alpha$-arylvinyl cations derived from the precursors of Tables 6.24 and 6.25 are long-lived and are remarkably stable. This raises the interesting question as to whether the vinylic solvolysis obeys the same reactivity–selectivity rule that is applicable to the solvolysis of saturated compounds. For saturated compounds, when the solvolysis is faster, the intermediate ion is more selective and, therefore, it is also more stable, according to the Hammond postulate [61]. The $\alpha$ values of Table 6.24 seem moderate or small compared with values for competition between azide ion and water [139, 140] or

TABLE 6.25

The $\alpha'$ Values for $ArC(X)=CR_1R_2$ in Aqueous Organic Solvents

| Ar | $R_1$ | $R_2$ | X | Solvent | Temp (°C) | $\alpha_{obs}{}^a$ | $\alpha'$ $(M^{-1})^b$ | Reference |
|----|-------|-------|---|---------|-----------|---------|---------|-----------|
| An | H | H | Br | 80% EtOH | 120 | | 0 | [2, 22, 56] |
| 9-Anthryl | H | H | Cl | 80% EtOH | 120 | 1.3–4.2 | 1.6–5.0 | [17] |
| | | | | 90% Me$_2$CO | 140 | 52 ± 8 | 260 | [17] |
| An | Me | Me | Br | 80% EtOH | 120 | | 0 | [22] |
| | | | | TFE | 35 | | 394 | [95b] |
| An | Me | Me | Cl | 80% EtOH | 120 | | 7.5 | [22] |
| An | Ph | Ph | I | 70% DMF | 130.5 | | 40 | [77] |
| Ph | Ph | Ph | I | 70% DMF | 189.5 | | 0 | [77] |
| An | Ph | Ph | OTs | 70% MeCO | 75 | | 9 | [6] |
| Ph | Ph | Ph | OTs | 70% Me$_2$CO | 120 | | 0 | [6] |
| An | An | An | Br | TFE | 90 | | 78 | [10] |
| Ph | Ph | Ph | Br | TFE | 140 | | 42 | [8] |
| Ph | Ph$^c$ | An | Br | TFE | 140 | 11 | 73 | [8] |
| Ph | An$^d$ | Ph | Br | TFE | 140 | 18 | 120 | [8] |
| Ph | An | An | Br | TFE | 140 | 0 | | [8] |
| An | Ph$^d$ | An | Br | 80% EtOH | 120 | | 3 | [55] |
| An | Anth$^e$ | | Br | 80% EtOH | 105 | | 158 | [92] |
| | | | | TFE | 35 | | 1200 | [141] |
| An | Xant$^f$ | | Br | 80% EtOH | 105 | | 96 | [93] |
| | | | | TFE | 35 | | 4200 | [93] |

$^a$ Apparent $\alpha$ values obtained from the rate decrease.
$^b$ After correction for rearrangement or elimination of the cation when needed, $\alpha' = k_x/k_{SOH}$.
$^c$ Trans isomer.
$^d$ Cis isomer.
$^e$ $R_1R_2C= = $ 9-anthronylidene.
$^f$ $R_1R_2C= = $ xanthenylidene.

chloride ion and water [136], but their significance becomes clear when it is realized that they are selectivity factors between the two ions rather than between an anion and a neutral molecule.

Fortunately, a more direct comparison is possible between some of the data of Table 6.25 and data derived from common ion rate depression for the solvolysis of highly reactive saturated halides. Such a comparison is given in Table 6.26 and, in spite of the slightly different conditions, it is clear that several $\alpha$-arylvinyl cations are highly selective compared with the *tert*-butyl cation and even with other sp$^2$-hybridized ions that are regarded as highly stable. This is especially remarkable for the anthronylidene-substituted cation **126a** [92], the selectivity of which is higher than those of the *tert*-butyl [133], the benzhydryl [133], and the $p,p'$-dimethylbenzhydryl cations [133] and approaches the selectivity of the triphenylmethyl cation [139]. Since the solvolysis rates of the $\alpha$-arylvinyl bromides of Table 6.26

**TABLE 6.26**

**Selectivities of Trigonal and Vinyl Cations**

| RX | Solvent | Temp (°C) | $\alpha' = k_X/k_{SOH}$ | Reference |
|---|---|---|---|---|
| $An_2CHCl$ | 85% $Me_2CO$ | 0 | 2300 | [136] |
| $AnC(Ph)CHCl$ | 85% $Me_2CO$ | 0 | 700 | [136] |
| 9-Anthryl-CH(Cl)Me | 90% $Me_2CO$ | 0 | 630 | [17, 142] |
| $Ph_3CCl$ | 85% $Me_2CO$ | $-34$ | 500 | [139] |
| $(p\text{-}MeC_6H_4)_2CHCl$ | 85% $Me_2CO$ | 0 | 69 | [133] |
| $Ph_2CHBr$ | 90% $Me_2CO$ | 50 | 60 | [133] |
| $Ph_2CHCl$ | 80% $Me_2CO$ | 25 | 12 | [133] |
| $t\text{-}BuBr$ | 90% $Me_2CO$ | 25 | 1–2 | [133] |
| 9-Anthryl-C(Cl)=CH$_2$ | 90% $Me_2CO$ | 140 | 260$^a$ | [17] |
| Anth=C(Br)An$^b$ | 80% EtOH | 105 | 158 | [92] |
| $Ph_2C=C=C(Cl)Ph$ | 90% EtOH | 25 | 10 | [85] |
| AnC(OTs)=CPh$_2$ | 70% $Me_2CO$ | 75 | 9 | [6] |

$^a$ Corrected for the elimination reaction.
$^b$ Anth, 9-anthronylidene.

are several orders of magnitude lower than those of the saturated bromides, it is clear that the same reactivity–selectivity rule does not apply for vinylic and saturated compounds together.

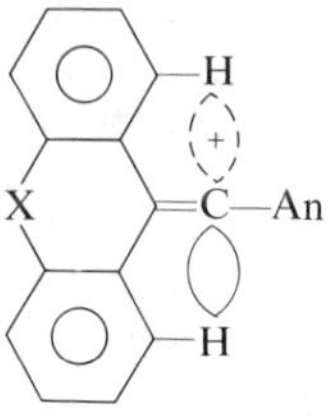

**126a**  X = C=O
**126b**  X = O

Several reasons can be suggested for this discrepancy and for the high selectivity of the vinylic compounds. It can be argued that the skeletons of the vinylic and most of the saturated compounds of Table 6.26 are sufficiently different, and comparison is unjustified. However, a comparison between α-(9-anthryl)vinyl chloride (**23**) [17] and α-(9-anthryl)ethyl chloride (**127**) [142], which show common ion rate depression within a run, gave α' values

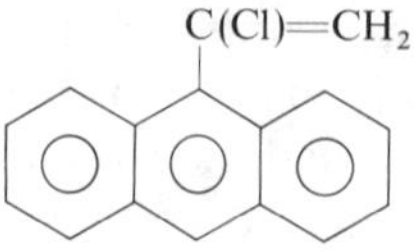

**23** 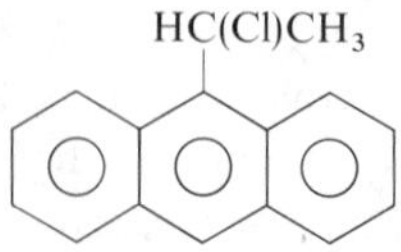                              **127**

of 630 for **127** at 0° and 260 for **23** at 140° in 90% acetone [17]. The value for **23** takes care of the elimination reaction to 9-ethynylanthracene, which comprises 80% of the reaction products. Hence, for structurally very close compounds substituted by a single α-aryl group, the saturated compound shows higher selectivity, but the small difference in selectivity does not reflect the $7 \times 10^4$-fold lower reactivity of **23** [17, 18].

The solvolyses of the less reactive vinylic compounds are conducted at a much higher temperature than are those of the saturated compounds. It was found that a change in the reaction temperature has a profound effect on the selectivity [143], but it is very unlikely that this effect contributes much to the selectivity of the vinyl cations.

Another possibility is either that the same reactivity–selectivity principle does not apply to the reaction of vinylic and saturated compounds together, or that the principle is not applicable at all to the reactions of carbenium ions with nucleophiles. Whereas selectivity is a property of the ion, its reactivity is indirectly evaluated from its rate of formation. Transition state differences are mainly responsible for the variation in the reactivity in the ion formation process within structurally related compounds. However, comparison of the rate of formation of a vinylic and a saturated compound should take into account the much lower ground state energy and the consequently slower solvolysis of the vinylic compound. The stability of the ions themselves may be similar. Consequently, the same reactivity–selectivity rule for both classes of compounds together is not expected to be the same [24], and the selectivity of the vinyl ions should be discussed only in relation to their electronic and steric effects.

The importance of the electronic effect is shown by the fact that the α' values for α-alkylvinyl cations are zero whereas they are appreciable for several α-arylvinyl cations. The large stabilization by an α-aryl group is sufficient to give higher α' values for the aryl-substituted cations of Table 6.26 than for the *tert*-butyl cation.

The dependence of the α' values on the nature of the α-aryl group is mechanistically very important. Ritchie showed that highly stable carbenium ions react with nucleophiles at relative rates that are strongly dependent on the nucleophile and on the solvent but independent of the nature of the

$$\log(k/k_0) = N_+ \tag{36}$$

cation [144]. His "constant selectivity" equation [Eq. (36)], where $N_+$ is a nucleophilic parameter independent of the cation and the $k$'s are the rate constants for the direct reaction of the cation with a nucleophile and with water [144], is in contrast to the reactivity–selectivity rule in solvolysis, which is based on competition experiments on solvolytically generated, less stable cations. Ritchie ascribes this discrepancy to product formation from different ionic species, e.g., free ions and ion pairs, which are formed in the solvolysis [144, 145]. The significance of the data of Tables 6.24 and 6.25 is that the common ion rate depression in the solvolysis of $\alpha$-arylvinyl cations indicates [see Eq. (37)] that almost all the products are obtained from capture of the *free* vinyl cation. If Eq. (36) applies to these cases, the $\alpha'$ values should be independent of the $\alpha$-aryl group but, if the reactivity–selectivity rule applies, $\alpha'$ should decrease as the $\sigma^+$ value of the substituent in the aryl group increases.

Most of the data of Tables 6.24 and 6.25 are consistent with the reactivity–selectivity principle and do not support Eq. (36). 1-Anisyl-2,2-diphenylvinyl iodide gives $\alpha'$ of 40 in 70% dimethylformamide [77], and the corresponding tosylate gives $\alpha'$ of 9 in 70% acetone [6]. In contrast, the 1-phenyl analogues, triphenylvinyl iodide and tosylate, show no common ion rate depression (i.e., $\alpha' = 0$) in the same solvents [6, 41]. The $\alpha'$ values of 9 and 0 for $\alpha$-(9-anthryl)vinyl cation [17] and the $\alpha$-anisylvinyl cation [22], respectively, lead to a similar conclusion. It may be argued that in the former cases the 1-anisyl derivatives give products from free ions and the 1-phenyl derivatives give products from ion pairs. Moreover, in the latter case, there is evidence that the $\alpha$-anisylvinyl cation gives products, at least partially, via ion pairs [56]. However, the solvolyses of the $\alpha$-anisyl and the $\alpha$-tolyl $\alpha$-bromoanisylideneanthrones in both AcOH and TFE give products from the free ions as deduced from the extensive common ion rate depression, but $\alpha$ and $\alpha'$ are lower for the *p*-methyl derivative [138].

On the other hand, the $\alpha'$ value of 78 for trianisylvinyl bromide in TFE buffered by 2,6-lutidine [10] is not much different from the $\alpha'$ of 42 for triphenylvinyl bromide [8], and a "low selectivity" rule is applicable for these systems, since trianisylvinyl bromide is 3000 times more reactive [8, 10].

A clearer answer to this problem as well as the recognition of a very important factor that contributes to the selectivity of the $\alpha$-arylvinyl cations comes from the study of the effect of the $\beta$ substituents. The solvolysis rates of most of the $\alpha$-anisyl-substituted systems of Tables 6.24 and 6.25 differ, but not by a large extent (Table 6.11). However, the $\alpha$ and $\alpha'$ values of these tables increase markedly with increased bulk of the $\beta$ substituent. The $\alpha'$ values for the ion **128** increase from 0 for $R_1 = R_2 = H$ [22, 56] to 158 for $R_1R_2C== = 9$-anthronylidene (**126a**) in 80% EtOH [3, 92], and from 394 for $R_1 = R_2 = Me$ [95b] to 1200 for **126a** in TFE [141] (Table 6.25). The increase in the $\alpha$ values is even more regular: from 0 for $R_1 = R_2 = H$,

$$An-\overset{+}{C}{=}C\overset{\displaystyle R_1}{\underset{\displaystyle R_2}{\big\langle}}$$

**128**

to 4.3 for $R_1 = R_2 =$ Me, to 15 to 21 for $R_1 = R_2 =$ Ph or An, to 32 for $R_1 R_2 C = =$ 9-anthronylidene (Table 6.24) [3].

Strong electron-withdrawing $\beta$ substituents such as bromine reduce appreciably the $\alpha$ values [31]. This is mainly an electronic effect of reducing the ion stability by electron withdrawal.

The absence of selectivity for the $\beta,\beta$-unsubstituted ion 128 ($R_1 = R_2 =$ H) indicates that the presence of an $\alpha$-anisyl group alone is insufficient for the observation of common ion rate depression and selectivity. Indeed, the rate constant for $\alpha$-bromo-$p$-methoxystyrene in AcOH/NaOAc increases slightly during the reaction, and this was ascribed to a positive salt effect by the formed NaBr [56]. However, the $\alpha$ value of zero for this system cannot be used as a guide to the relative reactivities of $Br^-$ and $OAc^-$ toward the free ion 128 ($R_1 = R_2 =$ H), since the product is probably derived from the ion pair rather than from the free ion (see below and Section V,B). The increase in $\alpha$ with the increased bulk of the $\beta$ substituents is reminiscent of the decrease in the $k_{OTs}/k_{Br}$ and $m$ values and the increased stereospecificity of the capture of the ion with the increased bulk of the $\beta$ substituent [3]. Space-filling models of the $\alpha$-arylvinyl cations reveal that, in spite of the partial relief of ground state strain associated with the $sp^2 \rightarrow sp$ hybridization change at $C_\alpha$, the cationic orbital is shielded by the $\beta$ substituents. Whereas such shielding is minimal in the $\beta,\beta$-unsubstituted ion, it increases with the bulk of the $\beta$ substituents. The models also show that, of all the compounds in Tables 6.24 and 6.25, the maximal shielding is achieved with the $\beta$-anthronylidene- and the $\beta$-xanthenylidene-substituted ions 126a and 126b, respectively. The enforced coplanarity of the $\beta$-aryl rings with the double bond places hydrogens 1 and 8 of these systems directly in the line of a perpendicular approach of a nucleophile to the empty orbital. Consequently, the steric hindrance for capture of the ion becomes severe, and the lifetime of the ion increases. The more polarizable nucleophile in the system, i.e., the bromide ion, which is capable of forming a bond from a larger distance by deforming its valence orbital, is more likely to capture the cation than the smaller acetate ion or the neutral acetic acid molecule. Although the effect of bulk alone favors the smaller acetate ion, the transition state for the cation–anion recombination is "early," and this factor is less important than the ability to form a bond from a distance. Hence, the steric hindrance, which becomes more important with the increase in the bulk of the $\beta$ substituent, accounts for the increased $\alpha$ values [3, 6, 36].

The absence of bulky substituents at the $\beta$ position does not necessarily rule out common ion rate depression. When the $\alpha$-aryl group is both a better stabilizer of a positive charge and bulkier than an $\alpha$-anisyl group, selectivity is observed. This is the case with the $\alpha$-anthrylvinyl cation 129,

**129**

which shows selectivity in 80% ethanol, 90% acetone, and AcOH [17]. Since the $\pi$ system of the $\alpha$-aryl group overlaps the vacant p orbital, hydrogens 1 and 8 are in a plane perpendicular to this orbital and the steric effects are lower than for the $\beta$-substituted system.

Consequently, even within the family of $\alpha$-arylvinyl cations, Ritchie's "constant selectivity" rule [144] does not apply, in spite of the fact that the products are formed mainly from the free vinyl cation. It is more difficult to assess the validity of the reactivity–selectivity principle, since due to ground state differences in the precursors the rates of formation of the ions do not necessarily reflect their stabilities. Within the limited family of compounds with $\alpha$-aryl groups of identical bulk, the rule seems to apply. For ions with different shielding of the cationic orbital, steric effects have a pronounced effect on the selectivity.

Since the shielding effect is constant, this explanation for the selectivity could be regarded as a "static" explanation. However, for triarylvinyl cations with three identical aryl groups, an alternative explanation is possible. If a rapid rearrangement of the aryl groups across the double bond is faster than capture by a nucleophile, this rapid movement can drive off nucleophile and solvent molecules and increase the lifetime of the cation [9]. This "dynamic" explanation for the selectivity is discussed and ruled out in Section VI.

The percentage of products formed from the free cation of Scheme 6.13 is given by Eq. (37), where $k_d$ is the depressed rate constant in the presence

$$\text{Percentage of products formed from free ions} = 100(1 - k_d/k_t^0) \quad (37)$$

of a certain concentration of $X^-$ [135]. Typically, the rate constant for most of the compounds of Table 6.24, and especially for the triarylvinyl halides, decreases less than ninefold, by the $X^-$ formed in the reaction or added externally, and less than 90% of the products are derived in most cases from the free $\alpha$-arylvinyl cations. This is a lower limit since no limit for the rate depression was observed in any of these cases and since the positive salt effect on $k_t$ was neglected.

Capture of the free cation by the leaving group and by the nucleophile competes frequently with other reactions of the ions, such as elimination, rearrangement, and cyclization. This affects the measured selectivity values since those ions that participate in the competing reactions are not available for the ion return process. Consequently, the apparent $\alpha$ and $\alpha'$ values ($\alpha_{obs}$) obtained in these cases should be divided by the fraction of ions involved in the return process; i.e., the real $\alpha$ values are higher than the $\alpha_{obs}$ values, and this correction was introduced in Table 6.25.

Three examples demonstrate this behavior. The solvolysis of $\alpha$-(9-anthryl)-vinyl chloride (23) in 90% acetone gives 80% alkyne 27 and 20% ketone 29 [Eq. (38)] [17]. Elimination is therefore four times faster than the solvolysis, and $\alpha_{obs}$ is five times smaller than $\alpha'$ (Table 6.25).

$$C_{14}H_9C(Cl)\!\!=\!\!CH_2 \xrightarrow{\ 90\% \ \text{acetone}\ } C_{14}H_9C\!\!\equiv\!\!CH + C_{14}H_9COMe \qquad (38)$$

$$\begin{array}{ccc} \textbf{23} & \textbf{27} & \textbf{29} \\ & (80\%) & (20\%) \end{array}$$

The solvolysis of the $\beta,\beta$-diaryl-$\alpha$-phenylvinyl halides [Eq. (39)] proceeds with $\beta$-aryl rearrangement across the double bond in TFE [8]. When Ar = Ar' = Ph, the rearrangement is degenerate and $\alpha_{obs}$ is identical with $\alpha'$. However, when Ar = Ph, Ar' = An or Ar = An, Ar' = Ph (compounds 88 and 89), 85% of the product is the rearranged 1-anisyl-2,2-diphenylvinyl trifluoroethyl ether (Scheme 6.10) [8]. The $\alpha_{obs}$ values are 11 to 18, but

$$ (39) $$

correction for the approximately sixfold higher rearrangement gives $\alpha' = 73$ to 120. It is clear that when the ion return constitutes only a small fraction of the overall reactions of the ion, the error in the $\alpha'$ value is high. An extreme example occurs when Ar = Ar' = An: only the anisyl rearranged products, E- and Z-1,2-dianisyl-2-phenylvinyl trifluoroethyl ethers, are formed and $\alpha_{obs}$ is zero [8]. In this and in the former case, return still takes place to both the unrearranged and the rearranged ions 130 and 131 [Eq. (40)], but since the rearrangements are 6 times and > 50 times faster than

$$ (40) $$

$$\begin{array}{cc} \textbf{130} & \textbf{131} \end{array}$$

capture of 130 when Ar = Ph and Ar = An, respectively, the ions 131 are captured preferentially or exclusively. This capture by bromide ion is not manifested in the $\alpha'$ value, since the formed $\alpha$-anisyl-$\beta,\beta$-diarylvinyl bromides solvolyze two or three orders of magnitude faster (Table 6.7) than the unrearranged $\alpha$-phenyl-substituted bromides, and the return to 131 and its reionization are hidden since they appear after the rate-determining step.

Finally, cyclization of the vinyl cation competes in many systems with external ion return. As discussed in Chapter 7, the reactions of 1,2-diaryl-2-(arylthio)vinyl 2,4,6-trinitrobenzenesulfonates lead mainly to cyclization to the corresponding aryl-substituted benzo[b]thiophenes [71, 146–148]. In acetone, cyclization competes with ion return [82], and it is remarkable that the lithium salt of 2,4,6-trinitrobenzenesulfonate ion, a very weak nucleophile, gives a small common ion rate depression for the reaction of E-1,2-ditolyl-2-(phenylthio)vinyl 2,4,6-trinitrobenzenesulfonate (68) [Eq. (41)]. The return by the free arenesulfonate ion is supported by incorporation of a $^{35}$S-labeled 2,4,6-trinitrobenzenesulfonate ion from the labeled lithium salt into the unreacted precursor. An "$\alpha'$" value of $3 \pm 1.3$ can be calculated

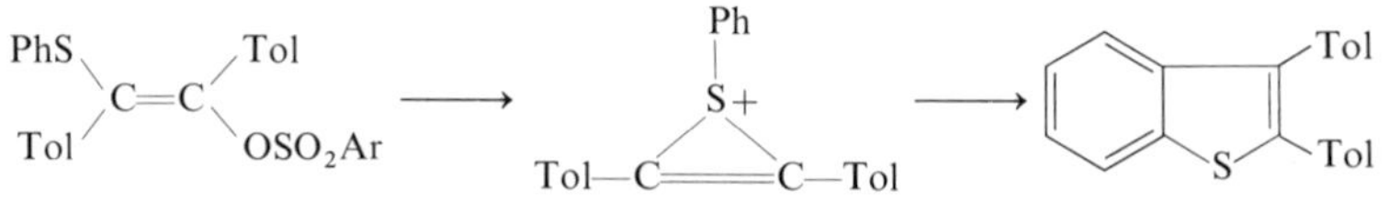

(41)

from the mass law effect for competition between return and cyclization rather than between capture by two nucleophiles. The common ion rate depression for this reaction disappears in more nucleophilic solvents such as AcOH, 9:1 acetone–methanol, and MeOH [82].

The importance of steric effects in the competition between capture and cyclization is shown by the work of Taniguchi and co-workers [54]. The reaction of 1-anisyl-2,2-bis(o-methoxyphenyl)vinyl bromide (61) or of the corresponding chloride in AcOH gave the benzo[b]furan 133 [Eq. (42)].

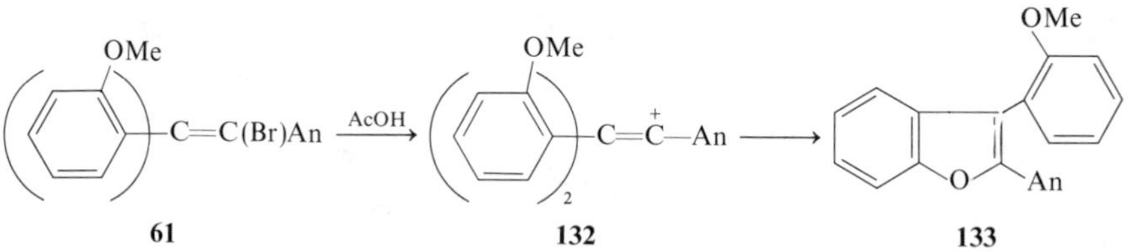

(42)

Common ion rate depression was not observed, although the intermediate ion 132 is very similar to the trianisylvinyl cation, which gives extensive common ion rate depression. It seems that steric effects reduce the rate of the capture of 132 to such an extent that the cyclization by the properly situated $\beta$-oxygen takes over. Indeed, even in the more nucleophilic 80% EtOH in the presence of good nucleophiles, such as $N_3^-$ or p-toluene thiolate ion, no capture product of the open ion 132 is observed and 133 is the exclusive product [54].

On the other hand, the $\alpha$-phenyl analogue **134** gave two benzo[$b$]furans (**136** and **137**) in a 91:9 ratio in AcOH/AgOAc and in a 93:7 ratio in 80% EtOH [Eq. (43)] [54]. Since the product **137** is formed by $\beta$-$o$-anisyl migration, the nucleophilic attack on the $\alpha$-carbon of **135** follows the order

$$(o\text{-}CH_3OC_6H_4)_2C{=}C(Br)C_6H_5 \xrightarrow[\text{AgOAc}]{\text{AcOH}} (o\text{-}CH_3OC_6H_4)_2C{=}\overset{+}{C}C_6H_5 \longrightarrow$$

$$\qquad\qquad\quad \textbf{134} \qquad\qquad\qquad\qquad\qquad\qquad \textbf{135}$$

$$(43)$$

$$\qquad\qquad\quad \textbf{136} \qquad\qquad\qquad\qquad \textbf{137}$$
$$\qquad\qquad\quad (91\%) \qquad\qquad\qquad\qquad (9\%)$$

internal nucleophilic attack (cyclization) > 2-$o$-methoxyphenyl group participation (migration) $\gg$ external nucleophilic attack (capture).

The delicate balance between the various processes suggests that the reactions of various systems in such solvents as TFE should be compared at the same concentrations of a nonnucleophilic base. The reaction of triphenylvinyl bromide in TFE buffered by NaOAc gave a lower $\alpha'$ value of 24 than in the presence of 2,6-lutidine ($\alpha' = 42$) [8], and triphenylvinyl acetate was among the products. Consequently, concurrent competition between the solvent, acetate ion, and bromide ion took place, and the apparent $\alpha'$ is composite.

Table 6.25 shows that common ion rate depression was observed with I$^-$ [77], Cl$^-$ [11, 17, 22, 50], Br$^-$ [8, 31, 36, 50, 55, 95], and OTs$^-$ [95] ions, and OMs$^-$ ion showed a small effect with the 1,2-dianisyl-2-phenylvinyl system [36]. Only with this system are there $\alpha$ and $\alpha'$ values for the reaction with several nucleophiles, and the $\alpha$ values give a scale of relative reactivities of nucleophiles, compared with AcO$^-$, toward the 1,2-dianisyl-2-phenylvinyl cation **85** (Table 6.27) [36]. Judged by other scattered data in other solvents, I$^-$ would be at the top of the table and 2,4,6-trinitrobenzenesulfonate ion would be at the bottom. The remarkable feature of Table 6.27 is that the order of relative reactivities toward **85** resembles that toward a saturated carbon atom [149], which reflects the combination of basicity, polarizability, and solvation effects. *A priori*, a cation should be "harder" than a saturated carbon and therefore more sensitive to basicity effects. However, the connection between the polarizability and the steric effects discussed above accounts for this behavior. The reactivity order of Table 6.27 would remain the same even if Eq. (35) rather than Eq. (34) better described the common ion rate depression in AcOH.

TABLE 6.27

Relative Reactivity of Nucleophiles
toward the Ion 85 in AcOH at 120°[a]

| Nucleophile | Relative reactivity |
|---|---|
| Br$^-$ | 45.5 |
| Cl$^-$ | 15.2 |
| AcO$^-$ | 1.0 |
| OMs$^-$ | 0.16 |
| AcOH | 0.0024 |

[a] From Rappoport and Apeloig [36].

The effect of solvent on the selectivity of the caption is twofold. First, a more nucleophilic solvent is prone to capture an earlier intermediate than the free ion, i.e., an ion pair, along the ionization–dissociation route. Unless this question is probed either by applying Eq. (37) to show that ≥95% of the products arise from the free ion or by applying the "special salt effect" [150], an erroneous α value will be obtained, since the relative capture rates of the free ion and the ion pair by the two nucleophiles are not necessarily identical. Second, the α values are dependent on solvent properties, such as the nucleophilicity and the dielectric constant.

For α-arylvinyl substrates in which the α and β substituents are not extremely bulky, the common ion rate depression in nucleophilic solvents, such as aqueous ethanol, aqueous acetone and aqueous dimethylformamide, is small. Even the addition of a large amount of the salt of the leaving group does not lead to a very large rate depression, although such an experiment introduces a large uncertainty in α' because the effect of the salt on the ionization is unknown. In these systems, products may well be formed from ion pairs (Section V,B), and comparison of the α' values is not recommended. However, whenever a common ion rate depression was observed in carboxylic acid media, $k_d$ decreased continuously on further addition of X$^-$, and for most systems ≥90% of the products were formed from the free ions [17, 36, 50, 55, 92]. Table 6.28 compares the α values and the corresponding percentages of return from the free ion 85 formed from the solvolysis of cis-1,2-dianisyl-2-phenylvinyl bromide in carboxylic acids [36, 51, 55]. In the low-dielectric pivalic acid, common ion rate depression was not observed and the products were obtained from the ion pair [51]. Since the nucleophilicities of AcOH and HCOOH are similar, it is assumed that the same is also true for the nucleophilicities of their conjugate bases. Hence, the α values are expected to increase with the dielectric constant, which increases the lifetime of R$^+$, and to decrease with the increased anion-solvating ability of the medium.

**TABLE 6.28**

**Relative Reactivity in Ionization, Ion-Pair Return, and External Ion Return of cis-1,2-Dianisyl-2-phenylvinyl Bromide (83)[a]**

| Solvent | $\varepsilon^b$ | $Y^c$ | $N^d$ | $k_{ion}{}^e$ | $1 - F^f$ | $\alpha$ | Return from $R^+$ (%) |
|---------|------|-------|-------|--------|---------|----|----------|
| 80% EtOH | 35.8 | 0.00 | 0.00 | 4.9 | 0.32 | Low | <1 |
| 1:1 AcOH–HCOOH | 32.3 | 0.76 | −2.05 | 52 | 0.46 | 2.6 | 84.5 |
| AcOH | 6.2 | −1.64 | −2.05 | 1.0 | 0.47 | 21 | 97.8 |
| 1:1 AcOH–Ac$_2$O | 15.0 | −2.47 | | 0.53 | 0.31 | 27 | 98.9 |
| Me$_3$CCOOH | 2.6 | −3.46 | | ∼0.1 | >0.99 | $g$ | Low$^g$ |

[a] From Lee et al. [35], Rappoport et al. [51], and Rappoport and Apeloig [55].
[b] Dielectric constant.
[c] Ionizing power.
[d] Nucleophilicity.
[e] At 120°.
[f] Fraction of ion-pair return.
[g] The free ion is not formed.

In carboxylic acid media the dielectric constant increases in the order 1:1 AcOH–HCOOH > 1:1 AcOH–Ac$_2$O > AcOH, and the anion-solvating ability via hydrogen bonding increases in the order 1:1 AcOH–Ac$_2$O < AcOH < 1:1 AcOH–HCOOH [55]. The lower $\alpha$ values in 1:1 AcOH–HCOOH [55] indicate that the increased differential solvation of Br$^-$ over AcO$^-$ more than compensates for the increase in the dielectric constant $\varepsilon$, and the higher $\alpha$ in 1:1 AcOH–Ac$_2$O is explained by the small change in solvation power compared with AcOH.

The low nucleophilicity ($N = -2.78$) [109] and the relatively high dielectric constant of CF$_3$CH$_2$OH ($\varepsilon = 26.67$ at 25°) [151] make it an ideal solvent for the observation of common ion rate depression. Table 6.25 shows that the $\alpha'$ values are higher than those in aqueous ethanol, and application of Eq. (37) shows that the products in this solvent are derived mainly or exclusively from the free vinyl cations. The high $\alpha'$ in pure TFE makes possible a systematic study of the solvent effect, and the $\alpha'$ values and the $k_1^0/k_1^{50}$ values for the solvolysis of 1-anisyl-2-methylpropen-1-yl bromide (**25**, X = Br), which are given in Table 6.29, are the most extensive data available on the effect of solvent on these values for a single substrate [95]. The decrease in the $\alpha'$ values with the increase in the water content of the medium was discussed in terms of several opposing effects. In mixed solvents $\alpha'$ should increase when the cation solvation is increased due to a longer lifetime of the cation. On the other hand, several factors reduce the $\alpha'$ values. These are an increase in anion solvation, which reduces the anion nucleophilicity (i.e., lower $k_{-1}$), an increase in the dielectric constant or in the ionic strength, which increases $k_2$ for the cation–neutral molecule (SOH)

**TABLE 6.29**

**Solvolysis of AnC(Br)=CMe$_2$ in Aqueous TFE at 35° [a]**

| % TFE (w/w) | $X_{H_2O}$ | $\varepsilon$ (35°) | $10^5 k_1^0$ (sec$^{-1}$) | $k_1^0/k_1^{50b}$ | $\alpha'$ (liter/mole) | $\alpha_{cor}$ (liter/mole) | $\alpha_{cal}$ (liter/mole) |
|---|---|---|---|---|---|---|---|
| 100 | 0.00 | 23.0 | 4.13 | 2.84 | 394 ± 42 | 373 | 360 |
| 97 | 0.15 | 25.0 | 3.70 | 2.30 | 225 ± 33 | 260 | 218 |
| 90 | 0.38 | 30.0 | 3.30 | 1.54 | 110 ± 5 | 153 | 80 |
| 80 | 0.58 | 36.5 | 3.40 | 1.12 | 27 ± 2 | 53 | 33 |
| 70 | 0.70 | 42.0 | 3.51 | 1.05 | 11 ± 3 | (18)[c] | 19 |
| 60 | 0.79 | | 3.63 | 1.04 | 9 ± 3 | | |
| 50 | 0.84 | 64.5 | 3.65 | 1.02 | 8 ± 1 | (8)[c] | 5.5 |

[a] From Rappoport and Kaspi [95b].
[b] Ratio of the integrated rate constant at 0% reaction to its value at 50% reaction.
[c] These values probably contain a contribution from reaction with the lyate ions.

reaction, a change from a less nucleophilic solvent to a more nucleophilic one or an increase in the concentration of the more nucleophilic solvent, and capture of $R^+$ by $SO^-$ rather than by SOH, which greatly increases the apparent $k_2$ value. Since water is a better cation solvator and TFE is a better anion solvator [152], $\alpha'$ should increase with an increase in the water content of the medium. However, not only is the dielectric constant of water higher than that of TFE, but water is a more nucleophilic solvent than TFE [109], and these two factors together should reduce $\alpha'$ at higher $X_{H_2O}$. The change in the nucleophilicity with the change in the nature and the concentrations of the capturing nucleophiles could be evaluated together independently, since the ratio of the vinyl trifluoroethyl ether to the ketone is given by Eq. (44), where $k_{H_2O}$ and $k_{TFE}$ are the rate constants for capture of $R^+$ by the solvent components, $k_2'$ is given by Eq. (45), and $k_2'/k_{TFE}$ is given by Eq. (46)

$$k_{H_2O}/k_{TFE} = [RCOCHMe_2][TFE]/[RCOCH_2CF_3][H_2O] \qquad (44)$$

$$k_2' = k_{TFE}[TFE] + k_{H_2O}[H_2O] \qquad (45)$$

$$k_2'/k_{TFE} = [TFE] + (k_{H_2O}/k_{TFE})[H_2O] \qquad (46)$$

[95]. Equations (45) and (46) are derived by assuming that the products are obtained exclusively from the free ion and that there is no specific solvation of $R^+$. The similar product distribution from the solvolysis of either 1-anisyl-2-methylpropen-1-yl bromide or tosylate in several solvent mixtures [95], as well as the $k_1^0/k_1^{50}$ values support the first assumption. By using the $k_{H_2O}/k_{TFE}$ values obtained from the observed product distribution and Eq. (44), the $k_2'/k_{TFE}$ values were calculated and used for correcting the $\alpha'$ values for the combined effects of change in the solvent concentration and nucleophilicity. As seen in Table 6.29, these $\alpha_{cor}$ values decrease less than the $\alpha'$ values, but they still decrease sevenfold from 100% to 80% TFE. This leaves the effect of the solvent on the $R^+ + SOH$ reaction as the main contribution to the decrease in the $\alpha'$ values with the increase in the dielectric constant. A plot of log $\alpha'$ versus $1/\varepsilon$, which is predicted to be linear for this reaction according to the Debye-Hückel equation, is indeed linear: log $\alpha = 65.3/\varepsilon' = 0.274$. The calculated $\alpha'$ values ($\alpha_{cal}$) of Table 6.29 show a reasonable fit with the observed values, suggesting that this effect, probably together with the other ones discussed above, accounts for the change in the $\alpha'$ values with the solvent [95]. Superimposed on these effects may be a decrease in the ion-pair return with increasing water content of the medium.

## B. Ion Pairs as Intermediates in Solvolysis and Isomerization

Scheme 6.13 is obviously a simplified solvolysis scheme since ion pairs probably play an important role in the solvolysis. The intervention of ion pairs in the solvolysis of α-alkylvinyl systems was inferred mainly from the nature of the products and the stereochemistry of the solvolysis of these systems [94, 119]. These tools are rarely useful for the more stable α-arylvinyl systems, since in many cases the product distribution and the stereochemistry of the reaction are determined at the free ion stage (Section IV,A). Kinetic methods, which were applied to several systems, indicated the importance of ion pairs in the overall solvolysis scheme and gave in some cases a quantitative estimation of the relative rate constants for the formation and return of some of the cationic intermediates.

Modena and Tonellato [82, 131] found that the salt effects on the solvolysis of $E$-1,2-ditolyl-2-(phenylthio)vinyl 2,4,6-trinitrobenzenesulfonate (**68**) [Eq. (41)] depend on the salt and the solvent. In acetone, lithium perchlorate gave a normal salt effect with $b = 2.3$, and lithium 2,4,6-trinitrobenzenesulfonate (LiTNBS) gave a common ion rate depression. However, in AcOH, LiTNBS gave only a normal salt effect with $b = 2$, but $LiClO_4$ showed a "special salt effect" with $k_{ext}/k_o = 1.96$ and $[LiClO_4]_{1/2} = 0.00035$ $M$, superimposed on a normal salt effect with $b = 3.9$. According to Winstein's extended solvolysis scheme, the special salt effect indicates that "solvent-separated" ion pairs are involved in an ion-pair return to the covalent substrate and that their capture by $LiClO_4$ directs them into product-forming intermediates [150]. The $k_{ext}/k_o$ value indicates that half of the ions reaching the solvent-separated ion-pair stage return to covalent material without forming product. However, it is not clear whether the product is formed from the free (bridged) ion or from the ion pair preceding it.

The most common means of determining the intermediacy of ion pairs during the solvolysis of saturated compounds is the comparison of the titrimetric rate constants $k_t$ and the polarimetric rate constant $k_\alpha$ for the loss of optical activity of one enantiomer of the substrate [132, 135b]. The cationic part of an ion-pair intermediate can lose its chirality and return to the covalent material without giving products, and the products will usually be racemic. Hence, $k_\alpha$ will be higher than $k_t$, and the gap between them can be used to calculate the extent of ion-pair return [132].

This method usually cannot be used for the achiral vinylic systems, although it can be used when a chirality that is present in the molecule as a whole is lost during the vinyl cation formation. An elegant use of this method was the recent comparison of $k_t$ and $k_\alpha$ for an optically active allene [153] (Chapter 5, Section IV,E). This technique was recently applied to the solvolysis of the optically active 9-(α-bromoanisylidene)-10-hydroxy-9,10-dihydroanthracene **138** [138]. In TFE [138a] and in 80% EtOH [138b] buffered by 2,6-lutidine the rate determining step is the C-Br heterolysis, and the reaction in TFE also shows an extensive common ion rate depression with $\alpha = 3205$ $M^{-1}$ [138a] whereas in 80% EtOH $\alpha = 65$ $M^{-1}$ [138b]. In

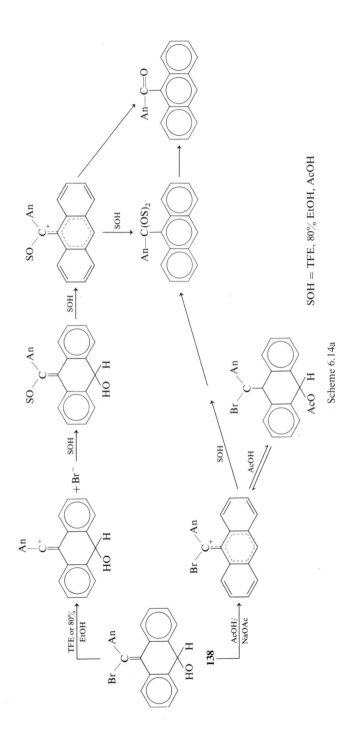

SOH = TFE, 80% EtOH, AcOH

Scheme 6.14a

both solvents $k_\alpha \sim k_t^\circ$ indicating that ion pair return with racemization is negligible.

In contrast, the loss of optical activity of **138** is much faster than the solvolysis in AcOH and in 1:1 AcOH—Ac$_2$O buffered by NaOAc: $k_\alpha/k_t = 136$. However, this is not due to an extensive ion pair return but to an initial ionization of the hydroxy group at the acidic media, followed by acetate capture at C-10 with the formation of the racemic acetate. Capture at C$_\alpha$ is much slower [138b]. The main mechanistic features of the two routes are given in Scheme 6.14a.

Rappoport and Apeloig suggested that another stereochemical probe could be applied and used the cis–trans isomerization of the precursor vinylic system, which is concurrent with the solvolysis, as a general tool for detecting ion-pair formation and for evaluating the extent of ion-pair return [36, 51, 55, 56, 130]. This approach is outlined in Scheme 6.14b [36].

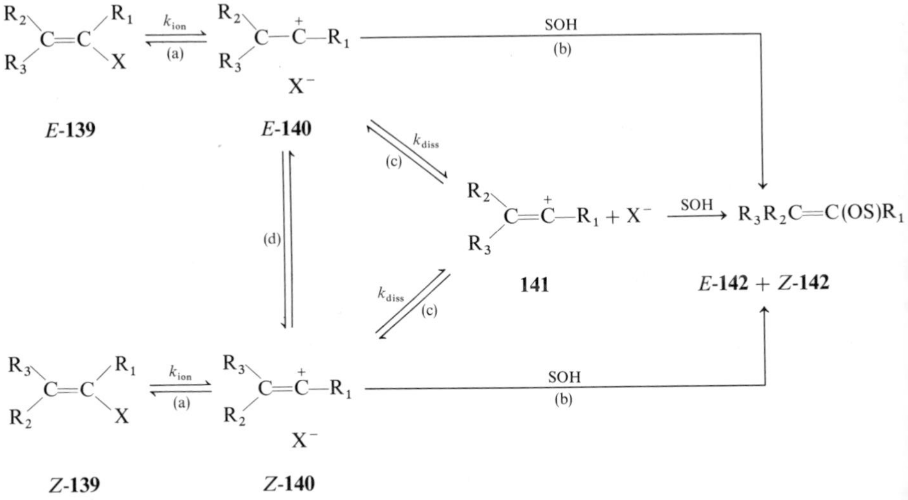

Scheme 6.14b

The first intermediate in the ionization ($k_{\mathrm{ion}}$) of the vinyl halide *E*-**139** is the ion pair *E*-**140**, which can participate in four reaction sequences: (a) return to *E*-**139**, which is hidden unless X$^-$ or R$^+$ is symmetric and ambident and return is to the isomeric labeled position; (b) reaction with the solvent to give the products *E*-**142** and *Z*-**142**, the ratio of which depends on the relative bulk of R$_2$ and R$_3$ and the extent of shielding of the cationic part by X$^-$; (c) dissociation ($k_{\mathrm{diss}}$) to X$^-$ and to the free ion **141**, which in turn can give products, return to the ion pairs *E*-**140** and *Z*-**140**, or be recaptured by X$^-$ to give both *E*-**139** and *Z*-**139**; and (d) isomerization to the isomeric ion pair Z-**140**, either by migration of X⁻ from one face of the cation to the other or by rotation of the ion along its axis. The two observable changes in the system are the formation of the products E-**142** and Z-**142** and a possible concurrent E-**139** ⇄ Z-**139** isomerization. By comparing the rate of the two processes, ion pairs could be detected and the distribution of E-**140**, Z-**140**, and **141** among the various routes could be determined. A simplified version of Scheme 6.14b was used in Chapter 5, where the stereochemical evidence for ion pairing was explained in terms of a faster solvent capture of either E-**140** or Z-**140** from the rear, as compared with their mutual interconversion [94, 119].

In spite of obvious similarities between Scheme 6.14b and the $k_\alpha - k_t$ probe in saturated systems, there are considerable differences between the use of the $d \rightarrow l$ racemization and the cis → trans isomerization as mechanistic tools [36]. The solvolysis rates of d-RX and l-RX are identical and the free trigonal ion is symmetric, so that the capture products by a nucleophile are racemized. In contrast, the solvolysis rates of E-**139** and Z-**139** may differ greatly (Table 6.11), and the capture rates of the unsymmetric ion **141** from both sides are also different. Since **141** is formed from both E-**139** and Z-**139**, excess retention is observed for one isomer and excess inversion for the other if products are formed from **141**. Moreover, nonsolvolytic addition–elimination isomerization routes are available for vinylic systems but not for saturated ones. Finally, isomerization takes place not only at the ion-pair stage, but also at the free ion stage. Hence, three prerequisites should be fulfilled before the cis-trans isomerization can be used as a mechanistic tool for ion-pair formation. First, isomerization routes that are not connected with the solvolysis (e.g., via addition–elimination) should be either ruled out or accounted for. Second, a system should be chosen in which the equilibrium constant for the E-**139**–Z-**139** pair is known and the difference in bulk of $R_2$ and $R_3$ is not too large; otherwise, one isomer will predominate so much in the mixture that its isomerization cannot be followed. Finally, isomerization via the free ion **141** should be either negligible or should be taken into account.

These prerequisites were not fulfilled for two systems in which cis–trans isomerization was observed during the reaction. The reaction profiles for the solvolysis–isomerization of E- and Z-1,2-dianisyl-2-chlorovinyl 2,4,6-trinitrobenzenesulfonates were recorded, but they were not used further for evaluating the extent of ion-pair return [84]. Isomerization of E- to Z-1-anisyl-1-bromopropenes (**53** → **52**, R = Me), in which the free-energy difference between the isomers is 1.65 kcal/mole, was observed during its solvolysis in 80% EtOH [90]. The intervention of two isomeric intimate ion pairs as the product-forming intermediates was suggested, but a more detailed analysis was not presented.

With two other systems, these prerequisites were fulfilled, and the method was applied. One was the $E$- and $Z$-$\alpha$-bromo-$\beta$-deuterio-$p$-methoxystyrenes **100** and **101**, respectively, in which the $Ad_E$–E route contributes little [22], if at all, to the solvolysis (Section II,B), the two $\beta$ substituents have essentially the same bulk, and common ion rate depression was not observed [56]. The other system was the 1,2-dianisyl-2-phenylvinyl system, in which the effects of the solvent and the leaving groups were extensively investigated [36, 51, 55, 130]. It was already shown that the solvolysis does not involve any $Ad_E$–E contribution, and the bulk of the $\beta$ substituents is nearly identical. However, the other prerequisites have to be dealt with in order to evaluate the ion-pairing phenomenon.

The acetolysis of *cis*- and *trans*-1,2-dianisyl-2-phenylvinyl bromides **83**, $X = Br$ (**83**-Br), and **84**, $X = Br$ (**84**-Br), respectively, and the *cis*-chloride **83**, $X = Cl$ (**83**-Cl), with and without added NaOAc shows an extensive common ion rate depression, and $k_t$ decreases extensively during the kinetic run or on addition of halide ion [36, 130]. By the use of Eq. (37), it was found that more than 94% of the products are derived from the free ion **85**. A concurrent extensive cis–trans isomerization of **83** to **84** or vice versa accompanies the solvolysis, and an equilibrium mixture of 54% cis acetate **86** (Nu = OAc) and 46% trans acetate **87** (Nu = OAc) is established during the solvolysis [36]. The reaction profile describing the concentration versus time changes for all the species is of a similar shape for the two isomers. The concentration of the starting halide decreases and that of the product acetates increases monotonously with time, whereas the concentration of the isomeric halide first increases and then decreases with the progress of the reaction.

A rate constant for the isomerization ($k_{\text{isom}}$) was defined in terms of Eq. (47), where $(\% \, \mathbf{84})_t$ and $(\% \, \mathbf{84})_\infty$ refer to the percentage of **84** in the RBr fraction at time $t$ and at infinity, respectively. It was found that $k_{\text{isom}}$ increased during the solvolysis, concurrently with the decrease in $k_t$.

$$k_{\text{isom}}^{83} = (2.3/t)\log(\% \, \mathbf{84})_\infty / [(\% \, \mathbf{84})_\infty - (\% \, \mathbf{84})_t] \qquad (47)$$

The appearance of common ion rate depression and the evidence for the linear cation **85** (Table 6.21) make it mandatory that the external ion return would lead to **83** $\rightleftarrows$ **84** isomerization, since capture of **85** from both sides should occur with equal probability. Moreover, from Table 6.21 it is clear that capture by $Br^-$ and $Cl^-$ would lead to an approximately 1:1 mixture of **83** and **84**. Scheme 6.9 for the vinyl bromides should then be modified to Scheme 6.15. When the $\alpha$ values were used to calculate the reaction profiles for all the stable species of Scheme 6.15, it was found that the experimental and the calculated points for the decrease in the precursor bromide and for the formation of **86**-OAc and **87**-OAc coincide (Fig. 6.2), but that more of

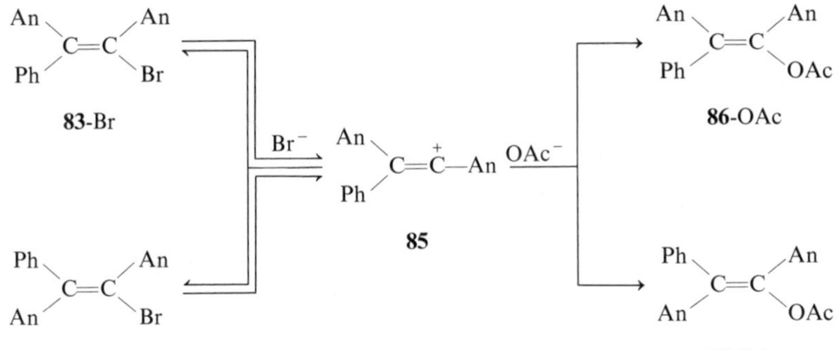

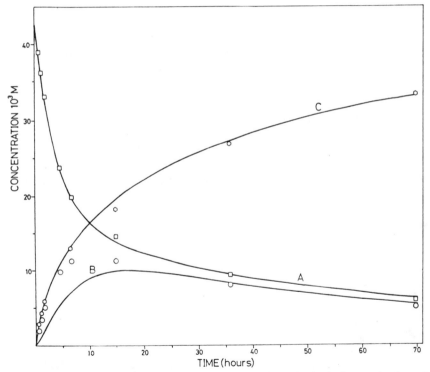

Scheme 6.15

**Fig. 6.2.** Concentration versus time profile for the solvolysis–isomerization of 0.044 $M$ **83**-Br with 0.087 $M$ NaOAc at 120.3°. The points are experimental [(○) **83**-Br (□) **84**-Br, (◔) **86**-OAc + **87**-OAc], and the lines [(A) for **83**-Br, (B) for **84**-Br, (C) for **86**-OAc + **87**-OAc] are theoretical and calculated by simulation of Scheme 6.14. [Reprinted by permission of the publisher from *J. Am. Chem. Soc.* **98**, 821 (1975).]

the isomeric bromide (e.g., **84**-Br from **83**-Br) is formed than is accounted for by the common ion rate depression. This extra isomerization should be ascribed either to an isomerization via a route that is irrelevant to the solvolysis or to the intervention of ion pairs the return of which proceeds with isomerization [36].

Although several nonheterolytic isomerization routes can be envisaged, the most likely one for electron-rich olefins such as $\alpha$-arylvinyl systems in an acidic medium is the $Ad_E$–E route [Eq. (48)], in which an electrophile, which

$$\underset{\substack{\\ \text{83—Br}}}{\overset{\text{An}}{\underset{\text{Ph}}{\diagdown}}C=C\overset{\text{An}}{\underset{\text{Br}}{\diagup}}} \underset{-H^+}{\overset{+H^+}{\rightleftharpoons}} \text{An} \cdots \underset{\substack{\text{An} \quad \text{Ph}}}{\overset{\text{H}}{\bigcirc}} \cdots \text{Br} \underset{\text{rotation}}{\rightleftharpoons} \text{An} \cdots \underset{\substack{\\ \text{H}}}{\overset{\text{Ph} \quad \text{An}}{\bigcirc}} \cdots \text{Br} \underset{+H^+}{\overset{-H^+}{\rightleftharpoons}} \underset{\substack{\\ \text{84—Br}}}{\overset{\text{Ph}}{\underset{\text{An}}{\diagdown}}C=C\overset{\text{An}}{\underset{\text{Br}}{\diagup}}}$$

(48)

adds to the double bond in the initial step, is eliminated after a free rotation in the cationic intermediate. There are ample precedents for this route [154]. This route was ruled out for the solvolysis of **83** and **84** in Section II,B. It should also be ruled out for the isomerization, since the solvent isotope effect $k_{isom}(AcOH)/k_{isom}(AcOD)$ is ca. 1.15 and since the $k_{isom}^0$ values ($k_{isom}$ extrapolated to zero reaction time) for X in *cis*-1,2-dianisyl-2-phenylvinyl-X are similar to the $k_t$ values and to the solvolysis values of saturated compounds and differ from the literature values for the relative rates for the $Ad_E$–E process (Table 6.30) [36]. Moreover, the isomerization of **83**-Br to **84**-Br is ten times faster in the presence of AgOAc than in the presence of

**TABLE 6.30**

Relative Reactivities[a] for the $Ad_E$ and the $S_N1$ Reaction[b]

| Group X | Relative $k_t$ (saturated RX) | Relative $k$ for $Ad_E$ | | | Relative $k_t$ for **83** | Relative $k_{isom}^0$ for **84** |
|---|---|---|---|---|---|---|
| | | A[c] | B[d] | C[e] | | |
| H | | 12 | 5000 | 11.4 | | 0.27 |
| Cl | 0.03 | 4.3 | 0.8 | 2.1 | 0.04 | 0.04 |
| Br | 1.0 | 1.0 | 1.0 | (1.0) | 1.0 | 1.0 |
| OAc | $<1.5 \times 10^{-5}$ | | | 314 | $<0.002$ | $<0.01$ |
| OMs | 85 | | | | 244 | 100 |

[a] Relative to Br.
[b] From Rappoport and Apeloig [36].
[c] Addition of $CF_3COOH$ to 2-X-propenes (Peterson and Bopp [57]).
[d] Bromination of $CH_2=C(X)CH_2Cl$ in 50% MeOH (Hooley and Williams [155]).
[e] Reactions of $AnC(X)=CH_2$ (Rappoport and Gal [22]).

NaOAc, whereas the isomerization of the corresponding ethylene **18**, which certainly proceeds via the $Ad_E$–E route, increases by only 15% with AgOAc. Hence, the isomerization of the vinyl bromide proceeds via an initial C—Br bond heterolysis.

The extra isomerization is therefore interpreted as being due to the formation of the ion-pair intermediates *E*-**143** and *Z*-**143** in addition to **85** along the solvolysis route.

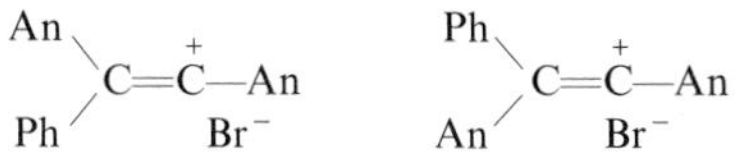

$$
\begin{array}{cc}
\underset{Ph}{\overset{An}{\diagup}}C\!\!=\!\!\overset{+}{C}\!\!-\!\!An & \underset{An}{\overset{Ph}{\diagup}}C\!\!=\!\!\overset{+}{C}\!\!-\!\!An \\
\qquad\quad Br^- & \qquad\quad Br^- \\
E\text{-}143 & Z\text{-}143
\end{array}
$$

Since the product formation is accounted for by **85** alone, these ion pairs are involved in the return and the dissociation processes, but not in the product-forming step. The rate of the interconversion of *E*-**143** and *Z*-**143** is unknown, and for all purposes the solvolysis scheme can be written as Scheme 6.16, in which **143** is a single ion pair replacing *E*-**143** and *Z*-**143**, and the superscripts refer to the species involved in the various specific reactions ($k_{Cl}$, $k_{Br}$, and $k_{OAc}$ are the rate constants for capture by $Cl^-$, $Br^-$, and $OAc^-$, respectively; $k_{ipr}$ and $k'_{ipr}$ are the rate constants for ion-pair return). Scheme 6.16 raises several questions. First, how should the ionization rate constant $k_{ion}$, which is higher than the $k_t$ of the simple Scheme 6.13, be evaluated? If $k_{ion}$ is known, the extent of ion-pair return can be estimated from the difference $k_{ion} - k_t{}^0$. Second, how can the involvement of the ion pair **143** in the reaction be independently verified, and how can its partition among the several reaction routes be calculated? Third, what is the exact nature of the ion pair and why is it not involved in the product-forming step?

$$
\begin{array}{c}
83 \underset{k_{ipr}^{143}}{\overset{k_{ion}^{83}}{\rightleftharpoons}} \\
\\
\quad\quad [R^+X^-] \underset{k_X^{85}}{\overset{k_{diss}^{143}}{\rightleftharpoons}} R^+ + X^- \overset{k_{OAc}^{85}}{\underset{OAc^-}{\longrightarrow}} 86\text{-OAc} + 87\text{-OAc} \\
\quad\quad\quad 143 \qquad\qquad 85 \\
\\
84 \underset{k_{ipr}'^{143}}{\overset{k_{ion}^{84}}{\rightleftharpoons}}
\end{array}
$$

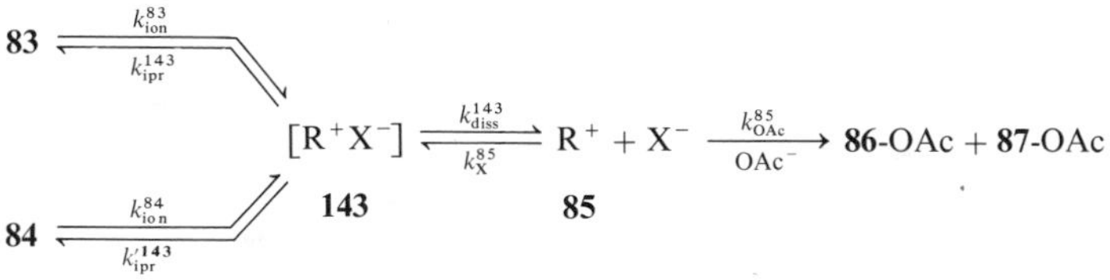

Scheme 6.16

The difficulty in evaluating $k_{ion}$ arises from the fact that it is given by the sum of two terms [Eq. (49)] [36]. The $k_t{}^0$ term accounts for the fraction of

$$k_{ion} = k_t{}^0 + k_{isom}^{143} \qquad (49)$$

ion pairs that dissociate further to **85** and end up as a product before return, and the $k_{isom}$ term accounts for the rest of the ion pairs, which return to covalent RX. However, $k_{isom}^{143}$ is not directly measured in a simple solvolysis experiment, since the measurable quantity is $k_{isom}$ (e.g., for 83) of Eq. (47), which is composed of the first-order constant $k_{isom}^{143}$ and the second-order rate constant for the bimolecular isomerization via reaction of 85 with Br$^-$ The latter constant is responsible for the increase in $k_{isom}$ during the reaction, since the [Br$^-$] term increases during the run. Four different methods were applied in order to overcome this difficulty [36].

According to Scheme 6.16, the reactions of the two intermediates can be decoupled if the free ion 85 will be completely captured by a nucleophile. The kinetics would then be much simplified, since a second-order capture of 85 with no common ion rate depression would be accompanied by a first-order isomerization via 143 alone. Attempts to capture 85 completely by increasing the acetate ion concentration failed for a practical reason: the extent of common ion rate depression is so high that the acetate concentrations required to overcome it were too high. However, when the solvolysis of 83-Br was conducted in the presence of high concentrations of LiCl, more than 90% of the products were the corresponding chlorides 83-Cl and 84-Cl in a 1:1 ratio. Significantly, although the return from 85 was practically excluded, the 83-Br ⇌ 84-Br isomerization was slower than in the absence of LiCl, but it still continued as a first-order process with a rate constant $k_{isom}^{143}$. By the use of a radioactive chloride ion, the capture rate of 85 ($k_{Cl}^{85}$) was determined. Since the two rate constants are independent and complementary, their sum together with the rate constant $k_t^{85}$ for the residual solvolysis should be equal to $k_{ion}$ (Table 6.31) [36].

**TABLE 6.31**

**The $k_{ion}$ Values Obtained by Different Methods for the Acetolysis of 83-Br in AcOH/NaOAc at 120.3°** [a]

| Method | $10^5 k_{ion}$ (sec$^{-1}$) | From |
|---|---|---|
| Chloride ion capture | 6.83 | $k_{Cl}^{85} + k_{isom}^{143} + k_t^{85}$ |
| Bromide ion capture | 7.23 | $k_{isom}^{85} + k_{isom}^{143} + k_t^{85}$ |
| Total cis content | 7.52 | $k_t + k_{isom}$ |

[a] From Rappoport and Apeloig [36].

Advantage can be taken of the large extent of common ion rate depression. In the presence of a large excess of bromide ion, the solvolysis is nearly completely suppressed and 143 and nearly all of 85 returns with isomerization to 83-Br and 84-Br. The sum of the rate constants for the major isomerization and the minor solvolysis should be identical to $k_{ion}$. Indeed, in the presence of a large excess of Bu$_4$NBr, $k_t$ accounts for less than 5% of the reaction, whereas a first-order isomerization is the main process observed [36].

A unique feature of the reactions of **83**-Br and **84**-Br is that the [**83**-Br]/ [**84**-Br] ratio of 54:46 at equilibrium is identical with the equilibrium ratio of the products, i.e., with [**86**-OAc]/[**87**-OAc]. It was shown that under these conditions the approach of the "total cis content" of the mixture, [**83**-Br] + [**86**-OAc], to this equilibrium value is a first-order process the rate constant of which is also $k_{ion}$. Moreover, $k_{ion}$ is then given by Eq. (50), where $k_t$ and $k_{isom}$ are the integrated rate constants for each kinetic point.

$$k_{ion} = k_{isom} + k_t \qquad (50)$$

The sum of these values, as obtained from the titrimetric and the isomer distribution data for each kinetic point, was indeed constant during a run. The basis for Eq. (50) is that the return of **85** to RX, which reduces $k_t$ within a run, gives a parallel increase in $k_{isom}^{85}$, so that their sum remains constant. Correction of the various rate constants for the normal salt effect gave the values of Table 6.31. The agreement between the $k_{ion}$ values, which are based on different processes, is good and strongly supports Scheme 6.16.

Simulation of Scheme 6.16 by the use of the best $k_{ion}$ value of Table 6.31, the best $\alpha$ value of Scheme 6.13 and Table 6.24, and the fraction of ion pairs that dissociate to the free ion **85** [$F$, Eq. (51)] as an adjustable parameter

$$F = k_{diss}^{143} /(k_{diss}^{143} + k_{ipr}^{143} + k_{ipr}'^{143}) \qquad (51)$$

gave an excellent fit between the experimental and the calculated reaction profiles (Fig. 6.3). A complete analysis of the partition of the two cationic intermediates is achieved by using the $\alpha$ values, the equilibrium constants for the **83**–**84** pair, and the fraction of ion-pair return $1 - F$. The values for **83**-Br and for **83**-Cl are given in Schemes 6.17 and 6.18, respectively. Since the partition of **85** between the forward and the reverse reactions depends on the Br$^-$ concentration, it was calculated for the arbitrary concentration of 1 $M$ Br$^-$ [36].

The remarkable feature of Schemes 6.17 and 6.18 is that at high concentration of Br$^-$ only a very small fraction of the initially formed cationic species ends up as the solvolysis product. For example, in the solvolysis of **83**-Br, nearly half of the cations return at the ion-pair stage, and 49% of the remaining 50%, which dissociate to the free ion, return by external ion return. Only 1% of the cationoid species gives the vinylic acetates [36].

The nature of the ion pair was investigated by carrying out the reaction in the presence of LiClO$_4$. The solvolysis rate is only slightly accelerated at LiClO$_4$ concentrations much higher than those usually required for a special salt effect [150], indicating that the ion pair is not a solvent-separated

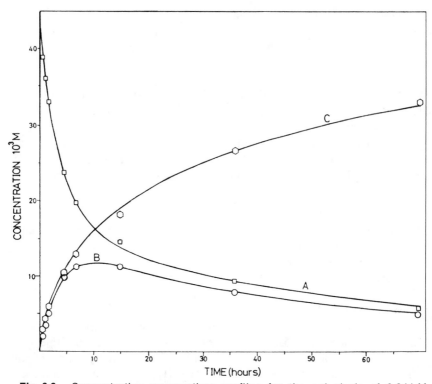

**Fig. 6.3.** Concentration versus time profiles for the solvolysis of 0.044 *M* **83**-Br with 0.087 *M* NaOAc in AcOH at 120.3°. The points are experimental [(○) **84**-Br, (□) **83**-Br, (◯) **86**-OAc + **87**-OAc], and the lines [(A) for **84**-Br, (B) for **83**-Br, (C) for **86**-OAc + **87**-OAc] are theoretical and calculated according to Scheme 6.17. [Reprinted by permission of the publisher from *J. Am. Chem. Soc.* **98**, 821 (1975).]

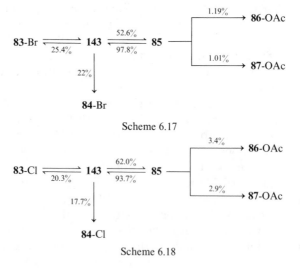

Scheme 6.17

Scheme 6.18

ion pair, but probably an intimate ion pair [135]. However, the isomerization rate increases 4.5 times with 0.03 $M$ LiClO$_4$, suggesting a lithium salt-promoted ionization and return from the ion pair **144**.

$$\begin{array}{c} An \\ \phantom{An} \diagdown \\ \phantom{AnAn} C{=}\overset{+}{C}{-}An \\ \phantom{An} \diagup \\ Ph \end{array}$$

$$ClO_4 {-} \overset{-}{Li} {-} Br$$

**144**

Behavior similar to that in AcOH was shown by **83**-Br, **84**-Br, and **83**-Cl in 1:1 AcOH–HCOOH and in 1:1 AcOH–Ac$_2$O [55]. The corresponding $\alpha$ and $1 - F$ values, which are given in Table 6.28, show a rather low sensitivity to the nature of the solvent. Whereas less ion-pair return is expected in a more dissociating and anion-solvating medium, this is not supported by the lower $1 - F$ value in 1:1 AcOH–Ac$_2$O than in 1:1 AcOH–HCOOH and by the similarity of the values for 1:1 AcOH–HCOOH and AcOH. Comparison with $1 - F$ values for saturated compounds is of little help since $1 - F$ (AcOH) values both higher [156] and lower [157] than $1 - F$ (HCOOH) values were reported. The low sensitivity of the $1 - F$ values to the change in the solvent supports a high degree of tightness in the ion pair.

The tightness of the ion pair and steric effects explain why products are not formed from the ion pairs in these media. A tight ion pair the cationic part of which is substituted by two $\beta$-aryl groups is highly shielded, from the front by the leaving group and from the rear by the $\beta$-aryl group, to approach of a solvent molecule in the plane of the vacant orbital. Both ion-pair return, which is a recombination of appropriately situated cation and anion, and dissociation, which is favored by a high dielectric constant, are apparently energetically less costly than the capture of the hindered ion pair by the solvent.

A dramatic change in the details of the process and in the partitioning of the ion-pair intermediate is observed when the dielectric constant of the carboxylic acid solvent becomes very low. In pivalic acid/sodium pivalate, a medium with a dielectric constant of ca. 2.6, the solvolyses of **25** (X = Br), trianisylvinyl bromide, **83**-Br, and **84**-Br proceed without common ion rate depression. Apparently, the dissociated ion is not formed, and products are fromed from the ion pair [51]. However, the **83**-Br $\rightleftarrows$ **84**-Br isomerization is much faster than the solvolysis, and $k_{isom}$ at 172.5° is 59 times higher than $k_t$ at 185.5°. The $1 - F$ value is therefore $\geq 0.99$; i.e., less than 1% of the initially formed ion pairs ends up as solvolysis products under conditions in which the free ion is not formed at all.

In this reaction, $k_t$ increases more than 10-fold and $k_{isom}$ only 1.5-fold in the presence of LiClO$_4$ [51]. This behavior is completely the opposite of that observed in AcOH [36] and is most likely due to a special salt effect [150] in which the products are formed from the solvent-separated ion pair. Nevertheless, the gap between $k_{ion}$ and $k_t$ is still large, and in the presence of 0.01 $M$ LiClO$_4$ only 11.5% of the ion pairs give products, whereas 88.5% return to covalent bromide. Whether return takes place only from a solvent-separated ion pair or also from an intimate ion pair is not known [51]. Consequently, when dissociation of the ion pair is not favored due to a low dielectric constant, the ion pair is still reluctant to give products, and ion-pair return becomes the dominant process.

A decrease in the dielectric constant causes a decrease in the ion-pair dissociation, but an increase in the bulk of the $\beta$ substituent should cause an increase in the ion-pair dissociation, since both the neutral precursor and the ion pair are sterically crowded and the crowding of the ion pair could be relieved by ion-pair dissociation. Only the $\beta$-anthronylidene derivative **62**, which is the record holder for common ion rate depression in AcOH, gives common ion rate depression also in pivalic acid, with $\alpha = 2.4$ (Table 6.24) [51]. In spite of the low $\alpha$, $\geq 80\%$ of the products are derived from the free vinyl cation **126a**. Since pivalate ion is larger than acetate ion and acetic acid probably forms better hydrogen bonds to the incipient bromide ion than does pivalic acid, the opposite is expected. The lower $\alpha$ value was ascribed to a smaller dissociation of NaBr in pivalic acid compared with that in AcOH; this results in a lower concentration of the free Br$^-$ ion, which is the active nucleophile in the return process [51].

High nucleophilicity of the solvent is another reason for product formation from ion pairs. As shown in Table 6.25, common ion rate depression is important in TFE but much less so in aqueous EtOH. The solvolysis rate of **83**-Br and **84**-Br in 80% EtOH does not decrease significantly within a run, but the rate is depressed by added Et$_4$NBr, giving $\alpha = 3$ liters/mole and showing that more than 21% of the products are derived from **85** [55]. Apparently, the much better nucleophilic solvent is probably able to capture even the ion pair, in spite of the higher dielectric constant. A first-order isomerization accompanies the solvolysis and should therefore proceed via the ion pair. The solvolysis in the presence of an extremely strong nucleophile, the benzyl thiolate ion, gives a 1 : 1 mixture of the 1,2-dianisyl-2-phenylvinylbenzyl thiolates **86** and **87** (Nu = PhCH$_2$S), but the isomerization rate decreases by only about one-third. It seems that most of the isomerization proceeds via an intimate ion pair, which is only partially capturable, whereas the solvolysis products are formed from the free ion and probably also from a solvent-separated ion pair [55].

The $1 - F$ value, calculated from the difference between $k_{ion}$ and $k_t$, is 0.32. This value does not differ much from the values in the other solvents, although it is lower than the value in 1 : 1 AcOH–HCOOH, a solvent that has a similar dielectric constant but much lower nucleophilicity, arguing again that some product is formed from the ion pairs in the nucleophilic solvent. Table 6.25 shows that the bulkier anthronylidene system **62** gives products in 80% EtOH mainly or exclusively from the free ion, as also observed in pivalic acid. Consequently, products are obtained from a more advanced intermediate in Winstein's solvolysis scheme either when the solvent becomes more dissociating and less nucleophilic or when the vinylic substituents become bulkier. This is summarized in Table 6.32.

**TABLE 6.32**

**Solvent and Structural Effects on the Products from AnC(Br)=CR$_1$R$_2$**

| Solvent | Dissociation power | Nucleophilicity | Products formed from free ion when R$_1$, R$_2$ are bulkier than | Reference |
|---|---|---|---|---|
| 80% EtOH | High | High | Ar | [55] |
| TFE | High | Very low | Me | [95] |
| AcOH | Moderate | Low | Me | [22] |
| Me$_3$CCOOH | Low | Low | Anthronylidene | [51] |

Extreme steric hindrance to capture of the free ion resulted in a preference for cyclization over external ion return [54], and ion pair return is also affected by this factor. Reaction of the two isomers of 1,2-dianisyl-2-(o-methoxyphenyl)vinyl bromides **145** and **146** in AcOH gave partial isomerization by internal reutrn but mainly cyclization [Eq. (52)], whereas no isomerization was observed in 80% EtOH [91].

The more hindered E- and Z-1-anisyl-2-(2,5-dimethoxyphenyl)-2-(o-methoxyphenyl)vinyl bromides (**148** and **149**, respectively) showed no common ion rate depression and no isomerization in either 80% EtOH or AcOH.

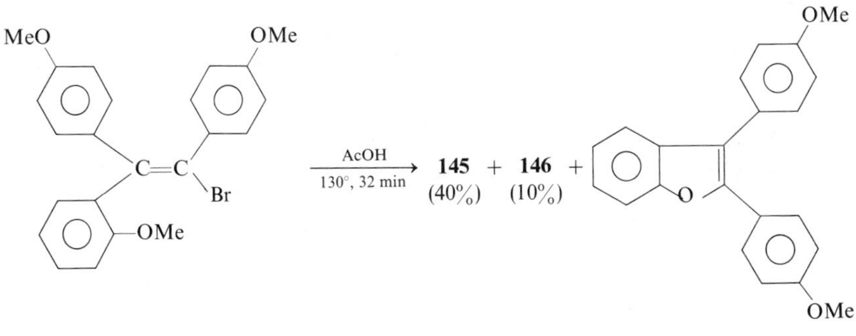

**145**

$$\xrightarrow[130°, \; 32 \; \text{min}]{\text{AcOH}} \quad \textbf{145} \; + \; \textbf{146} \; +$$
$$(40\%) \quad (10\%)$$

**147**
(50%)

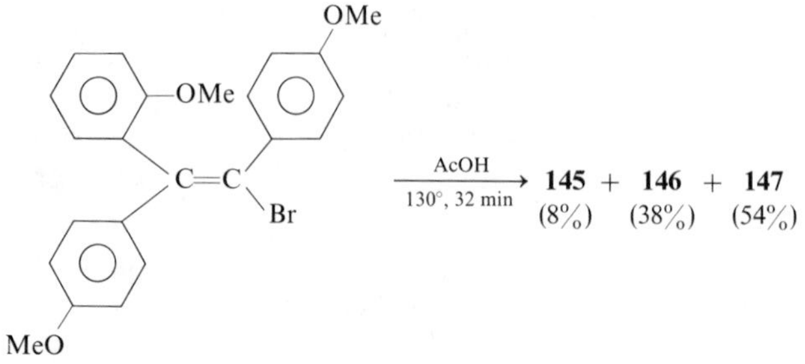

**146**

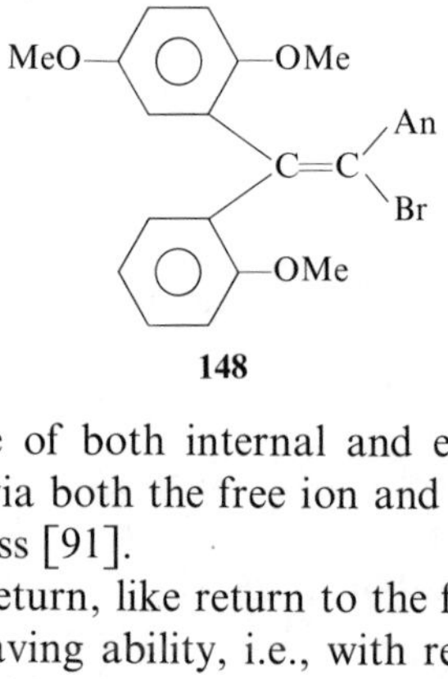

**148** 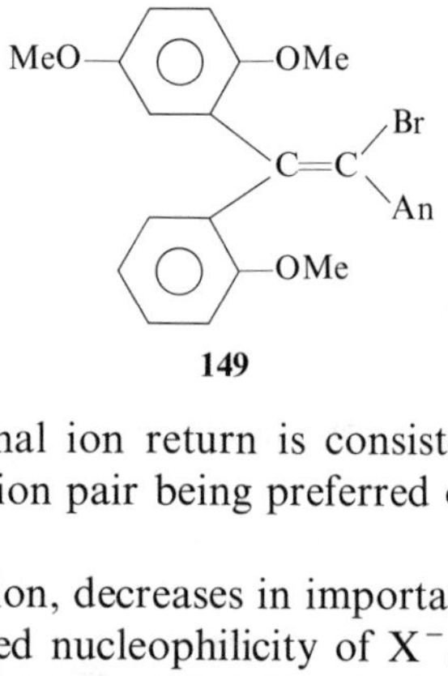                              **149**

The absence of both internal and external ion return is consistent with cyclization via both the free ion and the ion pair being preferred over any return process [91].

Ion-pair return, like return to the free ion, decreases in importance with increased leaving ability, i.e., with reduced nucleophilicity of $X^-$. Consequently, the kinetics become simpler, and $k_{ion}$ and $1 - F$ are calculated in

**TABLE 6.33**

**Behavior of $k_t$ and $k_{isom}$ for AnC(Pn)=C(X)An**

| $k_t$ | $k_{isom}$ | $k_{ion} =$ | Example[a] | Reference |
|---|---|---|---|---|
| Decreases[b] | Increases[b] | $k_t + k_{isom}$ | RBr/AcOH | [36] |
| Constant[b] | Constant[b] | $k_t + k_{isom}$ | RBr/80% EtOH | [55] |
| Decreases[c] | Constant | $k_t^0 + k_{isom}$ | RBr/80% EtOH | [55] |
| Constant | Constant | $k_t + k_{isom}$ | ROMs/AcOH | [36] |
| Constant | $0^d$ | $\sim k_t$ | ROMs/80% EtOH | [55] |
| $0^e$ | Constant | $\sim k_{isom}$ | RBr/Me$_3$CCOOH | [51] |

[a] R = AnC(Ph)=C(An)—.
[b] Within a run.
[c] By added $X^-$.
[d] Compared with $k_t$.
[e] Compared with $k_{isom}$.

several cases directly from $k_{isom}$ and $k_t$ without any extrapolations. This diversity of kinetic behavior is shown in Table 6.33.

The solvolysis of the vinyl mesylates 83-OMs and 84-OMs in AcOH/NaOAc gave a lower extent of both common ion rate depression and isomerization, but more than 28% of the products were formed from the free ion 85 [36]. However, when bromide ion was added to the solvolyzing mixture, the products were mainly 83-Br and 84-Br. The ([83-Br] + [84-Br])/([86-OAc] + [87-OAc]) ratio formed in this experiment should be identical with the $\alpha$ of Table 6.24 if the product-forming intermediate is 85. The $\alpha$ value of 17 obtained in this way is reasonably close to that calculated from the common ion rate depression. The partitioning of the intermediates is summarized in Scheme 6.19 [36].

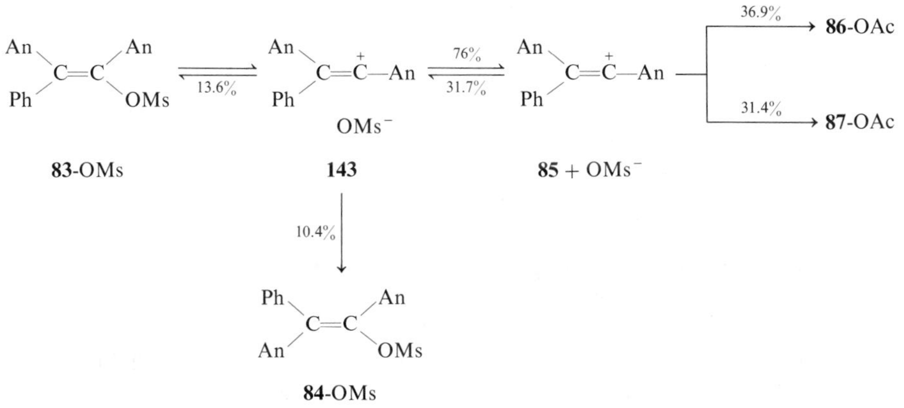

Scheme 6.19

The solvolysis–isomerization of the 1,2-dianisyl-2-phenylvinyl system gives a rare opportunity to compare the ion-pair return and the external ion return as a function of the leaving group in the same system. Table 6.34 shows

**TABLE 6.34**

**Ion-Pair Return and External Return in the Reaction of**
**cis-1,2-Dianisyl-2-phenylvinyl-X in AcOH/NaOAc$^a$**

| X | Temp (°C) | $10^5 k_{ion}$ (sec$^{-1}$) | $1 - F$ | $\alpha$ | Return from R$^+$ (%)$^b$ |
|---|---|---|---|---|---|
| Br | 120.3 | 7.52 | 0.47 | 21 | 97.8 |
| Cl | 120.3 | 0.31 | 0.38 | 5.7 | 93.7 |
| OMs | 75.1 | 13.6 | 0.24 | 0.16 | 31.7 |

$^a$ From Rappoport and Apeloig [36].
$^b$ For [X$^-$] = [OAc$^-$].

that the $\alpha$ values and the $1 - F$ values follow the same order: $Br > Cl > OMs$. As with the solvent effect, the sensitivity of the $\alpha$ values to the structural change is much higher than that of the $1 - F$ values. The order of the $1 - F$ values resembles those for the norbornyl system $[Br(0.96) > Cl(0.91)]$ [135b, 156c] and for the cyclobutyl system $[Cl(0.43) > OMs(0.20)]$ [158]. The $\alpha$ and $1 - F$ values are inversely related to the nucleophilicity of the anions $X^-$ in AcOH and to their solvation: the less solvated anion is more nucleophilic. The much lower sensitivity of the $1 - F$ values indicates that solvation of the anionic part of the ion pair is reduced compared with that of the free ion.

A weakly nucleophilic anion $X^-$ is unable to compete with a highly nucleophilic solvent. Consequently, the solvolysis of **83**-OMs and **84**-OMs in 80% EtOH proceeds without common ion rate depression and is not accompanied by a cis–trans isomerization [55].

The solvolysis–isomerization probe was used much less extensively for the reactions of $E$- and $Z$-$\alpha$-bromo-$\beta$-deuterio-$p$-methoxystyrenes (**100** and **101** respectively) [56]. Common ion rate depression is not a problem with these compounds, and $k_{ion}$ is again the sum of $k_t$ and $k_{isom}$. In 50% EtOH, isomerization was not observed. In AcOH, a **100** $\rightleftarrows$ **101** isomerization accompanied the solvolysis (Scheme 6.20), and $1 - F$ was 0.63 or 0.68 after correction for a possible isotope effect; i.e., approximately two-thirds of the ion pairs return to the covalent bromide. The absence of $\beta$-hydrogen exchange in AcOH indicates that an $Ad_E$–E contribution to the isomerization is small, if any exists at all. The nature of the products is discussed in Section II,B.

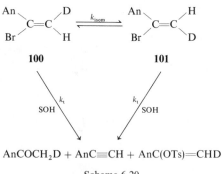

$$AnCOCH_2D + AnC{\equiv}CH + AnC(OTs){=}CHD$$

Scheme 6.20

These results are in interesting contrast to those for the 1,2-dianisyl-2-phenylvinyl system. The sterically unhindered ion pair **150** is captured by the nucleophilic 80% EtOH with preference over both dissociation and ion-pair return. In the less nucleophilic AcOH, ion-pair return becomes a

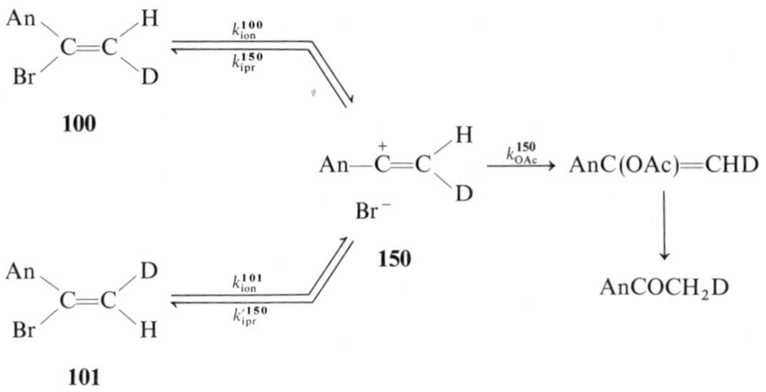

An, H / C=C / Br, D

**100**

$k_{ion}^{100}$
$k_{ipr}^{150}$

$$An-\overset{+}{C}=C\overset{H}{\underset{D}{\diagdown}} \quad \xrightarrow{k_{OAc}^{150}} \quad AnC(OAc)=CHD$$
$$Br^-$$

**150**

$\downarrow$

AnCOCH$_2$D

An, D / C=C / Br, H

**101**

$k_{ion}^{101}$
$k_{ipr}'^{150}$

Scheme 6.21

dominant process. In terms of Scheme 6.21, which describes the reaction [56], $k_{ipr}^{150} + k_{ipr}'^{150} > k_{AcO}^{150}$.

That the $1 - F$ value of 0.63 for **150** exceeds the value of 0.47 for the more crowded **143** at first seems surprising. However, the $1 - F$ values for the two systems measure different processes. Ion-pair return and dissociation compete for **143**, but ion-pair return, solvent capture, and probably some dissociation compete for **150**. Apparently, the capture of **150** is not sufficient to compensate for the sterically enhanced dissociation and for the sterically inhibited return of **143**, making the return more important for **150**.

## C. Product Distributions from the Cationic Intermediates

The observed products of vinylic solvolysis depend on the system and the solvent investigated. When a hydrogen substitutes a carbon adjacent to the postively charged α-carbon of the vinyl cation, partitioning of the intermediate between an S$_N$1 route and an E1 route takes place. Several products can be obtained for each of these routes. The E1 elimination can lead to both alkynes and allenes. The S$_N$1 reaction in solvent mixtures leads to substitution products from reaction with the solvent components. Rearranged and nonrearranged products can be obtained by capture of the intermediate ion by nucleophiles at different sites.

Some of the initially formed products are unstable in the reaction medium and are converted to other products during or after the reaction. For example, the alkyne formed in a carboxylic acid medium can add the solvent, giving the substitution rather than the elimination product of the vinylic precursor [17, 22] (Section II,B). Vinyl esters are hydrolyzed to the corresponding ketones [22, 60], especially when the β substituents are not bulky. On the other hand, hydrolysis is slow if the system is crowded, and the

1-anisyl 2,2-disubstituted vinyl pivalates formed under the pivalolysis conditions were stable toward hydrolysis [51]. Vinyl ethers and vinyl trifluoroethyl ethers are unstable toward hydrolysis to the corresponding ketones in a neutral or even a weakly basic medium if the $\beta$ substituents are not bulky enough [5, 95b], but hydrolysis of $\alpha$-phenyl-$\beta,\beta$-diarylvinyl trifluoroethyl ethers to the corresponding ketones proceeds with difficulty [8].

The partioning between the $S_N1$ and the E1 routes will be discussed first, and the selectivity of the cation toward the solvent components will be discussed later. The formation of rearranged products is discussed in Section VI.

In the cases in which it can be shown that the elimination product does not arise from an E2 reaction, the substitution/elimination ratio depends on the structure of the vinyl cation, on the solvent, and on the basicity of the medium.

Since an alkyne is sterically much less crowded than a substituted ethylene, it is expected that an increase in the bulk of the substituents on the vinyl cation will increase the elimination/substitution ratio. Indeed, under nearly identical conditions the percentage of the alkyne 151 [Eq. 53] in basic 80% EtOH is 11% from $\alpha$-bromo-$p$-methoxystyrene (21, R = H) [2, 73], 20%

$$\text{ArC(Br)=CHR} \longrightarrow \text{Ar}\overset{+}{\text{C}}\text{=CHR} \longrightarrow \text{ArC≡CR} \tag{53}$$

$$151$$

from $\alpha$-(9-anthryl)vinyl chloride (23, R = 9-anthryl) [17], 24% from $\alpha$-bromo-$p$-methoxy-$\beta$-methylstyrene (53, R = Me) [90], 43% from $\alpha$-bromo-$\beta$-$tert$-butyl-$p$-methoxystyrene (52, R = $t$-Bu) [73], and 63% from $E$-$\alpha$-bromo-$p,p'$-dimethoxystilbene (43, R = An) [31]. Likewise, the percentage of the alkyne formed from $\alpha$-arylvinyl bromides and tosylates increases from 10% for the $p$-methyl derivative 48b and 30% for the 2,4-dimethyl derivative 48c to 35% for the 2,4,6-trimethyl derivative 48d in aqueous ethanol buffered by sodium acetate [15].

The pH of the solution and the nature of the base have a profound effect on the elimination/substitution ratio, as shown by the work of Grob and Pfaendler on the solvolysis of $p$-amino-$\alpha$-bromostyrene (15) [46]. In aqueous buffer at pH 3.8 to 13.0 the rate constant for the hydrolysis of 15 is independent of the pH. However, the percentage of $p$-aminophenylacetylene (17) increases from 16% at pH 3.9 to 85% at pH 13.1 at the expense of the $S_N1$ product, $p$-aminoacetophenone (16) (Table 6.35). The percentage of the alkyne also increases with an increase in the concentration of a phosphate buffer, and at pH 8.6 it increases from 33% to 76% when the buffer concentration increases from 0.001 $M$ to 0.01 $M$. Consequently, the elimination step from the vinyl cation seems to be general base-catalyzed and is more sensitive

TABLE 6.35

Effect of pH on the Rates and Products of Solvolysis of
p-Amino-α-bromostyrene (15) in 50% Dioxane (v/v) at 20°

| pH | Buffer | $10^3 k_1$ (sec$^{-1}$) | pH | 16 (%) | 17 (%) |
|----|--------|------|------|--------|--------|
| 0.2 | | <0.03 | | | |
| 1.2 | | 0.15 | | | |
| 2.3 | | 1.4 | | | |
| 3.8 | Citrate | 4.1 | 3.9 | 84 | 16 |
| 5.6 | Citrate | 4.4 | 6.3 | 81 | 19 |
| 6.6 | Citrate | 4.1 | | | |
| 7.6 | Citrate | 4.1 | 8.0 | 75 | 25 |
| 8.6 | Phosphate | 4.3 | 8.7 | 65 | 35 |
| 9.6 | Borate | 4.3 | | | |
| 10.6 | Borate | 4.1 | 10.7 | 44 | 56 |
| 11.6 | Borate | 4.1 | | | |
| 12.1 | Phosphate | 4.4 | | | |
| 13.0 | Phosphate | 4.3 | 13.1 | 15 | 85 |

to the nature of the base than is the capture step. The large response to the nature of the base in the fast step of the process is rare and again suggests a relatively long lived intermediate.

Similarly, in the solvolysis of Z-α-bromo-p-methoxy-β-methylstyrene (52, R = Me), the percentage of the anisylmethylacetylene in the product increases from 15% with 1 equivalent of Et$_3$N to 29% with 5 equivalents of Et$_3$N. A smaller effect was found for the E isomer 53 (R = Me) [90].

The dependence on the nature of the base is exemplified by the first-order reactions of E-α-bromo-β-tert-butyl-p-methoxystyrene (53, R = t-Bu), in which elimination to anisyl-tert-butylacetylene in 50% EtOH accounts for 25% of the product with NaOAc but for 43% of the product with NaOAc [73].

In a binary mixture of nucleophilic solvents, the products are derived from both solvent components. Ketones and ethyl ethers are formed in aqueous ethanol, but a systematic study of their ratio has not been conducted, although it depends on the solvent composition, the nature and the concentration of the base, the temperature, the presence of silver salt, and the bulk of the β substituents. For example, the β,β-unsubstituted α-bromo- and α-chloro-p-methoxystyrenes 21 and 22 gave only the ketone and the alkyne in 80% EtOH, and no vinyl ether was observed [22]. The β,β-dimethyl derivative 25 (X = Br) gave both the ether and the ketone [Eq. (54)] in 80% EtOH

$$R_1R_2C{=}C(Br)An \xrightarrow{\text{EtOH-H}_2\text{O}} R_1R_2CHCAn + R_1R_2C{=}C(OEt)An \qquad (54)$$

with the carbonyl group (O) shown above the C of CHCAn.

in a 1.2 ratio [55]. The mono-$\beta$-methyl derivatives gave the corresponding alkyne, ether, and ketone, and the ether/ketone ratio was dependent on the base concentration. The $E$ isomer **53** (R = Me) gave 40% of the ketone and 35% of the ether with 1 equivalent of $Et_3N$, but the ether/ketone ratio increased progressively with an increase in the base concentration, and with 5 equivalents of $Et_3N$ the ratio was 2. This behavior is more pronounced for the $Z$ isomer **52** (R = Me); only the ketone was observed in the presence of 1 equivalent of $Et_3N$, whereas the ether/ketone ratio was 0.87 in the presence of 5 equivalents of the base [90]. The $\beta,\beta$-diphenyl derivative **9** (Ar = An; X = Br) gave a 1:1 ratio of the ether to the ketone in refluxing basic 80% EtOH, but the ether was the major product in the presence of silver carbonate [5, 30]. The increase in the ether/ketone ratio with the increased mole fraction of ethanol in aqueous ethanol was observed for the $\beta$-fluorenyl-idene derivative **10** (Ar = An; X = Br); the ratios were 1.08, 2.33, and 11.5 in 70, 80, and 90% ethanol, respectively [30]. Nevertheless, the $\alpha$-phenyl-substituted derivatives **88** and **89** and their $\beta,\beta$-diphenyl and dianisyl analogues gave only the ketones in 60% EtOH [8].

Some of these results are easily understood if part or all of the initially fromed vinyl ether hydrolyzes to the ketone during the reaction. This hydrolysis becomes significant when the $\beta$ substituents are small (e.g., for **21** and **22**) or when a high reaction temperature or a long reaction time is required, as is the case with the $\alpha$-phenyl derivatives or with **52** (R = Me). Since ethanol is more nucleophilic than water [109], more ether is expected to be formed initially when the concentrations of the solvent components are taken into account. This was observed in those cases in which the $\beta$ substituents are bulky enough so that the hydrolysis is relatively slow. In 80% EtOH the [EtOH]/[$H_2O$] ratio is 1.23, but the ether/ketone ratios are 2.33 and 4.6 [92] for the $\beta$-fluorenylidene (**10**) and $\beta$-anthronylidene (**62**) derivatives, respectively.

More systematic data on the product distribution are available for the solvolysis of 1-anisyl-2-methylpropen-1-yl tosylate (**75**) and its *ortho*-methoxy analogue (**69**) in aqueous trifluoroethanol [95b]. The data for **69** are given in Table 6.20 and show a progressive decrease in the percentage of the ether with the increase in the water content of the medium. Hydrolysis of the vinyl trifluoroethyl ether **71** to the ketone **70** takes place even in weakly basic media, and the values in Table 6.20 were obtained under conditions in which the hydrolysis was unimportant. The ratios of capture of the vinyl cation by the solvent components ($k_{H_2O}/k_{TFE}$) were calculated by using Eq. (44) and were reasonably constant in the region of 80 to 100% TFE. The values were lower for **63**, e.g., $k_{H_2O}/k_{TFE} = 0.30$ in 90% TFE [95b]. The ratios for both compounds were lower at higher molar percentages of water. Since water is a better nucleophile than TFE [109], it is surprising that the values are lower than unity. This may be due to the intervention of ion pairs in these reactions (Section V,A), but a specific solvation of the ion and collapse of the solvation shell may be an alternative explanation.

## VI. β-ARYL REARRANGEMENT IN TRIARYLVINYL SYSTEMS

Neighboring-group participation and rearrangement of heteroatoms from the β position during the solvolysis of α-arylvinyl systems, as well as β-hydrogen and β-aryl rearrangements during the solvolysis of α-arylvinyl systems, are discussed in Chapter 7. This section is concerned with the most extensively studied group of rearrangements, which are β-aryl migrations across the double bond in solvolytically generated triarylvinyl cations. Two types of rearrangements have been observed. When the product ion is more stable than the precursor ion, the rearrangement is nondegenerate whereas, when the stability of the product ion is identical to that of the parent ion, the rearrangement is degenerate. These rearrangements are facilitated by the shorter $C_\alpha$—$C_\beta$ bond as compared with the same bond in the corresponding saturated analogues. Models of triarylvinyl cations indeed show that the β-aryl group is in close vicinity to the potential migration terminus.

The intermediate or the transition state of the β-aryl rearrangement is a substituted vinylidenephenonium ion. The energy of the parent ion at different geometries was calculated and compared to that of the nonbridged analogues. Early calculations by Hoffmann and co-workers [159] suggested that the bridged ion should be stable. More recent MINDO/2 calculations showed that the charge delocalization in the vinylidenephenonium ion is lower than in the corresponding saturated ion [160]. These calculations also suggested that structure **152b**, with a planar geometry at the bridging carbon, is only 15.5 kcal/mole less stable than structure **152a**, in which the bridging carbon has the normal tetrahedral geometry. EHT calculations gave a difference of 25 kcal/mole [161], but more recent *ab initio* calculations by Collins and co-workers at the RHF/STO-3G level [162] gave a much higher difference, **152a** being 102 kcal/mole more stable than **152b**. By means of similar calculations, **152a** was also compared with the nonbridged structures. The bridged structure was found to be less stable than the α-phenylvinyl cation by 30.7 kcal/mole but more stable than the planar β-phenylvinyl cation

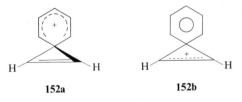

152a          152b

by 9.6 kcal/mole as well as the perpendicular $\beta$-phenylvinyl cation by 15.4 kcal/mole [87]. Although these values cannot be extrapolated to the tri-arylvinyl cations, they indicate that $\beta$-aryl rearrangements in these ions are feasible and that anchimeric assistance by the $\beta$-aryl group may be observed.

$\beta$-Aryl rearrangements across the double bonds in triarylvinyl cations [Eq. (55)] raise several questions about the mechanism of reaction, some of which are interconnected [8, 9, 163, 164]. (1) What is the driving force for the

$$Ar_3Ar_2C{=}C(Br)Ar_1 \longrightarrow Ar_3Ar_2C{=}\overset{+}{C}{-}Ar_1 \longrightarrow Ar_1Ar_2C{=}\overset{+}{C}{-}Ar_3 + Ar_1Ar_3C{=}\overset{+}{C}{-}Ar_2$$

$$(55)$$

rearrangement, especially when the rearrangement is degenerate? (2) Does the rearranging group participate in the transition state of the heterolysis or does the initial ionization give an unrearranged vinyl cation? (3) What is the role of free ions and ion pairs in the rearrangements? (4) What is the stereochemistry of the rearrangement? (5) Does the migratory aptitude of the migrating groups follow the order established for rearrangements in saturated compounds? (6) What are the relative rates for the rearrangement and the capture of the ion by the solvent and the leaving group? (7) What are the effects of the nonmigrating groups at the migration origin and at the migration terminus on the relative extent of the rearrangement? (8) What is the effect of solvent on these relative reactivities? (9) Can a rapid, reversible $\beta$-aryl rearrangement be the reason for the high selectivity of triarylvinyl cations? These questions were answered by a study of triarylvinyl systems substituted by all the combinations of anisyl and phenyl groups [8, 9, 51, 62, 64, 65, 163–167] and were recently extended by the study of several tolyl-substituted systems [35, 52, 168].

Eight different substrates (**153–160**) comprise all the combinations of phenyl and anisyl groups in triarylvinyl bromides. Two of these (**153** and **160**) have three identical aryl groups, and the other six compounds contain two pairs of geometrical isomers. Four of the substrates are $\alpha$-anisyl-substituted and four are $\alpha$-phenyl-substituted. These substrates can be in-volved in eight different rearrangements [Eqs. (56–63)]. Degenerate re-arrangements are indicated by the labeling of one of the aryl groups with an asterisk.

The initial ionizations lead to six different ions (**161–166**), which then rearrange to eight rearranged ions (**161\*–166\***, **162a**, and **165a**). Rearranged ions are labeled with an asterisk or with the suffix a.

Four of the rearrangements are $\beta$-phenyl rearrangements, and four are $\beta$-anisyl rearrangements. Two rearrangements [Eqs. (60) and (62)] lead from an $\alpha$-anisylvinyl cation to the less stable $\alpha$-phenylvinyl cation, and two rearrangements [Eqs. (57) and (59)] lead from an $\alpha$-phenylvinyl cation to an $\alpha$-anisylvinyl cation. The other four rearrangements are degenerate.

Rearrangements that lead from the more stable α-anisylvinyl cations **164** and **165** to the less stable α-phenylvinyl cations **162a** and **163\*** were not observed. This is in contrast to the deamination of the saturated analogous amines, in which the 1-anisyl-2,2-diphenylethyl cation gave 0.9% rearrangement to the 2-anisyl-1,2-diphenylethyl cation [169]. A similar small extent of rearrangement may escape detection in the vinylic systems.

$$
\begin{array}{c}
\underset{Ph}{\overset{Ph^*}{\diagdown}}C{=}C\underset{Br}{\overset{Ph}{\diagup}} \longrightarrow \underset{Ph}{\overset{Ph^*}{\diagdown}}C{=}\overset{+}{C}{-}Ph \overset{\sim Ph}{\rightleftharpoons} Ph^*{-}\overset{+}{C}{=}C\underset{Ph}{\overset{Ph}{\diagup}} \xrightarrow{SOH} products
\end{array}
\quad (56)
$$

**153**  **161**  **161\***

$$
\begin{array}{l}
\underset{An}{\overset{Ph^*}{\diagdown}}C{=}C\underset{Br}{\overset{Ph}{\diagup}} \\
\textbf{154} \\
\underset{Ph^*}{\overset{An}{\diagdown}}C{=}C\underset{Br}{\overset{Ph}{\diagup}} \\
\textbf{155}
\end{array}
\Bigg\} \rightarrow
\underset{An}{\overset{Ph^*}{\diagdown}}C{=}\overset{+}{C}{-}Ph \;
\textbf{162}
$$

$$
\xrightarrow{\sim Ph^*} An{-}\overset{+}{C}{=}C\underset{Ph}{\overset{Ph^*}{\diagup}} \xrightarrow{SOH} products \quad \textbf{164\*} \quad (57)
$$

$$
\xrightarrow{\sim An} Ph^*{-}\overset{+}{C}{=}C\underset{An}{\overset{Ph}{\diagup}} \xrightarrow{SOH} products \quad \textbf{162\*} \quad (58)
$$

$$
\underset{An}{\overset{An}{\diagdown}}C{=}C\underset{Br}{\overset{Ph}{\diagup}} \longrightarrow \underset{An}{\overset{An}{\diagdown}}C{=}\overset{+}{C}{-}Ph \xrightarrow{\sim An} An{-}\overset{+}{C}{=}C\underset{An}{\overset{Ph}{\diagup}} \xrightarrow{SOH} products \quad (59)
$$

**156**  **163**  **165a**

$$
\underset{Ph}{\overset{Ph}{\diagdown}}C{=}C\underset{Br}{\overset{An}{\diagup}} \longrightarrow \underset{Ph}{\overset{Ph}{\diagdown}}C{=}\overset{+}{C}{-}An \xrightarrow{\sim Ph} Ph{-}\overset{+}{C}{=}C\underset{An}{\overset{Ph}{\diagup}} \xrightarrow{SOH} products \quad (60)
$$

**157**  **164**  **162a**

$$
\begin{array}{l}
\underset{Ph}{\overset{An^*}{\diagdown}}C{=}C\underset{Br}{\overset{An}{\diagup}} \\
\textbf{158} \\
\underset{An^*}{\overset{Ph}{\diagdown}}C{=}C\underset{Br}{\overset{An}{\diagup}} \\
\textbf{159}
\end{array}
\Bigg\} \rightarrow
\underset{Ph}{\overset{An^*}{\diagdown}}C{=}\overset{+}{C}{-}An \;
\textbf{165}
$$

$$
\xrightarrow{\sim Ph} An^*{-}\overset{+}{C}{=}C\underset{Ph}{\overset{An}{\diagup}} \xrightarrow{SOH} products \quad \textbf{165\*} \quad (61)
$$

$$
\xrightarrow{\sim An} Ph{-}\overset{+}{C}{=}C\underset{An^*}{\overset{An}{\diagup}} \xrightarrow{SOH} products \quad \textbf{163\*} \quad (62)
$$

$$
\underset{An}{\overset{An^*}{\diagdown}}C{=}C\underset{Br}{\overset{An}{\diagup}} \longrightarrow \underset{An}{\overset{An^*}{\diagdown}}C{=}\overset{+}{C}{-}An \overset{\sim An}{\rightleftharpoons} An^*{-}\overset{+}{C}{=}C\underset{An}{\overset{An}{\diagup}} \xrightarrow{SOH} products \quad (63)
$$

**160**  **166**  **166\***

The most facile rearrangement among reactions (56) through (63) is reaction (59). The $\beta$-anisyl rearrangement across the double bond of the ion **163** was always faster than capture of the ion by an external nucleophile, including one of the best, the p-toluenethiolate ion. Solvolysis of **156** in AgOAc/AcOH [64], in 60% EtOH [8], in Me₃CCOOAg/Me₃CCOOH [51], and in TFE or TFE containing 2 molar equivalents of sodium p-toluenethiolate [8] gave the corresponding solvolysis–rearrangement products, the distribution of which is given in Scheme 6.22. The acetolyses showed no common ion rate depression, so that the rearrangement of **163** is faster than its capture by a bromide ion. Scheme 6.22 shows that all the products have (or are derived from) the rearranged 1,2-dianisyl-2-phenylvinyl

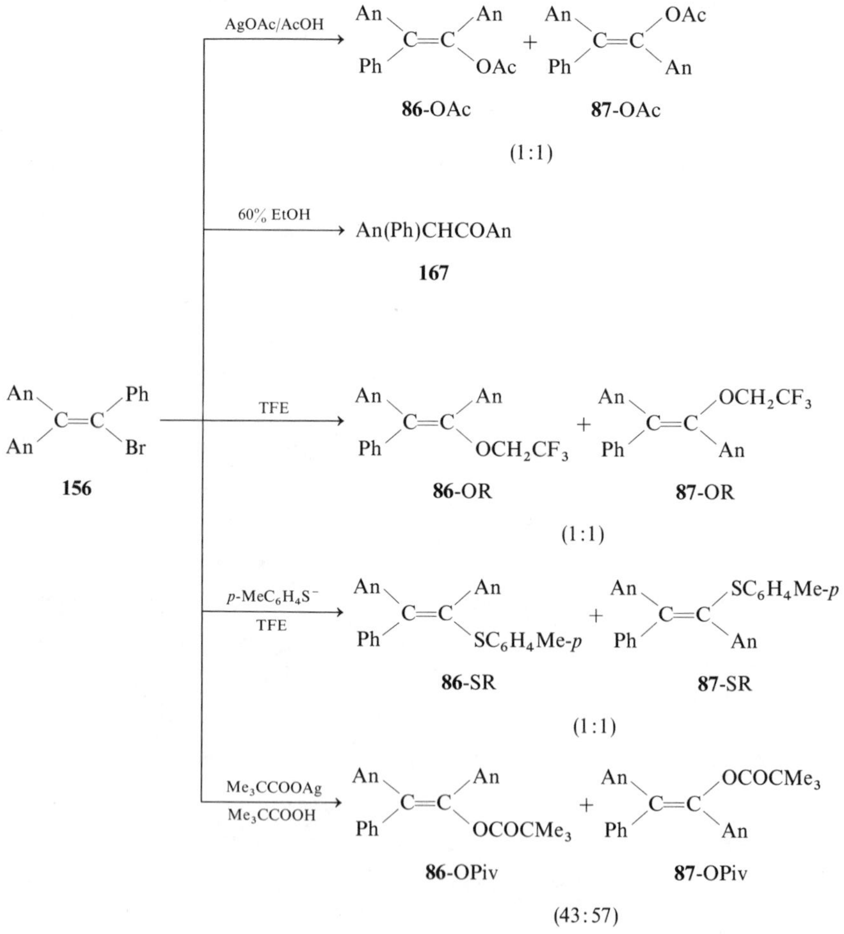

Scheme 6.22

skeleton. There is no indication, even from the nmr spectrum of the crude product, of the formation of any solvolysis product with the 2,2-dianisyl-1-phenylvinyl skeleton. The stereochemistry is lost in 60% EtOH, and only the ketone **167** is formed [8]. On the other hand, 1:1 mixtures of the cis and the trans isomers were obtained with AcOH [64], TFE [8], and $p$-MeC$_6$H$_4$S$^-$ [8] as the nucleophiles. One isomer predominates only in the low-dielectric pivalic acid, the *trans*-pivalate **87**-OPiv consisting of 57% of the vinyl pivalate mixture [51].

The kinetic evidence (Section III,B) indicates that the rate enhancement achieved by replacing a β-phenyl by a β-anisyl group is moderate [8]. In order to investigate the possibility of participation, the appropriate substrate for comparison with **156** is **154** with a phenyl trans to the leaving group. The rate enhancement factors $k_{156}/k_{154}$ are 2.6 and 3.7 in TFE at 120° and 140°, respectively, and 2.2 in 60% EtOH [8]. These rate factors are very similar to those found when a β-phenyl cis to the leaving group of **153** is replaced by a β-anisyl (Table 6.8), and this similarity rules out anchimeric assistance by the β-anisyl group. However, as discussed below, caution should be exercised when rate enhancement is used in triarylvinyl systems to probe anchimeric assistance.

Since even a seemingly small rate enhancement may correspond to a relatively large extent of anchimeric assistance [170], the Schleyer–Lancelot method was applied (see Chapter 7, Section III,C). β-Anisyl participation should take place from the rear and give the bridged ion **168**, the opening of which by a nucleophile should result in the 2,2-dianisyl-1-phenylvinyl derivative **169** and the cis product **86**. In contrast, capture of the open ion **165a** should give the product distribution listed in Table 6.21, i.e., an approximately 1:1 mixture of **86** and **87** (Scheme 6.23).

The fact that the product distributions in the reactions with TFE and $p$-MeC$_6$H$_4$S$^-$ are identical, within experimental error, with those obtained from the free ion **165a** indicate that this ion is the product-forming intermediate. By the Schleyer–Lancelot rate–product correlation, β-anisyl participation in the transition state is ruled out.

The product distributions in carboxylic acid media [51, 64] differ slightly from that for the solvolysis of $E$- and $Z$-1,2-dianisyl-2-phenylvinyl bromides (**158** and **159**, respectively) [36, 51, 118]. The acetolysis of the latter gives a 54:46 mixture of **86**-OAc and **87**-OAc [36], whereas the rearrangement gives a 50:50 mixture [64]. The pivalolysis of **158** and **159** gives a 50:50 mixture of **86**-OPiv and **87**-OPiv, whereas the rearrangement of **156** gives a 43:57 mixture [51]. Although the error in the distribution of the vinyl acetates from the rearrangement may be sufficient to give an identical distribution in both reactions, it is clear that, in the rearrangement of **156** in pivalic acid, there is a small inversion component leading to a preferential

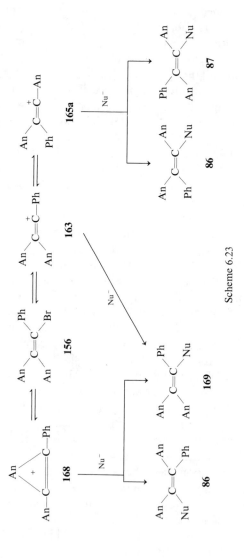

Scheme 6.23

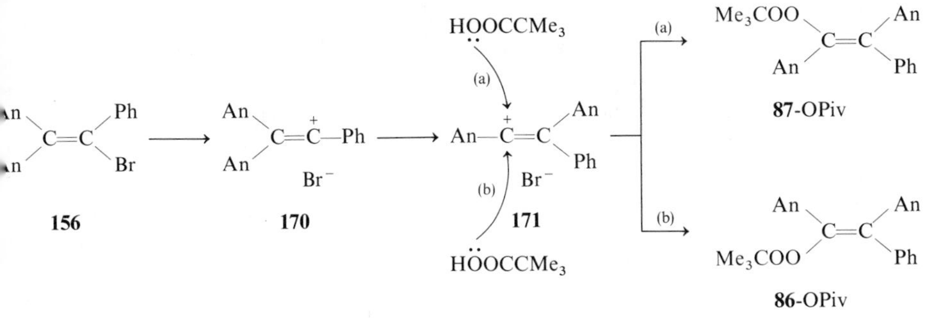

Scheme 6.24

$$85 \rightarrow 86\text{-OPiv} + 87\text{-OPiv}$$

$$(1:1)$$

formation of **87-OPiv**. This preference cannot be due to β-anisyl participation, since Scheme 6.23 predicts preferential formation of **86-OPiv** in this case. On the other hand, the stereochemical result is accounted for if the pivalolysis proceeds via the ion pairs (Scheme 6.24) [51].

The initial ionization gives the ion pair **170**, and preferential migration of the less hindered β-anisyl group would give the rearranged ion pair **171**. Capture of **171** by pivalic acid from its least hindered side [route (a) in Scheme 6.24] would give more of **87-OPiv**, as found, whereas dissociation to the free ion **85** (unlabeled **165**) or even to a more advanced ion pair would result in a 1:1 mixture of **86-OPiv** and **87-OPiv**. Since the kinetics indicate (Section V,B) that the 50:50 distribution from the solvolysis of **158** and **159** is due to capture of an ion pair [51], the inversion in the rearrangement argues strongly that the ion pair **171** is tighter than the product-forming ion pair in the solvolysis of **158** and **159**, i.e., that **171** is an intimate ion pair. Although capture of **171** is indicated by the product distribution, concurrent capture of a more dissociated cationic intermediate is not excluded [51]. The fact that the rearrangement in the pivalolysis proceeds via the ion pair while the other rearrangements apparently proceed via a more dissociated species is consistent with the low dielectric constant of pivalic acid, as discussed in Section V,B.

The rearrangement reactions of the 2-anisyl-1,2-diphenylvinyl bromides **154** and **155** [Eqs. (57) and (58)] are more complex [163, 164]. There are two isomeric precursors, which may be involved in two rearrangement reactions. Migration of a β-phenyl group across the double bond of the ion **162** would give the more stable ion **164***, whereas a β-anisyl migration would be degenerate and lead to **162***. Therefore, this system is suitable for elucidating the stereochemistry of the rearrangement, the nature of the intermediate, and the problem of the relative migratory aptitudes of different $\beta$-aryl groups.

Solvolysis of the unlabeled compounds, i.e., **88** and **89** is discussed in Section IV, and Scheme 6.10 gives the product distribution in TFE [8]. The main product (85%) from both isomers in this solvent was the phenyl-rearranged ether **90**, whereas the 15% skeletally unrearranged vinyl tri-fluoroethyl ethers consist of a 1:1 mixture of the $Z$ and $E$ isomers **91** and **92**, respectively [8]. The reaction of **89** with $p$-toluenethiolate ion in TFE gave 30% of a 1:1 mixture of the unrearranged $Z$- and $E$-vinyl $p$-toluene thiolates **93** and **94**, respectively, and 70% of the rearranged vinyl $p$-toluene thiolate **172** [Eq. (64)] [8].

$$
\begin{array}{c}
\underset{\text{An}}{\overset{\text{Ph}}{>}}C{=}C\underset{\text{Ph}}{\overset{\text{Br}}{<}} \xrightarrow{\ p\text{-MeC}_6\text{H}_4\text{S}^-\ } \underset{\text{An}}{\overset{\text{Ph}}{>}}C{=}C\underset{\text{SC}_6\text{H}_4\text{Me-}p}{\overset{\text{Ph}}{<}} + \underset{\text{An}}{\overset{\text{Ph}}{>}}C{=}C\underset{\text{Ph}}{\overset{\text{SC}_6\text{H}_4\text{Me-}p}{<}} \\[2mm]
\textbf{89} \qquad\qquad\qquad \textbf{93} \qquad\qquad\qquad \textbf{94} \\
\qquad\qquad\qquad (15\%) \qquad\qquad\qquad (15\%)
\end{array}
$$

$$
+ \ \underset{\text{Ph}}{\overset{\text{Ph}}{>}}C{=}C\underset{\text{SC}_6\text{H}_4\text{Me-}p}{\overset{\text{An}}{<}}
$$

$$
\textbf{172}
$$
$$
(70\%)
$$

$$\tag{64}$$

In 60% EtOH the two unlabeled compounds gave an identical mixture of the two ketones. The skeletally unrearranged **173** was formed in 95% yield, and the phenyl-rearranged ketone **174** was formed in 5% yield [Eq. (65)] [8].

$$
\underset{\textbf{88 and 89}}{\text{AnC(Ph)}{=}\text{C(Br)Ph}} \xrightarrow{60\%\ \text{EtOH}} \underset{\textbf{173}}{\text{An(Ph)CHCOPh}} + \underset{\textbf{174}}{\text{Ph}_2\text{CHCOAn}} \tag{65}
$$

Acetolysis of the $E$ isomer **89** in AcOH containing AgOAc gave only the unrearranged acetates **175** and **176** in a 45:55 ratio [Eq. (66)]. The phenyl-rearranged acetate was not observed [64].

$$
\underset{\text{Ph}}{\overset{\text{An}}{>}}C{=}C\underset{\text{Br}}{\overset{\text{Ph}}{<}} \xrightarrow[\text{AcOH}]{\text{AgOAc}} \underset{\text{Ph}}{\overset{\text{An}}{>}}C{=}C\underset{\text{OAc}}{\overset{\text{Ph}}{<}} + \underset{\text{Ph}}{\overset{\text{An}}{>}}C{=}C\underset{\text{Ph}}{\overset{\text{OAc}}{<}} \tag{66}
$$
$$
\textbf{89} \qquad\qquad \textbf{175} \qquad\qquad \textbf{176}
$$

The detection of a degenerate $\beta$-anisyl rearrangement requires labeling of one of the phenyl groups of **88** and **89**, and labeled **154** and **155** were ob-tained by using a $\beta$-pentadeuteriophenyl group (Ph*). The extent of $\beta$-anisyl rearrangement was determined by the combination of a mass spectral and an nmr method from the extent of scrambling of the label in the ketones in aqueous EtOH and in the vinyl acetates in AcOH [164]. The mass spectral method is especially suitable for this purpose, since the cleavage of the labeled ketones gives both a benzoyl and a benzhydryl fragment, making it possible to make an internal consistency check for the reliability of the results.

The extent of β-anisyl rearrangement was determined only in solvents in which the extent of β-phenyl rearrangement was very small or negligible. The data are given in Table 6.36. Two methods were described in the literature for presenting the extent of a degenerate rearrangement. Lee and co-workers [35, 52, 62, 65, 165–167] used the percentage of actual migration from $C_\beta$ to $C_\alpha$ so that a complete rearrangement of the labeled β group amounts to 50% migration. Rappoport and co-workers [9, 163, 164] gave the percentage of the rearrangement reaction when a complete statistical scrambling of the label between $C_\alpha$ and $C_\beta$ amounts to 100% reaction. Since different labeling methods were used, the latter, more general presentation is adopted in this section. The extent of β-anisyl rearrangement was found to be very high (90 ± 2%) in 60% EtOH and in AcOH, and it was still appreciable, although lower, in 80% EtOH [164].

**TABLE 6.36**

**β-Anisyl Rearrangement in the Solvolysis of 154 and 155[a]**

| Substrate | Solvent/base | Rearrangement[b] (%) |
|---|---|---|
| 155 | 60% EtOH/2,6-lutidine | 88.6 ± 1.4 |
| 3:7 155–154 | 60% EtOH/2,6-lutidine | 90.7 ± 0.5 |
| 155 | 80% EtOH/2,6-lutidine | 73 ± 3 |
| 155 | AcOH/AgOAc | 92.7 ± 1.8 |

[a] From Rappoport et al. [164].
[b] Based on an equal distribution of the label between the α and β positions at infinity.

The fact that the extent of both β-phenyl and β-anisyl rearrangements starting from the pure isomer 155 is identical with that starting from a mixture of 154 and 155 points to a common species as the product-forming intermediate in the solvents studied. The approximately 1:1/EZ product ratios of the vinyl trifluoroethyl ethers, the vinyl p-toluenethiolates, and the vinyl acetates suggest that the intermediate involved in the formation of the rearranged products is the free linear ion 162 [8, 164]. A scheme identical to Scheme 6.23 shows that an anisyl-bridged ion gives more of the E-2-anisyl-1,2-diphenylvinyl-Nu (Nu = $OCH_2CF_3$, OAc, $SC_6H_4Me$-p), contrary to what was observed. The almost identical solvolysis rates of 88 and 89 in 60% EtOH and in TFE (Table 6.8) [8] rule out both β-anisyl and β-phenyl participation in the heterolysis.

In $\beta$-aryl rearrangements to a cationic terminus in saturated systems, the migratory aptitude of a $\beta$-$p$-anisyl group generally is superior to that of a $\beta$-phenyl group [171]. This is because the charge dispersal ability of an anisyl in the transition state **177** is better than that of a phenyl in the transition state **178**. The $\beta$-anisyl/$\beta$-phenyl migratory ratios are ca. 6 to 500 for solvolysis and related reactions [171a–e] and much lower (1.2–2.0) for the deamination reaction [171f]. *A priori*, $\beta$-anisyl/$\beta$-phenyl ratios, which are obtained in vinylic systems by internal competition of these groups [Eq. (67)], can differ from those for saturated systems and may even be lower

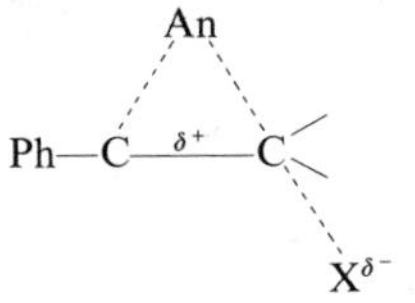

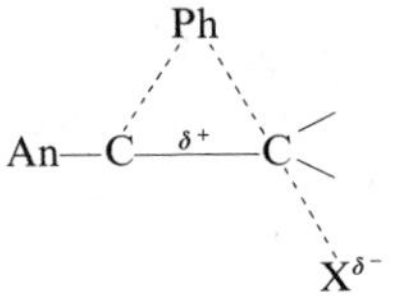

than unity. The reason for this is that two opposing effects should be considered for $\beta$-aryl rearrangements across the double bond. The stabilization of transition state **50** by the bridging anisyl group exceeds the stabilization achieved for transition state **51** by the bridging phenyl group. However, migration of anisyl to form the bridged species **50** is accompanied by a loss of $\pi(\beta\text{-An})$–$\pi(C{=}C)$ conjugation, whereas a $\pi(\beta\text{-Ph})$–$\pi(C{=}C)$ conjugation is lost during the formation of **51**. Hence, the $\beta$-anisyl/$\beta$-phenyl migration ratio is composed of the differential transition state versus ground state stabilization by the migrating group. The anisyl group stabilizes both states better than a phenyl group, and the ground state effect may reduce the ratios and even reverse them in an extreme case.

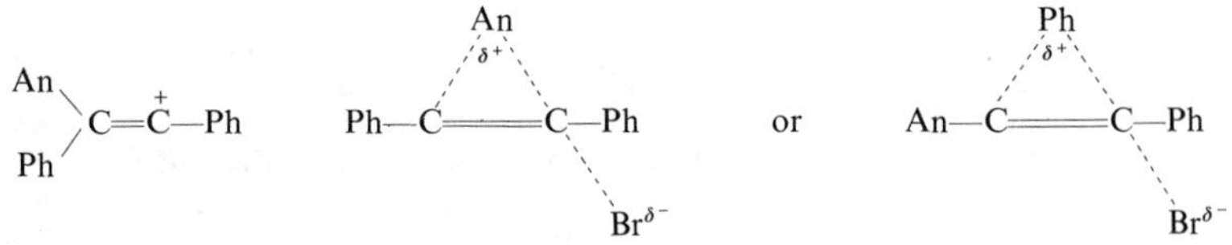

The relative migratory aptitudes and the relative capture rates by the solvent are calculated with reference to Scheme 6.25, which demonstrates the behavior in 60% EtOH. The first-order rate constants for the ionization of **155** and **154** are $k_{ion}$ and $k'_{ion}$, and the first-order rate constants for anisyl and phenyl migrations in the ion **162** are $k_{r(An)}$ and $k_{r(Ph)}$, respectively. The pseudo-first-order constants for capture of the $\alpha$-phenylvinyl cations (**162** and **162***) and of the $\alpha$-anisylvinyl cation **164*** are $k_{SOH}$ and $k'_{SOH}$,

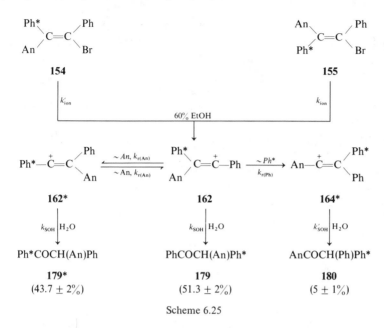

Scheme 6.25

respectively. Since the solvolysis of 1-anisyl-2,2-diphenylvinyl bromide gives only unrearranged products, the **162 → 164\*** process is regarded as irreversible. When the very small isotope effects on these constants are neglected, a steady-state treatment of the cationoid intermediates of Scheme 6.25 gives the relationships between the product distribution and the rate constants described by Eqs. (68) through (70). The data of Scheme 6.25 can be combined

$$[180]/[179] = k_{r(Ph)}/k_{SOH} \qquad (68)$$

$$[179]/[179^*] = 1 + (k_{SOH}/k_{r(An)}) \qquad (69)$$

$$[179^*] \cdot [179] [180]([179] - [179^*]) = k_{r(An)}/k_{r(Ph)} \qquad (70)$$

with the data of Table 6.36 to give the reactivity order $k_{r(An)}/k_{SOH}/k_{r(Ph)}$ of 42.4:10.6:1 in 60% EtOH [164]. In AcOH, β-phenyl rearrangement was not observed and, if an upper limit for $k_{r(Ph)}$ is assumed, $k_{r(An)}/k_{SOH}/k_{r(Ph)} =$ 67:10.5:1. As seen from Eq. (70), a very small extent of phenyl rearrangement or a very large extent of β-anisyl rearrangement will result in a large error in the $k_{r(An)}/k_{r(Ph)}$ ratios. This is why only a tentative relative reactivity of $k_{r(An)}/k_{r(Ph)}/k_{SOH} = > 100:11.3:1$ is obtained in TFE, where the β-anisyl re-arrangement is complete [8, 164]. A recent study of the 1-[13]C-labeled isomer **154′** in TFE gave the reactivity order $k_{r(An)}/k_{r(Ph)}/k_{SOH} = > 297:3:1$, and the products were the trifluoroethyl ethers, whereas the recovered bromide was a mixture of the four isomers **154′**, **155′**, **154\*′**, and **155\*′**, which were formed in equal amounts [91c].

$$
\begin{array}{cccc}
\underset{An}{\overset{Ph}{\diagdown}}C{=}^{13}C\underset{Br}{\overset{Ph}{\diagup}} &
\underset{Ph}{\overset{An}{\diagdown}}C{=}^{13}C\underset{Br}{\overset{Ph}{\diagup}} &
\underset{An}{\overset{Ph}{\diagdown}}{}^{13}C{=}C\underset{Br}{\overset{Ph}{\diagup}} &
\underset{Ph}{\overset{An}{\diagdown}}{}^{13}C{=}C\underset{Br}{\overset{Ph}{\diagup}} \\
\textbf{154′} & \textbf{155′} & \textbf{154*′} & \textbf{155*′}
\end{array}
$$

Consequently, the $k_{r(An)}/k_{r(Ph)}$ ratios of $\geq 11$ to $\geq 67$ are much higher than unity and do not show a significant contribution from a loss of ground state $\pi(\beta\text{-Ar})-\pi(C{=}C)$ conjugation when migrations of the two groups originate from the same ground state. A similar value of 164 could be obtained by comparing the extent of phenyl migration in the triphenylvinyl cation **161** [165] with the extent of anisyl migration in the ion **162** in AcOH [164]. In both cases, the migrations are between phenyl-substituted migration origin and terminus, indicating that the $k_{r(An)}/k_{r(Ph)}$ ratios are little affected by the identity of the nonmigrating $\alpha$ and $\beta$ substituents. The small influence of the $\pi(\beta\text{-Ar})-\pi(C{=}C)$ conjugation probably reflects the geometry of the substrates, since the aryl groups are already twisted from the plane of the double bond by mutual steric interactions in the ground state. Nevertheless, this effect should be borne in mind when $\beta$-aryl participation in vinylic systems is ruled out by the similar reactivities of cis and trans isomers.

Another degenerate $\beta$-anisyl migration is that in the trianisylvinyl cation **166** [Eq. (63)], in which the migration origin and terminus are anisyl-substituted. This rearrangement was studied both by Rappoport and co-workers [9] and by Lee and co-workers [62, 167] in several solvents, and different leaving groups and labeling techniques were used. In one labeling method, trianisylvinyl bromide was labeled by a $p$-trideuteriomethoxyphenyl group-($p$-$CD_3OC_6H_4$, An\*) at one of the $\beta$ positions, and the extent of rearrangement was measured by nmr, a 1:1:1 distribution of the An\* group at the three vinylic positions of the product being taken as 100% rearrangement [9]. Another labeling method used a $\beta$-$^{13}$C-enriched vinylic carbon, and $^{13}$C nmr or $^1$H satellite nmr was used for the analysis, a 1:1 scrambling of the label at the vinylic carbons amounting to 100% rearrangement [62]. The rearrangement of [2-$^{14}$C]trianisylvinyl phenyltriazene was followed by radioactive counting [167], and the infinity was taken similarly to that in the rearrangement of [2-$^{13}$C]trianisylvinyl bromide [62]. The extent of the $\beta$-anisyl rearrangement (Table 6.37) is highly solvent dependent: the rearrangement is complete in TFE [9] but is not observed at all in pivalic acid [51].

Kinetic evidence is not available, but there are two types of clear, strong evidence for the intermediacy of a free linear vinyl cation as the intermediate in the $\beta$-anisyl rearrangement in TFE and in AcOH. First, addition

# TABLE 6.37

## Degenerate β-Anisyl Rearrangements in $An_2C=\overset{+}{C}-An$

| Precursor | Solvent/base | Temp(°C) | Rearrangement[a] (%) | $k_{SOH}/k_{r(An)}$ | Reference |
|---|---|---|---|---|---|
| $AnC(An^*)=C(Br)An$[b] | 60% EtOH/NaOAc | 120 | 11.5 ± 2.0 | 23.1 ± 4.6[i] | [9] |
| | AcOH/NaOAc | 120 | 35 ± 2 | 5.6 ± 0.6[i] | [9] |
| | AcOH/NaOAc/Bu₄NBr | 120 | 4 ± 1.5[c] | | [9] |
| | Me₃CCOOH/Me₃CCOONa | 160 | 0 | >100[i] | [9] |
| | TFE/2,6-lutidine | 120 | 100 | <0.08[i] | [9] |
| | TFE/2,6-lutidine/Bu₄NBr[d] | 120 | 0 ± 3[e] | | [9] |
| | MeCN/2,6-lutidine | 120 | 26.4[f] | | [9] |
| | CF₃COOH/CF₃COONa | 50 | 100[g] | | [9] |
| $An_2{}^{13}C=C(Br)An$ | AcOH/AgOAc | Reflux | 40 ± 2 | 6.0 ± 0.6 | [62] |
| | CF₃COOH/CF₃COOAg | rt | 100 | | [62] |
| $An_2{}^{14}C=C(An)-N=N-NHPh$ | AcOH | rt | 76.2 ± 0.6 | 1.28 ± 0.02 | [167] |
| | AcOH/NaOAc[h] | rt | 33.6 ± 0.4 | 7.90 ± 0.14 | [167] |

[a] Complete scrambling of the α- and β-carbons or of the three anisyl groups between the α and β positions amounts to 100% rearrangement.

[b] $An^* = p\text{-}CD_3OC_6H_4$.

[c] 0.076 M Bu₄NBr and 0.011 M NaOAc.

[d] 3.6 M Bu₄NBr.

[e] In the recovered RBr.

[f] After 50 hr.

[g] Determined by mass spectra.

[h] 0.276 M NaOAc.

[i] Value corrected for error in Ref. [9].

of external bromide ion to **160** suppresses the rearrangement almost completely in TFE. The rearrangement is suppressed ninefold in AcOH, but it is likely that at higher bromide ion concentration the rearrangement would also be completely suppressed. Since external bromide ion captures the free trianisylvinyl cation **166** (Section V,A), this cation should be involved in the rearrangement [9]. Second, the extent of rearrangement in AcOH was studied by two labeling techniques, one that scrambles the three anisyl groups (Scheme 6.26) [9] and another that scrambles the α- and β-carbons (Scheme 6.27) [62]. Both experiments together amount to a "double-labeling" experiment of the type used by Collins and Bonner to establish the involvement of an "open" cation in the reactions forming the 1,2,2-triphenylethyl cation [172].

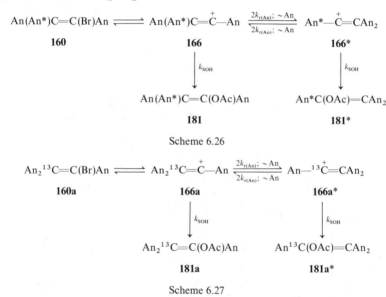

Scheme 6.26

Scheme 6.27

Equations (71) and (72) give the relationship between the extent of rearrangement and the $k_{SOH}/k_{r(An)}$ ratios, which are derived from these schemes when isotope effects and capture by Br⁻ (which does not affect the product

$$[181]/[181^*] = [2 + (k_{SOH}/k_{r(An)})] \qquad (71)$$

$$[181a]/[181a^*] = 1 + (k_{(SOH)}/k_{r(An)}) \qquad (72)$$

distribution at infinity) are neglected. In AcOH/AgOAc, $[181a]/[181a^*] = 4.0 \pm 0.25$; i.e., from Scheme 6.26, $k_{SOH}/k_{r(An)} = 6.0 \pm 0.6$. In AcOH/NaOAc, $[181]/[181^*] = 7.6 \pm 0.49$; i.e., from Scheme 6.27, $k_{SOH}/k_{r(An)} = 5.6 \pm 0.6$. The identity, within experimental error, of the two values for compounds labeled differently indicates that the assumption involved in deriving Eqs. (71) and (72), i.e., that the free ion is the intermediate, is correct [9].

that the assumption involved in deriving Eqs. (71) and (72), i.e., that the free ion is the intermediate, is correct [9].

The effect of solvent on the extent of the rearrangement is best analyzed in terms of the $k_{SOH}/k_{r(An)}$ ratios (Table 6.37) derived by equations analogous to (71) and (72). Two solvent properties are relevant in this respect: the dissociation power and the nucleophilicity. At higher dissociation power, as measured by the dielectric constant $\varepsilon$, the lifetime of the cation will be longer and $k_{r(An)}$ will be higher. The $k_{SOH}$ value will be directly proportional to the nucleophilicity of the solvent, as measured by Schleyer's $N$ values [109]. The lowest $k_{SOO}/k_{r(An)}$ value (0.08) is obtained in TFE, a highly dissociating but weakly nucleophilic solvent [109]. A higher value is obtained in the moderately dissociating but relatively weakly nucleophilic AcOH, and the highγst value is obtained in the most dissociating but also the most nucleophilic [109] 60% EtOH. Although the extents of rearrangement in the poorly dissociating and relatively weakly nucleophilic pivalic acid $k_{SOH}/k_{r(An)} > 100$) and in the highly dissociating and weakly nucleophilic trifluoroacetic acid ($k_{SOH}/k_{r(An)} < 0.08$) fit the above generalization, these two extreme behaviors occur for different reasons. The kinetics in pivalic acid indicate that the free ion is not involved to a large extent in the solvolysis [51], and Scheme 6.26 is not applicable. It seems that the lifetime of the ion-pair intermediate is too short to permit an appreciable rearrangement before return to covalent material or collapse with the solvent shell. Apparently, rearrangement via ion pairs in this solvent takes place only when the driving force of the rearrangement is high, as in the reaction of **156** (Scheme 6.24).

Complete rearrangement was observed in trifluoroacetic acid in the presence of either silver trifluoroacetate [62] or sodium trifluoroacetate [9]. In the latter case, the triarylethanones formed were analyzed by the less reliable mass spectral method, but the more reliable $^{13}C$ analysis gave a similar result. It was suggested by Oka and Lee [62] that the low nucleophilicity of the solvent is responsible for the large extent of rearrangement. However, Houminer et al. [9], who observed the complete rearrangement in the product trianisylethanone in $CF_3COOH$, also observed a large extent of rearrangement ($73 \pm 5\%$) in AcOH–HCOOH. However, they also found that the less reactive triarylethanone **182** rearranges partially to its isomer **182*** in this medium [Eq. (73)].

$$\text{An*CH(Ph)}\overset{O}{\overset{\|}{-}}\text{C}\text{—An} \xrightarrow{\text{1:1 AcOH–HCOOH}} \text{AnCH(Ph)}\overset{O}{\overset{\|}{-}}\text{C}\text{—An*} \qquad (73)$$

**182**                               **182***

Since the trifluoroacetolysis of **160** should give the two ketones **183** and **183*** [Eq. (74)], which are more prone to such an isomerization, the large

$$An^*(An)C{=}C(Br)An \xrightarrow{CF_3COOH} An^*CH(An){-}\overset{\displaystyle O}{\overset{\|}{C}}{-}An + An_2CH{-}\overset{\displaystyle O}{\overset{\|}{C}}{-}An^* \quad (74)$$

160                                              183                    183*

extent of rearrangement may be partly or completely due to postisomerization. The trifluoroacetolysis with $AgOCOCF_3$ gives trianisylvinyl trifluoroacetate, but rearrangement via initial protonation of either the vinyl bromide or the trifluoroacetate can be envisaged (Section II,A). Such protonation fits the development of a purple color when 160 is dissolved in $CF_3COOH$ [62].

The selectivity of triarylvinyl cations is discussed in Section V,A in terms of steric crowding of the cationic orbital by the $\beta$ substituents. However, a rapid dynamic equilibrium between degenerate ions may be an alternative explanation for the selectivity [9]. This possibility should be considered when three conditions are fulfilled. (1) The system should be symmetric so that migration of either of the $\beta$-aryl groups has equal probability. (2) The rearrangement should be fast and reversible. (3) The capture of the ion by the leaving group should be slower than the migration of the aryl group. The first condition is fulfilled only for triarylvinyl systems with three identical aryl groups. Of all these systems, the second condition of a complete rearrangement via an initial ionization has been observed so far only for the

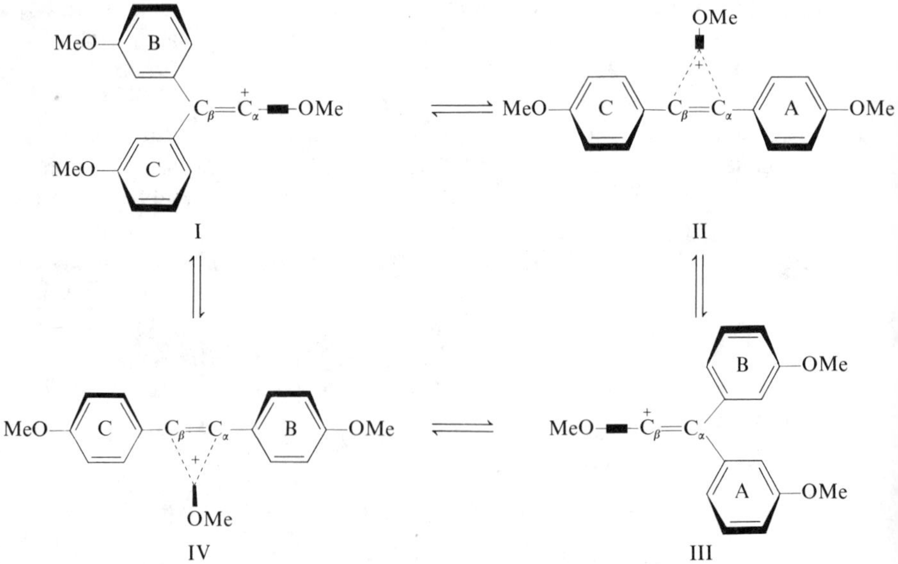

Scheme 6.28

trianisylvinyl cation in TFE [9]. The details of this rearrangement are demonstrated schematically in Scheme 6.28, which is a top view of the system, where A, B, and C specify the three anisyl groups.

Before the rearrangement, the α-anisyl group A of I is perpendicular to the plane of the double bond, whereas the two β groups B and C are somewhat twisted from this plane. Migration of group B involves movement in this plane but also a rotation to a perpendicular plane, since the bridging carbon becomes tetrahedral on bridging. The situation is less clear with groups A and C. Both groups move clockwise in plane, A away from and C toward a linear $C_\alpha$–$C_\beta$–C arrangement. Simultaneously, both groups also rotate, but the angle of rotation is unknown and is a result of a compromise between the tendency for a planar $\pi(Ar)$–$\pi(C=C)$ arrangement and a planar $\pi(Ar)$–$p(C^+)$ arrangement. The structure of ion II is based on the assumption that the $\pi(Ar)$–$\pi(C=C)$ interaction is dominant. Regardless of the angle of twist, the α,β-diarylvinylidenephenonium ion II is symmetric, and rings A and C are in a plane perpendicular to that of group B.

The rearrangement is completed by movement of B to the α-carbon. Since A and B are now conjugated with $C_\alpha$, they have to rotate by ca. 60°, whereas group C, which is now conjugated to the p orbital on $C_\beta$, has to move and rotate by nearly 90°. In the II → III transition, both clockwise phased movement and rotation take place. Group A can now rearrange, and the III → IV transition results in movements and rotations identical with those of the I → II process but on the opposite face of the double bond. Hence, the overall I ⇄ IV process involves a spherical, three-dimensional, coordinated movement of the three aryl groups. If this process drives solvent molecules and bromide ions from the vicinity of $C_\alpha$ and $C_\beta$, a term similar to that applied by Winstein for the suggested rapid 1.6 ⇄ 2.6 degenerate shift in the norbornyl system [173] may be used, and the process will be called a "three-dimensional windshield wiper effect." Such an effect might presumably increase the lifetime and consequently the selectivity of the ion.

The solvolysis product in TFE is completely rearranged, i.e., $k_{r(A)} \gg k_{SOH}$. The windshield wiper effect might contribute to the selectivity only if the third condition is fulfilled, i.e., when the rearrangement is also faster than capture of **166** by bromide ion ($k_{Br}$) and $k_{r(An)} \gg k_{Br}[Br^-]$ at appreciable bromide ion concentrations. Table 6.38 shows the conditions under which the ionization of **160** to the ion **166** leads to a high yield of the scrambled ether in the absence of added bromide ion. When bromide ion is added, **166** is captured preferentially by bromide ion rather than by the solvent. An increase in the concentration of bromide ion is accompanied by a decrease in the amount of the trianisylvinyl trifluoroethyl ether and in the extent of the rearrangement in the recovered trianisylvinyl bromide. At the high

**TABLE 6.38**

**Rearrangement of 160 in TFE/2,6-Lutidine at 120°** [a]

| Reaction time min | [Bu$_4$NBr] ($M$) | Yield (%) of An$_2$C=C(OCH$_2$CF$_3$)An | Rearrangement in An$_2$C=C(X)An (%) | |
|---|---|---|---|---|
| | | | X = Br[b] | X = OCH$_2$CF$_3$ |
| 480 | | 100 | | 100 ± 5 |
| 14 | | 60 | ≥ 80 ± 15 | > 95 |
| 14 | 0.8 | ≤ 15 | 56 ± 3 | |
| 14 | 3.6 | 0 | 0 ± 3 | |

[a] From Houminer *et al.* [9].
[b] In the RBr fraction that participated in the reaction.

concentration of 3.6 $M$ Br$^-$, the recovered bromide is unrearranged. Hence, $k_{Br}[Br^-] \gg k_{r(An)} > k_{SOH}$, and the selectivity of the ion is not due to the windshield wiper effect [9].

The nature of the intermediate in the degenerate $\beta$-anisyl rearrangement in the trianisylvinyl cation was also investigated by generating the ion from the corresponding phenyltriazene (**184**) in AcOH [167]. In this case, the leaving group is nitrogen, and a counterion is not present. The extent of rearrangement (Scheme 6.29) was higher (76.2%) than for the solvolysis of

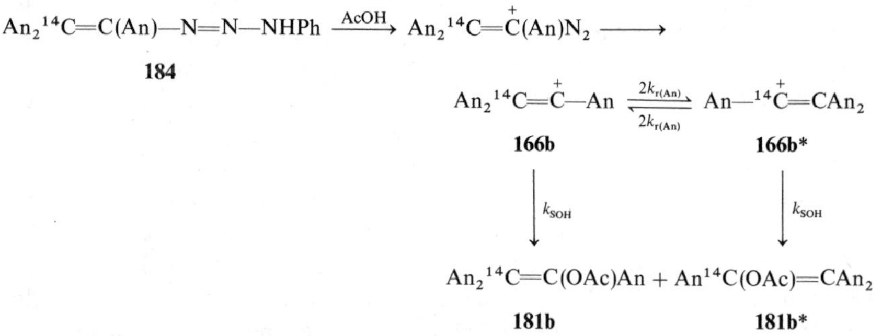

$$An_2{}^{14}C=C(An)-N=N-NHPh \xrightarrow{\text{AcOH}} An_2{}^{14}C=\overset{+}{C}(An)N_2 \longrightarrow$$

**184**

$$An_2{}^{14}C=\overset{+}{C}-An \underset{2k_{r(An)}}{\overset{2k_{r(An)}}{\rightleftharpoons}} An-{}^{14}\overset{+}{C}=CAn_2$$

**166b**                          **166b***

$$\downarrow k_{SOH} \qquad\qquad\qquad \downarrow k_{SOH}$$

$$An_2{}^{14}C=C(OAc)An + An{}^{14}C(OAc)=CAn_2$$

**181b**                          **181b***

Scheme 6.29

trianisylvinyl bromide (Table 6.37) and, from Eq. (72) $k_{SOH}/k_{r(An)} = 1.28$. The lower value in the reaction of the triazene in the unbuffered acid is not surprising since $k_{SOH}$ relates to capture by AcOH, whereas $k_{SOH}$ in AcOH/ NaOAc relates to capture by both AcO$^-$ and AcOH. Indeed, the rearrangement was suppressed to 33.6% in the presence of 0.273 $M$ NaOAc ($k_{SOH}/ k_{r(An)} = 7.90$), indicating the intermediacy of a free ion, which is capable of being captured before the rearrangement [167]. A $k_{AcO^-}/k_{AcOH}$ ratio of 372 is calculated from these values, suggesting that the capture of even a "hot" and therefore less selective trianisylvinyl cation from the triazene is predominantly by the acetate ion, as deduced from the common ion rate depression (Section V,A).

The rate of ionization of triarylvinyl bromides in nonsolvolytic media cannot be followed by titrimetry, and its measurement requires a different process, which also initiates in the heterolysis. For an asymmetric substrate, the cis–trans isomerization can serve as such a process and the **83**-Br ⇄ **84**-Br isomerization has a rate constant $k_{isom}$ of $2 \times 10^{-5}$ sec$^{-1}$ in acetonitrile at 120° [10]. For a substrate with three identical aryl groups, e.g., **160**, the rate of a β-aryl rearrangement, if any occurs, gives a lower limit for the ionization rate. The rate constant for anisyl migration for **160** is $9.5 \times 10^{-7}$ sec$^{-1}$ in acetonitrile at 120° [9], i.e., 20 times lower than $k_{isom}$ for **83**-Br, although **160** solvolyzes slightly faster than **83**-Br in protic solvents [5, 11, 36, 174]. Consequently, the cis–trans isomerization may require a less dissociated species than the rearrangement. This is supported by the results in pivalic acid, in which the **83**-Br ⇄ **84**-Br isomerization occurs at the ion pair stage, whereas no rearrangement takes place during the solvolysis of **160** [51]. A lower activation energy for the isomerization than for the rearrangement is reasonable since isomerization proceeds either by rotation of Br$^-$ around the cation or by rotation of the cation around its C=C axis, followed by cation–anion recombination, whereas the rearrangement involves a bond cleavage–bond formation sequence. Hence, the cis–trans isomerization is a superior method for estimating a lower limit for the ionization rate constant in aprotic solvents.

The extent of β-phenyl rearrangements is usually smaller than that of β-anisyl rearrangements. Lee and co-workers [65, 165, 167b] used $^{13}$C- and $^{14}$C-labeled substrates and observed a degenerate β-phenyl rearrangement during the solvolysis of triphenylvinyl triflate [165] and bromide [65, 167b] in various solvents [Eq. (56)]. The extents of rearrangement in AcOH were the same regardless of whether triflate or bromide was the leaving group, but deamination of the labeled phenyltriazene showed no β-phenyl rearrangement [167a] (Table 6.39). The vinyl triflate gave no rearrangement in the nucleophilic aqueous dioxane, and in carboxylic acids the extent of rearrangement was only slightly higher in formic acid (15.4%) than in AcOH (13.3%) but much higher in trifluoroacetic acid (52.0%) [165]. The decrease in the corresponding $k_{SOH}/k_{r(Ph)}$ values, which are derived from an equation analogous to (72), follows the decrease in the nucleophilicity of the solvents [$N$(AcOH) $= -2.05$; $N$(HCOOH) $= -2.05$; $N$(CF$_3$COOH) $= -4.74$] [109]. The reactions were also insensitive to the presence of a 1.1 molar equivalent

**TABLE 6.39**

**Degenerate $\beta$-Phenyl Rearrangement in $Ph_2C\overset{+}{=}C\!\!-\!\!Ph$**

| Precursor | Solvent/base | Temp (°C) | Rearrangement[a] (%) | | $k_{SOH}/k_{\tau(Ph)}$ | Reference |
|---|---|---|---|---|---|---|
| | | | In the product | In the recovered RX | | |
| $Ph_2{}^{14}C\!=\!C(Ph)OTf$ | 60% Dioxane/NaOH | Reflux | 0 | | | [165] |
| | AcOH | Reflux | 13.3 ± 0.3 | | 26.0 | [165] |
| | AcOH/NaOAc | Reflux | 13.6 ± 0.4 | | 25.4 | [165] |
| | 97% HCOOH | Reflux | 15.4 ± 0.2 | | 22.0 | [165] |
| | 97% HCOOH/HCOONa | Reflux | 15.3 ± 0.1 | | 22.0 | [165] |
| | $CF_3COOH$ | 40 | 52.0 ± 0.4 | | 3.68 | [165] |
| | $CF_3COOH/CF_3COONa$ | 40 | 55.9 ± 1.3 | | 3.16 | [165] |
| $Ph_2{}^{13}C\!=\!C(Ph)Br$ | AcOH/AgOAc | Reflux | 13.6 ± 1.8 | | 25.4 | [65] |
| $Ph_2{}^{14}C\!=\!C(Ph)Br$ | 90% AcOH | 150 | 36.8 ± 0.2 | 0 | 6.8 | [167b] |
| | 80% AcOH | 150 | 33.4 ± 0.4 | 0 | 8.0 | [167b] |
| | 70% AcOH | 150 | 30.4 ± 1.3 | 1 | 9.2 | [167b] |
| | 70% AcOH/NaOAc | 150 | 30.7 ± 0.9 | 0.6 | 9.0 | [167b] |
| | 70% AcOH/NaBr | 150 | 30.7 ± 0.7 | 0 | 9.0 | [167b] |
| | 60% AcOH | 150 | 25.5 ± 0.7 | | 11.8 | [167b] |

| Substrate | Solvent/Conditions | Temp. (°C) | | | | Ref. |
|---|---|---|---|---|---|---|
| $\text{Ph}_2{}^{14}\text{C=C(Ph)Br}$ | 50% AcOH | 150 | 19.3 ± 0.3 | | 16.6 | [167b] |
| | TFE/2,6-lutidine | 150 | 61[b] | 40.4[c] | 2.6 | [167b] |
| | TFE/2,6-lutidine | 150 | 78.8[d] | 60 ± 0.6[b] | 1.02 | [167b] |
| | TFE/2,6-lutidine | 150 | 90[e] | 80.6[d] | 0.44 | [167b] |
| | TFE/2,6-lutidine/Et$_4$NBr[f] | 150 | 60.8[b] | 39.8[c] | | [167b] |
| | TFE/2,6-lutidine/Et$_4$NBr[f] | 150 | 78.6[d] | 60.4[b] | | [167b] |
| | TFE/2,6-lutidine/Et$_4$NBr[f] | 150 | 88[e] | 58.2[b,g] | | [167b] |
| | TFE/2,6-lutidine/Et$_4$NBr[f] | 150 | 86.4[g] | 80.8[c] | | [167b] |
| $\text{Ph}_2{}^{14}\text{C=C(Ph)Br}$ | AcOH/AgOAc | Reflux | 13.1 ± 0.5 | | 26.0 | [65] |
| $\text{Ph}_2{}^{14}\text{C=C(Ph)—N=N—NHPh}$ | AcOH | rt | 0 | | | [167a] |
| | AcOH/NaOAc | rt | 0 | | | [167a] |
| | CF$_3$COOH | rt | 0 | | | [167a] |
| | 90% Me$_2$CO/HClO$_4$ | rt | 0 | | | [167a] |
| | TFE/HClO$_4$ | rt | 0 | | | [167b] |

[a] Complete scrambling of the α- and β-carbons amounts to 100% rearrangement.
[b] Reaction time, 48 hr.
[c] Reaction time, 24 hr.
[d] Reaction time, 96 hr.
[e] Reaction time, 480 hr.
[f] [Et$_4$NBr] = 0.19 $M$.
[g] [Et$_4$NBr] = 0.38 $M$.

of the sodium carboxylate. The products were the vinylic esters in AcOH and $CF_3COOH$ and the triphenylethanones **185** and **185*** in 97% HCOOH [Eq. (75)] [165].

$$Ph_2*C{=}C(X)Ph \quad \xrightarrow{\quad\quad}$$
$$X = Br, OTf$$

$$\xrightarrow{AcOH, CF_3COOH} Ph_2*C{=}C(OCOR)Ph + Ph_2C{=}*C(OCOR)Ph$$
$$R = Me, CF_3$$

$$\xrightarrow[\text{AcOH–H}_2\text{O}]{97\% \text{ HCOOH or}} Ph_2*CHCOPh + Ph_2CH*COPh$$
$$\qquad\qquad\qquad\quad \textbf{185} \qquad\qquad\quad \textbf{185*}$$

(75)

The extent of the degenerate $\beta$-phenyl rearrangement is more sensitive to the ionizing power of the solvent in the saturated systems. The values for the solvolysis–rearrangement of the saturated analogue 2-phenylethyl tosylate (**186**) [Eq. (76)] or its $1,1\text{-}d_2$ derivative are 12, 90, and 100% in AcOH/NaOAc, HCOOH/HCOONa, and $CF_3COOH/CF_3COONa$, respectively [175].

$$PhCH_2{}^{14}CH_2OTs \xrightarrow{RCOOH} Ph^{14}CH_2CH_2OCOR + PhCH_2{}^{14}CH_2OCOR \quad (76)$$
$$\textbf{186}$$

The importance of solvent nucleophilicity is also shown by the recent work of Lee and Ko on the solvolysis–rearrangement of $[2\text{-}{}^{14}C]$triphenylvinyl bromide in aqueous acetic acid. The extent of rearrangement in the product ketones **185** and **185*** [Eq. (75)] increases regularly from 19.3% in 50% AcOH to 36.8% in the less nucleophilic, less aqueous 90% AcOH (Table 6.39) [167b]. However, these results are in clear contrast to the smaller extents of rearrangement (13.3–13.6%) of the bromide in AcOH/AgOAc [65] or of the triflate in unbuffered AcOH [165], but the reason for this is not yet clear.

The solvent dependency of the extent of rearrangement could be explained in two ways [165]. The ionization may involve $\beta$-phenyl participation in the rate-determining step for the heterolysis, the importance of which increases with an increase in the solvent polarity. Alternatively, the reaction may proceed via a nonbridged species, rearrangement competing with capture by the solvent and the more nucleophilic solvent giving a higher $k_{SOH}$ value and a smaller extent of rearrangement. The insensitivity of the extent of rearrangement in 70% AcOH to either added NaBr or added NaOAc [167b] and in pure AcOH to an added metal acetate, in contrast to its decrease in the solvolyses of **160** and **184** described above, suggested that the rearrangement proceeds via the ion pairs **187** and **187*** (Scheme 6.30) [165]. In this case, $k_{r(Ph)}$ would be little affected by a change in solvent and the extents of rearrangement would be determined by the $k_{SOH}$ values, as observed.

Although the possibility of ion-pair return is embodied in Scheme 6.30 (see $k_{ipr}$), it is clear that such a process in the nucleophilic 70 to 90% AcOH

$$Ph_2*C{=}C(X)Ph \underset{k_{ipr}}{\overset{k_{ion}}{\rightleftharpoons}} \underset{X^-}{Ph_2*C{=}\overset{+}{C}{-}Ph} \underset{2k_{r(Ph)}:\, \sim Ph}{\overset{2k_{r(Ph)}:\, \sim Ph}{\rightleftharpoons}} \underset{X^-}{Ph_2C{=}*\overset{+}{C}{-}Ph} \underset{k_{ion}}{\overset{k_{ipr}}{\rightleftharpoons}} Ph_2C{=}*C(X)Ph$$

$$\quad 153' \qquad\qquad\quad 187 \qquad\qquad\qquad\qquad 187* \qquad\qquad 153*'$$

$$k_{SOH}\downarrow \qquad\qquad\qquad\qquad k_{SOH}\downarrow$$

$$Ph_2*C{=}C(OS)Ph \qquad Ph_2C{=}*C(OS)Ph$$

Scheme 6.30

is not competitive with either product formation or with the rearrangement since the recovered triphenylvinyl bromide in these media was $\leq 1\%$ rearranged [167b].

However, although reactions via ion pairs might be more likely for an α-phenyl than for an α-anisylvinyl cationic species, the ion-pair mechanism in carboxylic acid media presents two difficulties. First, triphenylvinyl bromide itself gives products from the free ion in TFE [8] (Section V,A) and, although the rearrangement may proceed via ion pairs [167b] as discussed below, it is highly probable that at least in HCOOH and CF$_3$COOH, the ionizing power of which is higher than that of TFE, products will also be formed from the free ion. Second, this conclusion requires further justification.

In view of these results, it is surprising that the reaction of [2-$^{14}$C]triphenylvinyl phenyltriazene (188) [Eq. (77)] gave no rearrangement in several

$$Ph_2{}^{14}C{=}C(Ph){-}N{=}N{-}NHPh \xrightarrow{\text{RCOOH}} Ph_2{}^{14}C{=}C(OCOR)Ph \qquad (77)$$

$$\mathbf{188}$$

solvents [167] (Table 6.39). Although a smaller extent of rearrangement is expected for the triphenylvinyl cation compared with the trianisylvinyl cation, these results are in contrast with the larger extent of β-anisyl rearrangement in the ion formed from trianisylvinyl phenyltriazene [167] than from trianisylvinyl bromide [9]. Since a counterion is not present, the products should be formed from the free ion. It was suggested that the lack of rearrangement confirms the conclusion that the β-phenyl rearrangement during the solvolysis proceeds at the ion-pair stage [167]. However, an alternative explanation is that a free "hot" ion is formed in the dediazoniation

**TABLE 6.40**

**Degenerate β-Phenyl Rearrangements in An(Ph)C=$\overset{+}{\text{C}}$—An**

| Precursor | Solvent/base | Temp (°C) | Rearrangement[a] (%) | $k_{SOH}/k_{(Ph)}$ | Reference |
|---|---|---|---|---|---|
| Ph(An*)C=C(Br)An[b] | 60% EtOH/2,6-lutidine | 125 | 5 ± 1 | 38 ± 9 | [9] |
| (5:1 E/Z) | 80% EtOH/2,6-lutidine | 115 | 4 ± 1.5 | 48 ± 20 | [9] |
|  | AcOH/NaOAc | 120 | <3 | ≥65 | [9] |
|  | AcOH/NaOAc/Bu$_4$NBr | 120 | <3 | ≥65 | [9] |
|  | Me$_3$CCOOH/Me$_3$CCOONa | 160 | 0 |  | [9] |
|  | TFE/2,6-lutidine | 120 | 53 ± 6 | 1.77 ± 0.4 | [9] |
|  | 1:1 AcOH–HCOOH/NaOAc | 95 | 68 ± 4[c] |  | [9] |
| (3:2 E/Z) | 1:1 AcOH–HCOOH/NaOAc | 95 | 56 ± 4[c] |  | [9] |
| E-Ph(An)[13]C=C(Br)An | AcOH/AgOAc | Reflux | 0 |  | [166] |
| Z-Ph(An)[13]C=C(Br)An | AcOH/AgOAc | Reflux | 0 |  | [166] |

[a] Complete scrambling of the α- and β-carbons or of the anisyl groups between the α and β positions amounts to 100% rearrangement.

[b] An* = $p$-CD$_3$OC$_6$H$_4$.

[c] Approximate value.

reaction, and it is therefore more reactive in its reaction with the solvent than is the solvolytically generated ion.

The last degenerate rearrangement of a β-phenyl group between anisyl-substituted migration origin and terminus [Eq. (61)] was investigated both with β-$^{13}$C-labeled [166] and with β-anisyl-labeled [9] 1,2-dianisyl-2-phenylvinyl bromides **158** and **159**. No rearrangement was observed in AcOH/NaOAc or AcOH/AgOAc or in pivalic acid [51] (Table 6.40). Some complications that accompanied the nmr and mass spectral analysis suggest that the small extent of rearrangement obtained in aqueous ethanol may contain a large error, but the relatively high value (53%) in TFE is authentic. Large extents of rearrangement were also obtained in a 1:1 AcOH–HCOOH mixture, but the products are the ketones **182** and **182*** [Eq. (73)], which interconvert in the reaction medium [9]. Consequently, the large extent of rearrangement cannot be unequivocally associated with β-phenyl rearrangement in ion **165**. These solvent effects resemble those observed for the rearrangement of trianisylvinyl bromide.

The results of Tables 6.36 through 6.40 together with the other data on rearrangements during the solvolysis of compounds **153** to **160** make possible a detailed analysis of the effects of the solvent, the migrating group, and the groups at the migration origin and terminus on the facility of these β-aryl migrations across the double bond [9, 166]. The transition states for the migrations can be schematically represented by structures **168** and **189** to **193**. Structures **189, 191, 192,** and **193** are those for the degenerate rearrangements, and **168** and **190** can be obtained in two ways, either when the migration origin is substituted by a phenyl or when it is substituted by an anisyl group.

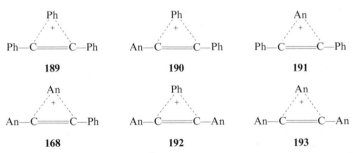

The relative extents of rearrangement in several solvents via these transition states are compared in Table 6.41 [9, 166] with similar data for saturated compounds [169, 176]. It is clear that TFE, which is a good rearrangement medium for saturated systems [177, 178], is the best of these solvents for promoting the β-aryl rearrangement. Its low nucleophilicity enables $k_r$ to greatly exceed $k_{SOH}$. Even degenerate rearrangements that are unfavored in other solvents proceed to an appreciable extent in this solvent. The other extreme is the low-dielectric pivalic acid, in which only the rearrangement with the largest driving force (via **168**) takes place. The abilities of AcOH and 60% EtOH to promote the different rearrangements appear to be roughly the same.

Table 6.41 also shows that the anisyl group stabilizes a positive charge better than a phenyl group, and for the same combination of migration origin and terminus (either phenyl- or anisyl-substituted) anisyl bridging is favored over phenyl bridging. Quantitative comparison is not possible in TFE, in which all the anisyl migrations are nearly complete. The $k_{r(An)}/k_{r(Ph)}$ values for the **192–193** pair are >22 in TFE and >11.6 in AcOH, in line with the high values obtained by the internal competition in the ion **162** [164]. The low value of 1.65 for 60% EtOH suggests that $k_{r(Ph)}$ for **165** in this solvent is erroneous.

Substitution at the migration origin and terminus shows similar effects. Migration to a phenyl-stabilized terminus is always preferred over migration to an anisyl-stabilized terminus (see reactions via **190** versus **192**, and **168** versus **193**). Migration from an anisyl-substituted origin to a phenyl-substituted terminus is therefore favored over the corresponding degenerate rearrangement (see reactions via **168** versus **191** and **193**, and **190** versus **192**).

**TABLE 6.41**

**Extents of β-Aryl Rearrangement in the Solvolysis of 1,2,2-Triarylvinyl Systems (A) and in the Deamination of 1,2,2-Triarylethylamines (B)**[a,b]

| Solvent | Transition state (A) | | | | | | Reference |
| | 189 | 190 | 191 | 168 | 192 | 193 | |
|---|---|---|---|---|---|---|---|
| TFE | 61[c] | 85 (0)[b] | 100[c] | 100 (0)[b] | 53 | 100 | [8, 9, 64, 164, 165, 167b] |
| AcOH | 13.4 | ~0 | 93 | 100 | <3 | 35 | [9, 11, 50, 64, 164–166] |
| 60% EtOH | | 5 | 89 | 100 | 5 | 11.5 | [8, 9, 64, 164] |
| Me₃CCOOH | | (0)[b] | | 100 | 0 | 0 | [51] |

| Solvent | Transition state (B) | | | | Reference |
| | 194 | 195 | 196 | 197 | |
|---|---|---|---|---|---|
| H₂O | 54 | (0.9)[b] | 90.8 | 33.4 | [169, 176] |

[a] From Houminer et al. [9].

[b] The migration origin is the carbon drawn on the left in structures **168** and **189–197**, except for values in parentheses, in which case the migration origin is the carbon drawn on the right.

[c] This value is for 35% reaction and is probably lower at lower reaction percentages.

In contrast, rearrangement from an α-anisyl-substituted vinyl cation to an α-phenyl-substituted cation was never observed, and transition states **190** and **168** are formed exclusively in the direction of the phenyl-substituted migration terminus. Consequently, anisyl migration from an anisyl-substituted migration origin to a phenyl-substituted migration terminus (via **168**) is the most favored process, and phenyl migration between two anisyl-substituted centers (via **192**) is the least favored.

Data for comparison with β-aryl rearrangements to an sp²-hybridized cationic center are very limited. The acetolysis of C-1-labeled 1,2,2-triphenyl-ethyl tosylate gave 39% net phenyl migration from $C_\alpha$ to $C_\beta$ [171c, 172a], giving a $k_{SOH}/k_{r(Ph)}$ ratio of 1.13 for migration via transition state **194**. The extent of migration is much smaller (6,7% net phenyl migration) in the vinylic analogue, for which $k_{SOH}/k_{r(Ph)} = 26$ [65, 165]. Whether this difference is dominated by a higher $k_{r(Ph)}$ or by a lower $k_{SOH}$ in the saturated system or whether it is due to an appreciable change of $k_r$ and $k_{SOH}$ in both systems is not known. More data are available from the work of Bonner and co-workers on the rearrangement that accompanies the deamination of 1,2,2-triarylethylamines in water, which proceeds via open ions [169, 176] and the transition states of which are **194** to **197**. There are three differences between the vinylic and the saturated systems (Table 6.41). (1) Phenyl migration in

Ph—CH————CH—Ph
**194**

An—CH————CH—An
**195**

An—CH————CH—Ph
**196**

An—CH————CH—Ph
**197**

the 1,2,2-triphenylethyl cation via **194** exceeds the anisyl migration in the 1,2,2-trianisylethyl cation via **195** [176] whereas the opposite is true with the vinylic analogues. (2) There is a small extent of phenyl migration to an anisyl-substituted terminus via **196** [169], but none was observed in the vinylic analogues, e.g., via **190**. (3) The anisyl migration in the 2,2-dianisyl-1-phenylethyl cation via **197** is extensive but not complete [169], whereas the vinylic analogue gave a complete rearrangement via **168**.

Bonner ascribed the greater extent of migration in the 1,2,2-triphenylethyl cation to a lower electrophilicity of the migration terminus in the 1,2,2-trianisylethyl cation due to a more extensive charge delocalization on the anisyl group [176]. However, this effect reduces both $k_{SOH}$ and $k_r$, and the situation observed by Bonner is apparently due to the fact that $k_r$ is more sensitive than $k_{SOH}$ to a change from an $\alpha$-phenyl- to an $\alpha$-anisyl-substituted cation. For the 1,2,2-triphenylethyl cation **198**, $k_{H_2O}/k_{r(Ph)} = 3.4$ and, for the 1,2,2-trianisylethyl cation **199**, $k_{H_2O}/k_{r(An)} = 8.0$. The reverse is true for the vinyl cations; $k_{AcOH}/k_{r(An)} = 5.6$ for **166**, and $k_{AcOH}/k_{r(Ph)} = 26$ for **161**. If a

$$Ph_2CH\overset{+}{-}CH-Ph \qquad An_2CH-\overset{+}{C}H-An$$

$$\textbf{198} \qquad\qquad\qquad \textbf{199}$$

migratory aptitude ratio $k_{r(An)}/k_{r(Ph)} > 20$ is used, then **166** is captured at least 4.4 times faster by AcOH than is **161**. This can be explained if the replacement of the nonmigrating $\beta$-phenyl by a $\beta$-anisyl group increases the extent of rearrangement in the vinylic system $\geq 4.4$-fold.

Differences (2) and (3) can be understood if the demand for charge dispersal by an anisyl group is greater in the vinylic system than in the saturated one. However, all the differences may well be due to the different solvent used. The possibility that the difference is due to the fact that the trigonal ions are formed by deamination and are less selective than the solvolytically generated vinyl cations is incompatible with the even larger difference between **161** and **166**, which are formed by deamination.

**TABLE 6.42**

**Relative Rate Constants for Several Processes Involving Triarylvinyl Cations**

| Ion | Solvent | Temp (°C) | $k_{Br}{}^a$ | $k_{SOH}$ | $k_{r(Ph)}$ | $k_{r(An)}$ | Reference |
|-----|---------|-----------|-----------|-----------|-------------|-------------|-----------|
| | | | | Relative rate constant | | | |
| **161** | TFE | 140 | 42 | 1 | $0.77^b$ | | [8, 167b] |
| **162** | TFE | 120 | 97 | 1 | 5.7 | | [8] |
| | 60% EtOH | 160 | | 1 | 0.053 | 4 | [8, 164] |
| | AcOH | 117 | | 1 | 0.053 | 6.3 | [164] |
| **163** | TFE | 140 | | 1 | | >50 | [8] |
| | 60% EtOH | 140 | | 1 | | >50 | [8] |
| | AcOH | 160 | | 1 | | >50 | [64] |
| **165** | TFE | 120 | | 1 | 0.56 | | [9] |
| | 80% EtOH | 115 | 3.1 | 1 | 0.021 | | [9, 55] |
| | AcOH | 120 | 2.1 | 1 | <0.01 | | [9, 36] |
| **166** | TFE | 120 | 78 | 1 | | 12.5 | [9, 10] |
| | AcOH | 120 | 18 | 1 | | 0.18 | [9, 50] |

$^a$ At 1 $M$ Br$^-$.

$^b$ Maximal value since the extent of rearrangement increases with time. The value relates to reaction within the ion pair and should not be compared with the other values in the table.

The above data for the various processes in which the free triarylvinyl cations are involved can be summarized in terms of relative rate constants compared with $k_{SOH}$ (Table 6.42). Comparison with the second-order constant for capture by $Br^-$ is done at a concentration of 1 $M$ $Br^-$. The capture by $Br^-$ is usually the most favored process except when the β-anisyl migration gives an anisyl-substituted transition state and an anisyl-stabilized product. β-Anisyl migration is usually faster than capture by the solvent except when the driving force for it is inherently low, whereas phenyl migration is slower than capture by the solvent except in a solvent of low nucleophilicity. The important conclusion is that these triarylvinyl cations have sufficiently long lifetimes so that they are able to participate in several processes before being quenched by the solvent.

Recent work investigated the rearrangements in tolyl-substituted systems [35, 52, 168]. The charge dispersal ability of the tolyl group is intermediate between those of the phenyl and the anisyl groups, and it is expected that the behavior of the tolyl-substituted systems will be intermediate between those of the phenyl- and the anisyl-substituted systems. This is the case with three different systems, and the results are given in Table 6.43.

β-Anisyl rearrangement of 2,2-dianisyl-1-tolylvinyl bromide (200) is complete in TFE and in AcOH and is extensive, but not complete, in the more nucleophilic 60% EtOH [Eq. (78)] [168]. The reaction probably proceeds

$$An_2C\!\!=\!\!C(Br)Tol \xrightarrow[\text{2,6-lutidine}]{60\% \text{ EtOH}} An_2CH\!-\!\overset{\overset{\displaystyle O}{\|}}{C}\!-\!Tol + AnCH(Tol)\!-\!\overset{\overset{\displaystyle O}{\|}}{C}\!-\!An \qquad (78)$$

$$\underset{\textbf{200}}{\phantom{An_2C}} \qquad\qquad\qquad \underset{\substack{\textbf{201}\\(14\%)}}{\phantom{An_2CH}} \qquad\qquad \underset{\substack{\textbf{202}\\(86\%)}}{\phantom{AnCH(Tol)}}$$

via the open ions **203** and **204**, since the products are nearly 1:1 mixtures of the E- and the Z-trifluoroethyl vinyl ethers **205** and **206** in TFE and the vinyl acetates **207** and **208** in AcOH. The same product distributions are obtained from E- and Z-1,2-dianisyl-2-tolylvinyl bromides **209** and **210** under the same conditions (Scheme 6.31) [168].

These results fit nicely with the 20% p-tolyl rearrangement in the 1-phenyl-2,2-ditolylvinyl cation in AcOH [180] and the complete β-anisyl rearrangement in the 2,2-dianisyl-1-phenylvinyl cation **163** [8, 64]. In a good rearrangement solvent, such as TFE, or even in a moderate one, such as AcOH, the driving force for the formation of an α-anisyl-stabilized vinyl cation via an anisyl-stabilized transition state is sufficient for a complete migration to a tolyl-substituted migration terminus. This process competes, rather effectively, with capture by the nucleophilic solvent in 60% EtOH.

The extent of β-tolyl rearrangement to a tolyl-substituted migration terminus was investigated by Lee and co-workers [52] for the labeled

# TABLE 6.43
## Rearrangements in Tolyl-Substituted Systems

| Precursor | Solvent/base | Temp (°C) | Migrating group | Rearrangement$^a$ (%) | $k_{SOH}/k_r$ | Reference |
|---|---|---|---|---|---|---|
| An$_2$C=C(Br)Tol **200** | 60% EtOH/2,6-lutidine | 140 | An | 86 ± 2 | 0.16 | [168] |
| | TFE/2,6-lutidine | 140 | An | 100 | <0.02 | [168] |
| | AcOH/NaOAc | 140 | An | 100 | <0.02 | [168] |
| Tol$_2$$^{13}$C=C(Br)Tol **211** | AcOH/AgOAc | Reflux | Tol | 27 ± 1 | 10.8$^d$ | [52] |
| | CF$_3$COOH | 100 | Tol | 86$^b$ | 0.66$^d$ | [52] |
| | CF$_3$COOH/CF$_3$COOAg | rt | Tol | 93 ± 1 | 0.30$^d$ | [52] |
| | CF$_3$COOH/CF$_3$COONa | 100 | Tol | 94 | 0.26$^d$ | [52] |
| E-Tol$^{13}$C(Ph)=C(Br)Tol **213** | AcOH/AgOAc | Reflux | Ph | 4.0 | 48 | [35] |
| | CF$_3$COOH/CF$_3$COOAg | rt | Ph | 70 | 0.86 | [35] |
| Z-Tol$^{13}$C(Ph)=C(Br)Tol **214** | AcOH/AgOAc | Reflux | Ph | 3.0 | 65 | [35] |
| | CF$_3$COOH/CF$_3$COOAg | rt | Ph | 68 | 0.94 | [35] |
| 1:1 E,Z-Tol$^{13}$C(Ph)=C(Br)Tol **213 and 214** | CF$_3$COOH/CF$_3$COOAg | rt | Ph | 70 | 0.86 | [35] |
| | CF$_3$COOH | 100 | Ph | 90$^c$ | | [35] |

$^a$ Complete scrambling of the $\alpha$- and $\beta$-carbons amounts to 100% rearrangement for the degenerate rearrangements.
$^b$ Value at two half-lives. The value increases with the increase in reaction time up to 100% rearrangement.
$^c$ Value at two and one-half half-lives. The value increases to 98% with time.
$^d$ Value corrected for error in ref. 52.

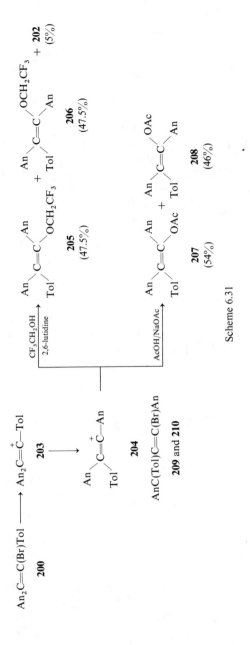

Scheme 6.31

$$\text{Tol}_2{}^{13}\text{C}=\text{C(Br)Tol} \longrightarrow \text{Tol}_2{}^{13}\text{C}=\overset{+}{\text{C}}-\text{Tol} \xrightarrow{\text{AcOH}}$$

$$\underset{\textbf{211}}{} \qquad\qquad \underset{\textbf{212}}{}$$

$$\text{Tol}_2{}^{13}\text{C}=\text{C(OAc)}-\text{Tol} + \text{Tol}_2\text{C}={}^{13}\text{C(OAc)Tol} \quad (79)$$

$$86.5\% \qquad\qquad 13.5\%$$

tri-$p$-tolylvinyl bromide **211** [Eq. (79)]. The extent of rearrangement in AcOH/AgOAc (27%) is intermediate between the values for triphenylvinyl bromide (13.6%) and trianisylvinyl bromide (40%) [9, 62]. The decrease in the $k_{\text{SOH}}/k_r$ values with the increase in the positive charge stabilizing ability of the aryl group is the outcome of opposing effects. Increased stabilization by the better migrating group will increase $k_r$, and reduced electrophilicity of the α-carbon will reduce $k_{\text{SOH}}$. However, these effects are apparently more than counterbalanced by the smaller demand for bridging in the more stable ion and the reduced electrophilicity of the migration terminus, which will decrease $k_r$.

A similar gradation of effects along the series An > Tol > Ph is shown by the very small, but still observable, extent (3–4%) of migration of the phenyl group between the tolyl-substituted migration origin and terminus of either $E$- or $Z$-2-phenyl-1,2-ditolyl-2-$^{13}$C-vinyl bromide (**213** or **214**, respectively) in AcOH [Eq. (80)][35]. The products were the acetates **216** and **217**, and

$$\text{Tol}-{}^{13}\text{C(Ph)}=\text{C(OAc)Tol} + \text{Tol(Ph)C}={}^{13}\text{C(OAc)Tol} \quad (80)$$

$$\underset{\textbf{216}}{} \qquad\qquad\qquad \underset{\textbf{217}}{}$$

each of them was a 1:1 $E/Z$ mixture. The increase in the $k_{\text{SOH}}/k_r$ values (Tables 6.39, 6.40, 6.43) from 25.4 for the triphenylvinyl cation **161** [165] to 57 ± 9 for the 2-phenyl-1,2-ditolylvinyl cation **215** [35] to ≥100 for the 1,2-dianisyl-2-phenylvinyl cation **165** [9, 166] is consistent with the decreased need for bridging (lower $k_r$), which more than compensates for the corresponding decrease in $k_{\text{SOH}}$, which is due to the more delocalized cation.

Comparison of the degenerate migrations in the ions **212** and **215**, in which the migration origin and terminus and presumably also the $k_{\text{SOH}}$ values are identical, gives a $k_{r(\text{Tol})}/k_{r(\text{Ph})}$ ratio of ca. 5.3 [35]. This ratio differs somewhat from the ratio of 15.7 found for the pinacol rearrangement in the saturated analogues [171a].

Much larger extents of rearrangement in the reactions of **211**, **213**, and **214** were found in trifluoroacetic acid, as expected for this solvent of low nucleophilicity (Table 6.43). However, the solvent isotope effect $k_{CF_3COOH}/k_{CF_3COOD}$ for **211** in unbuffered acid was 2.6, and in the presence of $CF_3COOAg$ it was 1.6 [52], whereas for unbuffered **214** or a 1:1 mixture of **213** and **214** the values were 3.4 and 3.9, respectively [35] (Table 6.4). These values strongly indicate that electrophilic addition of a proton from the solvent to the double bond is faster than the C—Br bond heterolysis in the unbuffered acid. The sequence of Scheme 6.32 accounts for the formation of both the rearranged and the unrearranged trifluoroacetates via initial formation of the trigonal ions **218** and **219**. A similar scheme can account for the rearrangements during the solvolyses of **213** and **214** in the absence of silver salts [35]. It was suggested that the much faster reaction in the presence of $AgOCOCF_3$ is initiated by an assisted C—Br bond heterolysis, so that the scrambling measures genuine rearrangement in the vinyl cation [35, 52]. Moreover, the identical extents of rearrangement in the presence of $AgOCOCF_3$ and $CF_3COONa$ were interpreted as indicating rearrangement via the vinyl cation also in the presence of $CF_3COONa$ [52]. However, the isotope effect in the latter case casts doubt on this explanation. Moreover, the trifluoroacetate mixture from the unbuffered acetolysis of **211** undergoes scrambling,

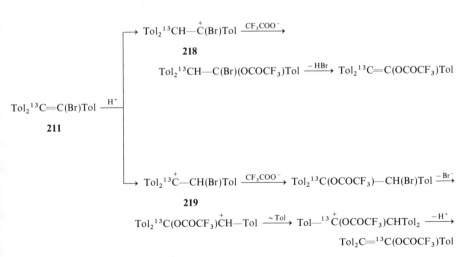

Scheme 6.32

and a sample that was only 86% scrambled after two half-lives is completely scrambled after seven half-lives [52]. The possibility was raised that this is due to formation, by an addition–elimination, of an unrearranged saturated vinyl trifluoroacetate, which subsequently ionizes to a species that undergoes a 1,2-tolyl shift. This is unlikely in view of the absence of an $S_N1$ route for the solvolysis of 1-anisyl-2-methylpropen-1-yl trifluoroacetate [32], and Scheme 6.32 seems more likely. Since trianisylvinyl bromide will undergo an even faster addition of $CF_3COOH$, the results in this solvent cannot be taken as representing unequivocally rearrangement via the open triarylvinyl cations.

An interesting comparison between the migration of $p$-anisyl and $o$-anisyl groups is given by the work of Taniguchi and co-workers [54]. The reaction of 2,2-bis($o$-methoxyphenyl)-1-phenylvinyl bromide (**134**) is discussed in Section V,A. The main product is the benzo[$b$]furan **136** with the unrearranged skeleton, whereas the benzo[$b$]furan **137** is obtained by an initial $o$-anisyl migration [Eq. (43)]. In basic 50% EtOH and in AcOH/AgOAc, **137** consists of 7 and 9% of the product, respectively, whereas only the unrearranged benzo[$b$]furans were obtained from the $\alpha$-anisyl analogue **61** [Eq. (42)] or from the $\alpha$-tolyl analogue.

Likewise, no rearrangement was observed in the reactions of $E$- and $Z$-1-anisyl-2-($o$-methoxyphenyl)-2-(2,5-dimethoxyphenyl)vinyl bromides **148** and **149**, respectively or their 1-phenyl analogues **220** and **221**, respectively [91]. Although two benzo[$b$]furans were obtained, both are substituted by the formerly $\alpha$-aryl group at the 2 position (Table 6.44), whereas a 2-$o$-anisyl-substituted benzo[$b$]furan is the expected product from rearrangement of either the $\beta$-$o$-anisyl or the $\beta$-2,5-dimethoxyphenyl group.

The preference for the benzo[$b$]furan formally obtained by attack on the $o$-methoxy group of the aryl trans to the leaving group suggests that the cyclization occurs earlier than in the free ion stage. This may be either via 5-$O$ participation or in the ion pair in which the $o$-methoxy of the $cis$-aryl group is shielded by the leaving group to attack by the positive charge. Hence, whereas $p$-anisyl rearrangement is complete for the ion **163** [8] and is 86% for the ion **203** [168] in aqueous ethanol, the extent of the $o$-anisyl

|         163          |         203          |         135          |         222          |

rearrangements in the analogous ions **135** and **222** is only 7 and 0%, respectively [54]. The $k_{SOH}/k_{r(An)}$ values for **163** and **203** are $\geq 400$ times lower than the $k_{cyclization}/k_{r(o\text{-}An)}$ value. Since the $k_{r(An)}$ and $k_{r(o\text{-}An)}$ ratios should not differ much, the capture of the cation by the internal OMe nucleophile is

**TABLE 6.44**

**Benzo[b]furans from the Reactions of Triarylvinyl Bromides**[a]

| Substrate | α-Ar | Solvent | Temp (°C) | Products (%) | |
|---|---|---|---|---|---|
| | | | | 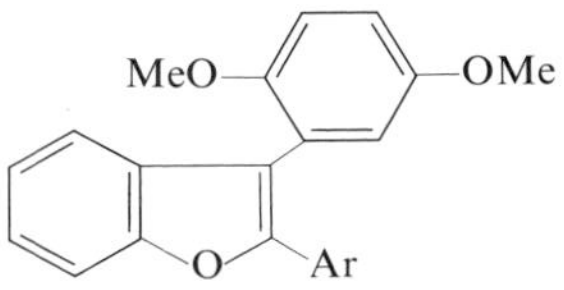 | 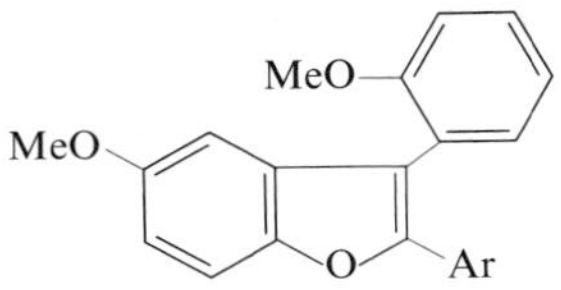 |
| 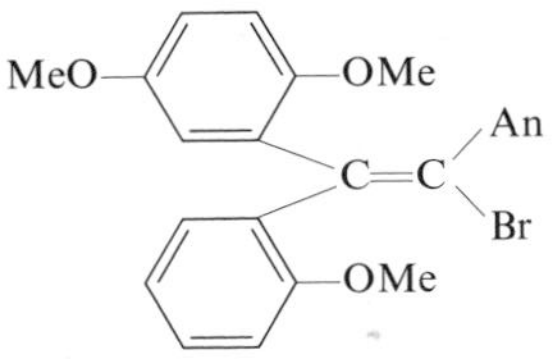 **148** | An | 80% EtOH<br>AcOH | 120<br>118 | 33<br>34 | 67<br>66 |
| 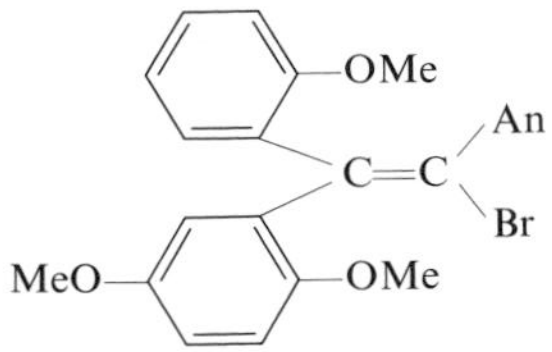 **149** | An | 80% EtOH<br>80% EtOH/Ag⁺<br>AcOH/NaOAc | 120<br>120<br>118 | 68<br>60<br>70 | 32<br>40<br>30 |

(continued)

**TABLE 6.44** (*Continued*)

| Substrate | α-Ar | Solvent | Temp (°C) | Products (%) | |
|---|---|---|---|---|---|
| 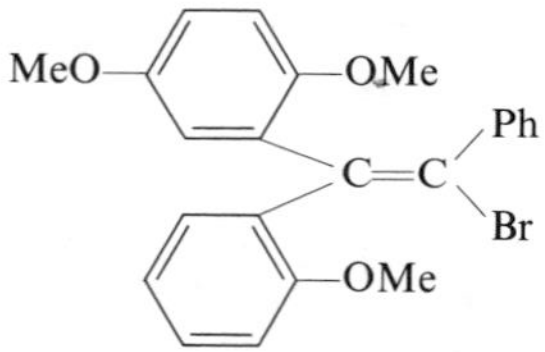 **220** | Ph | AcOH/Ag⁺ | 118 | 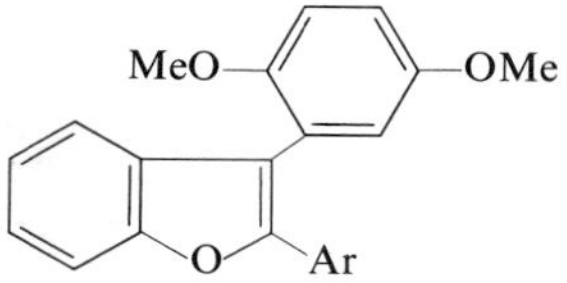 36 | 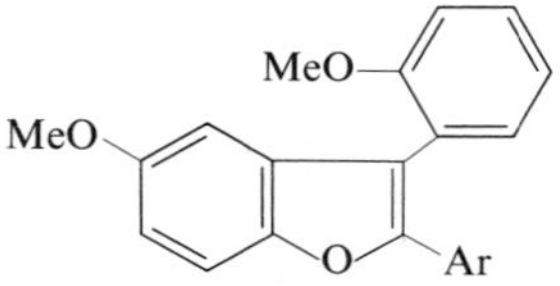 64 |
| 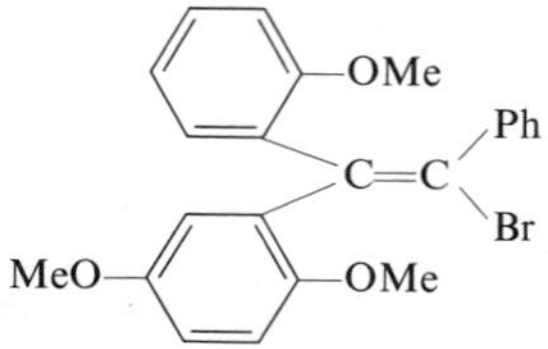 **221** | Ph | 50% EtOH | 160 | 74 | 26 |
| | | AcOH/Ag⁺ | 118 | 62 | 38 |
| | | 80% EtOH/Ag⁺ | 150 | 64 | 36 |

$^a$ From [91].

430

**TABLE 6.45**

**Intermediates in β-Aryl Rearrangements across the Double Bond**

| Compound | Solvent | Migrating group | Cationic intermediate | Evidence | Reference |
|---|---|---|---|---|---|
| E-ArC(Me)=C(OTf)Me | 60% EtOH, 97% TFE | Ar | Bridged | Stereochemistry, isotope effects, kinetics, substituent effects | [79, 181] |
| Z-ArC(Me)=C(OTf)Me | 97% TFE | Ar | Ion pair | Kinetics, substituent effects, stereochemistry | [79, 181] |
| Ph$_2$C=C(X)Ph (X = Br, OTf) | RCOOH | Ph | Ion pair | No capture by RCOO$^-$ | [65, 165] |
|  | TFE | Ph | Ion pair | No capture by Br$^-$, ion pair return | [167b] |
|  |  |  | Free ion? | Independence of X |  |
| An$_2$C=C(Br)Ph | Me$_3$CCOOH | An | Ion pair | Stereochemistry | [51] |
| An$_2$C=C(Br)Ph | AcOH, TFE | An | Free ion | Stereochemistry, kinetics | [8, 64] |
| An(Ph)C=C(Br)Ph | TFE, aq. EtOH | Ph, An | Free ion | Stereochemistry | [8, 164] |
|  |  |  | Free ion? | Only partial capture by ArS$^-$ |  |
| Tol$_2$C=C(N=NNHPh)Ph | AcOH | Tol | Free ion | Capture by AcO$^-$ | [180] |
| An$_2$C=C(Br)Tol | AcOH, TFE | An | Free ion | Stereochemistry | [168] |
| Tol$_2$C=C(Br)Tol | AcOH | Tol | Free ion? |  | [52] |
| TolC(Ph)=C(Br)Tol | AcOH | Ph | Free ion | Stereochemistry | [35] |
| An$_2$C=C(Br)An | AcOH, TFE | An | Free ion | Capture by Br$^-$ | [9] |
| An$_2$C=C(N=NNHPh)An | AcOH | An | Free ion | Capture by AcO$^-$ | [167] |

two or three orders of magnitude faster than capture by external analogous nucleophiles, such as EtOH or $H_2O$. Entropy considerations of the juxta-positions of the positively charged carbon and the methoxy group are mainly responsible for this difference.

The extensive study of rearrangements discussed above makes it possible to draw conclusions concerning the structural dependence of the nature of the cationoid intermediates in rearrangements across the double bond [9, 35]. The intermediates and the evidence for their intermediacy are given in Table 6.45, which shows that the nature of the $\alpha$ substituent is one of the important factors determining the nature of the intermediate. Convincing evidence for a bridged transition state or intermediate exists only for the $\alpha$-methyl-substituted system, as discussed in Chapter 7, with an aryl group at a position trans to the leaving group. Ion pairs are probably involved in the rearrangement of the geometric isomers of these compounds [79, 181]. Bridging by a phenyl group or an ion-pair intermediate was suggested for the $\beta$-phenyl rearrangement of triphenylvinyl bromide and triflate [65, 165], but further evidence, especially concerning the capture of the intermediate at higher concentrations of $RCOO^-$, is required. Strong evidence for the intermediacy of ion pairs exists for the rearrangement of 2,2-dianisyl-1-phenylvinyl bromide in the low-dielectric pivalic acid [51]. For some other $\alpha$-phenyl-substituted systems, the rearrangement is faster than the capture, even by powerful nucleophiles, and, although this may be taken as evidence for the intervention of ion pairs [164], this is not supported by the kinetics or the stereochemistry of the products, which suggest a nonbridged transition state and a free ion as an intermediate. Long-lived free ions that are capable of capture before rearrangement are the intermediates in the solvolysis–rearrangement of systems substituted by $\alpha$-tolyl [35, 52, 168, 180] and $\alpha$-anisyl [9] groups. However, it should be realized that the stereochemical evidence is related to the product-forming reactions of the rearranged ion, whereas the initially formed species before the rearrangement can still be an ion pair.

## VII. NONSOLVOLYTIC METHODS FOR GENERATING $\alpha$-ARYLVINYL CATIONS FROM VINYLIC PRECURSORS

The generation of $\alpha$-arylvinyl cations by electrophilic addition to aryl-acetylenes and by solvolysis is discussed at length in Chapter 3 and in this chapter. Several additional routes, which are related to the solvolysis and which have occasionally been used to generate $\alpha$-arylvinyl cations, are mentioned briefly in this section.

## A. Fragmentation

Fragmentation reactions were investigated by Grob and co-workers [182–185] as a possible source of α-arylvinyl cations. The first suggestion [1] for heterolytic formation of a vinyl cation was for the thermal decarboxylation in water of the potassium salts of cis and trans $\alpha,\beta$-unsaturated $\beta$-bromo acids **223** and **224**, including the derivatives of acrylic, crotonic and cinnamic acids [182]. The sole products from the cis $(E)$ series were the corresponding alkynes [Eq. (81)], whereas the trans $(Z)$ isomers gave both the alkynes and ketones [Eq. (82)]. Electron-donating substituents increased the reaction rates in both series, and the cis salts reacted faster than their trans isomers, e.g., when R = Ph, $k_{223}/k_{224} \sim 90$. The reactions are much faster than the solvolyses of the corresponding α-bromostyrenes [2].

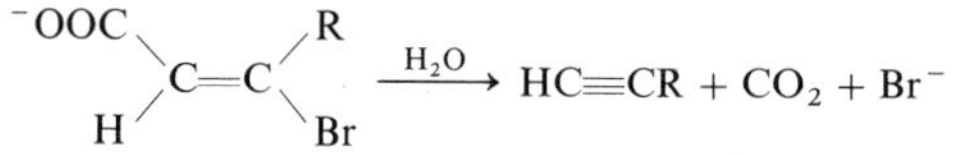

$$\underset{H}{\overset{^-OOC}{>}}C=C\underset{Br}{\overset{R}{<}} \xrightarrow{H_2O} HC\equiv CR + CO_2 + Br^- \quad (81)$$

**223**

$$\underset{H}{\overset{^-OOC}{>}}C=C\underset{R}{\overset{Br}{<}} \xrightarrow{H_2O} HC\equiv CR + RCOCH_3 + CO_2 + Br^- \quad (82)$$

**224**

It was suggested that the cis salts decarboxylate in a concerted one-step mechanism via transition state **225** without the intermediacy of vinyl cations. However, a two-step mechanism via a rate-determining formation of the zwitterion **226**, which may be internally solvated by the carboxylate group as in **227**, was suggested for the trans isomers. Loss of $CO_2$ leads to the alkyne, whereas a reaction with water followed by decarboxylation leads to the ketone.

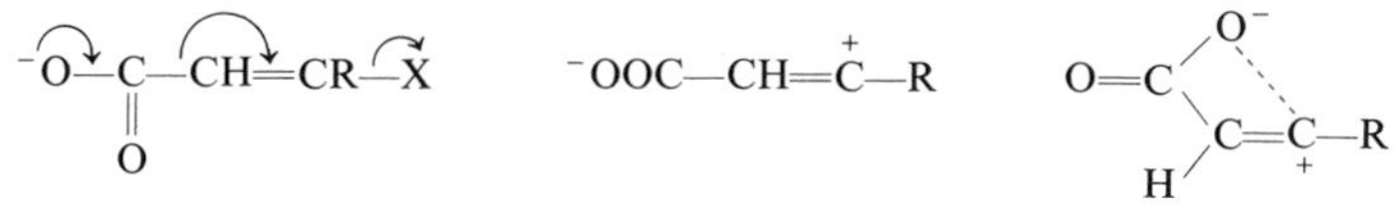

$$^-O-\overset{\overset{||}{C}}{\underset{O}{C}}-CH=CR-X \qquad ^-OOC-CH=\overset{+}{C}-R \qquad O=C\overset{O^-}{\underset{H}{<}}\overset{}{C}=\overset{+}{C}-R$$

**225**                        **226**                        **227**

A more clear-cut example of the formation of vinyl cations by fragmentation was given by Grob and Wenk [184]. Acyclic *anti*-methyl vinyl ketoximes in which the two π systems are orthogonal can undergo π-3-assisted Beckmann rearrangement via a substituted azirine cation [183]. However, when planarity is enforced on the system, the π-3 participation is suppressed due to the geometric constraints, and fragmentation leading to a vinyl cation is observed, provided that the cation is stabilized by an electron-donating α-aryl group.

**TABLE 6.46**

**Rates and Products for the Reactions of the Tosylates 228**[a]

| Ar | $10^6 k_f$ at 90° | **234** (%) | **231 + 232** (%) |
|---|---|---|---|
| Ph | — | $83^b$ | — |
| An | 2.72 | $61^b$ | $11^b$ |
| $p\text{-Me}_2\text{NC}_6\text{H}_4$ | 5330 | $10^c$ | $63^c$ |

[a] From Grob and Wenk [184].
[b] At 110°.
[c] At 80°.

The reaction of 2-aryl-3-methyl-2-cyclopenten-1-one oxime tosylates **228** in 80% EtOH gave the hydrolysis products **234** and the E- and the Z-enol ethers **231** and **232**, respectively, in a 1:1 ratio [184]. The product distributions and the fragmentation rate constants $k_f$ are given in Table 6.46. It is suggested that **231** and **232** are obtained from the vinyl cation **230**, which is generated by the fragmentation of **228** via the transition state **229**. A competing reaction of 4-aryl participation gives the phenonium ion **233**, which is hydrolyzed to the ketone **234** (Scheme 6.33). Both processes are enhanced by electron-donating aryl groups, but the fragmentation is apparently more sensitive to the substituent effect since **228** when Ar = Ph gives no fragmentation, **228** when Ar = An gives only minor fragmentation, and only when Ar = $p\text{-Me}_2\text{NC}_6\text{H}_4$ does the fragmentation lead to the main product. Likewise, fragmentation does not take place at all when the two double bonds of the *anti*-vinyl ketoxime derivatives are not perpendicular [183].

Vinyl cations were also generated in the partial fragmentation of the vinyl bromides *E*-**235** and *Z*-**235**. The solvolysis of both isomers in 80% EtOH leads to unsubstituted arylacetylenes by fragmentation and to substituted arylacetylenes by elimination of a proton and to the capture products by water and ethanol [Eq. (35)] [185]. The product distribution depends on the nature of the substituent in the aryl group. The solvolysis is accompanied by an *E*-**235** ⇌ *Z*-**235** isomerization.

The substituent effect supports a carbenium ion mechanism, since electron-donating substituents accelerate the reaction considerably. The relative rates in the *E* series are *E*-**235a** (636), *E*-**235b** (7.6), *E*-**235c** (1), and those in the *Z*-series are *Z*-**235a** (245), *Z*-**235b** (4), *Z*-**235c** (1). The $k_{E\text{-}235}/k_{Z\text{-}235}$ ratios for the methoxy and methyl substituents (14 and 10.5, respectively) are higher than the value of 8.3 for the *p*-methoxy-$\beta$-methyl analogues **53** and

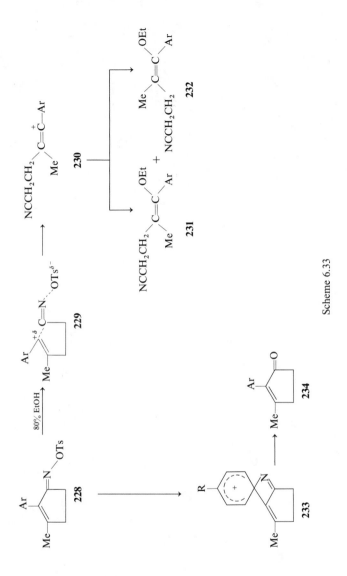

Scheme 6.33

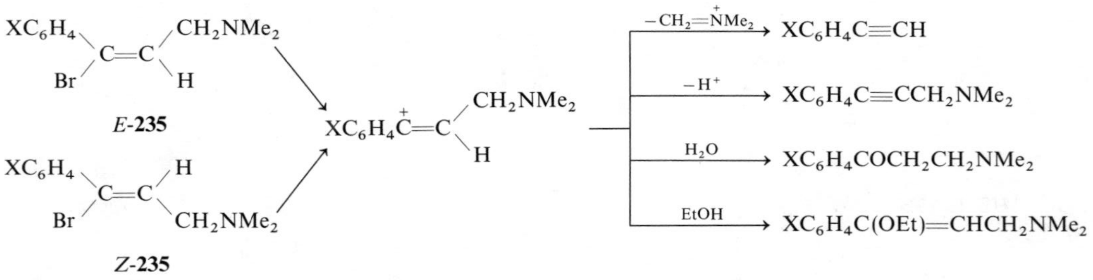

$$
\begin{array}{c}
\underset{\text{Br}}{\text{XC}_6\text{H}_4}\!\!>\!\!\text{C}\!=\!\text{C}\!<\!\!\underset{\text{H}}{\text{CH}_2\text{NMe}_2} \\
E\text{-}235
\end{array}
\qquad
\begin{array}{c}
\underset{\text{Br}}{\text{XC}_6\text{H}_4}\!\!>\!\!\text{C}\!=\!\text{C}\!<\!\!\underset{\text{CH}_2\text{NMe}_2}{\text{H}} \\
Z\text{-}235
\end{array}
$$

$$\text{XC}_6\overset{+}{\text{H}}_4\text{C}\!=\!\text{C}\!<\!\!\underset{\text{H}}{\text{CH}_2\text{NMe}_2}$$

$$\xrightarrow{-\,\text{CH}_2=\overset{+}{\text{NMe}}_2} \quad \text{XC}_6\text{H}_4\text{C}\!\equiv\!\text{CH}$$

$$\xrightarrow{-\,\text{H}^+} \quad \text{XC}_6\text{H}_4\text{C}\!\equiv\!\text{CCH}_2\text{NMe}_2$$

$$\xrightarrow{\text{H}_2\text{O}} \quad \text{XC}_6\text{H}_4\text{COCH}_2\text{CH}_2\text{NMe}_2$$

$$\xrightarrow{\text{EtOH}} \quad \text{XC}_6\text{H}_4\text{C(OEt)}\!=\!\text{CHCH}_2\text{NMe}_2$$

$$(83)$$

a   X = $p$-MeO
b   X = $p$-Me
c   X = H

**52** (R = Me) (Section III,B, Table 6.11), whereas the ratio for the unsubstituted compounds is 5.5. A similar explanation of ground state $\pi(Ar)$–$\pi(C=C)$ deconjugation holds also in this case.

An interesting result is that E-**235a** solvolyzes approximately three times faster than **53** (R = Me). Since $CH_2NMe_2$ should be inductively more electron withdrawing than methyl, the opposite is expected. A possible explanation for the weak acceleration may be a contribution to a C–C hyperconjugation by the dimethylamino group in the transition state.

The kinetic and product evidence for a vinyl cation intermediate indicates that the $Me_2NCH_2$ group is too weak an electrofugal group to direct the reaction to a concerted fragmentation route [185].

## B. Photolysis

Irradiation of a vinylic halide can lead to an initial heterolytic cleavage of the carbon–halogen bond. The vinyl radical formed in this way can give the corresponding vinyl cation by electron transfer to the halogen atom. Irradiation of the 2,2-di-p-anisyl, 2,2-di-m-anisyl, and 2,2-di-o-anisylvinyl bromides **236a** through **236c** gave mainly the diarylacetylenes **240a** through **240c** in 50 to 80% yield. The 2,2-di-o-anisyl derivative gave the 2- and 3-o-anisylbenzo[b]furans **241** and **242**, respectively, and irradiation of the 2,2-di-o-thioanisylvinyl bromide **236d** gave the 3-o-thioanisylbenzo[b]-thiophen **243** [186]. These transformations were accounted for by Scheme 6.34, in which the initially formed vinyl radical (**237**) was converted to a $\beta,\beta$-diarylvinyl cation (**238**) by electron transfer. The ion gave either cyclization similar to that discussed previously for $\beta$-o-anisylvinyl cations or rearrangement to the $\alpha,\beta$-diarylvinyl cation **239**, which then gave either elimination to **240** or cyclization to **241**. Products derived from the radical intermediate are omitted from Scheme 6.34, and it should be noted that except for the nature of the products other evidence for the intermediacy of vinyl cations is absent. Moreover, a direct cyclization of the radical **237** to **242** and **243** is also possible.

## C. Oxidative Decarboxylation

Oxidation of cinnamic acid and its p-chloro and p-nitro derivatives with lead tetraacetate gave the $\beta$-methylstyrenes as the exclusive or the main products via a vinyl radical intermediate. However, the decarboxylation of p-methoxycinnamic acid gave mainly cis- and trans-$\beta$-acetoxy-p-methoxy-styrenes **246** together with smaller amounts of 1,1,2-triacetoxy-2-anisyl-ethane **247** and p-methoxyphenacyl acetate **248**. A radical chain process involving an initial formation of the anisylvinyl radical **244**, followed by

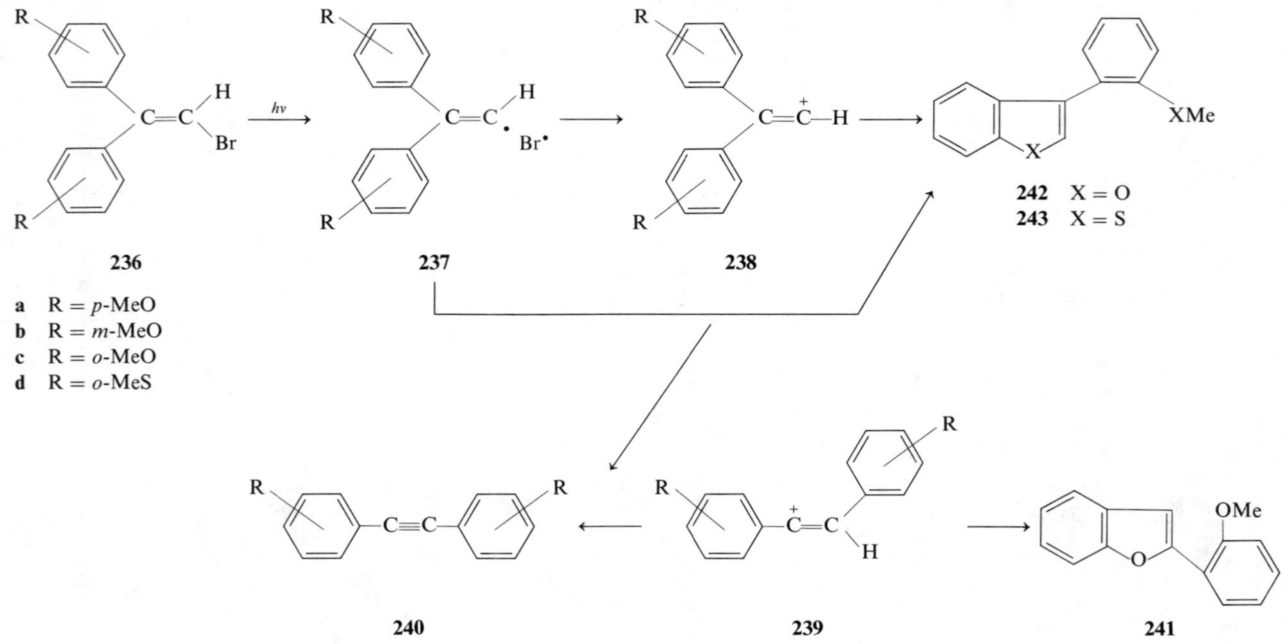

Scheme 6.34

oxidation of **244** to the corresponding cation **245** by Pb(IV) in an electron transfer process (Scheme 6.35), was suggested [187].

$$AnCH=CHCOOPb(IV) \longrightarrow \underset{\textbf{244}}{AnCH=CH^{\cdot}} + CO_2 + Pb(III)$$

$$AnCH=CHCOOPb(III) \longrightarrow \underset{\textbf{244}}{AnCH=CH^{\cdot}} + CO_2 + Pb(II)$$

$$AnCH=CH^{\cdot} + Pb(IV) \longrightarrow \underset{\textbf{245}}{AnCH=\overset{+}{C}H} + Pb(III) + AcO^{-}$$

$$AnCH=\overset{+}{C}H \xrightarrow{AcO^{-}} \underset{\textbf{246}}{AnCH=CHOAc} \xrightarrow[\text{few steps}]{Pb(OAc)_4} \underset{\textbf{247}}{AnCH(OAc)CH(OAc)_2} + \underset{\textbf{248}}{AnCOCH_2OAc}$$

Scheme 6.35

The intervention of an intermediate vinyl cation is supported by the lead tetraacetate oxidations of 3,3-diphenylacrylic acid and 3,3-dianisylacrylic acid **249**. Enol acetates **253** and **254** and dianisylacetylene **255** were formed only in the oxidation of the latter. The formation of the rearranged products **254** and **255** can be accounted for only by the intermediacy of the rearranged vinyl cation **252**, as shown in Scheme 6.36 [187], since 1,2-aryl migration does not occur in 2-arylvinyl radicals (such as **250**) (Section VI gives many examples of such rearrangements in vinyl cations) and such rearrangement is highly likely for the primary cation **251**.

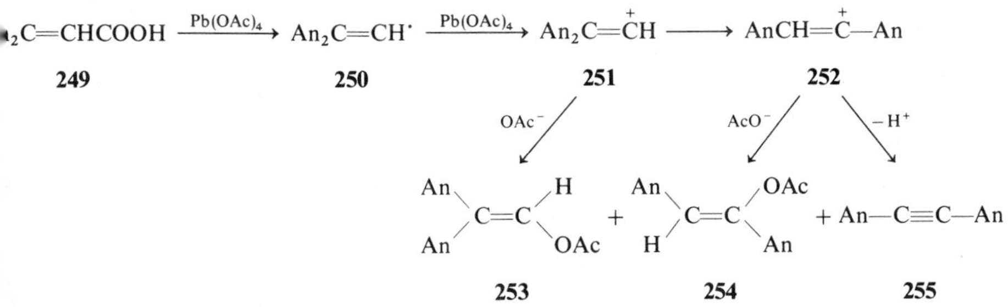

$$\underset{\textbf{249}}{_2C=CHCOOH} \xrightarrow{Pb(OAc)_4} \underset{\textbf{250}}{An_2C=CH^{\cdot}} \xrightarrow{Pb(OAc)_4} \underset{\textbf{251}}{An_2C=\overset{+}{C}H} \longrightarrow \underset{\textbf{252}}{AnCH=\overset{+}{C}-An}$$

Scheme 6.36

## D. Electroxidation

Electrooxidation of *cis*-1-(1-morpholino)-1,2-diphenylethylene (**256**) in aqueous *tert*-butanol with LiClO$_4$ as a supporting electrolyte, a platinum cathode, and a graphite anode gives deoxybenzoin **257** (26–31%), benzoin **258** (20–34%), and benzil **259** in 13 to 16% yield and morpholinium perchlorate. The same products were obtained from the deuterated compound, except that instead of deoxybenzoin a mixture of deoxybenzoin-α-*d*

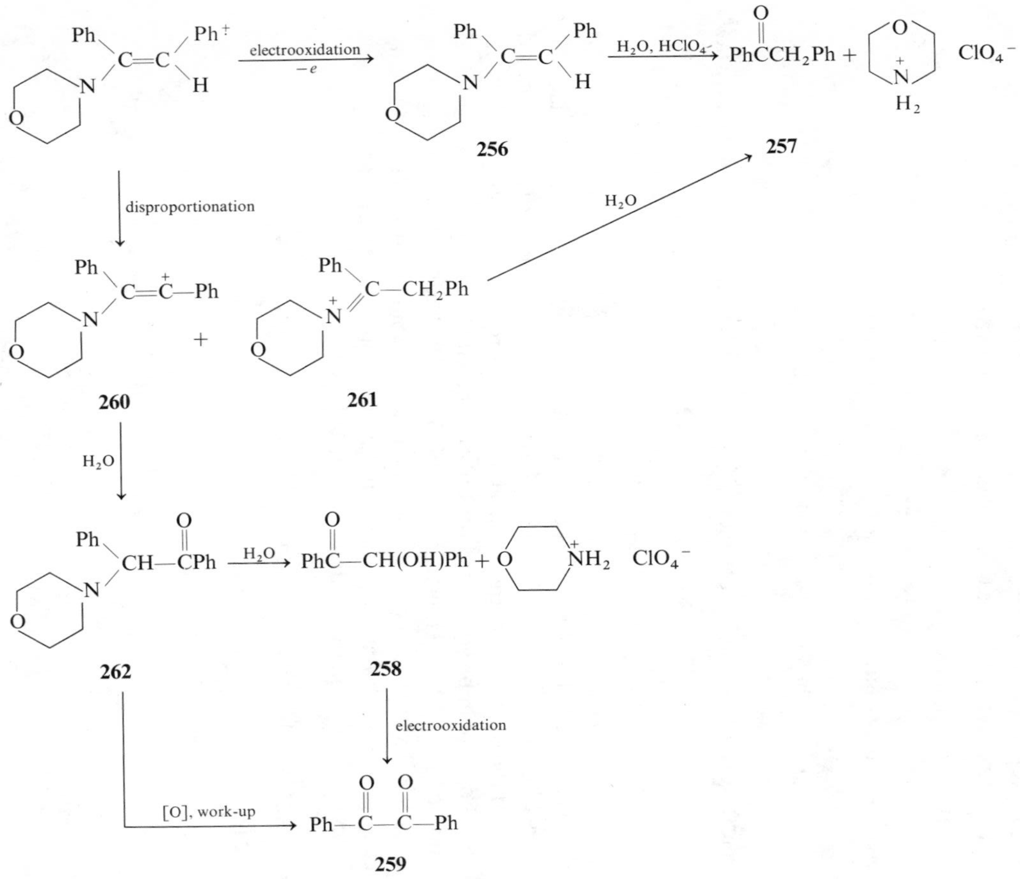

Ph — C(N-morpholine)=C(Ph)(H)  **256**

electrooxidation −e

$H_2O$, $HClO_4$

$PhCCH_2Ph$ + O (morpholinium, +N $H_2$) $ClO_4^-$

**257**

$H_2O$

disproportionation

**260**  +  **261**

$H_2O$

$O=C(Ph)—CH(morpholine)$  →  $PhC(=O)—CH(OH)Ph$ + O (morpholinium $\overset{+}{N}H_2$) $ClO_4^-$

**262**   **258**

[O], work-up   electrooxidation

$Ph—C(=O)—C(=O)—Ph$

**259**

Scheme 6.17

and deoxybenzoin-$\alpha,\alpha$-$d_2$ was formed in a 4:1 ratio [188]. The suggested reaction sequence for **256** (Scheme 6.37), which is supported by control experiments, involves the formation of **258** by solvolysis via the vinyl cation **260** and formation of **259** either by electrooxidation of **258** or during the work-up of the morpholino ketone **262**. Deoxybenzoin and its α-deuterio derivatives are the hydrolysis products of **256** or **256**-2-$d$, but the formation of deoxybenzoin-$\alpha,\alpha$-$d_2$ is a result of the electrooxidation reaction and indicates that the vinyl cation **260** is formed by disproportionation of the initially formed cation radical, which also gives **261**, the precursor of **257**. Some of **260** may also be formed by further oxidation of the cation radical, followed by loss of a proton.

## E. Mass Spectrometry

Huang and Lessard [189] also obtained the halogen analogues of **256**, i.e., **263** and **264**, as a mixture of the cis and trans isomers. The products of their solvolysis in water, probably via the ion **260**, were benzoin **258** and benzil **259** [Eq. (84)]. The same products were obtained in the presence of silver nitrate but in a much faster reaction. No further details on the solvolysis reaction were given.

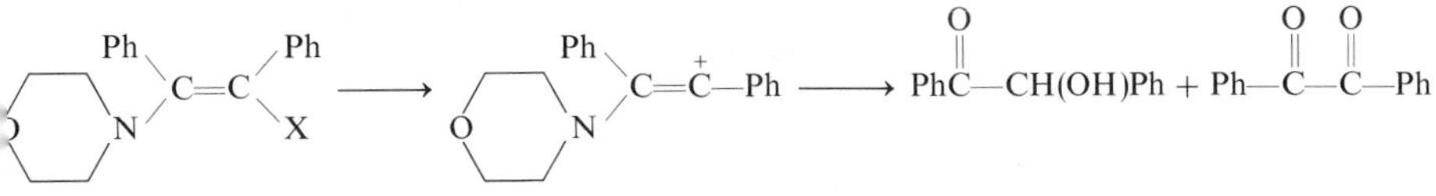

cis and trans            **260**            **258**            **259**

**263**   X = Br
**264**   X = Cl

(84)

The mass spectra of **256**, **263**, and **264** gave a fragment with $m/e = 264$, which was either the base peak or one close to it in intensity. Since this is the $m/e$ for the vinyl cation **260**, this is probably the main ion formed on fragmentation. Most of the other abundant ions in the spectra are derived from further fragmentations of **260**. From a study of the relative intensities of the peaks, it was suggested that the loss of bromine from the molecular ion of **263** is easier than the loss of chlorine from the molecular ion of **264**, which in turn is more facile than the loss of hydrogen from the molecular ion of **256** [189].

Mass spectral analysis of many of the α-arylvinyl bromides, chlorides, and tosylates discussed in the preceding sections on solvolysis gave a fragment with $m/e$ corresponding to the vinyl cation as one of the important fragments in their mass spectra [190]. It is noteworthy that the mass spectra of a pair of the geometric isomers of $\alpha$-aryl-$\beta,\beta$-disubstituted vinyl halides were usually very similar [190], except when a $\beta$-hydrogen was present [31] and elimination of HX, either thermal or during the fragmentation, was more facile with one isomer. Therefore, it is likely that the same linear vinyl cation is formed from both precursors and this ion itself is the precursor for further fragmentations.

## REFERENCES

1. C. A. Grob, *Bull. Soc. Chim. Fr.* p. 1360 (1960).
2. C. A. Grob and G. Cseh, *Helv. Chim. Acta* **47**, 194 (1964).
3. Z. Rappoport, *Acc. Chem. Res.* **9**, 265 (1976).
4. W. J. Hehre, L. Radom, and J. A. Pople, *J. Am. Chem. Soc.* **94**, 1496 (1972).
5. Z. Rappoport and A. Gal, *J. Am. Chem. Soc.* **91**, 5246 (1969); Z. Rappoport, A. Gal, and Y. Apeloig, *Isr. J. Chem.* **6**, 16p (1968).
6. Z. Rappoport and J. Kaspi (a) *J. Am. Chem. Soc.* **92**, 3220 (1970); (b) *J. Chem. Soc., Perkin Trans. 2* p. 1102 (1972).
7. P. von R. Schleyer, J. L. Fry, L. K. M. Lam, and C. J. Lancelot, *J. Am. Chem. Soc.* **92**, 2542 (1970).
8. Z. Rappoport and Y. Houminer, *J. Chem. Soc., Perkin Trans. 2* p. 1506 (1973).
9. Y. Houminer, E. Noy, and Z. Rappoport, *J. Am. Chem. Soc.* **98**, 5632 (1976).
10. Z. Rappoport and Y. Apeloig, unpublished results.
11. Z. Rappoport and A. Gal, *Tetrahedron Lett.* p. 3233 (1970).
12. P. E. Peterson and J. M. Indelicato, *J. Am. Chem. Soc.* **91**, 6194 (1969).
13. T. W. Bentley and P. von R. Schleyer, *Adv. Phys. Org. Chem.* **14**, 1 (1977).
14. V. J. Shiner, Jr., W. E. Buddenbaum, B. L. Murr, and G. Lamaty, *J. Am. Chem. Soc.* **90**, 418 (1968).
15. K. Yates and J. J. Périé, *J. Org. Chem.* **39**, 1902 (1974).
16. J. C. Charlton and E. D. Hughes, *J. Chem. Soc.* p. 850 (1956).
17. Z. Rappoport, P. Schulman, and M. Thuval (Shoolman), *J. Am. Chem. Soc.*, **100**, 7041 (1978).
18. L. Verbit and E. Berliner, *J. Am. Chem. Soc.* **86**, 3307 (1964).
19. W. M. Jones and D. D. Maness (a) *J. Am. Chem. Soc.* **91**, 4314 (1969); (b) **92**, 5457 (1970).
20. K. Yates, G. H. Schmid, T. W. Regulski, D. G. Garratt, H. W. Leung, and R. McDonald, *J. Am. Chem. Soc.* **95**, 160 (1973).
21. F. Marcuzzi and G. Melloni, *Tetrahedron Lett.* p. 2771 (1975).
22. Z. Rappoport and A. Gal, *J. Chem. Soc., Perkin Trans. 2* p. 301 (1973).
23. G. Modena, F. Rivetti, G. Scorrano, and U. Tonellato, *J. Am. Chem. Soc.* **99**, 3392 (1977).
24. Z. Rappoport, J. Kaspi, and Y. Apeloig, *J. Am. Chem. Soc.* **96**, 2612 (1974).
25. M. Hanack and T. Bässler, *J. Am. Chem. Soc.* **91**, 2117 (1969).
26. M. Hanack, T. Bässler, W. Eymann, W. E. Heyd, and R. Kopp, *J. Am. Chem. Soc.* **96**, 6686 (1974).
27. C. A. Grob and R. Spaar (a) *Tetrahedron Lett.* p. 1439 (1969); (b) *Helv. Chim. Acta* **53**, 2119 (1970).
28. S. Patai and Z. Rappoport, *in* "The Chemistry of Alkenes" (S. Patai, ed.), Vol. 1, p. 469. Wiley (Interscience), New York, 1964.

29. (a) Z. Rappoport, *Adv. Phys. Org. Chem.* **7**, 1 (1969); (b) G. Modena, *Acc. Chem. Res.* **4**, 73 (1971).
30. Z. Rappoport and A. Gal, *J. Org. Chem.* **37**, 1174 (1972).
31. Z. Rappoport and M. Atidia (a) *Tetrahedron Lett.* p. 4085 (1970); (b) *J. Chem. Soc., Perkin Trans. 2* p. 2316 (1972).
32. Z. Rappoport and J. Kaspi, *Isr. J. Chem.* **6**, 989 (1974).
33. J. Kaspi, Ph.D. Thesis, Hebrew Univ., Jerusalem, 1975.
34. Z. Rappoport and J. Kaspi, *Tetrahedron Lett.* p. 4039 (1971).
35. C. C. Lee, A. J. Paine, and E. C. F. Ko, *J. Am. Chem. Soc.* **99**, 7267 (1977).
36. Z. Rappoport and Y. Apeloig, *J. Am. Chem. Soc.* **97**, 821 (1975).
37. Z. Rappoport and R. Ta-Shma, *J. Chem. Soc. (B)* p. 871, 1461 (1971).
38. (a) E. Grunwald and S. Winstein, *J. Am. Chem. Soc.* **70**, 846 (1948); (b) S. Winstein, E. Grunwald, and H. W. Jones, *J. Am. Chem. Soc.* **73**, 2700 (1951); (c) S. Winstein, A. H. Fainberg, and E. Grunwald, *J. Am. Chem. Soc.* **79**, 4146 (1957).
39. (a) L. P. Hammett, "Physical Organic Chemistry," McGraw-Hill, New York, 1940; (b) H. H. Jaffe, *Chem. Rev.* **53**, 191 (1953).
40. Z. Rappoport and D. Ladkani, *Chem. Scr.* **5**, 124 (1974).
41. D. J. Cram, "Fundamentals of Carbanion Chemistry," Chaps. 1 and 2. Academic Press, New York, 1965; K. Bowden, A. F. Cockerill, and J. R. Gilbert, *J. Chem. Soc. (B)* p. 179 (1970).
42. Z. Rappoport and N. Pross, unpublished results.
43. J. L. Derocque, F. B. Sundermann, N. Youssif, and M. Hanack, *Justus Liebigs Ann. Chem.* p. 419 (1973).
44. P. E. Peterson and J. M. Indelicato, *J. Am. Chem. Soc.* **90**, 6515 (1968).
45. W. M. Schubert and G. W. Barfknecht, *J. Am. Chem. Soc.* **92**, 207 (1970).
46. C. A. Grob and H. R. Pfaendler, *Helv. Chim. Acta* **54**, 2060 (1971).
47. (a) D. C. Green, Ph.D. Thesis, Univ. of Washington, Seattle, 1974; W. M. Schubert, unpublished results; (b) G. W. Barfknecht, Ph.D. Thesis, Univ. of Washington, Seattle, 1970; (c) W. M. Schubert, personal communication.
48. Z. Rappoport, T. Bässler, and M. Hanack, *J. Am. Chem. Soc.* **92**, 4985 (1970).
49. (a) T. S. C. C. Huang and E. R. Thornton, *J. Am. Chem. Soc.* **98**, 1542 (1976); (b) P. M. Laughton and R. E. Robertson, *in* "Solute–Solvent Interactions" (J. F. Coetzee and C. D. Ritchie, eds.), Chap. 7. Dekker, New York, 1969.
50. A. Gal, Ph.D. Thesis, Hebrew Univ., Jerusalem, 1972.
51. Z. Rappoport, I. Schnabel, and P. Greenzaid, *J. Am. Chem. Soc.* **98**, 7726 (1976).
52. C. C. Lee, A. J. Paine, and E. C. F. Ko, *Can. J. Chem.* **55**, 2310 (1977).
53. R. J. Hargrove, T. E. Dueber, and P. J. Stang, *Chem. Commun.* p. 1614 (1970).
54. T. Sonoda, S. Kobayashi, and H. Taniguchi, *Bull. Chem. Soc. Jpn.* **49**, 2560 (1976).
55. Z. Rappoport and Y. Apeloig, *J. Am. Chem. Soc.* **97**, 836 (1975).
56. Z. Rappoport and Y. Apeloig, *J. Am. Chem. Soc.* **96**, 6428 (1974).
57. P. E. Peterson and R. I. Bopp, *J. Am. Chem. Soc.* **89**, 1283 (1967).
58. O. Exner, *in* "Advances in Linear Free Energy Relationships" (N. B. Chapman and J. Shorter, eds.), Chap. 1, p. 37. Plenum, New York, 1972.
59. W. E. Nelson and J. A. V. Butler, *J. Chem. Soc.* p. 957 (1938); J. C. Hornel and J. A. V. Butler, *J. Chem. Soc.* p. 1361 (1936).
60. S. A. Sherrod and R. G. Bergman, *J. Am. Chem. Soc.* **93**, 1925 (1971).
61. G. S. Hammond, *J. Am. Chem. Soc.* **77**, 334 (1955).
62. M. Oka and C. C. Lee, *Can. J. Chem.* **53**, 320 (1975).
63. C. C. Lee and P. J. Smith, *Can. J. Chem.* **54**, 3038 (1976).
64. Z. Rappoport, A. Gal, and Y. Houminer, *Tetrahedron Lett.* p. 641 (1973).

65. F. H. A. Rummens, R. D. Green, A. J. Cessna, M. Oka, and C. C. Lee, *Can. J. Chem.* **53**, 314 (1975).
66. D. Kaufman and L. L. Miller, *J. Org. Chem.* **34**, 1495 (1969).
67. T. Moeller, "Qualitative Analysis," 1st Ed., p. 255. McGraw-Hill, New York, 1958; T. E. Stevens, *J. Org. Chem.* **25**, 1658 (1960).
68. N. Frydman, R. Bixon, M. Sprecher, and Y. Mazur, *Chem. Commun.* p. 1044 (1969).
69. G. Capozzi, G. Melloni, and G. Modena, *J. Chem. Soc. (C)* p. 2625 (1970).
70. D. S. Noyce and R. M. Pollack, *J. Am. Chem. Soc.* **91**, 119 (1969).
71. G. Capozzi, G. Melloni, G. Modena, and M. Piscitelli, *Tetrahedron Lett.* p. 4039 (1968).
72. P. J. Stang and R. Summerville, *J. Am. Chem. Soc.* **91**, 4600 (1969).
73. Z. Rappoport, A. Pross, and Y. Apeloig, *Tetrahedron Lett.* p. 2015 (1973).
74. S. J. Cristol and R. J. Bly, Jr., *J. Am. Chem. Soc.* **83**, 4027 (1961).
75. H. C. Brown and Y. Okamoto, *J. Am. Chem. Soc.* **80**, 4979 (1958).
76. P. J. Stang, R. J. Hargrove, and T. E. Dueber, *J. Chem. Soc., Perkin Trans. 2* p. 1486 (1977).
77. L. L. Miller and D. A. Kaufman, *J. Am. Chem. Soc.* **90**, 7282 (1968).
78. T. Sonoda, M. Kawakami, T. Ikeda, S. Kobayashi, and H. Taniguchi, *Chem. Commun.* p. 612 (1976).
79. P. J. Stang and T. E. Dueber, *J. Am. Chem. Soc.* **99**, 2602 (1977).
80. G. Capozzi, G. Modena, and U. Tonellato, *J. Chem. Soc. (B)* p. 1700 (1971).
81. (a) G. Modena and U. Tonellato, *Boll. Sci. Fac. Chim. Ind. Bologna* **27**, 373 (1969); (b) G. Modena and U. Tonellato, *J. Chem. Soc. (B)* p. 1569 (1971).
82. G. Modena and U. Tonellato, *J. Chem. Soc. (B)* p. 374 (1971).
83. J. Salaun and M. Hanack, *J. Org. Chem.* **40**, 1994 (1975).
84. (a) P. Bassi and U. Tonellato, *Gazz. Chim. Ital.* **102**, 387 (1972); (b) P. Bassi and U. Tonellato, *J. Chem. Soc., Perkin Trans. 2* p. 1283 (1974).
85. M. D. Schiavelli, S. C. Hixon, H. W. Moran, and C. J. Boswell, *J. Am. Chem. Soc.* **93**, 6989 (1971).
86. D. S. Noyce and M. D. Schiavelli, *J. Am. Chem. Soc.* **90**, 1020 (1968).
87. Y. Apeloig and P. von R. Schleyer, unpublished results.
88. Y. Apeloig, P. von R. Schleyer, and J. A. Pople, *J. Org. Chem.* **42**, 3004 (1977).
89. A. Burighel, G. Modena, and U. Tonellato, *J. Chem. Soc., Perkin Trans. 2* p. 2026 (1972).
90. C. A. Grob and R. Nussbaumer, *Helv. Chim. Acta* **54**, 2528 (1971).
91. (a) T. Sonoda, S. Kobayashi, and H. Taniguchi, unpublished results; (b) T. Sonoda, Ph.D. Thesis, Kyushu Univ., Fukuoka, Japan, 1977; (c) H. Taniguchi and S. Kobayashi, unpublished results.
92. Y. Apeloig, Ph.D. Thesis, Hebrew Univ., Jerusalem, 1974.
93. D. Tsidoni, M.Sc. Thesis, Hebrew Univ., Jerusalem, 1974.
94. (a) R. H. Summerville and P. von R. Schleyer, *J. Am. Chem. Soc.* **94**, 3629 (1972); (b) R. H. Summerville, C. A. Senkler, P. von R. Schleyer, T. E. Dueber, and P. J. Stang, *J. Am. Chem. Soc.* **96**, 1100 (1974); (c) R. H. Summerville and P. von R. Schleyer, *J. Am. Chem. Soc.* **96**, 1110 (1974).
95. Z. Rappoport and J. Kaspi (a) *J. Am. Chem. Soc.* **96**, 586 (1974); (b) **96**, 4518 (1974).
96. A. H. Fainberg and S. Winstein, *J. Am. Chem. Soc.* **79**, 1597, 1602 (1957).
97. (a) L. Demeny, *Recl. Trav. Chim. Pays-Bas* **50**, 60 (1931); (b) R. E. Robertson, *Can. J. Chem.* **31**, 589 (1953).
98. D. N. Kevill, K. C. Kolwyck, D. M. Shold, and C. B. Kim, *J. Am. Chem. Soc.* **95**, 6022 (1973).
99. L. R. Subramanian and M. Hanack, *Chem. Ber.* **105**, 1465 (1972).
100. R. K. Crossland, W. E. Wells, and V. J. Shiner, Jr., *J. Am. Chem. Soc.* **93**, 4217 (1971).

101. T. M. Su, W. Sliwinski, and P. von R. Schleyer, *J. Am. Chem. Soc.* **91**, 5386 (1969).
102. H. M. R. Hoffmann (a) *J. Chem. Soc.* p. 6753 (1965); (b) p. 6762 (1965).
103. Y. Inomoto, R. E. Robertson, and G. Sarkis, *Can. J. Chem.* **47**, 4599 (1969); C. A. Grob, K. Kostka, and F. Kuhnen, *Helv. Chim. Acta* **53**, 608 (1970).
104. R. C. Bingham and P. von R. Schleyer, *J. Am. Chem. Soc.* **93**, 3189 (1971).
105. E. D. Hughes, *Trans. Faraday Soc.* **34**, 185 (1938); **37**, 603 (1941).
106. L. Radom, P. C. Hariharan, J. A. Pople, and P. von R. Schleyer, *J. Am. Chem. Soc.* **95**, 6531 (1973).
107. (a) V. J. Nowlan and T. T. Tidwell, *Acc. Chem. Res.* **10**, 252 (1977); (b) M. Kaftory, unpublished results.
108. (a) P. von R. Schleyer, W. F. Sliwinski, G. W. Van Dine, U. Schöllokopf, J. Paust, and K. Fellenberger, *J. Am. Chem. Soc.* **94**, 125 (1972); (b) W. F. Sliwinski, T. M. Su, and P. von R. Schleyer, *J. Am. Chem. Soc.* **94**, 133 (1972).
109. F. L. Schadt, T. W. Bentley, and P. von R. Schleyer, *J. Am. Chem. Soc.* **98**, 7667 (1976).
110. T. W. Bentley, F. L. Schadt, and P. von R. Schleyer, *J. Am. Chem. Soc.* **94**, 992 (1972).
111. T. W. Bentley and P. von R. Schleyer, *J. Am. Chem. Soc.* **98**, 7658 (1976).
112. P. D. Bartlett and T. T. Tidwell, *J. Am. Chem. Soc.* **90**, 4421 (1968).
113. G. Modena, U. Tonellato, and F. Naso, *Chem. Commun.* p. 1363 (1968).
114. V. J. Shiner, Jr., W. Dowd, R. D. Fisher, S. R. Hartshorn, M. A. Kessick, L. Milakofsky, and M. W. Rapp, *J. Am. Chem. Soc.* **91**, 4838 (1969).
115. D. E. Sunko and I. Szele, *Tetrahedron Lett.* p. 3617 (1972).
116. D. E. Sunko, I. Szele, and M. Tomić, *Tetrahedron Lett.* p. 1827 (1972).
117. K. Dimroth, C. Reichardt, T. Siepmann, and F. Bohlmann, *Justus Liebigs Ann. Chem.* **661**, 1 (1963).
118. Z. Rappoport and Y. Apeloig (a) *Isr. J. Chem.* **7**, 34p (1969); (b) *J. Am. Chem. Soc.* **91**, 6734 (1969).
119. (a) T. C. Clarke, D. R. Kelsey, and R. G. Bergman, *J. Am. Chem. Soc.* **94**, 3626 (1972); (b) T. C. Clarke and R. G. Bergman, *J. Am. Chem. Soc.* **94**, 3627 (1972).
120. R. Maroni, G. Melloni, and G. Modena, *Chem. Commun.* p. 857 (1972); *J. Chem. Soc., Perkin Trans. 1* p. 2491 (1973).
121. F. Marcuzzi, G. Melloni, and G. Modena, *Tetrahedron Lett.* p. 413 (1974).
122. F. Marcuzzi and G. Melloni, *Gazz. Chim. Ital.* **105**, 495 (1975).
123. R. Maroni, G. Melloni, and G. Modena, *J. Chem. Soc., Perkin Trans. 1* p. 353 (1974).
124. F. Marcuzzi and G. Melloni, *J. Am. Chem. Soc.* **98**, 3295 (1976).
125. G. F. P. Kernaghan and H. M. R. Hoffmann, *J. Am. Chem. Soc.* **92**, 6988 (1970).
126. H. Weiner and R. A. Sneen, *J. Am. Chem. Soc.* **87**, 287 (1965).
127. D. R. Kelsey and R. G. Bergman, *J. Am. Chem. Soc.* **92**, 228 (1970); **93**, 1941 (1971).
128. Z. Rappoport, Y. Apeloig, and Y. Spiegel, unpublished results.
129. I. L. Reich and H. J. Reich, *J. Am. Chem. Soc.* **96**, 2654 (1974); I. L. Reich, C. L. Haile, and H. J. Reich, *J. Org. Chem.* **43**, 2403 (1978).
130. (a) Z. Rappoport and Y. Apeloig, *Tetrahedron Lett.* p. 1845 (1970); (b) Z. Rappoport and Y. Apeloig, *Abstr., IUPAC Conf. Phy. Org. Chem., 2nd, Noordwijkerhout* p. 38 (1974).
131. G. Modena and U. Tonellato, *Chem. Commun.* p. 1676 (1968).
132. (a) A. H. Fainberg and S. Winstein, *J. Am. Chem. Soc.* **78**, 2763 (1956); (b) D. J. Raber, J. M. Harris, and P. von R. Schleyer, *in* "Ions and Ion Pairs in Organic Reactions" (M. Szwarc, ed.), Vol. 2, p. 247. Wiley (Interscience), New York, 1974.
133. L. C. Batman, M. G. Church, E. D. Hughes, C. K. Ingold, and N. A. Taher, *J. Chem. Soc.* p. 979 (1940).
134. C. K. Ingold, "Structure and Mechanism in Organic Chemistry," 2nd Ed., p. 483. Cornell Univ. Press, Ithaca, New York, 1969.

135. (a) S. Winstein, E. Clippinger, A. H. Fainberg, R. Heck, and G. C. Robinson, *J. Am. Chem. Soc.* **78**, 328 (1956); (b) S. Winstein, B. Appel, R. Baker, and A. Diaz, *Chem. Soc., Spec. Publ.* No. 19, p. 109 (1965).
136. T. H. Bailey, J. R. Fox, E. Jackson, G. Kohnstam, and A. Queen, *Chem. Commun.* p. 122 (1966).
137. G. Lodder, F. I. M. van Ginkel, and E. R. Hartman, unpublished results.
138. (a) Z. Rappoport and J. Greenblatt, *J. Am. Chem. Soc.* **101**, 1343 (1979); (b) **101**, 3967 (1979).
139. C. G. Swain, C. B. Scott, and K. H. Lohmann, *J. Am. Chem. Soc.* **75**, 136 (1953).
140. (a) R. A. Sneen, J. V. Carter, and P. S. Kay, *J. Am. Chem. Soc.* **88**, 2594 (1966); (b) D. J. Raber, J. M. Harris, R. E. Hall, and P. von R. Schleyer, *J. Am. Chem. Soc.* **93**, 4821 (1971).
141. J. Kaspi and Z. Rappoport, unpublished results.
142. N. Shieh, Ph.D. Thesis, Bryn Mawr College, Bryn Mawr, Pennsylvania, 1957.
143. B. Giese, *Angew. Chem., Int. Ed. Engl.* **16**, 125 (1977).
144. (a) C. D. Ritchie and P. O. I. Virtanen, *J. Am. Chem. Soc.* **94**, 4966 (1972); (b) C. D. Ritchie, *Acc. Chem. Res.* **5**, 348 (1972).
145. C. D. Ritchie, *J. Am. Chem. Soc.* **93**, 7324 (1971).
146. G. Capozzi, G. Melloni, and G. Modena, *J. Chem. Soc. (C)* p. 2621 (1970).
147. G. Modena and U. Tonellato, *J. Chem. Soc. (B)* p. 381 (1971).
148. G. Melloni and G. Modena, *J. Chem. Soc., Perkin Trans. 1* p. 218 (1972).
149. C. G. Swain and C. B. Scott, *J. Am. Chem. Soc.* **75**, 141 (1953).
150. S. Winstein, E. Clippinger, A. H. Fainberg, and G. C. Robinson, *J. Am. Chem. Soc.* **76**, 2597 (1954); S. Winstein, P. E. Klinedinst, Jr., and G. C. Robinson, *J. Am. Chem. Soc.* **83**, 885 (1961).
151. L. M. Mukherjee and E. Grunwald, *J. Phys. Chem.* **62**, 1311 (1958).
152. D. F. Evans, J. A. Nadas, and M. A. Matesich, *J. Phys. Chem.* **75**, 1708 (1971).
153. D. Scheffel, P. J. Abbott, G. J. Fitzpatrick, and M. D. Schiavelli, *J. Am. Chem. Soc.* **99**, 3769 (1977).
154. D. S. Noyce, W. A. Pryor, and P. A. King, *J. Am. Chem. Soc.* **81**, 5423 (1959); D. S. Noyce, G. L. Woo, and M. J. Jorgenson, *J. Am. Chem. Soc.* **83**, 1160 (1961); D. S. Noyce and M. J. Jorgenson, *J. Am. Chem. Soc.* **83**, 2525 (1961); D. S. Noyce, D. R. Hatter, and F. B. Miles, *J. Am. Chem. Soc.* **90**, 4633 (1968); R. C. Fahey and H. J. Schneider, *J. Am. Chem. Soc.* **92**, 6885 (1970).
155. S. R. Hooley and D. L. H. Williams, *J. Chem. Soc., Perkin Trans. 2* p. 1053 (1973).
156. (a) D. J. Raber, J. M. Harris, and P. von R. Schleyer, *J. Am. Chem. Soc.* **93**, 4829 (1971); (b) I. L. Reich, A. F. Diaz, and S. Winstein, *J. Am. Chem. Soc.* **94**, 2256 (1972); (c) J. P. Hardy, A. Ceccon, A. F. Diaz, and S. Winstein, *J. Am. Chem. Soc.* **94**, 1356 (1972).
157. A. F. Diaz, I. Lazdins, and S. Winstein, *J. Am. Chem. Soc.* **90**, 1904 (1968).
158. D. D. Roberts, *J. Org. Chem.* **37**, 1510 (1972).
159. R. Hoffmann, A. Imamure, and G. D. Zeiss, *J. Am. Chem. Soc.* **89**, 5215 (1967).
160. W. W. Schoeller, *Chem. Commun.* p. 872 (1974).
161. R. Hoffmann, R. W. Alder, and C. F. Wilcox, Jr., *J. Am. Chem. Soc.* **92**, 4992 (1970).
162. J. B. Collins, J. D. Dill, E. D. Jemmis, Y. Apeloig, P. von R. Schleyer, R. Seeger, and J. A. Pople, *J. Am. Chem. Soc.* **98**, 5419 (1976).
163. Y. Houminer, E. Noy, and Z. Rappoport, *Abstr., IUPAC Congr., 25th, Jerusalem* p.11 (1975).
164. Z. Rappoport, E. Noy, and Y. Houminer, *J. Am. Chem. Soc.* **98**, 2238 (1976).
165. C. C. Lee, A. J. Cessna, B. A. Davis, and M. Oka, *Can. J. Chem.* **52**, 2679 (1974).
166. C. C. Lee and M. Oka, *Can. J. Chem.* **54**, 604 (1976).

167. (a) C. C. Lee and E. C. F. Ko, *Can. J. Chem.* **54**, 3041 (1976); (b) **56**, 2459 (1978).
168. Z. Rappoport, Y. Houminer, and M. Aviv, unpublished results.
169. M. J. McCall, J. M. Townsend, and W. A. Bonner, *J. Am. Chem. Soc.* **97**, 2743 (1975).
170. C. J. Lancelot, D. J. Cram, and P. von R. Schleyer, *in* "Carbonium Ions" (G. A. Olah and P. von R. Schleyer, eds.), Vol. 3, Chap. 27. Wiley (Interscience), New York, 1972.
171. (a) G. W. Wheland, "Advanced Organic Chemistry," 3rd Ed., p. 593. Wiley, New York, 1960; (b) W. E. Bachman and F. H. Moser, *J. Am. Chem. Soc.* **54**, 1124 (1932); (c) W. E. Bachman and J. W. Ferguson, *J. Am. Chem. Soc.* **56**, 2081 (1934); (d) J. G. Burr, Jr., and L. S. Ciereszko, *J. Am. Chem. Soc.* **74**, 5426 (1952); (e) W. E. McEwen, M. Gilliland, and B. I. Spaar, *J. Am. Chem. Soc.* **72**, 3212 (1950); R. F. Tietz and W. E. McEwen, *J. Am. Chem. Soc.* **77**, 4007 (1955); (f) D. Y. Curtin and M. C. Crew, *J. Am. Chem. Soc.* **76**, 3719 (1954); L. S. Ciereszko and J. G. Burr, *J. Am. Chem. Soc.* **74**, 5431 (1952); J. D. Roberts and C. M. Regan, *J. Am. Chem. Soc.* **75**, 2069 (1953).
172. (a) C. J. Collins and W. A. Bonner, *J. Am. Chem. Soc.* **77**, 92 (1955); (b) W. A. Bonner and C. J. Collins, *J. Am. Chem. Soc.* **78**, 5587 (1956).
173. S. Winstein, *J. Am. Chem. Soc.* **87**, 381 (1965).
174. Z. Rappoport and Y. Apeloig, *Tetrahedron Lett.* p. 1817 (1970).
175. (a) C. C. Lee, G. P. Slater, and J. W. T. Spinks, *Can. J. Chem.* **35**, 1417 (1957); (b) J. E. Nordlander and W. G. Deadman, *J. Am. Chem. Soc.* **90**, 1590 (1968).
176. W. A. Bonner and T. A. Putkey, *J. Org. Chem.* **27**, 2348 (1962).
177. W. S. Trahanovsky and M. P. Doyle, *Tetrahedron Lett.* p. 2155 (1968); D. D. Roberts, *J. Org. Chem.* **36**, 1913 (1971); D. S. Noyce, R. L. Castenson, and D. A. Meyers, *J. Org. Chem.* **37**, 4222 (1972); D. S. Noyce and R. L. Castenson, *J. Am. Chem. Soc.* **95**, 1247 (1973).
178. F. L. Schadt and P. von R. Schleyer, *J. Am. Chem. Soc.* **95**, 7860 (1973).
179. W. A. Bonner and C. J. Collins, *J. Am. Chem. Soc.* **75**, 5372 (1953).
180. W. M. Jones and F. W. Miller, *J. Am. Chem. Soc.* **89**, 1960 (1967).
181. P. J. Stang and T. E. Dueber, *J. Am. Chem. Soc.* **95**, 2683, 2686 (1973).
182. C. A. Grob, J. Csapilla, and G. Cseh, *Helv. Chim. Acta* **47**, 1590 (1964).
183. C. A. Grob and P. Wenk, *Tetrahedron Lett.* p. 4191 (1976).
184. C. A. Grob and P. Wenk, *Tetrahedron Lett.* p. 4195 (1976).
185. C. A. Grob, unpublished results.
186. T. Suzuki, T. Sonoda, S. Kobayashi, and H. Taniguchi, *Chem. Commun.* p. 180 (1976).
187. B. Danieli, F. Ronchetti, and G. Russo, *Chem. Ind. (London)* p. 1067 (1973).
188. S. J. Huang and E. T. Hsu, *Tetrahedron Lett.* p. 1385 (1971).
189. S. J. Huang and M. V. Lessard, *J. Am. Chem. Soc.* **90**, 2432 (1968).
190. Z. Rappoport, Y. Houminer, Y. Apeloig, A. Pross, E. Noy, and H. Schwarz, unpublished results.

# 7

# REARRANGEMENT OF
# VINYL CATIONS

## I. GENERAL CONSIDERATIONS

Perhaps the most characteristic behavior of carbenium ions is their great propensity for rearrangements. These rearrangements occur by a multitude of pathways, such as hydride shifts, alkyl or aryl shifts, and Wagner–Meerwein and cyclopropylcarbinyl rearrangements. The driving force for carbenium ion rearrangements is generally the formation of a more stable intermediate ion from a less stable precursor. This may occur with concomitant ion formation and migration, and hence with anchimeric assistance, or in several discrete stages.

Vinyl cation rearrangements can be classified into two broad categories: (1) migration to the double bond (**1a–1b**) and (2) migration across the double bond (**2a–2b**). Each of these broad areas can be subdivided on the basis of

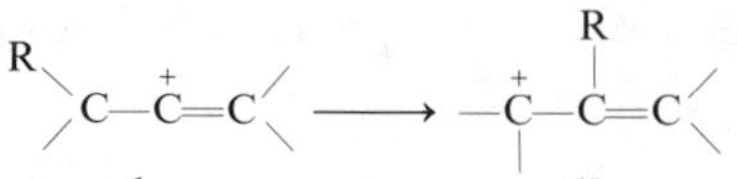

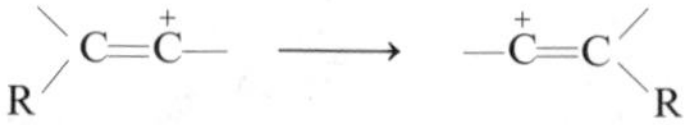

the nature of the migrating group into hydrogen, alkyl, aryl, or heteroatom shifts. As with saturated carbenium ions, rearrangement in vinyl cations is accompanied by the formation of a more stable ion from a less stable progenitor. This chapter deals in detail with each type of vinyl cation rearrangement, except for rearrangements of triarylvinyl cations, which are discussed in Chapter 6, Section VI.

## II. MIGRATION TO THE DOUBLE BOND

In a formalistic sense, migrations to the double bond should result in an allylic cation (**3b**) and hence should occur with considerable facility. However, closer examination of the mechanism of such migrations, as shown below, indicates that the initially formed intermediate must be a perpendicular allyl cation (**3b**), which must first undergo rotation about the C-2–C-3 bond in order to result in a stable allylic ion (**3c**) (Scheme 7.1).

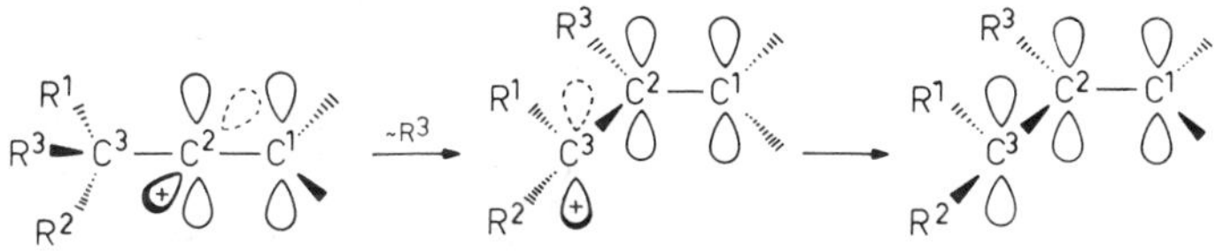

Scheme 7.1

Since, as discussed in Chapter 2, the parent linear allyl cation is some 40 kcal/mole more stable than the perpendicular species due to inductive destabilization of the perpendicular ion, the initially formed intermediate of 1,2 migrations to the double bond might be *less* stable than the precursor vinyl cation. In fact, no 1,2 migrations to the double bond are known in vinyl cations in which the resultant allylic ion is primary (i.e., $R^1 = R^2 = H$ in **3c** and in precursor **3a**). Of course, additional stabilization may be provided to the rearranged ion by the remaining alkyl or aryl substituents ($R^1 = R^2 =$ alkyl or aryl) after the migration. Indeed, as will be seen, such migrations occur most readily when the resultant carbenium ion (**3c**) is not only allylic but tertiary.

## A. Hydride Shifts to the Double Bond

Hydrogen shifts are known [1] to occur very readily in carbenium ions, frequently with an activation barrier of less than 3 to 4 kcal/mole [2]. 1,2-Hydrogen shifts to the double bond in vinyl cations occur much less readily, due in part to the above-mentioned instability of the perpendicular allyl cation and in part to the lower inherent stability of vinyl cations compared to alkyl cations and hence the more ready intrusion of other reactions.

# TABLE 7.1
## Solvolysis Products of Vinyl Triflate 4

| Solvent,[a] $CF_3CH_2OH/H_2O$ | Products (%) | | | | |
|---|---|---|---|---|---|
| | $(CH_3)_2CHC{\equiv}CH$ (5) | $(CH_3)_2C{=}C{=}CH_2$ (6) | $ROC(CH_3)_2CH{=}CH_2$ (7 $R = CF_3CH_2$) | $(CH_3)_2CHC(OR){=}CH_2$ (8 $R = CF_3CH_2$) | $(CH_3)_2CHCCH_3$ (O) (9) |
| 100 | 15.7 | 8.8 | 16.8 | 51.6 | 7.1 |
| 80 | 9.2 | 34.3 | 13.0 | 29.3 | 14.2 |
| 60 | 22.8 | 30.3 | 6.8 | 26.3 | 13.8 |

[a] Percent $CF_3CH_2OH$.

In other words, due to the selectivity of vinyl cations compared to alkyl cations, such reactions as proton loss and collapse with nucleophiles or solvent compete more favorably with rearrangements. Because of this, hydride shifts, and rearrangements in general, in vinyl cations as in alkyl cations are strongly dependent on both structure (i.e., substitution) and solvent polarity and nucleophilicity, as well as temperature.

Solvolysis of 3-methyl-1-buten-2-yl triflate **4** at 80° in buffered media gave, as first observed by Jäckel and Hanack [3, 4], a mixture of products, as shown in Table 7.1.

Although the majority of products is seen to arise from unrearranged ion **10**, 7 to 17% of the products arise from ion **11**, the result of a 1,2-hydride shift to the double bond, as shown in Scheme 7.2. As expected, the extent of hydride shift, as measured by the percentage of **7**, increases with reduced nucleophilicity of the solvent and reaches its maximum in the least nucleophilic solvent, absolute TFE. It is noteworthy that, despite the fact that the rearranged ion **11** is a tertiary carbonium ion, only 7 to 17% of the products arises via this path and the majority via the unrearranged vinyl cation **10**. As pointed out above, this is due to the destabilized perpendicular allyl cation initially formed in the rearrangement as well as the competing reactions of **10**.

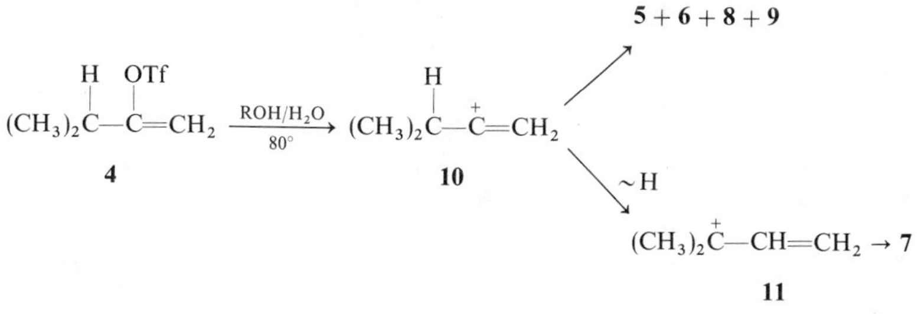

Scheme 7.2

When competing reactions, such as elimination or solvent capture, are hidden or impossible, hydride shifts become considerably more important and are sometimes the exclusive reaction pathway. This is shown by protonation of a series of alkynes with $FSO_3H-SbF_5$ in $SO_2$ or $SO_2Cl$ solution under stable ion conditions [5]. Since magic acid is the least nucleophilic system known, the only competing reactions of the vinyl cations resulting from protonation were cycloaddition and oligomerization with excess alkyne. At $-78°$ essentially only oligomerizations were observed [5]. However, cycloadditions as well as oligomerizations are bimolecular and hydride shifts are unimolecular, and since bimolecular reactions have more negative activation entropies the unimolecular hydride shifts should be favored at higher temperatures. Indeed, at $-20°$ virtually complete hydride shift and allyl cation formation with very little oligomerization were observed, as shown in Scheme 7.3.

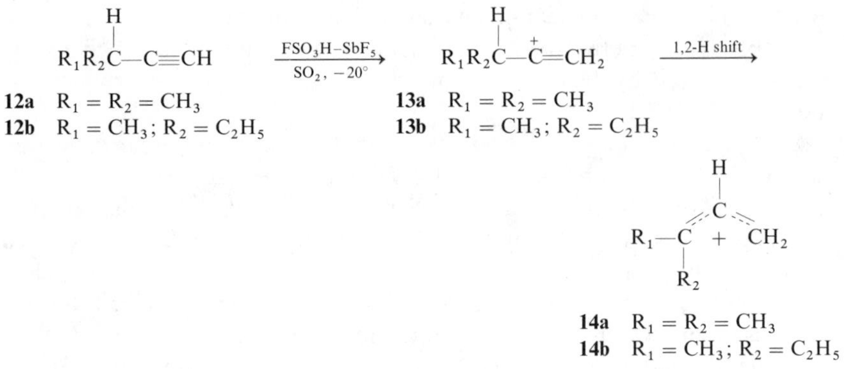

$$R_1R_2\overset{\overset{H}{|}}{C}\!-\!C\!\equiv\!CH \xrightarrow[\text{SO}_2,\,-20°]{\text{FSO}_3\text{H}-\text{SbF}_5} R_1R_2\overset{\overset{H}{|}}{C}\!-\!\overset{+}{C}\!=\!CH_2 \xrightarrow{\text{1,2-H shift}}$$

**12a** $R_1 = R_2 = CH_3$          **13a** $R_1 = R_2 = CH_3$
**12b** $R_1 = CH_3; R_2 = C_2H_5$    **13b** $R_1 = CH_3; R_2 = C_2H_5$

$$R_1\!-\!\overset{\overset{R_2}{|}}{C}\!\! + \overset{\overset{H}{|}}{\overset{\,\,C}{\diagdown}}CH_2$$

**14a** $R_1 = R_2 = CH_3$
**14b** $R_1 = CH_3; R_2 = C_2H_5$

Scheme 7.3

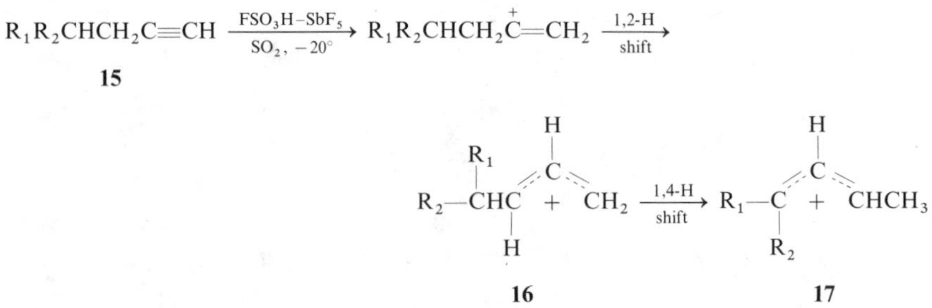

$$R_1R_2CHCH_2C\!\equiv\!CH \xrightarrow[\text{SO}_2,\,-20°]{\text{FSO}_3\text{H}-\text{SbF}_5} R_1R_2CHCH_2\overset{+}{C}\!=\!CH_2 \xrightarrow[\text{shift}]{\text{1,2-H}}$$

**15**

$$R_2\!-\!\overset{\overset{R_1}{|}}{\underset{\underset{H}{|}}{C}}HC + \overset{C}{\diagdown}CH_2 \xrightarrow[\text{shift}]{\text{1,4-H}} R_1\!-\!\overset{\overset{H}{|}}{\underset{\underset{R_2}{|}}{C}} + \overset{C}{\diagdown}CHCH_3$$

**16**                  **17**

Scheme 7.4

A similar protonation of 1-alkynes **15** unbranched at C-3 gave initially a vinyl cation, which likewise rearranged to the allyl cation **16** but then underwent an additional 1,4-hydride shift to give ions **17**, as shown in Scheme 7.4.

Protonation of dialkylacetylenes **18** resulted in allylic ions **20**, again via a 1,2-hydride shift from the intermediate vinyl cations **19**, as shown in Scheme 7.5. In all instances, the product allylic ions were characterized by means of their $^1$H and $^{13}$C nmr spectra [5].

Finally, an interesting 1,5 intramolecular hydride shift to a vinyl cation was observed [6] in the addition of an acylonium ion to alkynes, as outlined in Scheme 7.6 (see also Chapter 3).

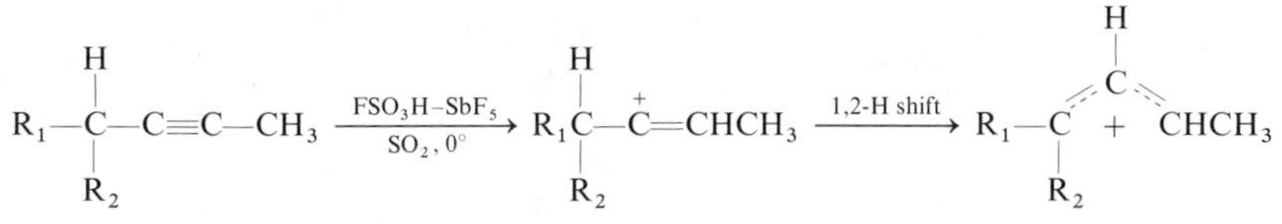

$$R_1-\underset{\underset{R_2}{|}}{\overset{\overset{H}{|}}{C}}-C\equiv C-CH_3 \xrightarrow[SO_2,\,0^\circ]{FSO_3H-SbF_5} R_1\underset{\underset{R_2}{|}}{\overset{\overset{H}{|}}{C}}-\overset{+}{C}=CHCH_3 \xrightarrow{1,2\text{-H shift}} R_1-\underset{\underset{R_2}{|}}{C}\overset{\overset{H}{|}}{\diagup}\overset{C}{\diagdown} + \quad CHCH_3$$

$$\textbf{18} \qquad\qquad\qquad \textbf{19} \qquad\qquad\qquad \textbf{20}$$

Scheme 7.5

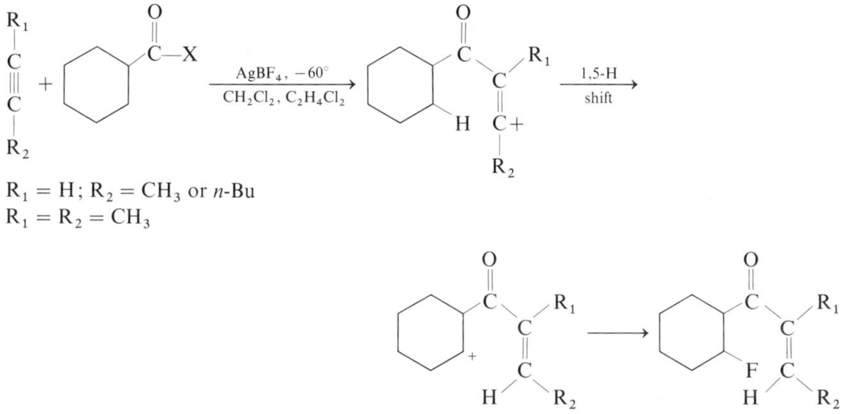

$$\underset{\underset{R_2}{|}}{\overset{\overset{R_1}{|}}{\underset{\|}{\overset{C}{C}}}} + \text{(cyclohexyl)C(=O)-X} \xrightarrow[CH_2Cl_2,\,C_2H_4Cl_2]{AgBF_4,\,-60^\circ} \text{(cyclohexyl ... } \overset{O}{\overset{\|}{C}}\diagup\overset{R_1}{\underset{\underset{\underset{R_2}{|}}{\overset{+}{C}}}{\underset{H}{C}}) \xrightarrow[\text{shift}]{1,5\text{-H}}$$

$$R_1 = H;\ R_2 = CH_3\ \text{or } n\text{-Bu}$$
$$R_1 = R_2 = CH_3$$

Scheme 7.6

## B. Alkyl Shifts to the Double Bond

Numerous alkyl shifts to the double bond of vinyl cations have been observed. Addition of HCl to *tert*-butylacetylene results [7] in a methyl shift, as shown in Scheme 7.7. At a 5:1 molar ratio of HCl to *tert*-butyl-acetylene, nearly 50% of the products derive from unrearranged vinyl cation **21** and approximately 50% from methyl-migrated ion **22**, after subsequent addition of a second mole of HCl to the initially formed products.

Similar results were obtained [8] when vinyl cation **21** was generated solvolytically from vinyl triflate **23**, as shown in Scheme 7.8. At 80° in 80% aqueous ethanol buffered with pyridine, 28% of the products were derived from unrearranged ion **21** and 47% of the products from rearranged ion **22**. The fact that triflate **23** reacts only three times faster than 2-propenyl triflate and there is only a 2:1 rearranged/unrearranged product ratio (rather than quantitative rearrangement) makes an anchimerically assisted synchronous ionization–migration unlikely and suggests a stepwise process for this reaction.

Quantitative methyl migration and allyl cation **24** formation was observed [5] upon addition of neat *tert*-butylacetylene to a solution of $FSO_3H-SbF_5$

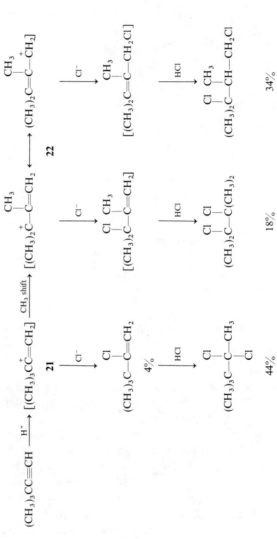

Scheme 7.7

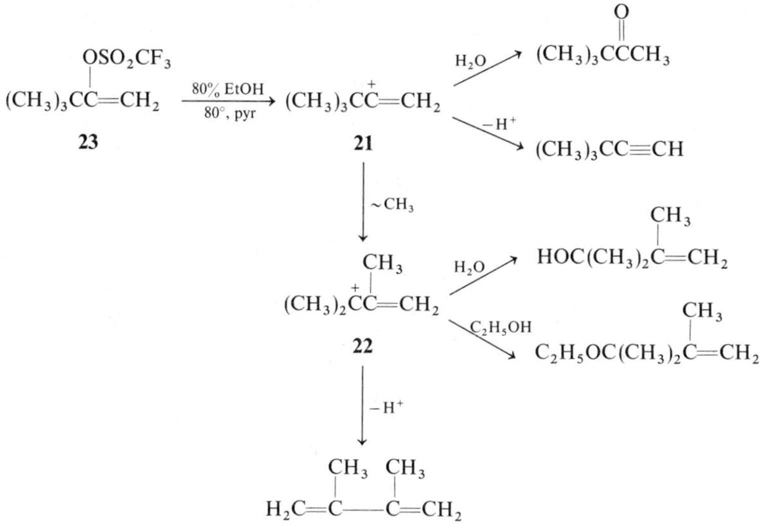

Scheme 7.8

in $SO_2$ at $-20°$:

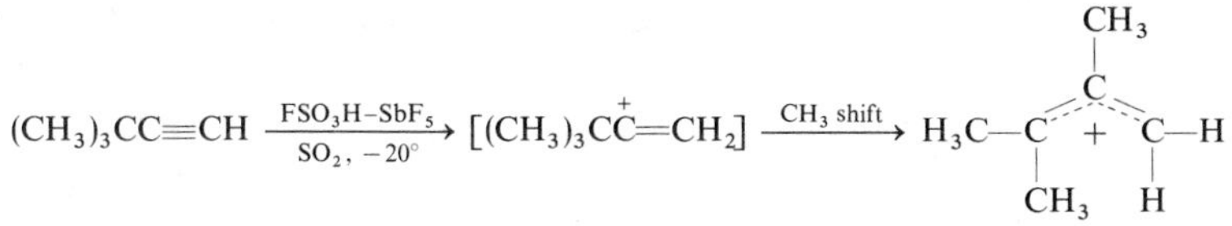

**21**        **24**

Ion **24** was identified by $^1H$ and $^{13}C$ nmr spectra and shown to be the result of kinetic rather than thermodynamic product control [5].

Quantitative rearrangement via methyl migration was also observed [5] in the protonation of di-*tert*-butylacetylene by $FSO_3H$–$SbF_5$ at $-78°$. However, the expected allyl cation **25** rearranges further via a postulated homoallylic ion (**26**), resulting in the observed [5] product mixture of *E*- and *Z*-**27**, as shown in Scheme 7.9.

Solvolysis of 1-adamantylvinyl triflate **28** also leads [9] to rearranged products, as shown in Scheme 7.10. The rearrangement is markedly solvent dependent. In 60% aqueous ethanol, the products are 90% adamantylacetylene **31** and 10% adamantyl methyl ketone **32**. In 90% aqueous $CF_3CH_2OH$ buffered with 2,6-lutidine, a less nucleophilic solvent, rearranged product predominates, with 69% homoadamantyl trifluoroethyl ether **33** $(R = CF_3CH_2)$ and only 25% acetylene **31** and 7% ketone **32** from the unrearranged ion **29**.

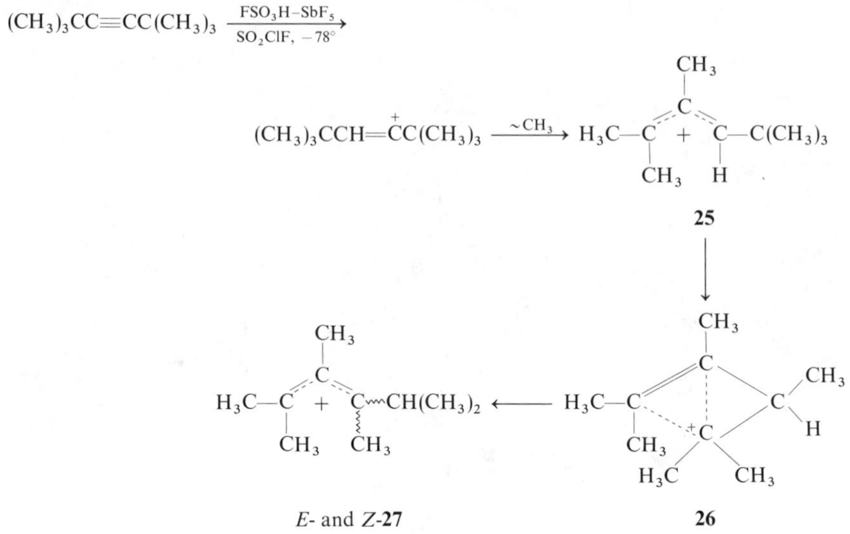

$$(CH_3)_3CC\equiv CC(CH_3)_3 \xrightarrow[SO_2ClF, \ -78°]{FSO_3H-SbF_5}$$

Scheme 7.9

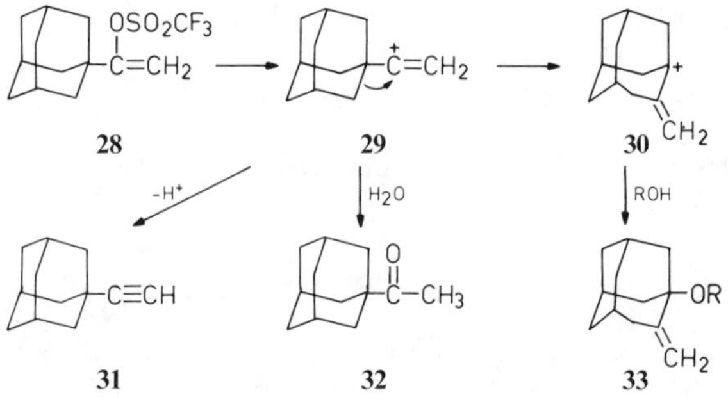

Scheme 7.10

The homoadamantyl product **33** is the result of a 1,2-alkyl shift in vinyl cation, **29** giving rise to ion **30** and then to product **33** [9]. Once again, it is not known whether ionization and rearrangement are concerted or nonconcerted. However, there must be some driving force for the rearrangement of **29** to **30**, since there is no rearranged product observed in a variety of solvents in the solvolysis of 1-adamantylcarbinyl brosylate, the saturated analogue of **28** [10].

Similar results were obtained in the addition of halide- or alcohol-derived adamantyl cation to acetylene in $H_2SO_4$ [11, 12]. However, it is extremely unlikely that the primary vinyl cation **34** is an intermediate in this reaction

due to its high energy, and more likely the reaction proceeds via the intermediacy of sulfates [13] (see Chapter 3 for details):

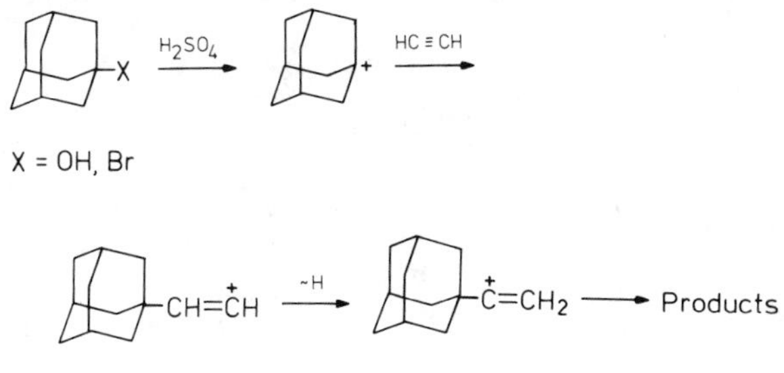

X = OH, Br

**34**

   Alkyl migrations to the double bond have also been observed in cyclic vinyl
cations. In particular, the majority of products arise via rearranged ion **36** in
the solvolysis [14a] of vinyl triflate **35**, as depicted in Scheme 7.11.

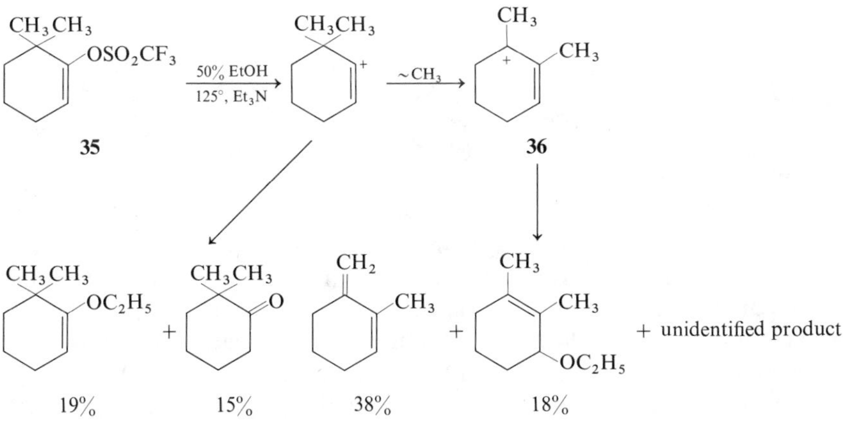

Scheme 7.11

   Similarly, solvolysis of spirovinyl triflate **37** gave [15] diene **39** as the sole
product via ion **38**, resulting from a 1,2-alkyl shift, as shown in Scheme 7.12.
The observation of only partially rearranged products and hence only partial
alkyl migration in the solvolysis of **35** and complete rearrangement in the
solvolysis of **37** strongly argues for, but does not prove, a concerted ionization migration in the reaction of **37** and a stepwise process in the solvolysis of
**35** [15]. Such a concerted, perhaps anchimerically assisted process in **37** is
likely, due to the favorable geometry of the migrating spiroalkyl group as
well as possible relief of strain in the spiro ring system of **37**.

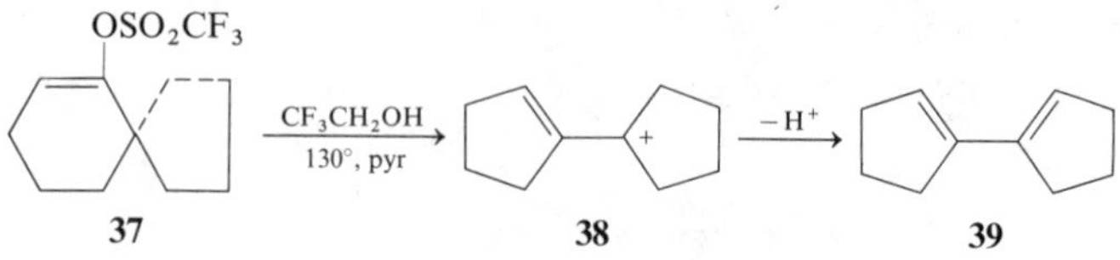

$OSO_2CF_3$

$\xrightarrow[130°,\ pyr]{CF_3CH_2OH}$    +    $\xrightarrow{-H^+}$

**37**        **38**        **39**

Scheme 7.12

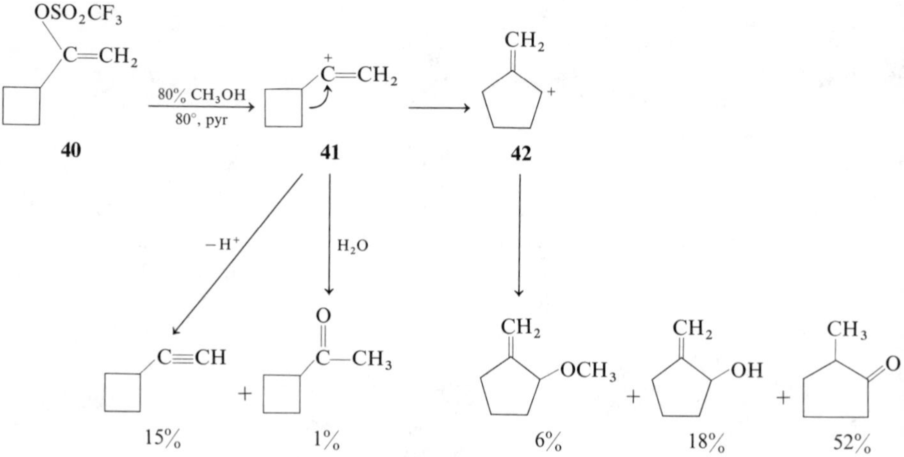

$OSO_2CF_3$

$C{=}CH_2$

$\xrightarrow[80°,\ pyr]{80\%\ CH_3OH}$    $\overset{+}{C}{=}CH_2$   $\longrightarrow$   $CH_2$

**40**        **41**        **42**

$-H^+$      $H_2O$

$C{\equiv}CH$    +    $\overset{O}{\overset{\|}{C}}{-}CH_3$      $CH_2$ ... $OCH_3$  +  $CH_2$ ... $OH$  +  $CH_3$ ... $O$

15%      1%      6%      18%      52%

Scheme 7.13

In all of the above cases of alkyl shifts to the double bond, the driving force for rearrangement was provided by the formation of a more stable tertiary carbenium ion upon migration. There is only one example known to date in which migration to the double bond results in the formation of a secondary carbenium ion. Solvolysis of vinyl triflate **40**, as indicated in Scheme 7.13, gives [16], besides 16% unrearranged products, 76% rearranged products derived from rearranged ion **42**. Presumably, there is an additional driving force for rearrangement of **41** to **42** by relief of the strain in the four-membered ring of **41**.

## C. Aryl Shift to the Double Bond

There is only one example of a potential aryl shift to the double bond involving a vinyl cation [17]. Bromination of 3-arylpropynes **43** in acetic acid resulted in mixtures of *cis*- and *trans*-1,2-dibromo-3-arylpropenes, but no solvent-incorporated bromoacetates, which are common in such brominations (see Chapter 3), were observed. The stereochemistry of the dibromides was predominantly trans, except for the *p*-methoxy isomer, which gave 83%

cis and 17% trans products. In the presence of 0.1 $M$ LiBr, all substrates, including the $p$-methoxy isomer, gave predominantly trans dibromides. Furthermore, the $p$-methoxy isomer was found to react 2.5 times faster than predicted from a Hammett plot. These results were accounted for by aryl participation and formation of a phenonium ion (44) in the case of the $p$-methoxy isomer:

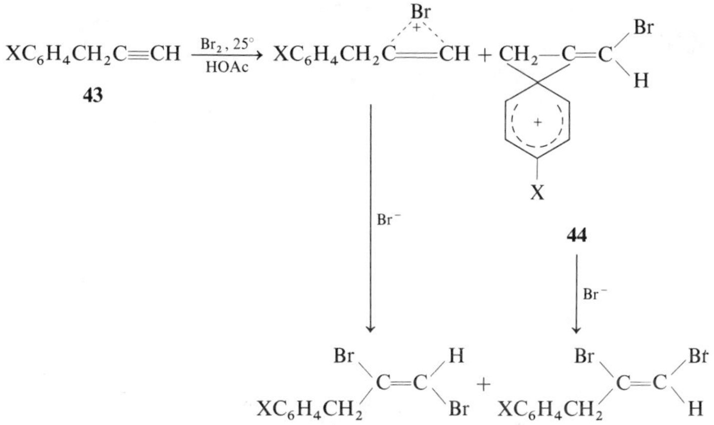

$X = p\text{-}CH_3O, p\text{-}CH_3, H, m\text{-}CF_3$

Interestingly, no 1,3-dibromo-2-arylpropene resulting from Br attack on C-3 of 44 was observed. Hence, although the product and kinetic data indeed suggest aryl participation in the formation of the initial vinyl cation, no actual phenyl migration to the double bond was detected [17].

## III. MIGRATIONS ACROSS THE DOUBLE BOND

### A. Hydride Shift

A number of hydride shifts across the double bond of a vinyl cation have been observed. In particular, solvolysis [9, 18] of $E$- and $Z$-cyclopropylvinyl triflates 45 gives 11 to 17% of cyclopropyl ethyl ketone 48 as a result of rearrangement of the intermediate propenyl cation 46 to ion 47, as shown in Scheme 7.14.

Hydride migration across the double bond was also observed in the solvolysis [19] of $E$- and $Z$-$\beta$-phenylvinyl triflates 49, as indicated in Scheme 7.15. The ratio of rearranged to unrearranged products was strongly solvent dependent, with 17% rearrangement in the less nucleophilic and more polar neat $CF_3CH_2OH$ and only 4% rearrangement in the more nucleophilic 60% aqueous $CF_3CH_2OH$.

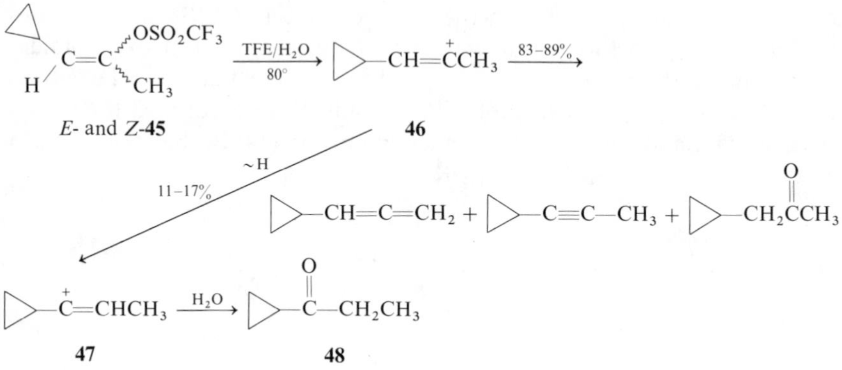

$$\text{(cyclopropyl)} \overset{H}{\underset{}{\diagup}} C = C \overset{OSO_2CF_3}{\underset{CH_3}{\diagdown}} \quad \xrightarrow[80°]{TFE/H_2O} \quad \text{(cyclopropyl)}-CH=\overset{+}{C}CH_3 \quad \xrightarrow{83-89\%}$$

$$E\text{- and }Z\text{-45} \qquad\qquad\qquad\qquad\qquad 46$$

$$\xrightarrow[11-17\%]{\sim H}$$

$$\text{(cyclopropyl)}-CH=C=CH_2 \;+\; \text{(cyclopropyl)}-C\equiv C-CH_3 \;+\; \text{(cyclopropyl)}-CH_2\overset{O}{\overset{\|}{C}}CH_3$$

$$\text{(cyclopropyl)}-\overset{+}{C}=CHCH_3 \quad \xrightarrow{H_2O} \quad \text{(cyclopropyl)}-\overset{O}{\overset{\|}{C}}-CH_2CH_3$$

$$47 \qquad\qquad\qquad\qquad 48$$

Scheme 7.14

$$\overset{C_6H_5}{\underset{H}{\diagup}} C = C \overset{OSO_2CF_3}{\underset{CH_3}{\diagdown}} \quad \xrightarrow[80°,\,pyr]{TFE/H_2O} \quad C_6H_5CH=\overset{+}{C}CH_3 \quad \xrightarrow{83-96\%}$$

$$E\text{- and }Z\text{-49} \qquad\qquad\qquad\qquad\qquad 50$$

$$\xrightarrow[4-17\%]{\sim H}$$

$$C_6H_5CH_2\overset{O}{\overset{\|}{C}}CH_3 \;+\; C_6H_5C\equiv CCH_3 \;+\; C_6H_5CH=C=CH_2$$

$$C_6H_5\overset{+}{C}=CHCH_3 \quad \xrightarrow{H_2O} \quad C_6H_5\overset{O}{\overset{\|}{C}}CH_2CH_3$$

$$51$$

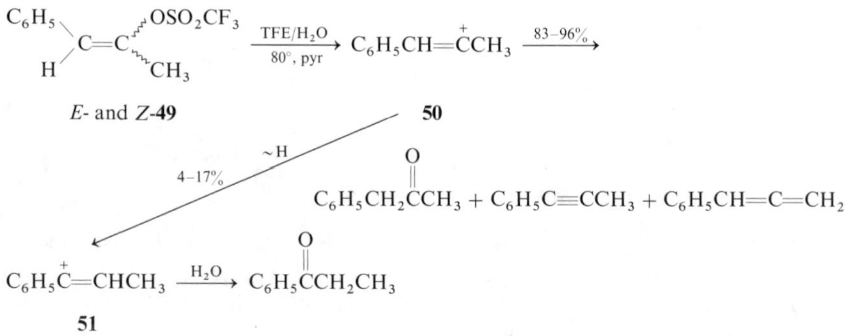

Scheme 7.15

The $Z$ isomer of **49** was found to react some three to ten times faster than the corresponding $E$-**49**, indicating the possible incursion of a synchronous E2 elimination. The slower reactivity of the $E$ isomer also rules out the intervention of a possible vinylidenephenonium ion (*vide infra*) in these solvolyses [19].

In both of the above cases, the driving force for rearrangement was the formation of a more stable vinyl cation from a less stable precursor: in the case of **45**, a cyclopropyl-stabilized ion and, in the case of **49**, an α-phenyl-stabilized ion. The fact that in comparable solvents under the same conditions there is more rearrangement in the case of cyclopropylvinyl triflate **45** than for phenylvinyl triflate **49** indicates the greater stabilizing effect of a cyclo-propane ring compared to a phenyl group, in agreement with the solvolytic observations, as discussed in Chapter 5.

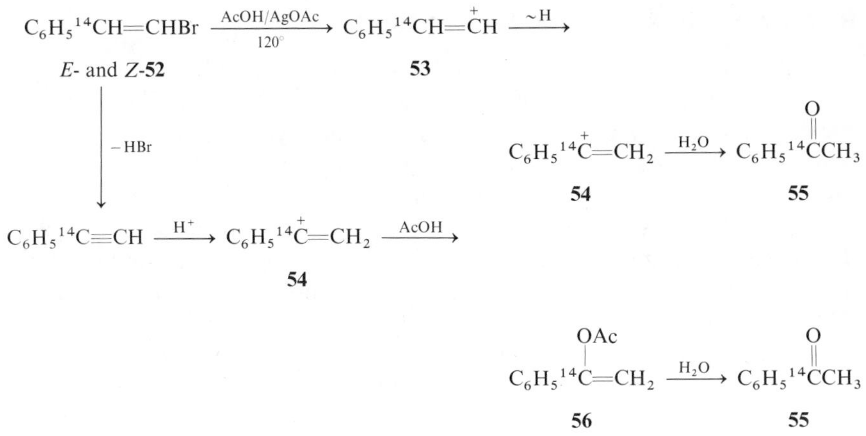

$$C_6H_5{}^{14}CH\!=\!CHBr \xrightarrow[120°]{AcOH/AgOAc} C_6H_5{}^{14}CH\!=\!\overset{+}{C}H \xrightarrow{\sim H}$$

E- and Z-**52**            **53**

$$\downarrow {\scriptstyle -HBr}$$

$$C_6H_5{}^{14}\overset{+}{C}\!=\!CH_2 \xrightarrow{H_2O} C_6H_5{}^{14}\overset{O}{\overset{\|}{C}}CH_3$$

         **54**           **55**

$$C_6H_5{}^{14}C\!\equiv\!CH \xrightarrow{H^+} C_6H_5{}^{14}\overset{+}{C}\!=\!CH_2 \xrightarrow{AcOH}$$

         **54**

$$C_6H_5{}^{14}\overset{OAc}{\overset{|}{C}}\!=\!CH_2 \xrightarrow{H_2O} C_6H_5{}^{14}\overset{O}{\overset{\|}{C}}CH_3$$

         **56**           **55**

Scheme 7.16

A hydride shift across the double bond was claimed [20] to occur in the reaction of E- and Z-bromostyrenes **52** with AgOAc in AcOH at 120°, as shown in Scheme 7.16. This conclusion was arrived at on the basis of the formation of acetophenone **55** as the major product along with very small amounts ($\sim 0.5\%$) of phenylacetylene and the essential independence of product formation as a function of the stereochemistry of the starting styryl bromide, as well as the absence of a reaction when AgOAc was replaced by NaOAc [20]. However, on the basis of the calculations and discussions in Chapter 2, the formation of a highly energetic primary vinyl cation such as **53** is very unlikely, and the observed results must have arisen by another pathway. One possibility would be a synchronous ionization hydrogen migration and hence direct formation of rearranged ion **54**, thereby bypassing the energetically unfavorable primary ion **53**. However, contrary to observations, such a pathway should be dependent on the stereochemistry of the starting halide. The solvent isotope effect $k_{AcOH}/k_{AcOD}$ of ca. 1.2 [20b] rules out an electrophilic addition–elimination mechanism. Rather, an elimination–addition mechanism with intermediate formation of phenylacetylene (Scheme 7.16) has been suggested for this reaction [19]. The rate of addition of AcOH to phenylacetylene under these conditions can be roughly estimated to be $2.7 \times 10^{-7} \sec^{-1}$ from the rate constant of $2.7 \times 10^{-4} \sec^{-1}$ for the addition of AcOH to $p$-$CH_3OC_6H_4C\equiv CH$ [21], together with a $\rho^+$ value of $-3.8$ for proton addition to arylacetylenes in water [22]. This value is of a comparable magnitude to the rate constant of $1.4 \times 10^{-7} \sec^{-1}$ estimated from the above data for the solvolysis reaction.

## B. Alkyl Shifts Across the Double Bond

To date, with one exception, alkyl shifts across the double bond of a vinyl cation have been observed only in cyclic systems. Deuterium labeling, as shown in Scheme 7.17, established that no methyl migration across the double bond occurs in the solvolysis of the simple vinyl triflate **57**. That is, only products derived from the unrearranged ion **58** and none from the rearranged ion **59** were observed under a variety of solvolytic conditions [9]. This is perhaps not surprising since alkyl migration across the double bond in simple alkylvinyl cations results in another simple alkylvinyl ion with no additional stabilization and hence no ready driving force. Furthermore, alkyl groups are not the best migrating groups in any event. This, of course, is in contrast to the triarylvinyl systems discussed in Chapter 6 (Section VI), in which, due to the longer lifetime of the more stable arylvinyl cations and the better migratory aptitude of aryl groups, rearrangements readily occur.

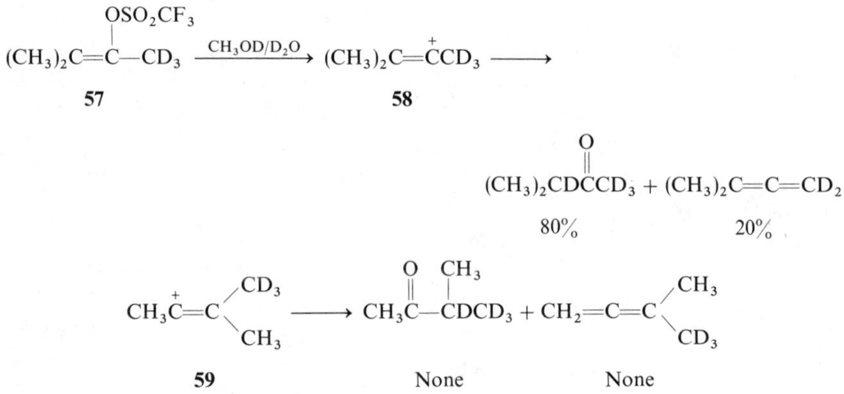

Scheme 7.17

However, an interesting double 1,2-methyl shift has been observed in the rearrangement of a vinyl cation to an allyl cation under super acid conditions [23]. As shown in Scheme 7.18, addition of 2-butyne to a solution of $(CD_3)_3CCl$ and $SbF_5$ in $SO_2$ at $-78°$ results in the formation of only ion **61**, with none of allyl ion **62**. The absence of ion **62** in the nmr spectrum rules out a *direct* 1,3-methyl shift in the initially formed vinyl cation **60** and argues for two consecutive methyl shifts, first across the double bond and then to the double bond, to give the product ion **61**. Such consecutive 1,2-methyl shifts in preference to a direct 1,3-methyl shift are also observed in trigonal carbenium ion chemistry [24].

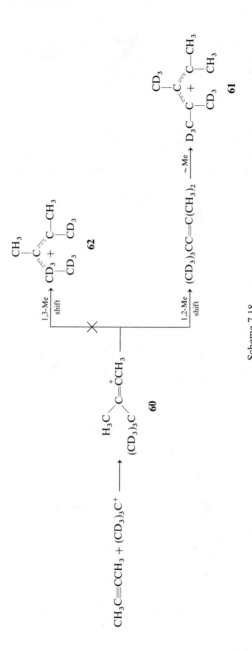

Scheme 7.18

A 1,2-methyl shift across the double bond of a vinyl cation was also claimed [25] to occur in the CO trapping of the 2-butenyl ion by bubbling a 2-butyne–CO mixture through a $FSO_3H$–$SbF_5$ solution in an nmr tube:

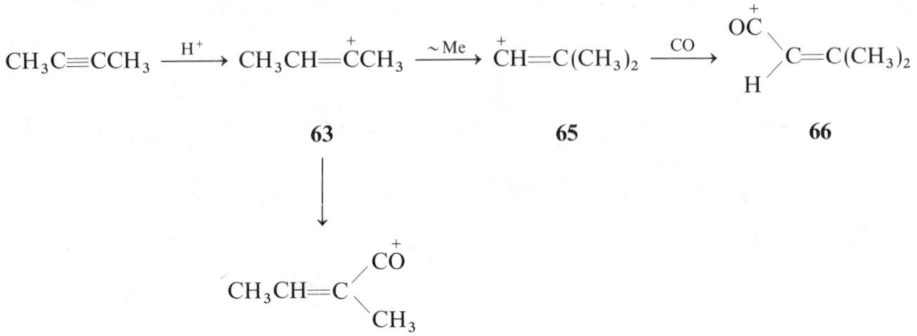

$$CH_3C\equiv CCH_3 \xrightarrow{H^+} CH_3CH\overset{+}{=}CCH_3 \xrightarrow{\sim Me} \overset{+}{CH}=C(CH_3)_2 \xrightarrow{CO} \underset{H}{\overset{\overset{+}{OC}}{\diagup}}C=C(CH_3)_2$$

$$\quad\quad\quad\quad\quad\quad\quad\quad \textbf{63} \quad\quad\quad\quad\quad\quad\quad \textbf{65} \quad\quad\quad\quad\quad\quad \textbf{66}$$

$$CH_3CH=C\underset{CH_3}{\overset{\overset{+}{CO}}{\diagup}}$$

$$\textbf{64}$$

Observation of product ion **66** was explained by a 1,2-methyl shift across the double bond in ion **63** to give **65** followed by trapping of **65** by CO. However, once again, such a migration involves formation of primary vinyl cation (**65**), which seems extremely unlikely on the basis of energetics even in super acid media, unless there is some unusual complexing by CO or $SbF_5$, and ion **66** must have formed by some other process.

Considerable driving force for alkyl migration across the double bond of a vinyl cation may exist in a cyclic system due to relief of ring strain. Indeed, a number of such migrations have been observed. The reactions of vinyl triflates **67** and **69** have been investigated in some detail by Hanack and Fuchs [14b], and the results are summarized in Table 7.2 and Scheme 7.19.

**TABLE 7.2**

**Solvolyses of Vinyl Triflates 67 and 69$^a$**

| Starting triflate | Reaction conditions | Product distribution (relative %) | | | | | | |
|---|---|---|---|---|---|---|---|---|
| | | 67 | 67a | 67b | 69 | 69a | 69b | 69 |
| 67 | 60% EtOH, 125°, 2,6-lutidine | — | — | 63 | — | — | 27 | <1 |
| | 80% TFE, 100° | — | 29 | 39 | — | 12 | 15 | 2 |
| | 97% TFE, 100° | — | 50 | 7 | — | 31 | 5 | 4 |
| | 100% TFE, 100° | — | 56 | 1 | — | 39 | <1.0 | 3 |
| 69 | 80% EtOH, 60° | 2 | <1.0 | — | 16$^b$ | 28 | 29 | 24 |
| | 80% TFE, 60° | 2 | 1 | <0.1 | — | 42 | 52 | 3 |
| | 97% TFE, 60° | 3 | 1 | <0.1 | — | 79 | 10 | 6 |
| | 100% TFE, 60° | 3 | 2 | — | — | 89 | 2 | 4 |

$^a$ From Hanack and Fuchs [14b].
$^b$ Unreacted starting triflate.

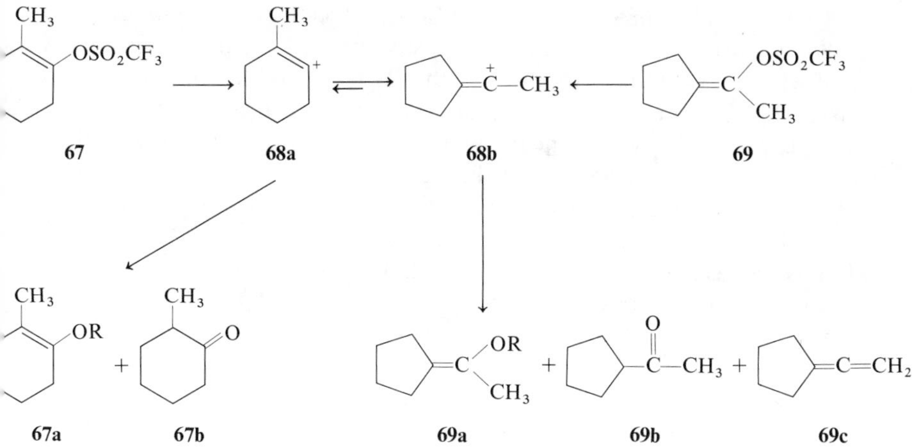

Scheme 7.19

It is evident from these data that vinyl triflates **67** and **69** interconvert via cations **68**. In media of low nucleophilicity, in particular ion **68a** undergoes rearrangement to **68b** and, to a considerably smaller extent, ion **68b** rearranges to **68a**. It is particularly significant that in the reaction of **69**, 2 to 3% of **67** was observed, presumably resulting from internal return via **68b** → **68a** → **67**. Due to the much faster reaction of **69** compared to that of **67**, no internal return to **69** from **67** could be detected.

Similarly, triflate **70** in aqueous ethanol gives both unrearranged and rearranged products via ions **71** and **72**, respectively:

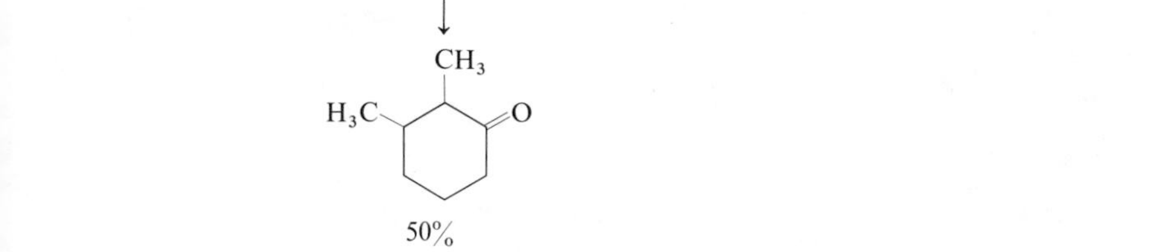

On the other hand, cyclohexenyl triflate gives only cyclohexanone upon solvolysis [14]. Hence, 1,2-alkyl shifts occur only if there is a substituent on the double bond. In such cases, rearrangement leads from a bent secondary vinyl cation, such as **68a**, to a more stable linear secondary vinyl cation, such as **68b**, whereas, in the absence of a substituent (i.e., the parent system), a 1,2-alkyl shift would lead to an energetically unfavorable "primary" vinyl cation. Hence, these results show that even a strained bent secondary vinyl cation is preferred over a linear but high-energy "primary" vinyl cation, showing the extreme instability of the latter species.

An interesting ring contraction via alkyl migration across the double bond, followed by ion-pair return, was observed in the trifluoroethanolysis of 4-homoadamanten-4-yl triflate. At 100° in TFE or 90% TFE buffered with pyridine, the ring-contracted primary vinyl triflate, in 95% yield, was isolated as the sole product [26a]:

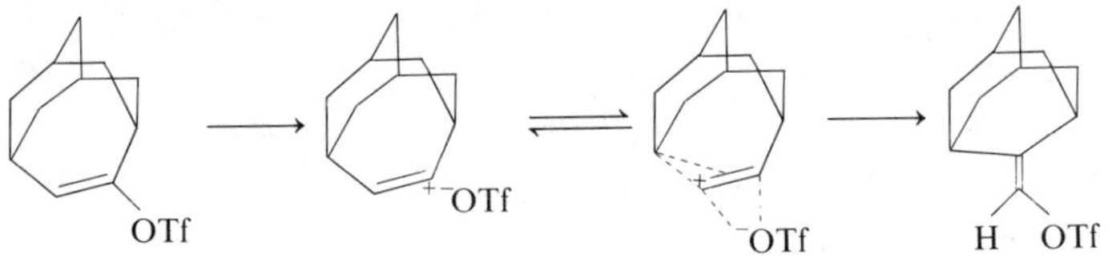

An intimate ion pair with a bridged structure, as shown, was invoked to explain these unusual results [26a].

Ring enlargement, rather than ring contraction, has also been observed [14] via alkyl migration in the solvolysis of cyclic vinyl triflate **73**:

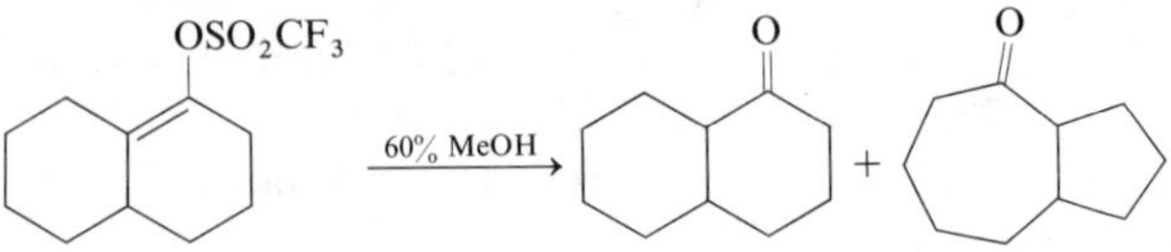

Similarly, reaction of triflate **74** results [16] in 50% ring-enlarged products, presumably via a 1,2-alkyl migration across the double bond in ion **75**, resulting in ion **76**:

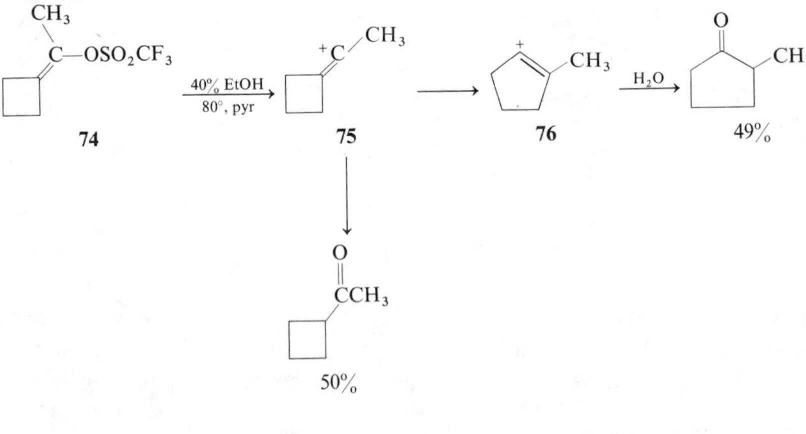

That ion **76** should form from **75** is interesting, for ion **76** should be a highly strained bent vinyl cation, whereas **75** can be linear. In fact, it has been shown [26b] that solvolysis of cyclopentenyl triflate does not proceed via a vinyl cation as previously reported [14] but rather by nucleophilic attack on sulfur and S–O bond cleavage, presumably due to the high energy of the strained five-membered cyclic vinyl cation (Chapter 5, Section V,B).

It is possible that, in order to avoid the highly strained cyclopentenyl ion **76**, vinyl triflate **74** undergoes a concerted intramolecular rearrangement to give cyclopentenyl triflate, which then reacts by nucleophilic attack on the sulfur and S–O cleavage to give the observed 2-methylcyclopentanone product.

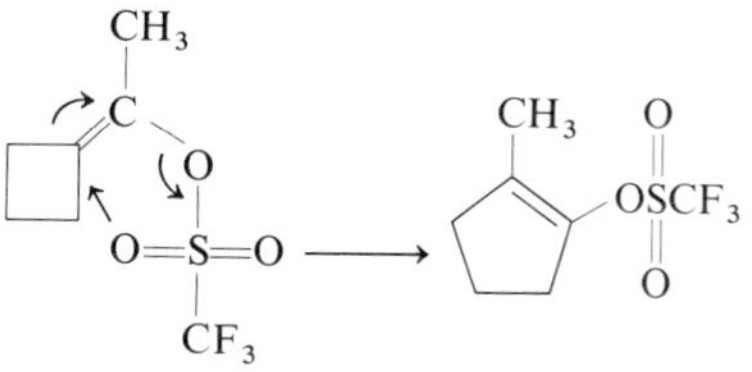

Ring-enlarged products were also observed [15] in the solvolysis of triflate **77**. However, in order to avoid an energetically unfavorable primary vinyl cation, ionization and alkyl shift and the attendant formation of the cycloheptenyl cation **78** are likely to be synchronous in this system.

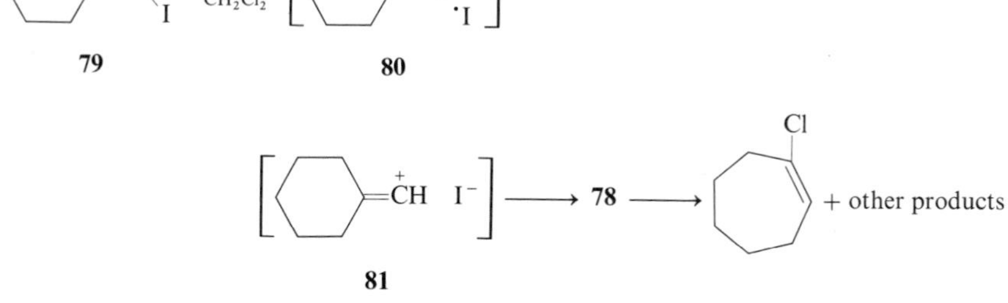

Ring-enlarged products have also been observed in the photolysis of vinyl iodide **79** in $CH_2Cl_2$ [27]. The initial radical (**80**) formed in the photolysis converts to a vinyl cation via an electron transfer process [27], giving **81**, which then undergoes an alkyl shift across the double bond, resulting in **78**.

This is followed by chloride abstraction from $CH_2Cl_2$ and formation of 1-chlorocycloheptene, as well as radical-derived products.

Considerable driving force for alkyl shift and ring-enlarged products also exists in the solvolysis of the cyclopropylidene system **82** due to the formation of the very stable nonclassical ion **83**. For further details on this system, see Chapter 5, Section IV,C.

**82**                **83**

## C. Aryl Migrations

$\beta$-Aryl migrations across the double bond in triarylvinyl systems have been extensively investigated by Rappoport and co-workers [28] and by Lee and co-workers [29]. The extent of rearrangement is strongly dependent on the nature of the aryl groups as well as solvent polarity and nucleophilicity [28, 29]. It is well established that these rearrangements proceed in the systems investigated (see Chapter 6, Section VI, for details) by way of open linear vinyl cations with little or no anchimeric assistance by the neighboring $\beta$-aryl group since $\alpha$-aryl vinyl cations are already strongly stabilized and would gain little by aryl bridging.

A different picture emerges in the solvolysis of $E$- and $Z$-3-aryl-2-buten-2-yl triflates $E$-**84** and $Z$-**85**. As shown in Table 7.3, deuterium label established that there is nearly 50% phenyl migration in the $E$ isomers **87** and substantial rearrangement as well in the $Z$ isomers **86** upon solvolysis in 60% aqueous ethanol [30]. Furthermore, reisolated partially reacted starting material indicated that no $E \rightleftarrows Z$ isomerization occurs under the reaction

**TABLE 7.3**

**Products of Reaction of Vinyl Triflates 86 and 87 in 60% EtOH**

| Starting triflate | $\underset{(\%)}{\overset{\displaystyle CD_3\ \ O}{\underset{\displaystyle \ }{C_6H_5CH{-}CCH_3}}}$ | $\underset{(\%)}{\overset{\displaystyle CH_3\ \ O}{\underset{\displaystyle \ }{C_6H_5CH{-}CCD_3}}}$ | $\underset{(\%)}{\overset{\displaystyle CH_3}{\underset{\displaystyle \ }{C_6H_5C{=}C{=}CD_2}}}$ | $\underset{(\%)}{\overset{\displaystyle CD_3}{\underset{\displaystyle \ }{C_6H_5C{=}C{=}CH_2}}}$ |
|---|---|---|---|---|
| **86a** | $34.5 \pm 0.4^a$ | $65.4 \pm 0.4$ | $72.7 \pm 0.6^a$ | $27.3 \pm 0.6$ |
| **86b** | $65.5 \pm 0.4^a$ | $34.5 \pm 0.4$ | — | — |
| **87a** | $47.6 \pm 0.3^a$ | $52.3 \pm 0.3$ | — | — |
| **87b** | $51.5 \pm 0.7^a$ | $48.6 \pm 0.7$ | $41.1 \pm 0.7^a$ | $58.9 \pm 0.79$ |

$^a$ Percent distribution of labeled ketones and allenes, respectively; not absolute yields.

conditions and no deuterium scrambling occurs in the $Z$ series **86**, but extensive scrambling occurs in the $E$ series **87**.

$$\underset{XC_6H_4}{\overset{H_3C}{>}}C=C\underset{CH_3}{\overset{OSO_2CF_3}{<}} \qquad \underset{H_3C}{\overset{XC_6H_4}{>}}C=C\underset{CH_3}{\overset{OSO_2CF_3}{<}}$$

*E*-84            *Z*-85

$$\underset{R_2}{\overset{C_6H_5}{>}}C=C\underset{R_1}{\overset{OSO_2CF_3}{<}} \qquad \underset{C_6H_5}{\overset{R_2}{>}}C=C\underset{R_1}{\overset{OSO_2CF_3}{<}}$$

**86a** $R_1 = CD_3; R_2 = CH_3$     **87a** $R_1 = CD_3; R_2 = CH_3$

**86b** $R_1 = CH_3; R_2 = CD_3$     **87b** $R_1 = CH_3; R_2 = CD_3$

In 97% aqueous $CF_3CH_2OH$, the rates of the $E$ isomers **84** are very much faster [31] than the rates of the $Z$ isomers **85**, with a $\rho$ value of $-3.76$ for **84** and only $-1.96$ for **85**, as shown in Fig. 7.1. Moreover, the rates of the $E$ isomers deviate significantly from a plot of $\log k$ versus $\sigma^*$ that includes the $Z$ isomers as well as simple alkylvinyl systems, as shown in Fig. 7.2. If the anchimerically unassisted Taft line is taken as a measure of the unassisted rates $k_u$, then, by the well-known Winstein [32] relationship, the deviation is a direct measure of the assisted $FK_\Delta$ rate constants summarized in Table 7.4 together with other relevant data.

Product studies [31] in anhydrous TFE at 85° established the formation of the two isomeric vinyl ethers **88** and **89** and the allene **90**. As the data

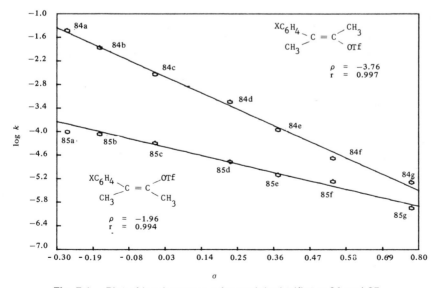

**Fig. 7.1.** Plot of $\log k$ versus $\sigma$ for arylvinyl triflates **84** and **85**.

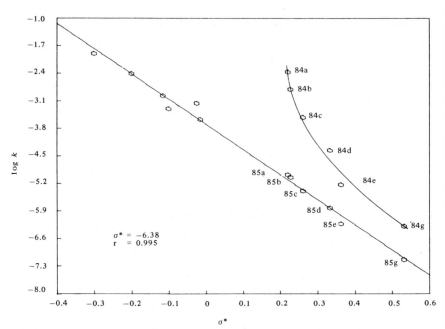

**Fig. 7.2.** Plot of log $k$ versus $\sigma^*$ for arylvinyl triflates **84** and **85** and some simple alkylvinyl triflates (see ref. 31).

**TABLE 7.4**

**Partitioning of Products and Reaction Rates of 84 in 97% TFE at 75°C**

| X | $k_u$ (sec$^{-1}$) | Rate enhancement, $k_t/k_u$ | $Fk_\Delta$ (sec$^{-1}$) | Assist. path (%), $(Fk_\Delta/k_t) \times 100$ | Yield of product et with retair geometry ( |
|---|---|---|---|---|---|
| $p$-CH$_3$O | $7.24 \times 10^{-6}$ | 599 | $4.33 \times 10^{-3}$ | 99.8 | — |
| $p$-CH$_3$ | $6.69 \times 10^{-6}$ | 235 | $1.56 \times 10^{-3}$ | 99.4 | 98.0 |
| H | $4.50 \times 10^{-6}$ | 68 | $3.00 \times 10^{-4}$ | 98.7 | 93.0 |
| $p$-Cl | $1.52 \times 10^{-6}$ | 29 | $4.26 \times 10^{-5}$ | 96.6 | — |
| $m$-Cl | $1.00 \times 10^{-6}$ | 6 | $4.89 \times 10^{-6}$ | 83.0 | — |
| $p$-NO$_2$ | $8.10 \times 10^{-8}$ | 6 | $4.33 \times 10^{-6}$ | 84.1 | 52.4 |

TABLE 7.5

**Products of Reaction of Vinyl Triflates 84 and 85 in Anhydrous TFE at 85°C**

| Triflate | Total % vinyl ethers | % 90 | Stereochemistry of vinyl ethers | |
| --- | --- | --- | --- | --- |
| | | | % 86 | % 87 |
| 84 (X = $p$-CH$_3$) | 98.2 ± 0.3 | 1.8 ± 0.3 | 99.85 ± 0.07 | 0.15 ± 0.07 |
| 84 (X = H) | 93.3 ± 1.0 | 6.7 ± 0.0 | 99.47 ± 0.05 | 0.53 ± 0.05 |
| 84 (X = $p$-NO$_2$) | 56.0 ± 3.0 | 44.0 ± 3.0 | 93.5 ± 1.0 | 6.5 ± 1.0 |
| 85 (X = $p$-CH$_3$) | 89.1 ± 0.4 | 10.9 ± 0.4 | 98.2 ± 0.1 | 1.80 ± 0.1 |
| 85 (X = H) | 76.7 ± 2.0 | 23.3 ± 2.0 | 96.4 ± 0.2 | 3.6 ± 0.2 |
| 85 (X = $p$-NO$_2$) | 40.1 ± 2.5 | 59.9 ± 2.5 | 79.4 ± 1.0 | 20.6 ± 1.0 |

in Table 7.5 indicate, a very high degree of stereoselectivity, with retained geometry, was observed in the solvolysis of the $E$ isomers 84 and a high degree of inversion was observed with the $Z$ isomers 85.

$$88 \qquad\qquad 89 \qquad\qquad 90$$

The different effects of substituents on the rates of reaction of the isomeric vinyl triflates 84 and 85, as well as the different product compositions, argue for a difference in the mechanism of reaction of the two triflates. The extensive phenyl migration across the double bond, the Hammett correlation of Fig. 7.1, the Taft correlation of Fig. 7.2, the high degree of product stereoselectivity, and the excellent Schleyer–Lancelot [33] correlation between kinetics and products all indicate the involvement of the vinylidenephenonium ion 93 in the solvolysis of the geometrically favorable $E$ isomers 84. The results are summarized and rationalized by the mechanism [31] outlined in Scheme 7.20. The observed internal return in the reaction of 84 requires the inital formation of an ion pair (91), which converts to 93 and then into products.

On the other hand, vinyl triflates 85 must initially ionize to an open linear vinyl cation pair (92) that can either lose a proton to give allene 90 or be preferentially captured by solvent on the side opposite the departing triflate, resulting in the inverted vinyl ether products 88 and 89. As the triflate gegenion moves away and the initial unfavorable geometry is lost, ion 92 may also convert to ion 93, accounting for the observed rearranged products and partially accounting for the stereoselectivity of the ether products.

However, it should be noted that, in the absence of detailed knowledge of the expected product distribution in the capture of the open ion 92, the

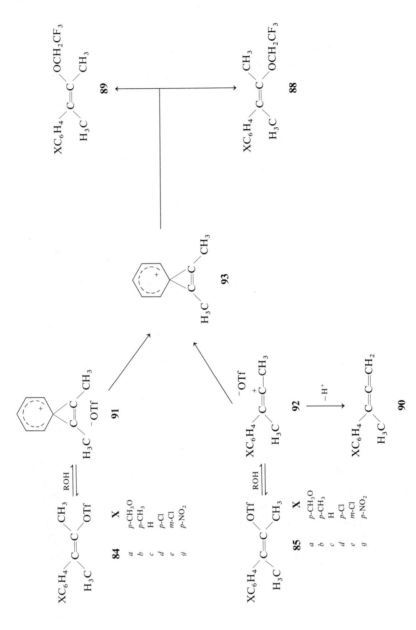

Scheme 7.20

TABLE 7.6

Deuterium Isotope Effects in the
Solvolyses of Vinyl Triflates 86
and 87 in 60% EtOH at 100°

| Compound | $k_H/k_D$ |
|----------|-----------|
| 86a | $1.47 \pm 0.06$ |
| 86b | $0.90 \pm 0.04$ |
| 87a | $1.16 \pm 0.02$ |
| 87b | $1.04 \pm 0.02$ |

stereoselectivity of the reaction of the $Z$ isomers **85** cannot give a quantitative estimate of the extent of leakage to the bridged ion **93**. The analogous ion **95** formed in the solvolysis of **94** in TFE gives an approximately 70:30 ratio of capture products, the capture product from the side of the methyl group predominating [34]. Nevertheless, in view of the analysis in Chapter 6, Section IV, and the similarity of the structures of **92** and **95**, leakage to the bridged ion **93** seems likely.

$$\underset{\textbf{94}}{\overset{\displaystyle H_3C}{\underset{\displaystyle p\text{-}O_2NC_6H_4}{>}}C=C\overset{\displaystyle An}{\underset{\displaystyle Cl}{<}}} \longrightarrow \underset{\textbf{95}}{\overset{\displaystyle H_3C}{\underset{\displaystyle p\text{-}O_2NC_6H_4}{>}}C=\overset{+}{C}-An} \longrightarrow$$

$$\overset{\displaystyle H_3C}{\underset{\displaystyle p\text{-}O_2NC_6H_4}{>}}C=C\overset{\displaystyle OCH_2CF_3}{\underset{\displaystyle An}{<}} + \overset{\displaystyle H_3C}{\underset{\displaystyle p\text{-}O_2NC_6H_4}{>}}C=C\overset{\displaystyle An}{\underset{\displaystyle OCH_2CF_3}{<}}$$

These considerations are confirmed by the kinetic deuterium isotope effects [35] summarized in Table 7.6. The $\beta$-deuterium isotope effect in **87a** is very much smaller than in **86a** (1.16 versus 1.47), indicating a lower residual charge on the $\alpha$-carbon of **87** than on that of **85** due to delocalization into the bridging phenyl group in ion **93**. The $\gamma$-deuterium isotope effect is inverse, as it should be [36], for **86b** but is normal for **87b**, again consistent with the formulation of vinylidenephenonium ion **93**, in which charge is delocalized into the $\beta$-carbon of the double bond.

Indeed, Hoffmann and co-workers [37] have shown by means of molecular orbital arguments that spiroarenes or bridged ions of type **93** are stable.

**96**

They are the unsaturated analogues of the well-known [38] phenonium ion **96**, which was first postulated by Cram [39] some three decades ago.

As discussed in Chapter 5, Section II, rearrangement and aryl migration across the double bond has also been observed in the decomposition of vinyldiazonium ions from diverse precursors.

## D. Heteroatom Migrations

Among the best migrating groups are neighboring heteroatoms with nonbonding lone-pair electrons [40]. Heteroatom rearrangements and participation in the solvolysis of vinylic systems have been extensively investigated by Modena and co-workers.

In particular, solvolysis of 1,2-diaryl-2-arylthiovinyl 2,4,6-trinitrobenzene-sulfonates (ROTNB) **97** in dichloromethane gave substitution products **99** in the presence of weak nucleophiles (Nu), such as methanol, HCl, and thiophenol, and cyclized 2,3-diarylbenzo[$b$]thiophenes in the absence of nucleophiles [41]. The cyclization, which takes place in the presence of anhydrous $BF_3$ or $Ag^+$ ions, is often accompanied by rearrangement, since the benzo[$b$]thiophene formed from **97** when X = $p$-Me, $p$-MeO, $p$-Cl, or $p$-Br is substituted by X at the 6 position meta to sulfur rather than at the expected 5 position, whereas when X = $m$-Cl or $m$-MeO no rearrangement takes place and X is in the expected 4 or 6 position of the product [42, 43]. Scheme 7.21 accounts for the formation of the two types of products (**100** and **101**) via the open ion **98**, although the cyclic thiirenium ion **102** may

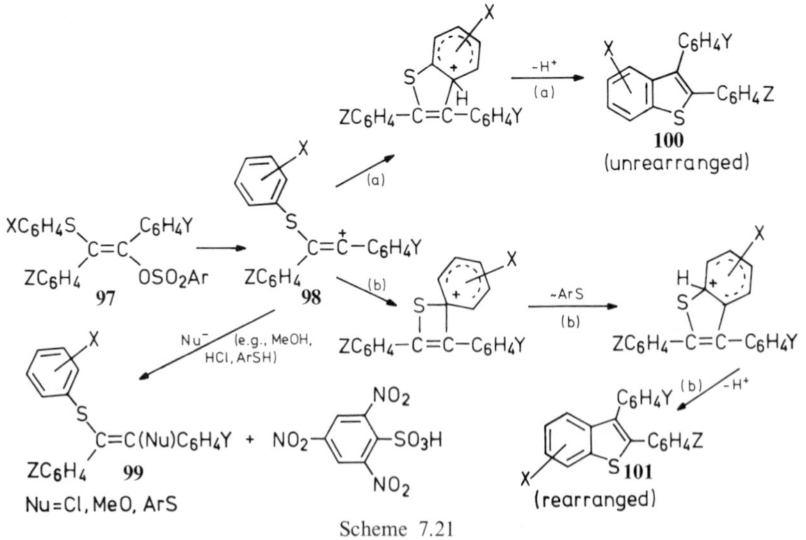

Scheme 7.21

$$
\begin{array}{c}
\text{C}_6\text{H}_4\text{X} \\
| \\
\overset{+}{\text{S}} \\
\diagup \diagdown \\
\text{C}{=}\text{C} \\
\diagup \qquad \diagdown \\
\text{ZC}_6\text{H}_4 \qquad \text{C}_6\text{H}_4\text{Y}
\end{array}
$$
**102**

also be an intermediate (see below). The unrearranged product **100** is formed by attack of the carbenium ion on the activated position ortho to sulfur (route a of Scheme 7.21), and the meta relationship of X and S is retained in the products. The rearranged benzo[*b*]thiophene **101** is formed by ipso attack, followed by a 1,2-sulfur migration (route b), and X and S are at positions meta to one another in the product. The balance between the two routes depends on the substituent. An ortho,para-directing X at the para position favors an ipso attack para to X, whereas the same group at the meta position directs the reaction to position ortho or para to X, i.e., to the unrearranged product [43]. An electron-withdrawing phenylsulfonyl substituent in **97** at a position either meta or para to the sulfur gives a low yield of the same 2,3-diphenyl-5-phenylsulfonylbenzo[*b*]thiophene. Consequently, attack of the vinyl cation takes place at a position meta to the substituent, i.e., at a position ortho to the sulfur when X = $p$-SO$_2$Ph and ipso to the sulfur, followed by 1,2-sulfur rearrangement when X = $m$-SO$_2$Ph [44].

Evidence for the S$_N$1 reaction is the first-order kinetics [45, 46], the rate increase with increasing solvent polarity [46, 47], the normal salt effect [47], the common ion rate depression [45, 47], the special salt effect [45], and the incorporation of trinitrobenzene[$^{35}$S]sulfonate ion from the medium (see Chapter 6, Section V) for **97** (Y = Z = Me; X = H) [45, 47]. The solvolysis rate of **103** was twice as fast as that of **97** (X = Y = Z = H). For X-substituted **97**, when Y = Z, a Yukawa–Tsuno type of equation [Eq. (1)] was obtained [45].

$$
\begin{array}{c}
\text{Ph} \qquad\qquad \text{OTNB} \\
\diagdown \qquad \diagup \\
\text{C}{=}\text{C} \\
\diagup \qquad \diagdown \\
\text{MeS} \qquad\qquad \text{Ph}
\end{array}
$$
**103**

$$\log k_{\text{X,Y,Z}} = -3.38 - 2.85(\sigma_\text{Y}^+ + 0.44\sigma_\text{Z} + 0.52\sigma_\text{X}) \tag{1}$$

The negative $\rho^+$ values and the dependence on the $\sigma^+$ values for substituents on the $\alpha$-aryl group (Chapter 6, Section III,A) indicate the formation of a positive charge, which partially resides on the $\alpha$-aryl group. The combination of a relatively low $\rho^+$ value of $-2.85$ for the $\alpha$-aryl substituents and relatively high $\rho$ value of $-1.45$ for the substituents on sulfur, which is higher than $\rho = -1.25$ for the substituents on the closer $\beta$-aryl group, indicates that more than inductive effects are transmitted by the substituents on sulfur [45]. Hence, these values support some sulfur participation in the transition state. The transition state geometry was therefore suggested to involve overlap between the incipient cationic orbital and one of the filled orbitals on the sulfur, without a severe bond angle distortion [45]. This is supported by stereochemical and kinetic data.

Stereochemical evidence for a symmetric intermediate comes from methanolysis studies in 4:1 acetone–methanol. The ester **104** gives the ether **105** in basic methanol [48, 49] [Eq. (2)] although, at equilibrium, **105** and its Z isomer are present in an approximately 1:1 ratio [48]. Solvolysis of either a 3.1:1 or a 1:2.1 mixture of the isomeric **106** and **107** gave the same 4.4:1 mixture of the ethers **108** and **109** [Eq. (3)] [48, 49]. Likewise, the isomer

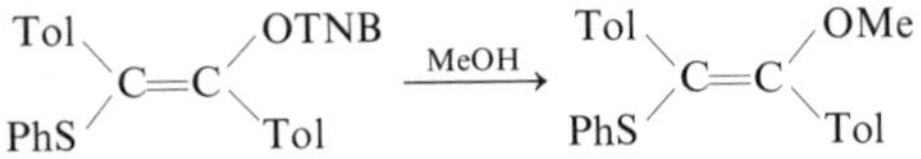

$$
\begin{array}{ccc}
\underset{PhS}{\overset{Tol}{>}}C{=}C\underset{Tol}{\overset{OTNB}{<}} & \xrightarrow{\text{MeOH}} & \underset{PhS}{\overset{Tol}{>}}C{=}C\underset{Tol}{\overset{OMe}{<}} \\
\mathbf{104} & & \mathbf{105}
\end{array}
\tag{2}
$$

$$
\begin{array}{ccc}
\underset{PhS}{\overset{Ph}{>}}C{=}C\underset{Tol}{\overset{OTNB}{<}} & & \underset{PhS}{\overset{Ph}{>}}C{=}C\underset{Tol}{\overset{OMe}{<}} \\
\mathbf{106} & \xrightarrow{\text{MeOH}} & \mathbf{108} \\
\underset{PhS}{\overset{Tol}{>}}C{=}C\underset{Ph}{\overset{OTNB}{<}} & & \underset{PhS}{\overset{Tol}{>}}C{=}C\underset{Ph}{\overset{OMe}{<}} \\
\mathbf{107} & & \mathbf{109}
\end{array}
\tag{3}
$$

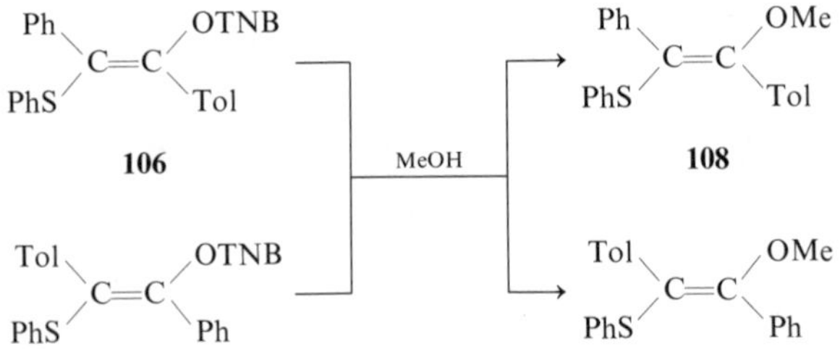

distribution in the cyclization reaction is also independent of the structure of the precursor, since either a 2.9:1 or a 1:4.1 mixture of **110** and **111** gives a 5:1 ratio of the benzo[b]thiophenes **112** and **113** [Eq. (4)] [48]. The high

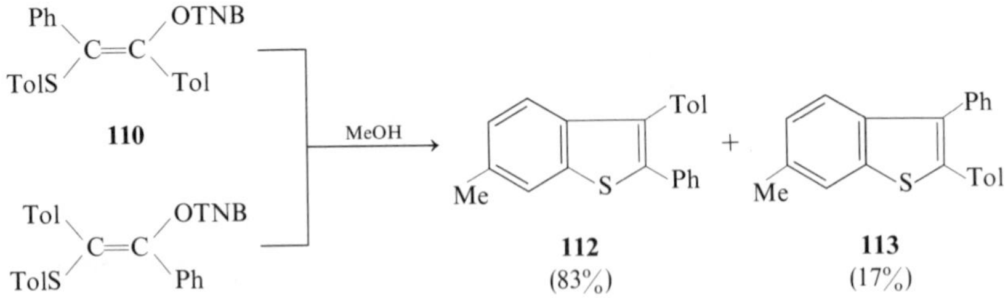

$$
\begin{array}{ccc}
\underset{TolS}{\overset{Ph}{>}}C{=}C\underset{Tol}{\overset{OTNB}{<}} & & \\
\mathbf{110} & \xrightarrow{\text{MeOH}} & \mathbf{112} \; (83\%) \;+\; \mathbf{113} \; (17\%) \\
\underset{TolS}{\overset{Tol}{>}}C{=}C\underset{Ph}{\overset{OTNB}{<}} & & \\
\mathbf{111} & &
\end{array}
\tag{4}
$$

stereospecificity and the identical product mixtures in reactions (3) and (4) are consistent only with a thiirenium ion intermediate (**102**) in which the sulfur serves as a configuration holder and ring opening leads to inversion or to an overall retention for the substitution process. However, as is true for other cases, it is difficult to distinguish between **102** and a mixture of the two rapidly equilibrating ions **114** and **115**, which give the same stereo-chemical results.

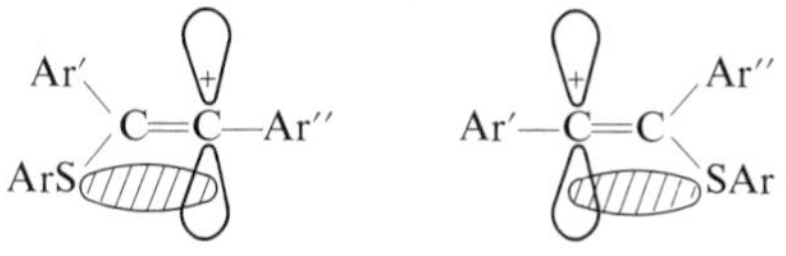

                    **114**                    **115**

A symmetric (or an average) structure for the intermediate is supported by $^{14}C$ scrambling experiments [49, 50]. The $\alpha$-$^{14}C$-labeled brosylates **116** cyclize with $BF_3$ in chloroform to the benzo[b]thiophene **117**, in which the $^{14}C$ is equally distributed at the 2 and 3 positions [Eq. (5)].

Further kinetic evidence for anchimeric assistance by the $\beta$-sulfur comes from comparison of the solvolysis rates of the sulfonate esters **118** and **119**. The reactivity ratios $k_{118}/k_{119}$, when R = 2,4,6-$(O_2N)_3C_6H_2$, are 20 to 22 in MeOH, in nitromethane, and in 19:1 $MeNO_2$–MeOH and 33 in AcOH,

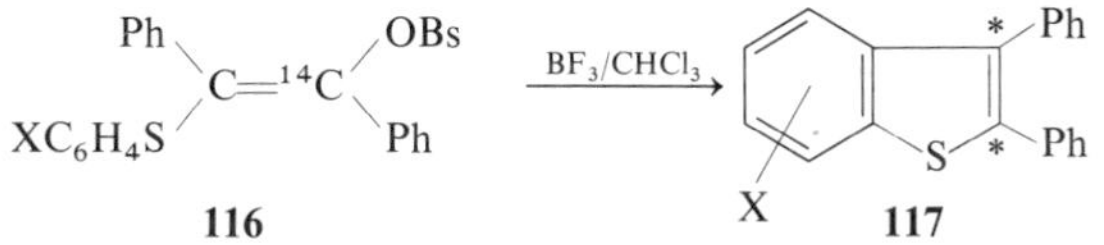

$$(5)$$

                **116**              X    **117**

X = H, p-MeO, m-Cl, m-MeO

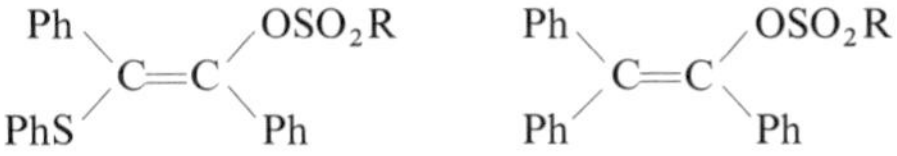

                **118**                    **119**

whereas the ratio is 15.6 in AcOH, when R = p-$MeC_6H_4$ [51]. Since inductive electron withdrawal of a PhS group is slightly higher than that of a phenyl group, a small rate retardation, i.e., $k_{118}/k_{119} < 1$, is expected on the basis of inductive effects alone. It was suggested that the different steric bulk of Ph and PhS does not contribute much to the rate difference and that the rate enhancement should therefore be ascribed to anchimeric assistance by the sulfur. The lower values compared with those of $10^3$ for sulfur assistance in the saturated $\beta$-arylthioethyl halides [52] could be due to the large leveling effect by the $\alpha$-aryl group. Moreover, overlap of the occupied orbitals of the $\beta$ group requires more severe bond distortion than in saturated systems, and the participation may be quite "late" along the reaction coordinate, although the ion may still be bridged [51]. The higher $k_{118}/k_{119}$ value in AcOH suggests a larger extent of participation in the acetolysis than in the methanolysis.

The extent of sulfur participation in the solvolysis of $\beta$-(arylthio)vinyl esters depends on the nature of the other $\beta$ substituent and increases in

**TABLE 7.7**

**Relative Solvolysis Rates of Compounds 120
in MeNO$_2$ at 25°[a]**

| R$_1$ | R$_2$ | Relative rate | $\rho^+$(Ar) | $\rho$(R$_2$ = ArS) |
|-------|-------|---------------|--------------|---------------------|
| H     | PhS   | 0.61          | −4.8         | −0.9[b]             |
| Ph    | Ph    | 1.0           |              |                     |
| Ph    | PhS   | 22            |              |                     |
| Me    | PhS   | 321           | −3.0         | −1.55[b]            |

[a] From Capozzi et al. [53].
[b] Ar = Ph.

the order H ≪ Ph < Me, as shown by the relative solvolysis rates of compounds **120**, which are given in Table 7.7 [53]. The highest negative $\rho^+$(Ar)
value is associated with the lowest negative $\rho$(ArS) value, in line with
decreasing sulfur participation with increasing charge dispersal by the α-aryl

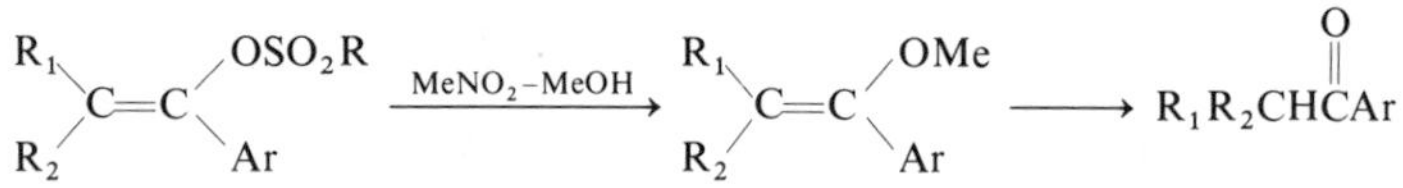

**120**

group. The similarity of the $\rho^+$(Ar) value for the β-H system with that for
the unassisted solvolysis of α-aryl-substituted systems indicates that anchimeric assistance by the β-sulfur is unimportant, although the stereochemistry suggests that the β-sulfur still acts as a configuration holder in the ion in a
"late" participation process. Hence, the transition state structure changes
from one near the linear ion for R$_1$ = H to one near the bridged ion for
R = Me. The effect of the β substituents is not easily understood in terms
of inductive or steric effects. However, it reflects their stabilizing effect on
the bridged form, as shown by the fact that the order is analogous to that
found for the effect of substituents on the stability of the corresponding
cyclopropenium ions [54]. These structural shifts indicate a small energy
difference between the open and the bridged ions, so that a minor structural
variation may change the geometry of the intermediate [53].

Anchimeric assistance and thiirenium ion formation have also been observed [55] in the solvolysis of dialkyl-β-thiovinyl sulfonates **121** and **122**,
with **121** reacting 3.8 × 10$^4$ and **122** 4.2 × 10$^3$ times faster, respectively, than

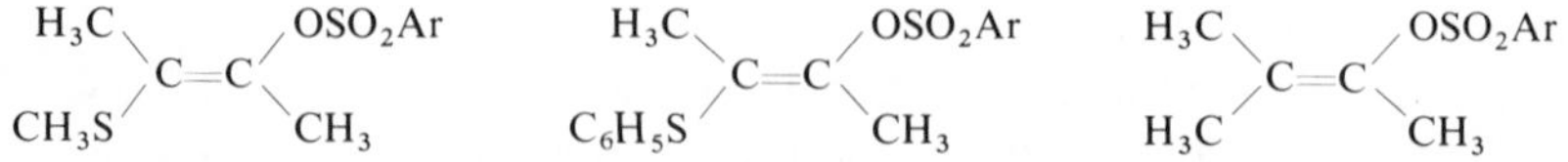

**121**                          **122**                          **123**

model compound **123** at 25°. The large anchimeric effects of $10^3$ to $10^4$ in the solvolysis of **121** and **122** relative to the factors of 20 to 330 in **120** suggest that the bridged thiirenium ion is much more stable than the open vinyl cation in the alkyl compared to the corresponding aryl systems. This is as one would expect, since α-aryl-stabilized vinyl cations need less stabilization by any neighboring group, including sulfur, than simple alkyl vinyl cations.

Other evidence for the intermediacy of thiirenium ions is discussed in Chapter 3, since the addition of substituted sulfenyl chlorides to acetylenes proceeds via the same cyclic intermediates. Modena and co-workers observed a direct inversion in the ring opening of the 1-methyl-2,3-di-*tert*-butylthiirenium ion **124** by chloride ion, which gives the inverted *E*-1,2-di-*tert*-butyl-2-(methylthio)vinyl chloride **125** [Eq. (6)] [56]. It is interesting that the transition state for this vinylic substitution reaction may involve the first example of a planar tetracoordinate carbon in a transition state [57].

$$t\text{-Bu}-\overset{\overset{\displaystyle Me}{\overset{\displaystyle |}{\underset{/\backslash}{S+}}}}{C=C}-\text{Bu-}t \quad + \text{ Cl}^- \longrightarrow \quad \overset{t\text{-Bu}}{\underset{\text{MeS}}{}}C=C\overset{\text{Cl}}{\underset{\text{Bu-}t}{}}$$

**124**           **125**

The calculations [58] related to the preference of the cyclic thiirenium ion over the noncyclic analogue are discussed in Chapter 2, and the nmr evidence [59] for the long-lived thiirenium ion is presented in Chapter 8.

In contrast to participation by β-sulfur, there is no kinetic or other evidence for participation by β-oxygen or β-nitrogen in the solvolysis of α-arylvinyl derivatives, although cyclization reactions have been observed. For example, solvolysis of the *E*-1,2-diaryl-2-aryloxyvinyl brosylates **126** in 80% EtOH proceeds at rates similar to those for triphenylvinyl sulfonates, but the reactions are accompanied by side reactions [60]. In dry EtOH, the solvolysis products are mainly benzil, phenol, and sulfinic acid derivatives and, since the substituent effects also differ from those of the $S_N1$ reaction of the thio analogues **97**, the possibility of homolytic reaction mechanism was raised [61]. When compounds **126** react with $BF_3$ in chloroform or in methylene chloride, the unrearranged benzo[*b*]furan **128** is obtained [Eq. (7)]. The open ion **127** is a probable intermediate, although a concerted process cannot be ruled out. Participation by the β-oxygen and intervention of an oxygen-bridged species are ruled out, since in this case both **126** when $Ar_1 = Ph$,

$$\overset{XC_6H_4O}{\underset{Ar_2}{}}C=C\overset{Ar_1}{\underset{OBs}{}} \longrightarrow \overset{XC_6H_4O}{\underset{Ar_2}{}}C=C\overset{+}{-}Ar_1 \longrightarrow \quad \text{(7)}$$

**126**        **127**        **128**

$Ar_2 = Tol$, $X = H$ and **126** when $Ar_1 = Tol$, $Ar_2 = Ph$, $X = H$ will give a mixture of **128** and **129** via a common intermediate, contrary to the formation of **128** alone [60].

It is therefore not surprising that participation by the $\beta$-acetoxy group of **130** was not observed [62], since the $\alpha$-aryl group is a better positive charge stabilizer.

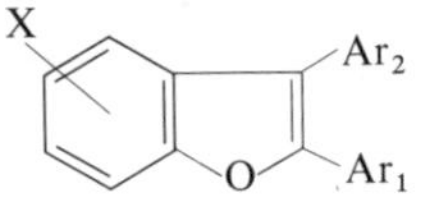

129 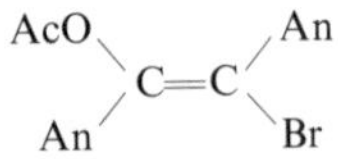                          130

The absence of O-3 participation is in line with the results of calculations (Chapter 2) which indicate that the bridged three-membered cyclic vinylic oxonium ion is much less stable than the open ion [63], in contrast with the relationship of the two species for the sulfur analogues [63, 64].

Participation by a more remote oxygen is *a priori* possible, since the solvolysis of 1-aryl-2,2-bis(*o*-methoxyphenyl)vinyl bromides **131** gives only substituted benzo[*b*]furans [65]. The reaction leading to **134** most likely involves the formation of the five-membered oxonium ion **133**, followed by nucleophilic dealkylation [Eq. (8)]. Anchimeric assistance at O-5 via participation by the *o*-methoxy group is ruled out by the rate data: **131** solvolyzes 1.7 times slower in 80% EtOH and 1.1 times slower in AcOH than trianisylvinyl bromide, and the $\rho^+$ value for a change of the $\alpha$-aryl group is $-3.8$ [65]. Moreover, *E*-1,2-dianisyl-2-(*o*-methoxyphenyl)vinyl bromide solvolyzes 0.74

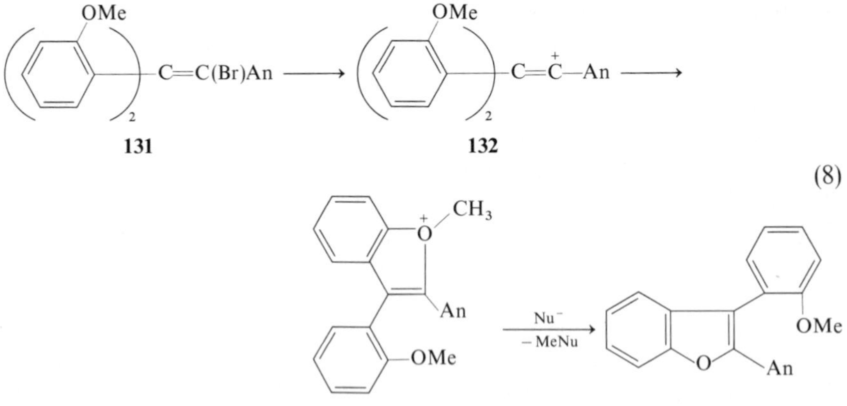

131                          132

(8)

133                          134

times slower than its *Z* isomer in 80% EtOH at 120°, indicating that the trans relationship of the bromine and the *o*-methoxyphenyl group does not accelerate the reaction [66].

Reaction of $E$-1,2-diaryl-2-[($N$-aryl)methylamino]vinyl brosylate **135** in methanol gave both the indole **136** and the substituted benzil **137** [Eq. (9)].

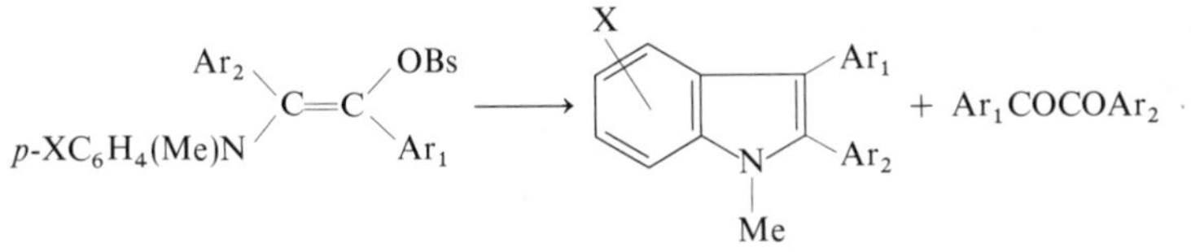

$$
\begin{array}{ccc}
\underset{p\text{-}XC_6H_4(Me)N}{\overset{Ar_2}{\diagdown}}C=C\overset{OBs}{\underset{Ar_1}{\diagup}} & \longrightarrow & \text{indole} \quad + \quad Ar_1COCOAr_2 \\
\textbf{135} & & \textbf{136} \qquad\qquad \textbf{137}
\end{array}
\tag{9}
$$

X = H, Me

Benzil itself was also formed from $E$-1,2-diphenyl-2-[($N$-phenyl)methyl-amino]vinyl chloride [67]. The cyclization also takes place in benzene in the presence of $AlCl_3$ and occurs without rearrangement. If it is assumed that the cyclization involves an intermediate vinyl cation, this ion is formed without any bridging by the nitrogen to give an azirinium ion. Moreover, since the solvolysis of **138** does not give any rearranged indole, $\beta$-nitrogen migration, even in the open ion **139** to form the more stable ion **140** [Eq. (10)],

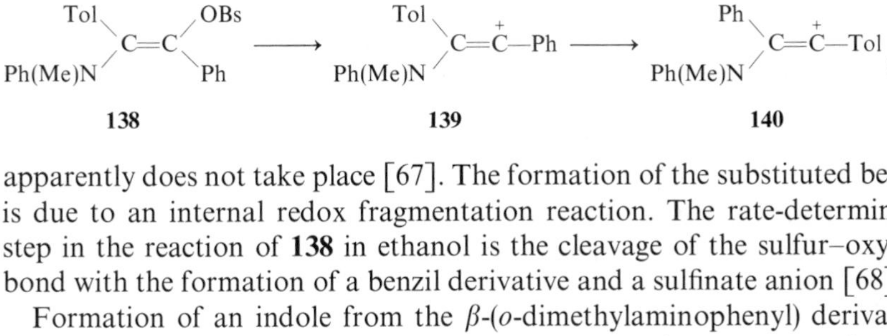

$$
\underset{Ph(Me)N}{\overset{Tol}{\diagdown}}C=C\overset{OBs}{\underset{Ph}{\diagup}} \longrightarrow \underset{Ph(Me)N}{\overset{Tol}{\diagdown}}C=\overset{+}{C}-Ph \longrightarrow \underset{Ph(Me)N}{\overset{Ph}{\diagdown}}C=\overset{+}{C}-Tol
\tag{10}
$$

$$
\textbf{138} \qquad\qquad\qquad \textbf{139} \qquad\qquad\qquad \textbf{140}
$$

apparently does not take place [67]. The formation of the substituted benzil is due to an internal redox fragmentation reaction. The rate-determining step in the reaction of **138** in ethanol is the cleavage of the sulfur–oxygen bond with the formation of a benzil derivative and a sulfinate anion [68].

Formation of an indole from the $\beta$-($o$-dimethylaminophenyl) derivative **141** was suggested to involve a reaction sequence analogous to reaction (8),

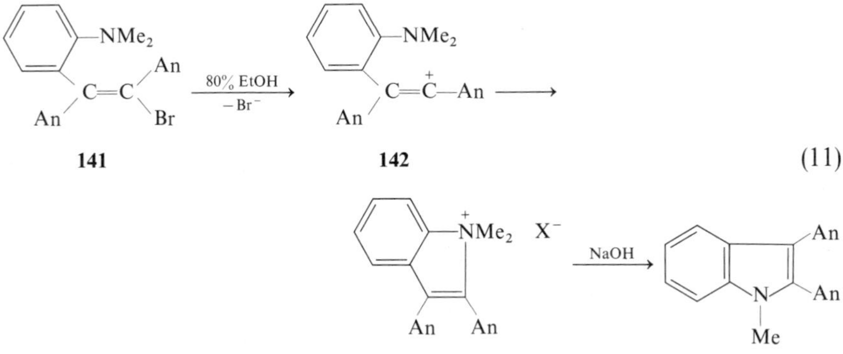

$$
\textbf{141} \quad \xrightarrow[-Br^-]{80\%\ EtOH} \quad \textbf{142} \quad \longrightarrow
\tag{11}
$$

$$
\textbf{143} \quad X = Br \quad \xrightarrow{NaOH} \quad \textbf{144}
$$

i.e., via initial formation of the open ion **142**, followed by reaction of the positively charged carbon with the nitrogen to form the onium salt **143**, which gives the indole **144** on dealkylation [Eq. (11)] [66]. The onium salt can be trapped as its perchlorate **143** ($X = ClO_4$) when the reaction is conducted with $AgClO_4$ in 70% EtOH, and on reaction with NaOH in 1:1 EtOH–THF it is converted to **144**.

In an earlier attempt to evaluate the possibility of $\beta$-bromine participation, the reactions of the Z- and E-$\alpha,\beta$-dibromo-$p,p'$-dimethoxystilbenes **145** and **146**, respectively, were compared [62]. In AcOH/AgOAc, only the E-monoacetate **130** was obtained from both compounds, but the products in AcOH/NaOAc at higher temperature were deoxyanisoin (**147**), 4,4'-dimethoxybenzoin acetate (**148**), anisil (**149**), and dianisylacetylene (**150**) [Eq. (12)], and geometric evidence for participation could not be obtained. The 2.5

$$AnC(Br){=}C(Br)An \xrightarrow[NaOAc]{AcOH/} AnCH_2COAn + AnCH(OAc)COAn + AnCOCOAn + AnC{\equiv}CAn$$

**145** and **146**            **147**          **148**          **149**        **150**

$$(12)$$

times faster solvolysis rate of the E isomer **146** compared with that of **145** cannot assist much in this respect in view of the differences in ground state energies. However, since the compounds are slower than the $\beta$-hydrogen analogoues, participation by the $\beta$-bromine was suggested to be negligible [62].

A more extensive study deals with the solvolysis of E- and Z-$\beta$-halogenovinyl 2,4,6-trinitrobenzenesulfonates **151** and **152**, respectively [69]. The kinetic data were given in Tables 6.7 and 6.12. The acetolysis of **151** ($Ar_1 =$ Ph, Tol; $Ar_2 =$ Ph; $X = I$) gave the retained acetate **151** ($X = OAc$) and

            **151**                    **152**                   **153**

diarylacetylene, whereas **151** ($Ar_1 = Ar_2 = Tol$; $X = I$) gave in nitromethane–methanol the retained ether **151** ($X = OMe$) as the main product (53%) in addition to 42% ditolylacetylene. Acetolysis of either of the bromo esters **151** and **152** ($Ar_1 = Ar_2 = Tol$; $X = Br$) gave a mixture of the acetates **151** ($X = OAc$) and **152** ($X = OAc$) in a 3.5:1 ratio, but $E \rightleftarrows Z$ isomerization of the bromo esters took place during the reaction. The less reactive esters and all the chloro esters **151** and **152** ($X = Cl$) gave only the $\beta$-halo ketones $Ar_2CHXCOAr_1$, except when $Ar_1 = Ar_2 = An$, in which case **151** ($X = OAc$) and **152** ($X = OAc$) were formed in a 1:3.4 ratio starting from either pure 152 (X = Cl) or from a 1 : 1.85 mixture of 151 and 152 (X = Cl). Extensive $E \rightleftarrows Z$ isomerization of the chloro esters took place during the reaction.

Table 6.12 demonstrated the strong retardation by $\beta$-bromine and $\beta$-chlorine and the rate enhancement by $\beta$-iodine over the $\beta$-phenyl-substituted compounds. It is clear that 151 (X = I) shows an anchimerically enhanced rate, whereas 151 (X = Cl, Br) gives no assisted reaction. The close reactivity of the cis and the trans chloro esters shows that anchimeric assistance is negligible, whereas the 10 to 15 times higher reactivity of 151 (X = Br) compared with 152 (X = Br) indicates a substantial assistance. Models show that the departure from planarity is higher for compounds 152, which should therefore be more reactive in the absence of participation. The reactivity ratio $k(151, X = Br)/k(152, X = Br)$ is therefore a lower value for the rate enhancement by a participating $\beta$-bromine. The results are supported by the stereochemistry, since with $\beta$-iodine retention of configuration by the configuration-holding iodine in the bridged ion 153 was observed.

In contrast, the acetolyses of 151 and 152 ($Ar_1 = Ar_2 = Tol$; X = Br and $Ar_1 = Ar_2 = An$; X = Cl) are nonspecific [69]. It was suggested that oriented ion pairs are formed from the bromo esters and that they may either return to their precursors or give solvated ions before collapse to the products. The cation formed from the trans bromo ester may consist of a rapid equilibrium of the bridged structure with a slightly more stable open form. Nucleophilic attack occurs on the open ion, and the steric effect of the bromine determines the low trans/cis ratio of the acetates. There is no evidence for bridging in the reaction of the chloro esters, but the high relative abundance of the cis acetate is not easily explained. In conclusion, the extent of bridging increases in the order $\beta$-I > $\beta$-Br > $\beta$-Cl.

A similar behavior was observed in the electrophilic addition of 2,4,6-trinitrobenzenesulfonyl hypohalites to diarylacetylenes. Only 151 was obtained when X = I, but a mixture of 151 and 152 was formed when X = Cl or Br [70]. A bridged species was suggested in the addition of the hypoiodite, whereas open ions are mainly involved when X = Cl or Br.

## REFERENCES

1. J. L. Fry and G. J. Karabatsos, in "Carbonium Ions" (G. A. Olah and P. von R. Schleyer, eds.), Vol. 2, p. 52. Wiley, New York.
2. L. A. Telkowski and M. Saunders, in "Dynamic Nuclear Magnetic Resonance Spectroscopy" (L. M. Jackman and F. A. Cotton, eds.), p. 523. Academic Press, New York, 1975.
3. K. P. Jäckel and M. Hanack, Tetrahedron Lett. p. 4295 (1975).
4. K. P. Jäckel and M. Hanack, Justus Liebigs Ann. Chem. p. 2305 (1975).
5. G. A. Olah and H. Mayr, J. Am. Chem. Soc. 98, 7333 (1976).
6. A. A. Schegolev, W. A. Smit, V. F. Kucherow, and R. Caple, J. Am. Chem. Soc. 97, 6604 (1975).

7. K. Griesbaum and Z. Rehman, *J. Am. Chem. Soc.* **92**, 1417 (1970).

8. A. G. Martinez, M. Hanack, R. H. Summerville, P. von R. Schleyer, and P. J. Stang, *Angew. Chem., Int. Ed. Engl.* **9**, 302 (1970).

9. M. A. Imhoff, R. H. Summerville, P. von R. Schleyer, A. G. Martinez, M. Hanack, T. E. Dueber, and P. J. Stang, *J. Am. Chem. Soc.* **92**, 3802 (1970).

10. See footnote 13 in reference 9.

11. T. Sasaki, S. Eguchi, and T. Tom, *Chem. Commun.* p. 780 (1968).

12. K. Bott, *Tetrahedron Lett.* p. 1747 (1969).

13. K. Bott, *Chem. Commun.* p. 1349 (1969).

14. (a) W. D. Pfeifer, C. A. Bahn, P. von R. Schleyer, S. Bocher, C. E. Harding, K. Hummel, M. Hanack, and P. J. Stang, *J. Am. Chem. Soc.* **93**, 1513 (1971); (b) M. Hanack and K. A. Fuchs, unpublished results.

15. P. J. Stang and T. E. Dueber, *Tetrahedron Lett.* p. 563 (1977).

16. M. Hanack, P. von R. Schleyer and A. G. Martinez, *An. Quim.* **70**, 941 (1974).

17. J. Pincock and C. Somawardhana, *Can. J. Chem.* **56**, 1164 (1978).

18. K. P. Jäckel and M. Hanack, *Tetrahedron Lett.* p. 1637 (1974).

19. K. P. Jäckel and M. Hanack, *Chem. Ber.* **110**, 199 (1977).

20. (a) C. C. Lee and E. C. F. Ko, *J. Org. Chem.* **40**, 2132 (1975); (b) C. C. Lee, personal communication.

21. Z. Rappoport and A. Gal, *J. Chem. Soc., Perkin Trans. 2* p. 301 (1973).

22. D. S. Noyce and M. D. Schiavelli, *J. Am. Chem. Soc.* **90**, 1020 (1968).

23. G. Capozzi, V. Lucchini, F. Marcuzzi, and G. Melloni, *Tetrahedron Lett.* p. 717 (1976).

24. G. A. Olah, *Angew. Chem., Int. Ed. Engl.* **12**, 173 (1973); D. M. Brouwer and H. Hogeveen, *Prog. Phys. Org. Chem.* **2**, 179 (1972).

25. H. Hogeveen and C. F. Roobeek, *Tetrahedron Lett.* p. 3343 (1971).

26. (a) R. Partch and D. Margosian, *J. Am. Chem. Soc.* **98**, 6746 (1976); (b) L. R. Subramanian and M. Hanack, *J. Org. Chem.* **42**, 174 (1977).

27. S. A. McNeely and P. J. Kropp, *J. Am. Chem. Soc.* **98**, 4319 (1976).

28. For a review and leading references, see Z. Rappoport, *Acc. Chem. Res.* **9**, 265 (1976).

29. C. C. Lee, A. J. Paine, and E. C. F. Ko, *Can. J. Chem.* **55**, 2310 (1977); *J. Am. Chem. Soc.* **99**, 7267 (1977); C. C. Lee and M. Oka, *Can. J. Chem.* **54**, 604 (1976); C. C. Lee and E. C. F. Ko, *Can. J. Chem.* **54**, 3041 (1976); M. Oka and C. C. Lee, *Can. J. Chem.* **55**, 320 (1975); F. H. A. Rummens, R. D. Green, A. J. Cessna, M. Oka, and C. C. Lee, *Can. J. Chem.* **53**, 314 (1975); C. C. Lee, A. J. Cessna, B. A. Davis, and M. Oka, *Can. J. Chem.* **52**, 2679 (1974).

30. P. J. Stang and T. E. Dueber, *J. Am. Chem. Soc.* **95**, 2683 (1973).

31. P. J. Stang and T. E. Dueber, *J. Am. Chem. Soc.* **99**, 2602 (1977).

32. A. Diaz, I. Lazdins, and S. Winstein, *J. Am. Chem. Soc.* **90**, 6546 (1968); E. F. Jenny and S. Winstein, *Helv. Chim. Acta* **41**, 807 (1958).

33. P. von R. Schleyer and C. J. Lancelot, *J. Am. Chem. Soc.* **91**, 4297 (1969).

34. Z. Rappoport and N. Pross, unpublished results.

35. P. J. Stang and T. E. Dueber, *J. Am. Chem. Soc.* **95**, 2686 (1973).

36. C. J. Collins and N. S. Bowmen, eds., "Isotope Effects in Chemical Reactions," Van Nostrand–Reinhold, New York, 1970.

37. R. Hoffmann, A. Imamura, and G. D. Zeiss, *J. Am. Chem. Soc.* **89**, 5215 (1967).

38. For a review and leading references, see C. J. Lancelot, D. J. Cram, and P. von R. Schleyer, *in* "Carbonium Ions" (G. A. Olah and P. von R. Schleyer, eds.), Vol. 3, p. 1347. Wiley (Interscience), New York, 1972.

39. D. J. Cram, *J. Am. Chem. Soc.* **70**, 4244 (1948); **71**, 3863, 3875 (1949); **74**, 2129, 2137, 2149 (1952).

40. T. H. Lowry and K. S. Richardson, "Mechanism and Theory in Organic Chemistry," Harper, New York, 1976.
41. G. Capozzi, A. di Bello, G. Melloni, and G. Modena, *Ric. Sci.* **39**, 267 (1969).
42. G. Capozzi, G. Melloni, G. Modena, and M. Piscitelli, *Tetrahedron Lett.* p. 4039 (1968).
43. G. Capozzi, G. Melloni, and G. Modena, *J. Chem. Soc.* (*C*) p. 2621 (1970).
44. G. Melloni and G. Modena, *J. Chem. Soc.*, *Perkin Trans. 1* p. 218 (1972).
45. G. Modena and U. Tonellato, *J. Chem. Soc.* (*B*) p. 374 (1971).
46. G. Modena, U. Tonellato, and F. Naso, *Chem. Commun.* p. 1363 (1968).
47. G. Modena and U. Tonellato, *Chem. Commun.* p. 1676 (1968).
48. G. Modena and U. Tonellato, *J. Chem. Soc.* (*B*) p. 381 (1971).
49. G. Capozzi, G. Melloni, G. Modena, and U. Tonellato, *Chem. Commun.* p. 1520 (1969).
50. G. Capozzi, G. Melloni, and G. Modena, *J. Chem. Soc.* (*C*) p. 3018 (1971).
51. (a) G. Modena and U. Tonellato, *Boll. Sci. Fac. Chim. Ind. Bologna* **27**, 373 (1969); (b) G. Modena and U. Tonellato, *J. Chem. Soc.* (*B*) p. 1569 (1971).
52. A. Streitwieser, Jr., "Solvolytic Displacement Reactions," pp. 108–110. McGraw-Hill, New York, 1962.
53. G. Capozzi, G. Modena, and U. Tonellato, *J. Chem. Soc.* (*B*) p. 1700 (1971).
54. R. Breslow, H. Hover, and H. W. Chang, *J. Am. Chem. Soc.* **84**, 3168 (1962); J. Ciabattoni and E. C. Nathan, III, *Tetrahedron Lett.* p. 4997 (1969).
55. A. Burighel, G. Modena, and U. Tonellato, *J. Chem. Soc.*, *Perkin Trans. 2* p. 2026 (1972); *Chem. Commun.* p. 1325 (1971).
56. G. Capozzi, V. Lucchini, G. Modena, and P. Scrimin, *Tetrahedron Lett.* p. 911 (1977).
57. Z. Rappoport, *Tetrahedron Lett.* p. 1073 (1978).
58. A. S. Denes, I. G. Csizmadia, and G. Modena, *Chem. Commun.* p. 8 (1972).
59. G. Capozzi, O. De Lucchi, V. Lucchini, and G. Modena, *Chem. Commun.* p. 248 (1975).
60. G. Capozzi and G. Modena, *J. Chem. Soc.*, *Perkin Trans. 1* p. 216 (1972).
61. U. Tonellato and G. Versini, *Gazz. Chim. Ital.* **105**, 1237 (1975).
62. Z. Rappoport and M. Atidia, *Tetrahedron Lett.* p. 4085 (1970); *J. Chem. Soc.*, *Perkin Trans. 2* p. 2316 (1972).
63. I. G. Csizmadia, F. Bernardi, V. Lucchini, and G. Modena, *J. Chem. Soc.*, *Perkin Trans. 2* p. 542 (1977).
64. I. G. Csizmadia, A. J. Duke, V. Lucchini, and G. Modena, *J. Chem. Soc.*, *Perkin Trans. 2* p. 1808 (1974).
65. T. Sonoda, S. Kobayashi, and H. Taniguchi, *Bull. Chem. Soc. Jpn.* **49**, 2560 (1976).
66. T. Sonoda, M. Kawakami, T. Ikeda, S. Kobayashi, and H. Taniguchi, *Chem. Commun.* p. 612 (1976).
67. G. Capozzi, G. Modena, and L. Ronzini, *J. Chem. Soc.*, *Perkin Trans. 1* p. 1136 (1972).
68. A. Burighel, G. Modena, and U. Tonellato, *J. Chem. Soc.*, *Perkin Trans. 2* p. 1021 (1973).
69. P. Bassi and U. Tonellato, *J. Chem. Soc.*, *Perkin Trans. 2* p. 1283 (1974); *Gazz. Chim. Ital.* **102**, 387 (1972).
70. P. Bassi and U. Tonellato, *J. Chem. Soc.*, *Perkin Trans. 1* p. 669 (1973).

# 8

# SPECTROSCOPIC EVIDENCE FOR VINYL CATIONS

The best evidence for any intermediate is, of course, actual isolation of the species involved. Many organic intermediates are much too reactive to be isolated even at low temperatures. However, modern spectroscopic techniques, and nuclear magnetic resonance methods in particular, in many instances make it possible to examine the structure of short-lived intermediates in solution without actual isolation. Such techniques have been very successfully applied in the general area of carbenium ions and are being used with increasing frequency in the case of vinyl cations.

## I. ALKYNYL CATIONS

Spectroscopic evidence for the existence of vinyl cations in solution was obtained for the first time by H. R. Richey, Jr., and his co-workers using alkynyl cations **1a** in their mesomeric form (**1b**) as models [1]. Alkynyl cations **1a** may be regarded as vinyl cations if the allenic resonance structure **1b** is important.

$$R_2\overset{+}{C}-C\equiv CR \longleftrightarrow R_2C=C=\overset{+}{C}R$$

**1a**          **1b**

Alkynyl alcohols **2** and **4** were used as precursors for the generation of the first alkynyl cations. Solutions of **2** and **4** in concentrated sulfuric acid gave the corresponding cations **3** and **5**, respectively, as observed by $^1$H nmr and uv spectroscopy [1, 2]. The precursor alcohols were recovered by quenching the sulfuric acid solutions with excess sodium hydroxide.

In a more detailed investigation [3, 4] of substituted alkynyl cations **7** by $^1$H nmr spectroscopy, the ions **7a** through **7e** were generated by dissolving the corresponding alcohols **6a** through **6e** in liquid $SO_2$ at $-50°$ and adding this solution to a mixture of $SbF_5$, $FSO_3H$, and $SO_2$ at $-70°$.

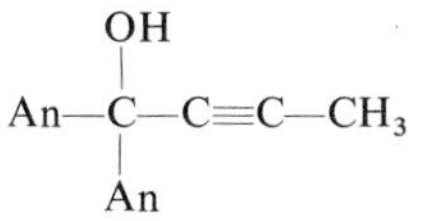

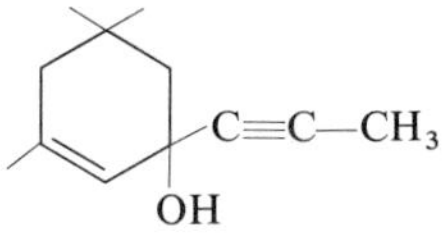

$$
\begin{array}{cc}
\text{An--C--C}\equiv\text{C--CH}_3 & \text{C}\equiv\text{C--CH}_3 \\
\mathbf{2} & \mathbf{4} \\
\end{array}
$$

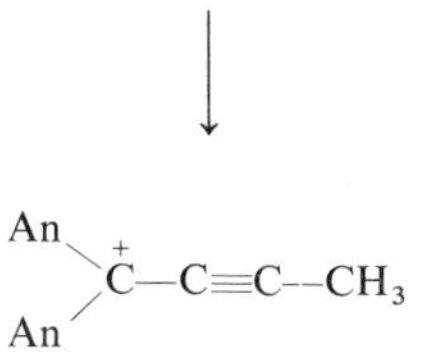

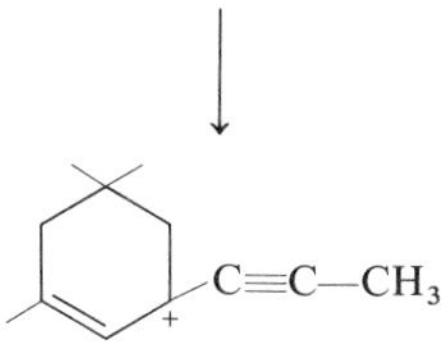

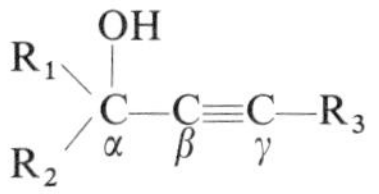

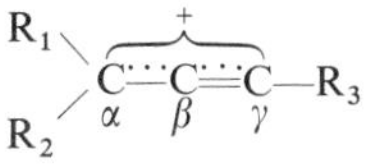

$$
\begin{array}{cc}
\mathbf{3} & \mathbf{5} \\
\end{array}
$$

| | |
|---|---|
| **6a** $R_1 = R_2 = R_3 = CH_3$ | **7a** $R_1 = R_2 = R_3 = CH_3$ |
| **6b** $R_1 = R_2 = Ph; R_3 = CH_3$ | **7b** $R_1 = R_2 = Ph; R_3 = CH_3$ |
| **6c** $R_1 = R_2 = CH_3; R_3 = Ph$ | **7c** $R_1 = R_2 = CH_3; R_3 = Ph$ |
| **6d** $R_1 = R_3 = Ph; R_2 = CH_3$ | **7d** $R_1 = R_3 = Ph; R_2 = CH_3$ |
| **6e** $R_1 = R_2 = R_3 = Ph$ | **7e** $R_1 = R_2 = R_3 = Ph$ |

The $^1H$ nmr spectra of ions **7** at $-60°$ showed large downfield shifts for the groups attached to the triple bond as well as for groups adjacent to the positive charge. The position of methyl groups directly attached to the charged carbon steadily moved downfield in the series of cations **7d**, **7c**, and **7a**. This is about the order expected on the basis of the predicted positive charge density at $C_\alpha$. Cation **7a** has no phenyl or other delocalizing groups on the alkynyl system and therefore would be expected to have the highest charge density, accounting for the methyl absorption being at the lowest field in the above series [4]. Delocalization of the positive charge into the phenyl group in **7c** resulted in a reduced charge density on the alkynyl cation system, accounting for the upfield position of the methyl signal in **7c** with respect to the corresponding methyl absorption in **7a**. On the basis of these $^1H$ nmr spectra, it was concluded that the mesomeric form **1b** is an important contributor to the structure of the alkynyl cations [3, 4].

In general, proton chemical shifts alone do not give sufficient information about the charge distribution in cations. Recently, the charge distributions in the alkynyl cations **7a** through **7e** were more accurately determined by $^{13}C$ nmr spectroscopy [5], the chemical shifts of the $C_\alpha$, $C_\beta$, and $C_\gamma$ carbon atoms in **7** being observed directly. The $^{13}C$ nmr data for the ions **7a** through

**7e** indicated significant contributions from the allenyl form **1b**. This was shown by a marked deshielding of $C_\alpha$ and $C_\gamma$ in the cations **7a** through **7e** relative to their precursor alcohols **6a** through **6e**. The deshielding for $C_\alpha$ was in the range of 204 ppm for **7a** to 112.3 ppm for **7e**, and that for $C_\gamma$ was 141.4 ppm in **7a** to 72.5 ppm in **7e** [5]. Aromatic substitution at $C_\alpha$ and $C_\gamma$ as in **7e** resulted in less deshielding of these carbons in the corresponding cations relative to their precursor alcohols, which is consistent with less positive charge at $C_\alpha$ and $C_\gamma$ due to its delocalization into the phenyl rings. In the case of **6a**, the relative contributions of the mesomeric forms **8a** and **8b** were estimated to be in a ratio of about 2:1 [5].

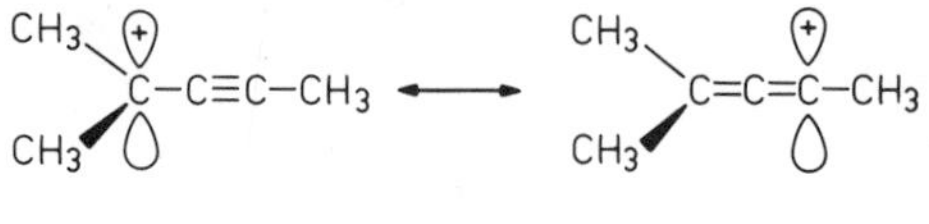

                   **8a**                     **8b**

Alkynyl ions substituted by the powerful electron-donating ferrocenyl group have also been investigated [6]. However, rather than alkynyl cation formation, primary alcohol **9a** was observed to give the ferrocenyl-stabilized vinyl cation **10a** by protonation in acetic acid. Tertiary alcohol **9b** underwent dehydroxylation to the mesomeric ion **10b**, which rapidly added $CF_3CO_2H$ to give the actually observed allylium ion **11**, as shown in Scheme 8.1. Only alcohol **9c** gave an alkynyl ion (**12**) of sufficient stability to be actually

$$FcC \equiv CCH_2OH \xrightarrow{H^+} Fc\overset{+}{C}=CHCH_2OH$$

**9a**   Fc = ferrocenyl                 **10a**

$$Fc'C \equiv CC(CH_3)_2OH \xrightarrow{H^+} Fc'C \equiv C-\overset{+}{C}(CH_3)_2 \longleftrightarrow Fc'\overset{+}{C}=C=C(CH_3)_2$$

**9b**   Fc' = 1',2-di-*tert*-butylferrocenyl             **10b**

                                             $\downarrow$ $CF_3COOH$

$$Fc'C \cdots \overset{+}{C}H \cdots C(CH_3)_2$$
$$\underset{OCOCF_3}{|}$$

                                             **11**

$$ArC \equiv CCH(OH)Fc \xrightarrow{H^+} ArC \equiv C-\overset{+}{C}HFc \longleftrightarrow Ar\overset{+}{C}=C=CHFc$$

**9c**   Fc = ferrocenyl                   **12a**                 **12b**

                            Scheme 8.1

observed [6]. The low chemical shift of the single nonaromatic proton (7.2 ppm) strongly suggests that the ion exists essentially in its alkynyl form **12a** with little or no contribution from the allenyl hybrid **12b**.

## II. VINYL CATIONS

The large stabilizing effect of a *p*-anisyl group was used for the first time by Hanack and co-workers [7] to attempt to generate a "free" vinyl cation and to observe its spectroscopic properties by $^1$H and $^{13}$C nmr spectroscopy. For this purpose, the *p*-anisylvinyl fluoride **13** was treated at low temperature with SbF$_5$ in SO$_2$ClF, and the $^1$H and $^{13}$C nmr spectra of the resulting solutions of **13** + SbF$_5$ were recorded [7]. In comparison with **13**, the $^1$H nmr spectrum of the reaction product **13** + SbF$_5$ showed peaks that were downfield-shifted between 0.50 and 1.6 ppm. The aromatic region of the spectrum was similar to that of the *p*-methoxybenzyl cation investigated under the same conditions [8] with regard to both the chemical shift of the protons and the nonequivalence of the two ortho and meta protons. In the $^{13}$C nmr spectrum of **13** + SbF$_5$, the $\beta$-methyl groups were shifted only

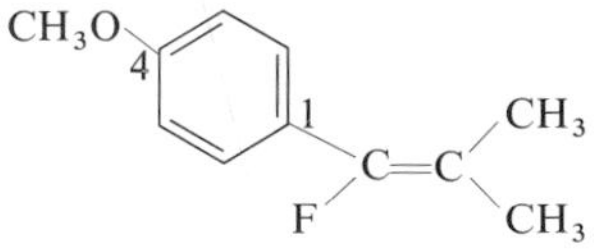

**13**

slightly downfield. An especially large downfield shift (30 ppm) was observed for C-4; again, this signal was shifted downfield by the same amount as the corresponding signal in the spectrum of the *p*-methoxybenzyl cation. The signals of the *o*- and *m*-carbon atoms were shifted downfield by about 20 and 30 ppm, respectively. Again, the two meta and ortho positions were not equivalent, presumably due to the frozen rotation of the *p*-methoxy group at low temperatures. Quenching of the solution of **13** + SbF$_5$ with CH$_3$OH–K$_2$CO$_3$ gave the isopropyl anisyl ketone.

Besides the anisylvinyl derivatives **13**, the aryl-substituted vinyl halides **14** through **19** (fluorides and chlorides) labeled in various positions with a high concentration of $^{13}$C were reacted with SbF$_5$ at low temperature in various nonnucleophilic solvents (SO$_2$ClF, CH$_2$Cl$_2$, SO$_2$) [9].

Although it has been reported that the phenylvinyl chloride **19** can be converted to the corresponding vinyl cation [10], a more recent systematic and thorough investigation of the reaction of SbF$_5$ with the vinyl fluorides and vinyl chlorides **13** through **19** by Siehl and Hanack [9] shows that arylvinyl cations are not formed with compounds **13** through **19** under these conditions.

$$
\begin{array}{cc}
\underset{F}{\overset{An}{>}}C\overset{\alpha}{=}\underset{H}{\overset{CH_3}{C<}} & \underset{Cl}{\overset{An}{>}}C=\underset{CH_3}{\overset{CH_3}{C<}} \\
\mathbf{14} & \mathbf{15}
\end{array}
$$

An–C(F)=C(CH₃)(H)   **14**

An–C(Cl)=C(CH₃)(CH₃)   **15**

An–C(Cl)=C(CH₃)(H)   **16**

Tol–C(X)=C(CH₃)(CH₃)   **17**  X = Cl, F

Tol–C(Cl)=C(CH₃)(H)   **18**

Ph–C(Cl)=C(CH₃)(CH₃)   **19**

The $^1$H nmr spectra showed without doubt arylcarbenium ion structures for the stable solutions of the vinyl halides + SbF$_5$. This was demonstrated by the chemical shifts as well as by the coupling constants and splitting patterns of the aromatic signals. The species resulting from interaction of **14** or **17** with SbF$_5$ showed fluorine couplings larger than the couplings in the starting vinyl fluorides and therefore characteristic of $\alpha$-fluorocarbenium ions. The spectra of the vinyl halides **16** and **17** + SbF$_5$, although symmetrically substituted by methyl at the $\beta$-carbon, exhibited nonequivalent aromatic protons. This is possible only if the benzylic $\alpha$-carbon in the corresponding cations is symmetrically substituted. Both effects are not in accordance with a vinyl cation structure for the reaction products of the vinyl halides with SbF$_5$. Vinyl cations would require the absence of a fluorine coupling and a symmetric structure for the species from **16** and **17**.

To obtain $^{13}$C nmr spectra, the vinyl halides, e.g., **13**, **14**, **16**, and **17**, labeled with $^{13}$C in high concentration in the $\alpha$ or $\beta$ position of the double bond were reacted with SbF$_5$ [9]. The chemical shifts of the $\alpha$-carbon atoms were found to be in the typical range of benzyl cations. The $\beta$-carbon atoms did not absorb in the range of sp$^2$ carbon atoms but at about 70 ppm, which is typical for sp$^3$ carbon atoms shifted to low fields. In the case of the vinyl fluorides **14** or **17** as precursors, the spectra of the solutions with SbF$_5$ showed $^{13}$C–$^{19}$F couplings typical for fluorocarbenium ions. The $^{13}$C$_\alpha$–$^{13}$C$_\beta$ coupling was found to be 40 to 47 Hz, which is in the range of a sp$^2$–sp$^3$ $^{13}$C coupling constant [9]. These results with vinyl halides **13** through **19** are in accordance with the formation of a carbenium ion that did not lose its halogen and no longer contained a double bond.

Since proton addition to the double bond was ruled out by control experiments, the species observed in the nmr were explained by an addition of $SbF_5$ to the double bond of the vinyl halides. Therefore, the initially formed $\pi$ complex of $SbF_5$ with the double bond is subsequently converted to a $\sigma$ complex, which then gives an $\alpha$-halocarbenium ion substituted by $SbF_4$ at the $\beta$ position. The structure of the addition complex was suggested to be **20**, in which the antimony tetrafluoride substituent interacts with the halogen in the $\alpha$ position, as well as with the excess $SbF_5$ in the solution [9].

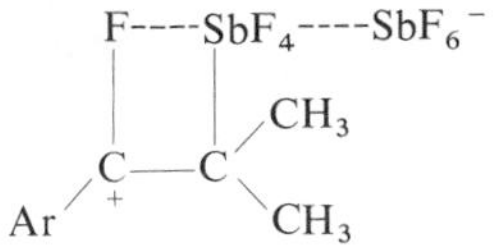

$$F\text{----}SbF_4\text{----}SbF_6{}^-$$

**20**

Protonation of alkynylferrocenes leads to species that were assigned vinyl cation structures from their $^1H$ nmr spectra [11]. Addition of $CF_3COOH$ to alkynes **21** gave green solutions, which quickly reacted further with solvent, forming trifluoroacetoxycarbenium ions. Upon hydrolysis with aqueous $NaHCO_3$, these ions gave ketones **23** in quantitative yields.

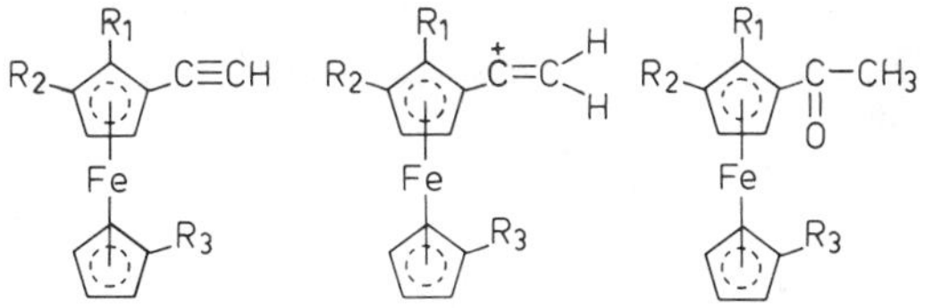

**21**          **22**          **23**

The $^1H$ nmr spectra of alkynes **21** substituted in the indicated positions with methyl or *tert*-butyl groups were recorded in cold trifluoroacetic acid. The initial signals in these spectra that were assigned to the corresponding vinyl cations **22** quickly decayed and were replaced by signals characteristic of addition products. Vinyl cation **22** was proposed as an intermediate due to the similarity of the $^1H$ nmr spectra with those of the related allenylferrocenyl ions **10** and also due to the fact that the vinyl protons appeared as an AB quartet with a relatively large coupling constant of 12 Hz. This would be in accordance with a linear geometry for the vinyl cation **22**. The lifetime of the vinyl cations was dependent on the substituents $R_1$, $R_2$, and $R_3$ in **22**. If $R_2 = R_3$ was *t*-Bu then, due to steric hindrance of solvent attack, the lifetime of **22** increased.

Substituted propynylferrocenes **24** and alkynylferrocenes of type **25** were also reacted with trifluoroacetic acid [11]. The protonation of **24** ($R_1 = R_3 = t$-Bu; $R_2 = H$) leads to a mixture of the stereoisomeric vinyl cations

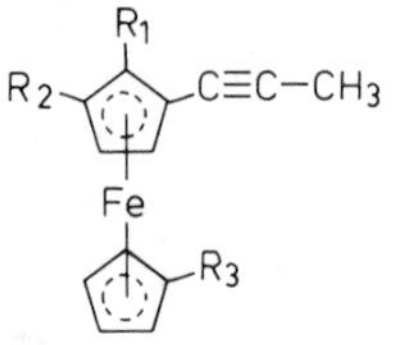

24 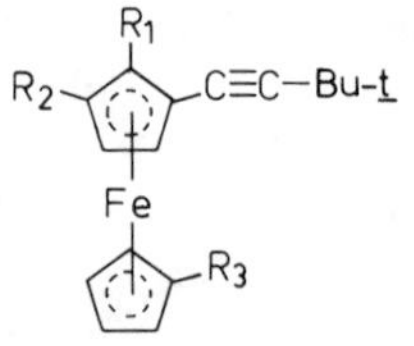                          25

**26** and **27** in a ratio of 2:3. At $-8°$, **26** and **27** underwent rotational inter-conversion to give an equilibrium mixture consisting predominantly (>95%) of the sterically less crowded cation **26**. Similarly, protonation of alkynyl-

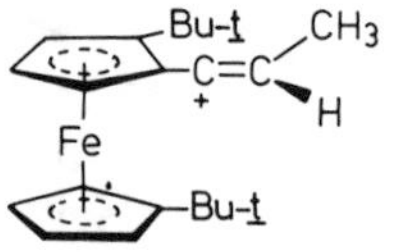

26 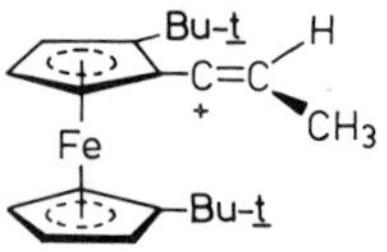                          27

ferrocenes **25** with different substituents $R_1$, $R_2$, and $R_3$ (H, $CH_3$, or $t$-Bu) with trifluoroacetic acid was suggested to give stable vinyl cations. The [1]H nmr spectra of these species showed the presence of a single stereoisomer in each case [11]. To date, the [13]C nmr spectra of these ferrocenylvinyl cations are not available.

## III. THIIRENIUM IONS

An example of the generation of a species quite similar to the vinyl cation is that of the thiirenium ion **30**, a long-lived bridged vinyl cation [12] (see also Chapters 3 and 7). The trimethylthiirenium ion **30** is quantitatively formed by reaction of bis-(methylthio)methylsulfonium hexachloroantimonate (**28**) with excess 2-butyne (**29**) at $-80°$ in liquid $SO_2$ [12]. The [1]H nmr spectrum

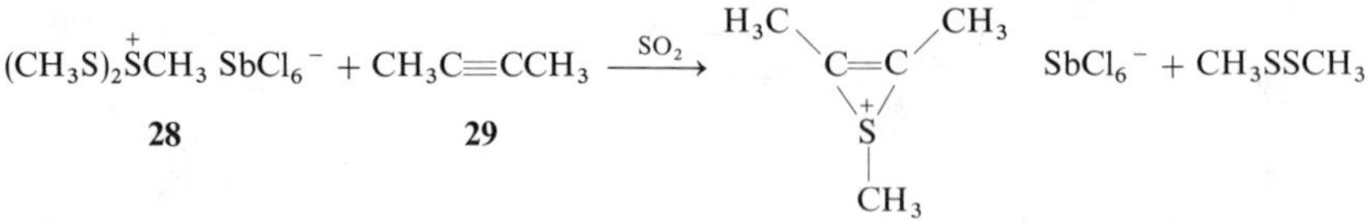

$$(CH_3S)_2\overset{+}{S}CH_3 \ SbCl_6^- + CH_3C≡CCH_3 \ \xrightarrow{SO_2} \ \begin{array}{c} H_3C \diagdown \ \ \diagup CH_3 \\ C=C \\ \diagdown \ \overset{+}{\diagup} \\ \underset{|}{S} \\ CH_3 \end{array} \ SbCl_6^- + CH_3SSCH_3$$

28                29                                        30

of **30** showed singlets at $\delta$ 2.51 and 2.77 in a ratio of 1:2 due to the $S-CH_3$ and $C-CH_3$ protons of the thiirenium ion **30**, respectively. Reaction of **28** with excess 3-hexyne under the same conditions gave a [1]H nmr spectrum consistent with the formation of a 1-methyl-2,3-diethylthiirenium ion [12].

The structure of the thiirenium ion **30** was confirmed by $^{13}$C nmr spectroscopy. The C–CH$_3$ and S–CH$_3$ shifts compare well with the reported shifts for the $\alpha$-CH$_3$ and S–CH$_3$ of the known pentamethylthiophenium ion. Ion **30** was obtained from methanesulfenyl chloride, antimony pentachloride, and 2-butyne (**29**) at $-80°$ in dichloromethane as a white, crystalline hexachloroantimonate, stable up to $-40°$ [13]. 1-Methyl-2,3-di-*tert*-butylthiirenium hexachloroantimonate (**31**), prepared from **28** and di-*tert*-butylacetylene in dichloromethane at $0°$, was stable even at room temperature.

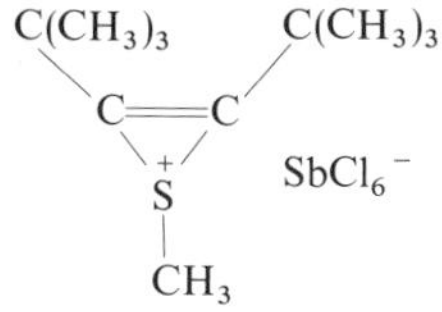

**31**

The $^1$H nmr spectrum of **31** (SO$_2$ solution at $-60°$ or CH$_2$Cl$_2$ solution at room temperatures) showed a singlet at $\delta$ 2.62 for the methyl group bonded to the sulfur atom and a singlet with $\delta$ 1.54 for the two equivalent *tert*-butyl groups. The $^{13}$C nmr spectrum in liquid SO$_2$ at $-50$ showed signals at $\delta$ 113.42 for the ring carbons, 33.12 for the quaternary carbons, 28.63 for the methyl carbon bonded on the sulfur, and 26.67 for the methyl carbons in the *tert*-butyl group. The $^1$H and $^{13}$C nmr spectra of **31** are in accordance with the given structure and are also in good agreement with the data for the trimethylthiirenium salt **30** [13].

It is evident from the foregoing that spectral observation of stabilized vinyl cations, such as alkynyl, $\alpha$-ferrocenyl, and thiirenium ions, is possible. The picture is less clear regarding the actual observation of $\alpha$-aryl and other vinyl cations, and more work is certainly required to answer the question as to whether such vinyl cations can also be observed under stable ion conditions using spectroscopic methods.

## REFERENCES

1. H. G. Richey, Jr., J. C. Phillips, and L. E. Rennick, *J. Am. Chem. Soc.* **87**, 1381 (1965).
2. H. Fischer and H. Fischer, *Chem. Ber.* **97**, 2959 (1964).
3. H. G. Richey, Jr., L. E. Rennick, A. S. Kushner, J. M. Richey, and J. C. Phillips, *J. Am. Chem. Soc.* **87**, 4017 (1965).
4. C. U. Pittman, Jr. and G. A. Olah, *J. Am. Chem. Soc.* **87**, 5632 (1965).
5. G. A. Olah, R. J. Spear, P. W. Westerman, and J. M. Denis, *J. Am. Chem. Soc.* **96**, 5855 (1974).
6. T. S. Abram and W. E. Watts, *J. Chem. Soc., Perkin Trans. 1* p. 1532 (1977).
7. H. U. Siehl, J. C. Carnahan, Jr., L. Eckes, and M. Hanack, *Angew. Chem., Int. Ed. Engl.* **13**, 675 (1974).

8. G. A. Olah, R. D. Porter, C. L. Jenell, and A. M. White, *J. Am. Chem. Soc.* **94**, 2044 (1972).
9. H. U. Siehl and M. Hanack, *J. Am. Chem. Soc.* (in press).
10. S. Masamune, M. Sakai, and K. Morio, *Can. J. Chem.* **53**, 784 (1975).
11. T. S. Abram and W. E. Watts, *J. Chem. Soc., Chem. Comm.* p. 857 (1974); *J. Chem. Soc., Perkin Trans. 1* p. 1522 (1977).
12. G. Capozzi, O. De Lucchi, V. Lucchini, and G. Modena, *J. Chem. Soc., Chem. Comm.* p. 248 (1975).
13. G. Capozzi, V. Lucchini, G. Modena, and P. Scrimin, *Tetrahedron Lett.* p. 911 (1977).

# 9

# MISCELLANEOUS AND CONCLUSIONS

## I. SYNTHETIC USES OF VINYL CATIONS

The previous chapters have discussed some reactions of vinyl cations leading to a variety of products. Undoubtedly, the most interesting and valuable use of vinyl cations in synthesis to date has been the work of Johnson and co-workers on biogenetic-type cyclizations and formation of steroids, including progesterone, as discussed in Chapter 3. Other examples include the conversion of vinyl halides or vinyl arenesulfonates to the corresponding vinyl ethers, acetates, and thiolates via $S_N 1$ processes, the formation of alkynes via E1 processes, and the formation of heterocycles such as benzothiophenes, benzofurans, and indoles by reaction of the vinyl cation with an ortho heteroatom on a $\beta$-aryl group.

Similar to the formation of benzothiophenes (Chapters 6 and 7), thiophene dioxides **2** were formed by cyclization via vinyl cations of E-2-arylsulfonyl-1,2-diphenylvinyl p-bromobenzenesulfonates **1** [1]:

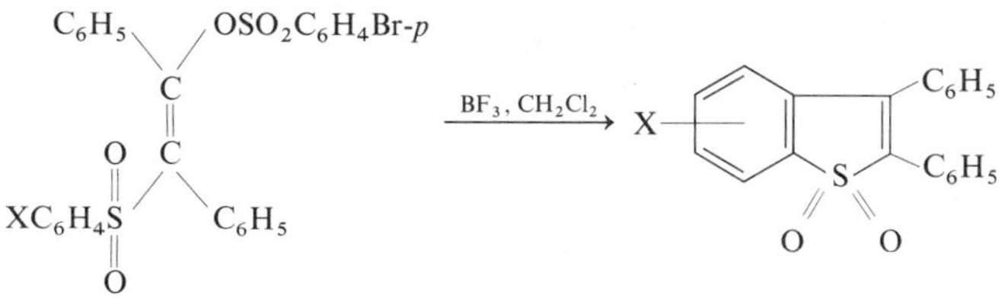

In all cases, a substituent initially para (or meta) to the sulfonyl group appears at a position para (or meta) to the new C—C bond. This was rationalized by an ipso attack followed by a 1,2-sulfur shift (analogous to Scheme

7.21) rather than alkylation at the ortho position, which is strongly de-activated by the sulfonyl group [1].

Intermolecular Friedel–Crafts type of alkylations with vinyl cations have also been observed. Alkynes and various mineral acids, both in the presence and absence of mercuric salts, have often been employed as precursors in aromatic alkylations [2]. Whether vinyl cations were actually involved in any of these reactions is difficult to ascertain from the available experimental data but, in view of what we know about vinyl cations today, they were most likely electrophiles, at least in some of these alkylations [2]. A similar situation exists with regard to the early reports on Friedel–Crafts type of alkylations with vinyl halides and various Lewis acids [3]. Alkylations have also been observed with unsaturated esters, such as Angelica lactone and vinyl acetates [4]. An exact interpretation of the mechanisms of these reactions is difficult not only for lack of appropriate data, particularly kinetics, but also due to complications arising from the use of heterogeneous catalysts and reaction conditions, the presence of acid and in some instances starting material, as well as product isomerizations.

Roberts and Abdel-Baset [5] found that $\alpha$-bromostyrene condensed rapidly with toluene in the presence of $Al_2Br_6$, resulting in 88% 1-phenyl-1-tolylethylene, as well as 4% phenylacetylene and 8% acetophenone:

$$C_6H_5C(Br)\!\!=\!\!CH_2 \xrightarrow[Al_2Br_6]{C_6H_5CH_3} p\text{-}CH_3C_6H_4(C_6H_5)C\!\!=\!\!CH_2 + C_6H_5C\!\!\equiv\!\!CH + C_6H_5\overset{\overset{\displaystyle O}{\displaystyle \|}}{C}CH_3$$

Alkylation via initial addition of a proton to form $C_6H_5\overset{+}{C}(Br)CH_3$ was ruled out since none of the expected 1,1-ditolyl-1-phenylethane was observed, thus making a vinyl cation mechanism likely but not required in this alkylation.

Stang and Anderson [6] reported an extensive investigation of aromatic alkylations with vinyl triflates. Alkylation of monosubstituted benzenes readily occurred, even in the presence of 2,6-di-*tert*-butyl-4-methylpyridine, a sterically hindered nonnucleophilic base, with vinyl triflates 3 and 4 in 50 to 85% yields. Lower yields of alkylated products were observed with

$$(C_6H_5)_2C\!\!=\!\!C(C_6H_5)OSO_2CF_3 \qquad (CH_3)_2C\!\!=\!\!C(C_6H_5)OSO_2CF_3$$

3                                           4

cyclooctenyl and cycloheptenyl triflates, and no reaction occurred with cyclohexenyl triflate. No alkylation was observed with trimethylvinyl triflate in refluxing benzene in the presence or absence of $BF_3$ at 80°, and only tar formed at 150°. In the case of cyclooctenyl, cycloheptenyl, and trimethylvinyl triflates, proton elimination from the incipient vinyl cation followed by dimerization and oligomerization competes with alkylation.

Alkylation of a series of monosubstituted benzenes with **4** gave a $\rho$ value of $-2.57$, one of the lowest observed in any electrophilic aromatic substitution and comparable to the $\rho$ value of $-2.4$ for ethylation [6]. Despite this low intermolecular selectivity, a very high intramolecular selectivity with exclusive formation of the ortho and para products and no meta (except for toluene) alkylation was observed. Competitive alkylation of $C_6D_6$ and $C_6H_6$ by **4** gave $k_H/k_D = 0.98 \pm 0.03$ [6]. These results were interpreted as indicating a vinyl cation as the intermediate electrophile with two distinct transition states for control of inter- and intramolecular selectivities [6]. This study represents the first unambiguous demonstration of the involvement of vinyl cations in Friedel–Crafts type of electrophilic aromatic substitution. Unfortunately, from a synthetic point of view, the reaction seems to be restricted to stabilized vinyl cations with no $\beta$-hydrogens, either on the double bond or on the adjacent saturated carbon.

## II. METAL-STABILIZED VINYL CATIONS

As mentioned in Chapter 2, theoretical calculations indicate that electron-rich transition metals may provide considerable stabilization to vinyl cations. Such complexes in fact are hybrids of two resonance forms, **5a** and **5b**. If the positive charge resides predominantly on the $\alpha$-carbon, as in **5a**, the

$$L_nM-\overset{+}{C}=C\diagup_{\diagdown} \longleftrightarrow L_n\overset{+}{M}-C=C\diagup_{\diagdown}$$

$$\textbf{5a} \qquad\qquad\qquad \textbf{5b}$$

species is a metal-stabilized vinyl cation. If the charge resides mostly on the metal, the species is in fact a positively charged alkylidene–carbene [7] complex. Recently, such complexes have been invoked as reaction intermediates and at least in one instance actually isolated as a stable crystalline salt.

Chisholm and co-workers [8, 9] reported the addition of HCl to a platinum(II) complex (**6**). Addition of dry HCl to **6** in $CDCl_3$ or $CD_2Cl_2$ occurred even at $-60°$. Addition of ammonia or trimethylamine to **7** in $CD_2Cl_2$ or $C_2D_6$ at $25°$ resulted in the formation of $R_3NHCl$ and complex **6**. The rate of HCl elimination from **7** was found to depend strongly on solvent polarity as measured by dielectric constant, with faster elimination in the more polar solvent. The rate of HCl elimination was found to be first order in complex **7** but zero order in the amine base [8]. Furthermore, complex **7** ($X = CH_2{=}CCl$) showed an unusually large $C_\alpha$—Cl bond of 1.809 Å, the C—Cl bond in vinyl chloride being only 1.728 Å. These data strongly suggest a Pt-stabilized vinyl cation (**8**) as the intermediate in the interconversion of **6** and **7**.

$$L_2XPt—C\equiv CH \qquad + HCl \longrightarrow L_2XPt—\underset{\underset{Cl}{|}}{C}=CH_2$$

$$\textbf{6} \qquad\qquad\qquad\qquad\qquad \textbf{7}$$

$$L = PMe_2C_6H_5$$
$$X = Cl, CH_2{=}CCl, HC\equiv C,$$

It is known (see Chapter 3) that HCl does not react with acetylene even at room temperature, so the addition of HCl to **6** at $-60°$ is possible only if a highly stable vinyl cation such as **8** is invoked as an intermediate. Likewise, alkylvinyl halides, let alone the parent vinyl chloride, do not react even under forcing solvolytic conditions (see Chapter 5). Even α-styryl bromide requires high temperature and aqueous polar solvents to undergo reaction via the phenyl-stabilized vinyl cation. Yet complex **7** undergoes unimolecular ionization to **8** at room temperature in much less polar solvents, such as benzene and methylene chloride. Although the relatively high $k_H/k_D \cong 3.0$ [8] for elimination from **7** may suggest the incursion of an E2 process in the formation of **6**, other data (*vide supra*) strongly indicate a unimolecular ionization and formation of **8**.

$$XL_2Pt—\overset{+}{C}=CH_2$$

$$\textbf{8}$$

A similar metal-stabilized complex was invoked by Davison and Solar [10] to explain the reactivity of cyclopentadienylcarbonyl iron–acetylides **9** with electrophiles. Reaction of iron-acetylide **9** with $HBF_3$–dimethyl ether complex in anhydrous methanol gave iron–acyl complex **11**. The same product was observed upon acid-catalyzed addition of water [10]:

$$[Fe]—C\equiv CPh + HX \longrightarrow [Fe]—\overset{+}{C}=CHPh \quad X^-$$

$$\textbf{9} \qquad\qquad\qquad\qquad \textbf{10}$$

$$[Fe]—\overset{+}{C}=CHPh \quad X^- \xrightarrow{H_2O} [Fe]—\underset{\underset{}{\overset{|}{OH}}}{C}=CHPh \longrightarrow [Fe]—\underset{\overset{||}{O}}{C}—CH_2Ph$$

$$\textbf{10} \qquad\qquad\qquad\qquad\qquad\qquad\qquad \textbf{11}$$

$$[Fe] = (n^5—C_5H_5)Fe(CO)_2 \xrightarrow{CH_3OH} [Fe]{\cdots}\overset{+}{C}{\overset{OCH_3}{\underset{CH_2Ph}{\big\langle}}} X^- \xrightarrow{CH_3OH} [Fe]—\underset{\overset{||}{O}}{C}—CH_2Ph + HX + (CH_3)_2C$$

$$\textbf{11}$$

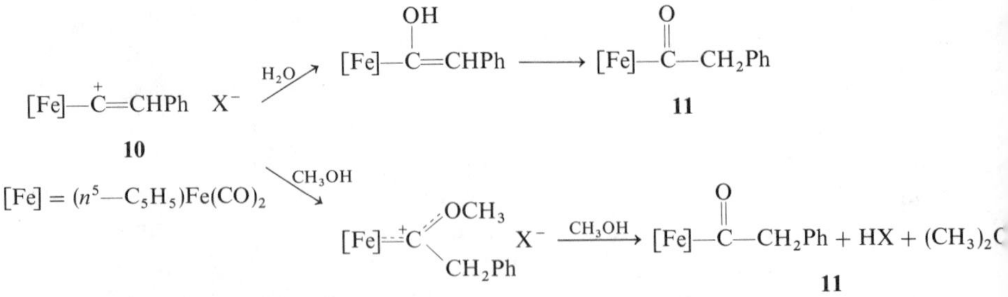

Both reactions probably proceed via the iron–vinyl cation complex **10**, as shown above. Reaction of complex **9** with $HBF_4 \cdot O(CH_3)_2$ in $CH_2Cl_2$ at $-78°$ gave a rapid reaction, which, upon addition of diethyl ether, precipitated crystalline complex **13** in 86% yield. This reaction was further evidence for the formation of complex **10**. In the absence of competing nucleophiles, complex **10** reacts with a second molecule of **9** to give **12**, which, upon cyclization, gave the observed product **13** [10]:

$$[Fe]-C \equiv CPh + HBF_4 \longrightarrow [Fe]-\overset{+}{C}=CHPh \ BF_4^- \overset{+9}{\longrightarrow}$$

**9**                            **10**

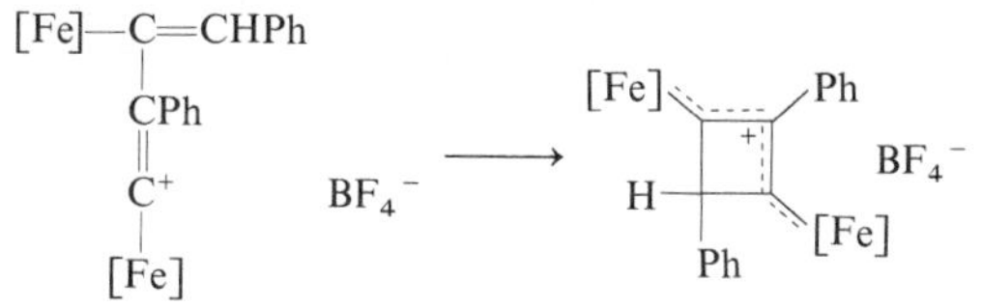

**12**                        **13**

Reaction of ynamines $RC \equiv CNR'_2$ with HCl or $HBF_4$ has been shown [11] to give analogous cyclic products:

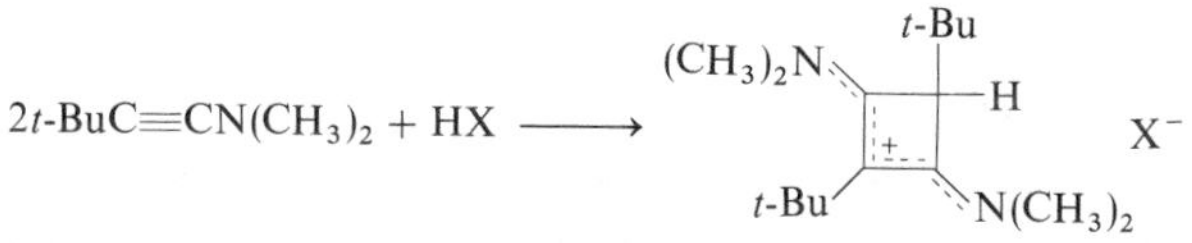

$$2t\text{-}BuC \equiv CN(CH_3)_2 + HX \longrightarrow$$

Similar results were observed with the acetylide complex **14**. With this complex the intermediate vinyl cation–iron complex **15** was found to be metastable. Reaction of complex **15** with methanol gave the stable crystalline alkoxycarbene complex **16**, and reaction with diethylamine resulted in deprotonation to starting material **14** [12]:

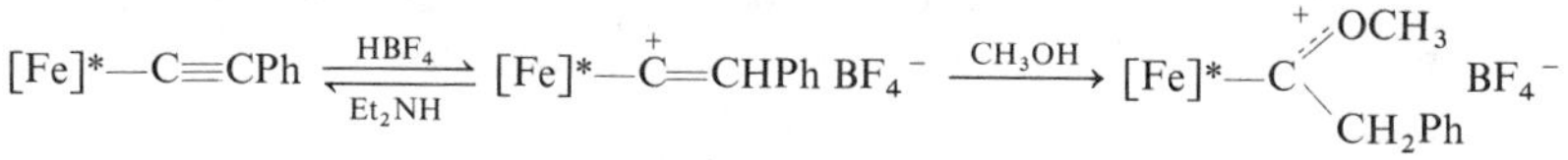

$$[Fe]^*-C \equiv CPh \underset{Et_2NH}{\overset{HBF_4}{\rightleftharpoons}} [Fe]^*-\overset{+}{C}=CHPh \ BF_4^- \overset{CH_3OH}{\longrightarrow} [Fe]^*-C\overset{\overset{+}{\diagup OCH_3}}{\diagdown CH_2Ph} \ BF_4^-$$

**14**                     **15**                     **16**

$[Fe]^* = (n^5\text{-}C_5H_5)Fe(CO)[P(C_6H_5)_3]$

Recently, a stable crystalline vinyl cation–iron complex (**17**) has actually been isolated and spectroscopically fully characterized [12].

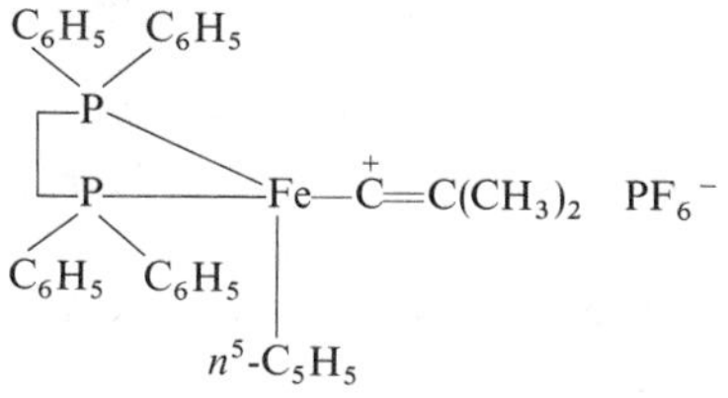

**17**

## III. SPECIES RELATED TO VINYL CATIONS

Besides vinyl and allenyl cations, other members of the family of unsaturated carbenium ions are acylium ions (**18**), nitrilium ions (**19**), aryl cations (**20**), and alkynyl cations (**21**). Both acylium [13] and nitrilium [14]

$$R-\overset{+}{C}=O \longleftrightarrow R-C\equiv\overset{+}{O} \qquad R_1\overset{+}{C}=\overset{..}{N}-R_2 \longleftrightarrow R_1C\equiv\overset{+}{N}-R_2$$

**18a**             **18b**           **19a**             **19b**

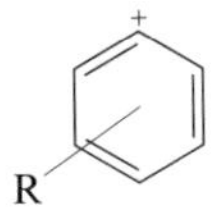

$$RC\equiv C^+$$

**20**         **21**

ions are well-established intermediates that derive a considerable degree of their stability from the stabilization of the positive charge by the adjacent heteroatom lone-pair electrons. In fact, spectroscopic evidence indicates that, in both acylium [15] and nitrilium [16] ions, the major resonance contributors are **18b** and **19b**, respectively, with the positive charge being predominantly on the respective heteroatoms. Hence, these species only vaguely resemble vinyl cations, and for this reason and the fact that they have been well covered in the literature [13–16], they will not be further discussed in this monograph.

### A. Aryl Cations

An excellent and extensive review of aryl cations has recently appeared [17]; hence, this monograph will treat only the essential features of the recent data on these intermediates. Gas-phase data derived from ion cyclotron resonance experiments [18] place the heat of formation of the parent phenyl cation $C_6H_5^+$ at $\Delta H^\circ = 270$ kcal/mole. This value, along with the $\Delta H_f^\circ = +19.8$ kcal/mole for benzene and the thermodynamic data of Table 2.1, can be used to derive relative stabilities of the phenyl cation and other carbenium ions by means of the isodesmic relationship of Eq. (1).

$$R^+ + C_6H_6 \longrightarrow C_6H_5^+ + R-H \tag{1}$$

These calculations indicate that the phenyl cation is 29 kcal/mole more stable than the methyl cation and 4 kcal/mole more stable than the parent vinyl cation. On the other hand, the phenyl cation is seen to be 11 kcal/mole less stable than the ethyl cation and 18 kcal/mole less stable than the 2-propenyl cation.

Theoretical calculations at the 4-31G level [19] indicate that for the phenyl cation the ground state is a highly distorted singlet $^1A_1$ with a $C\overset{+}{C}C$ bond angle of 145°. There are two low triplet states resembling the benzene geometry, the $^3B_1$ and $^3A_2$, with a singlet triplet splitting $^1A_1 \rightarrow {}^3B_1$ of approximately 20 kcal/mole. Calculations further indicate [20] that for the ground state species the positive charge is dispersed through the $\sigma$ framework, and hence it should be stabilized by $\sigma$ donors in the direction $o > m > > p$. In the triplet state $^3B_1$, the charge is delocalized throughout the $\pi$ system, and hence this state is better stabilized by $\pi$ donors and in the direction $p \sim o > m$. In fact, these calculations also indicate [20] that, with $NH_2$ and OH as substituents, there is a crossover to a triplet ground state compared to the singlet ground state of the parent phenyl cation.

Experimentally, the best evidence for the formation of aryl cations comes from dediazoniation of arenediazonium salts [21]. Swain and co-workers reported a careful and elegant investigation of the dediazoniation of benzene-diazonium tetrafluoroborate (22) under a variety of conditions. In the absence of strong base, reducing agent, and light, first-order rates with respect to diazonium salt were observed for the decomposition of 22 [22]. No deuterium incorporation was observed in the decomposition of 22 in deuterated solvents. This observation, along with the observation of only unrearranged products in the dediazoniation of substituted diazonium salts, rules out any type of benzyne mechanism for these decompositions [22]. The high activation entropy of + 10.5 eu, the very low selectivity for externally added weak nucleophiles, and the first-order kinetics strongly suggest a unimolecular decomposition of 22 and formation of a singlet phenyl cation [22].

The observed aromatic deuterium isotope effects nicely confirmed [23] this conclusion. The experimentally observed isotope effects $k_H/k_D$ of 1.74 ± 0.05 for the 2,3,4,5,6-$d_6$ diazonium salt, 1.52 ± 0.03 for the 2,4,6-$d_3$ isomer, 1.19 ± 0.02 for the 3,5-$d_2$ compound, and 1.02 ± 0.01 for the 4-$d_1$ benzenediazonium tetrafluoroborate suggest an intermediate with a high degree of charge development in the transition state. The very large $k_H/k_D$ of 1.22 ± 0.01 for *each* ortho deuterium is a particularly persuasive argument for the formation of a singlet phenyl cation. This isotope effect is as large as the $\beta$-deuterium isotope effects observed in the formation of vinyl cations discussed in Chapter 5, Section V,A,3. In a singlet phenyl cation the bulk of the positive charge is constrained in a sp²-hybridized orbital perpendicular to the $\pi$ system and hence in the plane of the C—H bonds. Therefore, similar to vinyl cations, the $o$-C—H bonds with a dihedral angle of 0° are ideally situated for hyperconjugative overlap and stabilization and account for the large observed deuterium isotope effects. The relatively large $k_H/k_D$ of 1.08 ± 0.01 for *each* meta deuterium and the smaller but significant $k_H/k_D$ of 1.02 ±

0.01 for the para deuterium confirm the theoretical calculations of charge delocalization via the $\sigma$ framework in a singlet phenyl cation.

The unusually large nitrogen $^{14}N/^{15}N$ isotope effects $k_{14}/k_{15}$ of 1.0384 $\pm$ 0.0010 and 1.0106 $\pm$ 0.0003 for the $\alpha$-N and $\beta$-N effects, respectively, further support [24] the formation of a highly charged intermediate with complete or nearly complete C—N bond breaking in the transition state in these dediazoniation reactions.

Further evidence for the formation of a phenyl cation in solution comes from the work of Zollinger and co-workers [25] on the dediazoniation of **22** in trifluoroethanol. Similar to Swain's results [22], Zollinger observed first-order kinetics and low selectivity in TFE. However, at 300 atm $N_2$ pressure at 25°, he also observed 2.5% $N_2$ incorporation into recovered, previously labeled benzenediazonium salt. Furthermore, at 320 atm CO, 5.2% of trifluoroethyl benzoate was observed in addition to the normal products [25]. These are highly significant results, for they indicate the formation of a high-energy intermediate that as a consequence is very unselective in its reactivity and hence reacts not only with CO but with the normally "inert" nitrogen. Furthermore, they indicate that under these conditions dediazoniation is reversible.

These results are best summarized by the mechanism in Scheme 9.1. Loss of $N_2$ from **22** leads to **23** via a unimolecular process that under certain conditions (high pressure) is reversible. Due to its high energy, ion **23** reacts nearly indiscriminately with solvents such as $H_2O$ or alcohol and added weak nucleophiles such as halide ions. Under high-pressure conditions it even reacts with CO, giving **26**, which subsequently gives the observed ester product **27**. The high energy and consequent low selectivity of aryl cations are understandable when one compares them to cyclic vinylc cations (Chapter 5, Section V,B). The phenyl cation itself should be at least as strained as the cyclohexenyl cation that forms only under forcing conditions. Furthermore, inductively and electronically, the phenyl cation should be even less stable than the cyclohexenyl cation.

$$C_6H_5OS$$
$$24$$

$$\xrightarrow{SOH}$$

$$C_6H_5N_2{}^+BF_4{}^- \underset{}{\overset{-N_2}{\rightleftharpoons}} C_6H_5{}^+ \xrightarrow{Nu^-} C_6H_5Nu$$

$$22 \qquad\qquad 23 \qquad\qquad 25$$

$$\overset{CO}{\searrow}$$

$$C_6H_5\overset{+}{C}O \xrightarrow{ROH} C_6H_5\overset{\displaystyle O}{\overset{\|}{C}}{-}OR$$

$$26 \qquad\qquad 27$$

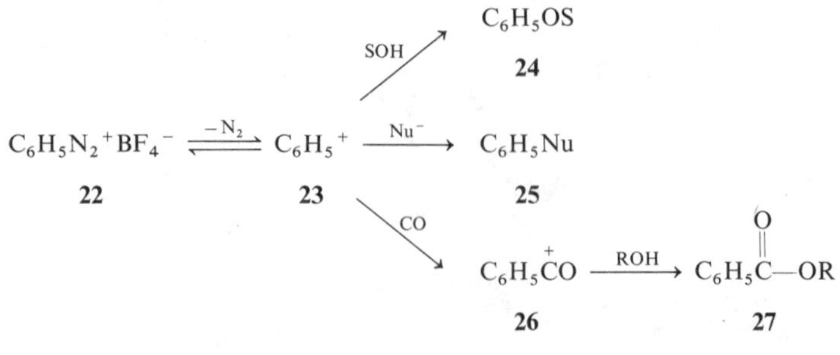

Scheme 9.1

Therefore, it is perhaps no surprise that, in contrast to dediazoniation, solvolysis of phenyl triflate does not result in the formation of a phenyl cation [26]. In fact, careful investigation of the solvolysis of a series of ring-substituted aryl triflates and nonaflates gave no evidence for aryl cation formation [27]. Rather, in nonnucleophilic solvents and in the absence of base as buffer, unreacted starting material was recovered whereas, in nucleophilic solvents or in the presence of base as buffer, nucleophilic attack on sulfur and S–O bond cleavage was observed [27].

## B. Alkynyl Cations

The gas-phase heat of formation of the parent alkynyl cation $HC_2^+$ has been estimated [28] to be $\Delta H_f = 399$ kcal/mole. The use of this value, along with the data in Table 2.1 and the isodesmic relationship of Eq. (2), allows

$$R^+ + C_2H_2 \longrightarrow C_2H^+ + R\text{—}H \qquad (2)$$

a comparison of the relative energies of various carbenium ions in relation to $HC_2^+$. Such a comparison shows that $HC_2^+$ is 66 kcal/mole *less* stable than the methyl cation, 90 kcal/mole *less* stable than the parent vinyl cation, and 106 kcal/mole *less* stable than the ethyl cation. Even if the value for the heat of formation of $HC_2^+$ were off by 15%, or some 60 kcal/mole, the parent alkynyl cation would still be only as stable as the methyl cation. Since there is no unambiguous evidence for the formation of a methyl cation in solution, it is unlikely that alkynyl cations will be seen in solution.

No data are available on the effect of substituents on the stability of alkynyl cations. However, since substituents can exert only an inductive effect in alkynyl cations, they would not be expected to provide sufficient stabilization to lower significantly the high energy of these intermediates.

Indeed, to date, no experimental evidence exists for the formation of alkynyl cations in solution [29]. Potential precursors, such as haloacetylenes **28**, do not give alkynyl cations but rather react by alternative pathways, such as by various nucleophilic addition–elimination processes [29]:

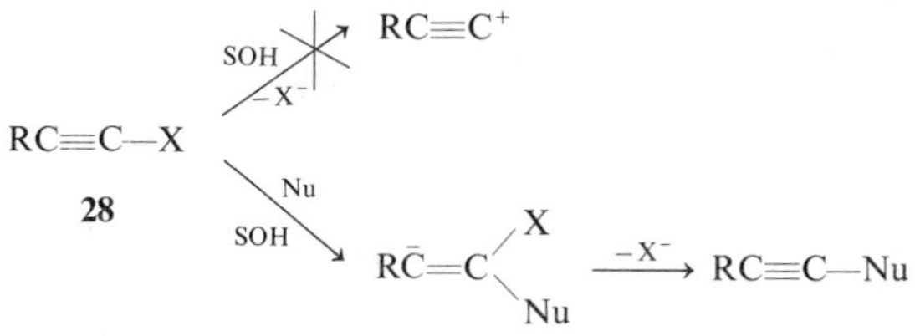

Indeed, as long as such alternative pathways of lower energy are available in solution, potential precursors, even alkynyl sulfonate esters $RC{\equiv}COSO_2R'$ (presently unknown compounds), will not form the high-energy alkynyl cation. Perhaps the only chance of generating an alkynyl cation in solution would be by means of alkynyldiazonium salts (29) as precursors, since nitrogen remains the best leaving group that exists, surpassing even the fluorosulfonates. However, to date such diazonium compounds (29) are unknown [30]. Attempts to generate alkynyl cations via an intermediate 29 have been made recently [31].

$$RC{\equiv}CN_2{}^+X^-$$

29

## IV. CONCLUSIONS AND PROGNOSIS

This monograph is evidence of the fact that, within the last 12 years, vinyl cations have emerged from relative obscurity, have been the subject of extensive and active research, and have become accepted as viable reaction intermediates. As the increasingly available gas-phase thermodynamic data discussed in Chapter 2 indicate, alkylvinyl cations are not as high in energy and therefore as unstable compared to trisubstituted carbenium ions, as previously believed. Specifically, the parent and the 2-propenyl cations are each only 15 kcal/mole less stable than their saturated counterparts, the ethyl and isopropyl cations, respectively. In other words, secondary vinyl cations, such as the 2-propenyl cation, are comparable in stability to primary carbenium ions, such as the n-propyl cation.

Theoretical calculations as well as experimental data indicate that the effects of substitution in vinyl cations are parallel but of somewhat greater magnitude than those in trisubstituted carbenium ions. There is little doubt from both theoretical and experimental results that vinyl cations prefer a linear geometry with an empty p orbital. Trigonally hybridized, bent, unsaturated cations, such as the small-ring $C_5$ and $C_6$ cycloalkenyl cations and the phenyl cation, are of considerably higher energy than their acyclic linear analogues. On the other hand, a nonclassical hydrogen-bridged structure cannot be ruled out for the parent vinyl cation. In fact, if the latest calculations are indeed correct, the exact structure of this ion will be difficult to establish. Such bridged structures are, however, unlikely for substituted vinyl cations, since the required two-electron three-center bonds are not as likely for the common substituent groups as for hydrogen.

The behavior of vinyl cations in solution, with some interesting differences in the case of the triarylvinyl systems, by and large parallel that of trisubstituted carbenium ions. They may be generated in solution by methods exactly analogous to the generation of trisubstituted carbenium ions by electrophilic addition to multiple bonds or solvolytic participation of such multiple bonds as well as by bond heterolyses of appropriate vinylic progenitors. Vinyl cations readily undergo a multitude of rearrangements and may in special cases be observed spectroscopically under strongly acidic conditions.

Arylvinyl cations substituted by bulky substituents show unexpected selectivities compared with trigonal cations, and their formation by bond heterolysis shows unusual solvent and leaving-group effects due to steric hindrance to solvation and ion capture, as well as steric acceleration of the bond heterolysis.

However, unlike the field of trisubstituted carbenium ions, in which there have been a plethora of investigations and results, the field of vinyl cations is but in its infancy or, at best, early childhood. Clearly, much more must be done in the gas phase, and reliable thermodynamic data for vinyl cations must be obtained. This is particularly true for such substituted systems as the arylvinyl and cyclopropylvinyl cations and the dienyl and allenyl cations. To date, no thermodynamic data exist, even for the simple phenylvinyl cation. The remaining questions and ambiguities in the nmr spectra of vinyl cations must be resolved and further work done on the spectroscopic characterization of these species.

The exact properties and behavior of highly hindered vinyl cations must be further explored. More information will undoubtedly be forthcoming in the area of metal-stabilized vinyl cations and their chemistry. Vinyl cations with potentially unusual properties, such as **30**, **31**, and **32**, might be profitably explored.

Although a limited amount of use has been made of vinyl cations analogous to trisubstituted carbenium ions in syntheses, they will no doubt be

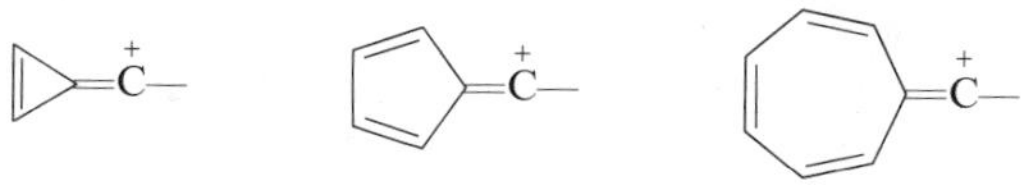

|     30     |     31     |     32     |

valuable synthetic intermediaries in the future. In fact, it is the potential of vinyl cations in syntheses that is presently most underexplored and hence the potentially most fertile territory in this field. Vinyl cations may be intermediates in polymerization and perhaps even in some as yet unknown biological processes.

## REFERENCES

1. G. Melloni and G. Modena, *J. Chem. Soc., Perkin Trans. 1* p. 1355 (1972).
2. R. Varet and V. J. Vienne, *C. R. Acad. Sci.* **104**, 1375 (1887); O. W. Cook and V. J. Chambers, *J. Am. Chem. Soc.* **43**, 334 (1921); J. Boeseken and A. A. Adler, *Recl. Trav. Chim. Pays-Bas* **48**, 474 (1929); J. S. Reichert and J. A. Nieuwland, *J. Am. Chem. Soc.* **45**, 3090 (1923); **50**,

2564 (1928); V. L. Vaiser, *Dokl. Akad. Nauk SSSR* **70**, 621 (1950); **84**, 71 (1952); **87**, 593 (1952); I. Iwai and T. Hiraoka, *Chem. Pharm. Bull.* **11**, 638 (1963); **14**, 262 (1966).

3. E. Demole, *Ber. Dtsch. Chem. Ges.* **12**, 2245 (1879); R. Anschütz, *Justus Liebigs Ann. Chem.* **235**, 150 (1886); L. Schmerling, J. P. West, and R. W. Welch, *J. Am. Chem. Soc.* **80**, 576 (1958); I. P. Tsukervanik and K. Y. Yuldashev, *Uzb. Khim. Zh.* p. 58 (1960); p. 40 (1961); *Zh. Obshch. Khim.* **31**, 858 (1961); **32**, 1293 (1963); **34**, 2647 (1964).

4. J. F. Eykman, *Chem. Weekbl.* **5**, 655 (1908); J. B. Niederl, R. A. Smith, and M. E. McGreal, *J. Am. Chem. Soc.* **53**, 3390 (1931); A. R. Buder, *J. Am. Chem. Soc.* **77**, 4155 (1955); V. V. Korshak, K. K. Samplavskaya, and A. I. Gershonovich, *J. Gen. Chem. USSR* **16**, 1065 (1946).

5. R. M. Roberts and M. B. Abdel-Baset, *J. Org. Chem.* **41**, 1698 (1976).

6. P. J. Stang and A. G. Anderson, *J. Am. Chem. Soc.* **100**, 1520 (1978); *Tetrahedron Lett.* p. 1485 (1977).

7. For reviews on alkylidene carbenes, see P. J. Stang, *Chem. Rev.* **78**, 383 (1978); *Acc. Chem. Res.* **11**, 107 (1978); H. D. Hartzler, in "Carbenes" (R. A. Moss and M. Jones, Jr., eds.), Vol. 2, p. 43. Wiley (Interscience), New York, 1975.

8. R. A. Bell and M. H. Chisholm, *Inorg. Chem.* **16**, 687 (1977).

9. R. A. Bell, M. H. Chisholm, D. A. Couch, and L. A. Rankel, *Inorg. Chem.* **16**, 677 (1977); M. H. Chisholm and H. C. Clark, *Acc. Chem. Res.* **6**, 202 (1973); *J. Am. Chem. Soc.* **94**, 1532 (1972).

10. A. Davison and J. P. Solar, *J. Organomet. Chem.* **155**, C8 (1978).

11. H. G. Viehe, R. Buijle, R. Fuks, R. Merènyi, and J. M. F. Oth, *Angew. Chem., Int. Ed. Engl.* **6**, 77 (1967).

12. A. Davison and J. P. Selegue, *J. Am. Chem. Soc.* **100**, 7763 (1978); see also, M. I. Bruce, A. G. Swincer, and R. C. Wallis, *J. Organomet. Chem.* **171**, C5 (1979); M. I. Bruce and R. C. Wallis, *ibid.*, **161**, Cl (1978); B. E. Roland, S. A. Fam, and R. P. Hughes, *ibid*, C29 (1979).

13. For reviews and leading references, see G. A. Olah, "Friedel Crafts and Related Reactions," Vols. 1–4. Wiley (Interscience), New York, 1963–1964.

14. For reviews and leading references, see K. Harada, in "The Chemistry of the Carbon–Nitrogen Double Bond" (S. Patai, ed.), Chap. 6, p. 225. Wiley (Interscience), New York, 1970; J. Morath and G. W. Stacy, in "The Chemistry of the Carbon-Nitrogen Double Bond" (S. Patai, ed.), Chap. 8, p. 327. Wiley (Interscience), New York, 1970; H. Ulrich, "The Chemistry of Imidoyl Halides." Plenum, New York, 1968; R. Ta-Shma and Z. Rappoport, *J. Am. Chem. Soc.* **99**, 1845 (1977); **98**, 8460 (1976); *J. Chem. Soc., Perkin Trans.* 2 p. 659 (1977); A. F. Hegarty, J. D. Cronin, and F. L. Scott, *J. Chem. Soc., Perkin Trans.* 2 p. 429 (1975); I. Ugi, F. Beck, and U. Fetzer, *Chem. Ber.* **95**, 126 (1952).

15. G. A. Olah, A. Germain, and A. M. White, in "Carbonium Ions" (G. A. Olah and P. von R. Schleyer, eds.), Vol. 5, Chap. 35. Wiley (Interscience), New York, 1976.

16. G. A. Olah and T. E. Kiovsky, *J. Am. Chem. Soc.* **90**, 4666 (1968); J. E. Gordon and G. C. Turrell, *J. Org. Chem.* **24**, 269 (1959).

17. Y. Apeloig and P. von R. Schleyer, submitted for publication.

18. T. B. McHahom, Ph.D. Thesis, Calif. Inst. of Technol., Pasadena, 1973.

19. J. D. Dill, P. von R. Schleyer, J. S. Binkley, R. Seeger, J. A. Pople, and E. Haselbach, *J. Am. Chem. Soc.* **98**, 5428 (1976).

20. J. D. Dill, P. von R. Schleyer, and J. A. Pople, *J. Am. Chem. Soc.* **99**, 1 (1977).

21. For review of early work and leading references, see H. Zollinger, *Angew. Chem., Int. Ed. Engl.* **17**, 141 (1978); *Acc. Chem. Res.* **6**, 335 (1973).

22. C. G. Swain, J. E. Sheats, and K. G. Harbison, *J. Am. Chem. Soc.* **97**, 783 (1975).

23. C. G. Swain, J. E. Sheats, D. G. Gorenstein, and K. G. Harbison, *J. Am. Chem. Soc.* **97**, 791 (1975).

24. C. G. Swain, J. E. Sheats, and K. G. Harbison, *J. Am. Chem. Soc.* **97**, 796 (1975).
25. R. G. Bergstrom, R. G. M. Landells, G. H. Wahl, Jr., and H. Zollinger, *J. Am. Chem. Soc.* **98**, 3301 (1976).
26. A. Streitwieser, Jr. and A. Dafforn, *Tetrahedron Lett.* p. 1435 (1976).
27. L. R. Subramanian, M. Hanack, L. W. K. Chang, M. A. Imhoff, P. von R. Schleyer, F. Effenberger, W. Kurtz, P. J. Stang, and T. E. Dueber, *J. Org. Chem.* **41**, 4099 (1976).
28. J. L. Franklin, J. G. Dillard, H. M. Rosenstock, J. T. Herron, and K. Draxl, "Ionization Potentials, Appearance Potentials and Heats of Formation of Gaseous Positive Ions," NSRDS-NBS-26. Natl. Bur. Stand., Washington, D.C., 1969.
29. S. I. Miller and J. I. Dickstein, *Acc. Chem. Res.* **9**, 358 (1976).
30. M. Regitz, "Diazoalkane." Verlag Chemie, Berlin, 1977.
31. M. Hanack and R. Helwig, unpublished results.

# INDEX

509